ANIMAL CELL TECHNOLOGY: BASIC & APPLIED ASPECTS

The Fifth International Meeting
of
Japanese Association
for
Animal Cell Technology

JAACT'92

JAACT'92 Organizing Committee

Chairperson: Shuichi KAMINOGAWA (The University of Tokyo, Japan)
Vice-Chairperson: Kazuo NAGAI (Tokyo Institute of Technology, Japan)
Secretary General: Shun'ichi DOSAKO (Snow Brand Milk Products Co., Ltd., Japan)
                 Akio AMETANI (The University of Tokyo, Japan)

Program Committee

David BARNES	Oregon State University, U.S.A.
Nobuo FUJIYOSHI	Vessel Research Laboratory Co., Ltd., Japan
Satoshi HACHIMURA	The University of Tokyo, Japan
Shuichi HASHIZUME	Morinaga Institute of Biological Science, Japan
Koji IKURA	Kyoto Institute of Technology, Japan
Hermann KATZINGER	University of Agriculture and Forestry, Austria
Kin'ichi KAWAMURA	Komatsugawa Chemical Engineering Co., Ltd., Japan
Yasuo KITAGAWA	Nagoya University, Japan
Jun-ichi KURISAKI	National Institute of Animal Industry, Japan
Anthony LUBINIECKI	SmithKline Beecham Pharmaceuticals, U.S.A.
Yoji MITSUI	Tsukuba University, Japan
Hiroki MURAKAMI	Kyushu University, Japan
Tadao OHNO	Riken Gene Bank, Japan
Ryuzo SASAKI	Kyoto University, Japan
Makoto SHIMIZU	The University of Shizuoka, Japan
Kazuki SHINOHARA	National Food Research Institute, Japan
Sanetaka SHIRAHATA	Kyushu University, Japan
Yukio SOGO	Snow Brand Milk Products Co., Ltd., Japan
Takamoto SUZUKI	Kirin Brewery Co., Ltd., Japan
Eiji SUZUKI	The University of Tokyo, Japan
Michiyuki TOKASHIKI	Teijin Co., Ltd., Japan

Animal Cell Technology: Basic & Applied Aspects

Volume 5

Proceedings of the Fifth International Meeting of the
Japanese Association for Animal Cell Technology,
Omiya, Japan, November 30–December 4, 1992

Edited by

S. KAMINOGAWA

A. AMETANI

and

S. HACHIMURA

*Department of Agricultural Chemistry,
The University of Tokyo, Japan*

SPRINGER–SCIENCE+BUSINESS MEDIA, B.V.

A C.I.P. Catalogue record for this book is available from the Library of Congress.

ISBN 978-0-7923-2477-5   ISBN 978-94-011-2044-9 (eBook)
DOI 10.1007/978-94-011-2044-9

Printed on acid-free paper

Contents

I. CELL CHARACTERIZATION

IV. ENHANCEMENT OF PRODUCTIVITY

VII. SERUM- AND PROTEIN-FREE CULTURE

VIII. CELL CULTURE ENGINEERING

IX. APPLICATION AND PRODUCTIVITY OF ANTIBODY

X. APPLICATION OF HUMAN MONOCLONAL ANTIBODY

XI. COMMERCIALIZATION OF ANIMAL CELL TECHNOLOGY PRODUCTS

Preface

Animal cell technology has been making tremendous progress. Originally this term reminded people of engineering for high density and large volume culture of animal cells. At present many fields of biological sciences are aiming at advance in animal cell technology. Cell culture engineering is aided not only with developments in apparatus, matrix, media, and computational analysis, but also with new biological procedures in gene and protein technology, cell biological resources and immunological methods. Results obtained with animal cell technology are applied to production of pharmaceuticals, diagnosis reagents and food endowed with physiological functions, and cell and gene therapy of animals and humans, and useful for elucidating scientific phenomena. It is also essential to establish methods of evaluation for functionality and safety of newly discovered molecules and cells. The progress in animal cell technology is supported by, and attributes in both of basic and applied sciences.

The proceedings of the Fifth International Meeting of the Japanese Association for Animal Cell Technology (JAACT) covers the subjects above mentioned. The articles in this book will help researchers in many fields to understand the current status and future trends in animal cell technology.

JAACT organized this Meeting and we express our gratitude to the members of JAACT. We gratefully acknowledge all the members of the organizing committee for their dedication in assuring the Meeting's success. For their valuable supports, we also thank the Japanese BioIndustry Association and Saitama Foundation for Culture and Industry.

The editors

SEARCH FOR EDIBLE PLANT CONSTITUENTS OF ANTI-TUMOR PROMOTION USING AN INHIBITION TEST OF EPSTEIN-BARR VIRUS ACTIVATION IN RAJI CELLS

Koichi KOSHIMIZU
Department of Food Science and Technology, Faculty of Agriculture,
Kyoto University, Kyoto 606, Japan

ABSTRACT

Tumor promoting inhibitory properties (anti-tumor promotion) of edible plants including vegetables, fruits and marine algae have been studied by a convenient in vitro assay designed to detect the inhibition of tumor promter-induced Epstein-Barr virus (EBV) activation in Raji cells. Some of the anti-tumor promoters were then isolated, however it was rare that the anti-tumor promoter occurring in each active plant extract was specified. Thus, the anti-tumor promoting activity of the crude extracts was suggested to be frequently enhanced or reduced with co-occurring factors acting additively, synergistically, or antagonistically. As an attempt to find strongly active anti-tumor promoters, tropical Zingiberaceae plants used as condiments and occasionally also used as local medicine, have been focused on. From one of such species, Languas galanga, 1'-acetoxychavicol acetate (ACA) was isolated as a potent EBV activation inhibitor. ACA was also proven to be a highly active anti-tumor promoter by an in vivo test on mouse skin. Thus, the tropical Zingiberaceae plants may be a promising source of anti-tumor promoting food constituents. Details of the in vitro assay method, screening results of edible plants as well as some anti-tumor promoters are reported.

INTRODUCTION

The two step process of chemical carcinogenesis by initiation and promotion, first proposed by Berenblum [1, 2], has recently been accepted to occur in a variety of cancers [3, 4]. Therefore, chemical inhibition of either process is possibly an effective means of cancer prevention. While inhibitors of tumor initiation has hitherto been extensively studied [5], less attention has been given to the inhibitors (anti-tumor promoters) of the promotion process. The author has investigated naturally occurring anti-tumor promoters using a convenient short-term in vitro method using B-lymphoblastoid Raji cells carrying the genome of the Epstein-Barr virus (EBV). This assay method first estimated for detection of tumor promoters by Ito et al. [6] is based on the fact that the EBV

1

S. Kaminogawa et al. (eds.), Animal Cell Technology: Basic & Applied Aspects, Vol. 5, 1–8.
© 1993 *Kluwer Academic Publishers.*

genome persisting in Raji cells is highly activated by the combination of a TPA (12-O-tetradecanoylphorbol-13-acetate) type tumor promoter and n-butyrate. Inhibitors of EBV activation were thus thought to possibly be anti-tumor promoters. Actually, most EBV activation inhibitors thus far isolated have been proven to inhibit tumor promotion when tested on mouse skin [7-10]. This assay system was, then, applied to evaluate anti-tumor promoting property of daily use food materials, particularly of edible plants, and to reveal their active constituents, because the daily intake of green and yellow vegetables has been reported to reduce the risk of cancers by epidemiological surveys [11]. In this report, details of the in vitro method of inhibition assay against EBV activation is first described. Possible anti-tumor promoting properties of edible plants as well as their active constituents are also reported.

MATERIALS and METHODS

Chemicals. TPA purchased from Funakoshi Chemicals (Tokyo, Japan) was used as a tumor promoter for in vitro and in vivo assays. Because of sample availability, HPA (12-O-hexadecanoylphorbol-13-acetate) and teleocidin B-4, isolated from a plant Sapium sebiferum [12] and an actinomycete Streptoverticillium blastmyceticum [13] in my laboratory, were also utilized as tumor promoters in place of TPA. 3-Oxoursolic acid used as a positive control for the inhibition of EBV activation was derived from ursolic acid as previously reported [7]. DMBA (7,12-dimethylbenz[a]anthracene) purchased from Nakarai Chemicals (Kyoto, Japan) was used as an initiator in anti-tumor promotion tests in vivo.

Inhibition test of EBV activation. Inhibition tests against EBV activation in Raji cells were basically conducted by the standard method reported previously [7], except for the promoters used and their amounts. In the cases of promoters TPA and HPA, the assay was carried out at a level of 40 ng/ml of the promoter and 352 μg/ml (4 mM) of sodium n-butyrate. When teleocidin B-4 at 20 ng/ml was used, the use of 264 μg/ml (3 mM) of sodium n-butyrate resulted in the highest EBV activation.

Screening tests for in vitro anti-tumor promoting activity. Fresh edible plants including vegetables and fruits were extracted with MeOH. After concentrated, each extract (200 μg) was assayed in a medium (1 ml) containing HPA (40 ng) as a promoter. Twenty-six MeOH extracts randomly selected were partitioned between EtOAc and water, and both parts (40 μg) thus obtained were assayed again. The inhibitory activity against early antigen (EA) induction of EBV was divided into 4 ranks as indicated in the TABLE. In the screening tests of marine algae, dried edible algae commercially obtained and non-edible fresh algae directly collected from the ocean were extracted with dichloromethane and MeOH, respectively. The MeOH extracts from fresh algae were re-extracted with EtOAc. The dichloromethane and EtOAc extracts (4 μg) were assayed for inhibitory activity of EBV-EA induction with teleocidin B-4 (20 ng). The activity was evaluated by the relative inhibitory rate (RIR) to the activity of 3-oxoursolic acid at 2 μg/ml, and divided into the following 4 ranks:

strongly active, RIR≥0.9; moderately active, 0.9>RIR≥0.7; weakly active, 0.7>RIR≥0.5; and inactive, 0.5>RIR. In the case of tropical Zingiberaceae plants, each EtOAc soluble part (10 μg) of the MeOH extracts was used for the assay with teleocidin B-4 (20 ng), and the activity was evaluated in the same way as the case of the screening test of edible plants.

In vivo anti-tumor promotion test. Anti-tumor promoting activity was tested by the standard initiation (DMBA)-promotion (TPA) protocol on mouse skin [9]. Each experiment was performed on 15 female ICR mice. One week after initiation with DMBA, each mouse was repetitively promoted by twice weekly application of TPA. The inhibitors were applied 1 hr before each TPA treatment. The anti-tumor promotion activity was evaluated by both the ratio of tumor-bearing mice and the average number of tumors with more than 1 mm in diameter per mouse after 15-20 weeks.

RESULTS and DISCUSSION

Tumor promoters such as TPA and teleocidins are known to cause several biological and biochemical responses, called pleiotropic effects [14]. Some of these responses have been utilized as short-term detection methods for tumor promoters, and anti-tumor promoters as well [15]. Among them, the author has used an inhibition test of tumor promoter-induced EBV activation in Raji cells for anti-tumor promoting activity in vitro. EBV is known as a virus closely associated with African Burkitt's lymphoma and nasophalyngeal carcinoma (NPC), endemic to southern China. Activation of this virus genome by TPA was first found by zur Hausen et al. [16]. Ito et al. found that the EBV genome in non-producer Raji cells was highly activated when TPA or analogous diterpene esters were applied in combination with n-butyrate [6]. This finding resulted in the estimation of a new assay method for tumor promoting activity. The method was afterwards adapted to anti-tumor promoting activity [7, 17]. The outline of the method is as follows; to 1 ml of a culture medium (RPMI 1640 containing 8% fetal calf serum, 200 units of penicillin, 250 μg of streptomycin and a known amount of sodium n-butyrate) including Raji cells ($5x10^5$) certain amounts of a tumor promoter and a test compound in 5 μl of DMSO are added. The cells are incubated at 37°C under 5% CO_2 atmosphere for 48 hr. Then, cell smears are prepared on glass slides, air dried, and fixed with acetone at room temperature for 10 min. The activated Raji cells expressing EBV-EA are stained with EA and viral capsid antigen (VCA) positive serum from a patient with NPC using an indirect immunofluorescent method [18]. The ratio of EA-induced cells in at least 500 cells is determined by fluorescence microscopical observation, and is compared with that in a control experiment without the test compound. The number of viable cells in the culture are also determined by the trypan-blue exclusion test for evaluation of toxicity of the test compound.

A total of 121 species of edible plants including herbage-, root- and fruit-vegetables, and fruits were tested for inhibition of EBV-EA induction to evaluate their anti-tumor promoting properties [19]. As shown in the TABLE, 14 plant extracts corresponding to 12% of the total test extracts showed strong activity, and 7 and 12 extracts corresponding to 6% and 10% of the total were respectively rated as being moderately to weakly active. Furthermore, when 26 inactive methanol extracts were randomly selected and purified by partition between ethyl acetate and water, 11 extracts were found to newly exhibit strong or moderate activities mainly in the ethyl acetate soluble part. Root-vegetables (taro root, edible burdock, Chinese yam, onion, ginger, lotus root and carrot) markedly had this tendency at a high rate. Thus, occurrence of anti-tumor promoters in a wide variety of edible plants was suggested. However, purification of the active constituent in each active extract did not always succeed to isolate specified constituents. Oleanolic acid, mokko lactone and arctic acid, and gingerol have been thus far identified as the active constituents of green perilla, edible burdock, and ginger, respectively [20]. Anti-tumor promoting activities of oleanolic acid [8] and mokko lactone [20] were confirmed by in vivo experiment on mouse skin.

TABLE. Edible plants inhibited EBV activation.

Inhibitory Activity[1,2,3]		
+ + +	+ +	+
Taro(shoot), curled lettuce	Green perilla,	Chinese mustard,
Field mustard, Cauliflower,	Japanese parsley,	Dittany of crete,
Japanese pepper(flower),	Table beet,	Bracken, Stone parsley,
Parsley, Ginger(sprout),	Raddish-cv.	Lily bulb, Chiboul,
Azuki bean, Avocado, Feijoa,	Koshin,	Eschallot, Mume,
Gingo nut, Japanese walnut,	Japanese chesnut,	Japanese yam,
Dwarf banana, Litch(skin)*	(skin)	Passion fruit,
	Sesame, Nectarine	Apple-cv. Fuji,
		Navel orange(skin)

[1] Activity was evaluated as follows: strongly active(+++), inhibition rate(IR)≥70%; moderately active(++), 70%>IR≥50%; weakly active(+), 50%>IR≥30%; and inactive(-), 30%>IR.

[2] See the reference for names of inactive plants (-) not shown here.

[3] Otherwise stated in parenthesis, data in the edible parts are shown.

* Weak activity (+) was found in the edible part of Litch.

Screening tests for possible anti-tumor promoting properties were also conducted in 36 species of marine algae [21], a part of which are important daily foods in Japan. In these screening tests, the activity was measured at reduced concentrations of both tumor promoter (teleocidin B-4 at 20 ng/ml) and test extracts (4 μg/ml), because most extracts were toxic to the Raji cells at high dosages. The condition used here might be stricter than that used for the tests of vegetables and fruits. Interestingly, strong inhibitory activity against EBV-EA induction was found only in the species of Phaeophyta, which includes several edible species such as wakame seaweed and sea tangles. The crude dichloromethane extract of wakame seaweed was then tested for in vivo anti-tumor promoting activity on mouse skin. The ICR mice at 7 weeks old were once initiated with DMBA (0.19 μmol/0.2 ml of acetone). One week after initiation, the mice were promoted with TPA (1.6 nmol/0.2 ml acetone) twice a week. In the inhibition test, the mice were treated with the dichloromethane extract (1 mg/0.2 ml acetone) 1 hr before each TPA treat-

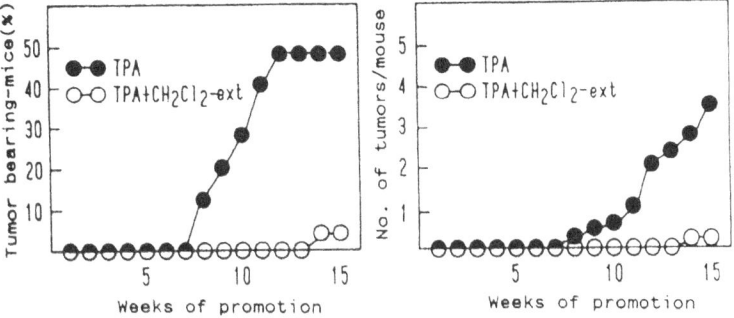

Fig. 1. Anti-tumor promoting activity of wakame seaweed extract.

ment. As shown in Fig. 1, the extract inhibited both the ratio of tumor-bearing mice and the number of tumors per mouse after 15 weeks [21]. An effort to purify the active constituent(s) was, however, not successful, because the strong inhibitory activity of the crude extract against EBV activation was not perfectly recovered in a specified fraction.

The above results strongly suggested that anti-tumor promoting activity of crude mixtures is enhanced or reduced with co-occurring factors acting additively, synergistically or antagonistically. This may reflect the pleiotropic effects of tumor promoters. Hence, anti-tumor promoting action may be positively controlled by the combination of a variety of constituents.

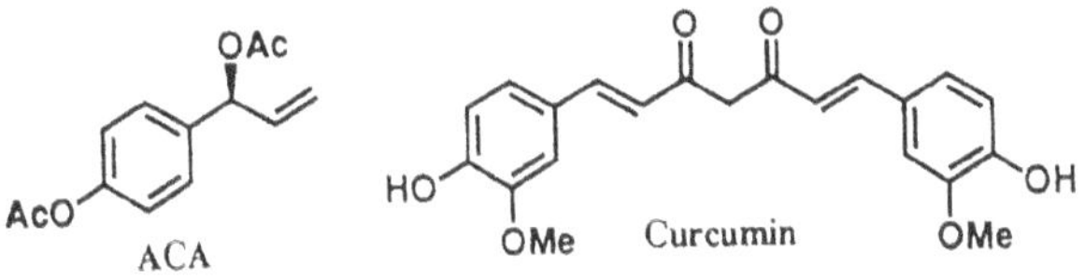

Fig. 2. Structures of ACA and curcumin.

Recently, several tropical Zingiberaceae plants used as condiments and occasionally also used as local medicine were found to possess significant inhibitory activities against EBV-EA induction. Among the plants tested, the activity of Languas galanga and Zingiber cassumunar were particularly remarkable. From L. galanga, (1'S)-1'-acetoxychavicol acetate (ACA, Fig. 2) was isolated as an EBV activation inhibitor [22]. ACA completely inhibited EBV-EA induction at a dosage of 100-fold molar equivalent of teleocidin B-4, and this activity was more than 10 times as potent as those of inhibitors thus far obtained from edible plants. As shown in Fig. 3, ACA at a dosage of 100-fold molar equivalent of TPA highly inhibited both the ratio of tumor-bearing mice and the number of tumors per mouse after 20 weeks. Significant inhibition was also found at an equimolar dosage of TPA, showing that ACA is one of the most potent anti-tumor promoters isolated from edible plants. On the other hand, the main inhibitor of _Z. cassumunar_ was confirmed to be curcumin (Fig. 2), whose anti-tumor promoting effect has been proven independently by Nishino et al. [23] and Huang et al. [24]. The tropical Zingiberaceae plants, thus, may be one of the most promising sources of anti-tumor promoting food constituents.

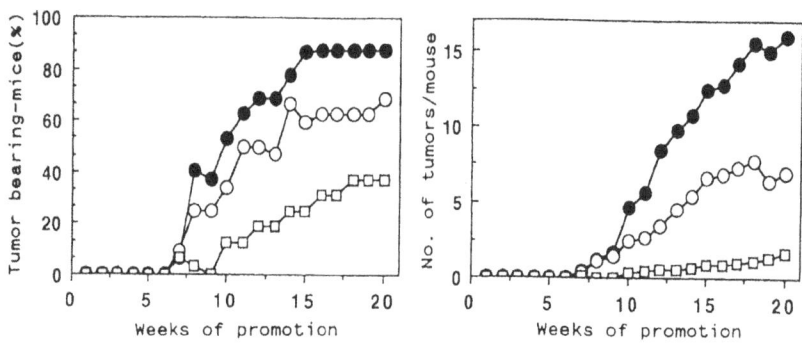

DMBA: 50 μg (0.19 μmol), TPA: 1 μg (1.6 nmol) –●–
ACA: 37.4 μg (160 nmol)   –□– (p < 0.001 in t-test)
374 ng (1.6 nmol)   –○– (p < 0.05 in t-test)

Fig. 3. Anti-tumor promoting activity of ACA.

The results obtained in these investigations may partly provide a chemical base for epidemiological surveys on the daily intake of vegetables and a reduction of the risk of cancer. Detailed study on the indication that the anti-tumor promoting action is enhanced by the combination of multiple food constituents would be necessary.

ACKNOWLEDGEMENTS

This study was partly supported by a Grant from the Ministry of Health and Welfare of Japan, and by a Grant-in-Aid for Scientific Research on Priority Areas from the Ministry of Education, Science and Culture of Japan. The author thanks Drs. Hajime Ohigashi, Harukuni Tokuda and Akira Murakami, Messrs. Akira Kondo, Yasushi Sakai and Shin Ohura, and Miss Kanoko Yamaguchi for their helpful collaborations.

REFERENCES

[1] Berenblum, I. (1941) 'Cocarcinogenic action of croton oil', Cancer Res., 1. 44-48.
[2] Berenblum, I. (1941) 'The mechanism of carcinogenesis, a study of the significance of cocarcinogenic action and related phenomena', Cancer Res., 1, 807-814.
[3] Wynder, E.L., Hoffman, D., McCoy, G.D., Cohen, L.A., Reddy, B.S.

(1978) 'Mechanisms of tumor promotion and cocarcinogenesis', in T.J. Slaga, A. Sivak, and R.K. Boutwell (eds.), Carcinogenesis-A Comprehensive Survey, Vol. 2, Raven Press, New York, pp. 11-48.
[4] Koshimizu, K. (1992) 'Plant constituents and their tumor inhibitory effects - Part III, inhibition of tumor promotion by plant constituents', Contemporary Health Digest, 7, Dec., 1-10 (in Japanese).
[5] Wattenberg, L.W. (1985) 'Chemoprevention of cancer', Cancer Res., 45, 1-8.
[6] Ito, Y., Yanase, S., Fujita, T., Harayama, T., and Imanaka, H. (1981) 'A short-term in vitro assay for tumor promoter substances using human lymphoblastoid cells latently infected with Epstein-Barr virus', Cancer Lett., 13, 29-37.
[7] Ohigashi, H., Takamura, H., Koshimizu, K., Tokuda, H., and Ito, Y. (1986) 'Search for possible antitumor promoters by inhibition of 12-O-tetradecanoylphorbol-13-acetate induced Epstein-Barr virus activation; ursolic acid and oleanolic acid from an anti-inflammatory Chinese medicinal plant, Glechoma hederacea L.', Cancer Lett., 30, 143-151.
[8] Tokuda, H., Ohigashi, H., Koshimizu, K., and Ito, Y. (1986) 'Inhibitory effects of ursolic and oleanolic acid on skin tumor promotion by 12-O-tetradecanoylphorbol-13-acetate', Cancer Lett., 33, 279-285.
[9] Murakami, A., Ohigashi, H., Jisaka, M.,, Hirota, M., Irie, R., and Koshimizu, K. (1991) 'Inhibitory effects of new types of biflavonoid-related polyphenols; lophirone A and lophiraic acid, on some tumor promoter-induced biological responses in vitro and in vivo', Cancer Lett., 58, 101-106.
[10] Murakami, A., Tanaka, S., Ohigashi, H., Hirota, M., Irie, R., Takeda, N., Tatematsu, A., Koshimizu, K. (1992) 'Chalcone tetramers, lophirachalcone and alatachalcone, from Lophira alata as possible anti-tumor promoter', Biosci. Biotech. Biochem., 56, 769-772.
[11] Hirayama, T. (1978) 'Diet and cancer', Nutr. Cancer, 1, 67-81.
[12] Ohigashi, H., Ohtsuka, T., Hirota, M., Koshimizu, K., Tokuda, H., and Ito, Y. (1983) 'Tigliane type diterpene-esters with Epstein-Barr virus inducing activity from Sapium sebiferum', Agric. Biol. Chem., 47, 1617-1622.
[13] Irie, K., Hirota, M., Hagiwara, N., Koshimizu, K., Hayashi, H., Murao, S., Tokuda, H., and Ito, Y. (1984) 'The Epstein-Barr virus early antigen inducing indole alkaloids, (-)-indolactam V and its related compounds, produced by Actinomycetes', Agric. Biol. Chem., 48, 1269-1274.
[14] Slaga, T.J., Fischer, S.M., Weeks, C.E., Nelson, K., Mamrack, M., Klein-Szanto, A.J.P. (1982) 'Cocarcinogenesis and biological effects of tumor promoters', in E. Hecker, N.E. Fusenig, W. Kunz, F. Marks and H.W. Thielmann (eds.), Carcinogenesis-A Comprehensive Survey, Vol. 7, Raven Press, New York, pp 19-34.
[15] Muto, Y., Ninomiya, M., and Fujiki, H. (1990) 'Present status of research on cancer chemoprevention in Japan', Jpn. J. Clin. Oncol., 20, 219-224.
[16] Zur Hausen, H., Bornkamm, G.W., Schmidt, R., and Hecker, E. (1979) 'Tumor initiators and promoters in the induction of Epstein-Barr virus', Proc. Natl. Acad. Sci. USA, 76, 782-785.
[17] Okamoto, H., Yoshida, D., and Mizusaki, S. (1983) 'Inhibition of 12-O-tetradecanoylphorbol-13-acetate-induced induction in Epstein-Barr

virus early antigen in Raji cells', Cancer Lett., 19, 47-53.

[18] Henle, G., and Henle, W. (1966) 'Immunofluorescence in cells derived from Burkitt's lymphoma', J. Bacteriol., 91, 1248-1256.

[19] Koshimizu, K., Ohigashi, H., Tokuda, H., Kondo, A., and Yamaguchi, K. (1988) 'Screening of edible plants against anti-tumor promoting activity', Cancer Lett., 39, 247-257.

[20] Koshimizu, K., and Ohigashi, H. (1990) 'Search for naturally occurring anti-tumor promoters by inhibition of tumor promoter-induced Epstein-Barr virus activation', in B.K. Kim, E.B. Lee, C.K. Kim and Y.N. Han (eds.), Advances in New Drug Development, The Pharmaceutical Society of Korea, Seoul, pp 438-447.

[21] Ohigashi, H., Sakai, Y., Yamaguchi, K., Umezaki, I., and Koshimizu, K. (1992) 'Possible anti-tumor promoting properties of marine algae and in vivo activity of wakame seaweed extract', Biosci. Biotech. Biochem., 56, 994-995.

[22] Kondo, A. Ohigashi, H., Murakami, A., Suratwadee, J., and Koshimizu, K. 'A potent inhibitor of tumor promoter-induced Epstein-Barr virus activation, 1'-acetoxychavicol acetate from Languas galanga, a traditional Thai condiment', Biosci. Biotech. Biochem., submitted.

[23] Nishino, H., Nishino, A., Takayasu, J., Hasegawa, T. (1987), 'Anti-tumor-promoting activity of curcumin, a major constituent of the food additive "turmeric yellow"', J. Kyoto Pref. Univ. Med., 96, 725-728.

[24] Huang, M.T., Smart, R.C., Wong, C.Q., and Conney, A.H. (1988) 'Inhibitory effect of curcumin, chlorogenic acid, caffeic acid, and ferulic acid on tumor promotion in mouse skin by 12-O-tetradecanoyl-phorbol-13-acetate', Cancer Res., 48, 5941-5946.

THE NATURE AND EFFECT OF CELL-CELL INTERACTIONS BETWEEN RECOMBINANT CHO CELLS PRODUCING HUMAN INTERFERON-GAMMA.

S.R. COPPEN, R. NEWSAM, A.J. BAINES, AND A.T. BULL.
Biological Laboratory,
University of Kent,
Canterbury,
Kent,
CT2 7NJ.
UK

C. CAULCOTT
Wellcome Foundation Ltd,
Beckenham,
Kent.
BR3 3BS
UK

ABSTRACT. The Chinese hamster ovary (CHO) cell line has emerged as one of the favoured mammalian cell lines for the production of recombinant human proteins. In large scale suspension culture, these cells have a tendency to form aggregates which can contain hundreds of cells. The structure of these clumps and factors influencing their formation are being investigated with the aim of determining the effect of cell aggregation on the production of the recombinant protein. Electron microscopy of the cell line CHO320 (which expresses human interferon gamma) showed the presence of specific junctions between the live cells in the aggregates. These junctions were not desmosomes, since a panel of antibodies to desmosomal components showed no reaction either by immunofluorescence or Western blotting. However, vinculin and cadherins were detected by Western blotting, suggesting that the junctions were most likely to be of the adherens type. Inhibition of adherens junction formation greatly affected the formation of the aggregates. The relationship between aggregate formation and the physiological state of the cultures has also been addressed; factors influencing aggregation were found to include cell growth rate and the rate of agitation of the cultures.

Introduction.

CHO cells are of great commercial importance in the production of recombinant mammalian proteins, especially for pharmaceutical purposes. The product should be homogeneous for such uses, but studies on the recombinant human interferon gamma produced by the CHO320 cell line have shown it to be heterogeneous, both in glycosylation and proteolytic processing (Curling, et al., 1990). This heterogeneity could be as a result of heterogeneity of the cells within the population or because a homogeneous population of cells is unable to give uniform processing due to nutrient limitation.

9

S. Kaminogawa et al. (eds.), Animal Cell Technology: Basic & Applied Aspects, Vol. 5, 9–18.
© 1993 *Kluwer Academic Publishers.*

CHO320 cells grown in suspension culture have been observed to form aggregates: these can contain many hundreds of cells. Aggregation means that the cells are heterogeneous in their environment, with variable access to oxygen and nutrients perhaps contributing to product heterogeneity. On the other hand, it is conceivable that the aggregation could make the cells more able to withstand the shear forces which occur in suspension cultures.

The aim of this study was to determine the effect of cell aggregation on the production of the recombinant protein. As a first step towards this, the structure of the aggregates and the factors influencing their formation have been investigated.

Materials and methods.

CELLS.

The CHO320 cell line was obtained from Wellcome Biotechnology Ltd (Wellcome Research Laboratories, Beckenham, Kent, UK). These were grown in serum free medium in 50 ml batch shake flask cultures. DUK and A2H were adherent CHO cell lines, grown in RPMI 1640 supplemented with 10% FBS and 2 mM L-glutamine and HT for the DUK cells. CHO-K1 cells were a suspension cell line, grown in the serum-free medium. The epithelial cell lines LLCPK1, OK and MDCK were obtained from the Research Councils Collection of cell lines, ECACC, PHLS, Porton Down, UK. They were grown in RPMI 1640 medium supplemented with 10% FBS and 2 mM L-glutamine.

QUANTITATION OF CELL AGGREGATES

Cells sampled from cultures were allowed to settle on poly-L-lysine coated slides and fixed with methanol at -20° C. The cells were then stained with 4',6-diamidino-2-phenylindole (DAPI) and the number of nuclei per particle counted under a fluorescence microscope. A cell aggregate is defined as a particle containing two or more nuclei.

INHIBITION OF ADHERENS JUNCTION FORMATION

It has been shown that vanadate in the presence of hydrogen peroxide inhibits phosphotyrosine phosphatases (Volberg, et al., 1992). This leads to an increase in the amount of phosphotyrosine associated with adherens junctions and causes them to dissociate. Cells were exposed to 2 mM H_2O_2 and 1 mM vanadate in culture medium for 15 minutes and then returned to normal medium. Samples were taken before treatment and at set time points after, for quantitation of the cell aggregates.

ANTIBODIES

R1PA rabbit polyclonal antibody to IFN-γ was made by immunising rabbits with purified recombinant IFN-γ from CHO320 cell culture supernatants. Guinea pig anti-sera to desmoplakin I + II and desmoglein were generously given by A. Magee (National Institute for Medical Research, Mill Hill, UK). The anti-pan-cadherin antibody (Geiger, et al., 1990) was a kind gift from B. Geiger (Weizmann Institute, Israel). Fluorescein labelled second antibodies were obtained from Sigma Chemical Company.

MEASUREMENT OF INTERFERON GAMMA CONCENTRATION

A sandwich ELISA method was used for the measurement of IFN-γ in cell culture supernatants (Curling, et al., 1990).

GEL ELECTROPHORESIS AND IMMUNOBLOTTING

Gel samples were prepared by adding five times concentrated sample buffer directly to the washed cells. Samples prepared from 10^5 cells were loaded in each lane and electrophoresed on 8% gels (Laemmli, 1970). The immunoblotting procedure was as described previously (Woods, et al., 1989), except that detection of antibody was done with enhanced chemiluminescence reagents (ECL kit, Amersham International, UK).

TRANSMISSION ELECTRON MICROSCOPY

Samples were fixed in 2.5% glutaraldehyde plus 2% paraformaldehyde in 0.1M cacodylate buffer, pH7.2, for one hour. They were washed in buffer and post-fixed in 1% osmium tetroxide for one hour, washed again and treated with 2% magnesium uranyl acetate overnight. Following this they were dehydrated through a graded series of ethanol and embedded in Spurr's resin. Silver/gold sections were cut and stained with uranyl acetate and lead citrate.

PREPARATION OF CRYOSECTIONS

Washed cells were fixed with methanol at -20°C and washed in 0.08M sodium phosphate buffer, pH7.3. The cells were then infiltrated with increasing concentrations of sucrose in 30% polyvinylpyrrolidone (PVP) up to 2.3M sucrose. Pelleted cells were mounted on notched specimen pins and immersed in liquid nitrogen. Sections (0.5μm) were cut using an ultramicrotome fitted with a cryo-chamber mantained at -95°C. Sections were placed on poly-L-lysine coated slides.

IMMUNOFLUORESCENCE

Sections were washed with PBS and incubated with the first antibody for 2 hours at room temperature in a moist chamber. The sections were then washed and incubated with the appropriate fluorochrome-conjugated second antibody for 2 hours and washed again. Sections were mounted in Mowiol containing p-phenylenediamine and using a Zeiss Universal microscope.

Results

INTERFERON GAMMA PURIFIED FROM CHO320 CELLS IS HETEROGENEOUS

Fig. 1 shows a Coomassie blue-stained 14% SDS/PAGE of IFN-γ purified from CHO320 culture supernatant. The preparation contains about 12 bands, grouped into three major classes, 15-17 kDa, 19-21 kDa, 23-27 kDa. These bands correspond respectively to non-, singly- and doubly-glycosylated IFN-γ (Curling, et al., 1990). Heterogeneity is apparent in each of the classes, which probably indicates that the protein is subject to variable glycosylation and proteolytic modification.

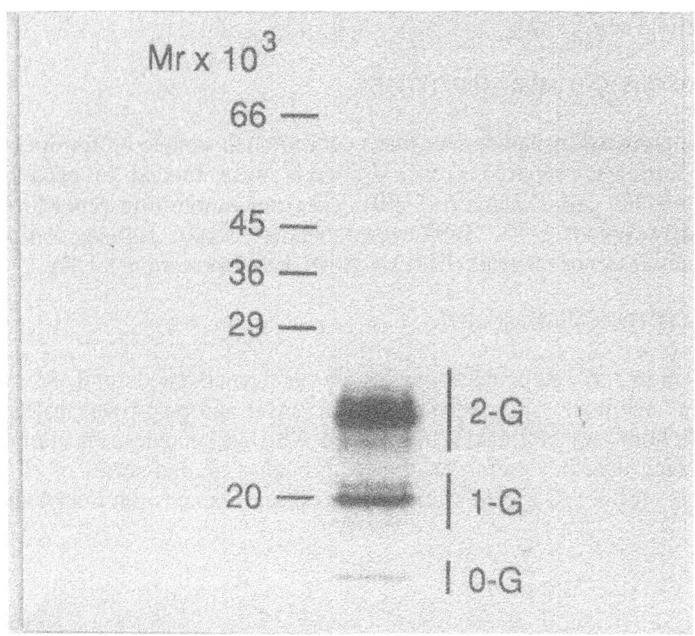

FIGURE 1. Heterogeneity of IFN-γ produced by CHO320 cells in suspension culture.

VARIATION OF INTRACELLULAR LEVELS OF INTERFERON GAMMA IN CHO320 CELLS

Fig. 2 shows that labelling of cryosections of CHO320 cells with R1PA gave a granular cytoplasmic labelling, characteristic of an endoplasmic reticulum label. All of the cells were stained but there was great variation in the intensity of staining between the cells. This was not due to uneven antibody penetration, since staining of control sections with other antibodies gave uniform intensity patterns.

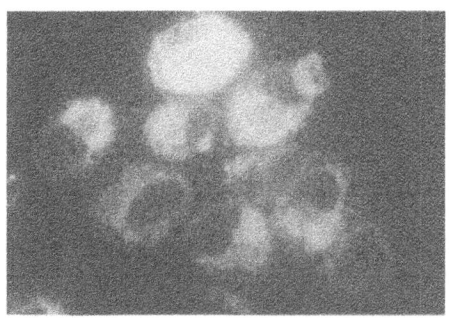

FIGURE 2. Heterogeneity of intracellular content of IFN-γ.

STRUCTURE OF THE AGGREGATES

Sections through the aggregates revealed that they consisted of a layer of live cells on a central core of dead cell material. (Fig. 3A). The layer of live cells can several cells deep. Between the live cells, specific junctions can be seen (Fig. 3B). These junctions could either be desmosomes or adherens type junctions as judged by their appearance under the EM.

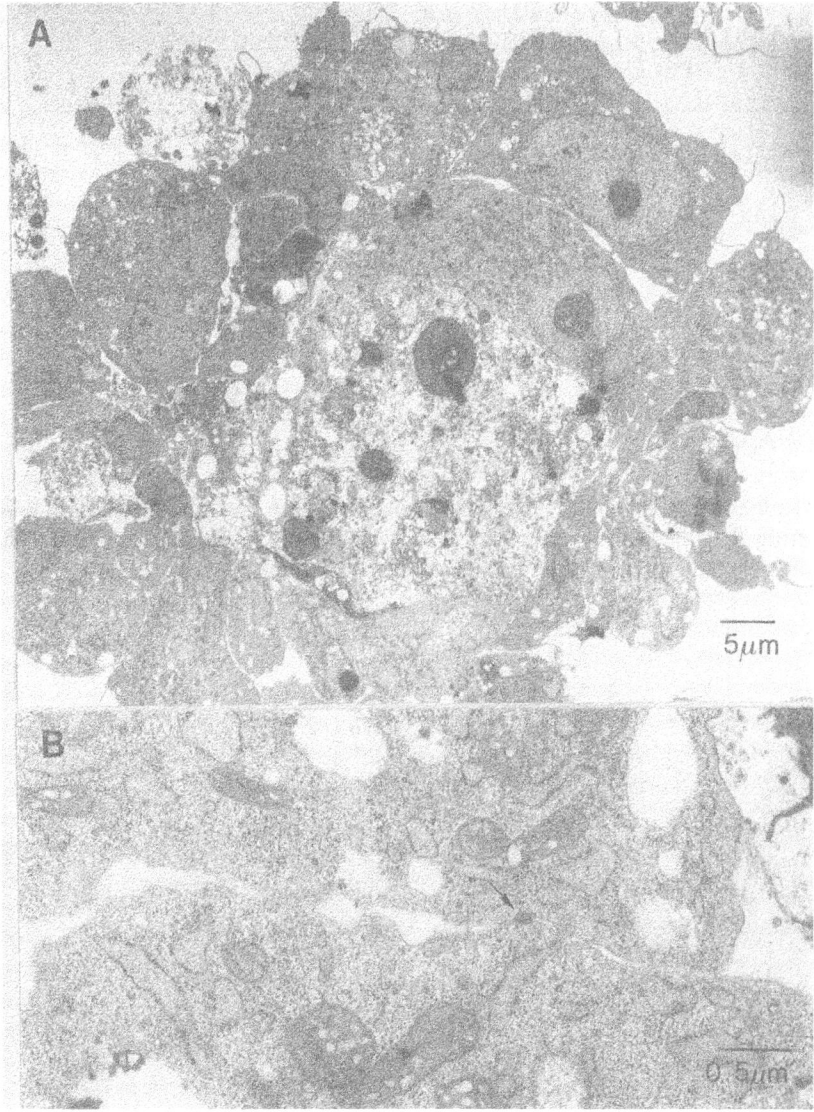

FIGURE 3. Electron micrograph to show the structure of a cell aggregate, a. Specific junctions can be seen between live cells, b.

NATURE OF THE JUNCTIONS

The junctions are unlikely to be desmosomes since desmoplakins I/II and desmoglein (major components of desmosomes), were not detected in immunoblotting although positive reactions were detected on the control cell lines LLCPK1, and MDCK, (Table 1).

TABLE 1. Occurrence of Junctional Components

Cell Line	Desmosomal		Adherens	
	DPI/II	DG1	Vinculin	Cadherins
MDCK	+	+	+	+
LLCPK1	+	+	+	+
DUK	-	-	+	+
A$_2$H	-		+	+
CHO-K1	-		+	+
CHO320	-	-	+	+

Vinculin and cadherins (components of adherens junctions) were present in all the cell lines tested, indicating that the junctions in the CHO320 cells are of the adherens type. The relative amounts of cadherins present in the different CHO cell lines was variable (Fg. 4).

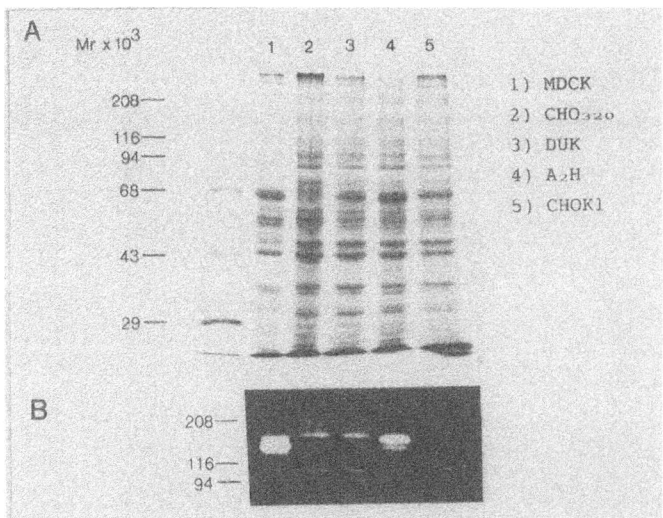

FIGURE 4. A. Cell homogenates were analysed by 8% SDS PAGE followed by Coomassie blue. B. Western blot of cell homogenates probed with pan-cadherin antibody.

INHIBITION OF ADHERENS JUNCTION FORMATION

The degree of aggregation was dramatically decreased upon treatment with the H_2O_2/vanadate, with the cells remaining viable during this time (Fig. 5). The degree of aggregation also decreased in the control culture , but was still at 30% after 24 hours. This indicates that adherens junctions are involved with holding the aggregates together, as the same treatment does not affect desmosomes.

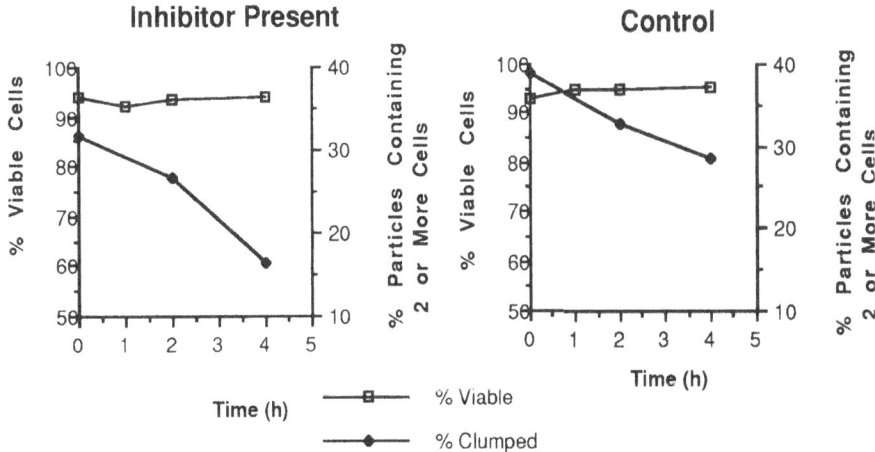

Figure 5. The effect of inhibition of adherens junctions.

TIME COURSE OF AGGREGATION

Initially the aggregation is independent of the viability of the cells, (Fig. 6). This corresponds to the lag phase of the culture. When the cells enter the exponential phase, aggregation reaches its maximal value and then declines as the cells begin to die. The phase of exponential growth is also the point of highest IFN-γ production. The dissintegration of the aggregates seems to lag slightly behind cell death. This lag time could represent the turn-over time of the proteins involved in the cell-cell interactions.

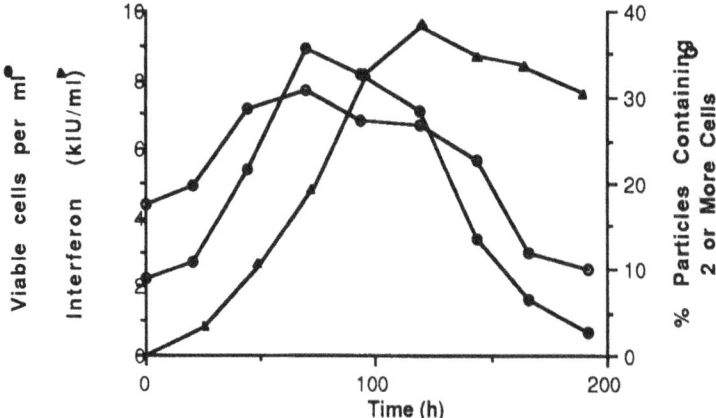

Figure 6. The relationship between cell viability, aggregation and interferon-gamma production within a batch culture.

IS CELL PROLIFERATION REQUIRED FOR AGGREGATION TO OCCUR?

In order to inhibit cell division cells were transfered to L-glutamine free medium. Since L-glutamine is the main donor of amino groups, nucleotide synthesis is arrested and hence so is DNA synthesis. The cells do, however, produce enough L-glutamine of their own for normal cell metabolism. The inhibition of cell division was almost 100% without loss of cell viability (Fig. 7a). The aggregation was not affected by the lack of cell division (Fig. 7b).

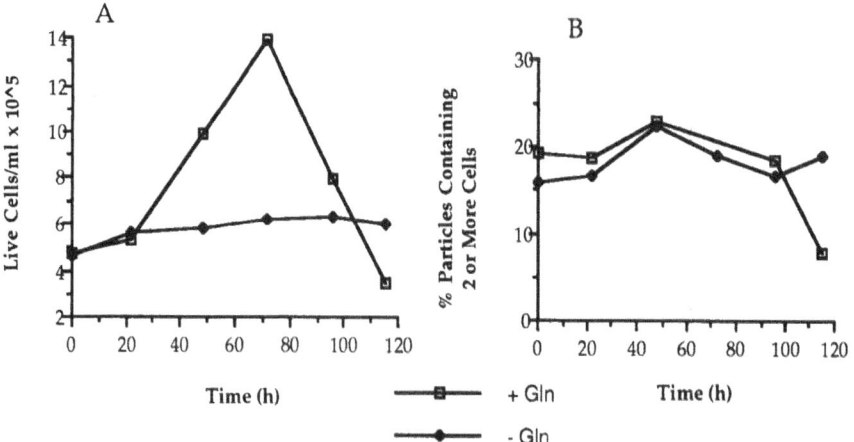

Figure 7. Cells are arrested by placing them in L-glutamine-free medium,A. The aggregation of the cells was not affected by this treatment,B.

THE EFFECT OF THE RATE OF AGITATION OF THE CULTURE ON AGGREGATION

Fig. 8 shows that the lower the shake speed, the greater the proportion of larger aggregates.

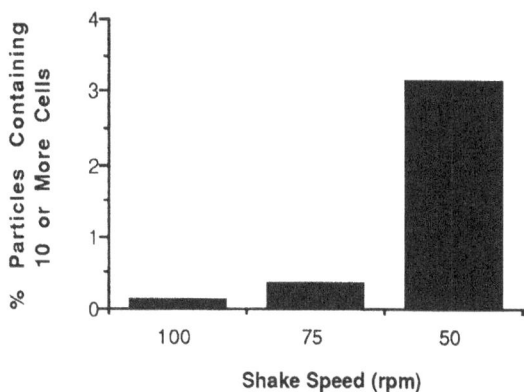

Figure 8. The effect of shake speed on the aggregation of the cells. The values represent the average percentage of particles containing ten or more cells in all the samples taken.

Discusssion

The heterogeneous nature of recombinant human interferon gamma produced by the CHO320 cell line could be as a result of the heterogeneity that occurs within the suspension culture. The cells have a tendency to form aggregates, as well as growing as single cells, which will cause variability in access to oxygen and nutrients. This may affect the production and proccessing of the recombinant protein. The aggregation could also be of an advantage, to help protect the cells from the shear forces that occur in suspension culture.

The aggregates consist of a dead mass surrounded by a layer of living cells, between which specific junctions can be seen. These junctions are most likely to be of the adherens type, due to the presence of vinculin and cadherins, though localization studies of these components has yet to be completed. Inhibition of adherens junction formation caused the aggregates to disintegrate, again implying the involvement of adherens junctions. Though, what is true for the CHO320 cell line may not be true for other recombinant CHO cell lines, since the expression of proteins may well vary between them. This is highlighted by the variable expression of the cadherins. The reason for this variability could be due to the different degrees of manipulation the different cell lines have undergone. The insertion of foreign fragments of DNA into the genome is very likely to alter the expression of neighbouring genes.

The formation of the aggregates can be related to the growth phase of a batch culture but is independent of cell division. One possible explanation, is that during the lag phase of a culture, many of the cells will be in G_0. The genes encoding the proteins involved with cell-cell interactions are not switched on in this phase, along with all the genes involved with cell division. As the cells begin to enter G_1, and go on to divide, these genes become activated, and agregates are formed. As the cells grown in the L-glutamine-free medium are

18

arrested in G_1, the genes are still being transcribed and the proteins necessary for cell-cell interactions are being produced; therefore the cells remain aggregated. The degree of aggregation is affected by the rate of agitation of the cultures as this will relate to the shear forces within the culture. The higher the rate of agitation the greater the shear forces which leads to smaller clumps. One interesting observation was that the proportion of single cells present in the cultures remained constant. This could be interpreted as evidence for the presence of two populations of cells within the culture. The aggregation of the cells is also likely to be affected by many other factors including medium composition.

Whether the elimination of aggregation will increase the fidelity and yield of the recombinant protein has still to be investigated.

Acknowledgements

We would like to thank the following for generous gifts of antibodies, cells and proteins: Anthony I. Magee at the National Institute for Medical Research, Mill Hill, UK, for the antibodies to DGi and DPI/II; Benjamin Geiger at the Weizmann Institute of Science, Israel, for the pan-cadherin antibody; Paul Hayter and Elizabeth Curling, University of Kent, for cell samples and purified interferon gamma.

S.R.C. is funded by a SERC-CASE award in association with The Wellcome Foundation Ltd, Beckenham, Kent, UK.

References

Curling, E. M. A., Hayter, P. M., Baines, A. J., Bull, A. T., Gull, K., Strange, P. G. and Jenkins, N. (1990) Recombinant human interferon-gamma: differences in glycosylation and processing lead to heterogeneity in batch culture. *Biochem. J.*, **272**, 333-337

Geiger, B., Volberg, T., Ginsberg, D., Bitzur, S. and Sabanay, I. (1990) Broad spectrum pan-cadherin antibodies, reactive with the C-terminal 24 amino acid residues of N-cadherin. *J. Cell Sci.*, **97**, 607-614

Laemmli, U. K. (1970) Cleavage of structural proteins during the assembly of the head of bacteriophage T4. *Nature*, **227**, 680-685

Volberg, T., Zick, Y., Dror, R., Sabany, I., Gilon, C., Levitzki, A. and Geiger, B. (1992) The effect of tyrosine specific protein phosphorylation on assembly of adherens-type junctions. *EMBO J*, **11**, 1733-1742

Woods, A., Sherwin, T., Sasse, R., McRae, T. H., Baines, A. J. and Gull, K. (1989) Definition of individual components within the cytoskeleton of Trypanosoma brucei by a library of monoclonal antibodies. *J. Cell Sci.*, **93**, 491-500

CHARACTERIZATION OF NEWLY ESTABLISHED CELLS WHICH PROVIDE AN ANIMAL MODEL FOR SPONTANEOUS METASTASIS.

[1]Takahiro Nomura, [2]Kazuo Ryoyama, [1]Gensaku Okada, [3]Sadaya Matano, [4]Haruhiko Tokuyama, [5]Isao Hori, [3]Shinobu Nakamura and [1]Tadanori Kameyama.
[1]Dept. Mol. Biol., [2]Dept. Exp. Ther., [4]Dept. Mol. Immunol.
Cancer Research Institute,
and [3]Dept. Third Intern. Med.
School of Medicine
Kanazawa Univ.
Kanazawa 920
and [5]Dept. Biol.
Kanazawa Medical University
Uchinada, Ishikawa 920-02
Japan

ABSTRACT. A novel cell line, r/mHM-SFME-1 (r/mHM-1) was established from ras/myc SFME cells transformed by human c-Ha-ras and mouse c-myc genes (SFME cells have been established from a Balb/c mouse embryo in a serum-free culture condition). This cell line was derived from a pulmonary metastasis developed in a Balb/c mouse which had been transplanted sub-cutaneously with pSV2-neo introduced ras/myc SFME cells. The r/mHM-1 cells had an ability to spontaneously metastasize into the lungs of syngeneic mice when injected subcutaneously, and survival of the mice which received the r/mHM-1 cells was significantly shorter than ones with ras/myc SFME cells. The r/mHM-1 cells grew slowly in vitro than their parental ras/myc SFME ones did, and produced dispersed colonies in agar whereas their parental ones produced packed ones. A urokinase type plasminogen activator activity was detected in the culture fluid in which the r/mHM-1 cells were cultured for 2 days, whereas the activity was not detected in those from the parent ones.

INTRODUCTION.

Our understanding about cancer has greatly progressed by the introduction of genetic engineering. One of the urgent requirements to control cancer is undoubtedly to accumulate knowledge about the mechanisms of metastasis and to control metastasis. Thus, this requirement needs development of new methods to detect micro-metastasis since it is apparent that micro-metastases in tissues are hardly detectable histologically. It seems to be much easier and exact if introduced genes of one species in tissues of another animal, that is,

19

S. Kaminogawa et al. (eds.), Animal Cell Technology: Basic & Applied Aspects, Vol. 5, 19–24.
© 1993 *Kluwer Academic Publishers.*

human genes in mouse tissues, are applied for detection of metastasis. In this context, the NIH3T3 cell is a good candidate and have been widely used since they are easily transformed by several human oncogenes (1, 2). However, this cell has no syngeneic mice so that micro-metastasis of the cell should be examined in an immuno deficient nude mice. This seems not to exactly represent the metastasis of human cancers and is also expensive.

We have recently established a novel cell line, r/mHM-SFME-1 (r/mHM-1) from ras/myc SFME cells transformed by human c-Ha-ras and mouse c-myc genes(3, 4, 5). The both cells are taken in syngeneic Balb/c mice and develop solid tumors, but only the r/mHM-1 cells spontaneously metastasize into the lung when injected subcutaneously into the back of mice. Thus, r/mHM-1 cells appear to be useful in analyzing mechanisms of metastasis in immuno competent host and developing new modalities to control micro-metastasis. In this report, we exhibit some of characters on the r/mHM-1 cell in comparison with its parental ras/myc SFME ones.

MATERIALS AND METHODS.

Cells and an animal: Transformed ras/myc SFME cells were provided by S. Shirahata of Kyushyu University courtesy of D. Barnes, Oregon State University. The cell was plated in plastic culture wares pre-coated with fibronectin and cultured at 37°C in a humidified CO_2 (5%) atmosphere by passaging every 6 days. The serum-free culture medium was F/D medium (1:1 mixture of Dulbecco's modified Eagle's medium and Ham's nutrient mixture F-12) supplemented with 10 nM sodium selenite, 10 μg/ml insulin (bovine), 10 μg/ml transferrin (bovine), 10 μg/ml high density lipoprotein (human) and 50 ng/ml epidermal growth factor (EGF, mouse submaxillary) (3, 6). The culture medium supplemented with FCS was the F/D medium containing only sodium selenite. The media and the supplements were prepared in pyrogen free ultra pure water. Six weeks old male and female Balb/c mice obtained from our own colony were used for animal experiments.

Estimation of cell numbers : Two methods were applied to estimate cell numbers. Firstly, cells were detached from plastic wares with trypsin and their numbers were counted with an electric particle counter. Secondly, when cells were cultured in 96-well microplates, the cell numbers were estimated by means of reduction of 3-(4,5-dimethylthioazol-2-yl)-2,5-diphenyltetrazolium bromide (MTT)(7).

Southern and Northern blotting : DNA was extracted from cultured cells or mouse tissues by proteinase-K followed by the phenol-chloroform treatment (8). After complete digestion with Ecoh-RI, 10 μg of DNA fragments were fractionated by electrophoresis and transfered on a Hybond N membrane (Amersham), and then hybridized with a probe (919 bp exon-1 fragments from pUCEJ6.6) (9, 10). Total cellular RNA was extracted using guanidium thiocyanate-phenol-chloroform (11). RNA fractionated by electrophoresis was transferred to a Hybond N membrane. The membrane was dried and UV-irradiated, and then hybridized with the same probe as cited above.

Tumorigenicity and histological examination : Six weeks old male and female mice received one million of r/mHM-1 or ras/myc SFME cells subcutaneously on their back. Lungs from three mice were examined histologically at 3 weeks after the inoculation The rest of the mice were offered to test their survival, and the test was terminated at 100 days after the inoculation. The subcutaneous solid tumors and their metastases in various organs, especially the lungs were examined histologically.

Soft agar method : A 0.4-0.6% double layered soft agar method with F/D medium supplemented 10% FCS was applied. Ten thousands cells were inoculated with 1 ml of 0.4% agar in the medium into ø 3 cm dish in which 3 ml of 0.6% agar in the medium was pre-plated and cultured as usual.

Detection of plasminogen activator : The radial caseinolysis method (12) was applied to detect activity of plasminogen activator. Briefly, 0.1M tris-HCl(pH 8.0) buffer solution containing 1.1% agarose, purified bovine plasminogen (0.04 C.U./ml) and 0.6% non-fat dry milk was poured and gelled on slide glasses. After gelling (1.5 mm thick) a small well (ø 3 mm) was made on the gel. The supernatant (10µl) from r/mHM-1 and ras/myc SFME cell culture was put into the well and the glasses were incubated for 48 hrs at 37°C. The plasminogen activator activity was measured by the lysis of non-fat dry milk. The amounts of the activities were estimated from lysis activities of purified high molecular weight urokinase.

RESULTS AND DISCUSSION.

Establishment of r/mHM-1 cell : The procedure for in vivo selection and in vitro establishment of r/mHM-1 cells are shown in [Plate 1]. A day after plating ras/myc SFME cells were transfected with pSV2-neo genes by the calcium-phosphate method (5). The cells were then cultured for 2 months with G418 (400µg/ml) in the serum-supplemented culture medium. After re-culturing in the serum-free culture medium, the cells were injected subcutaneously into the back of Balb/c mice. All of the mice which had received

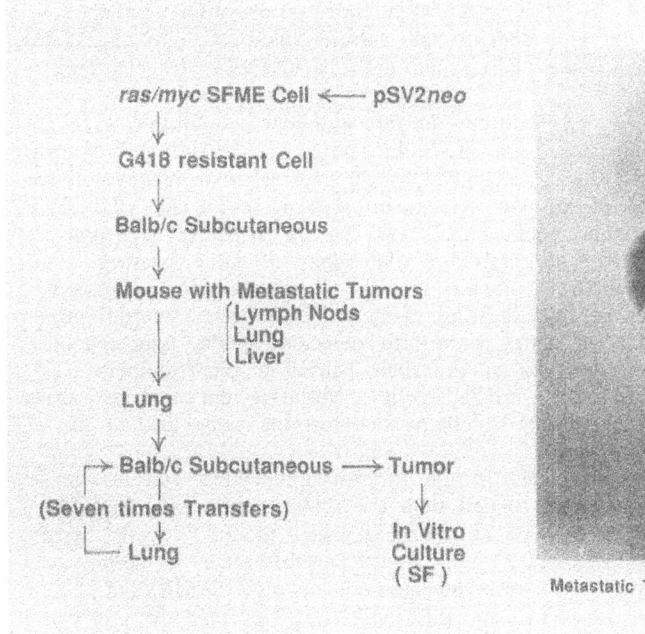
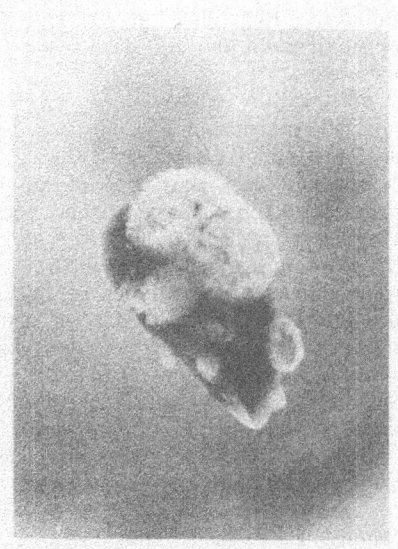

Metastatic Tumors in Lung at 3rd Transfer

[Plate 1]

each of four independently transfected cells developed solid tumors at the injected sites within 2 months. We found a mouse which had metastases in subaxillary and submaxillary lymph nodes. Clearly visible metastasis was also observable in the lung and liver. Freshly excised metastases were transplanted subcutaneously into the back of Balb/c mice . The metastases grew as subcutaneous masses and metastasized into the lungs. Such lung metastases were serially transplanted into the back of Balb/c mice. The subcutaneous tumor developed from the 7th lung metastases transplant were sterilely excised and cultured in the serum-free culture medium. The tumor cell grew in vitro and produced colonies when was plated in soft agar. Each colony was cultured in the serum-free culture medium and one of them was designated as r/mHM-SFME-1 (r/mHM-1). The r/mHM-1 cells no longer were resistant against G418.

Next, we examined whether exogeneously inserted human ras genes were detectable in r/mHM-1 cells. Eco-RI fragments of DNA from r/mHM-1 and ras/myc SFME cells were analyzed by Southern blotting and showed eight bands which were comparable between the both cells.

<u>Characterization of r/mHM-1 cell</u>: r/mHM-1 cells grew both in the serum-free and in serum-supplemented culture medium as their parental ras/myc SFME cells did. r/mHM-1 cells, however, exhibited slower growth than did their parental cells; the former grew continuously until day 9 in the serum-supplemented culture condition whereas the later reached maximum at day 4. The same tendency was also observed in the serum-free culture condition. Growth profiles of the two cells were quite different; r/mHM-1 cell tended to aggregate each other in the culture whereas ras/myc SFME cell did not. The aggregates were observed in both serum-free and serum-supplemented cultures. Regardless of the cell density and culture situations, the aggregates were observed by 3-4 days after plating and did not disappear by changing the media.

One million of r/mHM-1 cells cultured in the serum-free condition were injected subcutaneously into the back of Balb/c mice. r/mHM-1 cell-bearing mice started to die at day 18 and succumbed by 30 days whereas ras/myc SFME cell-bearing ones started to die at day 41 and survived much longer. All dead mice had solid tumors on their backs, and sexes did not affect the survivals. Metastases were histologically observable in the lungs of mice bearing r/mHM-1 cells at the fourth week of the inoculation whereas no metastases were detected in the lungs of mice bearing ras/myc SFME cell. In addition, the introduced human c-Ha-ras genes were also detected in the lung metastases at 28 days after the inoculation and their pattern of the Southern blotting was the same as that of r/mHM-1 cells. Northern blot analysis shows that expression of the ras in r/mHM-1 cells was almost the same as that of their parental cells.

Ten thousands of cells suspended in the serum-supplemented culture medium containing 0.4% agar were plated onto the 0.6% bottom agar in ø 3 cm dish and cultured. Colonies over 75 µm in diameter (about 70 cells) were counted at 3 weeks after the plating. No significant difference in colony formation efficiency in agar was observed between ras/myc SFME and r/mHM-1 cells. However, a significant difference was observed in morphology of the colonies; that is, the colonies of ras/myc SFME cells were spherical whereas those of r/mHM-1 ones were dispersed. This difference was further confirmed by an electron microscopy. The colonies of ras/myc SFME cells had smooth surfaces and were made of packed cells with some necrotic portions. On the other hand, cells in the colonies of r/mHM-1 cells were loosely associated and scattered. The cells in the former colonies tightly adhered each other whereas ones in the later colonies loosely attached. r/mHM-1 and ras/myc SFME cells were cultured for 2 days in the serumfree culture media. The culture fluids were offered for the detection of urokinase type plasminogen activator (U-PA) activity. Significant activity was detected in the culture fluid from the r/mHM-1 cells whereas it was negligible in one from ras/myc SFME cells.

There is a report that transfection of neo^r gene into rat mammary cancer cells does not potentiate their metastatic ability(13). On the other hand, Yoshikura et al. reported the alteration of cell surface carbohydrates by the neo^r gene transfection (14, 15). There is a possibility that the adhesion ability of ras/myc SFME cells may be altered by the introduction of neo^r gene. This could explain the behavior of the r/mHM-1 cells in soft agar and also in vivo. However, at present this is unknown and needs further experimentation. Meanwhile, an in vitro invasion assay using extra cellular matrix matrigel and type IV collagene is now on going.

Finally, the r/mHM-1 cell appears to provide a useful tool to analyze metastatic processes in immuno competent animals and to develop new modalities to control micro-metastasis.

ACKNOWLEDGEMENTS   We are grateful to Dr. Sanetaka Shirahata (Associate Professor, Graduate School, Genetic Resources Technology, Kyushyu University, Fukuoka) and Dr. David W. Barnes (Professor, Dept. Biochemistry and Biophysics, Environmental Health Science Center, Oregon State University, Corvallis, OR, U.S.A.) for the gift of SFME and ras/myc SFME cells and lots of useful suggestions on the cells and their culture methods.

This work was supported in part by a Grant-in-Aid from the Ministry of Education, Science and Culture of Japan (No.03671181).

REFERENCES
1) Shih, C. and Weinberg, R. A. 'Isolation of a transforming sequence from a human bladder carcinoma cell line', Cell, 29, 161-169, (1982).
2) Sakamoto, H., Mori, M., Taira, M., Yoshida, T., Matsukawa, S., Shimizu, K., Sekiguchi, M., Terada, M. and Sugimura, T. 'Transforming gene from human stomach cancers and a noncancerous portion of stomach mucosa', Proc. Natl. Acad. Sci. USA, 83, 3997-4001, (1986).
3) Loo, D. T., Fuquay, J. I., Rawson, C. L. and Barnes, D. W. 'Extended culture of mouse embryo cells without senescence: Inhibition by serum', Science, 236, 200-202 (1987).
4) Rawson, C. L., Shirahata, S., Collodi, P., Natsuno, T. and Barnes, D. W. 'Oncogene transformation frequency of nonsenescent SFME cells is increased by c-myc', Oncogene, 6, 487-489 (1991).
5) Shirahata, S., Rawson, C. L., Loo, D. T., Chang, Y. J. and Barnes, D. W. 'Ras and neu oncogenes reverse serum inhibition and epidermal growth factor dependence of serum-free mouse embryo cells', J. Celluar Physiol., 144, 69-76, (1990).
6) Loo, D. T., Rawson, C. L., Helmrich, A. and Barnes, D. W. 'Serum-free mouse embryo cells: Growth response in vitro', J. Cellular Physiol., 139, 484-491, (1989).
7) Carmichael, J., Degraff, W. G., Gazdar, A. F., Minna, J. D. and Mitchell, J. B. 'Evaluation of a tetrazolium-based semiautomated colorimetric assay: Assessment of chemosensitivity testing', Cancer Research., 47, 936-942 (1987).

24

8) Sambrook, J., Fritsh, E. F. and Maniatis, T. "Molecular cloning, A laboratory manual" Second Edition. pp. 7.26-7.52 (1989) Cold Spring Harbor Laboratory Press, New York.

9) Capon, D. J., Chen, E. Y., Levinson, A. D., Seeburg, P. H. and Goeddel, D. V. 'Complete nucleotide sequences of the T24 human bladder carcinoma oncogene and its normal homologue', Nature, 302, 33-37 (1983).

10) Feinberg, A. P. and Vogelstein, B. 'Addendum of "A technique for radiolabeling DNA restriction endonuclease fragments to high specific activity"', Analytical Biochem., 137, 266-267 (1984).

11) Chomczynski, P. and Sacchi, N. 'Single-step method of RNA isolation by acid guanidium thiocyanate-phenol-chloroform extraction', Analytical Biochem., 162, 156-159 (1987).

12) Saksela, O. 'Radial caseinolysis in agarose: A simple method for detection of plasminogen activator in the presence of inhibitory substances and serum', Analyt. Biochem., 111, 276-282, (1981).

13) Ichikawa, T., Kyprianou N. and Isaacs, J. T. 'Genetic instability and the acquisition of metastatic ability by rat mammary cancer cells following v-H-ras oncogene transfection', Cancer Res., 50, 6349-6357, (1990).

14) Yoshikura, H. ' Suppression of focus formation by bovine papilloma virus-transformed cells by contact with non-transformed cells: Involvement of sugar(s) and phosphorylation', Inter. J. Cancer, 44, 885-891, (1989).

15) Naito, A., Kitamura, Y., Sudoh, K. and Yoshikura, H. 'In vitro "progression" of bovine papillomavirus-transformed cells: Loss of contact sensitivity after multiple rounds of selection', Inter. J. Cancer, 48, 889-894, (1991).

BOVINE MAMMARY EPITHELIAL CELLS: MORPHOLOGY, PROTEIN SYNTHESIS AND GENE TRANSFECTION

T. MATSUDA, J. AHN, Y. MIZUNO,
T. ADACHI, N. AOKI and R. NAKAMURA
Department of Food Science and Technology,
School of Agricultural Sciences, Nagoya University,
Chikusa-ku, Nagoya 464-01, Japan.

ABSTRACT. Bovine mammary epithelial cells (BMEC) were isolated as "organoids" from a mammary gland of a Holstein cow by collagenase/pronase dissociation and low speed centrifugation. When BMEC were cultured on porous membrane or porous plastic sheet coated with extracellular matrixes such as "Matrigel", they assembled together and showed dramatic morphological changes, i.e. construction of nodule-like, alveolar and tubular structures. When BMEC were cultured in the presence of ^{35}S-Met, several radio-labeled proteins corresponding to bovine milk proteins were detected in the BMEC culture supernatant. Synthesis and secretion of the protein corresponding to $\alpha s1$-casein were remarkably increased by prolactin, hydrocortisone and insulin. Furthermore, $\alpha s1$-, β- and κ-caseins were immunologically detected in the culture supernatant by immuno-blotting analysis. As a preliminary experiment for gene manipulation of BMEC, a model gene containing LacZ was introduced into BMEC by the calcium phosphate precipitation method. β-Galactosidase activity was detected in some cells by activity-staining with X-gal, though transfection efficiency was not very high.

1. Introduction

Mammary epithelial cells synthesize a large amount of protein, fat and sugar (lactose), and such a milk synthesis, as well as cell division and differentiation are regulated by a variety of factors including peptide and steroid hormones, cell-substratum, and cell-cell interactions [1,2]. The molecular mechanism of developmentally and hormonally regulated expression of milk protein genes has been investigated mostly on primary culture and some cell lines of murine mammary epithelial cells [3-5]. The expression of casein genes has been reported to be stimulated by the synergistic actions of insulin, glucocorticoid and prolactin [6].

It is now possible to change milk composition or to produce entirely foreign proteins by genetic engineering such as transgenic techniques [7]. Therefore, mammary epithelial cells or the mammary gland might have a potential ability as an efficient bio-reactor for the production of valuable proteins and peptides. In the present study, bovine mammary epithelial cells were prepared from a lactating cow, and their morphogenesis, milk protein synthesis and gene transfection were investigated as a first step for the genetic modification of milk proteins and the synthesis of foreign proteins by bovine mammary epithelial cells.

25

S. Kaminogawa et al. (eds.), Animal Cell Technology: Basic & Applied Aspects, Vol. 5, 25–31.

2. Materials and Methods

2.1. CELL PREPARATION AND CULTURE

Bovine mammary epithelial cells (BMEC) were isolated from the mammary gland of a lactating Holstein cow obtained from a local slaughterhouse. The mammary tissue was minced with a surgical knife, and the tissue pieces were dissociated in Hank's balanced salt solution (BSS) containing 0.05% collagenase (Type I, SIGMA) with gentle shaking at 37°C for 14h. After filtration with a stainlesssteel mesh to remove undissociated tissues and debris, the cells were collected by low speed centrifugation at 80xg (700 rpm) for 3 min. The cells were further dissociated in Hank's BSS containing 0.05% pronase (SIGMA) with gentle shaking at 37°C for 30 min. The cells were collected by centrifugation at 80xg for 1 min, and washed three times with Hank's BSS.

The BMEC preparation was suspended in Dulbecco's modified Eagle's medium (DMEM) containing 10% fetal bovine serum, penicillin at 10 unit/ml, streptomycin at 50 µg/ml and fungizon (GIBCO) at 50 ng/ml (basal medium), and cultured on plastic dishes at 37°C in 95% air and 5% CO_2. The cells were cultured also on collagen-coated microporous membrane, Transwell (Coster) and on reconstituted basement membrane from EHS (Matrigel, Collaborative Research Incorporated) [8]. For hormonal induction of milk protein synthesis and extracellular matrix-dependent morphological changes, the cells were cultured in the presence of insulin (5 µg/ml;SIGMA), hydrocortisone(1µg/ml; SIGMA) and ovine prolactin (4µg/ml;SIGMA) as indicated.

2.2. RADIOISOTOPIC LABELING AND PROTEIN ANALYSES

Cultures in 24-well plates were washed once with methionine-free MEM and incubated in 200 µl of methionine-free MEM containing 0.5 mCi (18.5Mbq)/ml [^{35}S]methionine at 37°C for 4 h under 5% CO_2 and 95% air. The radio-labeled proteins secreted into the culture medium were collected by trichloroacetic acid precipitation.

The labeled proteins were separated by sodium dodecyl sulfate (SDS) polyacrylamide gel electrophoresis according to Laemmli [9], and the gel sheets were treated with EN^{3}HANCE, dried and exposed to X-ray film at -80°C.

2.3. IMMUNOBLOTTING ANALYSES OF MILK PROTEIN SYNTHESES

The BMEC culture supernatants were dialyzed against distilled water. The precipitated proteins including caseins were collected by centrifugation and used for immunoblotting analysis. The proteins were separated by the SDS-polyacrylamide gel electrophoresis and transferred electrophoretically onto polyvinylidene difluoride (PVDF) membrane (Immobilon-P, Millipore) [10]. The membrane was blocked with 3% bovine serum albumin (Fraction V, SIGMA) in phosphate buffered saline (PBS), and then incubated with rabbit antisera raised against αs1-, β- and κ-caseins. After washing with PBS containing 0.05% Tween-20, the membrane was incubated with peroxidase-coupled anti-rabbit IgG (Cappel). The protein bands reactive to the specific antibody were visualized by activity staining for peroxidase using 4-chloro-1-naphthol (Bio-Rad Laboratories).

2.4. GENE TRANSFECTION

BMEC cultured on plastic dishes (6 cm in diameter) were transfected with 1µg of DNA per dish using calcium phosphate precipitation procedure described by Chen and Okayama [11]. The DNA used for transfection was kindly provided by Dr. M. Tanaka (National Institute of Neuroscience) through Dr. Y. Kitagawa (Nagoya University). It is a model gene pENL containing promoter of human elongation factor Iα (EF-Iα) and LacZ gene with nuclear translocation signal [12]. After the cells were fixed with glutaraldehyde, the expression of LacZ gene was visualized by β-galactosidase activity staining using X-gal.

3. Results and Discussion

3.1. PRIMARY CULTURE OF BMEC

BMEC were effectively prepared from fresh mammary glands by the collagenase/pronase dissociation method and low speed centrifugation. The phase-contrast microscopic observation of the cells is shown in Fig. 1. Right after the preparation the cells did not appear singly but as masses of several cells, so-called "organoids" (Fig. 1A). About 4 to 5 days were required for the cells to start adhering, spreading and growing on plastic dishes (Fig. 1B). The cells grew well in the basal medium, and looked like typical epithelial cells (Fig.1C). The initial preparation was contaminated with fibroblast-like cells, though removed effectively by low speed centrifugation during several passages.

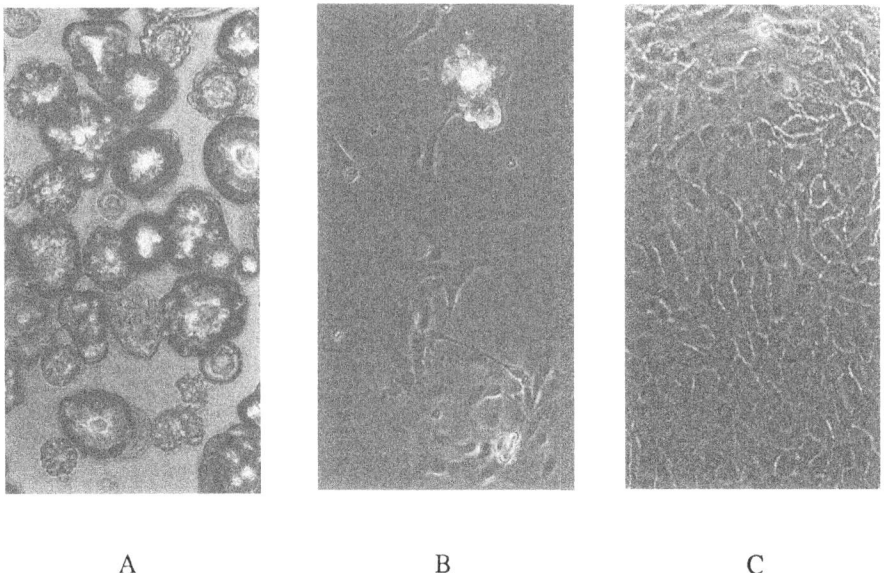

A   B   C

Figure 1. Bovine mammary epithelial cells cultured on plastic dishes

3.2 CULTURE CONDITION-DEPENDENT MOLPHOGENESIS

Since homogeneous BMEC were obtained, the cells were cultured under various different conditions and their morphological changes were observed. When BMEC were cultured on collagen-coated microporous membrane, Transwell, they formed continuous monolayer on the membrane, and the cells had more cubic form than the ones on plastic dishes did (data not shown).

BMEC showed dramatic morphological changes when cultured on the reconstituted basement membrane, Matrigel. As shown in Fig. 2A, the cells cultured on Matrigel-coated microporous plastic sheet (Cell-culture insert, Falcon) assembled together and constructed nodule-like, alveolar and tubular structures. Furthermore, an alternative morphological change was observed when the cells were cultured on thick gel layer of Matrigel formed on the microporous membrane. The cells formed domes with many dendritic structures, which grew into the gel matrixes (Fig. 2B). Thus, BMEC showed various morphogenesis depending on their culture conditions such as extracellular matrixes.

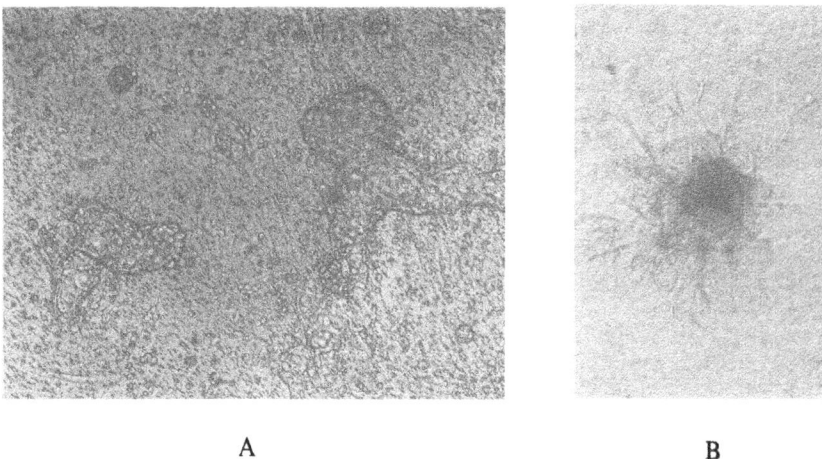

A   B

Figure 2. Morphogenesis of BMEC cultured on Matrigel-coated porous plastic sheet (A) and on a thick Matrigel layer (B)

3.3. PROTEIN SYNTHESIS AND SECRETION

Proteins synthesized and secreted by cultured BMEC were analyzed by radio-tracer experiments. The cells were cultured with three lactogenic hormones on plastic dishes, Matrigel-coated plastic dishes or thick Matrigel layer. Then, the proteins synthesized were radio-labeled and analyzed by gel electrophoresis. The cells cultured on plastic dishes , as well as those on Matrigel, secreted many proteins into the culture medium (data not shown). Although the proteins secreted by the plastic dish culture were slightly less than those of the cultures with Matrigel, BMEC were found to synthesize and secrete various proteins even in the absence of the reconstituted basement membrane. Hence, the plastic dish cultures were used for the following experiments.

To investigate the effect of lactogenic hormones on the BMEC protein synthesis, the cells were cultured with or without the three hormones, prolactin, hydrocortisone and insulin, and proteins secreted into the culture medium were analyzed. Fig. 3 shows the autoradiogram of proteins synthesized and secreted by the cultured BMEC. The synthesis of several proteins was induced or enhanced by the presence of both prolactin and hydrocortisone. Especially, the synthesis of about 32kDa-protein was remarkably induced by the two lactogenic hormones, though no induction was observed by adding either one alone.

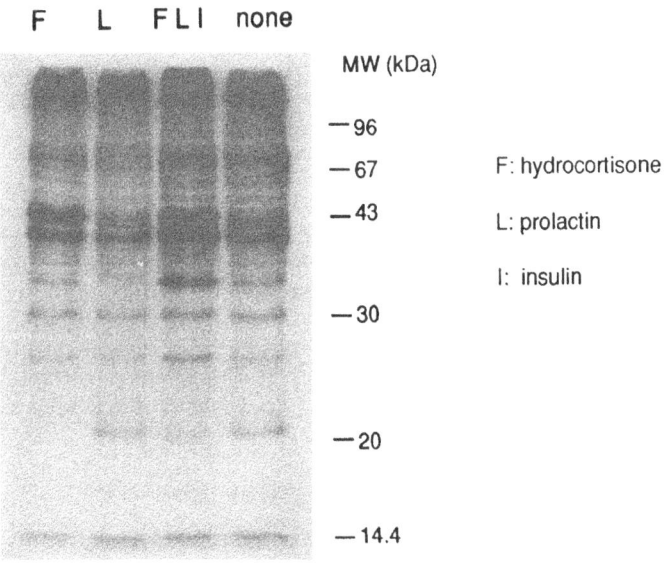

Figure 3. Protein synthesis and secretion by BMEC cultured in the presence of hydrocortisone, prolactin and insulin.

Since the hormone induced 32kDa-protein showed electrophoretic mobility similar to a major milk protein, αs1-casein, the secretion of milk proteins by the cultured BMEC was examined by immunoblotting using specific antibodies. As shown in Fig. 4, not only αs1-casein but also β- and κ-caseins were clearly detected in the BMEC culture supernatant. The major milk whey proteins such as β-lactoglobulin and α-lactalbumin, however, were not detected. Thus, the major milk proteins, caseins, were surely synthesized and secreted by the cultured BMEC.

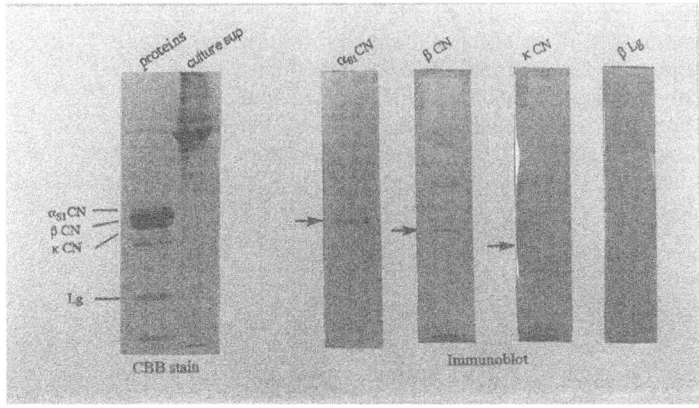

Figure 4.   Immunological detection of milk caseins synthesized and secreted by BMEC

3.4.   GENE TRANSFECTION INTO BMEC

Gene transfection into BMEC by the calcium phosphate method was done as a preliminary experiment for the expression of foreign genes in BMEC using milk protein gene promoters.   A model gene, pENL was transfected and the transfected cells were detected by the expressed β-galactosidase.   As shown in Fig. 5, β-galactosidase expressed and translocated into the nucleus was clearly observed for several cells by the activity staining.   However, the transfection efficiency of BMEC was rather lower than that of COS cells used as a positive transfection control.   Gene transfection by electroporation and preparation of milk protein gene promotors are now in progress.

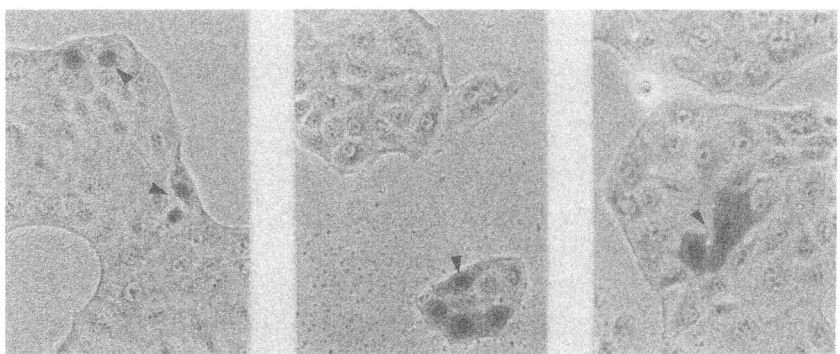

Figure 5.   Gene transfection into BMEC by the calcium phosphate method.

4. Acknowledgements

The authors gratefully acknowledge Mr. S. Tanaka (Cosmo Meat Inc.) for his kind arrangement to obtain fresh bovine mammary tissue. We are also grateful to Drs Y. Kitagawa, K. Miki (Nagoya University) and M. Tanaka (National Institute of Neuroscience) for the plasmid pENL, and to Dr. H. Otani (Sinshu University) for the anti-casein sera.

5. References

[1]   Mehta, N. M., Ganguly, N., Gangly, R. and Banerjee, M. R. (1980) 'Hormonal mammary gland of the mouse', J. Biol. Chem. 255, 4430-4434.

[2]   Li, M. L., Aggeler, J., Farson, D. A., Hatier, C., Hassell, J. and Bissell, M. J. (1987) 'Influence of a reconstituted basement membrane and its components on casein gene expression and secretion in mouse mammary epithelial cells', Proc. Natl. Acad. Sci. USA 84, 136-140.

[3]   Blum, J. L., Zeigler, M. E. and Wicha, M. S. (1987) 'Regulation of rat mammary gene expression by extracellular matrix components', Exp. Cell Res. 173, 322-340.

[4]   Yoshimura, M. and Oka, M. (1990) 'Transfection of β-casein chimeric gene and hormonal induction of its expression in primary murine mammary epithelial cells', Proc. Natl. Acad. Sci. USA 87, 3670-3674.

[5]   Schmidhauser, C., Bissell, M. J., Myers, C. A. and Casperson, G. (1990) 'Extracellular matrix and hormones transcriptionally regulate bovine β-casein 5'sequences in stably transfected mouse mammary cells', Proc. Natl. Acad. Sci. USA 87, 9118-9122.

[6]   Guyette, W. A., Matusik, R. J. and Rosen, J. M. (1979) 'Prolactin-mediated transcriptional and post-transcriptional control of casein gene expression', Cell 17, 1013-1023.

[7]   Simons, J. P., McClenaghan, M. and Clark, A. J. (1987) 'Alteration of the quality of milk by expression of sheep β-lactoglobulin in transgenic mice', Nature 328, 530-532.

[8]   Kleinman, H. K., MaGarvey, M. L., Hassel, J. R., Star, V. L., Cannon, F. B., Laurie, G. W. and Martin, G. R. (1986) 'Basement membrane complexes with biological activity', Biochemistry 225, 312-318.

[9]   Laemmli, U. K. (1970) ' Cleavage of structural proteins during the assembly of the head of bacteriophageT4', Nature 227, 680-685.

[10]   Herper, D. R., Liu, K. M. and Kangro, H. O. (1986) 'The effect of staining on the immunoreactivity of nitrocellulose bound proteins', Anal. Biochem. 157, 270-274.

[11]   Chen, C. and Okayama, H. (1987) 'High- efficiency transfomation of mammalian cells by plasmid DNA', Mol. Cell Biol. 7, 2745-2752.

[12]   Kim, D. W., Vetsuki, T., Kaziro, Y., Yamaguchi, N. and Sugano, S. (1990) 'Use of the human elongation factor 1α promoter as a versatile and efficient expression system', Gene 91, 217-223.

GROWTH AND DIFFERENTIATION OF A SMALL INTESTINAL EPITHELIAL CELL LINE IEC-6

AKIO AMETANI, SATOSHI HACHIMURA, AKEMI IMAOKA,
YOSHIO YAMAMOTO, HO-KEUN YI and SHUICHI KAMINOGAWA
Department of Agricultural Chemistry
The University of Tokyo
Bunkyo-ku, Tokyo 113
Japan

ABSTRACT. A small intestinal epithelial cell line, IEC-6, was analyzed for its growth and differentiation pattern *in vitro*. After IEC-6 cells had reached confluence, alkaline phosphatase (ALP), a differentiation marker, was introduced and its activity increased thereafter. Histochemical staining of the cells indicated that most of the ALP-positive cells did not proliferate. These results indicate that the cells differentiated during culture, which accompanied the end of proliferation. Furthermore, we found that the cells which became detached from the cultured plates had undergone apoptosis (programmed cell death). The differentiation of IEC-6 cells in the culture closely resembled the *in vivo* differentiation of small intestinal epithelial cells.

1. Introduction

Epithelial cells located along the intestinal crypt villus axis represent a continuous differentiation system. Crypt cells, which rapidly proliferate, move on to the villus tip, where the cells become terminally differentiated, this differentiation being characterized by changes in enzyme activity, transport ability and cell morphology (1,2). However, the mechanisms for regulating the proliferation and differentiation of intestinal epithelial cells remain ill-defined. The establishment of *in vitro* models is important for the further development of such studies. Cell lines such as HT-29 (3-5) and Caco-2 (6, 7) display typical features of intestinal cell differentiation; however, these cells are paradoxically carcinoma cell lines, and it has been difficult to induce differentiation *in vitro* with cell lines established from normal tissues. In the present study, we show that IEC-6 cells, established by Quaroni *et al.* from the rat small intestine (8), can be induced to differentiate in a normal culture. The differentiation of IEC-6 cells in the culture closely resembled the *in vivo* differentiation of small intestinal epithelial cells. This system should be useful for examining the regulation of cell differentiation and cell death in the small intestine. Furthermore, IEC-6 cells, being derived from the small intestine, may be of value in screening various biochemicals from food resources.

2. Material and Methods

2.1. Cell culture

S. Kaminogawa et al. (eds.), Animal Cell Technology: Basic & Applied Aspects, Vol. 5, 33–40.
© 1993 *Kluwer Academic Publishers.*

IEC-6 cells, purchased from ATCC (American Type Culture Collection), were maintained in Dulbecco's modified Eagle's medium (DMEM, Nissui) containing 5% FCS and 10 μg/ml of insulin (Collaborative Research). The cells were passaged at a seed density of 3 x 10^5 cells/100 mm plastic dish (Falcon) at 3- or 4-day intervals. The cells were removed from the plates with 0.05% trypsin/0.02% EDTA/PBS. For assays, cells were cultured in 60-mm plastic dishes (seed density of 1 x 10^5 cells), 96-well plates (3 x 10^3 cells), or 24-well plates (7 x 10^4 cells).

2.2. MTT assay

Cells were cultured in 96-well plates. A 5 mg/ml MTT/PBS solution was added to each well, and the plates were cultured for 6 hours. A 0.04 N HCl-isopropanol solution was added (100 μl/well), and the absorbance at 570 nm was measured.

2.3. Measurement of ALP activity

Cells were washed with PBS, before the ALP activity and protein content were measured. To measure ALP activity, a 1.25 mg/ml p-nitrophenyl phosphate solution in a 1 M diethanolamine buffer was added before incubating for 30 min at 37°C. The enzyme reaction was terminated by adding 10 μl of 1 N NaOH to the wells. The absorbance at 405 nm was measured, and the concentration of the product was determined against a standard curve drawn for p-nitrophenol. The protein content in each well was determined by using a protein staining solution (Bio-rad), the cells being solubilized by incubating with a solubilizing agent (180 mg urea, 39 mg 2-mercaptoethanol, 300 μg SDS and 20 mg EDTA/10 ml). The cell solution was mixed with the staining agent, and the absorbance at 595 nm was measured. The protein content was calculated by comparison with a standard curve drawn for BSA. The ALP activity of each culture is expressed as U/g protein; one unit of ALP activity being defined as that producing 1 μmol of p-nitrophenol per min.

2.4. Scanning electron microscopy

After being washed extensively with Hank's balanced salt solution, the cells were fixed with 2.5% glutaraldehyde in 0.1 M phosphate buffer (pH 7.4), dehydrated by passage through ethanol and t-butyl alcohol, and lyophilized. Mounting was performed with a sputter coater. The specimen was examined using a Hitachi S-800 scanning electron microscope.

2.5. ALP activity staining

Cells were histochemically stained for ALP activity by an ALP activity staining kit (Sigma). The cells were fixed with 10% formalin/PBS, and staining was performed according to the manufacturer's instructions, except that the incubation was conducted for 30 min at 37°C.

2.6. Immuncytochemical staining of proliferating cells

Proliferating cells were stained by a cell proliferation detection kit (Amersham) to detect those cells which had incorporated bromo-deoxyuridine (BdU) with an anti-BdU monoclonal antibody. Staining was performed according to the manufacturer's instructions, except that the cells were fixed with 99.5% ethanol.

2.7. Analysis of DNA degradation

An analysis of DNA degradation was performed according to the method of Neiman *et al* (9). Briefly, cells were suspended in an RSB buffer (10 mM Tris-HCl/10 mM NaCl/5 mM MgCl₂ at pH 7.4) mixed with an equal volume of a lysis buffer (1% SDS/20 μmol Tris-HCl/100 mM KCl/10 mM EDTA/ 16 mM dithiothreitol at pH 7.4) and proteinase K to a final concentration of 0.5 mg/ml for at least 3 hr at 55°C. After adding ammonium acetate, nucleic acids were precipitated with isopropyl alcohol. The precipitate was resolved in 10 mM Tris-HCl/1 mM EDTA, digested with 150 μg/ml of ribonuclease A and subsequently with proteinase K. Aliquots of this digest were analyzed by agarose-gel electrophoresis (0.7% SeaKem/1.1% NuSieve, FMC).

3. Results

The activity of ALP, a brush border membrane enzyme, was examined for IEC-6 cells cultured in DMEM with 5% FCS. Figure 1 shows that ALP was expressed in the cells on day 6, and increased thereafter. The growth curve for the IEC-6 culture is shown in Fig. 2. The increase in cell number, as detected by MTT assay, stopped on day 5, indicating that the cells had reached confluence. These results demonstrate that ALP activity was induced just after the IEC-6 cells had reached confluence. The cultures were histochemically stained for ALP activity, and it was confirmed that ALP-positive cells could be detected in the confluent cultures, but were not present in the earlier cultures (data not shown). We note that the cells expressing ALP activity formed colonies in areas of high cell density. In addition to the increase in ALP activity, there were morphological changes in the confluent cells; they developed apical brush border microvilli, which was revealed by scanning electron micrography (Fig. 3). From these results, we consider that IEC-6 differentiated in culture.

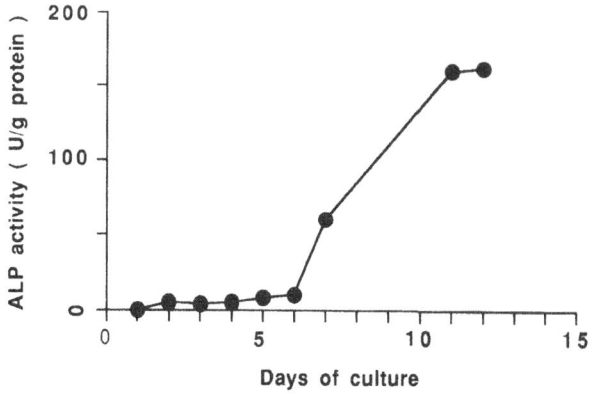

Fig. 1 ALP activity in IEC-6 culture

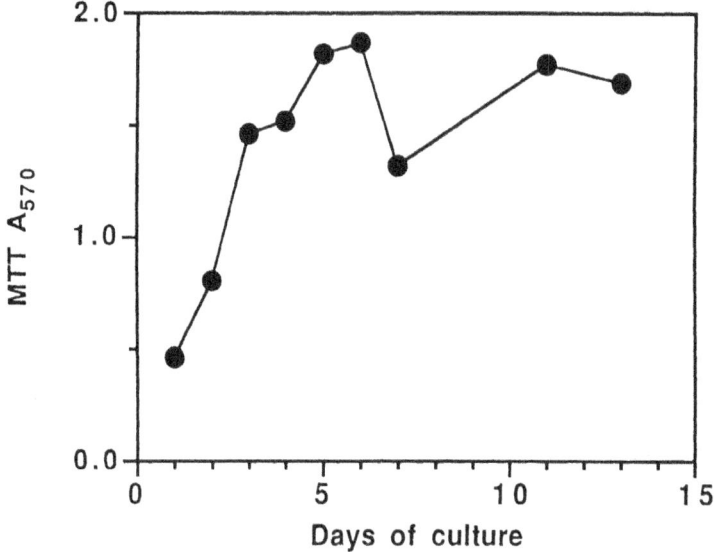

Fig. 2 Growth curve for IEC-6 culture

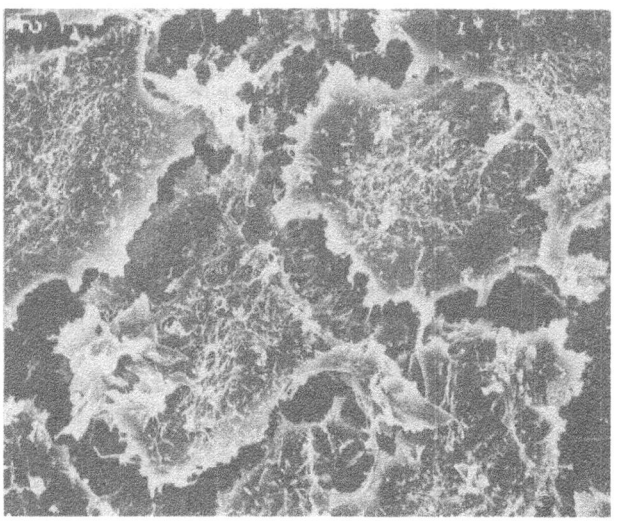

Fig. 3 Scanning electron micrograph (3000x) of IEC-6 cells cultured for 20 days

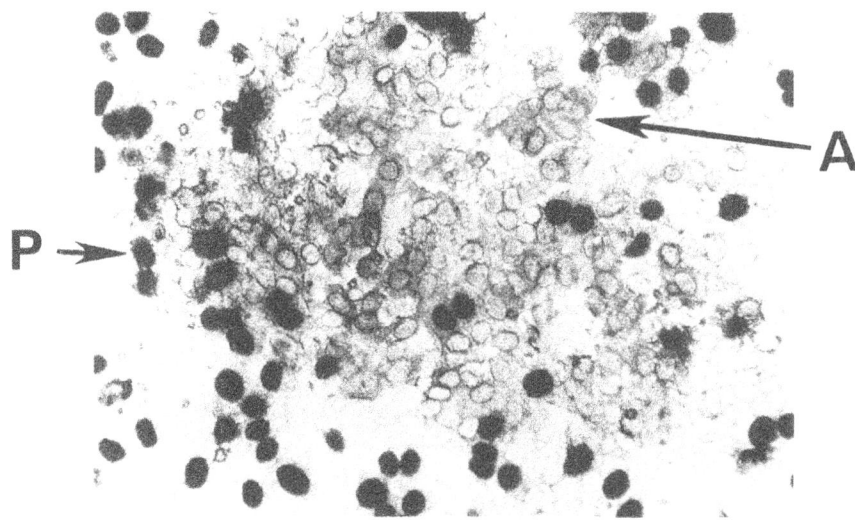

Fig. 4 Cytochemical staining of proliferating (P) and ALP-expressing (A) cells of confluent IEC-6 cells

Table 1 ALP expression on adhesive and non-adhesive IEC-6 cells derived from confluent cells

Cells	ALP activity (U/g protein)
Confluent cells	40.11
Adhesive cells	9.63
Non-adhesive cells	93.13

As cell differentiation generally accompanies termination of cell growth, we investigated whether such cells expressing ALP were proliferating or not. Confluent cells (day 13 of the culture) were double-stained for ALP activity and BdU incorporation. It is shown in Fig. 4 that most of the ALP-positive cells did not proliferate. Next, after finding cells which had become detached from the plate in confluent cultures of IEC-6, we checked whether such cells had differentiated and ceased to proliferate. Confluent cells (cultured for 20 days) were trypsinized and replated to test their adhesiveness and ALP activity. Table 1 shows that the cells which failed to re-adhere possessed a high level of ALP activity, while those cells that were adsorbed to the plates expressed only a low level of ALP. These results indicate that cells expressing ALP showed poor proliferation and adherence to the dishes, while the ALP-negative cells retained their adhesive properties and could be cultured further.

38

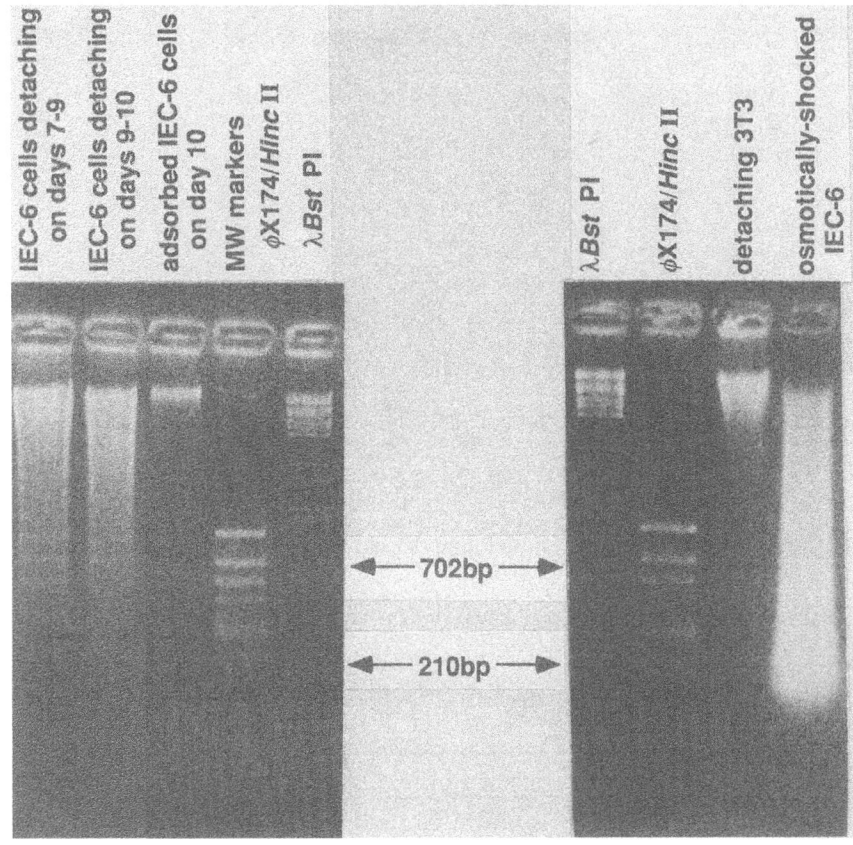

IEC-6 cells detaching on days 7-9
IEC-6 cells detaching on days 9-10
adsorbed IEC-6 cells on day 10
MW markers
φX174/Hinc II
λBst PI

λBst PI
φX174/Hinc II
detaching 3T3
osmotically-shocked IEC-6

←— 702bp —→

←— 210bp —→

Fig. 5 Agarose gel electrophoresis of DNA from IEC-6 cells which had detached from culture

Furthermore, agarose gel electrophoresis of DNA from these detached, floating cells revealed fragmentation of chromosomal DNA (Fig. 5). The size of the bands was a multiple of approximately 200 bp, indicating that these cells had undergone apoptosis (programmed cell death). The DNA ladder could not be seen in 3T3 fibroblast cells which had become detached during culture, nor was it seen in osmotically-shocked IEC-6 (Fig. 5), demonstrating that the DNA fragmentation was specific to the IEC-6 cells which had reached confluence.

4. Discussion

In this study, a small intestinal epithelial cell line, IEC-6, from newborn rat was analyzed for growth and differentiation pattern. This cell line maintains growth ability and phenotypes as undifferentiated epithelial cells in passage cultures (8). We cultured the cells in a single dish, exchanging a medium of DMEM containing 5% FCS at 2-day intervals. The cells during the 14-day culture period showed several characteristic changes. The results indicate that colonies of cells having ALP activity and microvilli structure appeared when the cell culture had reached confluence and stopped growing. These cells in the colonies formed multilayers. Even 20 days after starting the culture, the number of colonies of ALP-positive cells was limited. The areas and number of clusters of ALP-positive cells can be considered to correspond to those of the cells with microvilli.

One of the isoforms of ALP on small intestinal epithelium is known to be expressed predominantly on absorptive enterocytes which have differentiated from stem cells and been translocated from crypt to villi (10). However, undifferentiated epithelia express much less ALP on their surface. Thus, ALP can be used as a differentiation marker for small intestinal epithelial cells. Microvilli structure is also found on absorptive enterocytes (1). From the two aspects of enzyme expression and morphological change, we have demonstrated that IEC-6 differentiated during culture. This differentiation did not spread to all cells of IEC-6 in a culture, but occurred with a limited number of cells.

Other cells than the differentiated ones in a confluent culture also contain epithelial cells which can differentiate thereafter. Cells recovered from a confluent culture were tested for their adhesiveness and ALP activity. The results (Table 1) indicate that most of the cells that expressed ALP activity were not adsorbed to the plates again, while most of the cells that did not express ALP were adsorbed. These latter adsorbed cells were able to express ALP activity after reaching confluence again (data not shown), indicating that undifferentiated epithelial cells in a confluent culture can differentiate in the next culture. These suggest that the adhesiveness of cells changed during differentiation, the differentiated cells losing their adhesiveness to plastics and being piled up in a colony. The double-staining experiments for ALP activation and DNA replication indicate at a single-cell level that the differentiation of IEC-6 cells occurred at the end of proliferation (Fig. 4). We can speculate that differentiation with a loss of adhesiveness and halted growth would occur unidirectionally. During culture after reaching confluence, some cells became detached and floated in the culture supernatant. A DNA analysis showed that apoptosis had been induced in these floating cells (Fig. 5), suggesting that the final stage of differentiation is cell death. It is demonstrable that differentiation of IEC-6 as a whole was unidirectional.

Our results allow us to demonstrate several sequential events during the differentiation of IEC-6, compared with those of the small intestinal epithelium. Undifferentiated IEC-6 cells grow well and correspond to proliferating epithelia at the crypt base. Clusters of cells at a high density appear in a confluent culture of IEC-6, and the cells in these clusters form multilayers. These cells express ALP activity and a microvilli structure. *In vivo* cells on a villus express brush border membrane-bound enzymes such as ALP and other digestive enzymes, and also a microvilli structure. ALP-positive IEC-6 cells end their proliferation and lose adhesiveness. Piling up of IEC-6 cells with changed adhesiveness into clusters is compatible with the *in vivo* transportation of cells from crypt to villus. At the last stage of differentiation, IEC-6 cells become detached from the clusters in the culture supernatant, resulting in apoptopic cell death. Differentiated absorptive enterocytes at the top of the villus are also finally detached. It is noticeable that differentation pattern of IEC-6 corresponds to that of the *in vivo* small intestinal epithelium. IEC-6 can be useful for assessing biological substances and food components inducing or inhibiting growth, differentiation and cell death.

5. Acknowledgements

We would like to thank Dr. Toshiaki Kimura (Snow Brand Milk Products, Co.) for scanning electron microscopic analysis.

6. References

1. Madara, J. L. and Trier, J. S. (1987) in L. R. Johnson (ed), Physiology of the Gastrointestinal Tract, Raven Press, New York, pp. 1209-1250.
2. Moog, F. (1979) in K. Elliot and J. Whelan (eds), Development of Mammalian Absorptive Process (Ciba Foundation Symposiun 70), Exerpta Medica, Amsterdam, pp. 31-44.
3. Pinto, M., Appay, M. D., Simon-Assmann, P., Chevalier, G., Dracopoli, N., Fogh, J., and Zweibaum A. (1982) Biol. Cell. 44, 193-196.
4. Zweibaum, A., Pinto, M., Chavelier, G., Dussaulx, E., and Rousset, M. (1985) J. Cell Physiol. 122, 21-29.
5. Wice, B. M., Trugnan, G., Pinto, M., Pousset, M., Chevalier, G., Dussaulx, E., Lacroix, B. and Zweibaum, A. (1985) J. Biol. Chem., 260, 139-146.
6. Rousset, M., Laburthe, M., Pinto, M., Chevalier, G., Rouyer-Fessard, C., Dussaulx, E., Trugnan, G., Boige, N., Brun, J. L. and Zweibaum, A. (1985) J. Cell. Physiol. 123, 177-185.
7. Pinto, M., Robine-Leon, S., Appay, M. D., Kedigner, M., Triadou, N., Dussaulx, E., Lacroix, B., Simon-Assmann, P., Haffen, K., Fogh, J., and Zweibaum, A. (1983) Biol. Cell. 47, 323-330.
8. Quaroni, A., Wands, J., Trelstad, R. L., and Isselbacher, K. J. (1979) J Cell Biol. 80, 248-265.
9. Neiman, P. E., Thomas, S. J., Loring, G. (1991) Proc. Natl. Acad. Sci. USA 88, 5857-5861.
10. Quaroni, A.and Isselbacher, K. J. (1985) Dev. Biol. 111, 267-279.

SUBUNIT ASSEMBLY OF LAMININ VARIANTS IN CULTURED ANIMAL CELLS

Hoon Jeon[1], Masaaki Ono[1], Chino Kumagai[1] and,
Yasuo Kitagawa[1,2]

1) Graduate School of Biochemical Regulation Nagoya University,
2) Nagoya University BioSciences Center,
Chikusa-ku, Nagoya-shi 464-01, Japan

ABSTRACT

Bovine aortic endothelial cells (BAEC) produce two variant forms of laminin with a subunit composition of AB_1B_2 and $A'B_1B_2$. Analyses of the intracellular assembly of these subunits revealed that the B_1B_2 dimer formed first, and that A or A' joined to form the AB_1B_2 or $A'B_1B_2$ trimer. Angiostatic steroids shifted the relative size of the A and A' monomer pool in BAEC, and competition between the A and A' subunits in joining the B_1B_2 dimer produced AB_1B_2 and $A'B_1B_2$ in different ratios. This result suggests that subunit replacement is the general mechanism for producing laminin variants by various cells for tissue morphogenesis. When laminin subunits in BAEC were cross-linked with dithio-*bis*-(succinimidylpropionate)(DSP) and immunoprecipitated with anti-laminin antiserum, monomeric A, A', B_1 and B_2 monomers and the B_1B_2 dimer migrated as extremely large molecules in sodium dodecyl sulfate gel electrophoresis under non-reducing conditions. When the crosslinking disulfide bonds were cleaved under reducing conditions, they migrated as the usual subunits. This result suggests that molecular chaperones were involved in the process of the assembly and replacement of laminin subunits.

1. INTRODUCTION

Basement membranes are a special type of extracellular matrix which form the protein sheets underlying epithelial and endothelial cells. While a large number of molecular components have been suggested, the bulk is formed by only a few proteins including laminin, collagen IV and proteoglycans. Basement membranes influence the shape, motility and differentiation of cells. They are thus involved in the formation of polarized epithelia and endothelia during morphogenesis and development. They also play a role in metastasis, as tumor cells need to pass the basement membranes to penetrate into tissue.

Laminins are the major glycoproteins in basement membranes, and exert a strong influence on cellular activity. A prototype laminin (Mr 900,000) was first purified from the Engelbreth-Holm-Swarm (EHS) tumor and consisted of three disulfide-linked subunits named A, B_1 and B_2. On electron micrographs, laminin has the shape of an asymmetric cross with terminal globular domains on all arms, and inner globules within the short arms [1]. A complete set of cDNA sequences is known for human [2], mouse [3] and

41

S. Kaminogawa et al. (eds.), Animal Cell Technology: Basic & Applied Aspects, Vol. 5, 41–48.
© 1993 Kluwer Academic Publishers.

drosophila [4] laminins. Furthermore, the sequence of one B₁ chain isoform (s-laminin) [5] and the partial sequence of one A chain isoform (merosin) [6] have been determined. Recently, an epithelium-specific protein, kalinin, has been characterized [7], and shown to be a component of the anchoring filaments adjacent to hemidesmosomes. Kalinin consists of three different disulfide-linked polypeptides which form a long rod with one large and three small globular domains at opposite termini. The cDNA sequence has now revealed that kalinin is a member of the laminin family containing only truncated short-arm regions. The A chain component of kalinin can be assembled with the prototype B₁ and B₂ chains to form a trimer of Mr 600,000 called K-laminin [8].

This sequence information together with corresponding structural studies have led to an accurate structural model of laminin [9]. In this model, it is clear that the long arm must be the site of assembly, as this is the only structure in laminin in which all three chains meet. Most of the amino acid sequence in the long-arm can be arranged into a repeating heptad with hydrophobic residues in the first and fourth positions. The positions fifth and seventh are frequently occupied by charged amino acids. This kind of sequence pattern is well known as the coiled-coil α-helical structure found in many fibrous proteins such as myosin [10]. Dissociation-association experiments on laminin long-arm fragments have shown that the assembly of different laminin chains was guided by the specific interaction of coiled-coil domains. The selectivity for forming a heterodimer and -trimer structure could be attributed to the selective distribution of charged residues in the position fifth and seventh of the heptad repeat.

In bovine aortic endothelial cells (BAEC), we found that two variant laminins were synthesized at the same time, and that two A chain homologues (A and A' subunits) were interchanged in the process of heterotrimer formation [11]. Together with the fact that apparently the same A chain homologue is found in kalinin as well as in K-laminin, this suggests that the mechanism for coiled-coil formation driven by heptad repeats is shared among many laminin variants to produce a variety of structures and functions. In this paper, we will describe the possible involvement of molecular chaperones in the process of laminin assembly.

2. MATERIALS AND METHODS

2.1 Cell culture

BAEC isolated as described by Gospodarowicz [12] was maintained in Dulbecco's modified Eagle's medium (DMEM) containing 10% fetal calf serum and antibiotics.

2.2 Labelling cells with [³⁵S]methionine

A Confluent culture of BAEC in 24-well plates was washed once with methionine-free Eagle's minimal essential medium (MEM) and incubated in 200 µl of methionine-free MEM containing 0.5 mCi/ml [³⁵S]methionine at 37°Ç for 4 h.

2.3 Cross-linking

Metabolically labelled BAEC cultures were rinsed with phosphate-buffered saline (PBS) and a 100-µl solution of 2 mM DSP in PBS was added. After incubating on ice for 30 min, the cultures were first rinsed with 2 mM glycine in PBS to block the DSP activity and then with PBS [13].

2.4 Immunoprecipitation and SDS-polyacrylamide gel electrophoresis
Immunoprecipitation of the laminin-related polypeptides and SDS-electrophoresis was carried out as described [11,14,15].

3. RESULTS

3.1 Production of two laminin variants in BAEC and shift of their relative levels during angiogenesis
The organization of endothelial cells as the lining of the vascular system is supported by basement membranes [16]. The important role of laminins has been assumed in angiogenesis. We found that cultured endothelial cells produced a laminin variant ($A'B_1B_2$) in addition to the prototype laminin (AB_1B_2) and that their relative levels were modulated with angiostatic steroids [11]. BAEC produced four laminin-related polypeptides, A, A', B_1 and B_2. In order to analyze the intracellular assembly of these laminin subunits, we took advantage of sodium dodecyl sulfate (SDS) gel electrophoresis under reducing and non-reducing conditions. Since the laminin subunits are linked to each other with disulfide bonds at their long-arm region, we could follow the process of assembly by a combination of

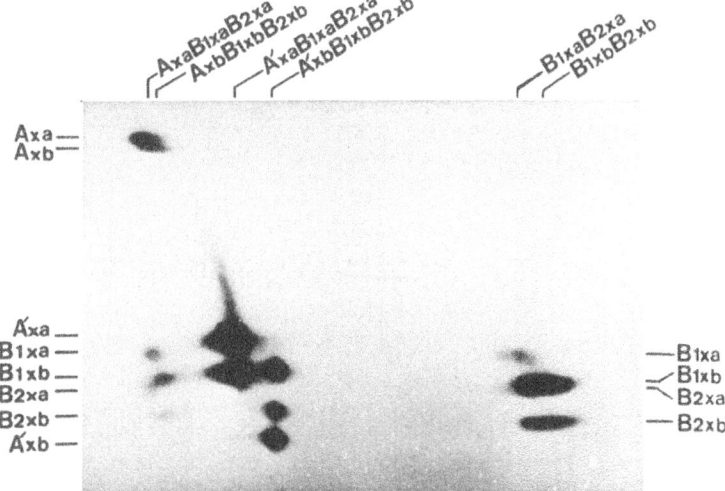

Fig. 1 Non-reducing and reducing two-dimensional SDS gel electrophoregram of laminin-related polypeptides in BAEC.
Radiolabelled cell lysate from BAEC was immunoprecipitated and electrophoresed from *left* to *right* under non-reducing conditions, and under reducing condition from *top* to *bottom*. The subscripts to A, A', B_1 and B_2 indicate the intracellular precursors for each subunit at different stages of N-glycan processing. It is important to notice that the glycan processing of each subunit in a same complex was synchronized.

non-reducing and reducing electrophoresis. The two-dimensional SDS electrophoregram shown in Fig. 1 was obtained by separating the radio-labelled laminin subunits form BAEC under non-reducing conditions in the first dimension (from left to right) and then under reducing conditions in the second dimension (from top to bottom). By this technique, we can find the monomeric subunits on the diagonal, and the subunits in a complex as spots on a vertical line. Such an analysis shows that the B_1B_2 dimer was formed first and then A joined it to form the AB_1B_2 trimer, or that A' joined to form the $A'B_1B_2$ trimer (Fig. 1). This mechanism shows that the replacement of subunits in the laminin complex by a corresponding variant can produce a variety of structures and functions of laminin (Fig. 2). $A'B_1B_2$ was the major product in endothelial cells under *normal* culture conditions where the cells were shifted to the angiogenic state due to transferring to culture. Angiostatic steroids such as medroxyprogesterone, on the other hand, suppressed the A' synthesis while stimulating the A synthesis in BAEC. Consequently, the major product of BAEC was converted to AB_1B_2. These results suggest that endothelial cells can produce different types of laminin during angiogenesis.

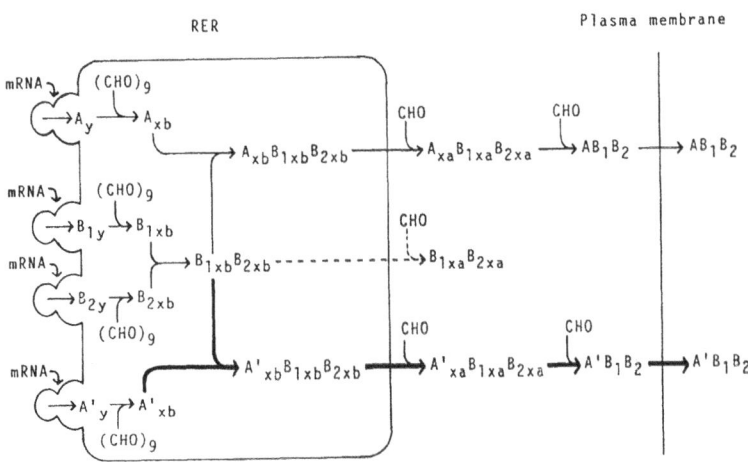

Fig. 2 Intracellular assembly of four laminin-related polypeptides into two laminin variants in endothelial cells.

A, A', B_1 and B_2 are the four laminin-related polypeptides found in endothelial cells. Subscripts *xa* and *xb* indicate stable intermediates at different stages of glycan processing. *y* indicates an unglycosylated form observed in the presence of tunicamycin. *(CHO)9* shows the high-mannose oligosaccharide chain. *CHO* shows terminally transferred monosaccharide, and *RER* shows rough endoplasmic reticulum.

3.2. Molecular chaperones involved in laminin assembly

In order to explore whether molecular chaperones were involved in the assembly and replacement of laminin subunits, we crosslinked such proteins to laminin subunits by incubating radiolabelled BAEC with DSP, and lysed the cells for immunoprecipitation. By this technique, we can detect proteins closely localized with laminin subunits. As the crosslinking reagent, we chose DSP [17], since this forms disulfide bonds reversibly cleavable under reducing conditions. As shown in Fig. 3A, intracellular crosslinking with DSP markedly reduced the amount of A, A', B_1 and B_2 monomers and the B_1B_2 dimers detected by SDS-electrophoresis under non-reducing conditions. Most of the radiolabelled bands migrated as polypeptides with extremely large sizes. By SDS-electrophoresis of the same samples under reducing conditions (Fig. 3B), on the other hand, the bands corresponding to all laminin subunits could be detected even after crosslinking. This result suggests that the monomeric and dimeric laminin subunits need to be associated with some proteins like chaperonin before their assembly into the AB_1B_2 or $A'B_1B_2$ trimer.

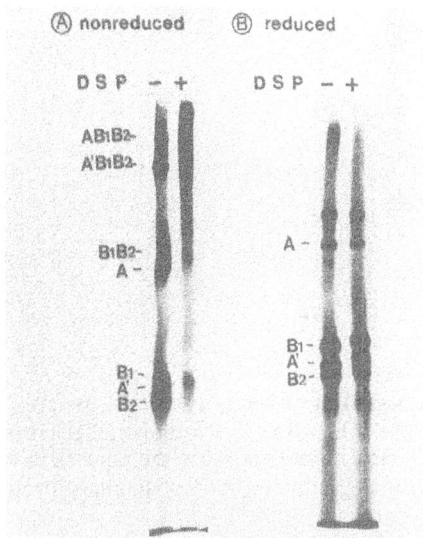

Fig. 3 Non-reducing and reducing SDS-electrophoresis of the radiolabelled laminin subunits in BAEC after crosslinking with DSP.
Metabolically radiolabelled BAEC cultures were crosslinked by incubating with (+) or without (-) 2 mM DSP for 30 min at 4 °C before the cells were lysed. The laminin subunits were immunoprecipitated together with crosslinked proteins and separated by SDS-electrophoresis under non-reducing (A) and reducing (B) conditions. A separating gel of 4% acrylamide was used.

In order to monitor the behavior of the laminin subunits, 4% acrylamide gel was used for the electrophoresis in Fig. 3. With such a soft gel, putative molecular chaperones crosslinked to the laminin subunits might have run off the gel. To detect smaller polypeptides, we therefore separated the same samples with 8% acrylamide gel (Fig. 4). In this electrophoresis, we could detect three major bands with estimated sizes of 80, 60 and 50 kDa, in addition to the bands of the laminin subunits. Since the emergence of these bands clearly depended on DSP crosslinking, these proteins are suggested to have been associated with the laminin subunits. The estimated sizes of these proteins are different from those of such well-characterized molecular chaperones as GRP94, GRP78 and HSP47 and their identity is open to further characterization.

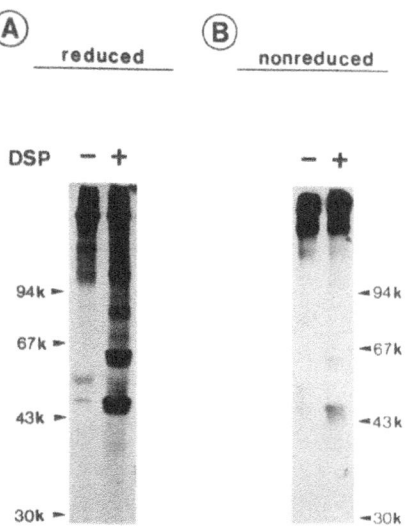

Fig. 4 Polypeptides in BAEC crosslinked to laminin subunits by DSP.

The radiolabelled lysate from BAEC with (+) or without (-) DSP treatment was separated in 8% acrylamide gel. The other details are the same as in Fig. 3. Arrow heads bside the lanes indicate the migration position of standard proteins with the corresponding sizes.

DISCUSSION

The present investigation has identified putative molecular chaperones which might be involved in the process of intracellular assembly and replacement of laminin subunits.   For selective recognition among the laminin subunits, interchain ionic and hydrophobic interactions due to the heptad repeat are the most important (Fig. 5). Laminin subunits help themselves to form such a structure based on the amino acid sequence in the long-arm region. This might suggest that no molecular chaperon would be needed for the assembly of laminins. On the other hand, we need to assume some sort of mechanism to neutralize the hydrophobic surface in the structure of monomeric and dimeric laminin subunits before assembly. This surface might eventually be neutralized when the assembly has been completed, but some chaperonin would be needed to neutralize it and keep the monomer and dimer soluble. It is worthwhile to notie that the crosslinking experiment particularly shifted the monomeric and dimeric subunits to larger complexes. In addition, our former analysis of intracellular laminin assembly in embryonal carcinoma F9 [14] and 3T3-L1 adipocytes [15] showed that laminin subunits were allowed to leave the rough endoplasmic reticulum (RER) for Golgi apparatus only after the trimer had been formed.   These facts suggest that the putative molecular chaperones might be important not only for preventing the aggregation of the laminin monomer and dimer, but also for arresting them in RER.   To prove the latter function, we need to demonstrate the presence of the KDEL sequence [18] at the N-terminal of the 80, 60 and 50 kDa polypeptides.

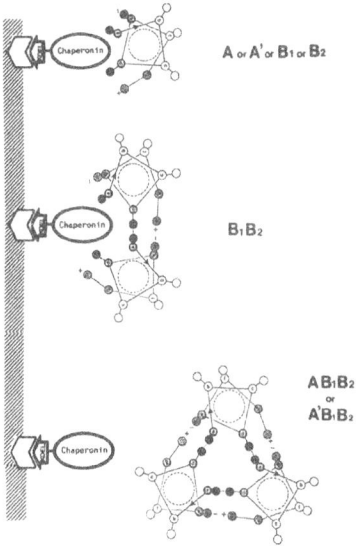

Fig. 5 Putative model for the role of molecular chaperones in laminin assembly.
    This figure shows the top view of the long-arm region of each laminin subunit. Due to the heptad repeat, hydrophobic amino acid (closed circles) forms a surface to interact with the molecular chaperones. The putative molecular chaperones might have a KDEL sequence at their N-terminal and interact with the specific receptor at the lumen of RER. See the text for more details.

    Figure 5 depicts our putative scenario for laminin assembly, which can be summarized as follows: 1) Each laminin subunit forms the structure of an α-helix having a hydrophobic surface along the axis of molecule with charged amino acid residues lined up on both sides.   2) Some types of molecular chaperone might interact with the laminin monomers at this hydrophobic surface and prevent them from aggregation.   3) If such molecular chaperones have the RER arresting signal (KDEL), the monomeric laminin subunits might be held in RER by the receptor recognizing this signal.   4) The B_1 and B_2 subunits assemble into the B_1B_2 dimer driven by interchain hydrophobic and ionic interactions.   5) Even after the B_1B_2 formation, the hydrophobic surface might remain to be neutralized, and the molecular chaperones might arrest the B_1B_2 dimer by interacting there. 6) Since no hydrophobic surface is remains after trimer formation, the trimer is allowed to leave RER for the Golgi apparatus.

REFERENCES

1) Engel, J., Odermatt, E., Engel, A., Madri, L. A., Furthmayr, H., Rohde, H. and Timpl, R. (1981) *J. Mol. Biochem.* **150**, 97-120.
2) Pikkarainen, T., Kallunki, T. and Tryggvason, K. (1988) *J. Biol. Chem.* **263**, 6751-6758.
3) Sasaki, M., Kleinman, H. K., Huber, H., Deutzmann, R. and Yamada, Y. (1988) *J. Biol. Cem.*, **263**, 16536-16544.
4) Chi, H,- C. and Hui, C. -F. (1989) *J. Biol. Chem.*, **264**, 1543-1550.
5) Ehrig, K., Leivo, I., Argraves, W.S ., Rouslahti, E. and Engvall, E. (1990) *Proc. Natl. Acad. Sci. U.S.A,,* **87**, 3264-3268.
6) Hunter, D. D., Shah, Y., Merlie, J.- P. and Sanes, J. R. (1989) *Nature,* **338**, 229-234
7) Rousselle, P., Lunstrum. G. P., Keene, D. R and Burgeson, R. E.(1991) *J. Cell. Biol.,* **114**, 567-576,
8) Marinkovich, M. P., Lunstrum, G. P., Keene, D. R. and Burgeson, R. E. (1992) *J. Cell Biol.,* **119**, 695-703.
9) Beck, K., Hunter, I. and Engel, J. (1990) *FFASEB J.,* **4**, 148-160.
10) McLachian, A. D. and Stewart, M.(1975) *J. Mol. Biol.,* **98**, 293-300.
11) Tokida, Y., Aratani, Y., Morita, A. and Kitagawa,Y. (1990) *J. Biol. Chem.* **265**, 18123-18129.
12) Gospodrowwicz, D., Moran, J., Braun, D. and Birdwell, C.(1976) *Proc. Natl. Acad. Sci. U.S.A,* ,**73**, 4120-4124.
13) Nakai, A., Satoh, M., Hirayoshi, K. and Nagata, K. (1992) *J. Cell Biol.,* **117**, 903-914.
14) Morita, A,, Sugimoto, E. and Kitagawa,Y. (1985) *Biochem. J.,* **229**, 259-264.
15) Aratani, Y. and Kitagawa,Y. (1988) *J. Biol. Chem.,* **263**, 16163-16169 .
16) Harbst, T.J., McCarthy, J. R., Tsilibary, E. C. and Furcht, I. T. (1988) *J. Cell Biol.,* **106**,1365-1373.
17) Roth, R. A. and Pierce, S. R.(1987) *Biochemistry,* **25**, 4179-4182.
18) Pelham, H. R. B.(1989) *Ann. Rev. Cell Biol.,* **5**, 1-23.

REGULATION OF MITOCHONDRIAL GENE EXPRESSION BY INTERFERONS

Hidetoshi Inagaki[1], Yuichi Matsushima[1], Mikiko Ohshima[1] and
Yasuo Kitagawa[1,2]

1) Graduate School of Biochemical Regulation, Nagoya University,
2) Nagoya University BioSciences Center
Chikusa-ku, Nagoya-shi 464-01, Japan

ABSTRACT

When interferon α or γ was added to a culture of HeLa cells, the steady-state levels of mitochondrial mRNAs and rRNAs were reduced. Interferon γ was more effective than α, and the reduction was more evident for mRNAs than rRNAs. In organello transcription with isolated mitochondria showed that the effect of the interferons was mainly due to reduced transcriptional activity. South-Western hybridization of a mitochondrial extract with the promoter sequence of the mitochondrial genome visualized a protein band of ca. 30 kDa with a size close to that of the transcription factor (mtTF1) reported by Clayton et al. Interferon γ reduced the signal of this band, suggesting that the reduced transcription was through regulation of the amount of mtTF1. A Northern blot analysis of total RNA extracted from HeLa cells with a cDNA fragment coding for mtTF1 showed reduced level of mRNA. A gel retardation assay with the promoter sequence, however, suggests that other factor(s) were also involved.

1. INTRODUCTION

Mitochondria are the main energy plant of most animal cells and are important for the metabolism of amino acids and lipids. For these functions, several hundreds of enzymes are localized in the mitochondrial structure. Most of them are synthesized outside mitochondria according to genetic information encoded in the nuclear genome and are then transported into the structure of mitochondria. On the other hand, mitochondria have their own genome which encodes 13 polypeptides of oxidative phosphorylation. This mitochondrial genome is a double-stranded circular DNA with ta size of ca. 17 kbs and encodes 22 tRNAs and 2 rRNAs in addition to mRNAs. The mitochondrial genome depends on the nuclear genome for most of machinery of its replication, transcription and translation, but it retains independence for coding the genetic information with a set of codons different from the universal codon. In this respect, the mitochondrion is a

49

S. Kaminogawa et al. (eds.), Animal Cell Technology: Basic & Applied Aspects, Vol. 5, 49–55.
© 1993 Kluwer Academic Publishers.

double-headed system controlled by both nuclear and mitochondrial genomes (Fig. 1). Thus, extracellular signals might have two access pathways to control the function of mitochondria. One is direct access to mitochondria, and we have shown that oxygen tension during the culture of HeLa cells could control mitochondrial genome transcription through this access [1]. The other access is through the nucleo-cytoplasmic system, by which the signals can indirectly control the mitochondrial genome by modulating the expression of the genes encoding machinery of mitochondrial genome replication, transcription and translation. We have shown that mitochondrial genome transcription depends on the growth stimulation of cultured cells, probably by the latter mechanism [2]. In this paper, we will show that interferons also govern mitochondrial gene transcription by controlling the level of trans-acting factors of the mitochondrial genome.

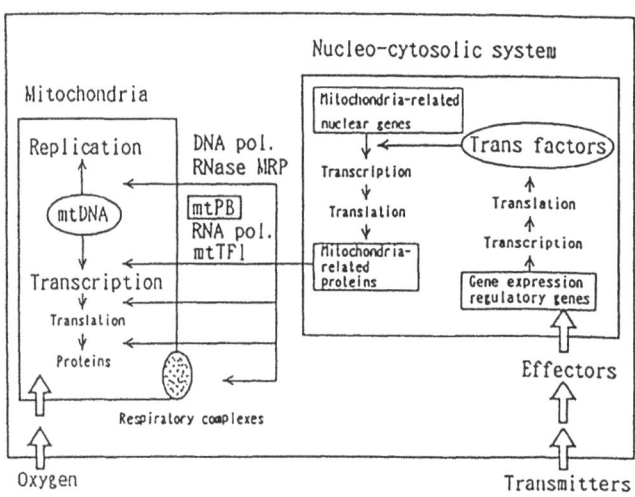

Fig. 1 Double-headed control of mitochondrial functions

Interferons affect a variety of cellular functions by regulating the expression of specific genes [3]. These include genes encoding the proteins involved in anti-viral response such as oligo 2.5 A synthetase, eIF-2 kinase and Mx. The enhanced expression of class I and II major histocompatibility genes is important in stimulating immune responses by interferons. The potent immunomodulatory effect of interferons through helper and effector T cell or B cell proliferation has also been reported reported. The ability of interferons to affect the proliferation of various tumor cells has led to wide interest in the use of interferons for treating neoplasia, and . T the expression of several oncogenes is inhibited by interferons. We will show here that expression of the mitochondrial genome is another target of interferons.

2. MATERIALS AND METHODS

2.1 Cell culture

HeLa cells were cultured in Eagle's minimal essential medium containing 10% calf serum on 150 mm dishes under humidified 95% air and 5% CO_2. The cells were treated with 1000 unit/ml of interferon α or γ for 18 h.

2.2 Preparation of hybridization probes

The PstI-HindIII fragment of human mitochondrial DNA was cloned into M13mp19 and used for a Northern blot analysis of the mRNAs encoding H'-ATPase subunits 6/6L, cytochrome c oxidase subunit 3, and NADH-dehydrogenase subunit 3 and 4/4L. The XbaI-KpnI fragment was cloned into pUC19 and used for detecting 12S and 18S rRNAs. pACT1.5 having a 1.5 kb inserted sequence of mouse β-actin, was used for standardizing the Northern blot analysis. [^{32}P]dCTP labeled probes were prepared by nick-translation.

2.3 Northern blot analysis

Electrophoresis of total cellular RNA (10-20 μg) on 1.0% agarose gel was done after denaturation [4]. RNA was trans-blotted to a nylon membrane and hybridized with radioactive probes as described in Ref. 5.

2.4 In organello transcription

The reaction mixture contained isolated mitochondria (10 mg protein/ml), 40 mM Tris-HCl (pH 7.5), 25 mM NaCl, 5 mM $MgCl_2$, 2 mg/ml bovine serum albumin, 20% glycerol, 10 mM Na_2HPO_4, 0.2 mg/ml creatine kinase, 2 mM ATP, 1 mM CTP, 1 mM GTP, 5 mM creatine phosphate and 4 mM pyruvate in a total volume of 50 μl. The reaction was started by adding radioactive UTP. To estimate the background radioactivity, 0.1 mg/ml ethidium bromide was added in control assays. An aliquot of the reaction mixture corresponding to 3-10 μl was mixed with 1 μl of 20% SDS at 10 min intervals, spotted on DE81 filter paper, and the radioactivity remaining on the filter after thorough washing was counted.

3. RESULTS AND DISCUSSION

3.1 Reduced level of mitochondrial mRNAs and rRNAs by interferons

When total RNA extracted from HeLa cells treated for 18 h with 1000 unit/ml of interferon α or γ was analyzed by Northern blot analysis with cloned mitochondrial DNA fragments encompassing H⁺-ATPase subunits 6/6L and NADH dehydrogenase 4/4L as the probes, the level of mRNAs was markedly reduced (Fig. 2, center panel). Compared to the signals observed with actin cDNA as a standard (Fig. 2, left panel), the effect of interferon γ seems to be stronger than that of α. The levels of 12S and 16S were also reduced by both interferon, the effect of interferon γ being stronger (Fig. 2, right panel). Probably due to larger mass and slower turnover rate of rRNAs, the effect of interferons was weaker on rRNAs than on mRNAs.

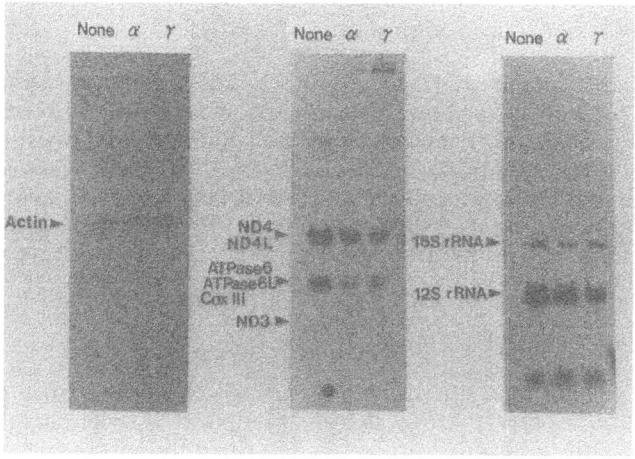

Fig. 2. Reduced level of mitochondrial mRNAs and rRNAs in HeLa cells treated with interferons α and γ.

HeLa cells were treated with none, 1000 unit/ml of interferon α or γ for 18 h and total cellular RNA was extracted. RNA corresponding to 10 μg was separated on 1.0% agarose gel, trans-blotted to a nylon membrane, and hybridized with radiolabelled *Pst*I -*Hind*III (center panel), *Xba*I -*Kpn*I (right panel) of the human mitochondrial DNA fragment, or pACT1.5 (left panel).

3.2 Reduced transcriptional activity in mitochondria isolated from HeLa cells treated with interferons

The reduced level of mitochondrial genome transcripts suggests a) that their turnover rate was accelerated by interferons, or b) that transcription of the mitochondrial genome was depressed by interferons. The results summarized in Fig. 3 clearly show that the reduction was mainly due to depressed transcription. In this analysis, we isolated mitochondria from

HeLa cells after treating with interferons and assayed the incorporation of radioactive UTP *in organello* into the RNA fragment polymerized long enough to be adsorbed to DE81 filter paper. As a negative control, we ran a parallel assay in the presence of an inhibitor of transcription, ethidium bromide. Taking the incorporation observed without the interferon treatment as 100%, interferon α and γ inhibited the transcription by 30 and 50%, respectively. Assuming the majority of transcripts to be rRNAs, it is reasonable that the relative effect of interferons α and γ on transcription (Fig. 3) was similar to that on the steady-state level of 12S and 16S rRNAs (Fig. 2). These results strongly suggest that interferons reduced the level of mitochondrial genome transcripts by depressing the intrinsic activity of transcription. This is in contrast to the hypoxic depression of the mitochondrial genome transcript [1], where no effect of hypoxic pretreatment of HeLa cells was detected by isolated mitochondria supplied with excess ATP.

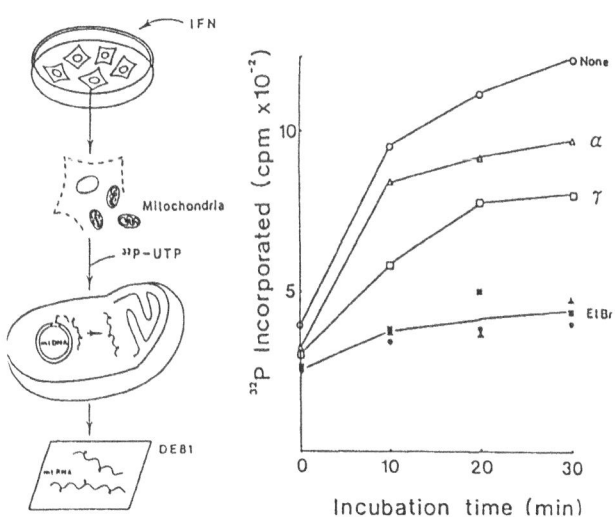

Fig. 3. *In organello* transcription with mitochondria isolated from HeLa cells after treatment with interferons α and γ

Mitochondria were isolated from HeLa cells treated with none (○,●) , interferon α (△,▲) or γ (□,■) as in Fig. 2. The *in organello* transcription activity of isolated mitochondria (10 mg/ml protein) was assayed in the absence (○,△,□) or presence (●,▲,■) of ethidium bromide.

3.3. Depressed gene expression of mtTF1 by interferons

The mammalian mitochondrial genome is transcribed as polycistronic mRNA from H- and L-strand promoters located in the D-loop region. For the precise initiation of transcription, transcription factors are important in addition to mitochondrial RNA polymerase. As one of such transcription factors, Clayton's group have purified a protein (mtTF1) with an approximate size of 27 kDa and isolated its cDNA clone as well [6]. mtTF1 has the ability to bind both H- and L-strand promoters, and stimulates transcription of the mitochondrial genome *in vitro* from the correct position. This suggests that regulation of mtTF1 gene expression might be an effective mechanism for interferons to control mitochondrial genome transcription.

In order to confirm this possibility, we took advantage of the technique called South-Western hybridization. For this, mitochondrial extract from HeLa cells treated with none, interferon α or γ was first separated by SDS-electrophoresis and the proteins were renatured during transferring to a membrane. After renaturating the proteins, the membrane was hybridized with a radiolabelled DNA fragment containing human mitochondrial H- and L-strand promoters. By this technique, we could detect a radioactive band with a size (30 kDa) close to that of mtTF1. The signal of this band was markedly reduced in the mitochondrial extract from HeLa cells treated with interferon γ, while interferon α did not cause the reduction. Based on the reported sequence of mtTF1 cDNA, we constructed a mtTF1 cDNA clone by the technique of reverse transcriptase polymerase chain reaction (RT-PCR) and estimated the level of mtTF1 mRNA in HeLa cells. A northern blot analysis suggested that interferon α and γ reduced the level by 50 and 70 %, respectively. These results strongly suggest that the effect of interferon on mitochondrial genome transcription was partially due to the depressed expression of the mtTF1 gene.

3.4 Other promoter binding factors might be involved in regulation of mitochondrial gene expression by interferons

In order to explore the other possible factors involved in transmitting the interferon effect, we ran a gel retardation assay of a mitochondrial extract from HeLa cells with radiolabelled H- or L-strand promoter sequences of the human mitochondrial genome. In this analysis, the H- and L-strand promoters gave one and three shifted bands, respectively. Since different genes are encoded on two strands, it is reasonable to detect different number of the binding proteins to H- and L-strand promoters. These proteins might be responsible for transcripting two strands with different efficiency. In addition, the L-strand promoter is important for producing RNA primer to replicate the mitochondrial genome which has its replication origin on the H-strand. We can reasonably assume that one of each band detected by the H- and L-strand promoters might correspond to mtTF1, and that the two additional bands detected by the L-strand promoter may correspond to factors controlling strand-specific transcription or the initiation of replication. When we compared the signals of the shifted bands observed in mitochondrial extracts from HeLa cells treated with none, interferon α or γ, all the bands detected by both promoters were weakened by interferons. Here again, interferon γ was more effective than α. These results suggest that other factors are involved in the mechanism for reduced transcription of the mitochondrial genome with interferons.

REFERENCES

1) Kadowaki, T. and Kitagawa, Y. (1991) *Exp. Cell Res.,* **192**, 243-247.
2) Kadowaki, T.,and Kitagawa, Y .(1988) *FEBS Lett.,* **233**, 51-56.
3) Revel, M. and Chebath, J. (1986) *TIBS,* **2**, 166-170.
4) McMaster, G. K. and Carmichael, G. G. (1077) *Proc. Natl. Acad.S ci.U.S.A.,* **74**, 4835-4838
5) Miki, K. and Kitagawa, Y. (1987) *J.Biochem.,* **102**, 385-392
6) Fisher, R. P., Parisi, M. A. and Clayton, D. A. (1989) *Gene & Dev.,* **3**, 2202-2217.

PRODUCTION OF THE HUMAN GROWTH HORMONE IN SERUM-FREE UC203 MEDIUM

Masahiro Mizuguchi,[1] Ken Matsumoto[1] and Kazukiyo Onodera[2]

1 Nissui Pharmaceutical Co. Ltd., Yuuki, Ibaraki, Japan
2 Department of Agricultural Chemistry, The University of Tokyo, Japan

ABSTRACT. We developed a serum-free medium from UC202 medium [1] that could support cell growth and enhance the production of the human growth hormone (hGH) coded by the introduced human gene. This developed UC203 medium was better than alpha-MEM [2] with 5% dialyzed fetal calf serum (DFCS) in terms of supporting cell growth and enhancing the production of hGH.

1. Materials and Methods

1-1. CELLS AND CELL CULTURE

CHO-K1 and its GT19 derivative cells were used in this study. UC202 medium (Nissui Pharmaceutical Co.) was used as the starting medium, and insulin, human transferrin (Holo type), fibronectin and catalase were used as growth factors.

1-2. CELL GROWTH EXPERIMENTS

The cells were harvested from the maintained medium by low-speed centrifugation and then washed with Dulbecco's PBS(-) (Nissui Pharmaceutical Co.). The cells were inoculated at 5×10^4 cells/5ml of viable cells into triplicate 60-mm plastic petri dishes (A/S Nunc, Kamstrup, Dk4000 Roskilde, Denmark) with the serum-free medium or conventional serum containing a medium, and incubated at 37°C in 5% CO_2.

The production of hGH was measured by the 2-site immuno-radiometric assay, authentic hGH being obtained from Pharmacia Co.
Osmotic pressure was measured by freezing-point thermodynamics with an Advanced DigiMatic® Osmometer (model 3D2, Advanced Instruments, USA).

2. Results

Many kinds of serum-free media have been developed and commercialized for the production of biologically active substances using cultured mammalian cells. They have often contained a large amount of proteins which hampered the isolation of useful substances.

Generally, the enrichment of each component has led to the enhanced production of biologically active substances. However, there are several exceptions to this rule.

S. Kaminogawa et al. (eds.), Animal Cell Technology: Basic & Applied Aspects, Vol. 5, 57–61.
© 1993 Kluwer Academic Publishers.

2-1. COMPONENTS OF AMINO ACIDS AND SUGARS

L-cysteine-HCl was essential for cell growth, but high concentrations inhibited the production of hGH (*Figure 1*). Judging from the residual amino acids during culture, L-arginine-HCl and L-lysine-HCl were consumed at the early stage of culture. The addition of a low concentration of L-arginine-HCl and L-lysine-HCl was effective for cell growth and hGH production (data not shown).

Among the sugars, mannose and D-(+)-glucose were effective. On the contrary, the addition of D-(-)-fructose and D-(+)-galactose was inhibitory. It was particularly evident that the concentration of D-(-)-fructose was critical for the production of hGH and cell growth in a dose-dependent manner (*Figure 2*).

Enrichment of the components resulted in a change in the osmotic pressure of the medium. Therefore, the osmotic pressure of the medium was adjusted by adding sodium chloride. The medium was buffered by using dihydroxyethylglycine (DHEG) at 1.8g/l (*TABLE 1*).

2-2. NEED FOR THE ADDITION OF NEW PROTEIN FACTORS

The UC203 medium required growth factors such as catalase, fibronectin, human transferrin and insulin. It was found that the addition of a new protein factor was effective for the production of hGH. Growth curves for CHO-K1 and GT19 in UC203 supplemented with the new protein factor are shown in *Figure 3*.

3. Conclusion

We chose a CHO cell line producing a high level of hGH as the model to find an appropriate chemically defined medium for industrial use. UC203 medium was better compared with alpha-MEM supplemented with 5% DFCS in terms of supporting both cell growth and enhancing the production of biologically active substances.

References

[1] Hata, J., Tamura, T., Yokoshima, S., Yamashita, S., Kabeno, S., Matsumoto, K. and Onodera, K. (1992) 'Chemically defined medium for the production of biologically active substances of CHO cells', Cytotechnology 10, 9 - 14.

[2] Stanners, C. P., Eliceiri, G. L. and Green, H. (1971) 'Two types of ribosome in mouse-hamster hybrid cells', Nature New Biology 230, 52 - 54.

TABLE 1. Chemically defined medium for the production of biologically active substances by CHO

Chemical component	mg/l	Chemical component	mg/l
Essential amino acids		*Purines and pyrimidines*	
L-Arginine-HCl	631.98	Hypoxanthine	2.00
L-Cysteine-HCl-H2O	135.25	Thymidine	0.40
L-Glutamine	584.00	*Other organic compounds*	
L-Histidine-HCl-H2O	209.50	Succinic acid	23.60
L-Isoleucine allo-free	262.50	Choline bitartrate	125.40
L-Leucine	262.50	D-(-)-Fructose	4000.00
L-Lysine-HCl	913.25	D-(+)-Mannose	1000.00
L-Methionine	14.90	N-Acetyl-D-glucosamine	220.00
L-Phenylalanine	33.00	Glutathion (reduced)	1.00
L-Threonine	47.60	i-Inositol	200.00
L-Tryptophan	10.20	Putrescine-2HCl	0.16
L-Tyrosine	36.20	Sodium pyruvate	110.00
L-Valine	46.90	*Major inorganic ions*	
Nonessential amino acids		Calcium Chloride	200.00
L-Alanine	25.00	Potassium Chloride	400.00
L-Asparagine-H2O	400.00	Sodium Chloride	4728.00
L-Aspartic acid	30.00	NaH2PO4-H2O	140.00
L-Glutamic acid	75.00	Magnesium sulfate	97.70
Glycine	50.00	*[Trace elements]*	
L-Proline	150.00	FeSO4-7H2O	0.80
L-Serine	200.00	ZnSO4-7H2O	1.76
Vitamins		*Buffers and indicators*	
D-Biotin	0.10	Sodium bicarbonate	2000.00
Folic acid	1.00	Dihydroxyethylglycine	1800.00
Lipoic acid	0.20	Phenol red-Na	5.00
Niacinamide	1.00	*Supplementation for clonal growth*	
Ca Pantothenate	1.00	Catalase	50.00
Pyridoxal-HCl	1.00	Fibronectin	5.00
Riboflavin	0.10	Human transferrin (Holo type)	10.00
Thiamine-HCl	1.00	Insulin	1.00
Vitamin B12	1.40	New protein factor	1000.00

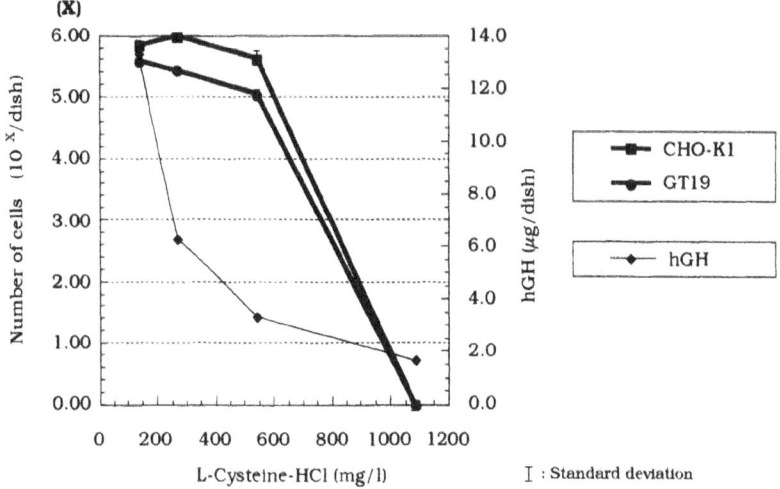

Figure 1. Effect of L-cysteine-HCl on GT19 and CHO-K1.

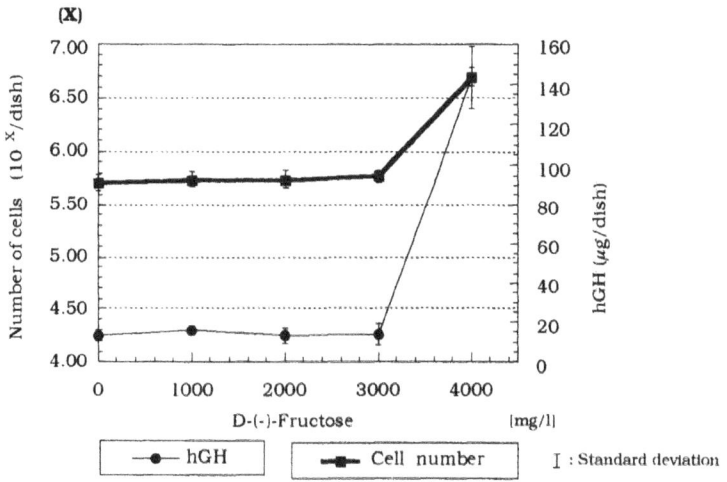

Figure 2. Effect of D-(-)-fructose on GT19.

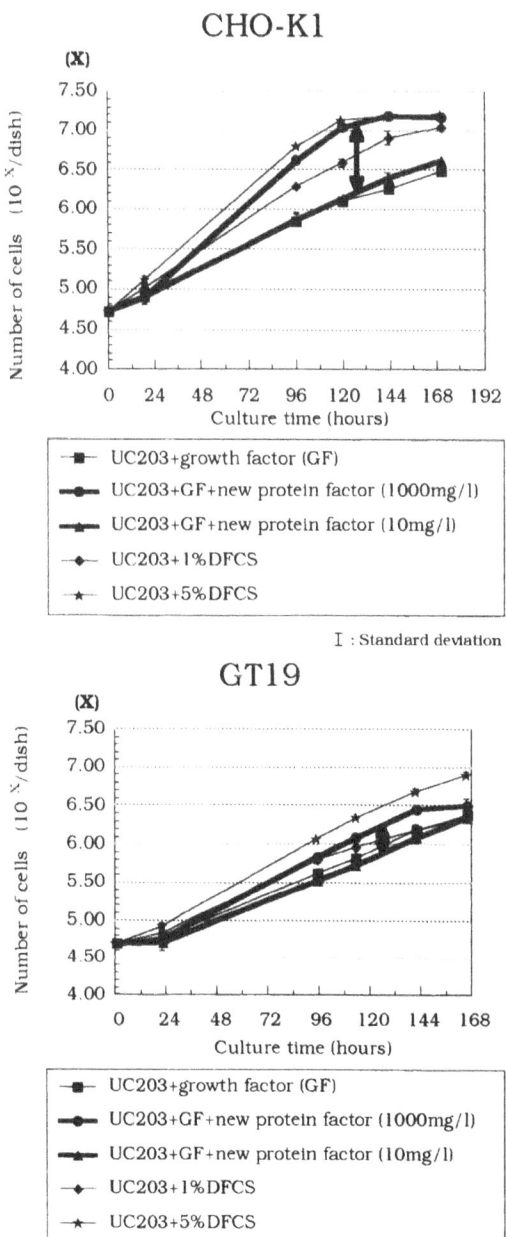

Figure 3. Growth curves for CHO-K1 and GT19 in UC203 supplemented with the new protein factor.

Stable Production of Pro-urokinase and its Derivative by Namalwa KJM-1 Cells
Adapted to a Serum-free Medium

Shinji Hosoi, Mitsuo Satoh, Hiromasa Miyaji, Tatsunari Nishi,
Tamio Mizukami, Mamoru Hasegawa, Seiga Itoh, and Tatsuya
Tamaoki
Tokyo Research Laboratories, Kyowa Hakko Kogyo Co. Ltd.
3-6-6 Asahi-machi, Machida-shi, Tokyo 194, Japan

Abstract
    We have previously reported that Namalwa KJM-1 cells adapted to a
serum-free medium and derived from Namalwa (B lymphoblastoid) cells were
more useful than CHO (Chinese hamster ovary) cells as the host cell line for
producing various recombinant proteins with a dhfr gene coamplification
method, because CHO cells secrete a cysteine endopeptidase (Satoh, 1990).
    An expression plasmid for pro-UK was constructed and introduced into
CHO and Namalwa KJM-1 cells, and cells resistant to methotrexate (MTX) were
then obtained. A Western blot analysis of pro-UK produced by Namalwa KJM-1
showed a single chain form, without the addition of protease inhibitors.
However, pro-UK produced by CHO was partly cleaved to a two-chain form that
was not activatable by plasmin, even with the addition of Aprotinin (10KIU/ml).
    Pro-UKS1 was designed as a thrombin-resistant derivative of pro-UK by
introducing sugar chains, Phe being substituted with Asn at the 164th amino
acid residue of pro-UK. An expression plasmid for pro-UKS1 was constructed
and introduced into Namalwa KJM-1 cells, and cells resistant to MTX were then
obtained. Among them, the highest pro-UKS1 producer (resistant to 500nM of
MTX), clone 41-8, was selected and further characterized. Clone 41-8 was
cultured in a serum-free (ITPSGF) medium, and under conventional conditions,
the maximal productivity of pro-UKS1 was about 10μg/ml/day. The addition of
glucose and tri-iodothyronine was effective for increasing the productivity, the
maximal productivity of pro-UKS1 being 67μg/ml/day. In this conditioned
medium, the content of pro-UKS1 was above 80% of the total protein.
    These results indicate that Namalwa KJM-1 cells are a useful host cell
line for producing recombinant proteins susceptible to proteases.

Introduction
    We have previously shown that a foreign gene expression system, using
recombinant Namalwa KJM-1 cells, and a perfusion culture system were
preferable for obtaining recombinant proteins in large quantities (Miyaji,1990a;

S. Kaminogawa et al. (eds.), Animal Cell Technology: Basic & Applied Aspects, Vol. 5, 63–70.
© 1993 Kluwer Academic Publishers.

Hosoi, 1991), and that the dhfr gene amplification method was available to this cell line (Miyaji,1990b; Hosoi, 1991).

In order to facilitate the purification of a target protein, it is important to avoid degrading of the protein. Thus, it is necessary to assay the protease activity in a serum-free conditioned medium of the host cells. We have reported that CHO secreted a cysteine endopeptidase in the conditioned medium (Satoh, 1990), and suggested that CHO was not suitable for the production of recombinant proteins susceptible to proteases. Pro-UK is a single-chain molecule as a pro-type of urokinase, and can be proteolytically converted by plasmin to a two-chain form of 54kDa molecular weight at amino acid positions 158 and 159 (Holmes, 1985). Besides this conversion, pro-UK would be cleaved by thrombin between Arg-Phe at amino acid positions 156 and 157. This form of pro-UK was inactive and had a 33kDa molecular weight by a Western blotting analysis under the reduced condition (Ichinose, 1986). Pro-UKS1 was designed as a thrombin-resistant derivative of pro-UK by introducing a glycosylation site, Phe being substituted with Asn at the 164th amino acid residue of pro-UK.

In this paper, we demonstrate the stable production of pro-urokinase and its derivative as examples of proteins susceptible to proteases by Namalwa KJM-1 cells adapted to a serum-free medium.

Materials and Methods

Cells and culture medium   Namalwa cells, a human lymphoblastoid cell line, were provided by Mr. F. Klein (Frederick Cancer Research Center, Frederick, Maryland, USA). They were adapted to a serum- and albumin-free RPMI-1640 medium supplemented with 4-(2-hydroxyethyl)-1-piperazineethanesulfonic acid (HEPES, 10mM), L-glutamine (4mM), penicillin (25U/ml), streptomycin (25μg/ml), insulin (3μg/ml), transferrin (5μg/ml), sodium pyruvate (5mM), sodium selenite (125nM), galactose (1mg/ml), and Pluronic F68 (1mg/ml); we call this the ITPSGF medium (Hosoi,1988).

Chemicals and materials   MTX was purchased from Sigma Chemicals Co., St. Louis, MO, USA., and RPMI-1640 medium was from Nissui Pharmaceuticals Co., Tokyo, Japan. Peptidyl-MCAs and 7-amino-4-methylcoumarylamide (AMC) were from Peptide Institute, Osaka, Japan. All other chemicals were obtained as previously described (Hosoi,1988; Miyaji,1990c).

Protease activity assay   The activities of proteases in the serum-free conditioned media were measured as described by Satoh (1990).

DNA manipulation   All enzymes were purchased from Takara Shuzo, Kyoto, Japan. Plasmid DNA preparation, DNA fragment purification, *E. coli* DNA polymerase I reaction, ligation, and the introduction of plasmid DNA into *E. coli* were done as described by Nishi (1984).

Construction of the expression vector of pro-UK and pro-UKS1
Expression plasmids pSE1UK1SEd1-3 (10.5kb, Fig. 1) and pMo1UKS1SEd1-5 (10.5kb, Fig. 2) were constructed. pSE1UK1SEd1-3 consisted of the following

four DNA fragments: (i) A 1.4kb HindIII-BamHI fragment, containing a part of 75 bp of the 5' non-coded region and a part of human pro-UK, which was cloned from human pharyngeal carcinoma cell line Detroit 562, using the Okayama-Berg cloning system previously described by Miyaji (1989). The amino acid sequence of pro-UK was identical with that previously reported (Holmes, 1985). The BamHI site was converted to a KpnI site by inserting synthetic DNA according to sequence A (68bp). (ii) A 1.6kb KpnI-ClaI (blunt) fragment containing rabbit β-globin RNA processing signals for splicing and polyadenylation from pSE1βd2-4 (Miyaji, 1990b). (iii) A 2.7kb XhoI (blunt)-ClaI fragment carrying a dhfr transcription unit from pSE1βd2-4. (iv) A 4.7kb ClaI-HindIII fragment containing the SV40 early promoter, an ampicillin-resistance gene, and a G418 gene from pSE1βd2-4.

(A) 5'- GATCCGCAGT CACACCAAGG AAGAGAATGG CCTGGCCCTC
      3'- GCGTCA GTGTGGTTCC TTCTCTTACC GGACCGGGAG
      TGAGGGTCCC CAGGGAGGAA ACGG GTAC -3'
      ACTCCCAGGG GTCCCTCCTT TGCC -5'

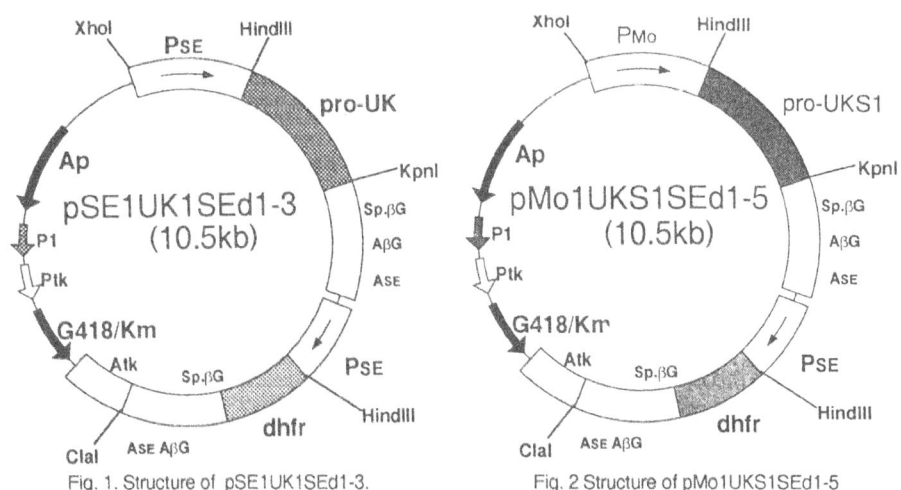

Fig. 1. Structure of pSE1UK1SEd1-3.   Fig. 2 Structure of pMo1UKS1SEd1-5

pMo1UKS1SEd1-5 (10.5kb: Fig. 2) consisted of the following three DNA fragments: (i) A 0.6kb ClaI-SmaI fragment containing Molony murine leukemia virus promoter derived from ECO-MOL LTR (Adachi, 1987). The ClaI site was converted to an XhoI site by inserting a synthetic DNA according to sequence 1. The SmaI site was converted to a HindIII site by inserting a synthetic DNA according to sequence 2. (ii) A 8.4kb XhoI-KpnI fragment from pSE1UK1SEd1-3. (iii) A 1.4kb HindIII-BamHI fragment from pSE1UK1SEd1-3, which was cloned between the HindIII and BamHI site of M13mp18 phage

DNA, the single-strand DNA being isolated. Phe was substituted with Asn at the 164th amino acid residue of pro-UK by site-directed mutagenesis, using a synthetic DNA primer according to sequence 3. The *Bam*HI site was converted to a *Kpn*I site by inserting a synthetic DNA according to sequence 4 (49bp).

(1) 5'- TCGAGGACC -3' (2) 5'- pCAAGCTTG- 3'
    3'- CCTGGGC -5' (3) 5'- GGGGAGAAAACACCACC -3'
(4) 5'- GATCCGCAGT CACACCAAGG AAGAGAATGG CCTGGCCCTC
    3'- GCGTCA GTGTGGTTCC TTCTCTTACC GGACCGGGAG
TAGAGGTAC -3'
ATCAC -5'

Electroporation An SSH-1 somatic hybridizer (Shimadzu Seisakusyo, Kyoto, Japan) and an SSH-C13 chamber (2mm electrode distance) were used, details of the procedures having been previously described by Miyaji (1990a).

Selection of MTX-resistant subclones Selection of the MTX-resistant subclones has been previously described by Hosoi (1991).

SDS-PAGE SDS-PAGE under reducing conditions was performed by the method of Laemmli (1970). Proteins on the gel were stained with Coomassie Brilliant Blue R-250 by the method of Burgess (1969). All samples were concentrated four times and then applied for SDS-PAGE.

Western blot analysis Culture supernatants were developed by SDS-PAGE and transferred to a nitrocellulose membrane by the method of Towbin (1979). The filter was washed and incubated with monoclonal antibody to urokinase (KM492). After being washed, the filter was incubated with rabbit anti-mouse IgG conjugated with horseradish peroxidase (HRP; DAKO Japan Co., Kyoto, Japan) and then developed by HRP colour reagent (Bio-Rad, Richmond, USA).

Measurement of pro-UK and pro-UKS1 by sandwich-type ELISA Pro-UK and pro-UKS1 were measured by sandwich-type ELISA, purified rabbit polyclonal anti-human pro-UK antibody (Japan Chemical Research) and purified KM492 being used. As a standard, purified pro-UK and pro-UKS1 obtained from a conditioned medium of clone 4-2 (Namalwa KJM-1-derived pro-UK producer) or from clone 41-8 (pro-UKS1) conditioned medium was used.

Results
Comparison of protease activity
    A comparison of protease activity in the serum-free conditioned medium between CHO and Namalwa KJM-1 cells is shown in Fig. 3. In the case of CHO, several kinds of protease activity were detected, these being specified mainly as trypsin-like cysteine endopeptidase. However, in the case of Namalwa KJM-1 cells, there was little endo-peptidase activity.

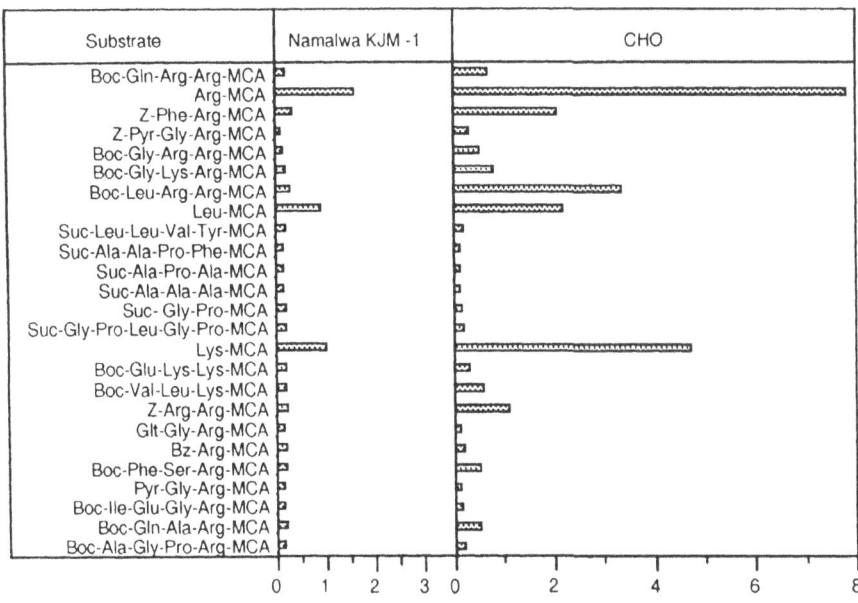

Activity (nmol/hr·ml)

Fig. 3. Protease activity in the serum-free conditioned medium.

Western blotting analysis of pro-UK

To confirm these findings, an expression plasmid for pro-UK, a model protein susceptible to proteases, was introduced to CHO and Namalwa KJM-1 cells.

After introducing pro-UK expression plasmid, pSE1UK1SEd1-3 (Fig. 1), to the CHO and Namalwa KJM-1 cell, cells resistant to G418 and MTX were obtained. Pro-UK produced by Namalwa KJM-1 and CHO was then analyzed by Western blotting (Fig. 4). Pro-UK produced by Namalwa KJM-1 showed a single-chain form without the addition of protease inhibitors. However, pro-UK produced by CHO was partly cleaved to an inactive two-chain form, even with the addition of Aprotinin (10KIU/ml). This degradation was inhibited by Leupeptin (data not shown).

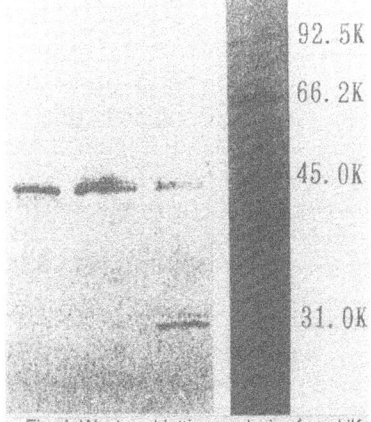

Fig. 4. Western-blotting analysis of pro-UK.

Productivity of pro-UKS1

An expression plasmid for pro-UKS1, pMo1UKS1SEd1-5 (Fig. 2), was introduced to Namalwa KJM-1, and cells resistant to G418 and MTX were obtained. Among them, the highest pro-UKS1 producer, clone 41-8, resistant to 500nM of MTX, was selected and further characterized. The productivity of pro-UKS1 by clone 41-8 under conventional conditions is shown in Fig. 5, the concentration of pro-UKS1 being about 30μg/ml.

Media optimization

The basal medium, RPMI-1640, contained 2.0g/l of glucose, which was enriched. Additional glucose (2.0g/l) had no effect on growth, but was effective for productivity (data not shown).

It has been reported that the Molony murine leukemia virus promoter region, which was used in pMO1UKS1SEd1-5, contained a thyroid hormone-responsible element (Sap, 1989), and it was predicted that tri-iodothyronine would be effective toward the productivity of pro-UKS1. In comparison with glucose enrichment control, 100nM of tri-iodothyronine had a greater effect on the productivity of pro-UKS1, the maximal productivity reaching 67μg/ml/day.

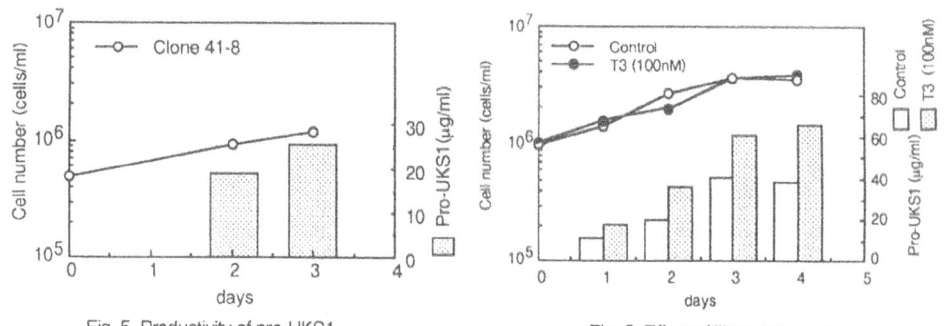

Fig. 5. Productivity of pro-UKS1.　　　　Fig. 6. Effect of Tri-iodothyronine.

An SDS-PAGE analysis of the conditioned medium (CM) was performed under reducing conditions (Fig. 7). 70μg/ml of purified pro-UKS1 showed a 57kDa smear band. In comparison with ITPSGF medium containing 5μg/ml of transferrin (75kDa, Tf), CM showed a pale band corresponding to Tf. CM containing 67μg/ml of pro-UKS1 showed a similar dense band corresponding to 70μg/ml of purified pro-UKS1 at 57kDa. From a densitometric analysis of this CM, the content of pro-UKS1 was above 80% of the total proteins.

Discussion

Namalwa KJM-1 cells adapted to a serum-free medium showed aminopeptidase activity and a small degree of endo-peptidase activity in the conditioned medium (Fig. 3). Pro-UK produced by Namalwa KJM-1 was mainly composed of a single-chain form, without the addition of protease inhibitors

(Fig. 4). In contrast, pro-UK produced by CHO was partly cleaved to an inactive two-chain form, even with the addition of Aprotinin (a serine protease inhibitor; 10KIU/ml). This degradation was inhibited by the addition of Leupeptin (data not shown). Therefore, in the case of a serum-free culture of CHO cells, appropriate protease inhibitors must be added for stable production of those proteins susceptible to proteases. Thus, Namalwa KJM-1 cells seem to be suitable for producing recombinant proteins susceptible to proteases.

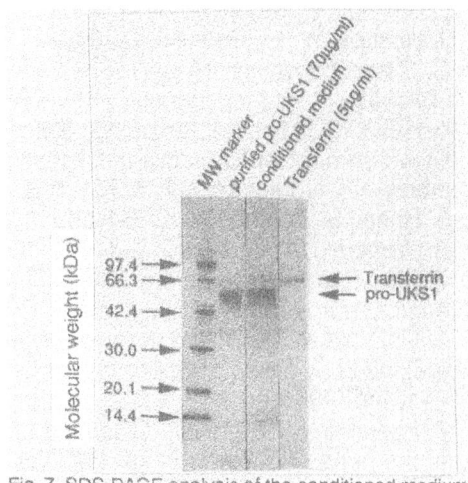

Fig. 7. SDS-PAGE analysis of the conditioned medium.

Glucose enrichment had no effect on growth, but was effective on the productivity of pro-UKS1; however, enrichment with serine and cysteine, and pH control were not effective for the productivity of pro-UKS1 (data not shown). These results are opposite of the results we have previously reported (Hosoi, 1991). Thus, it was necessary to optimize the culture conditions for each product and each clone.

As it has been reported that the Molony murine leukemia virus promoter region, which was used in pMO1UKS1SEd1-5, contained a thyroid hormone responsible element (Sap, 1989), it was predicted that tri-iodothyronine would be effective for the productivity of pro-UKS1. In comparison to glucose enrichment control, 100nM of tri-iodothyronine was more effective for the productivity of pro-UKS1 (Fig. 6), the highest productivity of pro-UKS1 reaching 67µg/ml/day. In this conditioned medium, the content of pro-UKS1 was above 80% of the total proteins (Fig. 7). Since the thyroid hormone and thyroid hormone-binding protein were present in a concentration of the nanomolar order in fetal calf serum, this enhancement of productivity could not be detected in the serum-containing medium.

These results indicate that Namalwa KJM-1 is a useful host cell line for producing recombinant proteins susceptible to proteases.

References
1. Adachi, A., Sakai, K., Kitamura, N., Nakanishi, S., Niwa, O., Matsuyama, M. and Ishimoto, A. (1984) "Characterization of the env gene and long terminal repeat of molecularly cloned friend mink cell focus-inducing virus DNA", J. Virol., 50, 813-821.
2. Burgess, R. R. (1969) "Separation and characterization of the subunits of

ribonucleic acid polymerase", J. Biol. Chem., 244, 6168-6176.
3. Holmes, W. E., Pennica, D., Blaber, M., Rey, M. W., Guenzler, W. A., Steffens, G. J. and Heyneker, H. L. (1985) "Cloning and expression of the gene for pro-urokinase in Escherichia coli", BIO/TECHNOLOGY, 3, 923-929.
4. Hosoi, S., Mioh, H., Anzai, C., Sato, S. and Fujiyoshi, N. (1988) Establishment of Namalwa cell lines which grow continuously in glutamine-free medium. Cytotechnology, 1, 151-158.
5. Hosoi, S., Murozumi, K., Sasaki, K., Miyaji, H., Satoh, M., Hasegawa, M., Itoh, S., Tamaoki, T. and Sato, S. (1991) "Optimization of cell culture conditions for G-CSF (granulocyte colony-stimulating factor) production by genetically engineered Namalwa KJM-1 cells", Cytotechnology, 7, 25-32.
6. Ichinose, A., Fujikawa, K. and Suyama, T. (1986) "The Activation of Pro-urokinase by Plasma Kallikrein and Its Inactivation by Thrombin", J. Biol. Chem., 261, 3486-3489.
7. Laemmli, U. K. (1970) "Cleavage of structural proteins during the assembly of the head of bacteriophage T4", Nature, 227, 680-685.
8. Miyaji, H., Nishi, T. and Itoh, S. (1989) "Expression of human lymphotoxin derivatives in Escherichia coli and comparison of their biological activity in vitro", Agric. Biol. Chem. 53, 277-279.
9. Miyaji, H., Mizukami, T., Hosoi, S., Sato, S., Fujiyoshi, N. and Itoh, S. (1990) "Expression of human beta-interferon in Namalwa cells which were adapted to serum-free medium", Cytotechnology, 3, 133-140.
10. Miyaji, H., Harada, N., Mizukami, T., Sato, S., Fujiyoshi, N. and Itoh, S. (1990b) "Efficient expression of human beta-interferon in Namalwa cells adapted to serum-free medium by a dhfr gene coamplification method", Cytotechnology, 4, 173-180.
11. Nishi, T., Saito, A., Oka, T., Itoh, S., Takaoka, C. and Taniguchi, T. (1984) "Construction of plasmid expression vectors carrying the Escherichia coli tryptophan promoter", Agric. Biol. Chem., 48, 669-675.
12. Sap, J., Munoz, A., Schmit, J., Stunnenberg, H. and Vennstrom, B. (1989) "Repression of transcription mediated at a thyroid hormone response element by the v-erb-A oncogene product", Nature, 340, 242-244.
13. Satoh, M., Hosoi, S. and Sato, S. (1990) "CHO (chinese hamster ovary) cells spontaneously secrete a cysteine endopeptidase", In Vitro Cell. Dev. Biol., 26, 1101-1104.
14. Towbin, H., Staehelin, T. and Gordon, J. (1979) "Electrophoretic transfer of proteins from polyacrylamide gels to nitrocellulose sheets", Proc. Natl. Acad. Sci. USA., 76, 4350-4354.

PRODUCTION AND CHARACTERIZATION OF RECOMBINANT SOLUBLE FORM ERYTHROPOIETIN RECEPTOR

Masaya Nagao, [1] Seiji Masuda, [1] Satoshi Abe, [1] Masatsugu Ueda, [2] and Ryuzo Sasaki [1]

[1]Department of Food Science and Technology, Faculty of Agriculture, Kyoto University, Kyoto 606, Japan and [2]Research Institute of Life Science, Snow Brand Milk Products Company Ltd., Tochigi 329–05, Japan

1. SUMMARY

A recombinant soluble form (sEPO–R) of erythropoietin (EPO) receptor (EPO–R) was produced by Chinese hamster ovary cells. One subclone, N14.2 could produce sEPO–R more than $40 \mu g/10^6$cells/day maximally. sEPO–R was isolated easily with EPO fixed gel in a high yield. Affinity of sEPO–R to EPO was determined by precipitating sEPO–R·radiolabeled EPO complex using anti EPO–R antibody and polyethylene glycol. The results showed a Kd of 13 nM which was much lower than those for cellular EPO–R. One $\underline{N}$–glycosylation site exists in sEPO–R but the glycosylation did not affect the binding affinity to EPO. A complex with a molecular size that corresponded to a 1:1 complex of EPO and sEPO–R was detected by gel filtration analysis.

2. INTRODUCTION

EPO, a major regulator of red blood cell production, supports survival of erythroid precursor cells and stimulates their proliferation and maturation through interaction with EPO–R (see Ref. 1 for review). Two cDNA clones encoding murine EPO–R have been isolated. One encodes the full–length receptor composed of total 507 amino acid residues including a 24 amino acid putative signal peptide, a 225 amino acid extracellular region, and a 236 amino acid cytoplasmic domain (2). A second cDNA encodes a sEPO–R, which consists of 24 amino acids in the putative signal sequence, 221 amino acids in the extracellular domain, and 20 amino acids newly added at the C–terminus (3). This cDNA is derived from the alternative splicing of mRNA transcribed from a single EPO–R gene and this product is able to bind EPO because the conditioned medium of COS cells transfected with the vector containing this cDNA is inhibitory on the binding of EPO to the cells expressing the membrane–bound EPO–R (3). The presence of sEPO–R in biologic fluids remains to be investigated. To study the physiological function of sEPO–R, powerful tools such as the purified sEPO–R and a detection method for sEPO–R with antibodies are necessary. Characterization of the complex between sEPO–R and

---

Abbreviations: EPO, erythropoietin; EPO–R, erythropoietin receptor; sEPO–R, soluble form of EPO–R; rHuEPO, recombinant human EPO; PBS, 10 mM phosphate buffered saline, pH 7.4; SDS, sodium dodecyl sulfate; MEM,α–minimum essential medium; FCS, fetal calf serum; dhfr, dihydrofolate reductase; MTX, methotrexate; PEG, polyethylene glycol; BSA, bovine serum albumin.

S. Kaminogawa et al. (eds.), Animal Cell Technology: Basic & Applied Aspects, Vol. 5, 71–77.

EPO may provide an important clue for understanding the mechanism of EPO action. The present work was undertaken to develop an efficient production system of murine recombinant sEPO-R and to characterize the ligand binding with the isolated sEPO-R.

3. MATERIALS AND METHODS

Expression and isolation of the wild-type and glycosylation-defective mutant sEPO-Rs. A cDNA of sEPO-R was constructed with three DNA fragments as described previously(4). Three fragments were ligated to yield the cDNA encoding sEPO-R that was composed of 24 amino acids in putative signal peptide, 221 amino acids in extracellular region of EPO-R, and 20 amino acids added at the C-terminus.

The expression plasmid of sEPO-R was constructed by ligating the HindIII-XbaI sEPO-R cDNA to pcDNAI·dhfr at HindIII and XbaI sites in multicloning sites as described previously(4). The plasmid for production of the N-glycosylation-defective mutant sEPO-R was constructed by the same procedures except that the DNA fragment for the N-terminal region in which Asn-51 was replaced by Gln was used. The plasmids were introduced into CHO dhfr⁻ cells(5) by the calcium phosphate method. Cells were cotransfected with pKSV10neo. The neo resistant cells were selected by culture in the presence of 400 µg/ml of G418. The dhfr⁺ clones with high production of sEPO-R were established with the amplification medium containing nucleoside-deprived MEM and 10% dialyzed FCS, increasing MTX from 25 to 100 nM. Production of sEPO-R in the culture media was monitored by measuring the inhibitory effect of the media on the radioiodinated rHuEPO binding to EPO-R on TSA8 cells(6). The culture medium of parental CHO cells had no effect on the binding.

Cells producing sEPO-R were maintained in the amplification medium. For production of sEPO-R, the cells were cultured to confluency and then the medium was replaced with the production medium, OPTI-MEM (GIBCO) containing 0.5% FCS. For isolation of sEPO-R, 90 ml of the culture medium was mixed in a tube containing 3 ml of rHuEPO-fixed CH Sepharose 4B gel (15 mg fixed EPO/ml gel) and the tube was rotated end over end for 12 h at 4°C. The tube content was put into a column and the column was washed thoroughly by PBS. Then sEPO-R was eluted by PBS containing 1.5 M MgCl₂. The eluted fractions containing sEPO-R were purified again with the EPO-fixed gel. Proteins were measured with a protein determination kit (Bio-Rad) using γ-globulins as a standard.

Antibody. The rabbit anti-N-terminal mouse EPO-R antiserum was prepared as described previously (7) using the peptide (APSPSLPDPKFESKAC) conjugated to keyhole limpet hemocyanin as an antigen. The anti-rHuEPO antiserum and the anti-sEPO-R antiserum were obtained by injection of rHuEPO and the isolated sEPO-R to rabbits, respectively.

Binding of radioiodinated rHuEPO to sEPO-R. The affinity of sEPO-R for rHuEPO(8) was estimated by precipitating the complex among sEPO-R, radiolabeled rHuEPO and anti-N-terminal EPO-R antibody by PEG in the presence of various concentration of unlabeled rHuEPO. The reaction mixture contained 15 pM radiolabeled rHuEPO, 3-200 nM sEPO-R, and 0.1% BSA in a total 300µl of PBS. After incubation of the mixture at 4°C overnight, 45µl of the rabbit anti-N-terminal EPO-R antiserum was added. The mixture was incubated again at 4°C overnight. To the mixture, 30µl of 1% γ-globulins and then 600µl of cold PBS containing 22.5% PEG6000 were added. The mixture was kept on ice for 10 min and then centrifuged at 4°C for 10 min at 12,000xg. The radioactivity associated with the precipitate was counted. Nonspecific binding was measured in the absence of sEPO-R or the antiserum. Both cases gave similar values of nonspecific binding.

Western blotting . For analyses with western blotting technique, proteins were fractionated with SDS–polyacrylamide gel electrophoresis and then electrotransferred to a nitrocellulose filter. The filter was incubated with the block ace (Snow Brand Milk Products) at 4°C overnight, with the rabbit anti–EPO antiserum for 1 h for detection of EPO or the rabbit anti–sEPO–R antiserum for detection of sEPO–R, and with the peroxidase conjugated goat anti–rabbit IgG antiserum (Capel) for 1h, in this order. The filter was soaked in an ECL–western blotting detection system buffer (Amersham) to detect the antigens on the filter by chemiluminescence.

Gel filtration analysis of EPO·sEPO–R complex. The size and stoichiometry of the complexes was established by separating mixture of EPO and sEPO–R(in ratios of 3:1, 2:1, 1:1, 1:2, 1:3) by gel filtration. The concentration of EPO was fixed at 18μM except at 2:1 and 3:1, where EPO was 36μM and 54μM, respectively. Protein mixtures in PBS(100μl) was incubated at either 4°C for 12h or 37°C for 1h, and analyzed by high performance liquid chromatography on a gel filteration column, TSK–G3000SW.

4. RESULTS

Fig. 1 shows the expression plasmid of sEPO–R and dhfr. Two CHO clones producing sEPO–R have been established. One clone, N14, produces the wild type sEPO–R in the presence of 25 nM MTX and the second clone, M19, produces the $\underline{N}$–glycosylation defective mutant sEPO–R in the presence of 2.5 nM MTX. In the mutant sEPO–R, Asn–51 responsible for $\underline{N}$–glycosylation of the wild type sEPO–R was replaced with Gln. Production of sEPO–R from both clone was detected by competitive binding assay. The culture media of these clones inhibited the binding of ^{125}I–rIIuEPO to EPO–R on TSA8 cells indicating that both the glycosylated and unglycosylated forms of sEPO–R are capable of binding with EPO(Fig. 2).

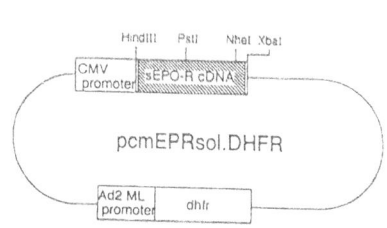

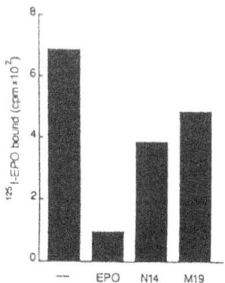

Fig. 1. **A Plasmid for expression of mouse sEPO–R.**The coding region of sEPO–R cDNA is hatched. Transcription of sEPO–R and dhfr is under control of the cytomegalovirus(CMV) promoter and adenovirus 2 major late(Ad2ML) promoter, respectively.

Fig. 2. **Competition assay of sEPO–R.** Each 60μl of conditioned medium of N14 and M19 was assayed. "–" represents the total counts in the absence of competitor and "EPO" represents the nonspecific counts in the presence of excess unlabeled rHuEPO as an competitor.

A high producing clone of sEPO–R, N14.2 was established after N14 was cultureed in the presence of 100 nM MTX and subcloned by limiting dilution. N14.2 produces the wild type sEPO–R, secreting 20~50μg sEPO–R /ml culture medium/day/10^{6}cells for at least 9 days in production medium(Fig. 3).

The wild–type sEPO–R in the culture medium of N14.2 was purified by binding to the EPO–fixed gel and elution with MgCl$_2$. The eluted sEPO–R was purified again with

74

the same procedure. An acidic condition (pH ~ 4.5) could elute sEPO–R from the EPO–fixed gel, but the recovered activity was about 30% of that with MgCl$_2$. The sEPO–R (1.2 mg) was isolated from 90 ml of the medium with 17–fold purification and 41% activity recovery. Fig. 4A shows SDS–polyacrylamide gel electrophoresis of proteins in the sEPO–R preparations. Two bands were detected in the purified sEPO–R; a major band had a size of 33 kDa, and a minor band with 29 kDa comprised about 5% of the total. The size of the minor band is little larger than Mr 26139 predicted from sEPO–R cDNA. As described below, the major band is the glycosylated form of sEPO–R and the minor band is the unglycosylated form. The EPO–fixed gel was quite effective for purification of sEPO–R; the sEPO–R preparation eluted from the first gel was almost pure. The glycosylated and unglycosylated forms could be separated by gel filtration with TSK–G3000SW (data not shown). Their molecular masses calculated from the elution positions with TSK–G3000SW were similar to those estimated from SDS–polyacrylamide gel electrophoresis. A very small amount (0.1% of the total) of the protein with a size equivalent to the homodimer was detected with TSK–G3000SW (data not shown). With the same purification procedures, the N–glycosylation defective sEPO–R was isolated.

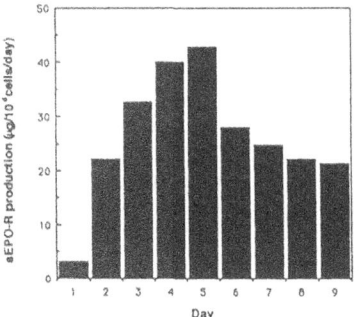

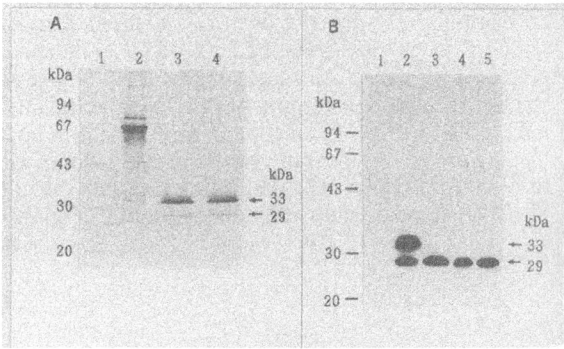

Fig. 3. **Time course of sEPO–R production** N14.2 was cultured to confluency and the medium was changed everyday. Conditioned medium was analysed by inhibitory binding assay using TSA8 cells(6).

Fig. 4. **SDS–PAGE and Western blotting analysis** of **sEPO–R** (A) 12% SDS–PAGE under reducing condition(lane 1 marker, lane 2. 1 μg proteins in the conditioned medium of N14.2, lane 3. 300 ng sEPO–R (first gel eluate), lane 4. 300 ng sEPO–R (second gel eluate). The gel was stained with silver. (B) Western blotting analysis with the rabbit anti–N–terminal EPO–R antiserum(lane 1. conditioned medium of parental CHO cells, lane 2. 10ng sEPO–R from N14.2, lane 3. 2ng sEPO–R N–glycanase treated sEPO–R from N14.2, lane 4. 2ng sEPO–R from N14.2 cultured with 1 μg/ml tunica–mycin, lane 5. 2 ng sEPO–R from M19.

Fig. 4B shows western blotting using the rabbit anti–N–terminal EPO–R antiserum to demonstrate that the minor component in the purified wild–type sEPO–R is an unglycosylated sEPO–R. The isolated wild–type sEPO–R was treated with N–glycanase for removal of the N–linked sugar (lane 3). The wild–type sEPO–R was isolated from the medium of N14.2 cells cultured in the presence of tunicamycin (lane 4). The glycosylation defective sEPO–R was isolated from the cultured medium of M19 cells (lane 5). They all migrated as a single band with 29 kDa, equivalent to the component with a smaller size in the isolated wild–type sEPO–R (lane 2). No bands were present in the sample prepared from the conditioned medium of nontransfected parental CHO cells (lane 1). Similar results were obtained when the conditioned media of parental CHO cells, N14.2 cells, and M19 cells were directly subjected to the western blotting analysis.

It was predicted from the cloned cDNA that the amino acid sequence of the N–terminal region of the mature EPO-R was APSPSLDPKFE(2). The sequence determined with the glycosylated form of wild–type sEPO-R was PSPSLDPKFE ; the N–terminal Ala predicted from cDNA had been removed. The same sequence missing N–terminal Ala was found with both the unglycosylated wild–type sEPO-R and the glycosylation defective mutant sEPO-R. The amino acid at the N–terminus of the transmembrane EPO-R remains unknown.

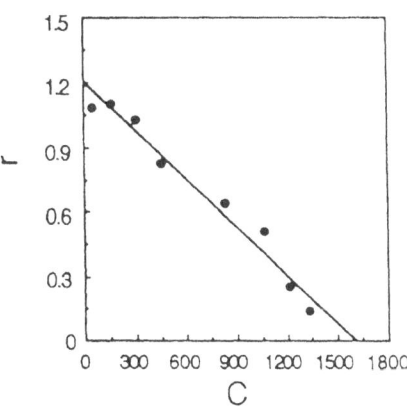

Fig. 5. **Binding of sEPO-R with EPO** Radiolabeled rHuEPO and glycosylated sEPO-R were kept at a constant 15 pM and 100 nM, respectively and the unlabeled rHuEPO was veried from 1–100 nM as a competitor. The ordinate, r, represents ratios of the bound radiolabeled rHuEPO to the free radioactive ligand, and the abscissa, C, represents ar/(1 +r) where a is the total EPO(ng/ml). Each point was the mean of sextuplicate assays. A Kd value of 13 nM was calculated from the slope of a linear Scatchard plot.

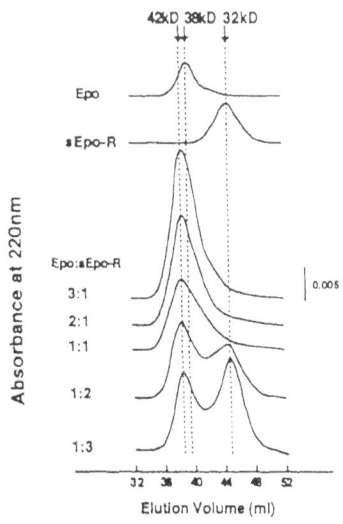

Fig. 6. Gel filtration analysis of complex formation between EPO and sEPO-R. Each mixture (100 μl) were chromatographed on a TSK-G3000SW gel filtration column (Toso). Peaks were monitored for absorbance at 220 nm.

Binding of sEPO-R to EPO was assayed by competitive inhibition by unlabeled EPO against the formation of the ternary complex among sEPO-R, ^{125}I-EPO and anti N–terminal EPO-R antibody. A Kd value of 13 nM was estimated from the slope of a straight line analyzed by Scatchard plot(Fig. 5). Anti N–terminal antibody binds sEPO-R and also EPO·sEPO-R complex without any effect on the interaction of sEPO-R with EPO. The affinity of sEPO-R for EPO was measured by two other methods and similar Kd values were estimated(4). Further, similar Kd values were obtained by using unglycosylated form of sEPO-R or by using unglycosylated or glycosylation defective mutant sEPO-R(data not shown). Asn–linked sugars of EPO affect the EPO binding to EPO-R; deglycosylation increases the affinity (6). Glycosylation of EPO-R appears to have no effects on the interaction with EPO.

The molecular sizes estimated by gel filtration were 38, 32, and 42kDa of EPO, sEPO-R, and the EPO·sEPO-R complex(Fig.6). The complex formation between different ratios of each component showed that the complex was formed in a ratio of 1:1.

5. DISCUSSION

In this paper, we described procedures for production and isolation of mouse sEPO–R and analyzed the interaction of the sEPO–R with rHuEPO. sEPO–R could be produced by subclone N14.2 more than $20\,\mu g/ml/day/10^6$ cells for 9 days. This high productivity was obtained by gene amplification using dhfr gene. In the course of the purification, sEPO–R was unstable at low pH(~ 4.5), so the elution of sEPO–R from EPO–fixed gel was performed with $MgCl_2$. Binding of sEPO–R with EPO showed a linear Scatchard plot, yielding a Kd value of 13 nM, which is much lower than those values of cellular EPO–R. Some erythroid cells including natural erythroid precursor cells show the biphasic Scatchard ligand–saturation curves with a Kd of 50 pM for high–affinity site and 500 pM for low–affinity sites, while others give linear plots with a Kd of 500 pM (see Ref. 9 for review). Although EPO–R cDNA was cloned from a mouse erythroid cell line with a single class of binding sites (Kd=240 pM), COS cells transfected with this cDNA displayed two classes of binding sites (Kd$_1$=30 pM, Kd$_2$=210 pM) (2). We infer the existance of other subunit of EPO–R which converts the affinity. Actually we have the result suggesting that the sugar chain of other subunit is crucial for the affinity when EPO–R was expressed on the surface of BHK cells(10). sEPO–R may be used as a good tool for seeking for other subunit.

Hormone–induced receptor oligomerization has been proposed as a signal transduction mechanism of the receptor family with a tyrosine kinase domain (11). More recently, a new cytokine receptor family that has no tyrosine kinase domain but has two characteristic conserved features (two pairs of cysteins and a Trp–Ser–X–Trp–Ser motif) in the extracellular region has been found (12). Oligomerization of the receptors in this family also appears to play a crucial role in construction of high–affinity sites, triggering the signal transduction pathway (13). EPO–R belongs to this new family, and oligomerization of EPO–R has been proposed (14–20) but has yet to be demonstrated. The ligand–induced homodimerization of a receptor has been demonstrated with a soluble form of growth hormone receptor (21). Our attempts to find the ligand–induced oligomerization using sEPO–R have been unsuccessful. However, this does not exclude the oligomerization of EPO–R on cell membrane, because a membrane–anchored form of EPO–R or an additional membrane component may be needed.

Soluble receptor forms have been found for many hematopoietic growth factors and also other hormone receptors(22). The physiological functions of these soluble receptors are currently not known but it is speculated that the soluble forms serve as feedback inhibitors by binding with the ligands, thereby preventing excess recruitment of the ligand–induced cells. Exploration of sEPO–R in serum and urine is currently in progress using mono– and polyclonal antibodies against sEPO–R.

6. REFERENCES

1. Krantz, S. B.(1991) 'Erythropoietin.' Blood 77, 419–434
2. D'Andrea, A. D., Lodish, H. F. and Wong, G. G. (1989) 'Expression cloning of the murine erythropoietin receptor.' Cell 57, 277–285.
3. Kuramochi, S., Ikawa, Y. and Todokoro, K. (1990) 'Characterization of murine erythropoietin receptor genes.' J. Mol. Biol. 216, 567–575.
4. Nagao, M., Masuda, S., Abe, S., Ueda, M. and Sasaki, R. (1992) 'Production and ligand binding characteristics of the soluble form of murine erythropoietin receptor.' Biochem. Biophys. Res. Commun. 188, 888–897
5. Urlaub, G. and Chasin, L. A. (1980) 'Isolation of chinese hamster cell mutants

deficient in dihydrofolate reductase activity.' Proc. Natl. Acad. Sci. USA 77, 4216–4220
6. Yamaguchi, K., Akai, K., Kawanishi, G., Ueda, M., Masuda, S. and Sasaki, R. (1991) 'Effects of site–directed removal of N–glycosylation sites in human erythropoietin on its production and biological properties.' J. Biol. Chem. 266, 20434–20439
7. Li, J.–P., D'Andrea, A. D., Lodish, H. F. and Baltimore,D. (1990) 'Activation of cell growth by binding of Friend spleen focus–forming virus gp55 glycoprotein to the erythropoietin receptor.' Nature 343, 762–764.
8. Goto, M., Akai, K., Murakami, A., Hashimoto, C., Tsuda, E., Ueda, M., Kawanishi, G., Takahashi, N., Ishimoto, A., Chiba, H. and Sasaki, R. (1988) 'Production of recombinant human erythropoietin in mammalian cells: Host–cell dependency of the biological activity of the cloned glycoprotein.' Bio/Technology 6, 67–71.
9. D'Andrea, A. D. and Zon, L. I. (1990) 'Erythropietin receptor: Subunit structure and activation.' J. Clin. Invest.86, 681–687.
10. Nagao, M., Matsumoto, S., Masuda, S. and Sasaki, R.(1993) 'Effect of tunicamycin treatment of ligand binding to the erythropoietin receptor: Conversion from two Classes of binding sites to single class.' Blood (in press)
11. Ullrich, A. and Schlessinger, J. (1990) 'Signal transduction by receptors with tyrosine kinase activity.' Cell 61, 203–212.
12. Bazan, J. F. (1990) 'Structural design and molecular evolution of a cytokine receptor superfamily.' Proc. Natl. Acad. Sci. USA 87, 6934–6938.
13. Nicola, N. A. and Metcalf, D. (1991) 'Subunit Promiscuity among Hemopoietic growth factor receptor.' Cell, 67, 1–4.
14. Sasaki, R., Yanagawa, S., Hitomi, K. and Chiba, H. (1987) 'Characterization of erythropoietin receptor of murine erythroid cells.' Eur. J. Biochem. 168, 43–48.
15. Hitomi, K., Fujita, K., Sasaki, R., Chiba, H., Okuno, Y., Ichiba, S., Takahashi, T. and Imura, H. (1988) 'Erythropoietin receptor of a human leukemic cell line with erythroid characteristics.' Biochem. Biophys. Res. Commun. 154, 902–909.
16. McCaffery, P. J., Frazer, J. K., Lin, F.–K. and Berridge, M. V. (1989) 'Subunit structure of the erythropoietin receptor.' J. Biol. Chem. 264, 10507–10512.
17. Mayeux, P., Casadevall, N., Lacombe, C., Muller, O. and Tambourin, P. (1990) 'Solubilization and hydrodinamic characteristics of the erythropoietin receptor : Evidence for a multimeric complex.' Eur. J. Biochem. 194, 271–278.
18. Mayeux, P., Lacombe, C., Casadevall, N., Chretien, S.,Dusanter, I., and Gisselbrecht, S. (1991) 'Structure of the murine erythropoietin receptor complex: Characterization of the erythropoietin cross–linked proteins.' J. Biol. Chem.266, 23380–23385.
19. Hosoi, T., Sawyer, S. T. and Krantz, S. B. (1991) 'Photoaffinity labeling of the erythropoietin receptor and its identification in a ligand–free form.' Biochemistry 30, 329–335.
20. Watowich, S. S., Yoshimura, A., Longmore, G. D., Hilton, D. J., Yoshimura, Y. and Lodish, H. F. (1992) 'Homodimerization and constitutive activation of the erythropoietin receptor.' Proc. Natl. Acad. Sci. USA 89, 2140–2144.
21. Cunningham, B. C., Ultsch, M., de Vos, A. M., Mulkerrin, M. G., Clauser, K. R., and Wells, J. A. 'Dimerization of the extracellular domain of the human growth hormone receptor by a single hormone molecule.' (1991) Science 254, 821–825.
22. Fernandez–Botran, R. (1991) 'Soluble cytokine receptors: their role in immunoregulation.' FASEB J. 5, 2567–2574

A ts-SV40 BASED NEW MAMMALIAN PLASMID VECTOR AND ITS COPY NUMBER CONTROL BY TEMPERATURE CHANGE

Hideyo Kirinaka, Shinji Iijima and Takeshi Kobayashi
Department of Biotechnology, School of Engineering, Nagoya University
Furo-cho, Chikusa-ku, Nagoya
JAPAN, 464-01

ABSTRACT. Simian virus 40 (SV40) DNA is generally used as a mammalian vector. However, stable transformants with SV40 based plasmids are difficult to obtain, since active propagation of SV40 DNA molecules kills cells. Therefore, we established a new SV40 based vector (pSVtsA), whose copy number could be controlled by temperature change. Results of Southern analyses of low molecular weight DNA indicated that the DNA propagated as circular molecules in transformants at semi-permissive temperature for virus DNA replication (37°C). When the temperature was shifted down from 37°C to permissive temperature (33°C), pSVtsA molecules propagated vigorously and copy number increased to 10^4/cells. Human erythropoietin cDNA was inserted into the downstream of SV40 late promoter of this plasmid and introduced into the simian cells. We could induce erythropoietin production by temperature shift down and erythropoietin production level increased more than 40-fold.

1. Introduction

Production of biological substances by recombinant DNA technology with mammalian cells are one of the most important procedures either for basic research or industrial production such as pharmaceutical industry. Different from bacterial recombinant DNA technology, a few plasmid vectors have been reported with mammalian cell system except for bovine papilloma virus based vectors (Law et al. (1981)). Therefore, we developed a plasmid vector based on SV40 in order to improve the production using mammalian cells.

SV40 infects productively to monkey and human derived cells and kills them (Tooze (1980)). We, therefore, tried to suppress virus genome replication using

S. Kaminogawa et al. (eds.), Animal Cell Technology: Basic & Applied Aspects, Vol. 5, 79–84.
© 1993 Kluwer Academic Publishers.

temperature sensitive large T antigen bearing mutant virus, since the large T antigen is essential for the initiation of virus replication. For this purpose, we chose tsA$_{640}$ mutant of SV40 which could replicate at 32°C but could not at 39°C (Kimura and Dulbecco (1973)). We cloned this virus genome and inserted human erythropoietin cDNA just downstream of the SV40 late promoter and showed effective erythropoietin production by changing the cultivation temperature.

2. Materials and Methods

2.1. VIRUS AND PLASMIDS

A mutated SV40 virus which showed a temperature dependent replication phenotype (tsA$_{640}$) was kindly gifted from Professor G. Kimura of Kyushu University (Kimura and Dulbecco (1973)). This virus was infected to simian CV-1 cells and propagated at 33°C. After 40 hours cultivation, virus particles were harvested and vilions were purified by equilibrium centrifugation. Virus DNA was extracted by the standard procedure (Hirt (1967)).

Recombinant plasmid pSVtsA was constructed by ligating *Bam*HI fragment of SV40 tsA$_{640}$ genome and the plasmid pMC1neo which has bacterial neo^r gene under the control of thymidine kinase gene promoter derived from herpes simplex virus (Figure 1).

pSVtsAcEP was constructed by inserting *Eco*RV-*Sal* I digested pSVLcEP fragment into pSVtsA at *Eco*RV-*Sal* I sites (Figure 2). pSVLcEP is a plasmid containing human erythropoietin cDNA (Jacobs et al. (1985)), which was kindly gifted from Snow Brand Milk Products Co. LTD.

2.2. CELL CULTURE AND TRANSFECTION

CV-1 cells were grown in minimum essential medium supplemented with 10% (vol./vol.) fetal bovine serum. Cells confluent in 100mm-diameter dishes were trypsinized, and were replated (1:10 dilution) 1 day before transfection. Two μg of plasmid DNA, with 20 μg calf thymus DNA as carrier was transfected by the calcium phosphate coprecipitation method (Graham and van der Eb (1973)). The DNA precipitates were incubated with the cells for 12 hours, then the fresh medium was added. For isolation of stable transformants, the cells were split by 1:15 into selective medium containing 400 μg/ml of G418(Gibco Co.) after 48h post infection. Individual G418-resistant CV-1 colonies were isolated approximately after 3 weeks.

To study erythropoietin production, isolated recombinant clones were seeded at various concentrations (between 1×10^5-1.6×10^6 cells per 100 mm dish) and were cultured for one day at 37°C and the temperature was then shifted down to

30 or 33°C. Samples were taken every day and the amounts of produced erythropoietin was determined by enzyme-linked immunosorbent assay (ELISA).

2.3. DNA BLOTTING AND HYBRIDIZATION

The low-molecular-weight DNA was isolated by the method of Hirt (Hirt (1967)). DNAs were analyzed by the method of Southern (Hames and Higgins (1985)) after electrophoresis on 0.8% agarose gels. After electrophoresis, DNAs in the gels were depurinated, denatured, and transferred onto nylon membranes (Hybond-N+, Amersham). ^{32}P-labeled probes were prepared by using a Random Primer DNA labeling kit (Takara Co.).

3. Results and Discussion

Monkey CV-1 cells were transfected with recombinant DNA derived from SV40 A gene (large T antigen) temperature sensitive mutant (tsA$_{640}$), by calcium phosphate coprecipitation method. Transfected cells were incubated at 33°C (permissive temperature for SV40 tsA$_{640}$ DNA replication) and 37°C (semipermissive temperature) for 48 hours. To select stable transformants, the cells were then split and medium was changed to selection medium. At 37°C, G418 resistant colonies were appeared after two weeks and 7 of them were picked up. On the other hands, cells could not form colonies at 33°C. By several trials, we isolated a few colonies at 33°C, but they soon died. Since SV40 tsA$_{640}$ replication is permissive at 33°C, vigorous plasmid propagation may give cytotoxic effects on the cells and probably killed cells.

To analyze the state of pSVtsA DNA in transformed cells, low molecular weight DNAs were isolated by the procedure of Hirt. From about 40% of clones, single bands were detected as a major band and the molecular size of the bands were the same by cutting with several different restriction enzymes. These results show that these G418 resistant clones contained circular form of SV40 derived plasmid at 37°C. The copy number of the plasmid was estimated to be more than 10^4 copy/cell by judging from dot blot analyses.

Human erythropoietin cDNA was inserted into temperature sensitive plasmid pSVtsA just downstream of SV40 late promoter. In this plasmid construct, exogenous erythropoietin gene is expressed under the control of SV40 late promoter. This erythropoietin expression vector was introduced into CV-1 cells by the calcium phosphate coprecipitation method and G418 resistant clones were obtained at 37°C after 2 weeks.

To study the inducible production of erythropoietin by changing temperature, cells were first cultured at 37°C and temperature was then shifted down at 33°C. As shown in Figure 3, erythropoietin production was fully induced by temperature change. Thirty degree gave higher production than that at 33°C. The produced erythropoietin concentration increased almost 40-folds by the temperature change from 37°C to 30°C.

These results show the usefulness of this SV40 based plasmid vector for production of other biological substances.

4. Acknowledgements

We thank Dr. G. Kimura of Kyushu University and Snow Brand Milk Products Co. LTD for kindly providing SV40 tsA mutant and human erythropoietin cDNA, respectively.

5. References

Graham, F. and van der Eb, A. (1973) 'A new technique for the assay of infectivity of human adenovirus 5 DNA', Virology 52, 456-457.

Hames, B. D. and Higgins, S. J. (1985) Nucleic acid hybridization-a practical approach, IRL press.

Hirt, B. (1967) 'Selective extraction of polyoma DNA from infected mouse cell cultures', J. Mol. Biol. 26, 356-369.

Jacobs, K., Shoemaker, C., Rudersdorf, R., Neill, S. D., Kaufman, R. J., Mufson, A., Seehra, J., Jones, S. S., Hewick, R., Fritsch, E. F., Kawakita, M., Shimizu, T. and Miyake, T. (1985) 'Isolation and characterization of genomic and cDNA clones of human erythropoietin', Nature 313, 806-810.

Kimura, G. and Dulbecco, R. (1973) 'A temperature-sensitive mutant of simian virus 40 affecting transforming ability', Virology 53, 529-534.

Law, M. F., Dowy, D. R., Dvoretsky, I. and Howley, P. M. (1981) 'Mouse cells transformed by bovine papilloma virus contain only extrachromosomal viral DNA sequences', Proc. Natl. Acad. Sci. USA 78, 2727-2731.

Tooze, J. (1980) The Molecular Biology of Tumor Viruses, 2nd edition, Cold Spring Harbor Laboratory, New York.

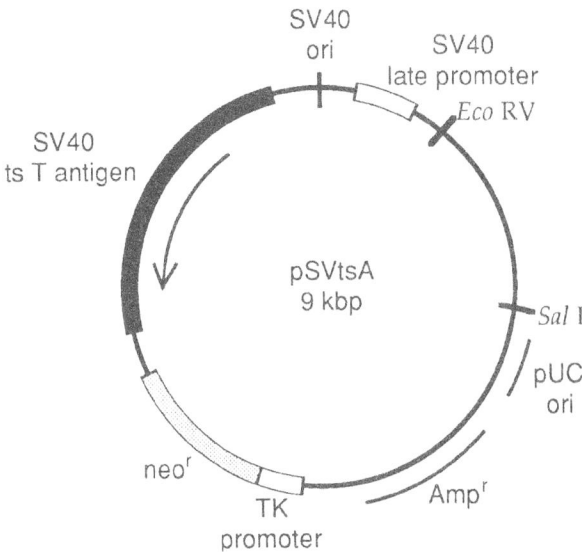

Figure 1. Construction of tsSV40 based vector pSVtsA

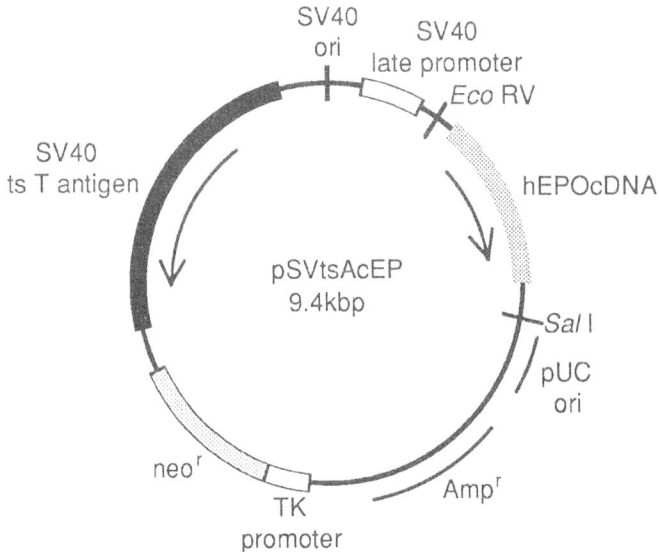

Figure 2. Construction of erythropoietin expression vector pAVtsAcEP.
Erythropoietin cDNA was inserted into the downstream of SV40 late promoter.

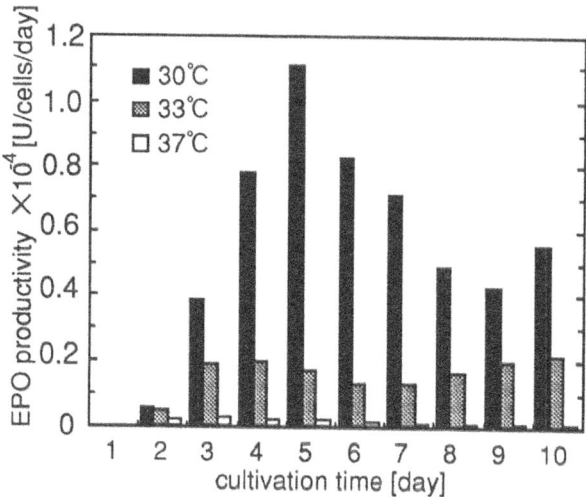

Figure 3. Erythropoietin productivity by pSVtsAcEP transfected cells after temperature shift down from 37°C to 30°C or 33°C. Initial cell density was 1x10⁵ cells/100mm dish.

SELECTION OF REGULATORY MUTANTS OF μ GENE EXPRESSION FROM MYELOMA CELLS TRANSFECTED WITH THE MODIFIED μ GENE

T. SHINDO*, H. UEDA, F. MAKISHIMA, E. SUZUKI, and H. NISHIMURA
Department of Chemical Engineering, Faculty of Engineering, University of Tokyo, 7-3-1 Hongo, Bunkyo-ku, Tokyo 113, Japan
*Engineering Research Laboratories, Research Institute, Kaneka Corporation, 1-8 Miyamae, Takasago, Hyogo 676, Japan

ABSTRACT. Cells of the murine immunoglobulin non-producing myeloma cell line could stably express the μ chain on the surface (μ_m) without a light chain by transfection of the modified μ gene that had the transmembrane sequence of the human epidermal growth factor receptor and that lacked the CH1 region. We aim to find regulatory mutants of the μ gene expression in the μ_m^- mutants that are selected from μ_m^+ cells harboring the modified μ gene on the Bovine papillomavirus vector as extrachromosomal DNA. We established the method of selectively killing μ_m^+ cells with ricin A-conjugated anti-μ antibody for efficient selection of the μ_m^- mutants.

1. Introduction

The transcription of the immunoglobulin (Ig) gene is regulated by the particular transcription factors in a plasma cell. Gene isolation of the transcription factors contributes to revealing the mechanism for the tissue-specific expression of the Ig gene. However, hardly any of these genes have been isolated yet, because the amounts of the transcription factors are so little and some of them make such strong complexes that their purification is very difficult. Thus, somatic cell mutants defective in the transcription factors are very useful. We can isolate their genes that can recover the transcription activity of the Ig gene in such regulatory mutants from the genome of normal cells. Many researchers have tried to obtain the regulatory mutants [1], but they have not yet succeeded in this. As the source for mutant selection, they used hybridoma cells that secreted hapten-specific IgM. The regulatory mutants are found as mutants that do not produce Ig. However, hybridoma cells have four copies of the genes of transcription factors as a result of cell fusion, whereas the active Ig gene is a single copy due to allelic exclusion. Accordingly, mutants suffering from defects in all four genes of the transcription factors, but not in the Ig gene, are very rare in Ig non-producing mutants, making it particularly difficult to obtain the regulatory mutants from hybridoma cells. Consequently, duploid myeloma cells are preferable for such purposes.

We used myeloma cells transfected with the μ gene for selecting the regulatory mutants of μ gene expression, and designed a strategy for efficiently selecting these mutants from

85

S. Kaminogawa et al. (eds.), Animal Cell Technology: Basic & Applied Aspects, Vol. 5, 85–91.

them. The strategy consists of the following steps. The first is the modification of the μ gene for surface μ expression (μ_m) without a light chain. From the $\mu_m{}^+$ cells treated with a chemical mutagen, the cells that do not produce μ can be easily selected as $\mu_m{}^-$ mutants. The second is the multiplication of the modified μ gene with the Bovine papillomavirus (BPV) vector. The $\mu_m{}^-$ mutants selected from the cells having multicopies of the μ gene are possibly the intended regulatory mutants. The last is the easier selection of the $\mu_m{}^-$ mutants by killing the $\mu_m{}^+$ cells with toxin-conjugated anti-μ antibody.

2. Materials and Methods

2.1. PLASMID

The construction of pSV-VμMeΔCH1 (Fig. 1)[2] and pCMVλ1 [3] has been described elsewhere, pSV-VμMeΔCH1 being constructed from pSV-Vμ1 [4]. pCMVλ1 can express the λ1 chain under control of the cytomegalovirus (CMV) promoter, and also has the neomycin-resistant gene as a selectable marker. The gene of the λ1 chain was cDNA cloned by PCR from J558L cells. pSV-VμMe was the same as pSV-VμMeΔCH1, except that the ClaI-XhoI fragment containing the *Eco*-gpt gene as a selectable marker and the SacI-SacI fragment containing the CH1 region remained in pSV-VμMe. BPV-VμMeΔCH1 (Fig. 3) was constructed from pSV-VμMeΔCH1 and BCMGSNeo [5]. The ClaI-XhoI fragment of pSV-VμMeΔCH1 containing the μ gene was inserted once into Bluescript at the polycloning site. The XhoI site and SmaI site of the recombinant Bluescript were modified to the NotI site and XhoI site, respectively, by using linker sequences. The fragment containing the μ gene and BCMGSNeo, whose XbaI site had been converted to the XhoI site with a linker and whose CMV promoter had been removed, were mutually connected at the NotI site and XhoI site.

2.2. CELL LINE AND TRANSFECTION

X63.653 is the murine Ig non-producing myeloma cell line. The cells were cultured in Dulbecco's Modified Eagle Medium (DMEM) supplemented with 10% FCS (DMEM/10% FCS), and transfected with plasmid by electroporation. The transfectants with plasmids having the neomycin-resistant gene were selected with 1 mg/ml of geneticin (G418; Sigma). The transfectants with pSV-VμMe were selected in the presence of 6 μg/ml of mycophenolic acid and 250 μg/ml of Xanthine.

2.3. MUTANT SELECTION WITH RICIN A-CONJUGATED ANTI-μ ANTIBODY

Ricin A (Sigma) was conjugated with goat anti-mouse IgM(μ) antibody (Cappel) by N-succimidyl-3-(2-pyridyldithio)-propionate (SPDP; Sigma), according to the E.Y. Laboratory procedure.

For mutant selection, cells were cultured with 3 nM of the ricin A-conjugated anti-μ antibody in DMEM containing 10% FCS and 50 mM lactose for five to six days, after having been cultured with 200-300 mg/ml of ethane methylsulfonate (Sigma) for mutation in DMEM/10% FCS for one day.

2.4. ANALYSIS

For a μ_m analysis by flow cytometry (EPICS; Coulter Electronics), the cells were incubated with FITC-conjugated anti-μ antibody. For an analysis of intracellular μ, the cells were fixed with ethanol before incubating with FITC-conjugated anti-μ antibody.

3. Results

3.1. μ_m EXPRESSION IN MYELOMA CELLS BY MODIFICATION OF THE μ GENE

Plasmid pSV-Vμ1 has the active μ gene of anti-Nitrophenacetyl (NP)-hapten antibody [4]. When it is transfected to the cells of murine myeloma cell line J558L, which only produce the λ1 chain, they secrete the anti-NP antibody. Two modifications of the μ gene in pSV-Vμ1 were made for μ_m expression in myeloma cells without a light chain. One involved replacing the μ tailpiece sequence with the transmembrane sequence of the human epidermal growth factor receptor (EGFR) to hold the μ chain on the cell surface. The other involved deleting the CH1 region to prevent the μ chain from being retained in the endoplasmic reticulum (ER) in the absence of the light chain.

We transfected resulting plasmid pSV-VμMeΔCH1 (Fig. 1) to X63.653 cells. Fig. 2 shows μ_m analyses by flow cytometry, Fig. 2A being for X63.653 cells. The peak at the left represents μ_m^- cells. Fig. 2B is for the transfectants of X63.653 with plasmid pSV-VμMe, which had a transmembrane sequence of EGFR instead of the μ tailpiece sequence, while still having the μ CH1 region. As shown, no transfectants could express μ_m. However, the transfectants expressed μ_m when additionally transfected with light chain expression vector pCMVλ1, as a new peak representing μ_m^+ cells can be observed at the right in Fig. 2C. On the other hand, part of the X63.653 cells transfected with pSV-VμMeΔCH1 expressed μ_m as shown in Fig. 2D. We could finally obtain some μ_m-expressing clones from the cells transfected with pSV-VμMeΔCH1 by limiting dilution after selecting with a cell sorter (Fig. 2E).

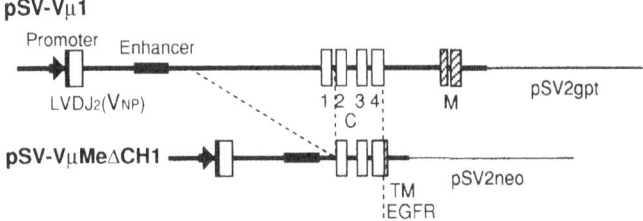

Fig. 1. Construction of pSV-VμMeΔCH1. Open boxes are exons for the μ chain of anti-NP antibody, corresponding to LVDJ(NP), CH1, CH2, CH3, and CH4. Hatched boxes are the transmembrane sequences of μ or EGFR. The thick arrow is the μ promoter sequence, and the thick line is the μ enhancer sequence. The thin lines are the vector sequences originating from pBR322 having the *Eco*-gpt gene or neomycin-resistant gene as a selectable marker.

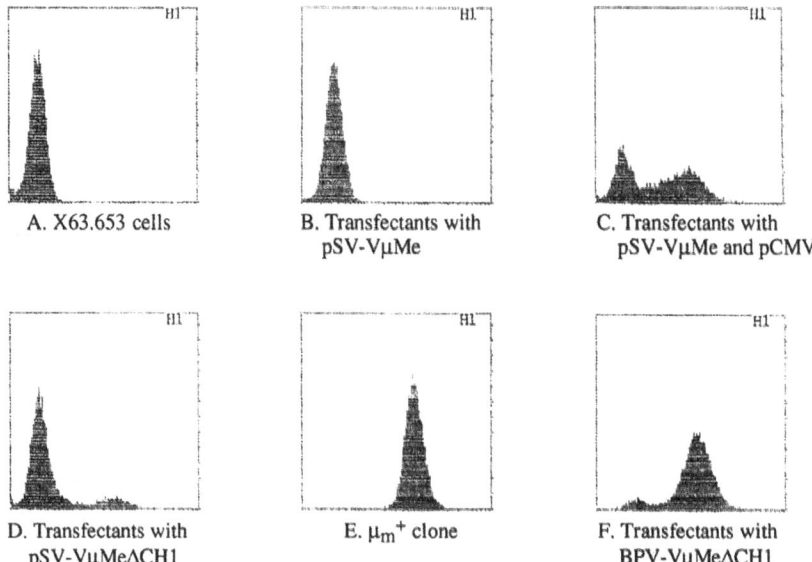

Fig. 2. Flow cytometric analyses of μ_m. The horizontal axis is the log of the fluorescence intensity, corresponding to the number of μ_m on a cell. The vertical axis is the relative cell number.

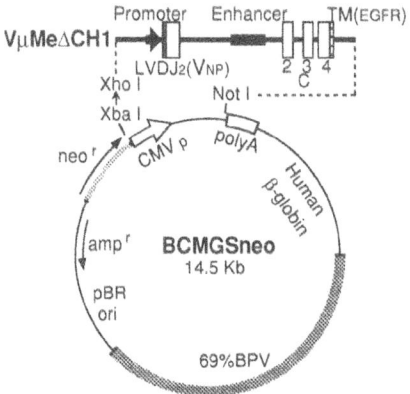

Fig. 3. Construction of BPV-VμMeΔCH1. cDNA integrated between the CMV promoter (open arrow) and the polyA site (open box) in the direction of the arrow can be transcribed under control of the CMV promoter. The dark box is the 69% BPV sequence. Thin arrows are the neomycin-resistant gene for selection in mammalian cells, and the ampicillin-resistant gene for selection in *E. coli* cells. The human ß-globin sequence located between the polyA site and BPV sequence is said to be helpful for replication of the vector as an episome.

3.2. MULTIPLICATION OF THE MODIFIED μ GENE IN MYELOMA CELLS

For multiplication of the modified μ gene, we used the BPV vector. It has been reported that BCMGSNeo propagated as an episome in X63.653 cells [5]. The 69% BPV sequence in the vector enables episomal replication, and we integrated the fragment containing the modified μ gene into BCMGSNeo at the insertion site for cDNA. We also deleted the CMV promoter from BCMGSNeo so that the μ gene was expressed only under the control of its own promoter and enhancer sequences (BPV-VμMeΔCH1; Fig. 3). When X63.653 cells were transfected with BPV-VμMeΔCH1, almost all the transfectants expressed μ_m (Fig. 2F).

3.3. μ_m^- MUTANT SELECTION WITH RICIN A-CONJUGATED ANTI-μ ANTIBODY

In order to examine the applicability of toxin-conjugated anti-μ antibody to the selection of μ_m^- mutants, we mutagenized the cells of a μ_m-expressing transfectant with pSV-VμMeΔCH1 with a chemical mutagen, and selected μ_m^- mutants from the mutagenized population with the conjugated antibody. The expression of μ_m and intracellular μ was analyzed by flow cytometry for the selected population (Fig. 4.). We could successfully obtain μ_m^- mutants. In the population of the μ_m^- mutants, there seem to have been μ-producing mutants in addition to μ-non-producing ones.

4. Discussion

The CH1 region of the Ig heavy chain is said to be the region where the light chain binds and where Ig heavy chain-binding protein (Bip), an ER protein, binds in the absence of the light chain [6]. Fig. 2 shows that the light chain was necessary for transporting the μ chain outside the cells and that the CH1 region was related to the retention of the μ chain in ER in the absence of the light chain. This supports the assembly-transport mechanism for the immunoglobulin molecule in ER. Besides, successful μ_m expression by using the transmembrane of EGFR instead of the μ transmembrane is consistent with the fact that, in a plasma cell, the intact μ gene cannot express surface IgM due to the absence of *mb-1*, which functions for the expression of surface IgM by associating with the μ_m transmembrane [7].

The μ gene which can express μ_m without the light chain and in which its intact promoter and enhancer sequences remain is effective for obtaining the regulatory mutants of μ gene expression, because the selection of μ_m^- mutants from myeloma cells transfected with the μ gene is more precise and easier than the selection of IgM non-secreting mutants from hybridoma cells by suicide selection [1], and we can do without considering a defect in the expression of the light chain.

Plasmid pSV-VμMeΔCH1 does not have the fragment required for replicating in mammalian cells. Hence, it seems that it was randomly cut in the cell, and its fragment was inserted into the genome at one point when transfected to a cell. This seems to be the reason why only a few transfectants expressed μ_m (Fig. 2D). Accordingly, this means that the μ_m^+ transfectant had a single copy of the μ gene. In contrast, almost all the transfectants with BPV-VμMeΔCH1 expressed μ$_m$ (Fig. 2F), suggesting that BPV-VμMeΔCH1 was not incorporated into the genome but replicated as an episome in the cell, resulting in the multiplied μ gene in the cell. We plan to use a clone with multicopies of the μ gene selected from the transfectants with BPV-VμMeΔCH1 as a resource for actual mutant selection.

The μ$_m$⁻ mutants obtained from a transfectant with pSV-VμMeΔCH1 (Fig. 4) could be classified into three types after further analyses (data not shown). The mutant without any expression of μ comprised about a half of the population of the μ$_m$⁻ mutants. We think that our system of μ$_m$⁻ mutant selection, using the toxin from myeloma cells, is applicable for obtaining regulatory mutants, because the proportion of the μ-non-producing mutants was sufficiently high. The remaining population consisted of the mutants accumulating μ in the cytoplasm and the ones secreting μ outside the cell.

We think that the three steps in our strategy for efficiently selecting the regulatory mutants of μ gene expression have been proved in this work, and that this system is ready to be applied to the selection of the intended regulatory mutants.

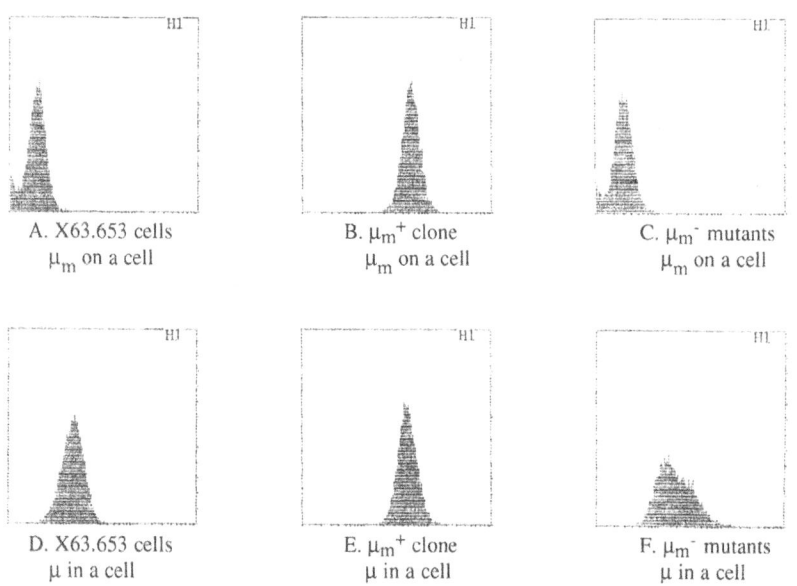

Fig. 4. Flow cytometric analyses of μ$_m$ (A-C) and intracellular μ (D-F). The horizontal axis corresponds to the number of μ on a cell (A-C) or in a cell (D-F). X63.653 cells (A and D) and a μ$_m$⁺ clone (B and E) were negative and positive controls, respectively.

5. Acknowledgements

We are indebted to Dr. M. S. Neuberger for kindly presenting plasmid pSV-Vμ1 and to Dr. H. Karasuyama for presenting plasmid BCMGSNeo.

6. References

1. Trimble, W. S., Baker, M. D., Boulianne, G. L., Murialdo, H., Hozumi, N., and Shulman, M. J. (1986) 'Analysis of hybridoma mutants defective in synthesis of immunoglobulin M', *Somat. Cell Mol. Genet.* **12**, 467-477.
2. Shindo, T., Ueda, H., Makishima, F., Suzuki, E., and Nishimura, H. (1992) 'Efficient selection of μ_m^- mutants from μ_m-expressing myeloma cells by treatment with ricin A-conjugated anti-μ antibody', *Somat. Cell Mol. Genet.* **18**, 553-558.
3. Ueda, H., Kikuchi, M., Shindo, T., and Nishimura, H. (1991) 'Direct cloning of immunoglobulin cDNA by polymerase chain reaction and its expression in myeloma and non-myeloma cells', *Proc. 24th Autumn Meeting Soc. Chem. Eng. Jpn.*, Nagoya, p. 73.
4. Neuberger, M. S. (1983) 'Expression and regulation of immunoglobulin heavy chain gene transfected into lymphoid cells', *EMBO J.* **2**, 1373-1378.
5. Karasuyama, H., and Melchers, F. (1988) 'Establishment of mouse cell lines which constitutively secrete large quantities of interleukin 2, 3, 4 or 5, using modified cDNA expression vectors', *Eur. J. Immunol.* **18**, 97-104.
6. Hendershot, L., Bole, D., Köhler, G., and Kearney, J. F. (1987) 'Assembly and secretion of heavy chains that do not associate posttranslationally with immunoglobulin heavy chain-binding protein', *J. Cell Biol.* **104**, 761-767.
7. Hombach, J., Tsubata, T., Leclercq, L., Stappert, H., and Reth, M. (1990) 'Molecular components of the B-cell antigen receptor complex of the IgM class', *Nature* **343**, 760-762.

PROCESS CONTROL OF BHK CELLS
BY NUCLEOTIDE POOL MEASUREMENT

Ulrich Valley, Thomas Ryll and Roland Wagner

Arbeitsgruppe Zellkulturtechnik, Gesellschaft für Biotechnologische Forschung m.b.H., GBF, Mascheroder Weg 1, D-38124 Braunschweig, Germany

ABSTRACT

Nucleotide pool measurements have been shown to be an attractive tool for the control of fermentation processes with recombinant mammalian cells. In particular, the NTP and U values expressed in the NTP/U-plot gives a direct access to the physiological state of the cells. A two times higher perfusion rate had to be adjusted for a suffucent supply to satisfy the physiological requirements of the cells than had been assumed by the use of a classical environmental control.

ABBREVIATIONS

ATP = adenosine triphosphate, CTP = cytidine triphosphate, DO = dissolved oxygen, GTP = guanosine triphosphate, HPLC = high performance liquid chromatography, IL-2 = interleukin 2, MEM = minimum essential medium, NCS = newborn calf serum, NTP = nucleotide triphosphates, PurTP = purine triphosphates, PyrTP = pyrimidine triphosphates, rpm = revolutions per minute, UDP-GalNAc = UDP-N-acetyl-galactosamine, UDP-GlcNAc = UDP-N-acetyl-glucosamine, UDP-GNAc = sum of UDP-GalNAc and UDP-GlcNAc, UTP = uridine triphosphate

INTRODUCTION

Parameters of intracellular analysis have been shown to be useful as a tool for control production processes based on *in vitro* cultivated hybridoma and recombinant animal cells. Nucleotides were found to present the best target as they reflect the exact physiological state of a culture (Ryll and Wagner, 1992). Following the progress of batch, perfused and chemostat cultures cell specific

S. Kaminogawa et al. (eds.), Animal Cell Technology: Basic & Applied Aspects, Vol. 5, 93–97.
© 1993 *Kluwer Academic Publishers.*

regularities were found which allowed the generation of three characteristic parameters: the nucleotide triphosphate ratio (NTP, eq. 1), the uridine ratio (U, eq. 2) and the combined ratio NTP/U (eq. 3).

Equations

$$NTP = \frac{PurTP}{PyrTP} = \frac{(ATP + GTP)}{(UTP + CTP)} \qquad (1)$$

$$U = \frac{UTP}{UDP\text{-}GNAc} \qquad (2)$$

$$\frac{NTP}{U} = \frac{UDP\text{-}GNAc \ (ATP + GTP)}{UTP \ (UTP + CTP)} \qquad (3)$$

These equations allow a direct description of the growth cycle by means of their specific values for every phase of the culture. The use of a specific function for the application on *in vitro* cultivations have been proposed with an NTP to U plot which combines the results obtained by the cell analysis and which offers a tool for the control and regulation of cell growth derived processes. We used it for the growth control of a perfused culture of BHK 21 cells.

MATERIALS AND METHODS

Cell line
The cell line BHK 21 pSVIL2, provided by GBF genetic engineering department, has been manipulated by genetic methods to produce human interleukin 2 constitutively under the control of the SV 40 promotor (Conradt *et al.*, 1989).

Culture conditions
Cells were cultivated in a stirred fermenter (2 dm^3 working volume) with bubble-free aeration (Wagner and Lehmann, 1988, Ryll *et al.*, 1990). The system was optionally equipped with continuous perfusion. DO was maintained at 40 % of air saturation (oxygen concentration: 2.7 mg dm^{-3}). The pH was controlled off-line and kept constant between 6.9 and 7.3. The stirrer speed was adjusted to 30 - 50 rpm. BHK cells were grown in suspension in a medium

formulated of a mixture of Iscove's MEM and Ham's F12 with 3.61 g dm^{-3} NaHCO$_3$ and 0.18 g dm^{-3} sodium pyruvate and supplemented with 5 % NCS.

Analysis of fermenter samples

Samples were analyzed as described previously (Ryll and Wagner, 1992). Glucose and lactate contents were determined with an YSI 27A glucose/lactate analyzer (Yellow Springs Instruments, Ohio). Free amino acids were quantified by means of a reversed phase HPLC system with pre-column derivatisation with o-phthaldialdehyde (Larsen and West, 1981). For the determination of the total cell number nuclei were fixed and stained with 0.1 % crystal violet in 0.1 mol dm^{-3} citrate and subsequently counted with a hemocytometer. The proportion of dead cells was estimated by trypan blue exclusion. Nucleotide pools were analyzed as described earlier (Ryll and Wagner, 1991).

RESULTS AND DISCUSSION

A strategy has been followed of maintaining an optimal exponential growth rate for the continuously perfused cultivation. The exact nucleotide ratios had to be determined as a prerequisite of a precise control. As previously shown for the used BHK cell line optimal NTP and U values for an exponential cell growth (log-phase) were found to 3.5 - 4.7 and 1.8 - 3 for NTP and U value, respectively (Ryll and Wagner, 1992). The process was started with a cell density of 2·10^5 ml^{-1} with cells from a continuously perfused cultivation. It had been controlled by adjusting the perfusion rate (nutrient supply) to the control parameters NTP and U which were not allowed to leave the 'log-box' of the NTP/U blot (Fig. 1b). If the parameter leave the defined area the perfusion rate was increased. As a result of exact control a constant growth rate could be maintained. No nutrient limitation had been detected by a simultaneous monitoring of glucose and amino acid concentrations. The perfusion rate had to be adjusted to up to 4 reactor volumes a day at a relatively low cell density of 3·10^6 ml^{-1} (Fig. 1a). When the perfusion rate had been adjusted by nutrient control it had been assumed that the maximum exponential growth had already been achieved with a perfusion rate of two reactor volumes a day at this cell density (Wagner et al., 1989).

These investigations show that the control of intracellular parameters offers direct access to informations about the real requirements of cell cultures. Although exponential growth can be achieved by environmental control cells are far away from their best physiology of exponential intracellular conditions. Best physiology however, can have a positive influence on the product integrity and can probably guarantee a homogeneous bulk.

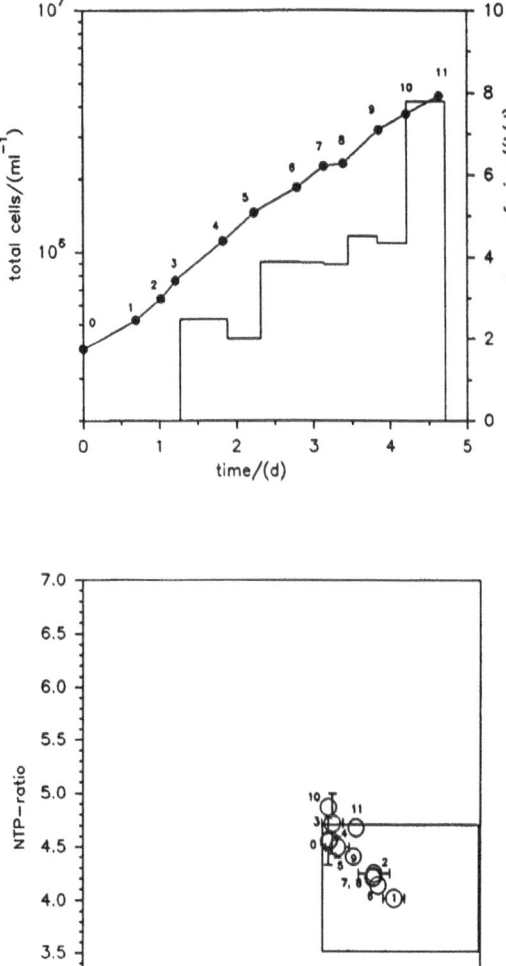

Fig 1.   Growth and perfusion rate (a) and NTP/U blot (b) of a continuously perfused cultivation of recombinant BHK cells

REFERENCES

Conradt, H.S., Nimtz, M., Dittmar, K.E.J., Lindenmaier, W., Hoppe, J. and Hauser, H. (1989) Expression of human interleukin-2 in recombinant baby hamster kidney, Ltk⁻ and Chinese hamster ovary cells. J. Biol. Chem. 264, 17368-17373.

Larsen, B.R. and West, F.G. (1981) A Method for Quantitative Amino Acid Analysis Using Precolumn o-Phthalaldehyde Derivation and High Performance Liquid Chromatography. J. Chromatogr. Sci. 19, 259-265

Ryll, T., Jäger, V. and Wagner, R. (1991) Intracellular concentration of ATP and other nucleotides during continuous cultivation of hybridoma cells. In: Production of Biologicals from Animal Cells in Culture (Spier, R.E., Griffiths, J.B., Meignier, B., eds.) Butterworths, UK, pp. 236-242

Ryll, T. Jäger, V. and Wagner, R. (1991) Variation of ratios and concentrations of nucleotide triphosphates and UDP-sugars during a perfused batch cultivation of hybridoma cells. In: Animal Cell Culture and Production of Biologicals (Sasaki, R., Ikura, K., eds.), Kluwer, Dordrecht, 307-317

Ryll, T., Lucki-Lange, M., Jäger, V. and Wagner, R. (1990) Production of recombinant human interleukin-2 with BHK cells in a hollow fibre and a stirred tank reactor with protein-free medium. J. Biotechnol. 14, 377-392

Ryll, T. and Wagner, R. (1991) Improved ion-pair high-performance liquid chromatographic method for the quantification of a wide variety of nucleotides and sugar-nucleotides in animal cells. J. Chromatogr. 570, 77-88

Ryll, T., Wagner, R. (1992) Intracellular Ribonucleotide Pools as a Tool for Monitoring the Physiological State of in vitro Cultivated Mammalian Cells during Production Processes.
Biotechnol. Bioeng. 40, 934-946

Wagner, R., Krafft, H., Lehmann, J. (1989) The Production of Human Interleukin 2 by Recombinant Mammalian Cells. In: Spier, R.E., Griffiths, J.B., Stephenne, J., Crooy, P.J. (eds.) ' Advances in Animal Cell Biology and Technology for Bioprocesses', Butterworths, Sevenoaks, pp. 374-377

Wagner, R. and Lehmann, J. 1988. The growth and productivity of recombinant animal cells in a bubble-free aeration system. Trends Biotechnol., 6, 101-104

PURIFICATION AND cDNA CLONING OF FLATFISH INTERFERON

T. Tamai[1], T. Noguchi[2], N. Sato[1], S. Kimura[1], S. Shirahata[2], and H. Murakami[2]

[1]*Taiyo Central R&D Institute, Taiyo Fishery Co. Ltd., 16-2 Wadai, Tsukuba-shi, Ibaraki 300-14, Japan.*
[2]*Graduate School of Genetic Resources Technology, Kyushu University, Fukuoka 812, Japan.*

ABSTRACT

Fish interferon (IFN) was found in cultured medium of a flatfish lymphocytes cell line immortalized by oncogene transfection. The IFN was purified by DEAE-Toyopearl and WGA agarose affinity chromatography from the cultured medium of the cell. The protein was a glycoprotein of about 16kDa and the cDNA which codes the protein was cloned. The cDNA encodes 138 amino acids including a signal peptide. The cDNA fused to the SRα promoter was shown to direct synthesis and secretion of active IFN in COS-1 cells.

INTRODUCTION

Infectious diseases pose a constant and highly costly threat to successful fish husbandry. So far there is no effective chemotherapy to control viral diseases. It is against this background that fish immunology will be perceived as potentially playing an important role in aquaculture not only for the prevention and therapy of infectious diseases but also on breeding of healthy fishes resistance for various diseases.

IFN is a general name of protein or glycoprotein of ca. 20KDa which shows antiviral activity via specific recepter (Demaeyer & Demaeyer-Guignard, 1988). Three types of IFN (α, β and γ) are distinguishable on the basis of different biological and biochemical properties (Stewart, 1980; Sano *et al.*, 1988; Hosoi *et al.*, 1988). Fish cells can secret IFN-α and β molecules in response to virus infection of fibroblast or epithelial cell lines either *in vitro* (Gravell & Malsberger, 1965; Sena & Rio, 1975; Okamoto *et al.*, 1983) or *in vivo* (de Kinkelin *et al.*, 1982). Secretion of fish IFN-γ has been also suggested (Dorson *et al.*, 1985; Graham & Secombes, 1990). However, complete purification and gene cloning of these fish IFNs have not been reported yet.

Because a small amount of IFN exhibits strong biological activity, it is difficult to purify IFN from fish body. Therefore first we immortalized flatfish leukocytes by oncogenes. We found the IFN activity in the cultured medium of the immortalized lymphocytes and purified the flatfish IFN by DEAE-Toyopearl and WGA agarose affinity column chromatography. The protein was a glycoprotein of about 16 KDa. We have also cloned an interferon cDNA encoding a polypeptide with 138 amino acids in length.

MATERIALS & METHODS

IFN preparation HL8 cells (2×10^5 cells/ml) were cultured in ERDF medium supplemented with 19 μg/ml insulin, 35 μg/ml transferrin, 20 μM ethanolamine and 25 nM

99

sodium selenium (ITES) (Murakami *et al.*, 1982). The cells were cultured at 15 °C in an atmosphere of 95% air/5% CO_2 for 24 h. The supernatant was used as an IFN preparation.

Antiviral activity test Carp epithelial (EPC) cells (8 x 10^5 cells/ml) were cultured in 96 well microplates, then the IFN preparation was added to the EPC cells. On the next day Hirame Rhabdovirus (HRV) was added to EPC cells with a multiplicity of infection (m.o.i.) of 5. On 7 days after transfection the cell viability was calculated by the amount of neutral red uptake which was measured by optical density using a microplate reader.

DEAE-Toyopearl chromatography A DEAE-Toyopearl 650M column (Toso, Tokyo)(1 cm x 20 cm; bed volume, 20ml) was equilibrated with 10 mM Tris-HCl buffer (pH 8.0). The spent medium was applied at a flow rate of 1 ml/min. The column was washed with 60ml of Tris-HCl buffer (pH 8.0), and the IFN was eluted using a linear gradient of NaCl (0-1M). Antiviral activity of each fraction was examined.

WGA agarose chromatography The WGA agarose column (Honen Oil Co. Ltd., Tokyo) was equilibrated with 10mM phosphate-buffered saline (PBS) (pH 7.2). The active fractions purified by DEAE-Toyopearl chromatography was adjusted to pH 7.2 and applied to the column at a flow rate of 30 ml/h. The column was then washed with 150 ml of PBS and IFN was eluted using a stepwise increase of N-acetyl glucosamine (GlucNAc). The antiviral activity of each fraction was assayed. The active fractions were pooled and stored at -20°C.

SDS-polyacrylamide gel electrophoresis (SDS-PAGE) The active fraction purified by WGA chromatography was dialyzed against PBS and lyophilized. After repeated dialysis and lyophilization, the sample was resuspended in sampling buffer (0.1M Tris-HCl, 4% SDS, 16% glycelin and 0.002% bromophenol blue and applied to a polyacrylamide gel containing a 15% acrylamide separating gel and a 3% stacking gel according to the Laemmli's method (Laemmli, 1970).

cDNA synthesis In order to prepare a cDNA library for the screening of IFN specific clone, we prepared and used partially purified mRNA from HL8 cells as template. After the stimulation of the cells by HRV, RNA was isolated using guanidinium thiocyanate and poly (A) RNA was purified with oligo dT column chromatography (Pharmacia). The mRNA was devided into 10 fractions by sucrose density gradient centrifugation. mRNA of each fraction was translated into recombinant products using rabbit reticulocyte lysate and antiviral activity of these products was examined. EPC cells (8 x 10^5 cells/ml) were cultured in 96 well microplates. Then the *in vitro* translated products were added to the culture medium. And antiviral activity was assayed using HRV (m.o.i. =1.0)

The mRNA screened by *in vitro* translation assay was reverse transcribed to cDNA. The cDNA was inserted into the HincII site of a cloning vector pBluescriptII and introduced into *E. coli* XL1 blue strain. Each plasmid from the library was isolated with alkaline SDS method (Ish-horowicz, D. *et al.*, 1981) and treated with proteinase K to eliminate contamination of RNase and transcribed into RNA using T3 and T7 polymerase

(Stratagene). This RNA was mixed with rabbit reticulocyte lysate and used for bioassay to screen the IFN specific cDNA.

Sequencing Dideoxynucleotide chain termination sequencing was carried out using T7 DNA polymerase modified (Sequenase) (United States Biochemical Corp., Cleveland, OH).

Expression Flatfish IFN cDNA insert which was cleaved by EcoRI and NotI from pBlue-IFN, was connected with EcoRI-PstI fragment of pcDLSRα which contains the SRα promoter which was one of the most potent promoter in mammalian cells. This plasmid was introduced into monkey cell line COS-1 by DEAE dextran method. The antiviral activity of the cultured medium of the COS-1 cells were assayed.

RESULTS

Antiviral activity of the cultured medium of HL8
HRV infection to EPC cells was completely inhibited by 6% supplement of the HL8 cultured medium (m.o.i of 5). As the amount of HL8 cultured medium became smaller, the viability of EPC cells became lower (Fig.1).
DEAE-Toyopearl chromatography
Dialyzed HL8 culture medium was applied to a DEAE-Toyopearl 650M column. Antiviral activity was detected in the fraction eluted with 500-600 mM NaCl (Fig.2). The active fraction was pooled and stored at -20°C.

Affinity chromatography
Affinity chromatography was performed using a column of WGA agarose as described in **Materials & Methods**. The column was washed with the adsorbing buffer and then developed by stepwise increase of GlucNAc. Antiviral activity was detected in the fraction eluted with 100mM GlucNAc (Fig.3). The active fraction was pooled and lyophilized.

SDS-PAGE
The active fraction was electrophoresed using SDS-polyacrylamide gel. As shown in fig.4, distinct band was detected in the location of 16 KDa protein. The extracted protein from the gel showed antiviral activity.

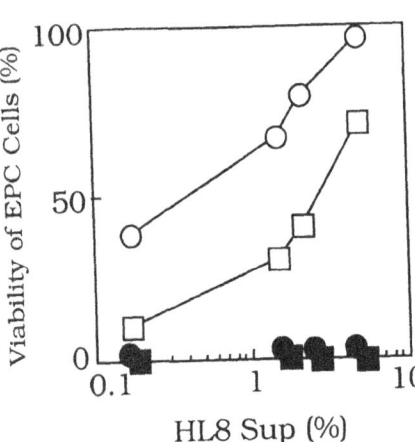

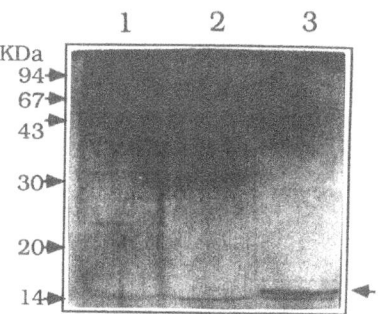

Elution Volume (ml)

HL8 Sup (%)

Fig. 1 Antiviral activity of the cultured medium of immortalized leukocyte on the HRV infection of EPC cells. An immortalized cell line, HL8 (2 x 10^5 cells/ml) was cultured in ITES-ERDF medium at 15°C for a week. Nontransformed leukocytes (8 x 10^{5}cells/ml) were cultured in 5% FCS-ERDF medium for a week. Varing concentration of the cultured medium was added to 0.1ml of EPC cell suspension (8 x 10^{5}cells/ml). On the next day, the cells were challenged by the various m.o.i. of HRV from 5 to 50. On 7th day, the viability of EPC cells were assayed by the uptake of neutral red. The viability of EPC cells (%) was calculated by the ratio of the optical density of the test well to that of control well containing uninfected cells. Open mark shows the culture media of HL8. Closed mark shows the culture media of nontransformed leukocytes. The m.o.i. of HRV:○ and ●, 5; ▢ and ▪ 50.

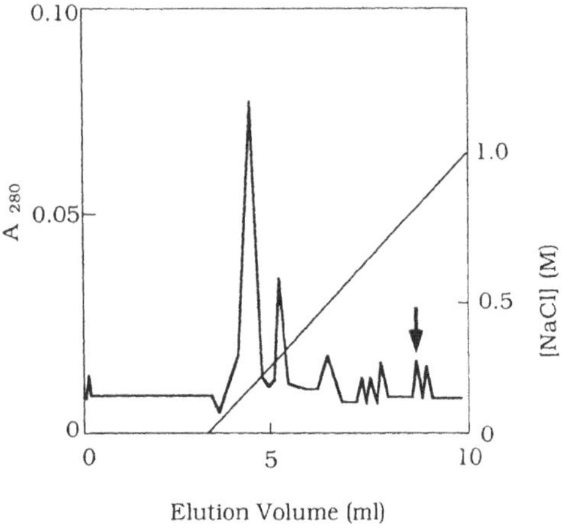

Fig. 2 Purification of flatfish IFN by a DEAE-Toyopearl 650M column chromatography. The HL8 cells (2 x 10^5 cells/ml) were cultivated in 90 mm plastic dishes in ITES-ERDF medium at 15°C for a week. The cultured medium (500ml) of the cells was applied to a column of DEAE-Toyopearl 650M resin (10mm I.D. x 20 cm) at a flow rate of 1ml/min. The IFN was eluted using a linear gradient of NaCl (0-1 M) and the antiviral ativity of each fraction was assayed using HRV (the m.o.i. of 5), as described in Fig. 1. Arrow showed the fraction in which the activity was detected.

Fig. 4 SDS-polyacrylamide gel electrophoregram of each fraction. The fraction of WGA agarose column chromatography (0.1 μg protein) dialized against PBS and lyophilized were applied to a SDS-polyacrylamide gel electrophoresis using a 15% separating gel. After electrophoresis, the gel was visualized by silver-staining. Lane 1, unadsorbed fraction; lane 2, peak 1; lane 3, peak 2. An arrow of the right side showed the upper protein band of lane 3 which exhibited the IFN activity, after extraction from the gel.

Cloning of flatfish IFN cDNA
mRNA prepared from HL8 was devided into 10 fractions using sucrose density gradient centrifigation. Each fraction's mRNA was translated and IFN activity of the resulted products was detected as described in Materials & Methods. mRNA of about 20S was found to encode the IFN. mRNA from the fraction was used to make a cDNA library. We screened the cDNA library by *in vitro* translation assay in order to identify an IFN-specific clone. Plasmids were prepared from 18 group of mixtures of 24 recombinant clones, and each plasmid group was transcribed and *in vitro* translated. One of the 18 groups gave a positive signal by this assay. Next we prepared 24 plasmids separately from the positive group and subjected them to the same assay. It appeared that the plasmid from a clone contained a cDNA specific for the IFN.

Sequencing and expression
The complete nucleotide sequence of the insert from pBlue-IFN is presented in Fig.5. The DNA sequence contains a single large open reading frame, the first ATG, which usually serves as initiation codon in eukaryotes is followed by 138 codons before a terminal codon TAA is found. The sequence analysis of flatfish IFN cDNA allowed us to predict the hitherto unknown primary structure of flatfish IFN which would consist of 138 amino acids. Flatfish IFN from HL8 cells is reportedly glycosilated *in vitro*(Tamai *et al.*, 1992). Potential glycosilation site (Asn-X-Ser) were present in deduced IFN sequence.

The flafish IFN cDNA were subcloned into a pcDLSRα 296 vector and introduced into COS-1 cells by DEAE dextran method. The cultured medium of the COS-1 cells inhibited HRV challenge to EPC cells completely, suggesting that active IFN was produced by the flatfish IFN cDNA.

DISCUSSION

The immortalized leukocytes secreted interleukin-2 (Tamai *et al.*, 1992). Here we report that the immortalized leukocytes also secreted IFN like protein, of which molecular weight was estimated to be about 16 KDa. Mammalian IFNs are known to have a broad molecular weight of about 15 -30KDa (Yip *et al.*, 1982). The flatfish IFN mature molecule deduced from the nucleotide sequence was composed of 108 amino acids which was result of cleavage by signalpeptidase and the deduced molecular size of 12 KDa was a little smaller than the purified flatfish IFN (about 16 KDa) from the cultured medium of the immortalized lymphocyte. The difference in the both sizes suggested that the glycosilation site had actually been glycosilated. In fact, the protein gave a positive glycosilation detection test. Although most mammalian IFNα do not appear to be extensively glycosilated and to have molecular weights range between 15KDa and 23KDa (Allen and Fantes, 1980), some subtypes of IFNα is suggested to have glycosilated moiety (Grab and Chadha, 1979).

The sequence of the cloned flatfish IFN DNA had 54% homology with that of human IFNαN. Lower homology of 26% was found between the sequence of flatfish IFN and human IFNβ DNA. The homology between the sequence of flatfish IFN cDNA and that

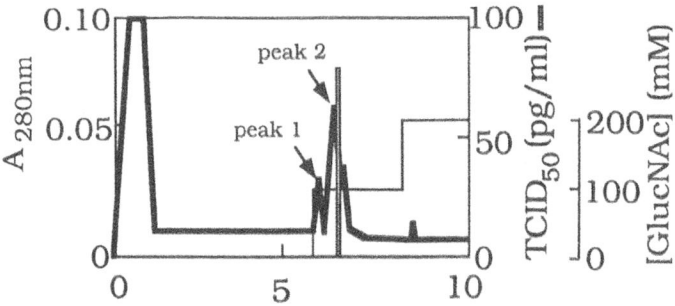

Fig. 3 Purification of flatfish IFN by a WGA agarose chromatography. The active fraction (1ml) of DEAE-Toyopearl column chromatography was applied to a column of WGA agarose (1ml of bed volume) at a flow rate of 30 ml/h. The adsorbed materials were eluted with a stepwise increase of N-acetyl glucosamine (0-100mM). The TCID$_{50}$ of antiviral activity of each fraction was assayed using HRV (m.o.i. = 1), as described in Fig. 1.

```
                                                        -12  TAGTCTACCTCC
         10        20        30        40        50        60        70        80        90
ATGATCAGAAGTACTAACTCCAACAAGTCAGATATATTGATGAATTGCCATCATCTGATAATCAGATATGATGATAATTCCGCTCCTTCT
MetIleArgSerThrAsnSerAsnLysSerAspIleLeuMetAsnCysHisHisLeuIleIleArgTyrAspAspAsnSerAlaProSer

        100       110       120       130       140       150       160       170       180
GGTGGTTCTTTGTTCCGCAAAATGATAATGTTACTCAAACTTTTAAAATTAATAACGTTCGGGCAACTACGAGTTGTCGAATTGTTTGTA
GlyGlySerLeuPheArgLysMetIleMetLeuLeuLysLeuLeuLysLeuIlethrPheGlyGlnLeuArgValValGluLeuPheVal

        190       200       210       220       230       240       250       260       270
AAGTCTAATACTTCTAAAACCTCAACAGTATTATCTATTGACGGCTCTAATCTAAATAGTTTGTTAGATGCTCCTAAAGATATTTTAGAT
LysSerAsnThrSerLysThrSerThrValLeuSerIleAspGlySerAsnLeuIleSerLeuLeuAspAlaProLysAspIleLeuAsp

        280       290       300       310       320       330       340       350       360
AAACCTTCCTGCAATTCCTTTCAACTGGATTTGGTGCTGGCCTCTAGTGCCTGGACACTGTTGACAGCGCGGCTGCTTAACTACCCCTAC
LysProSerCysAsnSerPheGlnLeuAspLeuLeuLeuAlaSerSerAlaTrpThrLeuLeuThrAlaArgLeuLeuAsnTyrProTyr

        370       380       390       400       410
CCTGCTGTTTTACTCTCTGCTGGTGTTGCTTCGGTAGTTTTAGTGCAAGTCCCATAAACTAATA
ProAlaValLeuLeuSerAlaGlyValAlaSerValValLeuValGlnValPro***
```

Fig.5 Nucleotide sequence of the flatfish cDNA and deduced amino acid sequence of the flatfish IFN. Glycosilation site was shown with underline.

of human IFNγ was low. cDNA cloning of the flatfish IFN opened the way to produce a large amount of recombinant fish IFN. The fish recombinant IFN is expected to be useful for prevention or therapies of various fish virus diseases.

Reference

1. Allen G & Fantes KH.(1980) Nature 287: 408-411.
2. De Klinken P, Dorson M & Hattenberger-Baudouy AM (1982) Dev. Comp. Immun. Suppl. 2: 167-174.
3. Demaeyer E. & Demaeyer-Guignard J. (1988) Interferon and Other regulatory Cytokines, John Wiley & Sons, New York.
4. Dorson M, Barde A & de Klinken P (1975) Ann. Microbial. (Inst. Pasteur) 126: 485-489.
5 Grab PM. & Chadha KC (1979) Biochemistry 18: 5782-5786.
6. Gravell M & Malsberger RG (1965) Ann. N.Y. Acad. Sci. 126: 555-565.
7. Graham D & Secombes CJ (1990) Fish Biol. 36: 563-573.
8. Hosoi K, Utsumi J, Kitagawa T, Shimizu H & Kobayashi S (1988) J. Interferon Res. 8: 375-384.
9. Ish-horowicz D & Burke JE (1981) Nucleic Acid Res. 9: 2989-2998.
10. Laemmli UK (1970) Nature 227: 680-685.
11. Murakami H, Masui H, Sato GH, Sueoka N, Chow TP & Kano-Sueoka T(1982) Proc. Natl. Acad. Sci., USA 79: 1158-1162.
12. Okamoto N, Shirakura T, Nagakura Y & Sano T (1983) Fish Pathology 18: 7-12.
13. Sano E, Okano K, Sawada R, Naruto M, Sudo T, Kamata K, Iizuka M & Kobayashi S (1988) Cell. Struct. Funct. 13: 143-160.
14. Stewart WE (1980) Nature 286: 110.
15. Tamai T., Sato N., Kimura S., Shirahata S. & Murakami H. (1992) Murakami H., Shirahata S. & Tachibana H. (Eds) Animal Cell Technology: Basic and Applied Aspects (pp509-514) Kluwer Academic Publishers, Dordrecht.
16. Yip YK., Barrowclough BS, Urban C & Vilcek J (1982) Science 215: 411-413.

DETECTION BY FLUORESCENCE IN SITU HYBRIDIZATION OF THE ERYTHROPOIETIN
GENE INTRODUCED INTO ANIMAL CELLS

K. OKUMURA, Y. HORIKITA, K. HISHIDA, H. TAGUCHI,
Y. SHIMABAYASHI, M. NAGAO,* AND R. SASAKI*
Laboratory of Biological Chemistry, Faculty of Bioresources,
Mie University, Tsu, Mie 514, Japan
*Department of Food Science and Technology, Faculty of
Agriculture, Kyoto University, Kyoto 606, Japan

ABSTRACT. The expression vectors of human erythropoietin (EPO), pZIP-
NeoSV(X)1-EPO, and the soluble form of the mouse EPO receptor (sEPO-R),
pcmEPR-sol.DHFR, were transfected to HepG2 and CHO-dhfr⁻ cells,
respectively, and analyzed with regard to their integration sites on the
chromosomes by using fluorescence in situ hybridization (FISH). The
integrated DNA was detected by FITC, and FISH images were taken by a
cooled CCD digital imaging system connected with a fluorescence
microscope. The integrated EPO gene was detected weakly at only one
specific site of a given chromosome. On the other hand, the integrated
sEPO-R gene gave intense signals because of gene amplification. The
integration site of the sEPO-R gene was the telomere region of a pair of
specific chromosomes. In interphase nuclei, this gave dispersed signals
in contrast to the integrated EPO gene, which gave singlets or doublets.

INTRODUCTION

The production of recombinant proteins in mammalian cells is very
important for clinical applications as well as for fundamental research.
In particular, the production of glycoproteins, which have carbohydrate
structures similar or preferably identical to naturally occurring ones,
has been widely promoted, because the carbohydrate chains can affect the
biological properties of proteins. On the other hand, this technique
provides a means of investigating the mechanisms for gene expression and
gene amplification. We are interested in the relationship between gene
expression and the DNA replication timing of the gene. Replication of
DNA in the animal cell genome takes place in a temporally ordered
fashion during the S phase of the cell cycle. In general, actively
transcribed genes in genomes replicate in the early S phase [1]. How is
a gene integrated into a genome? As the first step to answer this
question, we tried to detect the genes introduced into animal cells by
using fluorescence in situ hybridization (FISH) and digital-image
microscopy. FISH is a powerful tool in biology and medicine [2,3]. The

S. Kaminogawa et al. (eds.), Animal Cell Technology: Basic & Applied Aspects, Vol. 5, 107–113.
© 1993 Kluwer Academic Publishers.

most direct approach for mapping DNA sequences is by FISH, with which we can visualize targeted DNA sequences both in metaphase chromosomes and in interphase nuclei. This methodology has reached a high standard thanks to improved hybridization techniques [4-6] and sensitive equipment for image capture and processing [7].

In this paper, we show the detection of a human EPO gene integrated into HepG2 cells, and a mouse sEPO-R gene into CHO cells, by FISH as examples. A cooled charge-coupled device (CCD) digital image analyzing system will be introduced, and the future direction of this work will also be discussed.

Materials and Methods

DNA PROBES AND CELL CULTURE. The construction of the EPO expression plasmid, pZIP-NeoSV(X)1-EPO, has been described elsewhere [8]. The transfection of this plasmid to HepG2 cells and the isolation of EPO expression cells, HepG2-EPO, were done as described in the same reference. HepG2-EPO cells were cultured in minimum essential medium (MEM, GIBCO) supplemented with 10% fetal calf serum (FCS, GIBCO). The construction of both the sEPO-R expression vector (pcmEPR-sol.DHFR) and sEPO-R producing cells (DXB11N14.2), and the other culture conditions have been described elsewhere [9].

CHROMOSOME PREPARATION. Metaphase chromosome spreads and interphase nuclei from both cells were prepared by standard techniques with some modifications. Briefly, exponentially growing cells were trypsinized and harvested after a colcemid treatment, and then the cells were subjected by hypotonic shock and methanol/acetic acid fixation, as described in ref. 10. In the case of the HepG2 cells, to get longer chromosome spreads, a slide immersed in 70% ethanol was burned just after dropping the fixed cell suspension. For in situ hybridization reactions, chromosomes were denatured in 70% (vol/vol) formamide/2X SSC (1X SSC is 0.15M sodium chloride/0.015M sodium citrate, pH 7) for 2 min at 70 °C, and then incubated in 70%, 90%, and 99.5% ice-cold ethanol (5 min each), before air drying.

PROBE LABELING. pZIP-NeoSV(X)1-EPO and pcmEPR-sol.DHFR probes were labeled by nick translation [11] with biotin-16-dUTP (Boehringer Mannheim). The critical size range for the DNA probe molecules (smaller than 500 bp and preferably 150-250 bp) was achieved by empirically varying the amount of DNase I (Takara Shuzo) in the nick translation reaction. Unincorporated nucleotides were separated from the probe DNA by centrifuging through 1-ml Sephadex G-50 columns in the presence of 0.1% SDS.

IN SITU HYBRIDIZATION. The in situ hybridization conditions were as described elsewhere [10]. Briefly, about fifty nanograms of labeled probes and 10μg of DNase-treated salmon sperm DNA were ethanol-precipitated together and redissolved in 10μl of a hybridization mixture [50% (vol/vol) formamide/2X SSC/10% dextran sulfate]. After being denatured at 75 °C for 5 min, the probe was hybridized onto the denatured specimen. After incubating overnight and subsequent posthybridization washing and blocking with bovine serum albumin, the probes were detected

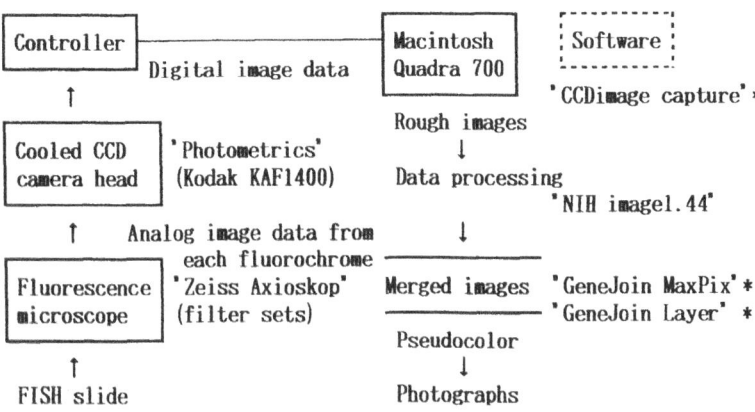

Figure 1. Schematic Diagram of the Cooled CCD Image Analyzing System.
The fluorescence microscope is directly connected with the CH250 camera
head, in which is installed the Kodak KAF1400 CCD camera. The
fluorescence signal passing through each filter from the FISH slide is
sent to the cooled CCD camera head. Analog image data are sent to the
controller, and then digital image data are transferred to the Macintosh
computer. Each FISH image is taken by using the software, 'CCD image
capture' and, if necessary, processed by 'image1.44' (by Wayne Rasband,
NIH). The images are merged and given pseudocolor by 'GeneJoin'. The
software marked (*) was programmed by Timothy Rand (Yale University).

by means of fluorescein isothiocyanate (FITC)-conjugated avidin
(Boehringer Mannheim). Unbounded fluorophores were removed by washing
and the slides were counterstained with DAPI (4,6-diamidino-2-phenol-
indole dihydrochloride, Sigma Co.). In the case of the specimens of
HepG2 cells, two micrograms of Cot1 DNA (BRL) were mixed with a probe,
and the repetitive DNA sequences were pre-annealed for 10 min at 37 °C
before hybridization.
IMAGING. Chromosome spreads and nuclei were imaged by fluorescence,
using a wide-field microscope [Zeiss Axioskop; 63x1.25 numerical
aperture Plan Neofluar oil-immersion objective equipped with a cooled
CCD camera (Photometrics CH250)]. CCD image acquisition and processing
employed an Apple Macintosh Quadra 700 running software developed by
Timothy Rand as shown in Fig. 1. The final images were photographed
from the computer monitor.

Results and Discussion

DETECTION OF THE EPO GENE BY FISH. Human genomic DNA clones usually
contain dispersed repetitive sequences such as the Alu family. When we

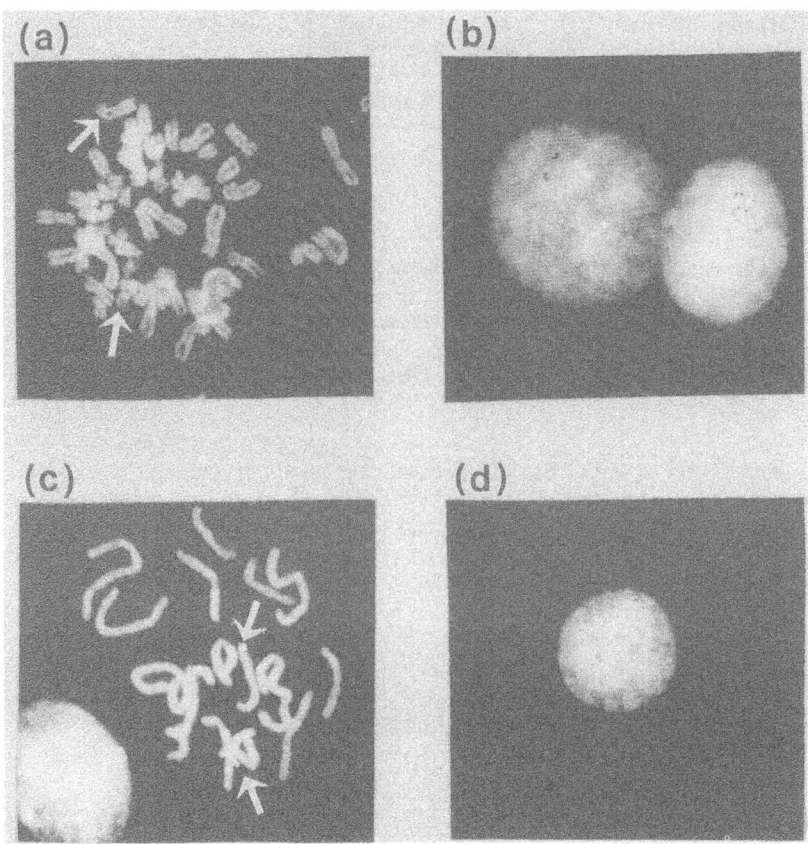

Figure 2. FISH Images from the Cooled CCD Image Analyzing System.
Typical images are shown of the metaphase spread (a) and interphase
nuclei (b) prepared from HepG2-EPO. Also shown are the amplified sEPO-R
gene on the metaphase (c) and interphase (d) from the sEPO-R gene-
transfected CHO cells. White arrows show signals. The specific signals
of FITC are shown, and the specimens were counterstained by DAPI. The
weak FITC signals in (a) and (b) were processed and intensified before
merging.

use them as probes of FISH on specimens derived from human cell lines,
total human genomic DNA or Cot1 DNA fraction must be mixed with labeled
probes as a competitor to suppress non-specific signals. The EPO
expression plasmid, pZIP-NeoSV(X)1-EPO, was directly labeled with
biotin-dUTP by nick translation, and mixed with Cot1 (2μg) and salmon
sperm (8μg) DNAs. After denaturation, the probe mixture was pre-
annealed for 10min at 37 °C and hybridized onto the denatured slides,
which were mounted as metaphase spreads and interphase nuclei prepared

from the HepG2 cells into which the EPO gene was introduced (HepG2-EPO). HepG2-EPO was confirmed to produce human EPO by the same method as that described in ref. 8 (data not shown). As shown in Fig. 2(a), only one set of FITC signals was detected on a given chromosome of short arm near the centromere. The endogenous EPO gene was not detected under these standard conditions because it was not detected in untransfected HepG2 cells. Although we haven't checked the number of integrated EPO genes per HepG2-EPO cell, it's possible to integrate multiple tandem copies of the exogenous EPO gene. In the interphase nuclei, singlet or doublet signals were observed as shown in Fig. 2(b). The FITC signals in Fig. 2(a) and (b) were originally weak and ambiguous, though they were processed and intensified by the computer. One of the most important points for mapping the specific DNAs by FISH is the quality of the specimens that can be prepared. The best way to get good metaphase spreads depends on the cell types, and we could only get very short and few chromosome spreads from HepG2 cells by using the standard method. We tried several other methods such as synchronizing the cells and using a DNA topoisomerase II inhibitor to get longer and more chromosomes. Although we haven't succeeded in this yet, it is necessary to identify the specific chromosome into which the EPO gene is integrated and map it with a banding method.

DETECTION OF THE sEPO-R GENE BY FISH. The sEPO-R expression vector was directly labeled by nick translation and mixed with only salmon DNA. As the sEPO-R gene is a mouse clone and the transfected cell was derived from hamster, no suppression is required. After denaturation at 75 °C for 5min, the probe mixture was placed on ice before applying to the denatured specimens. As shown in Fig. 2(c), intense signals were observed because of gene amplification under the culture conditions in the presence of methotrexate. The integration site of the sEPO-R gene was the telomere region of a pair of specific chromosomes. This amplified gene gave dispersed signals in the interphase nuclei as shown in Fig. 2(d), in contrast to the case of HepG2-EPO (Fig. 2(b)).

The introduction of an image analyzing system may be a powerful tool in the field of cytogenetics. In particular, a cooled CCD camera has very high sensitivity. The system introduced here (Fig. 1) was originally set up at the David C. Ward Laboratory (Yale University), and it is the latest version. To get an image from a FISH slide takes only 0.1 to just a few seconds. Once rough images have been captured and saved in the computer, they can be processed whenever necessary. The software named 'GeneJoin' (constructed by Timothy Rand, Yale University) has 8 windows, allowing eight different images to be merged and given individual pseudocolors.

We have already proposed a method for determining the DNA replication timing in the cell cycle of a gene by FISH [12]. In general, actively transcribed genes replicate in the early S phase of the cell cycle. Exogenous genes are integrated randomly into animal chromosomes, and some of them may be transcribed actively, while others may not. As we usually select the cells which have sufficient activity of the integrated genes at the screening step, they can be expected to have a replication timing in the early S phase. On the other hand, the chromosomal region of R positive bands replicates earlier than that of R negative ones [13]. Integrated EPO genes may replicate in the early S phase and might have been possibly located on the R positive bands. Unfortunately, the FISH signals in this experiment have been too weak so far to score the ratio of nuclei which gave doublets. Further experiments for preparing longer chromosome spreads, mapping by Alu-PCR banding [14] and amplifying the signals are in progress.

ACKNOWLEDGEMENTS

We thank T. Rand and D. C. Ward (Yale University School of Medicine) for providing the software, and H. Takanari (Mie University School of Medicine) for using the CCD equipment. This work was supported in part by a Grant-in-Aid for Scientific Research from the Ministry of Education, Science, and Culture of Japan.

REFERENCES

1. Holmquist, G. P. (1987) 'Role of replication time in the control of tissue-specific gene expression,' Am. J. Hum. Genet. 40, 151-173.
2. Lichter, P. and Ward, D. C. (1990) 'Is non-isotopic in situ hybridization finally coming of age?' Nature 345, 93-95.
3. Trask, B. J. (1991) 'Fluorescence in situ hybridization: application in cytogenetics and gene mapping,' Trends in Genet. 7, 149-154.
4. Lawrence, J. B., Villhave, C. A., and Singer, R. H. (1988) 'Interphase chromatin and chromosome gene mapping by fluorescence detection of in situ hybridization reveals the presence and orientation of two closely integrated copies of EBV in a human lymphoblastoid cell line,' Cell 52, 51-61.
5. Pinkel, D., Landegent, J., Collins, C., Fuscoe, J., Segraves, R., Lucas, J., and Gray, J. W. (1988) 'Fluorescence in situ hybridization with human chromosome-specific libraries: Detection of trisomy 21 and translocations of chromosome 4,' Proc. Natl. Acad. Sci. U.S.A. 85, 9138-9142.
6. Lichter, P., Cremer, T., Borden, J., Manuelidis, L., and Ward, D. C. (1988) 'Delineation of individual human chromosomes in metaphase and interphase cells by in situ suppression hybridization using recombinant DNA libraries,' Hum. Genet. 80, 224-234.
7. Jovin, T. M. and Arndt-Jovin, D. J. (1989) 'Luminescence Digital Imaging Microscopy,' Ann. Rev. Biophys. Chem. 18, 271-308.
8. Goto, M., Akai, K., Murakami, A., Hashimoto, C., Tsuda, E., Ueda, M., Kawanishi, G., Takahashi, N., Ishimoto, A., Chiba, H., and Sasaki, R. (1988) 'Production of recombinant human erythropoietin in mammalian cells: host-cell dependency of the biological activity of the cloned glycoprotein,' Bio/Technology 6, 67-71.
9. Nagao, M., Masuda, S., Abe, S., Ueda, M., and Sasaki, R. (1992) 'Production and ligand-binding characteristics of the soluble form of murine erythropoietin receptor,' Biochem. Biophys. Res. Commun. 188, 888-897.
10. Cremer, T., Lichter, P., Borden, J., Ward, D. C., and Manuelidis, L.

(1988) 'Detection of chromosome aberrations in metaphase and interphase tumor cells by in situ hybridization using chromosome-specific library probes,' Hum. Genet. 80, 235-246.

11. Lichter, P., Chang Tang, C.-J., Call, K., Hermanson, G., Evans, G. A., Houseman, D., and Ward, D. C. (1990) 'High-resolution mapping of human chromosome 11 by in situ hybridization with cosmid clones', Science 247, 64-69.

12. Sara, S., Okumura, K., Ward, D. C., and Cedar, H. (1992) 'Delineation of DNA replication time zones by fluorescence in situ hybridization,' EMBO J. 11, 1217-1225.

13. Hand, R. (1978) 'Eukaryotic DNA: organization of the genome for replication,' Cell 15, 317-325.

14. Baldini, A. and Ward, D. C. (1991) 'In situ hybridization banding of human chromosomes with Alu-PCR products: a simultaneous karyotype for gene mapping studies,' Genomics 9, 770-774.

HIGH-LEVEL EXPRESSION OF RECOMBINANT BOVINE β-LACTOGLOBULIN IN INSECT CELLS

K. MIZUMACHI, J. KURISAKI,* and N. M. TSUJI
Department of Animal Products, National Institute of Animal Industry
Tsukuba Norindanchi P. O. Box 5, Ibaraki 305, Japan

ABSTRACT A cDNA coding for bovine β-lactoglobulin (β-LG) was obtained from a cDNA library constructed in λgt11 with poly(A)$^+$ RNAs from mammary gland. The cDNA, although lacking 49 nucleotides of the 5' end, was cloned into baculovirus transfer vector pVL1392 and inserted into the genome of the *Autographa californica* nuclear polyhedrosis virus (AcNPV) by homologous recombination. Cloned recombinant AcNPV was obtained by plaque purification, which was carried out by 3 rounds of visual screening of infected *Spodoptera frugiperda* cells (Sf9). For the production of recombinant β-LG, the cells were infected with the cloned recombinant AcNPV at a multiplicity of infection of 10 or 20. The level of recombinant β-LG in the supernatant was determined by sandwich ELISA. At 72 to 96 hours post-infection, secreted recombinant β-LG reached its maximum level. Correct processing of recombinant β-LG in insect cells was suggested, since the sequence of the N-terminal 20 residues was identical to that of natural β-LG. Preparative-scale production was performed by a suspension culture, the yield of recombinant β-LG achieved being 5 mg per 1 liter of the culture.

1. Introduction

The functional properties of food proteins are greatly influenced by their primary, secondary and tertiary structures. The application of protein engineering to food proteins should, therefore, provide useful information for improving their properties. Bovine β-lactoglobulin (β-LG) is one of the major milk proteins and is known to be causative of allergy. It would be a significant advantage for patients allergic to milk if the allergenicity of β-LG could be reduced by modification of the structure. A baculovirus expression system has recently been applied to the production of various recombinant proteins [1] and is recognized for its ability to express functional products in a high yield. In the present study, a high-level expression system for bovine β-LG was established in insect cells by using a baculovirus vector.

2. Materials and Methods

2.1 cDNA CLONING

Total RNA was isolated from the lactating mammary gland of a Holstein cow by the guanidinium thiocyanate procedure [2]. Double-stranded cDNA was synthesized from poly(A)$^+$ RNA, enriched by oligo(dT)-cellulose column chromatography, and subcloned into λgt11 with *Eco*RI adaptors (cDNA cloning system-λgt11, Amersham Japan, Tokyo). The cDNA library was

---

*To whom all correspondence should be addressed.

S. Kaminogawa et al. (eds.), Animal Cell Technology: Basic & Applied Aspects, Vol. 5, 115–121.
© *1993 Kluwer Academic Publishers.*

screened for β-LG cDNA with mouse anti-β-LG antibodies. DNA fragments prepared from positive clones were subcloned into pUC18, and a Southern analysis was performed with a synthetic oligonucleotide probe (5'-GCAGCCATGAAGTGCCTCCTGCTT-3') corresponding to the N-terminal signal peptide sequence of β-LG [3]. As a result, positive clone pBLG11 was isolated. The DNA insert in pBLG11 was sequenced by the dideoxy-termination method [4], using a model 373A DNA sequencer (Applied Biosystems Japan, Tokyo). General genetic engineering techniques, including the isolation of bacteriophage, plasmid and single-stranded DNA, enzymatic reactions and hybridization experiments, were performed according to Sambrook *et al.* [5]. Oligonucleotides were synthesized by a DNA synthesizer (Gene Assembler Plus, Pharmacia Biotech, Tokyo).

2.2 EXPRESSION OF RECOMBINANT PROTEIN

Sf9 *Spodoptera frugiperda* insect cells (Invitrogen, San Diego, CA) were maintained at 27°C in supplemented Grace's insect medium containing 10% fetal bovine serum (FCS) and gentamycin (50 μg/ml). The cells were grown in tissue culture flasks to a density of 4×10^5 cells/ml or in spinner flasks to a density of $1-2 \times 10^6$ cells/ml.

The *Bam*HI fragment from pBLG11 containing the cloned cDNA for β-LG was inserted into the *Bam*HI site of transfer vector pVL1392 (Invitrogen). The resulting plasmid, pVLG101, was cotransfected with wild-type AcNPV DNA into Sf9 cells by homologous recombination according to Summers and Smith [6]. The recombinant virus was obtained by plaque purification, which was carried out by 3 rounds of visual screening of the infected Sf9 cell monolayers with a 0.75% agarose overlay. Southern analysis of DNA from the purified recombinant baculovirus was performed with the synthetic oligonucleotide probe described in 2.1.

The cells were seeded into flasks at density of 1.2×10^5 /cm^2 and infected for one hour with the recombinant virus at a multiplicity of 10 or 20 plaque-forming units per cell (pfu/cell). The cells were cultured further for 24, 48 and 96 hours in supplemented Grace's medium containing 10% FCS, and the supernatants were collected for analyses of the product.

2.3 PREPARATIVE ISOLATION OF RECOMBINANT PROTEIN

Sf9 cells cultured in spinner flasks were collected by centrifugation and incubated at 10^7 cells/ ml for one hour with the recombinant baculovirus at a multiplicity of 10 pfu/cell. After 24 hours of incubation in Grace's medium containing 10% FCS without supplements, the cells were incubated for 72 hours under serum-free condition. The supernatant was collected by centrifugation and lyophilized after dialysis against water at 4°C. The lyophilized product was subjected to column chromatography on TSKgel DEAE-Toyopearl 650S (Tosoh, Tokyo), using Waters 650 (Nihon Millipore, Tokyo). The proteins were eluted by a concave gradient from 0 to 0.5 M NaCl in a 0.05 M imidazole-HCl buffer (pH 6.4). The fractions containing recombinant β-LG were pooled, dialyzed against water and lyophilized.

2.4 PROTEIN ANALYSES

The recombinant β-LG in the culture supernatant was determined by sandwich ELISA, using the anti-β-LG monoclonal antibody for plate coating and rabbit polyclonal antibodies for detecting the captured antigens. The second antibodies were further reacted with anti-rabbit IgG conjugated with alkaline phosphatase. The activity was measured at 405 nm by a Titertek Multiskan MCC/340 (Flow Laboratories, McLean, VA), using *p*-nitrophenylphosphate as a substrate.

SDS-PAGE and native PAGE were performed by the modified method of Laemmli [7]. Western blotting was carried out with a nitrocellulose (BA83, Schleicher & Schuell, Dassel, Germany) or PVDF (Immobilon P^{SQ}, Nihon Millipore) membrane. Mouse polyclonal anti-β-LG antibodies were used for immunological staining. Proteins electroblotted on the PVDF membrane were sequenced by a model 477A protein/peptide sequencer (Applied Biosystems Japan).

3. Results

3.1 CLONING OF THE RECOMBINANT BACULOVIRUS TO PRODUCE β-LG

The size of the *Eco*RI insert in pBLG11, a positive clone to the hybridization probe, was about 800 base-pairs, which is close to the size of the reported cDNA of β-LG [3]. The sequence of the cloned cDNA matched 50-348 of the known cDNA sequence of β-LG (genetic variant A), indicating that the 49 oligonucleotides 5' end of the β-LG cDNA was missing (Figure 1). The rest of the cDNA sequence was not determined in the present study. Initiation codon ATG of β-LG cDNA was incidentally preserved in the cloned cDNA, owing to the *Eco*RI adaptor used for cloning, but next codon AAG was replaced by GAG.

Transfer vector plasmid pVLG101 was constructed with the cloned cDNA for β-LG as shown in Figure 1. After cotransfecting with pVLG101 and wild-type AcNPV DNA, four clones of the recombinant baculovirus were finally isolated from occlusion-negative plaques by plaque purification. By Southern analysis, all the clones were proved to contain cDNA for β-LG. The subsequent experiments were carried out with recombinant baculovirus AcLG101.

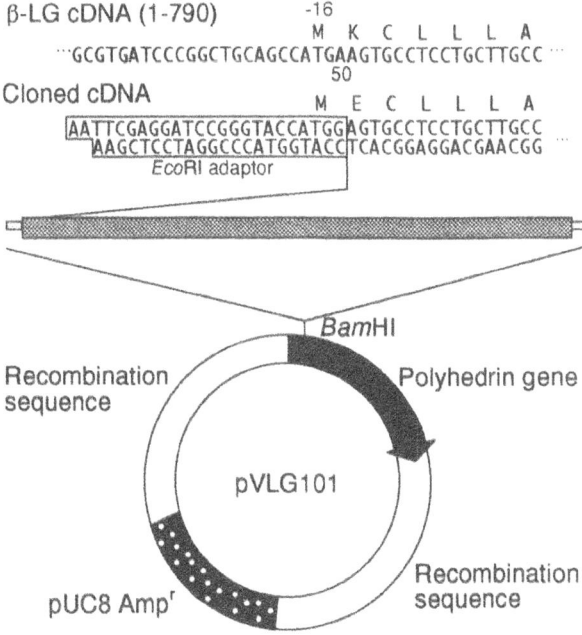

Figure 1. Construction of transfer vector pVLG101 to make a recombinant baculovirus. A *Bam*HI fragment was excised from the cloned cDNA and inserted into the *Bam*HI site of transfer vector pVL1392.

3.2 EXPRESSION AND ISOLATION OF RECOMBINANT β-LG

As shown in the SDS-PAGE patterns (Figure 2A), Sf9 cells infected with AcLG101 secreted a protein with a molecular mass of 18,000 daltons (lane 5), which is similar to that of natural β-LG, while the uninfected Sf9 cells (lane 3) and the cells infected with wild-type AcNPV (lane 4) did not produce such a protein. Mouse anti-β-LG antibodies reacted specifically with the recombinant protein (Figure 2B). The N-terminal sequence of the 20 residues of the protein was Leu-Ile-Val-Thr-Gln-Thr-Met-Lys-Gly-Leu-Asp-Ile-Gln-Lys-Val-Ala-Gly-Thr-Trp-Tyr, which is identical to that of natural bovine β-LG [8]. These results suggest that the product of Sf9 cells infected with AcLG101 was recombinant β-LG. Figure 3 shows the secretion of recombinant β-LG by Sf9 cells at various times following their infection with AcLG101. Although the level of secreted recombinant β-LG up to 48 hours post-infection was marginal, the production reached a relatively high level at 72 and 96 hours post-infection. The multiplicity of infection, 10 or 20, did not affect the level of recombinant β-LG production.

After chromatography on a DEAE-column of the lyophilized product from spinner flasks, the recombinant β-LG was still contaminated with bovine serum albumin and some other proteins, even though the culture medium was changed to serum-free Grace's medium at 24 hours post-infection (Figure 4, lane 6). The recombinant β-LG was purified by rechromatography as shown in Figure 4 (lane 7). The yield of the recombinant β-LG was 5 mg per liter of the culture.

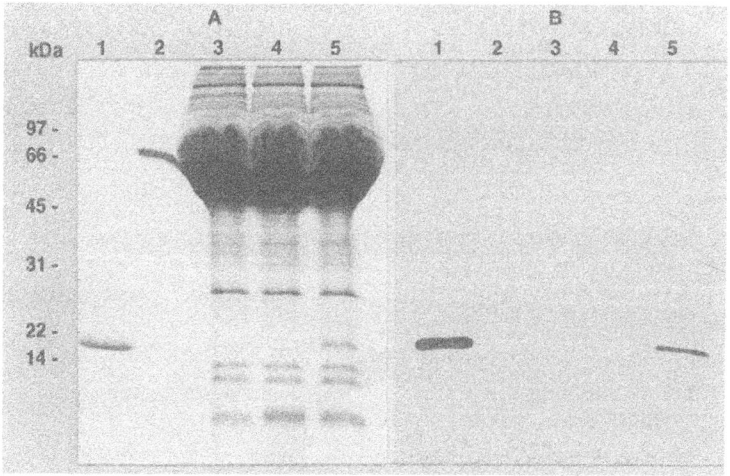

Figure 2. Expression of bovine β-LG by the recombinant baculovirus. (A) SDS-PAGE electrophoretograms of the culture supernatants of Sf9 cells. The supernatant was dialyzed , lyophilized and subjected to electrophoresis. Proteins were stained by Coomassie brilliant blue. (B) Immunoblot analysis of the culture supernatants. The proteins from the SDS-PAGE separation were transferred to a nitrocellulose membrane and reacted with mouse anti-β-LG antibodies. The bound antibodies were visualized by peroxidase-conjugated anti-mouse IgG (goat) and 3,3'-diaminobenzidine as a substrate. lane 1, bovine β-LG (1μg); lane 2, bovine serum albumin (1μg); lane 3, the culture supernatant of uninfected Sf9 cells; lane 4, the culture supernatant of Sf9 cells infected with wild-type AcNPV; lane 5, the culture supernatant of Sf9 cells infected with AcLG101.

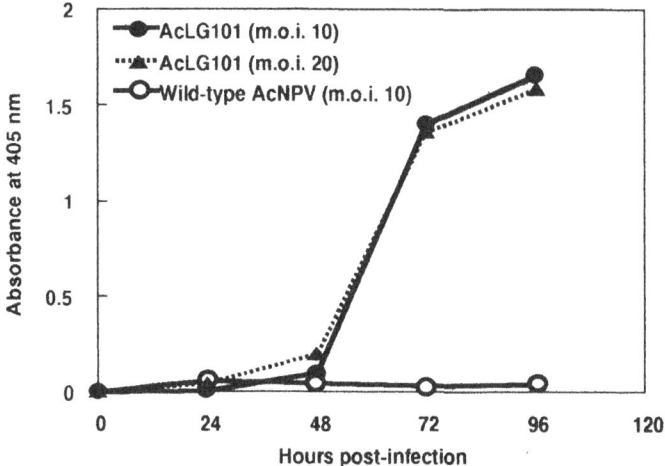

Figure 3. Secretion of β-LG from Sf9 cells after their infection with AcLG101. Sf9 cells $(1.2 \times 10^5/cm^2)$ were infected with AcLG101 at a multiplicity of infection of 10 or 20. At 24, 48, 72 and 96 hours post-infection, the supernatant was collected, and the concentration of β-LG was measured by sandwich ELISA.

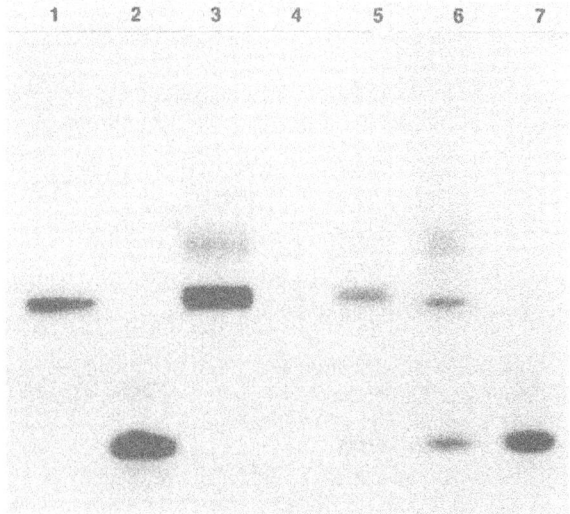

Figure 4. Native PAGE analysis of the recombinant β-LG produced on a preparative scale. Lane 1, BSA (3 µg); lane 2, native β-LG (1 µg); lane 3, the culture supernatant of Sf9 cells infected with wild-type AcNPV (5 µl); lane 4, the culture supernatant of Sf9 cells infected with AcLG101 in a spinner flask (5 µl); lane 5, the lyophilized supernatant of lane 4 (1 µg); lane 6, the fraction isolated by DEAE chromatography (1 µg); lane 7, the fraction from DEAE rechromatography (1 µg).

4. Discussion

Recombinant β-LG was produced by Sf9 cells infected with AcLG101, as judged from the sequence and the size of the cDNA cloned for β-LG and several properties of the product such as the approximate molecular weight, the immunoreactivity to anti-β-LG, and the N-terminal sequence. Correct processing was also suggested, since no difference was found in the sequences of the N-terminal 20 residues between recombinant and native β-LG. The recombinant β-LG would have Glu in place of Lys in native β-LG at the second residue (-15) of the signal peptide. This difference did not have any effect on the processing of the signal peptide, although it might have affected the efficiency of gene expression. A multiplicity of infection of 10 and harvest of the culture supernatant at 96 hours post-infection were the optimum conditions for producing the recombinant protein in the present study. Under such conditions, the level of the recombinant β-LG secreted by the insect cells was high enough to allow preparative isolation of the product. The purification of the recombinant β-LG was easily accomplished by rechromatography in a DEAE column. In order to avoid contamination by serum-derived proteins and cell damage by centrifugation, the application of a serum-free medium such as Sf-900 throughout the culture might have greatly simplified the procedures. Since Smith *et al.* [9] and Maeda *et al.* [10] have developed a baculovirus expression system, a variety of recombinant proteins have been produced by the system, and the levels of production in the medium ranged from 1 to 200 mg/l. The expression level in the present study, 5 mg/l after purification, is in the reported range and higher than that with the yeast expression system [11].

A high-level expression system for β-LG was thus established in the present study. The system should be useful for improving the functional properties of β-LG by protein engineering.

Acknowledgements

We thank Dr. M. Ugaki and Dr. H. Hirano for their cooperation in the sequencing of DNA and peptides, respectively. We also thank Dr. R. Irie, Mr. T. Okamoto and Mr. Y. Fujita for helpful discussion.

References

[1] Luckow, V. A. and Summers, M. D. (1988) 'Trends in the development of baculovirus expression vectors', Biotechnology 6, 47-55.

[2] Okayama, H., Kawaichi, M., Brownstein, M., Lee, F., Yokota, T., and Arai, K. (1987) 'High-efficiency cloning of full-length cDNA', in R. Wu and L. Grossman (eds.), Methods in Enzymology 154, Academic Press, San Diego, pp. 3-28.

[3] Alexander, L. J., Hayes, G., Pearse, M. J., Beattie, C. W., Stewart, A. F., Willis, I. M., and Mackinlay, A. G. (1989) 'Complete sequence of the bovine β-lactoglobulin cDNA', Nucl. Acids Res. 17, 6739.

[4] Sanger, F., Nicklen, S., and Coulson, A. R. (1977) 'DNA sequencing with chain-terminating inhibitors', Proc. Natl. Acad. Sci. U.S.A. 74, 5463-5467.

[5] Sambrook, J., Fritsch, E. F., and Maniatis, T (1989) 'Molecular Cloning: A Laboratory Manual', Cold Spring Harbor Laboratory, Cold Spring Harbor, New York.

[6] Summers, M. D. and Smith, G. E. (1987) 'A manual of methods for baculovirus vectors and insect cell culture procedures', Texas Agricultural Experiment Station Bulletin No. 1555.

[7] Laemmli, U. K. (1970) 'Cleavage of structural proteins during the assembly of the head of bacteriophage T4', Nature 227, 680-685.

[8] Braunitzer, G., Chen, R., Schrank, B., and Stangl, A. (1972) 'Automatische Sequenzanalyse eines Proteins (β-Lactoglobulin AB)', Hoppe-Seyler's Z. Physiol. Chem. 353, 832-834.

[9] Smith, G. E., Summers, M. D., and Fraser, M. J. (1983) 'Production of human beta interferon in insect cells infected with a baculovirus expression vector', Mol. Cell. Biol. 3, 2156-2165.

[10] Maeda, S., Kawai, T., Obinata, M., Chika, T., Horiuchi, T., Maekawa, K., Nakasuji, K., Saeki, Y., Sato, Y., Yamada, K., and Furusawa, M. (1984) 'Characteristics of human interferon-α produced by a gene transferred by a baculovirus vector in the silkworm, *Bombyx mori*', Proc. Japan Acad. 60 (ser. B), 423-426.

[11] Totsuka, M., Katakura, Y., Shimizu, M., Kumagai, I., Miura, K., and Kaminogawa, S. (1990) 'Expression and secretion of bovine β-lactoglobulin in *Saccharomyces cerevisiae*', Agric. Biol. Chem. 54, 3111-3116.

MASS PRODUCTION AND PURIFICATION OF RECOMBINANT FLATFISH INTERFERON

T. Tamai[1], T. Noguchi[2], K. Tsujimura[2], N. Sato[1], S. Kimura[1], S. Shirahata[2] and H. Murakami[2]

[1]Taiyo Central R&D Institute, Taiyo Fishery Co. Ltd, 16-2 Wadai, Tsukuba-shi, Ibaraki 300-42, Japan
[2]Graduate School of Genetic Resources Technology, Kyushu University, 6-10-1 Hakozaki, Higashi-ku, Fukuoka 812, Japan

ABSTRACT
Recombinant baby humster kidney cells (BHK-21) producing flatfish interferon (F-IFN) was cultured on the macroporous microcarrier to accomplish high density cell culture. Recombinant F-IFN was purified by WGA affinity chromatography as described by Tamai et al. (1992). To establish the detection system of F-IFN, we established 7 mouse monoclonal antibodies (MAbs) reactive to F-IFN. The MAb was applied to purify F-IFN fom the cultured medium of BHK cells. The preventive efficacy of recombinant F-IFN on the infected fish was investigated in vivo. F-IFN was effective even in a small amount of 10μg/kg body.

INTRODUCTION
Over the last 20 years, aquaculture has grown into a very important industry in many part of the world. High density culture of fishes in farm is giving a danger for fishes to suffer from feral diseases by bacteria or viruses. There is no effective chemotherapy for control of virus diseases (Ellis, 1988). It is against this background that fish lymphokines have been percieved as potentially playing an important role in aquaculture. Flatfish lymphocyte cell line HL8 immortalized by oncogene transfection was found to secret IFN-like antiviral protein, of which molecular size was estimated to be 16kDa (Tamai et al., 1992). cDNA cloning of the IFN revealed that this molecule is composed of 108 amino acid(Tamai et al., 1993). In this paper large quantity of the IFN secreted by recombinant BHK cells was purified by a single step chromatography using MAb. F-IFN made it possible to cure the infected fishes in vivo.

MATERIALS AND METHODS

Cultivation of recombinant BHK cells Recombinant BHK was produced by transfecting of BHK-21 cells with a plasmid pSR αIFN. The recombinant BHK cells were seeded in ERDF media supplemented with 5% FCS at the cell density of 1×10^5 cells/ml and incubated at 37°C in 5% CO_2 humidity for 6 days. The culture supernatant containing recombinant F-IFN was pooled.

S. Kaminogawa et al. (eds.), Animal Cell Technology: Basic & Applied Aspects, Vol. 5, 123–127.
© 1993 *Kluwer Academic Publishers.*

The recombinant BHK cells were cultured also in serum-free ERDF medium supplemented with 10 μg/ml insulin, 25 μg/ml transferrin, 20 μM ethanolamine and 25 nM selenite (ITES)(Murakami *et al.*, 1982) at 37°C in 5% CO_2 humidity for 6 days.

Cultivation on macroporous microcarrier   The recombinant BHK cells (7 x 10^6 cells/ml) was incubated with macroporous microcarrier (Asahikasei Chem. Co. Ltd.) with the density of 10^2 beads/cell for 6 hr at 37°C in 5% CO_2 humidity.   The cell number on the beads was measured by the amount of crystal violet uptake which was measured by optical density.

WGA agarose chromatography   A WGA agarose column (Honen Oil Co Ltd., Tokyo) was equilibrated with 10mM phosphate-buffered saline (PBS) (pH 7.2).   The recombinant BHK cell culture (50ml) was adjusted to pH 7.2 and applied to the column at a flow rate of 30 ml/h.   The column was then washed with 150 ml of PBS and antiviral protein was eluted using a 100mM of N-Acetyl glucosamine.

Preparation of MAbs   The purified F-IFN was intraperitoneally injected to immunize mice three times.   On the third day after the final immunization, splenocytes were isolated and fused with HAT-sensitive mouse myeloma cell line, P3-X53-Ag8U1 cells by the PEG method (Eto *et al.*, 1991).   On the third day after fusion cells were cultivated in HAT medium.   The antibody production was determined by enzyme linked immuno solvent assay (ELISA).   Examined cells were cloned using limiting dilution twice and established as antibody producing hybridomas.   These cells could be maintained in serum-free medium supplemented with ITES.   Approximately 30% of total protein was antibody.   IgG type of MAb, FIG was purified with protein A chromatography (0.4 x 1cm) (Pharmacia, England).   FIG was fixed using an activated CNBr Sepharose kit (Pharmacia, England).   Typing of MAb light chain and subclass of IgG was done using a typing kit (Tago Co. Ltd.).   Specific reactivity of MAbs to F-IFN, mouse IFN and IFNγ were determined by an ELISA.

Purification of F-IFN using MAB   F-IFN was purified from the cultured medium of the BHK cells using the MAb column chromatography.   The cultured medium of the BHK cells was applied to MAb column (0.4 x 1cm) chromatography at a flow rate of 1ml/ min.   The adsorbed IFN was eluted with 0.1M glycin-HCl (pH 3.0).   The purified F-IFN was solved in the phosphate buffered saline (PBS) to give the final concentration of 200μg/ml

Antiviral effect in vivo   The preventive effect of F-IFN against infected fish *in vivo* was investigated.   The mean body weight of flatfish was 2mg.   The flatfish were treated with 1μg of recombinant IFN intraveneously.   On the next day these flatfish were injected with Hirame Rhabdovirus (HRV) suspension (10^2 $TCID_{50}$) amplified using carp epithelial (EPC) cells (Kamei *et al.*, 1989).   As control, 10 flatfish were injected with only PBS.   These fish were kept in a tank supplied with fresh sea water continuously at 11°C, and the fish viability was observed for 10 days following virus injection.

RESULTS

Cultivation of F-IFN producing BHK cells
The seeded BHK cells producing F-IFN (1 x 10^5 cells /ml) grew to the cell density of 1 x 10^6 cells/ml 6 days after cultivation, in 5% FCS-ERDF medium at 37°C in 5% CO_2

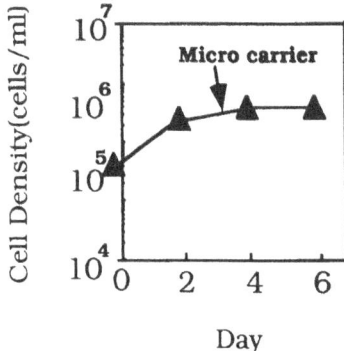

Fig. 1 BHK cell growth. IFN producing BHK cells were cultured in 3% FCS-ERDF medium in 5% CO₂ humidity at 37°C for 6 days. The BHK cells (1x10⁵ cells/ml) were mixed with macroporous microcarrier (1x10⁷ particles/ml) and cultured for 24 hours. The cell number was measured by crystal violet method.

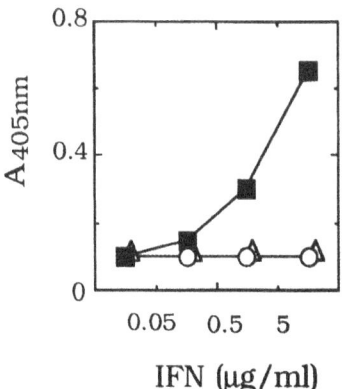

Fig. 2 Reactivities of FIG-4 against various kinds of IFNs. Crossreactivities of MAb (FIG-4) against flatfish IFN (■), human IFNγ (Δ) and mouse IFN (O) were detected with ELISA method. Various concentration of IFN was attached on the bottom of 96 well microplates, and 100ng/ml FIG-4 was reacted. Peroxidase labbeled antibody against mouse IgG was added and reacted with H₂O₂ and and ABTS.

humidity. However the recombinant BHK cells couldn't grow in serum-free ERDF medium supplemented with ITES. To accomplish high density cell cultivation, the cells were attached on the surface of macroporous microcarrier and seeded with the cell density of 1×10^5 cells/ml. The cell density reached over 2×10^6 cells/ml after 6 days(Fig.1).

Establishment and characterization of MAbs against F-IFN
Immunized mouse lymphocytes with F-IFN were fused with P3-X63-Ag8U1 cells. After two fusion experiments, seven hybridomas producing MAb reactive to F-IFN were produced. They were composed of two IgG1s (FIG-4 and FIG-11) and five IgMs (FIM-1, FIM-2, FIM-3, FIM-5 and FIM-12). All MAbs had λ light chains. These six MAbs showed different affinities to the F-IFN. Among them FIG4 and FIM1 showed the highest reactivity.
Figure 2 showed the reactivity of FIG-4. FIG4 showed OD value of 0.35 with 8μg/ml of F-IFN and the OD values were decreased as the amounts of F-IFN were decreased. FIG4 showed no cross reactivity with human IFNγ or mouse IFN. FIG-4 was purified using protein A column chromatography and fixed to the activated CNBr sephacel. The MAb-sephacel column (0.4 x 1cm) was constructed.

Purification of F-IFN using MAb column chromatography
F-IFN was purified from the cultured medium of the BHK cells using MAb column chromatography (Fig.3). The peak eluted with 0.1M glycin-HCl (pH3.0) had revealed to have antiviral activity. This fraction inhibited the HRV infection on carp epithelial (EPC) cells (Tamai *et al.*, 1992). The purified F-IFN was disolved in PBS to give the final concentration of 200 μg/ml.

Therapy
Over 60% of flatfish treated with F-IFN could survive even on 10 days after virus challenge. The control fish were all dead on the 7th day(Fig.4). Quite strong antiviral activity was exhibited not only *in vitro* but also *in vivo*.

DISCUSSION

BHK cells producing recombinant F-IFN were cultured on the macroporous microcarrier successfully which enabled us to prepare a large amount of F-IFN. To purify F-IFN from the cultured medium of the BHK cells we established seven hybridomas producing MAbs against F-IFN. All the MAbs reacted specifically with F-IFN without cross-reactivity against human IFNγ or mouse IFN, which were applicable to the detection of F-IFN. One of these MAbs ,FIG4, having the neutralizing activity may be useful to examine the functional segment of fish IFN molecule. We could purify F-IFN from the cultured medium of the recombinant BHK cells with a single step of chromatography. This MAb may be used for the preparation of high purified F-IFN.
The *in vivo* antiviral activity of recombinant F-IFN was evaluated against experimental infection with HRV on flatfish. The preventive effect against virus infection *in vivo* was detected at a low level of 10 μg F-IFN/kg body. We found the viral inhibitory effect on trout by F-IFN administrated orally *in vivo*. F-IFN might be useful against fish virus on varing fish in aquaculture. The manner to treat F-IFN easily into fish should be

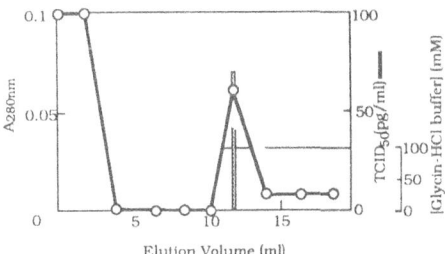

Fig. 3 IgG type MAb (FIG-4) purified using protein A column chromatography was fixed to activated CNBr sephacel, and a column (bed volume 1ml) was constructed. The spent medium (10ml) of the recombinant BHK cells was applied and eluted with 0.1M glycin-HCl buffer (pH 3.0). The $TCID_{50}$ of antiviral activity of each fraction was assayed using HRV (the m.o.i. of 1.0) infection to EPC cells (8×10^5 cells/ml) under the condotion of 15°C, 5% CO_2 humidity.

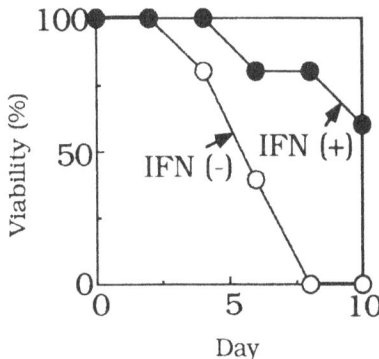

Fig.4 Antiviral activity of flatfish IFN *in vivo.* One μg of flatfish IFN was injected intravenouslyinto living flatfish weighing about 2 gram. On the next day Hirame Rhabdovirus (HRV) was challenged with 100 $TCID_{50}$. The viability of fishes was examined every other day.

developed,. Oral administration, soaking or showering system might open the way to the effective curing system in aquaculture.

REFERENCE

1. Ellis, AE. (1988)Fish Vaccination, pp.1-19, Academic Press, London.
2. Eto, N., Yamada, K., Koga, A., Shirahata, S. & Murakami, H. (1991) Cytotechnology 6: 13-21.
3. Kamei Y., Yoshimizu M. & Kimura T. (1987) Fish Pathol 22: 147-152.
4 Murakami, H., Masui, H., Sato, GH., Sueoka, N., Chow, TP. & Kano-Sueoka, T. (1982) Proc. Natl. Acad. Sci. USA, 79: 1158-1162.
5. Tamai, T., Shirahata, S., Sato, N., Kimura, S., Nonaka, M. & Murakami, H. (1992) Cytotechnology, in press.
6. Tamai, T., Shirahata, S., Sato, N., Kimura, S. & Murakami, H. (1993) B.B.A., in press.

STIMULATION OF PROTEIN DEGRADATION BY TUMOR NECROSIS FACTOR α IN CULTURED MUSCLE CELLS

K. YAGASAKI, Y. SAKAZAKI, Y. KIDA, Y. MIURA and R. FUNABIKI
Department of Applied Biological Science, Tokyo Noko University,
Fuchu, Tokyo 183, Japan

ABSTRACT. The action of tumor necrosis factor α (TNF)/cachectin on the degradation of intracellular proteins was studied in cultured L6 myotubes. Cellular proteins were preliminarily labeled with [^{3}H]tyrosine for either 18 hr (long-lived proteins, LLP) or 2 hr (short-lived proteins, SLP). Radiolabeled tyrosine released into the acid-soluble fraction of each experimental medium was used as an index of protein degradation. The degradation of LLP increased as the fetal calf serum (FCS) concentration was decreased from 10% to zero in the releasing medium. TNF exerted no infulence on LLP degradation at any TNF concentration (0.2-20 ng/ml) nor at any FCS concentration (0-10%), nor over any action period (1-18 hr) of the cytokine. The degradation of SLP was also increased when the releasing medium was low in FCS. However, TNF was found to stimulate SLP degradation at concentrations of 0.2, 2 and 20 ng/ml when 10% FCS was contained in the releasing medium. These results suggest that TNF may stimulate the degradation of SLP in L6 myotubes in a direct manner when the basal rate of protein turnover in the myotubes is low. They also suggest that TNF cannot directly exert any influence on LLP degradation, even when the basal rate of protein turnover in the cells is low.

1. Introduction

Tumor necrosis factor α (TNF)/cachectin has been reported to affect lipid metabolism [1]. In addition, this cytokine has recently been reported to stimulate muscle protein degradation when administered in vivo to animals, this being considered as a cause for muscle wasting during infection [2, 3]. However, it is still unclear whether or not TNF stimulates protein degradation by acting directly on muscle tissues. The present study was therefore conducted to clarify this aspect, using cultured muscle cells.

2. Materials and Methods

Stock cultures of L6 myoblasts [4] were maintained in Dulbecco's modified Eagle's medium (DMEM) supplemented with 10% fetal calf serum (FCS) and antibiotics (10% FCS/

S. Kaminogawa et al. (eds.), Animal Cell Technology: Basic & Applied Aspects, Vol. 5, 129–133.

DMEM) under an atmosphere of 5% CO_2/95% humidified air at 37°C, as described previously [5]. The prefused cells (5 X 10^4 cells/well) were subcultured in 24-well plates and grown for 11 or 12 days in 0.4 ml of 10% FCS/DMEM to form myotubes [5]. The medium was renewed every 3 days.

Proteins in the myotubes were prelabeled with L-[ring-3,5-^{3}H] tyrosine (50 Ci/mmol, ARC). Cellular proteins are generally classified into two types according to their turnover rates [6], that is, long-lived proteins (LLP) and short-lived proteins (SLP). To measure LLP degradation [7], the 11-day-old cells received 0.4 ml of 1% or 10% FCS/DMEM containing 0.4 μCi of [^{3}H]tyrosine, and were cultured for 18 hr (prelabeling). The cells were then washed twice with phosphate-buffered saline (PBS) containing 2 mM of nonlabeled tyrosine, and cultured in 0.4 ml of 0% or 10% FCS/DMEM containing 2 mM of nonlabeled tyrosine for 3 hr unless otherwise noted (chasing). The chasing medium was discarded to remove the SLP-derived [^{3}H]tyrosine. The myotubes were then cultured in 0.4 ml of DMEM containing 2 mM of nonlabeled tyrosine and various concentrations of FCS and TNF (murine recombinant, GIBCO BRL), usually for 4 hr, to release the LLP-derived [^{3}H]tyrosine (releasing). To measure SLP degradation, the 12-day-old cells received 0.4 ml of 1% or 10% FCS/DMEM containing 2 μCi of [^{3}H]tyrosine, and were cultured for 2 hr (prelabeling). After washing twice with PBS containing 2 mM of nonlabeled tyrosine, the cells were cultured for 1 hr in 0.4 ml of 1% or 10% FCS/DMEM containing excess (2 mM) nonlabeled tyrosine and various concentrations of TNF.

At the end of the experiments, the releasing medium was removed, and the cells were washed twice with PBS containing 2 mM of nonlabeled tyrosine. The washed cells were dissolved in 0.5 ml of 2 mM SDS in 0.2 N NaOH at 37 °C for 30 min, 0.1 ml of which was transferred into a vial so that the radioactivity could be counted after adding 4 ml of a nonione-toluene (NT) scintillator and 0.1 ml of 23.3% trichloroacetic acid (TCA) (DPMcell). The radioactivity in 0.1 ml of the releasing medium was also counted after adding 4 ml of the NT scintillator and 0.1 ml of 20% TCA (DPMmedium). To 0.2 ml of the releasing medium were added 0.05 ml of bovine serum albumin (2 mg/ml) and 0.2 ml of 22.5% TCA, the contents being mixed, stood on ice for 1 hr, and centrifuged. The supernatant (0.2 ml) was transfered to a vial and the radioactivity was counted after adding 4 ml of the NT scintillator (DPMsup). The results are expressed as the percentage degradation according to the following equation: (DPMsup/DPMcell+DPMmedium)X100. Statistical analyses were done by using Duncan's multiple-range test [8].

3. Results

3.1 EFFECT ON LLP DEGRADATION

The effect of TNF on LLP degradation was examined under various experimental conditions. As shown in Table 1, the degradation of LLP was increased as the FCS concentration was decreased from 10% to zero in the releasing medium. Various concentrations of TNF(0.2, 2, 20 ng/ml), as compared with the corresponding controls without TNF, exerted no significant influence on LLP degradation under any FCS conditions.

In the next experiment, the TNF action on LLP degradation was examined with 10% FCS

Table 1. Effects of FCS concentration during the prelabeling, chasing and releasing periods, and of TNF concentration on the degradation of long-lived proteins in L6 myotubes

FCS concentration (%)			Protein degradation (%)			
Prelabeling	Chasing	Releasing	TNF concentration (ng/ml)			
(18 hr)	(3 hr)	(4 hr)	0	0.2	2	20
1	0	0	12.5±1.1	13.2±0.8	13.7±0.4	13.3±0.4
1	0	1	11.6±0.4	11.4±0.5	11.2±0.2	11.1±0.4
10	0	10	7.3±0.1	7.7±0.3	7.6±0.2	7.5±0.3
10	10	10	5.8±0.2	6.0±0.1	6.1±0.2	6.1±0.2

Each value represents the mean ± standard error for five to six assays.

Table 2. Effects of releasing time and TNF concentration on the degradation of long-lived proteins in L6 myotubes

Releasing time (hr)	Protein degradation (%)			
	TNF concentration (ng/ml)			
	0	0.2	2	20
4	5.8±0.2	6.0±0.1	6.1±0.2	6.1±0.2
6	8.5±0.2	8.8±0.2	8.8±0.1	8.8±0.2
18	23.9±0.7	24.4±0.5	23.7±0.3	24.6±0.3

Each value represents the mean ± standard error for five to six assays.
The prelabeling and chasing times were 18 hr and 3 hr, respectively.
The concentration of FCS was 10% throughout the experiments.

Table 3. Effects of chasing and releasing times and of TNF concentration on the degradation of long-lived proteins in L6 myotubes

Chasing time (hr)	Releasing time (hr)	Protein degradation (%)			
		TNF concentration (ng/ml)			
		0	0.2	2	20
3	1	1.65±0.12	1.60±0.08	1.70±0.21	1.60±0.05
3	6	7.75±0.24	8.18±0.27	8.20±0.33	7.73±0.20
15	1	1.52±0.08	1.46±0.06	1.75±0.10	1.56±0.04
15	6	6.87±0.13	6.64±0.19	6.77±0.25	6.68±0.15

Each value represents the mean ± standard error for five to six assays.
The prelabeling time was 18 hr and the concentration of FCS was 10% throughout the experiments.

throughout the experiment by changing the releasing time (acting time)(Table 2). TNF had no affect on the degradation of LLP at any concentration.

The TNF effect on LLP degradation was further examined by changing the chasing (3 or 15 hr) and releasing (1 or 6 hr) times (Table 3). The degradation of LLP was increased according to the releasing time, but not the chasing time. Under both time conditions, TNF had no significant effect on LLP degradation in the myotubes.

3.2 EFFECT ON SLP DEGRADATION

The effect of TNF on the degradation of SLP was evaluated under various FCS conditions (Table 4). The degradation of SLP, like that of LLP, was also increased when the prelabeling or releasing medium was low in FCS. In the protein catabolic state, TNF did not stimulate the further degradation of SLP. However, the cytokine enhanced SLP degradation in the presence of sufficient (10%) FCS in the releasing medium.

Table 4. Effect of TNF on the degradation of short-lived proteins in L6 myotubes at various FCS concentrations

FCS concentration (%)		Protein degradation (%)			
		TNF concentration (ng/ml)			
Prelabeling (hr)	Releasing (hr)	0	0.2	2	20
1	1	26.7 ± 1.0	26.0 ± 0.5	26.8 ± 0.6	26.5 ± 0.6
10	1	20.4 ± 0.5	19.4 ± 0.6	20.4 ± 0.7	20.2 ± 0.4
10	10	18.2 ± 0.7^a	20.4 ± 0.5^b	20.6 ± 0.3^b	21.2 ± 0.2^b

Each value represents the mean ± standard error for five to six assays.
Values not sharing a common superscript letter are significantly different at P < 0.05 by Duncan's multiple-range test.

4. Discussion

The results obtained here suggest that TNF cannot directly stimulate the degradation of LLP in L6 myotubes. The stimulatory action of TNF on LLP degradation in muscle tissues, if any in vivo, may be mediated in an indirect manner. In contrast, TNF was found to directly stimulate SLP degradation in L6 myotubes when the cells were not in the protein catabolic state.

Protein degradation in eukaryotic cells occurs through distinct nonlysosomal and lysosomal pathways [6]. There seem to be two types of receptor for TNF [9]. It is therefore interesting to know which type of TNF receptor and which pathway for protein degradation are involved in the stimulatory action of TNF on SLP degradation in myotubes. In our separate experiment, inhibition of phospholipase A_2 and C was found to eliminate the stimulatory action of TNF on SLP degradation (data not shown). Thus, there is a possibility that a product(s) of phospholipase A_2 or C pathway may be involved in the signaling cascade leading from the binding of TNF and the receptor to the protein degradation site in L6 myotubes. Further studies are needed to clarify this aspect.

5. References

[1] Grunfeld, C., Gulli, R., Moser, A. H., Gavin, L. A., and Feingold, K. A. (1989) 'Effect of tumor necrosis factor administration in vivo on lipoprotein lipase activity in various tissues of the rat,' J. Lipid Res. 30, 579-585.

[2] Flores, E. A., Bistrian, B. R., Pomposelli, J. J., Dinarello, C. A., Blackburn, G. L., and Istfan, N. W. (1989) 'Infusion of tumor necrosis factor/cachectin promote muscle catabolism in the rat,' J. Clin. Invest. 83, 1614-1622.

[3] Goodman, M. N. (1991) 'Tumor necrosis factor induces skeletal muscle protein breakdown in rats,' Am. J. Physiol. 260, E727-E730.

[4] Yaffe, D. (1968) 'Retention of differentiation potentialities during prolonged cultivation of myogenic cells,' Proc. Natl. Acad. Sci. U.S.A. 61, 477-483.

[5] Yagasaki, K., Saito, K., Yamaguchi, M., and Funabiki, R. (1991) 'Involvement of arachidonic acid metabolism in insulin-stimulated protein synthesis in cultured L6 myocytes,' Agric. Biol. Chem. 55, 1449-1453.

[6] Goldberg, A. L., and Rock, K. L. (1992) 'Proteolysis, proteasomes and antigen presentation,' Nature 357, 375-379.

[7] Ballard, F. J. (1987) 'Regulation of intracellular protein breakdown with special reference to cultured cells,' in H. Glaumann and F. J. Ballard (eds.), Lysosomes: Their role in protein breakdown, Academic Press, London, pp. 285-318.

[8] Duncan, D. B. (1955) 'Multiple range and multiple F tests,' Biometrics, 11, 1-42.

[9] Ostade, X. V., Vandenabeele, P., Everaerdt, B., Loetscher, H., Gentz, R., Brockhaus, M., Lesslauer, W., Tavernier, J., Brouckaert, P., and Fiers, W. (1993) 'Human TNF mutants with selective activity on the p55 receptor,' Nature 361, 266-269.

FROM IN VITRO CELL CULTURE TO IN VIVO ANIMAL STUDY: DEMONSTRATION THAT EPIDERMAL GROWTH FACTOR (EGF) IS A POTENT INHIBITOR OF ADIPOSE DIFFERENTIATION.

GINETTE SERRERO.
W.Alton Jones Cell Science Center, Inc.
10 Old Barn Road.
Lake Placid, NY 12946.

ABSTRACT. Obesity which is characterized by abnormal adipose tissue development is a first degree public health hazard in industrialized countries as it is accompanied by a high incidence of cardiovascular disease, diabetes and elevated morbidity. One important aspect in the study of adipose tissue development is to investigate the hormonal control of adipose differentiation. In particular, if negative regulators of differentiation can be identified, than they can be used to develop a therapeutic approach towards the control of excessive adipose tissue development in human subjects. The data presented here describe the identification of epidermal growth factor as a potent inhibitor of adipose differentiation both in vitro and in vivo. Moreover, evidence are presented that genetically obese ob/ob mice present an impaired level of EGF thus demonstrating for the first time the existence of a correlation between low level of the adipose differentiation inhibitor EGF and excessive adipose tissue development .

1. Introduction

One of the most significant progress in cell culture techniques in the past 15 years has been the demonstration that the serum supplementation in cell culture medium can be replaced by an adequate combination of hormones, growth factors, attachments factors and vitamins. In particular, establishment of serum-free, defined medium techniques has made it possible to perform primary culture of functional cells in conditions where they can either undergo differentiation or maintain their differentiated functions. This in turn has allowed investigators to make significant progress in the investigation of the hormonal control of differentiation program and of the maintenance of differentiated functions in a well defined environment.

Thirty years ago, Sato and his collaborators were able to isolate from hormone-dependent tumors several cell lines which maintained in culture their differentiation properties (Sato, 1964). This led the way to the development of an in vitro cellular approach to endocrinology where isolated systems can be studied and their hormonal requirement for maintenance of tissue functions are investigated (McKeehan et al, 1987). Since the ultimate goal is to understand the physiological integrated control of tissue functions, it is then necessary to apply the information obtained with an in vitro model system to in vivo physiological situations. The conjunction of the culture of either immortalized cell lines or of primary cells with the use of defined culture conditions has allowed to establish such a correlation between in vitro and in vivo physiology.

We would like to present in this paper an example of such a correlation with our work on the hormonal control of adipose differentiation. In particular, an experimental approach is presented here to investigate the effect of epidermal growth factor (EGF) as an inhibitor of adipose differentiation in vitro

135

S. Kaminogawa et al. (eds.), Animal Cell Technology: Basic & Applied Aspects, Vol. 5, 135–141.
© 1993 *Kluwer Academic Publishers.*

and in vivo.. The discovery that EGF acts as a differentiation inhibitor comes from our work with the insulin-independent variant 1246-3A cell line isolated from the adipogenic cell line 1246. We have found that 1246-3A cells have become differentiation-deficient and tumorigenic in vivo. It was demonstrated that the culture medium from 1246-3A cells could block adipose differentiation of the parent cell line 1246 (Serrero, 1986). By biochemical characterization, it was shown that adipose differentiation inhibitory activity co-eluted with TGF-α (EGF)-like and TGF-β like polypeptides present in the cukture medium of the insulin-independent variant (Yamada and Serrero, 1989) and that pure TGF-α and TGF-β could inhibit differentiation of the 1246 cells (Serrero, 1986). Since TGF-β was produced by the 1246-3A cells as a latent precursor which was biologically inactive, it was concluded that the adipose differentiation inhibitory activity was mainly due to the presence of the TGF-α like (EGF-like) polypeptides. In addition, unlike TGF-β that blocked both adipose and muscle differentiation of 1246 cell derivatives, EGF (or TGF-α) only blocked adipose differentiation and not muscle differentiation of the teratoma-derived cell line. This result suggested a specific role of EGF (TGF-α) on adipose differentiation. Based on these results, it is important to determine whether EGF or its counterpart TGF-α can act as physiological regulator of adipose differentiation. The present paper provides a strategy to investigate this question. Several types of experiments were carried out: to determine the effect of EGF on the differentiation of adipocyte precursors freshly isolated fat pads in primary culture; to determine whether EGF can delay adipose tissue development in vivo; to investigate the status of EGF in animal model of obesity.

2. Material and methods

All methods used in this paper have been described in detail elsewhere. Primary culture of adipocyte precursors is described in Serrero (1987). In vivo animal studies were performed as in Serrero and Mills (1991). Northern blot analysis of prepro-EGF mRNA expression was done according to Serrero et al (1993).

3. Results and discussion

3.1. MODEL SYSTEM TO STUDY ADIPOSE DIFFERENTIATION IN VITRO AND IN VIVO.

Adipose differentiation is routinely studied with adipogenic cell lines. These cell lines can differentiate into cells that present all the characteristics of mature adipocytes. However, since the purpose is to understand physiological regulation of differentiation in vivo, it is necessary to develop a strategy in order to determine if the information obtained in vitro also pertain to physiological in vivo conditions. Such a strategy has been developed in my laboratory and is outlined in figure 1. The purpose of using such a strategy is to understand the physiological regulation of adipose differentiation with a final goal of establishing an understanding of normal adipose tissue development and possibly developing a mean of therapeutic intervention for abnormal adipose tissue development such as obesity.

The important points of the systems used are the followings:

a- the adipogenic cell line called 1246 which we have characterized is derived from a mouse C3H teratoma. This cell line is bipotential as it can differentiate either into adipocytes or into muscle cells. Moreover, these cells can proliferate and undergo adipose differentiation in defined medium (Serrero and Khoo, 1982). The defined medium consists of DME-F12 nutrient medium (1:1 mixture) supplemented with fibronectin (2 μg/ml), insulin (10 μg/ml), transferrin (10 μg/ml) and fibroblast growth factor, FGF (10 ng/ml). This medium will be referred as 4F medium. When the cells reach confluency at day 4, differentiation is triggered by treating the cells for 48 hrs with dex-mix consisting of dexamethasone ($2x10^{-7}$ M) and isobutylmethylxanthine ($2x10^{-4}$ M). Cells are harvested at day 14.

Other adipogenic cell lines such as 3T3-L1, 3T3-F442A and Ob17 cells can proliferate in a more complex medium but cannot differentiate unless 1-2% fetal bovine serum is added to the medium when the cells reach confluency (Serrero et al, 1979; Gaillard et al, 1984).

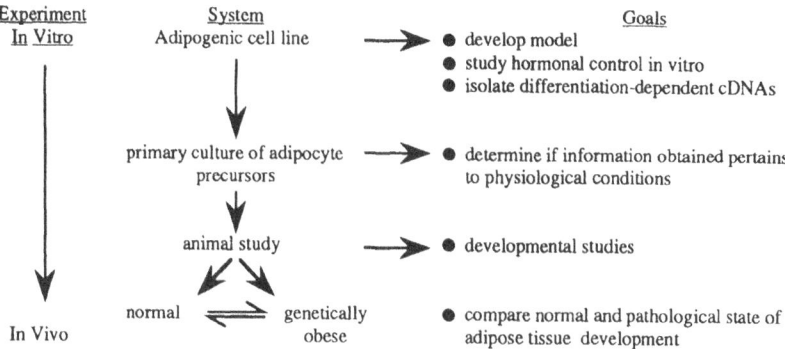

PHYSIOLOGICAL REGULATION OF ADIPOSE TISSUE DEVELOPMENT AT CELLULAR AND MOLECULAR LEVELS

Figure 1

b- The defined medium developed for the adipogenic cell line 1246 can also support the proliferation and differentiation of adipocyte precursors freshly isolated from inguinal fat pads of 48 hrs-old rats (Serrero, 1987). In the 4F medium, the adipocyte precursors proliferate and differentiate with a higher frequency and at a higher rate than the cell line. After 8 days in culture, 80-90% of the cells have undergone differentiation. Interestingly, the dex-mix treatment is not necessary to trigger differentiation in the primary culture. Thus, this primary culture is very adequate to investigate events associated to physiological conditions of adipose differentiation.

c- Animal models can be used to investigate adipose tissue development in vivo. For early developmental studies, one can use either newborn mice or rats. Right after birth, inguinal fat pads consist mostly of undifferentiated adipocyte precursors. During the first 10 days of life, a burst of differentiation takes place in the inguinal fat pads. Thus, inguinal fat pads of newborn animals is a good source of precursors for primary culture (see above) and is adequate to study the events related to early phases of adipose development in vivo. For comparing normal and abnormal adipose tissue development, several strains of genetically obese mice are available (Herberg and Coleman, 1977). In addition, several models of acquired obesity are available. The two models that we are using in the laboratory are the mice where obesity has been induced by injection of monosodium glutamate (Olney, 1969) and the mice rendered obese by cafeteria-diet.

3.2. EGF INHIBITS ADIPOCYTE PRECURSORS DIFFERENTIATION IN PRIMARY CULTURE.

As mentioned above, adipocyte precursors freshly isolated from inguinal fat pads of 48 hrs old rats can be cultivated in 4F medium, After 8 days the cells have differentiated and express all the markers specific of adipocytes. On a routine basis, differentiation is followed by measuring the increase in glycerol-3-phosphate dehydrogenase (G3PDH) specific activity since the assay is very easy, non-isotopic and since the increase of activity dirung differentiation in primary culture varies from 600 to 2,000 fold. If EGF is added to the 4F medium, increase of G3PDH specific activity is inhibited (figure 2). In contrast to the cells cultivated in 4F medium only, the cells cultivated in the presence of EGF remain fibroblastic and they do not accumulate triglycerides. Adipocyte differentiation is a cascade of event characterized by the induction of expression of early markers of differentiation such as the 50 kDa adipose differentiation related protein (ADRP), lipoprotein lipase (LPL) and fatty acid binding protein

(FAB) and of late markers of differentiation such as G3PDH and triglyceride accumulation. We have shown that EGF can also inhibit the expression of mRNA expression for ADRP, LPL and FAB (data not shown) suggesting that EGF acts on the early events of adipose differentiation.

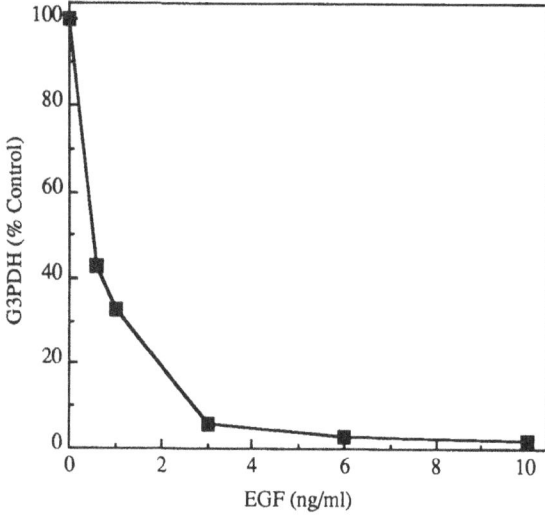

Figure 2. Effect of EGF on the differentiation of rat adipocyte precursors. Adipocyte precursors were cultivated in 4F defined medium (Serrero, 1987). EGF was added at day 1 and maintained throughout the experiment. Cell were harvested at day 6 to measure G3PDH activity. Values are expressed as % of control corresponding to G3PDH specific activity in cells cultivated in 4F medium in the absence of EGF (1750 mU/mg protein).

3.3. EGF DELAYS ADIPOSE TISSUE DEVELOPMENT IN VIVO.

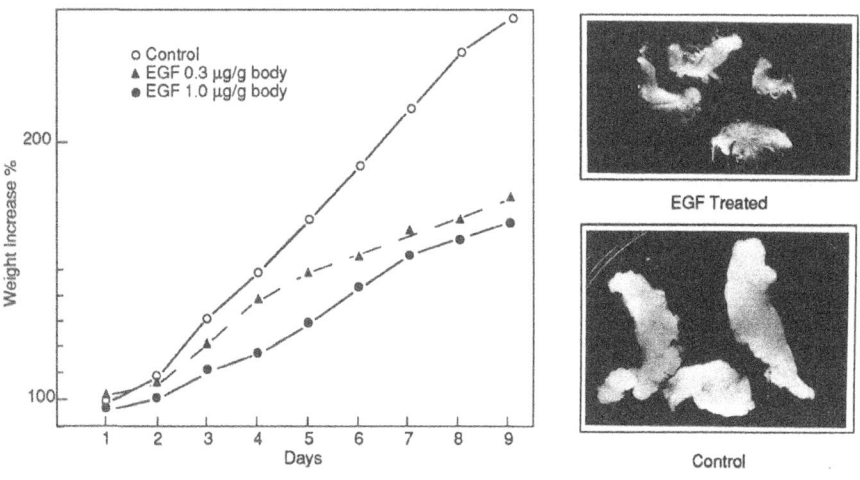

Figure 3A. Effect of subcutaneous injection of increasing concentrations of EGF on body weight of newborn NBR rats starting 24 hrs after birth.

Figure 3B. Photograph of inguinal fat pads from control animals (left) and from animals injected for 10 days with EGF at 1 µg/g body weight

Effect of EGF on in vivo adipose tissue development was investigated by performing daily subcutaneous injection of EGF (from 0.1 to 1 µg/g body weight) to newborn rats for 9 days, starting 24 hr after birth. Control animals were injected with saline only. Body weight was monitored daily. The rationale for the experimental conditions are the following: inguinal fat pads are the first developed. it can easily be dissected and its presence is independent of the sex of the animals; were done experiments with newborn animals since at birth the majority of the cells in the fat pads are not yet differentiated and since differentiation take place during the first 9-10days of life. Thus injecting EGF during this time frame would allow us to determine whether EGF can delay early phases of adipose tissue development. The results have been described in detail elsewhere (Serrero and Mills, 1991). Figure 3 shows that injection of EGF delays the weight gain of newborn rats. In addition, fat pads of EGF treated animals were not as develpoped as fats pads of control rats. In addition, fat pad weight was decreased by EGF injection (Serrero and Mills, 1991). Fat pads were digested with collagenase in order to separate the undifferentiated precursors and differentiated adipocytes. Number of precursors was counted and degree of differentiation of the fat pads was determine by measuring the amount of triglycerides accumulated in the fat pads of control and EGF treated animals. Same results were obtained if EGF was injected into newborn mice. Interestingly, the effect of EGF on reducing weight gains in the animals is long lasting since we have shown that animals treated with EGF only for the first 10 days of life will remain leaner than their littermates injected with saline only even 35 days later. It is not known at this time whether EGF can induce weight reduction if injected into adult animals. These experiments are currently underway in our laboratory.

3.4. Impaired EGF level in genetically obese ob/ob mice when compared to normal littermates.

The results presented above indicate that EGF is a potent inhibitor of adipose differentiation both in vitro and in vivo. As mentioned above, several models of genetic obesity have been isolated (Herberg and Coleman, 1977). In our studies, we used the ob/ob model in order to investigate the status of EGF in ob/ob mice when compared to their normal littermates. The obese mutation in mice is corresponds to a single autosomal recessive gene located on chromosome 6. Adult ob/ob mice are characterized by marked obesity of the hypertrophic-hyperplastic type, hyperphagia, transient hyperglycemia and markedly elevated plasma insulin. EGF is a 6 kDa polypeptide which is present in large amounts in submaxillary glands of mice (Cohen, 1962). EGF is synthesized as a 128 kDa prepro-EGF in several tissues (Gray et al, 1983). The highest level of expression is found in submaxillary gland followed by kidney. However, processing of prepro-EGF into the 6 kDa form does not occur in kidney. Since so far the majority of the biological activity of EGF in vivo is attributed to submaxillary gland EGF (Tsutsumi et al, 1986), we measured the level of submaxillary gland EGF in obese (ob/ob) and normal male littermates by radioimmunoassay (Serrero et al, 1993). Since EGF produced in the submaxillary gland is also found in plasma, we also determine the level of plasma EGF in obese and normal littermates although the exact role of circulating EGF is not clear. In addition, we determined by northern blot analysis, the level of expression of submaxillary gland prepro-EGF mRNA in obese and normal littermates. These experiments were carried out with mice at 5 weeks of age when metabolic perturbations are not yet very marked and at 16 weeks of age when the ob/ob mice have all the characteristics of severe obesity.

As shown in figure 4A, the level of EGF in the submaxillary gland extract of 5 weeks old ob/ob mice was 6% of the level found in the submaxillary gland of normal littermates. As the mice became older, the level of EGF increased in both control and obese mice. However, the difference of EGF level between obese and control mice was still apparent (38% of that found in control mice). The level of plasma EGF in obese mice was 27% at 5 weeks and 37% at 16 weeks of the level found in normal age matched littermates respectively (figure 4A).

140

Expression of submaxillary gland prepro-EGF mRNA was hardly detectable in obese mice but very high in normal littermates at 5 weeks or 16 weeks (figure 4B).

A

Age weeks	Phenotype	Body weight g	EGF Concentration	
			Submax gland µg/mg tissue	Plasma pg/ml
5	Control	22.3 ± 0.7	0.67 ± 0.06	203 ± 5.5
	Obese	29.8 ± 0.3	0.04 ± 0.01	56.5 ± 8.4
16	Control	28.3 ± 0.8	25.2 ± 2.5	433 ± 61
	Obese	54.6 ± 1.8	9.8 ± 2.5	162 ± 68

Figure 4A: Comparison of EGF level in submaxillary gland extract and in plasma of obese male mice and their normal littermates. EGF level was measured by radio-immunoassay as described in Serrero et al, (1993). Each experimental group contained 4 to 6 animals. Experiments were repeated three times. Values are given as mean ± SE.

B

Prepro EGF mRNA →

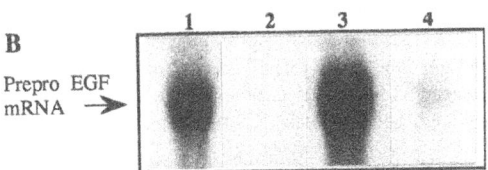

Figure 4B: Prepro-EGF mRNA expression in submaxillary gland of 5 weeks and 16 weeks old control and obese male mice. Northern blot analysis was performed as described in Serrero et al, (1993) with 10 µg of total RNA from the submaxilary gland of 5 weeks old control (lane 1) and obese mice (lane 2), 16 weeks old control (lane 3) and obese mice (lane 4).

Although the experiments described here were performed with male mice, decrease in EGF level was also found in female obese mice when compared to their sex matched littermates. In addition, impaired EGF level is already detected as early as 15 days in obese mice, suggesting that the decrease in EGF level may occur before the onset of metabolic perturbations in obese mice..

Based on these results and on the fact that submaxillary gland EGF is the major source of biologically active EGF in mice (Tsutsumi et al, 1986) , it is possible to speculate that abnormal tissue development observed in obese mice is at least partially correlated with an abnormal adipose tissue development. Further detailed in vivo studies with normal and obese mice are necessary in order to determine whether EGF can delay or decrease abnormal adipose tissue development in obese animals. Based on the results obtained so far, it is tempting to speculate that EGF or derivatives could be used as a mean to either control or monitor obesity in human subjects. In any case, the data presented here point out to the importance of in vitro studies using cell culture in defined conditions to discover factors affecting differentiation and to obtain data on their potential physiological importance that would warrant in vivo studies.

4. Acknowledgements.

This study is dedicated to Dr. Gordon Sato for his teaching and for his inspiring ideas, which are at the basis of this work. The author wishes to thank Dianne Mills, Nancy Lepak and Steve Goodrich for their technical support, Marina LaDuke for the illustrations and Valerie Oliver for preparing this manuscript. This work was supported by grants from the National Institute of Health and from the Council for Tobacco Research.

5. References

Cohen, S. (1962) Isolation of a mouse submaxillary gland protein accelerating incisor eruption and eyelid opening in the newborn animal. J. Biol. Chem. 237: 1555-1562.

Gaillard, D., R. Negrel, G. Serrero, C. Cermolacce and G. Ailhaud (1984) Growth of pre-adipocyte cell lines and cell strains from rodents in serum-free hormone-supplemented medium. In Vitro Cell. Devel. Biol. 20:79-88.

Gray, A., T.J. Dull and A. Ullrich (1983) Nucleotide sequence of epidermal growth factor cDNA predicts a 128,000 molecular weight precursor protein. Neture 303: 722-725.

Herberg, L. and D.L. Coleman (1977) Laboratory animals exhibiting obesity and diabetes syndromes. Metabolism 26: 59-99.

McKeehan, W.L., D. Barnes, L. Reid, E. Standbridge, H. Murakami and G.H. Sato (1990) Frontiers in mammalian cell culture. In Vitro Cell. Dev. Biol. 26: 9-23.

Olney, J. (1969) Brain lesions, obesity and other disturbances in mice treaedwith monosodium glutamate. Science 164: 719-721.

Sato, G.H. (1964) Use of dispersed animal cell cultures in the study of specialized physiologic processes. In: Methods in Medical Research, 10: 323-326 (H.N. Eisen, editor-in-chief), Chicago Year Book Medical Publishers, Inc.

Serrero, G. (1986) Endocrine and autocrine control of growth and differentiation of teratoma-derived cell lines. In: Cellular Endocrinology. Hormonal Control of Embryonic and Cellular Differentiation, (G. Serrero and J. Hayashi, eds.), Alan R. Liss, Inc., NY, pp. 191-204.

Serrero, G. (1987) Epidermal growth factor inhibits the differentiation of adipocyte precursors in primary culture. Biochem. Biophys. Res. Commun. 146: 194-202.

Serrero, G. and J.C. Khoo (1982) An in vitro model to study adipose differentiation in serum-free medium. Anal. Biochem. 120:351-359.

Serrero, G. and D. Mills (1991) Physiological role of epidermal growth factor on adipose tissue development in vivo. Proc. Natl. Acad. Sci. USA 88: 3912-3916.

Serrero, G., D. McClure and G. Sato (1979) Growth of mouse 3T3 fibroblasts in serum-free hormone-supplemented media. In: Hormones and Cell Culture, Cold Spring Harbor Conferences on Cell Proliferation. (G. Sato and R. Ross, eds.), Vol. 6, pp. 523-530.

Serrero, G., Lepak, N.M., Hayashi, J. and Goodrich, S.P (1993) Impaired epidermal growth factor production in genetically obese ob/ob mice. Am. J. Phys., in press.

Tsutsumi, O.H., H. Kurachi and T. Oka (1986) A physiological role for epidermal growth factor in male reproductive function. Science 233: 975-977.

Yamada, Y. and G. Serrero (1989) Characterization of transforming growth factors produced by the insulin-independent teratoma-derived cell line 1246-3A. J. Cell. Physiol. 140:254-263.

CO-OPERATIVE ACTION OF CYSTEINE AND SOME HORMONES ON BILE ACID SYNTHESIS IN PRIMARY
CULTURED RAT HEPATOCYTES

Y. MIURA, K. ISHIHARA, Y-H. ITO, Y-T. ITO, K. YAGASAKI, and R. FUNABIKI
Department of Applied Biological Science, Tokyo Noko University,
3-5-8 Saiwaicho, Fuchu, Tokyo 183, Japan

ABSTRACT. The effect of cysteine on bile acid synthesis in primary cultured rat
hepatocytes was investigated. Cysteine was found to increase bile acid synthesis at a
concentration of 5 mM cysteine. This effect of cysteine was not removed by the addition
of DL-buthionine-sulfoximine (BSO), a specific inhibitor of gluthathione (GSH)
biosynthesis. Cysteine was also found to modulate the thyroid hormone action on bile
acid synthesis. These results suggest that cysteine may stimulate bile acid synthesis in
a direct manner and not by increasing the production of GSH, and act co-operatively with
the thyroid hormone on bile acid synthesis in primary cultured rat hepatocytes.

1. Introduction

It is well known that tumor-bearing often causes a disorder of lipid metabolism. We
have already reported that rats transplanted with an ascites hepatoma line of AH109A
showed endogenous hypercholesteremia and that this hypercholesteremia could be reduced
by feeding a 20% casein diet supplemented with 1.2% cystine [1]. The serum cholesterol
level is known to be regulated by cholesterol absorption across the intestine, by
choleserol synthesis and by catabolism in the liver. Serum cholesterol is converted to
bile acids that are conjugated with taurine or glycine in the liver and excreted into
the duodenum. We have reported that the excretion of fecal bile acids was reduced by
tumor-bearing and increased by feeding a 20% casein diet supplemented with 1.2%
cystine [2]. Since the conversion of cholesterol to bile acids is known to take place
exclusively in the liver, we have therefore employed primary cultured rat hepatocytes to
investigate in vitro the action of cysteine (in place of cystine) on bile acid
synthesis, and found that excess cysteine in the medium augmented bile acid synthesis in
hepatocytes from both normal and hepatoma-bearing rats (unpublished observations). From
these results, we concluded that excess cysteine in the medium could augment bile acid
synthesis irrespective of the condition of the rats from which hepatocytes were
prepared, and that cysteine itself had the ability to augment bile acid synthesis in
primary cultured rat hepatocytes.
The present study was made to clarify the mechanism(s) for cysteine action on bile

S. Kaminogawa et al. (eds.), Animal Cell Technology: Basic & Applied Aspects, Vol. 5, 143–148.
© 1993 Kluwer Academic Publishers.

acid synthesis, using primary cultured hepatocytes from normal rats.

2. Materials and Methods

2.1. PREPARATION OF PRIMARY CULTURED RAT HEPATOCYTES

Hepatocytes were isolated from male Donryu rats weighing 180-250 g by the in situ collagenase perfusion method described by Seglen [3]. Cells were innoculated to collagen-coated 12-well culture plates (Corning, N.Y.) at a density of 1.54×10^5 cells/ cm^2 and cultured in a DM-160 culture medium supplemented with 10% fetal calf serum (FCS) (JRH Biosciences, Lenexa, KS) and antibiotics for 24 hr.

2.2. MEASUREMENT OF BILE ACID SYNTHESIS IN PRIMARY CULTURED RAT HEPATOCYTES

After a 24-hr culture in the presence of 10% FCS, the cells were cultured for another 24 hr in the presence of 0.3 μCi of [4-^{14}C] cholesterol (51.0 mCi/mmol, New England Nuclear) that had been previously conjugated with FCS (final concentration of 5%) with or without cysteine or some hormones. [^{14}C] bile acids converted from [^{14}C] cholesterol were extracted from the conditioned medium and the radioactivity was counted [4, 10]. After collecting the medium, the cells were lysed by 0.1% SDS, and the cellular concentration of DNA was measured fluorometrically by the method of Brunk et al. [5], using calf thymus DNA (Sigma, St. Louis, MO) as a standard.

2.3. STATISTICAL ANALYSIS

A statistical analysis was done by using Student's t test, a P value of < 0.05 being considered statistically significant.

3. Results and Discussion

3.1. EFFECT OF CYSTEINE ON BILE ACID SYNTHESIS

The effect of different concentrations of cysteine on bile acid synthesis was examined in primary cultured rat hepatocytes (Figure 1). Cysteine stimulated bile acid synthesis at a concentration of 5 mM cysteine.

Thus, in the following experiments, we employed a concentration of 5 mM cysteine to investigate the mode of action of cysteine. Normal DM-160 medium containing 0.5 mM cysteine was used as a control.

3.2. EFFECT OF THE INHIBITOR OF GLUTHATHIONE BIOSYNTHESIS ON CYSTEINE ACTION

A rate-limiting enzyme of bile acid synthesis is cholesterol 7α-hydroxylase, which has been reported to be activated by the reduced form of gluthathione (GSH) in a cell-free system [6]. Cysteine is a precursor of GSH, and we therefore examined the possibility that cysteine could regulate bile acid synthesis by increasing the GSH production in hepatocytes. The intracellular concentration of GSH was increased when

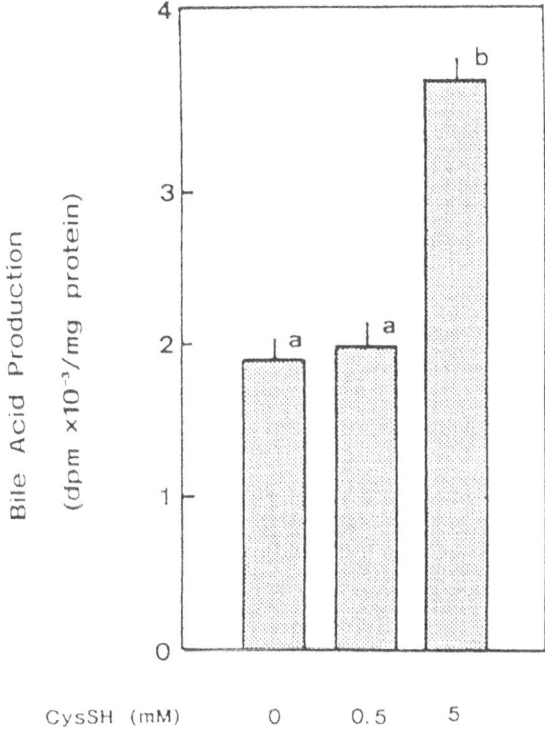

Figure 1. Effect of cysteine on bile acid synthesis in primary cultured rat hepatocytes. Values not sharing a common letter are significantly different at P < 0.05.

cysteine was present in the medium, while DL-buthionine-sulfoximine (BSO), a specific inhibitor of gluthathione biosynthesis, completely inhibited intracellular gluthathione biosynthesis. However, BSO failed to eliminate the stimulatory effect of 5 mM cysteine on bile acid synthesis (data not shown). From these results, we conclude that cysteine may stimulate bile acid synthesis in an unknown direct manner and not by increasing the production of GSH.

As described in the materials and methods section, the values obtained in this study represent the results of the overall process from the uptake of cholesterol by cells through the LDL receptor to the secretion of converted bile acids into the medium. The ligand-binding domain of the LDL receptor is known to have cysteine-rich regions. So one possible mode of action of cysteine is thought to be regulation of the ligand-binding activity of the LDL receptor.

3.3. CO-OPERATIVE EFFECT OF CYSTEINE WITH SOME HORMONES ON BILE ACID SYNTHESIS

Some hormones are known to regulate bile acid synthesis in the liver or hepatocytes, the effects of glucocorticoid and the thyroid hormone having been most extensively studied. Glucocorticoid is thought to increase bile acid synthesis by enhancing the expression of cholesterol 7 α-hydroxylase gene, and so is the thyroid hormone [7]. The thyroid hormone has been reported to activate not only cholesterol 7α-hydroxylase, but also other enzymes that participate in bile acid synthesis [8]. Insulin had been thought to have no or little effect on bile acid synthesis, but Hylemon et al. [7] have recently reported that, under the continuous presence of insulin in a serum-free medium of primary cultured rat hepatocytes, glucocorticoid and the thyroid hormone could induce cholesterol 7 α-hydroxylase gene expression to the level of that in intact liver.

We next investigated whether or not cysteine would modulate the effect of these hormones on bile acid synthesis (Figure 2). In the presence of insulin (1.4 μM) alone or of insulin and dexamethasone (0.1 μM) in combination, 5 mM cysteine stimulated bile acid synthesis to about 150% that of 0.5 mM cysteine. However, in the presence of insulin and T_3 (10 nM) in combination or of insulin, T_3 and dexamethasone in combination, 5 mM cysteine stimulated bile acid synthesis to about 250% that of 0.5 mM cysteine.

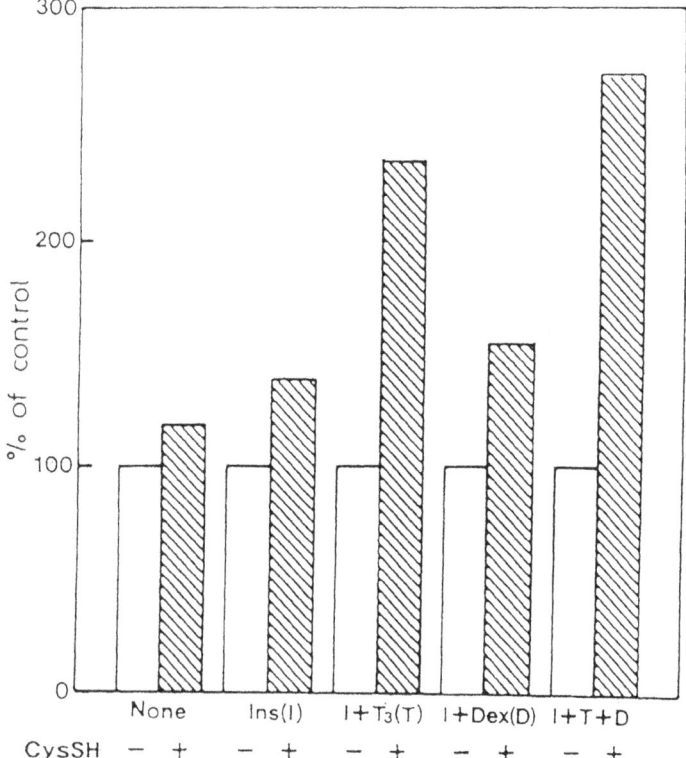

Figure 2. Co-operative action of cysteine and some hormones on bile acid
synthesis in primary cultured rat hepatocytes. The concentration
of each hormone is described in the text.

From these results, we conclude that cysteine might mainly modulate T_3 action on bile acid synthesis. T_3 is well known to show its action on gene expression after binding to its nuclear receptor. Yamamoto et al. have recently reported that 2-mercaptoethanol could stabilize the T_3 receptor by regulating the intracellular oxidoreductive condition in primary cultured rat hepatocytes [9]. It is likely that cysteine may modulate T_3 action on bile acid synthesis by stabilizing the T_3 receptor.

4. Conclusion

We have demonstrated that cysteine augmented bile acid synthesis without the involvement of newly synthesized GSH, and that this amino acid was capable of modulating the thyroid hormone action on bile acid synthesis in primary cultured rat hepatocytes. One possible mode for these actions of cysteine may be in regulating the intracellular oxidoreductive conditions. However, other possibilities, including direct activation of cholesterol 7 α-hydroxylase by cysteine, cannot be ruled out. Further extensive studies will clarify the detailed modes of cysteine action.

5. References

1) Yagasaki, K., Machida, M., and Funabiki, R. (1986) 'Effects of dietary methionine, cystine, and glycine on endogenous hypercholesteremia in hepatoma-bearing rats, J. Nutr. Sci. Vitaminol. 32, 643-651.

2) Ishihara, K., Yagasaki, K., Okushi, N., and Funabiki, R. (1991) 'Effects of dietary sulfur amino acids and glycine on fecal steroid excretion in hepatoma-bearing rats,' Agric. Biol. Chem. 55, 853-854.

3) Seglen, P. O. (1976) 'Preparation of isolated rat liver cells,' Methods Cell Biol. 13, 29-83.

4) Hylemon, P. B., Gurley, E. C., Kubaska, W. M., Whitehead, T. R., Guzelian, P. S., and Vlahcevic, Z. R. (1985) 'Suitability of primary cultures of adult rat hepatocytes for studies of cholesterol and bile acid metabolism,' J. Biol. Chem. 260, 1015-1019.

5) Brunk, C. K., Jones, K. C., and James, T. W. (1979) 'Assay for nanogram quantities of DNA in cellular homogenates,' Anal. Biochem. 92, 497-500.

6) Danielsson, H., Kalles, I., and Wikvall, K. (1984) 'Regulation of hydroxylations in biosynthesis of bile acids. Isolation of a protein from rat liver cytosol stimulating reconstituted cholesterol 7 α-hydroxylase activity,' J. Biol. Chem. 259, 4258-4262.

7) Hylemon, P. B., Gurley, E. C., Stravitz, R. T., Litz, J. S., Pandak, W. M., Chiang, J. Y. L., and Vlahcevic, Z. R. (1992) 'Hormonal regulation of cholesterol 7 α-hydroxylase mRNA levels and transcriptional activity in primary rat hepatocytes cultures,' J. Biol. Chem. 267, 16866-16871.

8) Myant, N. B., and Mitropoulos, K. A. (1977) 'Cholesterol 7 α-hydroxylase,' J. Lipid. Res. 18, 135-153.

9) Yamamoto, N., Inoue, A., Takahashi, K. P., Li, Q., Nakamura, H., Tagami, T., Sasaki,

S., Imura, H., and Morisawa, S. (1992) 'Loss of thyroid hormone receptor activity in primary cultured rat hepatocytes is reversed by 2-mercaptoethanol,' Biochem. J. 281, 669-673.

10) Yagasaki, K., Machida-Takehana, M., and Funabiki, R. (1990) 'Effects of dietary methionine and glycine on serum lipoprotein profiles and fecal sterol excretion in normal and hepatoma-bearing rats,' J. Nutr. Sci. Vitaminol. 36, 45-54.

SUPPRESSION OF KILLER T CELL INDUCTION BY PRODIGIOSIN 25-C

JUNJI MAGAE, TAKAO KATAOKA, MASAHIRO YAMASHITA, RYOHEI F.
TSUJI*, MI-HEON LEE, MAKARI YAMASAKI* AND KAZUO NAGAI
Department of Bioengineering, Tokyo Institute of Technology,
4259 Nagatsuta-cho, Midori-ku, Yokohama 227, Japan, and
*Department of Agricultural Chemistry, The University of Tokyo,
1-1-1 Yayoi, Bunkyo-ku, Tokyo 113, Japan.

ABSTRACT. Prodigiosin 25-C (PrG) suppresses induction of killer T cells
without inhibiting functions of B cells and helper T cells. To explore
the mechanism of the specific immunosuppression of the antibiotic, we
treated naive mice with PrG and evaluated functions of T cells and
macrophages in the spleen by mixed lymphocyte reaction (MLR). Our
results indicated that PrG inhibits allogeneic major histocompatibility
antigen complex (MHC)-specific stimulator function of macrophages but
not response of T cells. Presentation of exogenous antigen was rather
enhanced. These results suggest that PrG inhibits induction of killer T
cells through selective inhibition of alloantigen presentation of
macrophages.

1. Introduction

PrG (Fig.1) is a red pigment produced by Streptomyces hiroshimensis
[1]. The antibiotic suppresses proliferative response of T cells induced
by concanavalin A (Con A) more effectively than that of B cells induced
by lipopolysaccharide (LPS) [1]. PrG inhibits Con A-induced
proliferative response of T cells without suppressing IL-2 production
and IL-2 receptor formation in vitro [2,3]. In addition, the antibiotic
specifically suppressed activity of killer T cells induced by the
immunization of allogeneic mastocytoma, but not induction of antibody
production against sheep red blood cells (SRBC), T cell-dependent
antigen, or Brucella abortus, T cell-independent antigen [2,3]. In
contrast, FK506 significantly suppresses induction of killer T cells and
antibody production against SRBC but not against Brucella abortus [3].
These results suggest that PrG specifically suppresses induction of
killer T cells without affecting functions of helper T cells and B
cells. Present work is directed at elucidating the mechanism of specific
immunosuppressive effect of PrG.

149

S. Kaminogawa et al. (eds.), Animal Cell Technology: Basic & Applied Aspects, Vol. 5, 149–154.
© 1993 Kluwer Academic Publishers.

2. Materials and Methods

2.1. CHEMICALS

PrG was isolated from culture broth of <u>Streptomyces hiroshimensis</u> as described previously [1]. Fluorescent isothiocyanate (FITC)-labelled anti-Lyt2 antibody and phycoerythrin (PE)-labelled anti-L3T4 antibody were purchased from Becton Dickinson Co. and PE-labelled anti-B220 antibody from Coulter electronics (Florida).

2.2. MICE

Mice (6-10wk old, female) were purchased from Japan Charles River Inc.. They were maintained with a commercial pellet diet and tap water.

2.3. CELL CULTURE

Cell cultures were carried out in RPMI1640 medium supplemented with 0.05mM 2-mercaptoethanol, 0.05mg/ml kanamycin, 0.008mg/ml tylosin tartrate and 10% FCS in 5% CO_2 atmosphere at $37^{\circ}C$. CTLL-2 and BK-1 [4] were maintained as described previously [5].

2.4. ANALYSIS OF LYMPHOCYTE POPULATION

Splenocytes (1×10^6 cells) were incubated with FITC-labelled anti-Lyt2, PE-labelled anti-L3T4 or PE-labelled anti-B220 and analyzed by flow cytometry (Epics-CS, Coulter Electronics, Florida) as described previously [3,6].

2.5. INDUCTION AND DETERMINATION OF MLR

Nylon-wool purified spleen T cells and spleen adherent cells (SAC) were prepared as described previously [3,5]. T cells (5×10^5 cells/well) were mixed with SAC (1×10^5 cells/well) and cultured for 72h. [^{3}H]thymidine (500nCi/well) was pulse-labelled for 4h prior to harvesting [5].

2.6. ANTIGEN PRESENTATION OF EXOGENOUS ANTIGEN

SAC (3×10^3 cells/well) from BALB/c mice were mixed with $CD4^+$ T lymphocyte clone BK-1 (1.5×10^4 cells/well) and the relevant antigen, key hole limpet hemocyanin (KLH) (0.01mg/ml). After 24h-cultivation, IL-2 activity of the supernatant was determined [3] by culturing IL-2 dependent cell line CTLL-2 (5×10^3 cells /well) in the presence of the supernatant. [^{3}H]thymidine was pulse-labelled 4h prior to harvesting.

FIGURE 1. Structure of PrG.

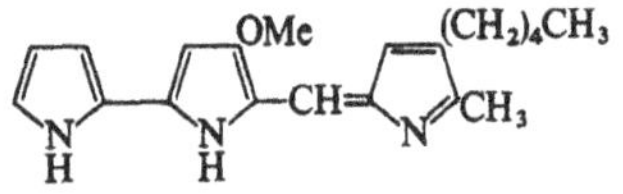

3. Results

3.1. ANALYSIS OF LYMPHOCYTE POPULATION

C57BL/6 mice were treated with PrG and the lymphocyte population of spleen was analyzed by flow cytometry (Table 1). The immunization with allogeneic P815 cells resulted in a decrease of B220[+] B cells and L3T4[+] T cells and an increase of Lyt2[+] T cells. This is in agreement with the fact that the mastocytoma only expresses class I MHC but not class II MHC. Treatment of the mice with PrG recovered the changes of lymphocyte population while the same treatment of non-immunized mice did not affect the population. This suggests that PrG has no selective toxicity against Lyt2[+] T cells and that the antibiotic inhibits induction phase of T cell activation.

TABLE 1. PrG induced decrease of B220[+] B cells and increase of Lyt2[+] T cells in spleen of P815-immunized mice but not of normal mice.

| antigen | positive cells (%) | | | |
| | normal | | P815-immunized | |
	control	PrG	control	PrG
B220	52.8	53.8	25.6	41.6
L3T4	26.2	33.2	12.5	18.5
Lyt2	15.5	17.5	28.6	12.9

Spleen cells from 3 C57BL/6 mice were pooled and stained by fluorescent antibodies. Mice were immunized i.p. with P815 (2x10[7]cells/head) on day 0. PrG (1mg/kg) was administered i.p. on days 0,3,5,8. Mice were sacrificed on day 10.

TABLE 2. PrG suppressed allogeneic MHC specific MLR by suppressing the antigen presentation of macrophages but not by suppressing response of T cells.

exp.	T cells	SAC	radioactivity (cpm)
1	normal	normal	$43,441\pm2,251$
	normal	PrG	$8,642\pm825$ (80)[*]
2	normal	normal	$72,189\pm3,225$
	PrG	normal	$71.051\pm3,260$ (2)

Cells prepared from spleen cells of 3 mice were combined (exp.1: T cells from BALB/c and SAC from C57BL/6, exp.2; T cells from C57BL/6 and SAC from BALB/c) and MLR was carried out as described in Materials and Methods. PrG (1mg/kg) was administered i.p. on day 0,3,5,8 and mice were sacrificed on day 10. Values, mean$\pm$SD. Parentheses, % of inhibition. [*]Statistically significant (p<0.001, t-test)

3.2 ANALYSIS OF T CELL AND SAC FUNCTIONS BY MLR

Induction of cytotoxic T cells requires macrophages for optimal activation. We further examined whether PrG inhibits function of T cells or macrophages using MLR (Table 2). When SAC from PrG-treated mice were used as antigen-presenting cells (APC), a marked reduction of proliferative response was observed, whereas T cells treated with PrG normally proliferated in response to allogeneic SAC. The result suggested that PrG inhibits function of APC but not of T cells in vivo.

3.3 EFFECT ON PRESENTATION OF EXOGENOUS ANTIGEN

The macrophage dysfunction observed above is controversial because antibody production against SRBC requires macrophages which present exogenous antigen to T cells. To examine the effect on presentation of exogenous antigen, we tested antigen presentation ability of macrophages from PrG-treated mice to BK-1, a type 1 helper T cell that recognizes KLH presented by syngeneic Ia$^+$ APC. BK-1 stimulated by KLH presented by PrG-treated SAC was found to produce even more IL-2 (Table 3). The result suggests that the dysfunction of PrG-treated APC is restricted to alloantigen presentation.

TABLE 3. Macrophages from PrG treated mice enhanced presentation of exogenous antigen.

SAC	KLH	radioactivity (cpm)	
		50% sup.	15% sup.
normal	+	515±106	153±20
	-	112±21	129±22
PrG	+	2,874±1,054*	420±135*
	-	158±61	109±18

SAC were prepared from spleens of 3 BALB/c mice and used as APC for stimulation of BK-1. IL-2 activity after 24h-cultivation was measured by proliferation of CTLL-2 in the presence of 50% or 15% of the supernatant. PrG (1mg/kg) was administered i.p. on days 0,2,4,6 and mice were sacrificed on day 10. Values, mean±SD.
*Statistically significant (p<0.05, t-test).

4. Discussion

Present results suggest that the immunosuppressive effect of PrG specific for killer T cells results from the suppression of antigen presentation specific for alloantigen. The results are in agreement with our recent observations that selective inhibition of Con A-induced suppression observed in vitro is due to synergistic toxicity of PrG and Con A but not due to selective suppression of T cell proliferation rather than B cell proliferation [5]. Furthermore, PrG also inhibits delayed-type hypersensitivity in efferent phase [7] in which inflammatory cells including macrophages play major roles [8]. Since similar synergy with Con A was observed with monovalent cation ionophores and inhibitors of vacuolar-type H^+-ATPase [9], it is assumed that inhibition of acidification of intracellular organella is a primary action of the antibiotic. Elevation of pH affects intracellular transport, glycosylation of proteins, and activity of the enzymes in organellas [10]. In fact, PrG increased binding of Con A to the cells [5]. It is, therefore, predicted that PrG inhibits activity of enzymes which participate in processing of antigen, intracellular transport of newly synthesized MHC products or processed peptide in APC.

T cells recognize allogeneic MHC as a complex of MHC product and processed antigenic peptide derived from endogenous proteins [11]. Endogenous antigens and exogenous antigens are processed and associate with MHC products by entirely different pathways [11,12]. PrG possibly inhibits the pathway of endogenous antigen specifically. Allogeneic MHC is presented to helper T cells in three different ways [13]; First, allogeneic MHC is incorporated and expressed by syngeneic class II positive macrophages. Second, $CD4^+$ T cells recognize allogeneic class II-bearing macrophages. Third, $CD8^+$ helper T cells recognize allogeneic class I MHC expressed on allogeneic cells. Because we use P815 as an allogeneic antigen which expresses only class I antigen, and because the first way is the same as processing pathway of exogenous antigen, PrG may suppress the third one selectively. However, we do not yet know whether these models are also applicable in the recognition of killer T cells.

Alternatively, it is possible that PrG inhibits proliferative response of T cells but not the production of IL-2 since alloantigen recognition was determined by the proliferative response of T cells in MLR while presentation of exogenous antigen was determined by IL-2 production. Further precise studies including IL-2 production in MLR reaction are essential to elucidate the mechanism of the selective suppression of PrG.

5. Acknowledgements

This work was in part supported by the Grant-in-Aid for the "Biodesign Research Program" from RIKEN and a Cancer Research Grant from the Ministry of Education, Science and Culture in Japan.

6. References

[1] Nakamura, A., Nagai, K., Ando, K. and Tamura, G. (1986) `Selective suppression by prodigiosin of the mitogenic response of murine splenocytes', J. Antibiot. 39, 1155-1159
[2] Nakamura, A., Magae, J., Tsuji, R. F., Yamasaki, M., and Nagai, K. (1989) `Suppression of cytotoxic T cell induction in vivo by prodigiosin 25-C', Transplant. 47, 1013-1016

[3] Tsuji, R, F., Yamamoto, M., Nakamura, A., Kataoka, T., Magae, J., Nagai, K., and Yamasaki, M. (1990) `Selective immunosuppression of prodigiosin 25-C and FK506 in the murine immune system', J. Antibiot. 43, 1293-1301

[4] Shinohara, N., Huang, Y., Muroyama, A. (1991) `Specific suppression of antibody responses by soluble protein-specific, class II-restricted cytotoxic T lymphocyte clones', Eur. J. Immunol. 21, 23-27

[5] Kataoka, T., Magae, J., Nariuchi, H., Yamasaki, M., and Nagai, K. (1992) `Enhancement by concanavalin A of the suppressive effect of prodigiosin 25-C on proliferation of murine splenocytes', J. Antibiot. 45, 1303-1312

[6] Kurisaki, T., Munemura, K., Kobayashi, K., Yamamoto, M., Hayashi, M., Nakamura, A., Magae, J., Kusakabe, H., Uramoto, M., Isono, K., Nagai, K., and Yamasaki, M. (1991) `E-15, a novel polysaccharide mitogen from Nocardia specific for B cells', Agric. Biol. Chem. 55, 2987-2991

[7] Tsuji, R. F., Magae, J., Yamashita, M., Nagai, K., and Yamasaki, M. (1992) `Immunomodulating properties of prodigiosin 25-C, an antibiotics which preferentially suppresses induction of cytotoxic T cells', J. Antibiot. 45, 1295-1302

[8] Askenase, P. W. (1992) `Delayed-type hypersensitivity recruitment of T cell subsets via antigen-specific non-IgE factors or IgE antibodies: relevance to asthna, autoimmunity and immune responses to tumor and parasites' Chem. Immunol. 54, 166-211

[9] Kataoka, T., Magae, J., Kasamo, K., Yamanishi, H., Endo, A., Yamasaki, M., and Nagai, K. (1992) `Effects of prodigiosin 25-C on cultured cell lines: Its similarity to monovalent polyether ionophores and vacuolar type H^+-ATPase inhibitors', J. Antibiot. 45, 1618-1625

[10] Barasch, J., and Al-Awqati, Q. (1992) `Chloride channels, Golgi pH and cystic fibrosis' Trends Cell Biol. 2, 35-37

[11] Monaco, J. J. (1992) `A molecular model of MHC class-I-restricted antigen processing' Immunol. Today 13, 173-179

[12] Neefjes, J. J., and Ploegh, H. L. (1992) `Intracellular transport of MHC class II molecules' Immunol. Today 13, 179-184

[13] Singer, A., Kruisbeek, A., and Andrysiak, P. M. (1984) `T cell-accessory cell interaction that initiates allospecific T lymphocyte responses. Existence of both Ia-restricted and Ia-unrestricted cellular interaction pathways' J. Immunol. 132, 2199-2209

IMMUNOPOTENTIATOR FROM *BIFIDOBACTERIUM ADOLESCENTIS*

J. LEE, A. AMETANI, Y. SATO, AND S. KAMINOGAWA
Department of Agricultural Chemistry
The University of Tokyo
1-1-1 Yayoi, Bunkyo-ku, Tokyo 113, Japan

ABSTRACT   Among the food microorganisms tested and screened, *Bifidobacterium adolescentis* showed especially strong mitogenic activity toward murine splenocytes and Peyer's patch cells, and enhanced the proliferation of ovomucoid-stimulated lymph node cells in an antigen-nonspecific manner. *B. adolescentis* also enhanced the production of anti-ovomucoid antibody by primed lymph node cells. A lipopolysaccharide inhibitor, polymyxin B, did not inhibit the mitogenic activity of *B. adolescentis* toward normal BALB/c splenocytes. A high molecular weight (>100 kDa) fraction from the proteinase K-treated cytoplasmic fraction of *B. adolescentis*, but not a low molecular weight fraction, retained the mitogenic activity. The mitogenic activity of *B. adolescentis* was also detected toward BALB/c nude (nu/nu) mouse splenocytes, as well as hetero (nu/+) and normal (+/+) mouse splenocytes. These results suggest that the high molecular weight immunopotentiator from *B. adolescentis* acted as a B cell mitogen, being different from lipopolysaccharide.

1. Introduction

An immunopotentiator has been defined as a substance that directly or indirectly enhances a particular immunological function, or modifies one or more components of the immunoregulatory network, immunopotentiators of bacterial origin being used in the field of vaccination and immunotherapy (4, 6). It has recently been reported that many food microorganisms such as lactic acid bacteria have a beneficial effect for preventing or treating intestinal disorders and for enhancing host immune responses (11, 12). In addition, *L. casei* (2, 10), *Lactococcus lactis* subsp. *cremoris* (7) and *Bifidobacterium infantis* (8, 13) have also been reported to have considerable antitumorial activity. A recent report has demonstrated that *B. breve* enhanced the anti-lipopolysaccharide (**LPS**) antibody production and proliferation of Peyer's patch (**PP**) cells (14).

At the present time, however, little is known about whether all lactic acid bacteria have such immunopotentiating activity, or whether specific species or strains do. Although several of the above-mentioned activities have been demonstrated, little has been elucidated about the molecular and cellular mechanisms by which these immunopotentiators exert their effect on various cellular components of the immune system.

In this study, several food microorganisms were screened for their mitogenic

155

S. Kaminogawa et al. (eds.), Animal Cell Technology: Basic & Applied Aspects, Vol. 5, 155–164.
© 1993 *Kluwer Academic Publishers.*

activity, using murine spleen and PP cells, in order to select strains with high immunopotentiating activity. The proliferation and antibody production of murine lymph node (**LN**) cells primed with hen egg ovomucoid (**OM**) were also investigated in this screening. In addition, we tried to isolate and characterize the immunopotentiating molecules and to clarify the molecular and cellular mechanisms for immunopotentiation. In characterizing immunopotentiators from these microorganisms, we are aiming to develop a physiologically functional food capable of maintaining the host defense system which would otherwise be lost with aging.

2. Materials and Methods

2.1. MICROORGANISMS

The microorganisms used in this study were purchased from Japan Bifidus Foundation (**JBF**; Tokyo, Japan) or American Type Culture Collection (**ATCC**; Rockville, MD) and isolated from cultured milk, gouda cheese, yogurt starter or kefir grain. After fermentation, the cells were harvested in a refrigerated centrifuge, washed three times with distilled water and lyophilized for storage. The lyophilized cells were resuspended in distilled water at i mg/ml and broken up with a Branson 200 cell disruptor (Danbury, CT) for 15 min under ice-cold conditions, before being lyophilized for the experiments. The harvested *B. adolescentis* (whole cells) were suspended in distilled water and treated twice with cell disruptor for 15 min. The homogenate was then centrifuged 2 times at $1,000 \times g$ for 20 min until the intact cells could be discarded. The sonically disrupted product was lyophilized and is designated as "sonicated cells". Part of the disrupted product was then ultracentrifuged at $70,000 \times g$ for 30 min at 5°C. The lyophilized supernatant and insoluble precipitates are designated as "cytoplasmic fraction" and "cell wall fraction," respectively.

2.2. MICE

Female, 6- to 8-week-old BALB/c mice were purchased from Charles River Japan (Atsugi, Japan). Female, 6- to 8-week-old BALB/c nude (nu/nu), hetero (nu/+) and normal (+/+) mice were from CLEA Japan (Tokyo, Japan).

2.3. ANTIGEN AND REAGENTS

OM was prepared from hen egg white as described by Fredericq and Deutsch (5) and then purified by DEAE-Sephacel (Pharmacia LKB, Uppsala, Sweden) anion-exchange chromatography. *Salmonella typhimurium* LPS, concanavalin A (**Con A**, Type 4) and polymyxin B were purchased from Sigma Chemical Co. (St. Louis, MO). Proteinase K was purchased from Merck (Darmstadt, Germany).

2.4. LN, SPLEEN AND PP CELL-PROLIFERATION ASSAY

LN cells from OM-primed mice, and spleen and PP cells from unprimed mice were dispersed to provide a single-cell suspension and plated on 96-well tissue culture plates at 4×10^5 LN cells/well, or 5×10^5 spleen and PP cells/well in RPMI 1640 medium containing 1% normal mouse serum and 50 μM 2-mercaptoethanol. To triplicate wells, samples at various doses for each microorganism, the mitogens (Con A and LPS) or the homologous antigen were added, and these cells were incubated at 37°C in 5% CO_2 for

72 h (LN cells) or 48 h (spleen and PP cells). All the cultures were pulsed with 1 μCi [^{3}H]thymidine (New England Nuclear, Boston, MA) and incubated for a further 20 h before harvesting. The incorporation of [^{3}H]thymidine was counted with a liquid scintillation analyzer (Packard, Downers Grove, IL) and is expressed as the mean number of counts per minute (**cpm**). The stimulation index (**SI**) was calculated as the mean cpm incorporated in the presence of a sample divided by the mean cpm incorporated in the absence of the sample.

2.5. *IN VITRO* ANTIBODY PRODUCTION

Two weeks after immunizing with 50 μg of OM, the draining LN cells were plated on 48-well tissue culture plates at 2×10^6 cells per well. Triplicate wells were challenged with various doses of each sample, the cells being incubated at 37°C in 5% CO_2. The upper 0.5 ml of the culture was then discarded, and 0.5 ml of fresh RPMI 1640 medium containing 1% normal mouse serum was added at the 3rd and 7th days of the culture period. After a further 4 days, the plates were centrifuged, and the culture supernatants were collected. The amount of anti-OM antibody in each supernatant was measured by an enzyme-linked immunosorbent assay (1).

2.6. PROTEINASE K TREATMENT AND ION-EXCHANGE CHROMATOGRAPHY

The cytoplasmic fraction from *B. adolescentis* (300 mg) was dissolved to 10 ml of a 0.05 M sodium phosphate buffer at pH 7.5, and 6 mg of proteinase K was added. After incubating for 14 h at 37°C while shaking, the mixture was separated by centrifugation through a membrane filter for a molecular weight of 50,000 (Ultrafree-MC; Millipore, Bedford, MA). The filtrate was then concentrated about five fold in a ultrafiltration system (Centricut 50; Kurabo Industries, Osaka, Japan).

The concentrated proteinase K-treated cytoplasmic fraction was applied to a column of DEAE-Sephacel that had been equilibrated with a 0.05 M sodium phosphate buffer at pH 7.5, and eluted with a linear gradient of NaCl concentration from 0 to 1.0 M. The flow rate was 2 ml/min and the eluate was collected in 5-ml portions. The absorbance was measured at 280 nm to detect protein and at 480 nm for saccharides after the phenol-sulfuric acid reaction (3). A 20-μl portion of each fraction was added to 5×10^5 spleen cells/well, and the proliferative response was determined as already mentioned. To investigate the protein and carbohydrate profiles, the active peak was separated by centrifugation through a membrane filter for a molecular weight of 100,000 (Ultrafree-MC).

SDS-PAGE was carried out in accordance with the procedure of Laemmli (9), using 12% polyacrylamide gel. Protein and carbohydrate patterns were visualized with a silver staining kit (2D-Silver stain II; Daiichi Pure Chemical Co., Tokyo, Japan) and by periodic acid-Schiff (**PAS**) staining (15).

3. Results and Discussion

3.1. SCREENING FOR THE IMMUNOPOTENTIATING ACTIVITY OF FOOD MICROORGANISMS

Mitogenicity in T cells and B cells was fully induced by Con A and LPS, respectively.

Five strains of *Bifidobacterium*, especially *B. adolescentis* and some particular strains of *L. casei*, had immunopotentiating activity toward the proliferation of murine LN, spleen and PP cells (Table 1). However, *Lactococcus, Streptococcus, Kluyveromyces*

Table 1. The results of screening for immunopotentiating activity of sonicated food microorganisms

| Samples | Proliferative Response | | | Antibody Production[c] (ng/ml) |
	LN[a]	Spleen[b] (cpm x 10^{-3})	PP[b]	
Control	1.9 ± 0.3	2.1 ± 0.1	0.5 ± 0.2	20 ± 2
Concanavalin A	44.5 ± 3.2	52.1 ± 2.1	12.9 ± 1.3	106 ± 20
Lipopolysaccharide	47.1 ± 4.2	31.0 ± 0.7	7.2 ± 1.0	391 ± 11
Ovomucoid	28.1 ± 3.7	-	-	387 ± 53
B. bifidum CHR	42.7 ± 1.6	26.3 ± 1.3	20.6 ± 2.0	433 ± 12
B. bifidum A234-4	36.5 ± 0.4	25.1 ± 0.2	23.1 ± 1.3	383 ± 48
B. longum M101-2	37.3 ± 3.2	27.4 ± 1.3	16.3 ± 1.0	385 ± 18
B. infantis I-10-5	26.5 ± 2.7	19.7 ± 1.5	14.0 ± 2.3	327 ± 30
B. adolescentis M101-4	54.1 ± 1.3	32.5 ± 0.5	27.8 ± 1.2	380 ± 7
L. casei subsp. *casei* R-C	6.3 ± 0.8	6.0 ± 3.9	1.7 ± 0.1	205 ± 23
L. casei subsp. *casei* Y	9.1 ± 1.1	5.5 ± 0.7	1.5 ± 0.3	210 ± 60
L. casei subsp. *casei* 3A-1	5.9 ± 0.5	3.4 ± 0.3	1.3 ± 0.4	173 ± 9
L. casei subsp. *casei* 1B-1	5.0 ± 1.2	24.0 ± 3.8	1.1 ± 0.4	252 ± 6
L. casei subsp. *casei* KEF-L	52.8 ± 3.3	22.0 ± 0.8	19.3 ± 1.4	154 ± 15
L. acidophilus CH2	2.6 ± 0.2	4.6 ± 0.4	1.7 ± 1.0	195 ± 18
L. acidophilus Wiesby	7.8 ± 0.6	3.7 ± 0.6	1.2 ± 0.5	172 ± 24
L. acidophilus L-08	25.9 ± 2.3	8.5 ± 0.8	3.7 ± 0.3	274 ± 11
L. acidophilus L-10	3.2 ± 0.4	3.7 ± 1.2	1.4 ± 0.4	77 ± 13
L. delbrueckii subsp. *bulgaricus* YOG10	6.0 ± 1.7	3.2 ± 0.7	1.0 ± 0.5	103 ± 13
L. delbrueckii subsp. *bulgaricus* YOG12	15.5 ± 1.1	6.7 ± 1.0	2.8 ± 0.6	240 ± 25
Lc. lactis subsp. *cremoris* H61	11.2 ± 0.9	2.1 ± 0.2	1.2 ± 0.2	106 ± 8
Lc. lactis subsp. *cremoris* 5621C	9.2 ± 1.4	4.1 ± 1.5	1.3 ± 0.4	186 ± 10
Lc. lactis subsp. *lactis* 527	3.0 ± 0.6	1.8 ± 0.1	1.7 ± 0.6	101 ± 10
S. thermophilus B3	10.2 ± 1.7	5.4 ± 1.0	2.9 ± 0.6	131 ± 11
K. lactis ATCC 124626	5.0 ± 1.1	3.9 ± 0.3	2.1 ± 0.4	108 ± 13
K. lactis IFO 1090	3.0 ± 0.8	4.1 ± 0.8	6.6 ± 2.0	118 ± 13
K. bulgaricus ATCC 16045	3.2 ± 0.8	4.9 ± 0.6	1.9 ± 0.9	88 ± 28
K. fragilis ATCC 56752	3.2 ± 0.9	8.1 ± 0.4	2.5 ± 1.1	86 ± 21
K. fragilis ATCC 36534	6.4 ± 1.7	3.1 ± 0.2	3.5 ± 0.5	267 ± 22
C. kefyr 200	4.1 ± 0.8	3.1 ± 1.3	2.9 ± 2.4	111 ± 21
C. holmii 205	3.3 ± 0.5	3.4 ± 0.4	3.1 ± 0.7	108 ± 21

[a] Growth of LN cells from OM-immunized mice was measured.

[b] Growth of splenocytes and PP cells from unimmunized mice was measured.

[c] OM-specific antibody production by LN cells from OM-immunized mice was estimated as the amount converted to that of anti-OM monoclonal antibody 23E5.

and *Candida* strains exhibited weak immunopotentiating activity or none at all. OM-dependent antibody synthesis of B cells was shown by adding the homologous antigen (OM) or LPS. Five strains of *Bifidobacterium* up-regulated the production of the anti-OM antibody of LN cells, but the other food microorganisms only stimulated it a little. Among the food microorganisms tested, *B. adolescentis* showed the strongest immunopotentiating activity in the various assays. Our results clearly indicate that the proliferative activation of unprimed spleen and PP, and of OM-primed LN cells, and the augmentation of anti-OM antibody production was possible for some, but not all the food microorganisms tested, and that the occurrence of immunopotentiating activity was strongly dependent on the properties of individual strains and a specific cell component, namely an immunopotentiator.

3.2. DISTRIBUTION OF THE IMMUNOPOTENTIATING COMPONENTS FROM *B. ADOLESCENTIS*

The characteristics of the immunopotentiating materials from microorganisms, and the molecular and cellular mechanisms for such immunopotentiation have not been clarified well, so that it is necessary to investigate these activities by using diverse immune tissues in order to cover the full immunopotentiating activity associated with an individual immune function. To characterize the immunopotentiator from *B. adolescentis*, which showed the strongest immunopotentiating activity, we investigated the distribution of activity in various fractions from *B. adolescentis* toward the proliferation of OM-stimulated LN cells, and toward unprimed spleen and PP cells (Fig. 1). The immunopotentiating activity of *B. adolescentis* was augmented by sonication in all assays

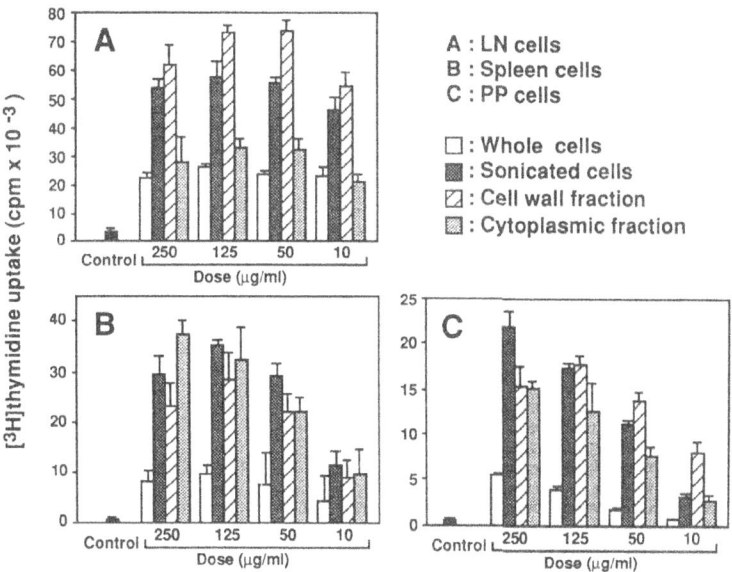

Fig. 1 The distribution of immunopotentiating components from *B. adolescentis* on proliferation of murine LN, spleen and PP cells.

160

using LN, spleen and PP cells, suggesting that the active component of *B. adolescentis* was disclosed by disrupting the cells. The soluble cytoplasmic and insoluble cell wall fractions were shown to have stimulating activity, but the response patterns for the LN, spleen and PP cells varied.

3.3. EFFECT OF POLYMYXIN B, AN LPS INHIBITOR

To determine whether or not the enhancement of proliferation by *B. adolescentis* was due to LPS, an LPS inhibitor, polymyxin B (1 and 5 µg/ml), was added to LN and spleen cells with LPS (2, 10 and 50 µg/ml) or with each fraction of *B. adolescentis* (250 µg/ml). The proliferation of OM-primed LN cells and unprimed spleen cells by LPS was inhibited in accordance with the addition of polymyxin B, but the enhancement of proliferation by whole cells, sonicated cells, and the cell wall and cytoplasmic fractions of *B. adolescentis* was not inhibited (Fig. 2). Therefore, the mitogenicity of the immunopotentiating component(s) from *B. adolescentis* was not due to LPS.

It is known that LPS is a highly productive source of immunomodulating molecules, but is also known to have high toxicity toward humans. It would be interesting to identify a safer immunopotentiator of bacterial origin. In this respect, we expect that the gram-positive food microorganism, *B. adolescentis*, could be a source for endowing food with physiological functions capable of augmenting the host defense against an infectious nonself such as toxic microorganisms, as well as against an aberrant self such as tumorial cells.

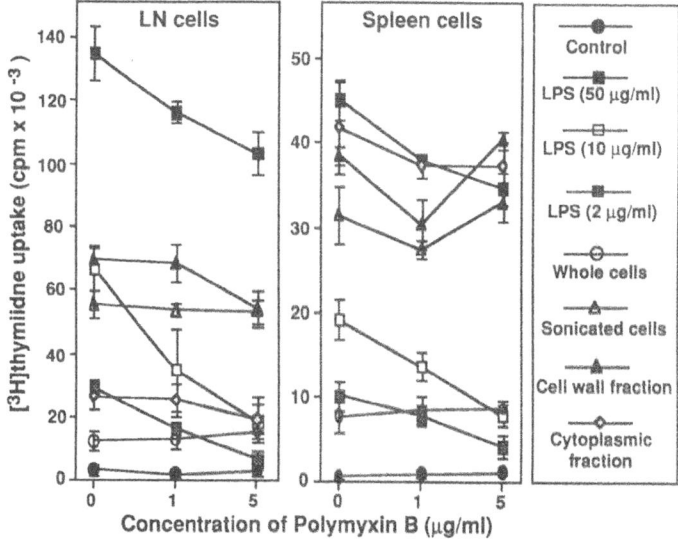

Fig. 2 The effect of polymyxin B on immunopotentiating activity of lipopolysaccharide and *B. adolescentis*.

3.4. CHARACTERIZATION AND ISOLATION OF THE IMMUNOPOTENTIATOR FROM THE CYTOPLASMIC FRACTION OF *B. ADOLESCENTIS*

To characterize the immunopotentiating materials in *B. adolescentis*, the proteinase K-treated cytoplasmic fraction from *B. adolescentis* was separated with a membrane filter for a molecular weight of 50,000, and the proliferative response and SDS-PAGE pattern were then investigated. The greater part of the proteins was excluded by the proteinase K treatment and filtration, but a 43 kDa protein band was resistant to proteinase K. The proliferation of spleen cells was proportional to the saccharide detected by the PAS staining method (Fig. 3).

The purification of the immunopotentiator from the proteinase K-treated cytoplasmic fraction was carried out by DEAE-Sephacel ion-exchange chromatography. The active peak was broad with several shoulders, but the top of the peak was located at a different position from the much larger protein and saccharide peaks (Fig. 4). These results suggest that this isolation method was suitable for removing the non-active materials from this active fraction.

Fraction No. 65, showing the highest activity, was separated with a molecular-sieving membrane. The higher molecular weight fraction from No. 65, of over 100,000, induced the growth of splenocytes. On the other hand, the lower molecular weight fraction did not have only proliferative activity (Fig. 5). These materials were also analyzed by SDS-PAGE. Both the higher and lower molecular weight fractions contained the 43 kDa protein band, but the lower molecular weight fraction did not contain only saccharide bands (data not shown).

These results indicate that the 43 kDa protein was not the active material and that

A: SDS-PAGE pattern

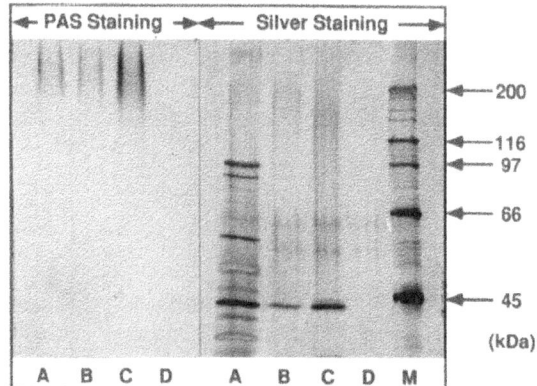

B: [3H]thymidine uptake (cpm x 10^-3)

Control	Pro K	A	B	C	D
4.1±0.2	7.5±0.5	43.1±1.7	42.5±3.2	67.0±2.3	5.7±0.4

Pro K; medium + proteinase K, A; cytoplasmic fraction, B; proteinase K treatment, C; M.W > 5 x 10^4, D; M.W < 5 x 10^4, M; M.W marker

Fig. 3 The SDS-PAGE pattern and proliferative response of proteinase K-treated cytoplasmic fraction from *B. adolescentis*.

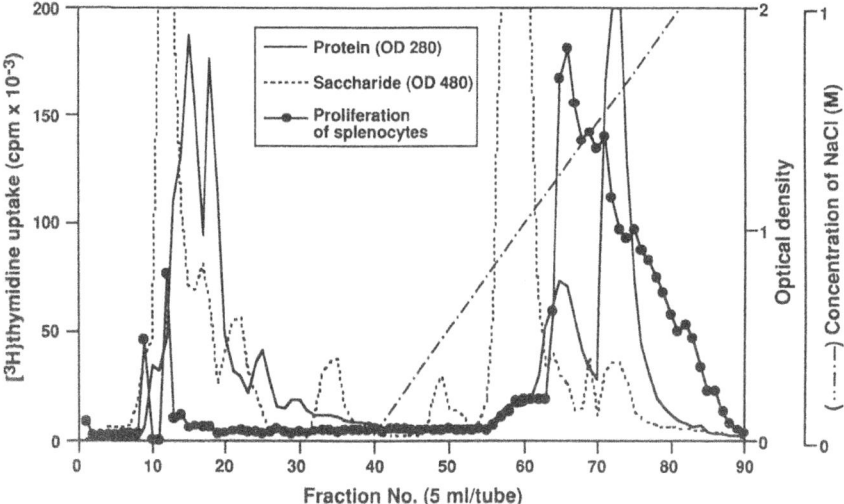

Fig. 4 The DEAE-Sephacel ion-exchange chromatography pattern of proteinase K-treated cytoplasmic fraction from *B. adolescentis*.

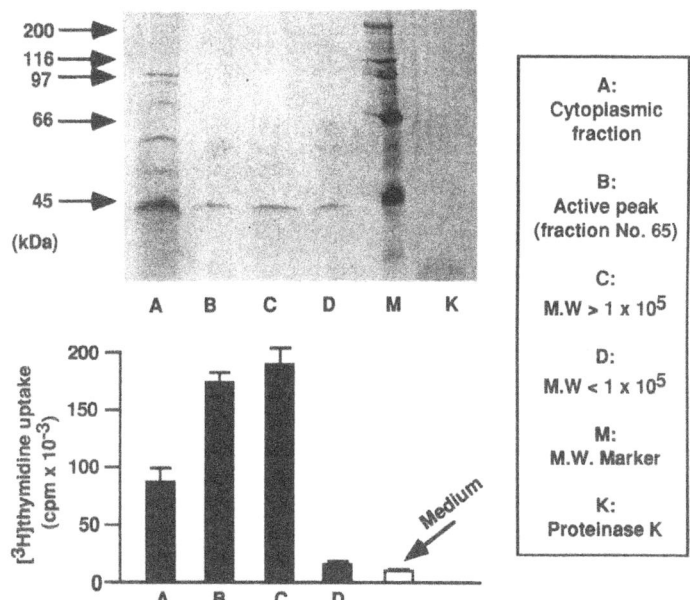

A:	Cytoplasmic fraction
B:	Active peak (fraction No. 65)
C:	M.W > 1 x 10^5
D:	M.W < 1 x 10^5
M:	M.W. Marker
K:	Proteinase K

Fig. 5 The SDS-PAGE pattern (silver staining) and proliferative response of immunopotentiating component from *B. adolescentis*.

the immunopotentiating material had a molecular weight higher than 100 kDa, and was probably a polysaccharide. We are now trying to isolate and characterize the immunopotentiating molecules from *B. adolescentis*.

3.5 TARGET CELLS FOR PROLIFERATION

To determine the target cells for proliferation, splenocytes from BALB/c athymic nu/nu mice, and from normal +/+ and littermate nu/+ mice were cultured with mitogens (Con A and LPS) and each fraction from *B. adolescentis*. The mitogenic activity of LPS occurred in both the cultures of nu/+ and nu/nu mice, but Con A did not induce any mitogenic activity in the cultures of nu/nu splenocytes that were deficient in mature T cells. Splenocytes from both the heterozygote and homozygote were activated with whole cells, sonicated cells, the cell wall fraction and cytoplasmic fraction of *B. adolescentis* (Fig. 6). Similar results were obtained in the experiments using normal and nude mice. These results indicate that mature T cells were not required for blastogenesis by *B. adolescentis*, and that the immunopotentiator from *B. adolescentis* was a potent mitogen to murine B cells.

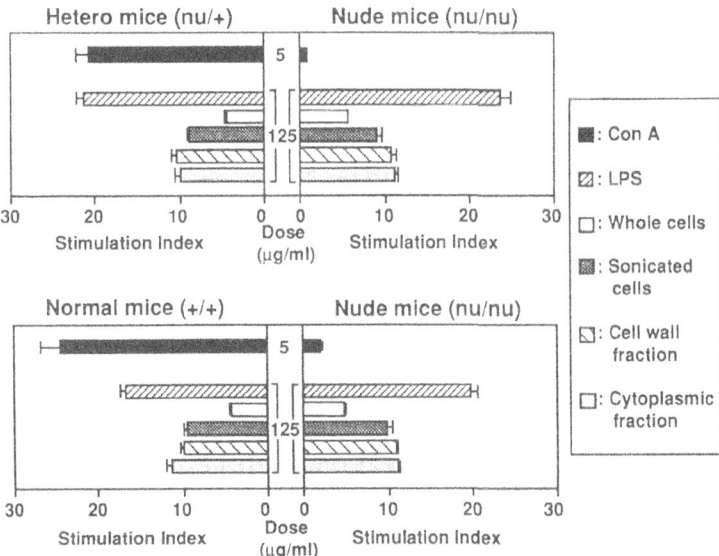

Fig. 6 The B cell mitogenic activity of *B. adolescentis* on splenocytes from BALB/c normal (+/+), hetero (nu/+) and nude (nu/nu) mice.

4. Concluding remarks

B. adolescentis was shown to contain high molecular weight polysaccharides which could strongly stimulate the proliferation and antibody production of B cells. We will further study the mechanism for B cell stimulation, and the immunopotentiating functionality and safety in the case of oral administration of this immunopotentiating molecule from *B. adolescentis*.

5. Reference

1. Ametani, A., Kaminogawa, S., Shimizu, M. and Yamauchi, K. (1987) 'Rapid screening of antigenically reactive fragments of αs_1-casein using HPLC and ELISA', J. Biochem. 102, 421-425.

2. Asano, M., Karasawa, E. and Takayama, T. (1986) 'Antitumor activity of *Lactobacillus casei* (LC 9018) against experimental mouse bladder tumor (MBT-2)', J. Urology 36, 719-721.

3. Dubois, M., Gilles, K. A., Hamilton, J. K., Rebers, P.A. and Smith, F. (1956) 'Colorimetric method for determination of sugars and related substances', Anal. Chem. 28, 350-356.

4. Fauci, A. S., Rosenberg, S. A., Sherwin, S. A, Dinarello, C. A, Longo, D. L. and Lane, H. C. (1987) 'Immunomodulators in clinical medicine', Ann. Int. Med. 106, 421-433.

5. Fredericq, E. and Deutsch, H. F. (1949) 'Studies on ovomucoid', J. Biol. Chem. 181, 499-510.

6. Gatenby, P. A. (1992) 'Immunopotentiation', in I. M. Roitt and P. J. Delves (eds.), Encyclopedia of immunology, Academic Press, London, pp. 847-852.

7. Kitazawa, H., Toba, T., Itoh, T., Kumano, N., Adachi, S. and Yamaguchi, T. (1991) 'Antitumoral activity of slime-forming, encapsulated *Lactococcus lactis* subsp. *cremoris* isolated from Scandinavian ropy sour milk "villi"', Anim. Sci. Technol. 62, 277-283.

8. Kohwi, Y., Hashimoto, Y. and Tamura, Z. (1982) 'Antitumor and immunological adjuvant effect of *Bifidobacterium infantis* in mice', Bifidobacteria Microflora 1, 61-68.

9. Laemmli, U.K. (1970) 'Cleavage of structural protein during the assembly of the head of bacteriophage T4', Nature 227, 680-685.

10. Matsuzaki, T., Yokokura, T. and Mutai, M. (1988) 'Antitumor effect of intrapleural administration of *Lactobacillus casei* in mice', Cancer Immunol. Immunother. 26, 209-214.

11. Perdigon, G., Alvarez, S. and Holgado, A. A. P. R. (1991) 'Immunoadjuvant activity of oral *Lactobacillus casei*: influence of dose on the secretory immune response and protective capacity in intestinal infections', J. Dairy Res. 58, 485-496.

12. Perdigon, G., Macias, M. E. N., Alvarez, S., Oliver, G. and Holgado, A. A. P. R. (1990) 'Prevention of gastrointestinal infection using immunobiological methods with milk fermented with *Lactobacillus casei* and *Lactobacillus acidophilus*', J. Dairy Res. 57, 255-264.

13. Sekine, K., Toida, T., Saito, M., Kuboyama, M., Kawashima, T. and Hashimoto, Y. (1985) 'A new morphologically characterized cell wall preparation (whole peptidoglycan) from *Bifidobacterium infantis* with a higher efficacy on the regression of an established tumor in mice', Cancer Res. 45, 1300-1307.

14. Yasui, H. and Ohwaki, M. (1991) 'Enhancement of immune response in Peyer's patch cells cultured with *Bifidobacterium breve*', J. Dairy Sci. 74, 1187-1195.

15. Zacharius, R. M., Zell, T. E., Morrison, J. H. and Woodlock, J. J. (1969) 'Glycoprotein staining following electrophoresis on acrylamide gels', Anal. Biochem. 30, 148-152.

IMMUNOSUPPRESSIVE PROPERTIES OF FK-506

Toru Kino, Hiroshi Hatanaka, Toshio Goto, and Masakuni Okuhara

Exploratory Research Laboratories, Fujisawa Pharmaceutical Co. Ltd.
5-2-3 Tokodai, Tsukuba, Ibaraki, Japan

ABSTRACT

FK-506, a novel immunosuppressant from Streptomyces tukubaensis, was a selective anti-T lymphocyte agent like ciclosporin (CS). This agent suppressed both cell-mediated and humoral immunity, without depressing the bone marrow. Futhermore, it was effective in a rat skin allograft model. These results provided a strong motivation to develop FK-506 for clinical application to organ transplantation.

Fig. 1 Structures of FK-506 and Ciclosporin

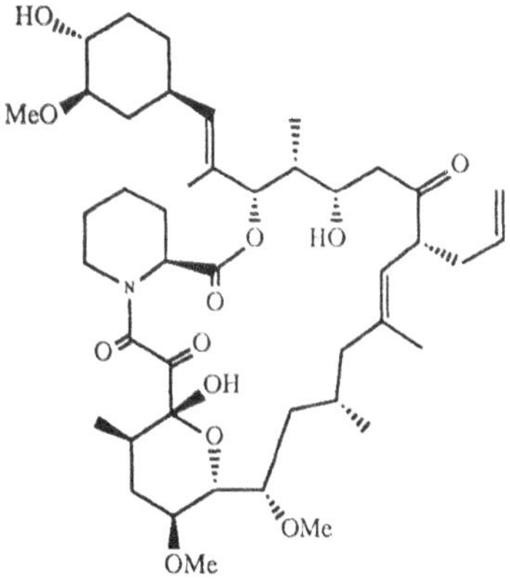

FK-506 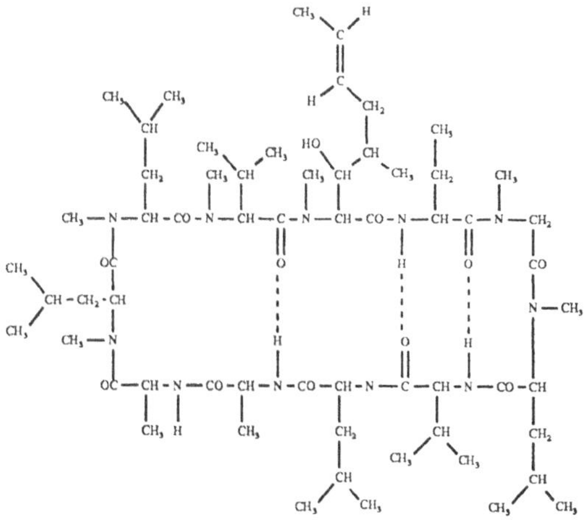   Ciclosporin

The remarkable successes in human organ transplantation are due in large part to the development of immunosuppressive drugs. Ciclosporin, a fungal metabolite, is

165

S. Kaminogawa et al. (eds.), Animal Cell Technology: Basic & Applied Aspects, Vol. 5, 165–169.
© 1993 Kluwer Academic Publishers. Printed in the Netherlands.

becoming the drug of first choice in transplantation surgery (Fig. 1). FK-506, a novel immunosuppressant from *Streptomyces tsukubaensis*, was discovered in 1984 in our laboratories and has undergone multicenter clinical trials for organ transplantations such as liver, kidney and heart (Fig. 1). The immunosuppressive properties of FK-506 *in vitro* and *in vivo* will be studied in this report.

Fig. 2   Schematic representation of MLR and allograft rejection

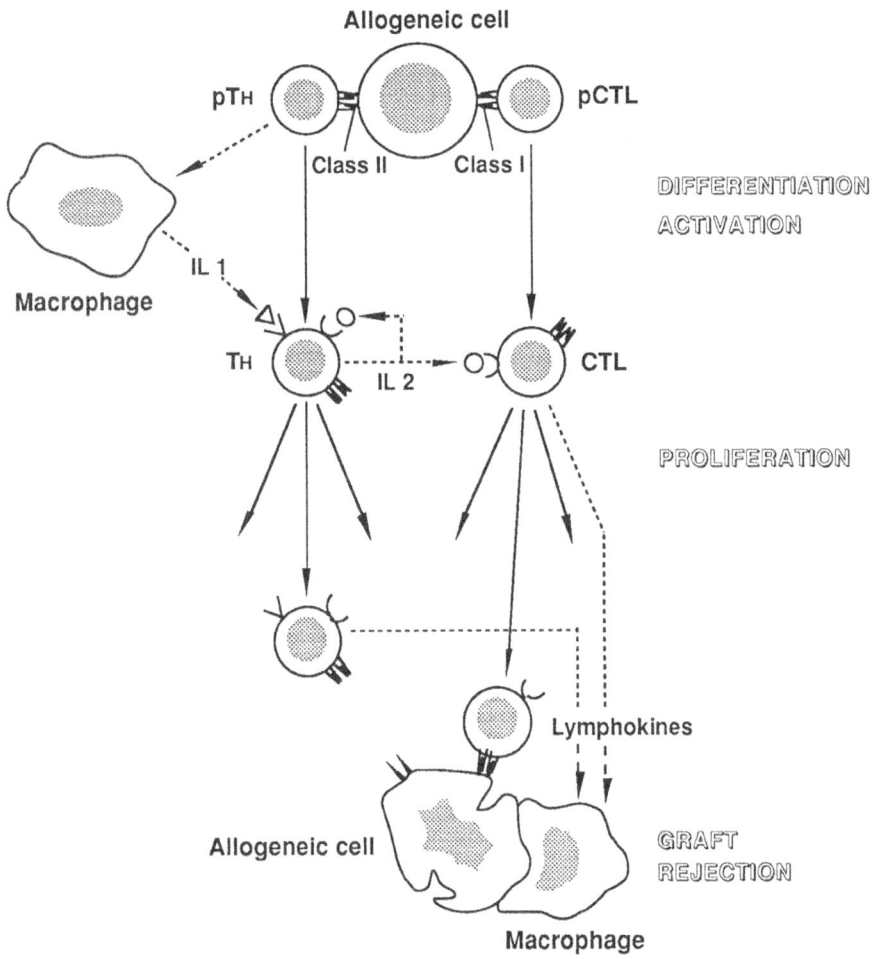

Fig. 2 shows the scheme for mixed lymphocyte reaction (MLR) and allograft rejection.   MLR is regarded as an *in vitro* correlate of allograft rejection and is widely used in graft -matching tests.   Interleukin 2 (IL2) is known to be a growth factor for T cells.   Cytotoxic T lymphocytes (CTL) receive IL2 and proliferate, before killing the allogeneic cells.   We therefore tested *in vitro* immunosuppressive properties of FK-506 by using MLR.

Table 1 shows the *in vitro* immunosuppressive effect of FK-506 to MLR.   FK-506 was highly effective in suppressing the proliferative response in murine and human MLRs.   Moreover, cytotoxic T cell generation, the production of IL2, and the expression of the IL2 receptor in MLR were also suppressed by FK-506.   The IC50 values for FK-506 and CS in all tests were around 0.1 nM and 10 nM, respectively.   However, FK-506 had no effect on bone marrow colony formation or on leukemic T cell proliferation.   These data show that FK-506 and CS share the unique similar biological properties, although FK-506 is more potent than CS[1].

Table 1. *In vitro* immunosuppressive activities

	IC50 (nM)	
	FK-506	CS
MLR	0.2-0.3	14-27
CTL induction	0.2	24
IL-2, γ-IFN production	0.1-0.3	10-32
IL-2R expression	0.1	10
Bone marrow	1400	800
Leukemic T cell	>3200	>1000

FK-506 is a selective anti-T lymphocyte agent like CS. FK-506 may prevent the clonal expansion of T cells and CTL generation, with inhibition of the production of IL2 and the expression of IL2R. These *in vitro* studies suggest that FK-506 would be an effective immunosuppressant that may prove useful in organ transplantation.

Table 2 shows the effect of FK-506 on skin allografting in rats. WKA recipients of F344 skin grafts were treated with FK-506 3.2 mg / kg 5 days a week for 2 weeks, and were given subsequent maintenance doses of 0.32 or 3.2 mg / kg intermittently twice a week for 120 days. All the animals retained healthy grafts as long as the treatment was continued[2].

These findings provided a strong motivation to develop FK-506 for clinical application to organ transplantation.

Table 2. Effect of FK-506 on rat skin transplantation

Treatment	n	Survival days	MST (range)
Not treated	8	5, 6, 6, 6, 6, 6, 7, 7	6.0 (5-7)
FK-506			
3.2→3.2 mg/kg	8	>120, >120, >120, >120, >120, >120, >120, >120	>120
3.2→0.32 mg/kg	8	>120, >120, >120, >120, >120, >120, >120, >120	>120

MST : median survival time (days)

Clinical trials were begun in March 1989 by Dr. Starzl at the University of Pittsburgh on liver transplantation. The wider evaluation of FK-506 in transplant recipients is now underway in multicenter, prospective, controlled trials in North America, Europe, and Japan.

The novel macrolide immunosuppressant FK-506 is a powerful and selective anti-T cell agent which has a similar mode of action to that of ciclosporin.

There are several points of interest in current FK-506 research[3].

1. After long-term clinical trials, the therapeutic window in the field of organ transplantation must be determined.
2. Even if long-term and controlled clinical studies are needed, therapeutic potential for treating auto-immune diseases can be expected.
3. There is still a need for new immunosuppressants with improved therapeutic indexes.
4. FK-506 is a useful probe for studying signal transduction and activation cascades in T cells. Several research groups have been investigating the mechanism for the immunosuppressive action of FK-506 at the cellular and molecular levels. The process of T cell activation has become one of the most popular tools in research on the molecular biology of cellular signal transduction.

References

1) Kino T., H. Hatanaka, S. Miyata, N. Inamura, M. Nishiyama, T. Yajima, T. Goto, M. Okuhara, M. Kohsaka, H. Aoki & T. Ochiai. 1987 J. Antibiot. 40(9): 1256-1265
2) Inamura N., K. Nakahara, T. Kino, T. Goto, H. Aoki, I. Yamaguchi, M. Kohsaka & T. Ochiai. 1988 Transplantation 45(1): 206-209
3) First International Congress on FK-506. 1991 Transplantation Proceedings 23(6) 2709-3380

ENHANCEMENT OF ANTIBODY PRODUCTION AND CELL PROLIFERATION BY MARINE BIOACTIVE SUBSTANCES

Z-L. Kong *1, G-T.Hung *1, K. Shinohara *2
*1 Department of Marine Food Science National Taiwan Ocean University, Keelung, Taiwan.
*2 National Food Research Institute, the Ministry of Agriculture, Forestry and Fisheries, Tsukuba, Japan.

It has long been suggested that some constitutes of foods has the physiological functions of modulating immune, nerve, hematopoiesis and endocrine systems. Techniques for serum-free mammalian cell culture assessment will be useful both for basic studies of cytobiology and for evaluation of physiologically active substances in vitro. In our previous study, some constitutes of vegetable and fruits are found to have significant physiological functions, such as antimutagenicity (1), antitumorgenicity (2), immunoglobulin secretion-promoting (3-5), and macrophage-activating (6) toward cultured cells. However, it is expected that the constitutes in marine source may also exhibit the potential effects on the cultured cells. The present report describes the immunoglobulin production and proliferation regulating in some kinds of shellfish extracts on serum-free cultured human-derived cell lines.

Materials and Methods

Cell lines and medium

In order to screen new biological active substances in shellfish, the human-derived cell lines used in these study were U-937, a histocytic lymphoma compared with U-M, a U-937 derived macrophage-like cell line (7). HL60, a promyelocytic lymphoma and two kinds of human hybridomas HB4C5 and SI102. These cell lines were obtained from American Type Culture Collection and Dr. Murakami (Kyushu University), and were cultured in enriched RDF (eRDF, Kyokuto) medium supplemented with 20mM HEPES buffer, 1.7 μ M insulin, 0.4 μ M iron-free human transferrin, 20 μ M ethanolamine, and 25 nM selenite. The cells at log phase were collected and replated at the cell density of 5×10^4 cells/ml for assay in 24-well microplates. The experimental plates were incubated at 37 ℃ for 3 days in a humidified 5% CO2 air atmosphere.

Preparation of hot water extracts of shellfish

Shellfish, such as abalone (*Italiotis discus*), Formosa squid (*Loligo formosana*),

S. Kaminogawa et al. (eds.), Animal Cell Technology: Basic & Applied Aspects, Vol. 5, 171–176.

cuttlefish (*Sepia esculenta*), shijimi (*Corbical maximum*), oyster *(Crassotrea gigas)*, hard clam (*Meretrix Jamarcki*) purchased from Keelung fish market. 300g fresh material were homogenized and then boiled in water for 20min around 100℃. After centrifugation and filtration, the supernatant was then concentrated by vacuumevaporator. The active fraction of concentratedaqueous extracts was obtained by 80% ammonium sulfate precipitation, then dialyzed against running water and PBS buffer to remove the low molecular weight water-soluble constituents and each dialyzate was lyophilized. The dialyzates was dissolved in phosphate buffered saline solution, and diluted in serum-free medium.

Cell proliferation assay
The effect of extract fractions on the cell growth was estimated by MTT (3-(4, 5-dimethyl-thiazol-2-yl)-2, 5-diphenyl tetrazolium bromide) assay. An water-insoluble formazan formed in the 96-well microplate was dissolved in HCl-isopropanol, and the absorbance at 550nm was measured in a microplate reader (Dynatech).

Determination of immunoglobulin M (IgM)
The cell lines used in this study were human hybridoma cell lines. The HB4C5 and SI102 cell lines are hybridoma producing monoclonal antibodies against lung cancer and breast cancer, respectively. Secreted monoclonal antibody (IgM) by cultured hybridoma cells was measured by Enzymed-Linked Immunosorbent Assay with peroxidase-labeled sheep anti-human IgM antibody.

Hydroxyapatite column chromatography
The active fraction (100mg) obtained by ammonium sulfate precipitation was then applied to a hydroxyapatite-gel column (2 X 30cm) pre-equilibrated with 20mM phosphate buffer (pH7.2). Non-absorbed fraction were wash out with the same buffer and eluted with PBS-0.5M NaCl, finally eluted with 0.5M PBS. The eluate was measured for absorbance at 235nm and the fraction eluted just around the peak top was dialyzed prior to being freeze-dried.

Butyl-TOYOPEARL 650 column chromatography
The active fraction (40mg dissolved in 40% ammonium sulfate 0.1M PBS buffer) obtained from hydroxyapatite column (Butyl-TOYOPEARL 650 size 1.5 X 25cm). Fraction were eluted with diluted ammonium sulfate, and monitorited by measuring the absorbance at 235nm.

Gel Electrophoresis
Electrophoresis was performed on a 10-20% gradient Sodium Dodecyl-polyacrylamide gel. Gels were stained with 0.25% Coomassie Brilliant Blue R-250.

Results

Effect of shellfish hot-water extracts on the morphological alteration of cultured cells

Nondialyzable fraction of shellfish hot-water extracts showed strong cell-aggluting activity on cultured U-M cells, hybridoma SI102 and HB4C5 cells but not on U-937 or HL60 cells. Fig.1 shows the binding of abalone fractions to specific cells.

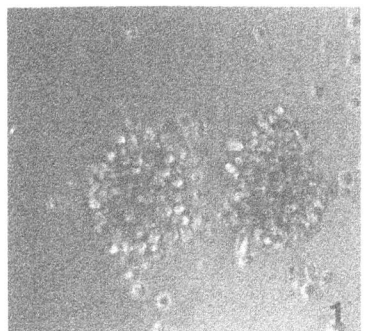

Fig. 1. Photomicrographys (X100) of Hybridoma SI102 Cells Cultured in Enriched RDF Medium Treated with Abalone Fraction For 12hr (1). The Cells (2) No Treatment as Control.

Chromatography of abalone fraction

Chromatogram of effective HA-fraction obtained by Hydroxyapatite column was shown in Fig.2-1, and principle purified by Butyl-Toyopearl 650 column was shown in Fig.2-2. SDS-PAGE proved that purified principle showed single band having a mol.wt. of 17,000 dalton.

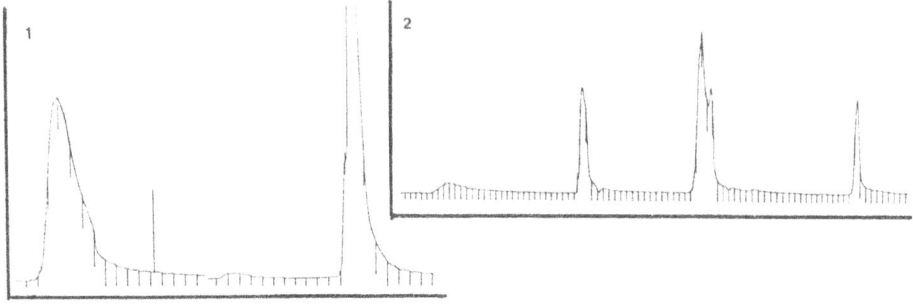

Fig. 2. Hydroxyapatite-gel (HA) Chromatography of Abalone AmmoniumSulfate

Fraction from Hot-Water Extract (1). Butyl-Toyopearl 650 Hydrophobic Chromatography of Active Abalone HA-Fraction. The Eluate ws Measured for Absorbance at 235nm (2).

Effect of the abalone fractions on the growth of HB4C5 cells

The results of a MTT assay of these above shellfish fractions on HB4C5 cells are shown in Fig.3. At low concentrations, the fraction of PW-PT2 were found to produce a higher absorbance than the control, suggesting that PW-PT2 may have growth-promoting activity on HB4C5 cells.

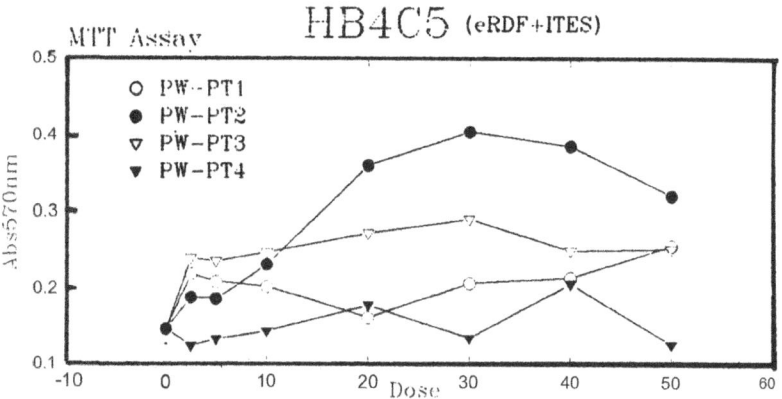

Fig. 3. MTT Assay. Dose-Response Curve of the Growth-Promoting Activity of Shellfish Fractions on Hybridoma HB4C5 Cells.

Effect of Abalone fractions on the IgM secretion of HB4C5 cells

The fractions of Abalone-BT fractions was examined. Fig.4 showedthe increase in the amounts of IgM after the HB4C5 cells were cultured with the fractions for 12 hr. The highest increase in the amount of IgM was observed in PW-PT3 fraction.

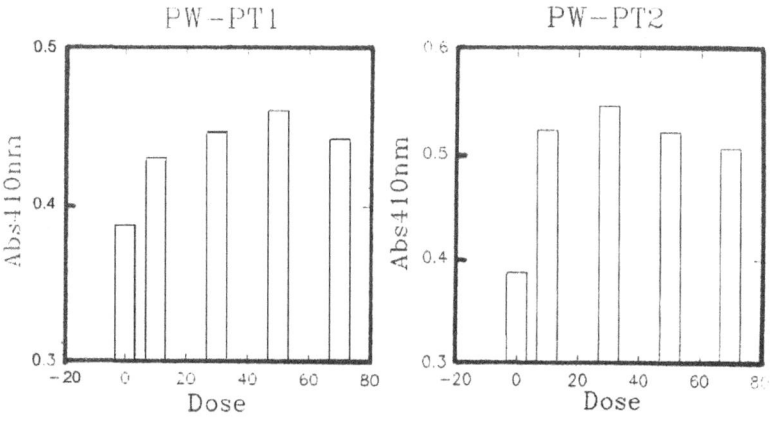

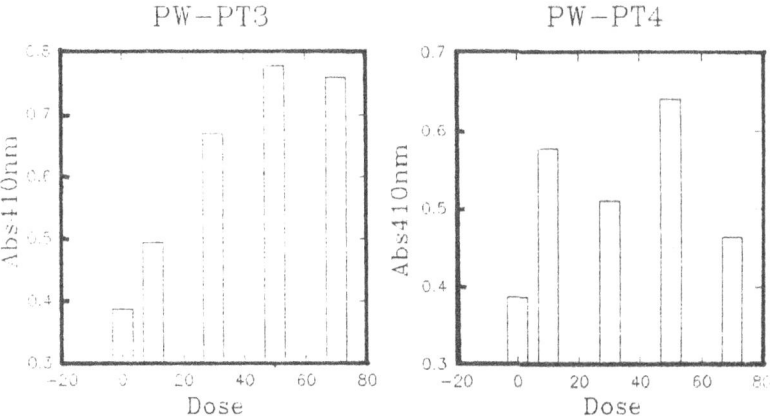

Fig. 4. Effect of Abalone-BT Fractions on IgM Secretion of Hybridoma HB4C5 Cells. The Cells (5X104 cells/ml) were Cultured with Fractions for 12hr, and the IgM in the Sepent Medium was measured by a ELISA Method.

Discussion

Marine animals are a potential source of molecules for pharmacology which are popular in traditional Chinese medicine. In this report, we elucidate the physiological functions of constitutes in marine resource, the effect of the fractions of hot-water extracts from shellfish toward several kinds of human-derived cells was done. Nondialyzable fraction of shellfish hot-water extracts showed strong cell-aggluting activity on cultured U-M cells, hybridoma SI102 and HB4C5 cells but not on U-937 or HL60 cells. Those results suggested that the binding of shellfish fractions to specific cells, indicating the existence of receptor on the surface of cells. In previous study (7), we found that mannose and N-acetylglucosamine residue are rich on the cell membrane of hybridoma and U-M cells, but rare on U-937 and HL60 cells. Furthermore, we found that abalone fractions showed significant enhancement of the production of monoclonal antibody in hybridomas and proliferating effect in cultured human cells. For characterization of bioactive substances, some lectin-like proteins were purified from those boiled shellfish soup, by ammonium sulfate precipitation, hydroxyapatite, butyl-hydrophobic and followed by gel filtration.On sodium dodecyl sulfate-polyacrylamide gel electrophoresis, it showed single protein band with mol.wt. of 17, 000 dalton in abalone and 22, 000 dalton in oyster respectively. And the principle was confirmed as glycoprotein by PAS stain. Further purification and characterization is now in progress. Partially purified fractions showed a cell-aggluting ability and induced enhancement of IgM secreting may have an important immunological potential. Moreover, glycoprotein molecular structure of principle fraction might be helpful to explain its cell specificity.

Literature cited

1) Z-L. Kong et at. Agric. Biol. Chem., 53, 2073-2079 (1989)

2) Z-L. Kong et al. Cytotechnology 7,113-119 (1991)

3) Y. Fuke et al. Nippon shokuhin kogyo Gakkaishi 39, 281-198 (1992)

4) Z-L. Kong et al. Nippon shokuhin kogyo Gakkaishi 39, 79-87 (1992)

5) Y. Fuke et al. Nippon shokuhin kogyo Gakkaishi 39, 193-196 (1992)

6) M. Miwa et al. Biochem. Biophy. Res. Commun.,30, 296-301 (1990)

7) Z-L. Kong et al. In Vitro Cell. Dev Biol., 26, 946-954 (1990)

Effect of a dialyzate of spinach on the chromosomal aberration induced
by mutagens in cultured Chinese hamster cells

I. SAKUMA(1), M. KOBORI(2), H. MORI(3), and K. SHINOHARA(2)
(1) Japan Food Research Laboratories, Tama, Tokyo 206, Japan
(2) National Food Research Institute, Ministry of Agriculture, Forestry
    and Fisheries, Tsukuba, Ibaraki 305, Japan
(3) Research Institute, Kagome Co. Ltd., Nishinasuno, Tochigi 329-27,
    Japan

ABSTRACT.   We examined the antimutagenic effect of a spinach extract on
the chromosomal aberration induced by mutagens in a cultured Chinese
hamster lung cell line (C H L).   The dialyzate of the spinach extract
suppressed the frequency and the total number of chromosomal aberrations
induced by such mutagens as hydrogen peroxide $(H_2 O_2)$, N-methyl-
N'-nitro-N-nitroso-guanidine (M N N G) and mitomycin C (M M C),
while it showed no effect on the clastogenicity of 4-nitroquinoline-1-
oxide (4 N Q O).   The dialyzate of the spinach extract completely inhibited the clastogenicity of $H_2 O_2$ at a concentration of 50 µg/ml.
The extract also suppressed the frequency of chromosomal aberration
induced by M N N G and M M C, although this suppression was not
total.

1. INTRODUCTION

Dietary inhibitors of mutagenesis have been investigated by short-term
bio-assays such as the Ames test, sister-chromatid exchange test, and
chromosomal aberration test [1].   A Chinese hamster lung cell line
(C H L) has been used for the chromosomal aberration test, which is
sensitive enough to detect the chromosomal aberration induced by many
mutagens [2].   This test has also been utilized to detect bio-anti-
mutagens and desmutagens.   Our previous studies have revealed that a
dialyzate of spinach showed an antimutagenic effect in the Ames
test [3].   In this present study, we examined the suppressing effect of
a spinach extract on the chromosomal aberration induced by direct-acting
mutagens such as hydrogen peroxide $(H_2 O_2)$, N-methyl-N'-nitro- N-
nitroso-guanidine (M N N G), mitomycin C (M M C) and 4-nitro-
quinoline-1-oxide (4 N Q O) in C H L cells.

S. Kaminogawa et al. (eds.), Animal Cell Technology: Basic & Applied Aspects, Vol. 5, 177–181.
© 1993 Kluwer Academic Publishers.

2. MATERIALS AND METHODS

2.1. Extraction of the spinach leaves
Spinach leaves were cut into small pieces and homogenized in 10 mM phosphate-buffered saline (P B S, pH 7.0) by a Waring blender. The homogenate was filtered through cotton cloth and centrifuged at 10,000 g for 30 min. The 20–80% saturated ammonium sulfate fractions of the supernatant were applied to a Sephadex G-25 column pre-equilibrated with P B S.

2.2. Cells and cell culture
A Chinese hamster lung cell line (C H L) was cultured in Eagle's M E M (Nissui Pharmaceutical Co., Japan) supplemented with 10% heat-inactivated new-born calf serum (Gibco). The doubling time of the cells was approximately 15 hrs.

2.3. Assay of the suppressive effect on chromosomal aberration
The cells were seeded at a density of 7×10^4 cells/well on 6-well plates and cultured for 24 hrs. The spinach extract was dissolved in P B S and added to the culture at several concentrations. At the same time, cells were treated with $H_2 O_2$, MNNG, MMC or 4NQO. After incubating for 24 hrs, the cells were treated with 0.2 µg/ml of colcemid for the last 2.5 hrs of incubation, and then fixed. Chromosome preparations made by an air-drying method were stained with a 1.5% Giemsa solution in a 1/150 M phosphate buffer (pH 6.9) for 15 min. One hundred well-spread metaphases were observed under an optical microscope.

a)   b)

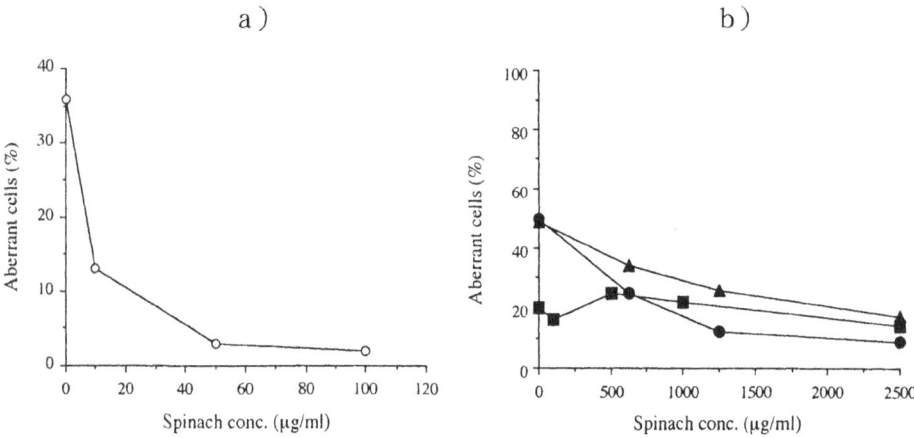

Fig. 1 Effect of different concentrations of the spinach extract on the induction of aberrant cells by 4.0 µg/ml of $H_2 O_2$ (a: ○), 0.25 µg/ml of MNNG (b: ●), 0.02 µg/ml of MMC (b: ▲) and 0.2 µg/ml of 4NQO (b: ■).

The numbers of chromatid and iso-chromatid gaps (g), chromatid breaks (ctb), chromatid exchanges (cte), chromosome breaks (csb) and chromosome exchanges (cse), and the frequency of cells with chromosome aberration were counted.

Table 1 Effect of the spinach extract on the frequency (%) and type of chromosomal aberration induced by H_2O_2, MNNG, MMC and 4NQO in CHL cells.

Mutagen	Spinach extract (μg/ml)	Aberrant cells (%)	Number of aberrations/100 cells					
			Total	g	ctb	cte	csb	cse
No treatment	0	2	2	0	1	0	1	0
	625	1	1	0	0	0	1	0
	1,250	3	4	0	0	0	4	0
	2,500	4	7	1	0	1	5	0
H_2O_2 (4.0μg/ml)	0	36	89	2	18	56	12	1
	10	13	20	1	2	12	4	1
	50	3	12	1	0	0	11	0
	100	2	2	1	0	0	1	0
MNNG (1.0μg/ml)	0	91	389	3	59	327	0	0
	625	88	278	3	63	206	3	3
	1,250	82	206	1	27	173	5	0
	2,500	69	130	2	27	99	2	0
(0.25μg/ml)	0	50	74	2	25	40	6	1
	625	25	37	2	10	21	3	1
	1,250	12	15	3	2	7	3	0
	2,500	9	10	0	7	3	0	0
MMC (0.02μg/ml)	0	49	87	2	20	62	1	2
	625	34	56	2	11	42	1	0
	1,250	26	35	3	9	23	0	0
	2,500	17	20	0	2	14	2	2
4NQO (0.2μg/ml)	0	20	60	0	5	51	2	2
	100	16	34	2	10	19	1	2
	500	25	83	2	6	75	0	0
	1,000	22	75	0	6	63	4	2
	2,500	14	37	0	1	31	4	1

3. RESULTS AND DISCUSSION

The effect of different concentrations of the spinach extract on the
induction of aberrant C H L cells by 4.0 µg/ml of $H_2 O_2$, 0.25 µg/ml
of M N N G, 0.02 µg/ml of M M C and 0.2 µg/ml of 4 N Q O was first
examined. As shown in Fig. 1, the spinach extract was found to suppress
the induction of aberrant cells by $H_2 O_2$, M N N G and M M C in a
dose-dependent manner, while no suppressive effect of the extract was
detected for 4 N Q O. The clastogenicity of $H_2 O_2$, which is one
of the active oxygen species, was completely inhibited by the extract at
a concentration of more than 50 µg/ml. It was also found that the
extract reduced the cytotoxicity of $H_2 O_2$ foward C H L cells (data
not shown). Compared with the case for $H_2 O_2$, the suppressive effect
of the spinach extract on M N N G and M M C was low, and a higher
concentration of the extract was needed to suppress their induction of
aberrant cells.
Table 1 shows the frequency (%) and type of chromosomal aberration
induced by $H_2 O_2$, M N N G, M M C and 4 N Q O, and the effect of
the spinach extract on their induction. As is evident from Table 1,
$H_2 O_2$ induced ctb, cte and csb, especially cte, in the C H L cells.
The induction of ctb and cte by $H_2 O_2$ was suppressed by adding
10 µg/ml of the spinach extract. M N N G and M M C induced ctb and
cte, and these types of aberration were also suppressed by the spinach
extract.

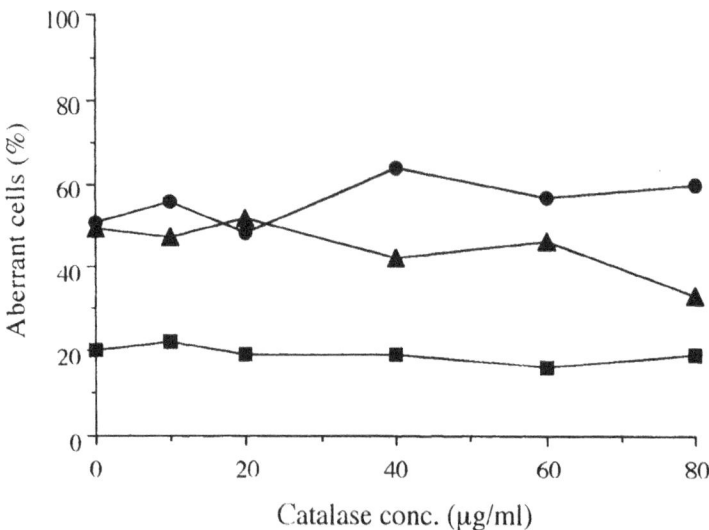

Fig. 2 Effect of catalase on the frequency (%) of aberrant cells
induced by 0.25 µg/ml of M N N G (●), 0.02 µg/ml of M M C (▲) and
0.2 µg/ml of 4 N Q O (■).

Spinach includes an amount of catalase, which is a catalyzing enzyme of the active oxygen species and is known to inhibit the clastogenicity and cytotoxicity of H_2O_2 [4], [5]. Therefore, catalase may be considered to play a key role in the suppressive effect of the spinach extract on the clastogenicity of MNNG, MMC and 4NQO. The effect of catalase on the frequency of aberrant cells induced by MNNG, MMC and 4NQO was to slightly reduce the frequency of aberrant cells caused by MMC, while no effect was observed for MNNG and 4NQO (Fig.2). While the reducing effect of the spinach extract on the number of chromosomal aberrations was not remarkable for any mutagens, these results suggest that the suppressive effect may be caused by high-molecular-weight substances existing in the spinach extract a part from catalase. Clarification of the active substances in the spinach extract is now being carried out.

REFERENCES

1. Hayatsu, H., Arimoto, S., and Negishi, T. (1988) "Dietary inhibitors of mutagenesis and carcinogenesis" , Mutat. Res. **202**, 429-446.
2. Ishidate, M., Jr., Sofuni, T., Yoshikawa, K., Hayashi, M., Nohmi, T., Sawada, M., and Matsuoka, A. (1984) "Primary mutagenicity screening of food additives currently used in Japan" , Fd. Chem. Toxicol. **22**, 623-636.
3. Shinohara, K., Kong, Z L., Fukuda, T., and Iino, K. (1991) "Desmutagenic actions of partially fractionated dialyzate of spinach on Trp-P-2" , Nippon Shokuhin Kogyo Gakkaishi **38** , 242-248.
4. Hirasawa, M., Gray, K. A., Shaw, R. W., and Knaff, D. B. (1987) "Spectroscopic properties of spinach catalase" , Biochim. Biophys. Acta. **911**, 37-44.
5. Oya, Y., Yamamoto, K., and Tonomura, A. (1986) "The biological activity of hydrogen peroxide I. Induction of chromosome-type aberrations susceptible to inhibition by scavengers of hydroxyl radicals in human embryonic fibroblasts " , Mutat. Res. **172**, 245-253.

SUPPRESSION OF MELANIN PRODUCTION IN A MOUSE B16 MELANOMA 4A5 CELL LINE BY A SPINACH EXTRACT

Yasuto SASAKI,* Masuko KOBORI,**Hironobu MORI, ***and Kazuki SHINOHARA **
*   Nitto Flour Milling Co. Ltd., Research Laboratory, 6-2-1 Tokai, Ohta-ku, Tokyo 143, Japan
**   National Food Research Institute, Ministry of Agriculture, Forestry and Fisheries, 2-1-2 Kannondai, Tsukuba 305, Japan
*** Kagome Co. Ltd., Research Laboratory, 17 Nishitomiyama, Nishinasuno, Tochigi 329-27, Japan

ABSTRACT. The effect of an extract of fresh and of freeze-dried spinach on the melanin production of a mouse melanoma cell line, B16 melanoma 4A5, was examined. Both the extracts of dried and fresh spinach suppressed the melanin production in B16 melanoma 4A5. The passage of the cells in the presence of 100 μg/ml of fresh spinach extract clearly reduced the melanin production. However, production was evoked again when the spinach extract was removed, and no cytotoxicity of the extract toward the cells was observed. No inhibition of mushroom tyrosinase by the extract was detected either. These results indicate that the reduction of melanogensis by spinach extract was not due to cytotoxicity toward the cells nor to the inhibition of tyrosinase activity in the cells.

INTRODUCTION

Epidermal melanin synthesis occurs in vivo within melanosomes in the melanocytes of cells via the oxidation of tyrosine by tyrosinase. In the melanosome, active tyrosinase is converted from protyrosinase that has accumulated in vesicles (premelanosome) formed in the cytoplasm at the first stage of melanin synthesis. The matrix of the premelanosomes is progressively obscured during melanin synthesis, resulting in the formation of fully melanized mature melanosomes. Many studies[1, 2, 3] have indicated that the amount of melanin in the epidermis was closely related to the activity of tyrosinase.

Melanin which governs skin and hair color protects the skin from harmful solar radiation. On the other hand, melanin precipitation in the epidermis causes such symptoms as freckles and liver spots. Consequently, it has became cosmetically and biochemically interesting to suppress epidermal melanin production. The establishment of a cell culture system of melanotic melanoma cells makes it possible to elucidate the phemomenon of melanogenesis.

Many studies have been carried out on factors that regulate the pigmentation and proliferation of many kinds of melanoma cells. Commonly used cell lines are the B16,[4] Harding-Passy,[5] and Cloudman strains[6] from mice; the Fortner (RPMI) strain[7] from hamsters; and many human strains such as the IIB-MEL cell line.[8] There are any number of agents that inhibit melanin production both in vitro and in vivo.

It has been reported that a low concentration of a thymidine analog, 5-bromodeoxyuridine,[9] and such pharmacologic agents[10] as retinoic acid, hexamethylene bisacetamide, sodium butyrate and dimethylsulfoxide suppressed the pigmentation of cultured melanoma cells. The inhibitory effect of arbutin

S. Kaminogawa et al. (eds.), Animal Cell Technology: Basic & Applied Aspects, Vol. 5, 183–188.

(hydroquinone - β-D-glucopyranoside), one of the hydroquinone derivatives, on melanogenesis in vitro and in vivo has also been investigated.[11]

In this study, we screened the substances that would suppress melanin production in a mouse B16 melanoma 4A5 cell line existing in several vegetable extracts.

MATERIALS AND METHODS

Materials

One hundred grams each of several freeze-dried vegetables (spinach, egg-plant, cabbege and broccoli) were homogenized in 200ml of 40% ethanol. After removing the precipitates by centrifugation, each concentrated aqueous extract was dialyzed against water and then lyophilized. In addition, one kilogram of fresh spinach (Okame) was homogenized in a phosphate-buffered saline solution (PBS, pH 7.4) and centrifuged. The protein - rich fraction of the extract was separated by salting out with 20 - 80% ammonium sulfate and then lyophilized after dializing against water. Each sample was solubilized in PBS and applied to a cell assay after sterilizing by filtration through a millipore filter (0.22 μm).

Cells and cell culture

B16 mouse melanoma 4A5 cells supplied by Riken Cell Bank were maintained in Dulbecco's modified Eagle's medium (DMEM) supplemented with 10% fetal calf serum (FCS) and 50 μg/ml of Kanamycin. The cells (1×10^5 cells/ml) were seeded in 6-well plates or 6cm-dia. dishes and cultured with or without the vegetable extracts in a CO_2 incubator under a humidified 5% CO_2-95% air atmosphere. After treating with 0.1% trypsin containing 0.005% EDTA, the cell numbers were counted with a hemocytometer. All experiments were carried out in duplicate.

Determination of the melanin content

Cells harvested from the dishes were sonicated in Solvable (E.I. du Pont de Nemours & Co.) and the absorbance at 400nm was measured. The melanin content was calculated from a standard curve for synthetic melanin, and the data are expressed as picograms per cell. Before harvesting, the pigmentation in the cells was also observed under a microscope.

RESULTS AND DISCUSSION

Effect of the freeze-dried vegetable extracts and fresh spinach extract on the proliferation and melanin production of a mouse B16 melanoma 4A5 cell line

The B16 melanoma 4A5 cells were cultured with different concentrations of the extracts of dried spinach, eggplant, cabbage and broccoli for 3 days. As shown in Fig. 1, the extract of dried spinach suppressed the melanin production most strongly among the vegetables, without any significant change in the proliferation of the cells. The extract of freeze-dried eggplant also suppressed it at a high dosage. The fresh spinach extract suppressed the melanin production in a dose-dependent manner more strongly than the freeze-dried sample without any effect on proliferation at under 250 μg/ml (Fig. 2).

The photographs in Fig. 3 indicate that pigmentation of the B16 melanoma 4A5 cells cultured with the fresh spinach extract (100 μg/ml) was clearly reduced in comparision with that of a control culture without the spinach extract.

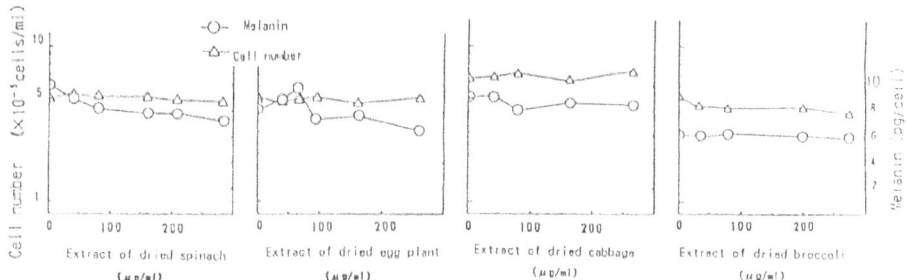

Figure 1. Effect of freeze-dried vegetable extracts on melanin production in
the mouse B16 melanoma 4A5 cell line.
The B16 melanoma 4A5 cells were cultured with different concentrations of the
dried vegetable extracts at 37°C for 3 days in DMEM supplemented with 10% FCS in
a humidified 5% CO₂-95% air atmosphere.

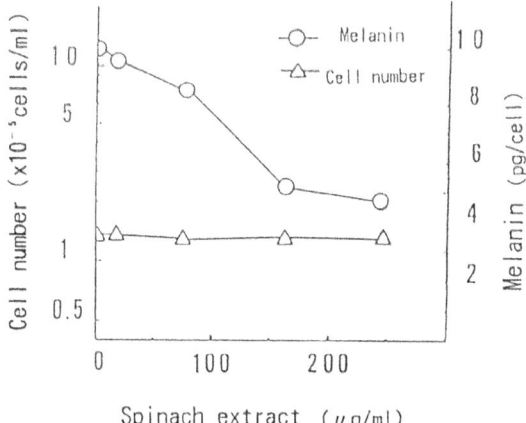

Figure 2. Effect of the fresh spinach extract (Okame) on melanin production in
the mouse B16 melanoma 4A5 cell line.
The experimental conditions were the same as those in Fig. 1.

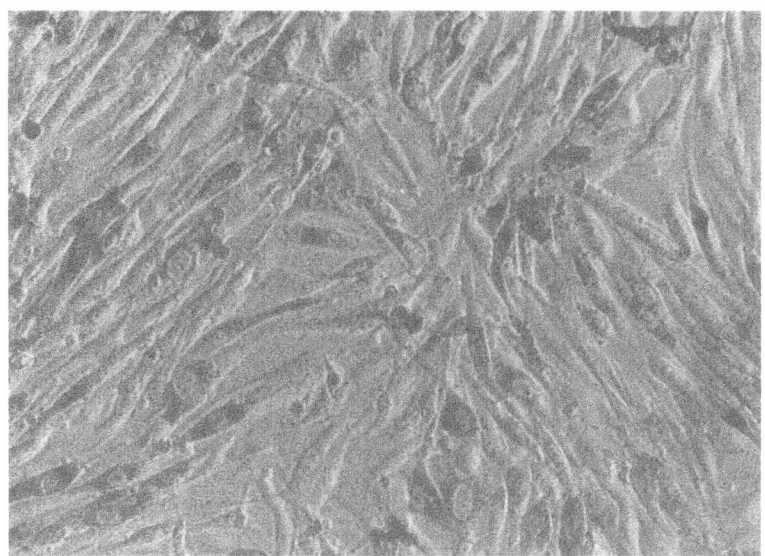

Control (PBS)

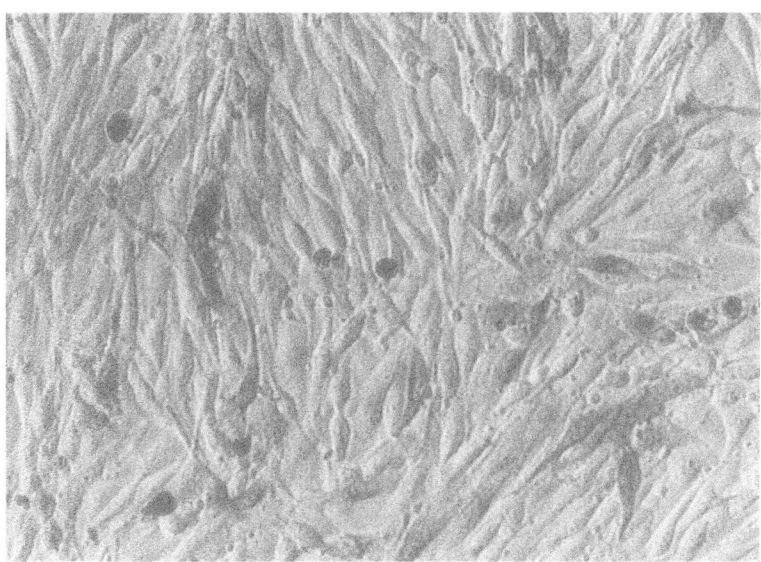

Fresh spinach (Okame; 100 μg/ml)

Figure 3. Photographs of the melanoma cells cultured with the fresh spinach extract (Okame ; ×100).

The B16 melanoma 4A5 cells were cultured with 100μg/ml of the fresh spinach extract (Okame) at 37 °C in DMEM supplemented with 10% FCS in a humidified 5% CO_2-95% air atmosphere for 3 days.

Melanin production of B16 melanoma 4A5 cells in the presence of the fresh spinach extract.

As shown in Fig. 4, the melanin production of proliferating B16 melanoma 4A5 cells was suppressed in a time-dependent manner in the presence of $100\mu g/ml$ of the fresh spinach extract. The passage of the cells in the presence of 100 $\mu g/ml$ of the fresh spinach extract continously reduced the melanin production. However, its production was evoked again when the spinach extract was removed (Fig. 5).

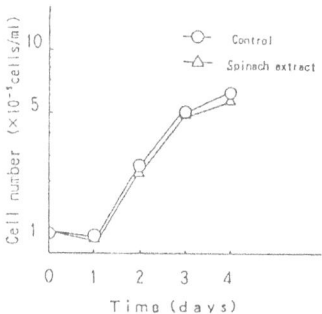
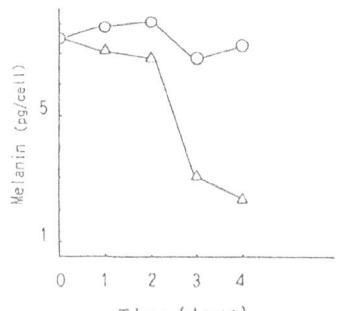

Figure 4. Effect of the fresh spinach extract (Okame) on the proliferation and melanin production of B16 melanoma 4A5.
The experimental conditions were the same as those in Fig. 3.

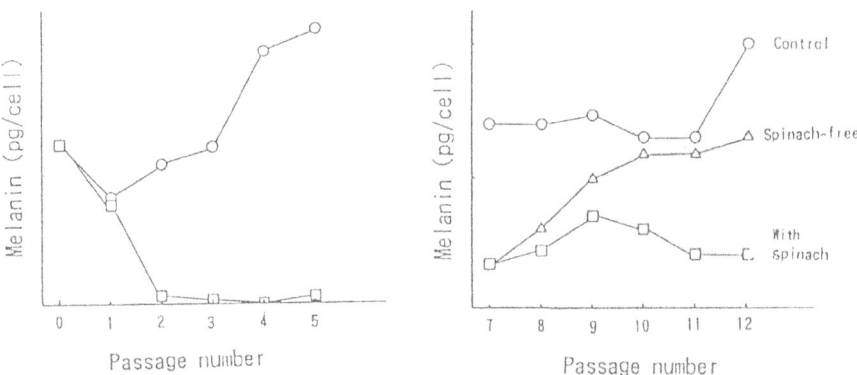

Figure 5. Melanin production of B16 melanoma 4A5 cells in the presence and absence of the fresh spinach extract (Okame).
B16 melanoma 4A5 was cultured with and without $100\mu g/ml$ of the fresh spinach extract (Okame) for several passages.

It was thus found that culturing highly pigmented mouse melanoma cells in the presence of the fresh spinach extract resulted in a complete loss of the melanin-producing system in the cells. The important process of suppressing melanin production by the spinach extract is considered to inhibit the first steps of melanin synthesis, which are 1) progressive reduction in tyrosinase activity, 2) inhibition of the formation of tyrosinase, and 3) reduction of quione derivatives. The fresh spinach extract did not inhibit mushroom tyrosinase activity in an in vitro test (data are not shown here). It has also been reported that a spinach extract had significant functions such as antioxidativity. These results imply that the suppression of melanogenesis by the spinach extract was not due to cytotoxicity and the inhibition of tyrosinase activity, and that its antioxidativity might be related to the suppression of melanogenesis in mouse melanoma cells.

REFERENCES

1) Oikawa, A., Nakayasu, M., and Nohara, M. (1973) 'Tyrosinase activities of cell-free extracts and living cells of cultured melanoma cells,' Develop. Biol. 30, 198-205.

2) Saeki, H. and Oikawa, A. (1978) 'Effects of pH and type of sugar in the medium on tyrosinase activity in cultured melanoma cells,' J. Cell. Physiol. 94, 139-146.

3) Saeki, H. and Oikawa, A. (1983) 'Stimulation of tyrosinase activity of cultured melanoma cells by lysosomotropic agents,' J. Cell. Physiol. 116, 93-97.

4) Hu, F. and Lensey, P.F. (1964) 'The isolation and cytology of two pigment cell strains from B16 mouse melanomas,' Cancer Res. 24, 1634-1643.

5) Harding, H.E. and Passy, R.D. (1933) J. Pathol. 33, 417.

6) Pawelek, J., Wong, G., Sansone, M., and Morowitz, J. (1973) 'Molecular control in mammalian pigmentation,' Yale J. Biol. Med. 46, 430-433.

7) Moore, G.E., Lehner, D.F., Kikuchi,Y., and Less, L.A. (1962) Science 154, 1186.

8) Guerra, L., Bover, L., and Mordoh, J. (1990) 'Differentiating effect of L-Tyrosine on the human melanoma cell IIB-MEL-J,' Exp. Cell Res. 188, 61-65.

9) Wrathwall, J.R., Oliver, C., Silagi, S., and Essner, E. (1973) 'Suppression of pigmentation in mouse melanoma cells by 5-Bromodeoxyuridine,' J. Cell Biol. 57, 406-423.

10) Orlow, S.J., Chakraborty, A.K., Boissy, R.E., and Pawelek, J.M. (1990) 'Inhibition of induced melanogenesis in Cloudman melanoma cells by four phenotypic modifiers,' Exp. Cell Res. 191, 209-218.

11) Akiu, S., Suzuki, Y., Fujinuma, Y., Asahara, T., and Fukuda, M. (1988) 'Inhibitiory effect of arbutin on melanogensis: biochemical study in cultured B16 melanoma cells and effect on the UV-induced pigmentation in human skin,' Proc. Jpn. Soc. Invest. Dermatol. 12, 138-139.

EFFECT OF SPINACH EXTRACTS ON THE PROLIFERATION AND ON SOME PHENOTYPIC CHARACTERISTICS OF HUMAN MYELOID LEUKEMIA CELL LINES

Masuko Kobori, Youichi Nishiba, and Kazuki Shinohara
National Food Research Institute,
Ministry Agriculture, Forestry and Fisheries,
2-1-2 Kannondai, Tsukuba, Ibaraki 305, Japan

ABSTRACT   The differentiation-inducing effect of spinach extracts on human myeloid leukemia cells was examined.   Non-dialyzable extracts of fresh and freeze-dried spinach leaves inhibited the proliferation of HL-60 promyelocytic leukemia cells, which are known to differentiate into granulocyte or monocyte/macrophage cells.   The extracts were also found to reduce nitroblue tetrazolium dye and induce the non-specific esterase activity of HL-60 cells.   These results suggest that the extracts induced monocytic differentiation of HL-60 cells.   The proliferation of K562 erythroleucemic cells and THP-1 monocytic leukemia cells was also inhibited by the freeze-dried spinach extract.   However, there was no induction of benzidine positive cells on K562 cells nor adherence of THP-1 cells, which indicate the differentiation into erythrocytes and monocyte/macrophages, respectively.

1. INTRODUCTION

Human myeloid leukemia cell lines are known to be differentiated in vitro by a variety of agents. In our studies on some physiological functions of vegetables and fruits, we have demonstrated that a non-dialyzable extract of freeze-dried spinach leaves induced morphological change in a human promyelocytic leukemia cell line, U-937, which is one of the characteristics of differentiation into macrophage cells[1].   To elucidate the functions of spinach on human myeloid leukemia cell lines and their differentiation, we have now examined the effect of non-dialyzable extracts of spinach on the proliferation and on some phenotypic characteristics of human myeloid cell lines such as the HL-60 human promyelocytic leukemia cell line[2], K562 human erythroleukemia cell line[3] and THP-1 human monocytic leukemia cell line[4].

2. MATERIALS AND METHODS

2.1. Preparation of the Non-dialyzable Extracts of Spinach

Freeze-dried spinach leaves and fresh leaves of Spinacia Oleacea cv. Okame were kindly provided by Asahi Food Industry Co. (Nagano, Japan)

189

S. Kaminogawa et al. (eds.), Animal Cell Technology: Basic & Applied Aspects, Vol. 5, 189–193.

and Takii Nursery Co. (Kyoto, Japan), respectively. The non-dialyzable extracts of spinach were then prepared. Freeze-dried spinach leaves were homogenized in 40% ethanol by a Waring blender and kept at $4^{\circ}C$ overnight. The homogenate was filtered through cotton cloth and centrifuged at 10,000 g for 30 min. The supernatant was concentrated by evaporation, dialyzed against running tap water and lyophilized. Fresh spinach leaves were cut into small pieces and homogenized in phosphate-buffered saline (PBS, pH 7.4) by a Waring blender. The homogenate was filtered through cotton cloth and centrifuged at 10,000 g for 30 min. The supernatant was fractionated with 20-80% saturated ammonium sulfate, and the precipitate was collected and dissolved in PBS.

2.2. Cells and Cell Culture

HL-60, K562 and THP-1 cells were provided by Japanese Cancer Research Resources Bank (JCRB). HL-60 cells were grown in RPMI1640 medium (Nissui Pharmaceutical Co., Tokyo, Japan) supplemented with 10% fetal bovine serum (FBS; Irvine Scientific, U.S.A.) in a 5% CO_2 humidified incubator. K562 cells and THP-1 cells were cultured in eRDF medium (Kyokuto Pharmaceutical Kogyo Co., Tokyo, Japan) supplemented with 10% FBS. The viable cells were counted with a hemocytemeter, the viability being determined by trypan blue exclusion.

2.3. Determination of Nitroblue Tetrazolium (NBT)-Reducing Activity

NBT reduction activity was measured according to the method of Collins et al.[5]. HL-60 cells (1×10^6 cells/ml) in eRDF medium were incubated for 30 min at $37^{\circ}C$ with an equal volume of 0.2% NBT dissolved in PBS containing $4 \times 10^{-6}M$ of freshly prepared TPA (12-O-tetradecanoylphorbol-13-acetate) and 20% FCS. The percentage of cells with formazan deposits was determined with a hemocytemeter.

2.4. Determination of Non-specific Esterase Activity

The nonspecific esterase activity of the HL-60 cells was determined by a histochemical assay as described by Li et al.[6], using alpha-naphtyl butyrate as a substrate.

2.5. Benzidine Staining

Benzidine-positive K562 cells were detected by staining with o-dianisidin[7].

3. RESULTS

Figure 1 shows the proliferation curves for HL-60 cells cultured with several concentrations of the non-dialyzable extract of freeze-dried spinach. The proliferation of HL-60 cells was found to be inhibited by the extract of 0.25-1.0 mg dry wt/ml of freeze-dried spinach. It was also found that 10-12% of HL-60 cells treated with 0.5 mg dry wt/ml of extract for 4 and 6 days were positive in NBT dye reduction activity, which is a marker for the differentiation of Hl-60 cells (data are not shown). The treatment of HL-60 cells with the extract induced about 16% of non-specific esterase-positive cells, which is a monocyte/macrophage-specific enzyme (Fig. 2). When the extract prepared from fresh spinach (Okame) was used, the proliferation of HL-60 cells was inhibited in a concentration range of 0.1-0.5 mg of protein/ml. The extract also produced HL-60 cells which reduced NBT dye and had non-specific esterase activity.

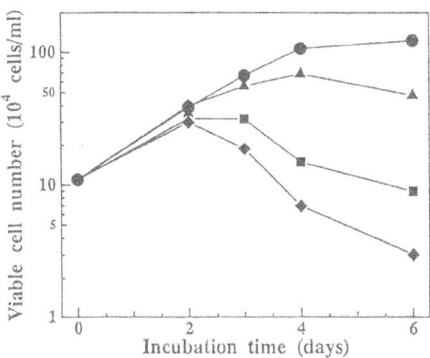

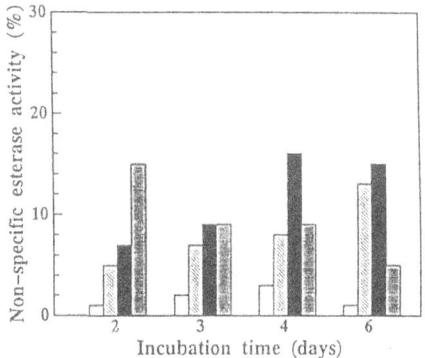

Fig. 1 Effect of the freeze-dried spinach extract on the proliferation of HL-60 cells

The cells were incubated in the absence (—●—) or presence of 250μg/ml (—▲—), 500μg/ml (—■—) and 1mg/ml (—◆—) of the freeze-dried spinach extract.

Fig. 2 Effect of the freeze-dried spinach extract on the non-specific esterase activity of HL-60 cells

The cells were cultured inn the absence (□) or presence of 250μg/ml (▨), 500μg/ml (■) and 1mg/ml (▦) of the freeze-dried spinach extract.

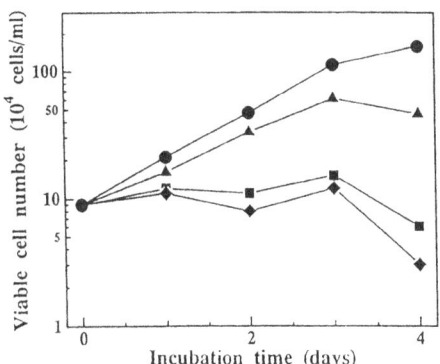

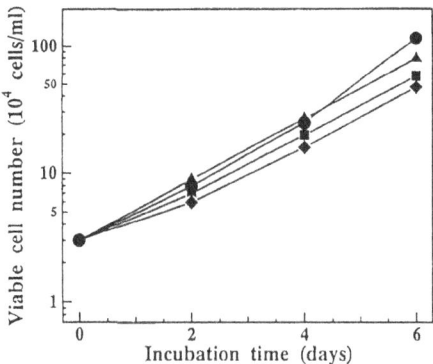

Fig. 3 Effect of the spinach extract on the proliferation of K562 cells

The cells were cultured in the absence (—●—) or presence of 250 μg/ml (—▲—), 500μg/ml (—■—) and 800μg/ml (—◆—) of the freeze-dried spinach extract.

Fig. 4 Effect of the spinach extract on the proliferation of THP-1 cells

The cells were cultured in the absence (—●—) or presence of 2μg/ml (—▲—), 8μg/ml (—■—) and 16μg/ml (—◆—) of the freeze-dried spinach extract.

The proliferation of K562 cells, which are known to be differentiated into erythrocytes, was also inhibited by 0.25-0.8 mg dry wt/ml of the freeze-dried spinach extract, similarly to the case of HL-60 cells (Fig. 3). However, no benzidine-positive cells containing hemoglobin were increased by exposure to the spinach extract exposure. THP-1 cells could not survive in the presence of more than 20 μg/ml of the freeze-dried spinach extract (Fig. 4). When the THP-1 cells were cultured with 2-16 μg/ml of the extract, no adherence of THP-1 cells to the plate was induced, which is an evident marker of the differentiation into macrophages.

3. Discussion

It is known that HL-60 human promyelocytic leukemia cells can be induced to differentiate into granulocytes, monocytes and macrophages by such agents as dimethyl sulphoxide (DMSO)[2] and TPA[8]. Most of the compounds which induce the differentiation of leukemia cells also inhibit the proliferation of the cells. The non-dialyzable extracts of freeze-dried and fresh (Okame) spinach at concentrations of 0.25-1.0 mg dry wt/ml inhibited the proliferation of HL-60 cells and induced NBT-positive cells. The reduction of NBT dye is a marker showing the differentiation of HL-60 cells into granulocytes or monocytes, and it has already been found in HL-60 cells differentiated into granulocytes and monocytes[9].

The extracts also increased the level of non-specific esterase activity, a monocyte/macrophage-specific enzyme. The extracts of freeze-dried and fresh spinach did't induce cell adherence, which is a typical function of HL-60 cells differentiated into macrophages. These results suggest that the spinach extracts induced the monocytic differentiation of HL-60 cells.

K562 cells and THP-1 cells can be induced to differentiate into erythrocytes and macrophages, respectively. However, the freeze-dried spinach extracts did't induce the differentiation of K562 and THP-1 cells, although they inhibited the proliferation of these cells. From these results, it is suggested that the spinach extracts induced the differentiaton of U-937 cells into macrophage cells, and the differentiation of HL-60 cells into monocytes, while they induced no differentiation of other myeloid leukemia cells such as K562 and THP-1.

References

1. Kong, Z-L., Murakami, H., and Shinohara, K. (1991) 'Effect of dialyzate fractions of spinach on growth of human-derived cells', Cytotechnology 7, 113.
2. Collins, S.J., Gallo, R.C., and Gallagher, R.E. (1977) 'Continuous growth and differentiation of human myeloid leukemia cells in suspension culture', Nature 270, 347.
3. Lozzio, C.B., and Lozzio, B.B. (1975) 'Human chronic myeloigenous leukemia cell line with positive Philadelphia chromosome', Blood 45,

321.
4. Tsuchiya, S., Yamabe, M., Yamaguchi, Y., Kobayashi, Y. Konno, T., and Tada, K. (1980) 'Establishment and characterization of a human acute monocytic leukemia cell line (THP-1)', Int J. Cancer 26, 171.
5. Collins, S.J., Ruscetti, F.W., Gallagher, R.E., and Gallo, R.C. (1979) 'Normal functional characteristics of cultured human promyelocytic leukemia cells after induction of differentiation by dimathyl-sulphoxide',J. Exp. Med. 149, 969.
6. Li, C.Y., Lam, K.W., and Yam, K.T., (1973) 'Esterase in human leukocytes', J. Histchem Cytochem. 21, 1.
7. Eto. Y., Tsuji, T., Takezawa, M., Takano, S., Yokogawa, Y., and Shibai, H. (1987) 'Purification and characterization of erythroid differentiation factor (EDF) isolated from human leukemia cell line THP-1', Biochem. Biophys. Res. Commun. 142, 1095.
8. Lotem, J., and Sachs, L. (1979) 'Regulation of normal differentiation in mouse and human myeloid leukemic cells by phorbol esters and the mechanism of tumor promotion', Proc.Natl.Acad.Sci.U.S.A. 76, 5158.
9. Collins, S. J. (1987) 'The HL-60 promyelocytic leukemia cell line: Proliferation, differentiation, and cellular oncogene expression', Blood 70, 1233.

SUSTAINED INDUCIBILITY OF CYTOCHROME P450 ACTIVITY IN RAT HEPATOCYTES CULTURED IN A SERUM-FREE MEDIUM

JULIET O. LOBO, ROXANNE L. SAMROCK, DAVID W. JAYME and
PAUL J. PRICE,
GIBCO BRL/Life Technologies, Inc.,
2086 Grand Island Blvd.
Grand Island, New York 14072 USA

ABSTRACT. Liver microsomal oxygenases are multicomponent enzyme systems which metabolize a wide variety of xenobiotics. A major component of the system is a group of enzymes collectively known as cytochrome P450 (CP450). A major limitation in the use of rodent hepatocyte cultures in toxicity testing and pharmacokinetic studies has been the rapid loss of phase 1 reactions catalyzed by the CP450-dependent mono-oxygenases. Using a sandwich matrix and a serum-free medium developed by GIBCO, total rat CP450 could be maintained for at least 9 days at 75-80% day "0" levels. Metabolic studies of the microsomal fraction of primary adult rat hepatocytes, measured by the conversion of 7-ethoxycoumarin to 7-hydroxycoumarin and of 3,4-benzo-[a]-pyrene to 3-hydroxybenzo-[a]-pyrene, demonstrated maintenance of activity over the same 9 days comparable to the "0" time controls.

1. Introduction

The ability to maintain cells in their natural phenotypic state *ex vivo* has immediate applications in both clinical and pharmacological laboratories. Cultures of functional hepatocytes on the proper matrix may allow for the development of an extracorporeal liver support system for the treatment of acute liver failure and, possibly, for the correction of numerous inborn errors of metabolism of liver origin. In the pharmaceutical laboratory, hepatocytes are playing a major role in understanding pharmacokinetics.

An important function of the liver; that is, the detoxification and metabolism of xenobiotics, is accomplished by a group of hemoproteins with differing substrate specificities (Guengerich, 1979). These enzymes are collectively known as the cytochrome P450 (CP450) system. A characteristic property of this enzyme system is its inducibility by many chemicals or lipophilic substances associated with enhanced microsomal oxidation and reduction. These inducers can be divided into three broad classes typified by drugs like phenobarbital, polycyclic hydrocarbons like benzo(a)pyrene and steroids such as testosterone.

A major limitation on the use of rodent hepatocytes in culture for pharmacokinetics and hepatotoxicity studies in traditional serum-supplemented culture has been the rapid loss (about 70% by day 2) (Bissell et al 1973, Guzelian et al 1977, Michalopoulos et al 1976) of CP450 catalyzed reactions. A number of attempts have been made to maintain activity of some CP450 isozymes and to improve expression of liver function. Choice of attachment matrix (Rojkind et

195

S. Kaminogawa et al. (eds.), Animal Cell Technology: Basic & Applied Aspects, Vol. 5, 195–201.
© 1993 *Kluwer Academic Publishers.*

al 1980), co-culture with other cell types (Begue et al 1984, Guigen-Guillouzo et al 1983) and modification of culture conditions (Edwards et al 1985, Paine et al 1979, Warren and Fry 1988) have been used with some success.

This study uses a serum-free growth medium for the culture of adult primary rat hepatocytes to investigate: a) the effect of different attachment matrices on morphology and growth of cells, and b) the effects of culture media on the long term maintenance and expression of selected CP450 isozymes. In monolayer cultures grown using either a collagen-1 sandwich or an EHS-collagen sandwich matrix, total rat CP450 could be maintained for 9 days at 75-80% day "0" levels. Cytochrome P450 isozymes (particularly P450c) in freshly isolated hepatocytes and in cultures attached in the presence and absence of serum were also analyzed using Western blots. Results correlated well with the metabolic activity measured by the conversion of 7-ethoxycoumarin to 7-hydroxycoumarin and benzo[a]pyrene to 3-hydroxybenzo[a]pyrene. These isoforms were not only maintained with time in culture, but demonstrated a substantial increase in activity over the same 9 days as compared to the "0" time controls.

2.   Materials and Methods

2.1.   PREPARATION OF ISOLATED HEPATOCYTES

Hepatocytes were isolated from adult male Sprague-Dawley rats by perfusing the liver *in situ* with a Ca-free buffer (Liver Perfusion Medium, GIBCO) followed by a Collagenase-Dispase L-15 digest medium (Liver Digest Medium, GIBCO). The liver was aseptically removed to a beaker containing cold transport media (Hepes buffered L-15 with BSA, Liver Transport Medium, GIBCO) and transferred to the cell culture laboratory. The hepatocytes were released by mincing and pipetting with a large bore pipette. The cell suspension was filtered through a 100 μm nylon mesh into a beaker placed on ice, sedimented by centrifugation at 50g for 5 minutes, resuspended and washed 2-3 times in cold wash medium (Williams Med E with BSA, Hepatocyte Wash Medium, GIBCO). Hepatocytes were purified by Percoll (Sigma, St.Louis, MO) density gradient separation and washed twice more before being resuspended in the attachment medium. This procedure gave a cell yield of about 2.5 x 10^8 cells with about 90-95% viability, as determined by trypan blue exclusion.

About 10 x 10^6 cells were plated in 150-cm^2 tissue culture flasks precoated with a Collagen 1 (Price 1975) matrix (12.5μg/cm^2) or flasks precoated with the EHS matrix (100 μg/cm^2 Matrigel, Collaborative Research Inc., Bedford, MA) and incubated in a humidified atmosphere of 5% CO$_2$ in air at 37°C. The initial attachment medium consisted of a hormone-supplemented modification of Williams Medium E (Hepatocyte Attachment Medium, GIBCO) containing 1.25μg/cm^2 and either 5% fetal bovine serum or 5μg/cm^2 fibronectin. Unattached cells were poured off 2 hours later and replaced with hepatocyte serum-free medium (Hepatocyte-SFM, GIBCO), a highly modified Chees medium. Cultures were refed with hepatocyte SFM at 24 hours and every 48 hours thereafter. Hexobarbital (0.5 mM) was added to cultures 48 hours before harvesting the cells.

2.2.   PREPARATION OF MICROSOMES

The medium was aspirated and the attached hepatocytes were rinsed with cold PBS. To each flask, 5 ml of ice-cold 100 mM KPO_4 buffer (pH 7.4) containing 1mM dithiothreitol was added, and the attached hepatocytes were scraped into the buffer. The cells from 1-2 flasks were pelleted by centrifugation at 1000 x g for 5 minutes. The pellet was resuspended in 5 ml of fresh buffer and cells disrupted with a sonicator (Heat Systems, Model W-225R). The nuclei and mitochondria were separated from the microsomes by centrifugation at 30,000 x g for 10 minutes. The microsomal pellet was obtained by centrifugation of the 30,000 x g supernatant fraction at 105,000 x g for 30 minutes. The microsomal pellet was either stored frozen at -70°C or was analyzed after being homogenized in a 40mM K PO_4 buffer (pH 7.4) containing protease inhibitors.

2.3.   ANALYTICAL PROCEDURES

Protein concentrations of the microsomal homogenates were determined by the commercially available Pierce BCA Protein Assay Reagent (Pierce Chemical Co., Rockford, IL) using BSA as a standard. Cytochrome P-450 concentrations were determined from the dithionite reduced CO-difference spectrum, using an extinction coefficient of $91mM^{-1} cm^{-1}$ (Omura and Sato, 1964). The spectra were recorded on a Shimadzu dual beam spectrophotometer (Shimadzu Scientific Instruments, Columbia, MD) equipped with baseline correction and peak wavelength indicator.

2.4.   METABOLIC STUDIES

Catalytic activity (deethylase and hydroxylase) of the cytochrome P450 isozymes was measured using two substrates. The 7-Ethoxycoumarin O-deethylase (ECOD) activity assay was based on procedures of Fry and Bridges (1980) and Greenlee and Poland (1978). The reaction mixture, in a total volume of 1.0 ml, contained 10mM Sodium Phosphate Buffer pH 7.4 with 1 mmol EDTA, 400 μmol NADPH, 500 μmol of 7-ethoxycoumarin (Aldrich; substrate freshly prepared in methanol) and 200 μg protein. The reaction was started by the addition of the enzyme and the mixture was incubated for 20 min at 37°C in air with vigorous shaking. The reaction was stopped by the addition of 1N HCl (150 μl). The fluorescent product 7-hydroxycoumarin was extracted with 2 ml chloroform by mixing for 20 seconds and 1 ml of the organic chloroform-phase was mixed with 2.5 ml 0.01M NaOH/1M NaCl for 20 seconds on a vortex mixer. Aqueous phase fluorescence was estimated using a Shimadzu spectrofluorometer with excitation and emission wavelengths of 370 and 455 nm, respectively.

The fluorometric assay for aryl hydroxylase (B-[a]-P) activity used 3,4-benzo[a]pyrene as substrate (Nebert and Gelboin, 1968). The reaction mixture, in a total volume of 1 ml, contained 100 μM Potassium phosphate buffer, pH 7.6, an NADPH-regenerating system containing 0.5 μmol NADP, 20 μM of glucose 6-phosphate, 3 μmol $MgCl_2$, 0.1 μmol EDTA, 1 Kornberg unit of glucose-6-phosphate dehydrogenase and 200 μg of protein. The reaction was initiated by addition of substrate (100 μmol made in acetone) in 50 μl aliquots and the mixture shaken at 37°C for 30 minutes, in air. The reaction was stopped by the addition of 1.0 ml cold acetone. Enzyme activities were determined in duplicate and compared to a blank to which acetone had been added prior to incubation. 3.5 ml of hexane was then added and the mixture incubated with shaking at 37°C for 10 minutes. A 1.0 ml aliquot of the organic phase was extracted with 3.0 ml of 1N NaOH. Extracted, hydroxylated benzo[a]pyrene in the alkali phase was determined spectrofluorometrically with excitation at 396 mμ and emission at 522 mμ. The samples were quantitated using a standard solution of 3-hydroxybenzo[a]pyrene made in NaOH.

2.5. IMMUNOLOGICAL STUDIES

Antibodies to three different isozymes of cytochrome P450 and to cytochrome P450 reductase were obtained from Oxygene, Dallas, TX. SDS-PAGE, performed according to the method of Laemmli (1970) were run overnight on a 10% reduced gel. Proteins were transferred to nitrocellulose sheets (Towbin et al 1979) using the Multiphor II Nova Blot (Pharmacia, Uppsala, Sweden). The non-specific binding sites on the nitrocellulose sheets were blocked for 1 hour at room temperature with 2% milk made in PBS containing 0.05% Tween 20. The sheets were incubated at 4°C overnight in the appropriate antibody. The blots were washed with PBS containing 0.05% Tween 20 and incubated for 1 hour at room temperature with a 1:100 dilution of horseradish-peroxidase-conjugated IgG. After several washes in PBS, the antibodies were detected by staining with a freshly prepared solution 20 mM 4 chloro-1-napthol, 0.03% H_2O_2 made in PBS containing 0.05% Tween 20 and 6% methanol. Staining was terminated by rinsing the sheets in water. The visualized sheets were dried and photographed over a fluorescent light box onto Polaroid high speed film.

3. Results

3.1. MORPHOLOGY AND CELL SURVIVAL

Using the cell isolation protocol as described, the hepatocytes attached to the collagen-coated or EHS-coated plates within 1-2 hours in attachment medium supplemented with 5% fetal bovine serum. Cell attachment improved by 20-40% (Table 1), as evidenced by the protein content per plate if the plates were precoated with EHS rather than collagen. After 2 hours, the culture medium containing unattached or dead cells was removed and replaced with Hepatocyte SFM containing collagen to form a sandwich matrix. Phase contrast microscopy of hepatocytes cultured for 24 hours on a collagen matrix revealed a number of flattened binucleated cells. A small number of those attached (nearly 5%) remained rounded and were presumed to be non-viable. Hepatocytes cultured on an EHS matrix revealed cells clumped together to form a mesh.

Table 1. Hepatocyte attachment efficiency using different media and substrata.

Medium	Matrix	Protein[*] (μg/flask)[**]
WE-FBS	EHS	420
H-SFM	EHS	1290
WE-FBS	collagen	210
H-SFM	collagen	970

[*] Value not corrected for matrix alone
[**] Average from multiple flasks

Table 2. CP450 activity retention in serum-free culture.

Day	pmol CP450/mg protein
0	245
3	225
8	200

3.2. CYTOCHROME P450 CONTENT AND ASSOCIATED MONO-OXYGENASE ACTIVITIES OF CULTURED HEPATOCYTES

Freshly isolated cells prepared from a single liver were cultured for 24 hours in Hepatocyte SFM to allow formation of a stable monolayer, and were then incubated for 120 hours prior to harvesting, in medium with or without Hexobarbital (HB). Treatment of cultures with HB resulted in an increase in protein content and in an increase in microsomal concentration of CP450 of 1.5-2-fold. Thus, we determined that treatment of cells with 2 mM of hexobarbital for at least 48 hours was a prerequisite for cytochrome P450 induction.

The level of CP450 in the microsomal fraction of fresh hepatocytes varied between experiments, with a range of 175.82-307.69 pmol/mg protein. Table 2 shows a minimal loss in total CP450 (pmol/mg protein) of 8.5% on day 3 and of less than 20% after 8 days in culture.

Cells cultured in Hepatocyte SFM showed an initial drop in both deethylase and hydroxylase activities (data not shown), followed by a restoration in 7-Ethoxycoumarin O-deethylation and benzyl-hydroxylation over time in culture. The increased 6 day level was maintained at 9 days (data not included). In the absence of hexobarbital both the deethylase and hydroxylase activities at 9 days were found to be lower than the 0h control. Hexobarbital added at 48h prior to harvesting the cells induced both activities substantially (3-4-fold), irrespective of the type of matrix the cells were plated on. In both instances the levels of activity of the enzymes was higher with the EHS matrix. The presence of serum in the medium inhibited the rise in activity of both enzymes, despite addition of hexobarbital 48h prior to harvest.

Table 3. Effect of culture medium and substratum on hexobarbital-induced CP450 enzyme activities.

Day	Condition	Hex Addition	ECOD nmol/min/mg protein	B(a)P nmol/min/mg protein
0	Primary Suspension	0	7.0	4.2
9	WE-FBS / Matrigel	+	3.2	4.6
	WE-FBS / Collagen	+	3.8	3.5
	H-SFM / Matrigel	-	4.7	2.3
	H-SFM / Matrigel	+	15.3	9.2
	H-SFM / Collagen	-	3.5	2.5
	H-SFM / Collagen	+	11.1	7.2

3.3. WESTERN BLOT ANALYSIS

Of the three CP450 apoenzymes (1A1, IIB1 and IIIA1 and reductase), 1A1 and IIIA1 were readily detected by Western blots using polyclonal rabbit antibody. We did not see expression of IIB1 under any of the conditions tested. Figure 2 shows a Western blots of microsomal enzymes isolated from hepatocytes cultured for 9 days in serum-free medium. While the expression of reductase was constant under most culture conditions tested, the hexobarbital-induced expression of 1A1 and IIIA1 were dependent on the medium and substratum composition.

Figure 1: Western blot analysis of P450 IIIA1, P450 IA1 and reductase from hepatocytes cultured for 9 days.

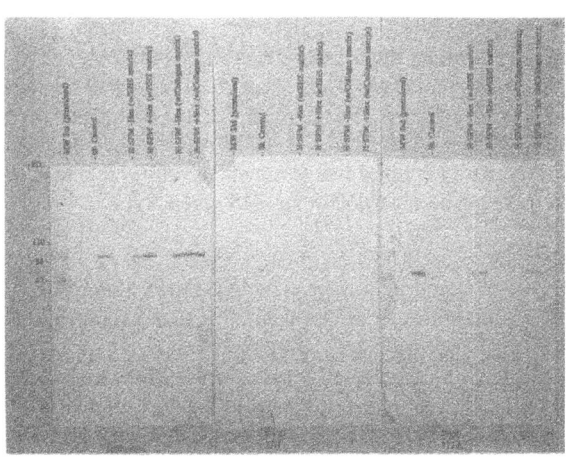

Cells were grown on an EHS or collagen matrix in Hepatocyte serum-free media. 50ug of sample was loaded per lane on 10% reduced SDS gels, run overnight at 10ma. The gels were blotted onto nitrocellulose membranes. The blots were stained for CP450 or for reductase (Oxygene, Dallas) following the recommendations of the supplier. Prestained molecular weight markers were run simultaneously. The blot was developed using 4-chloro-napthol as the peroxidase substrate.

4.   Discussion

Primary monolayers of adult rat hepatocytes may be cultured under serum-free conditions and retain CP450 inducibility at near normal levels for 8-10 days. We have also demonstrated increased attachment of the cells in a serum-free environment. While there was a large variation in the CP450 content and mono-oxygenase activities for isolated hepatocyte cultures prepared from different rats, we show almost complete retention of CP450 activity during the first 72 hours with 75% of the activity still retained by day 8. A higher level of CP450 inducibility was maintained when collagen rather than the extracellular matrix EHS as substratum. The choice of collagen was also important: cell growth was poorer in collagen IV and V than collagen I.

We examined the metabolic activity of cultured hepatocytes using two substrates commonly used to measure xenobiotic metabolic activity. ECOD and B-[a]-P activities (linked to several P450s) increased in our serum-free media and remained elevated for at least 8 days.

Long-term maintenance of functional adult hepatocytes *in vitro* without loss of enzymatic activity should be a valuable tool for studies on metabolic processing of drugs and carcinogens *in vitro*, and may allow for the development of an extracorporeal liver support system for the treatment of acute liver failure. As human hepatocytes exhibit more stable metabolizing system than adult rat cultures, future studies with human hepatocytes utilizing this sandwich matrix and Hepatocyte Serum-Free Medium may show even greater retention of activity.

5. References

Begue, J.M., Guguen-Guillouzo, C., Pasdeloup, N., Guillouzo, A.(1984). Prolonged maintenance of active cytochrome P-450 in adult rat hepatocytes co-cultured with another liver cell type. *Hepatology* 4:839-842.

Bissell, D.M., Hammaker, L.E., and Meyer, U.A. (1973). Parenchymal cells from adult liver in nonproliferating monolayer culture. *J Cell Biol.* 59:722-734.

Edwards, A.M., Glistak, M.L., Lucas, C.M., and Wilson, P.A. (1985). 7-Ethoxycoumarin Deethylase activity as a convenient measure of liver drug metabolizing enzymes: Regulation in cultured rat hepatocytes. *Biochem. Pharmacol.* 33:1537-1546.

Fry, J.R., and Bridges, J.W. (1980). The metabolism of 7-Ethoxycoumarin in primary maintenance cultures of adult rat hepatocytes. *Naunyn-Schmiedeberg's Arch Pharmacol.* 311:85-90.

Greenlee, W.F., and Poland, A. (1978). An improved assay of 7-Ethoxycoumarin O-Deethylase activity: Induction of hepatic enzyme activity in C57BL//6J and DBA/2J mice by Phenobarbital, 3-Methylcholanthrene and 2,3,7,8-Tetrachlorodibenzo-p-Dioxin. *J Pharmacol. Exptl. Therap.* 205:596-605.

Guengerich, F.P. (1979). Isolation and purification of cytochrome P-450, and the existence of multiple forms. *Pharmacol. Therap.* 6:99-121.

Guguen-Guillouzo, C., Clement, B., Baffet, G., Beaumont, C., Morel-Chany, E., Glaise D., Guillouzo, A. (1983). Maintenance and reversibility of active Albumin secretion by adult rat hepatocytes co-cultured with another liver epithelial cell type. *Exp. Cell. Res.* 143:47-54.

Guzelian, P.S., Bissell, D.M., Meyer, U.A. (1977) Drug metabolism in adult rat hepatocytes in primary monolayer culture. *Gastroentrology* 72:1232-1236.

Laemmli, U.K. (1970). Cleavage of structural protein during the assembly of the head of bacteriophage T4. *Nature* 227:680-685.

Michalopoulos, G., Sattler, C.A., Sattler, G.L., and Pitot, H.C. (1976) Cytochrome P-450 induction by phenobarbital and methylcholanthrene in primary cultures of hepatocytes. *Science* 193:907-909.

Nebert, D.W., and Gelboin, H.V. (1968). Substrate-inducible microsomal Aryl Hydroxylase in mammalian cell culture. *J Biol. Chem.* 243:6242-6249.

Omura, T., and Sato, R. (1964). The Carbon-monoxide binding pigment of liver microsomes. I. Evidence for its hemoprotein nature. *J Biol. Chem.* 239:2370-2378.

Paine, A.J., Williams, L.J., and Legg, R.F.(1979). Apparent maintenance of cytochrome P-450 by Nicotinamide in primary cultures of rat hepatocytes. *Life Sciences* 24:2185-2192.

Price, P.J.(1975). Preparation And use of rat-tail Collagen. *TCA Manual* 1:43-44.

Rojkind, M., Gatmaitan, Z., Mackensen, S., Giambrone, M., Ponce, P., and Reid, L.M. (1980). Connective Tissue Biomatrix: Its isolation and utilization for long-term cultures of normal rat hepatocytes. *J Cell Biol.* 87:255-263.

Towbin, H., Staehlin, T., and Gordon, J. (1979). Electrophoretic Transfer of Proteins From Polyacrylamide Gels To Nitrocellulose Sheets: Procedure and Some Applications. *Proc Natl Acad Sci.* 76:4350-4354.

Warren, M., and Fry, J.R.(1988). Influence of Medium Composition on 7-Alkoxycoumarin O-Dealkylase Activities of Rat Hepatocytes in Primary Maintenance Culture. *Xenobiotica* 18:973-981.

BIOSYNTHESIS OF PROTEIN PRODUCTS BY ANIMAL CELLS - ARE GROWTH AND NON-GROWTH ASSOCIATED CONCEPTS VALID OR USEFUL?

J.P. BARFORD[1], P.J. PHILLIPS[1], C.P. MARQUIS[1] and C. HARBOUR[2]
[1] Departments of Chemical Engineering and [2] Infectious Diseases
University of Sydney
N.S.W. 2006
Australia

ABSTRACT. The application of simple relationships describing substrate uptake and product formation derived from microbial systems, to the growth of hybridoma cells is considered unlikely to assist in the understanding of antibody formation and hence maximising antibody yield. Inadequacies in the current approach to the study of the kinetics of growth of hybridoma cells and antibody production are described and suggestions made as to which approach may be employed to assist in our further understanding of the process.

Introduction

In the study of the kinetics of growth and product formation in microbial systems, there has been the attempt to define simple mathematical relationships which describe the relationship between variables such as specific growth rate, specific substrate uptake rate and specific product formation rate. The simplest of these are the relationships:

$$Q_s = \alpha \mu + \beta$$

and $\quad Q_p = \gamma Q_s$

where $\quad Q_s$ = specific substrate uptake rate (mmol substrate/g biomass/h)

$\quad Q_p$ = specific product formation rate (mmol product/g biomass/h)

and $\quad \alpha, \beta$ and γ are constants.

This relationship (for continuous culture) is shown in Figure 1. Inherent in such a relationship are a number of assumptions, not all of which are widely accepted. These include:

(i) β is constant and is generally referred to as the maintenance requirement. This relationship assumes that maintenance requirements are met by a fixed amount of the incoming substrate, independent of the growth rate. This implies a fixed amount of energy derived from the substrate and, consequently, that the catabolic pathway or pathways and their stoichiometry, with respect to ATP formation (i.e. energetics), remains fixed.

S. Kaminogawa et al. (eds.), Animal Cell Technology: Basic & Applied Aspects, Vol. 5, 203–213.
© 1993 Kluwer Academic Publishers.

(ii) There is a direct relationship between the amount of substrate consumed and the amount of product produced, and that this is a fixed ratio. This means that there is one pathway of fixed stoichiometry relating substrate uptake to product formation. This would normally be associated with either:

(a) A single catabolic (energy producing) pathway resulting in the production of an end product of metabolism from the substrate. (The possibility of multiple catabolic pathways with a fixed ratio between the pathways under all conditions is also possible, but is not normally considered as a realistic metabolic condition).

or

(b) An anabolic (biosynthetic) product in which the production of the product is directly proportional to the amount of substrate apportioned to the anabolic section of the cell. In such cases, there is also the assumption that the proportion of the substrate apportioned to the catabolic and anabolic sections of the cell remains fixed. These concepts are illustrated in Figure 2(a) and 2(b).

(iii) The biomass is homogeneous. When looking at microbial systems, it is of importance to appreciate the extent of any applicability of this simple description. Three examples may be used to demonstrate this:

      (a) Ethanol formation (by high density yeast systems)
      (b) Aerobic Yeast Metabolism (considering ethanol as the end product)
and   (c) Amino Acid formation.

When looking at high density yeast systems for producing alcohol, the aim is to achieve a situation where the biomass density is so high that the growth rate is negligible and hence most substrate is transformed into the product. Under this condition, $Q_s \approx \beta$ and $Q_p = \gamma \ Q_s$ and the model is adequate. Essentially the product is the end product of catabolism only and, as such, has a direct (stoichiometric) relationship with the amount of substrate taken up and also is the only energy source for cell growth.

When aerobic yeast metabolism is considered, a number of limitations of the simple model are illustrated. Figure 3 shows (schematically) the biomass, substrate and product profiles associated with the aerobic growth of *S.cerevisiae*. Firstly, there is more than one catabolic pathway. The two catabolic pathways (respiration and fermentation) do not have a fixed ratio and hence fixed stoichiometry, it varies with the growth rate. This results in a non-linear relationship between growth rate and specific substrate uptake and specific product formation rate. As a result, any mathematical description of such a growth profile must include details of more than one catabolic pathway. Such descriptions have been reported elsewhere (1,2).

The control of the relative extent of the two catabolic pathways becomes a unique feature of the metabolism of *S.cerevisiae* and cannot be represented by a simple growth and non-growth associated description. **To understand this control, it is necessary to confront the issue of the origin of anabolic (biosynthetic) intermediates. By this, it is meant that the origin of biosynthetic intermediates and their relationship to the energy producing pathways must be** examined. This requires a much more sophisticated analysis than afforded by the simple growth and non-growth associated description.

Figure 4 illustrates how such an analysis was included in a model description of the cell containing more than one energy producing (catabolic) pathway, viz the aerobic growth of *S.cerevisiae*. When the production of amino acids by microorganisms is considered, any useful description must now confront the issue of a product which involves both the use of catabolic and anabolic pathways to achieve its manufacture (the catabolic pathways provide both energy and building blocks for the biosynthesis of the desired product by the anabolic pathway). In our previous descriptions (Figures 2(a) and 2(b)), we gave a simplistic description where substrate was either used for energy production (catabolic) or biosynthetic (anabolic) use and that the desired product was either catabolic in origin or anabolic in origin. Whilst this makes the growth and non-growth associated description broadly applicable to biosynthetic products, it does so in a manner which is practically too simplistic for the most basic of biosynthetic products. In reality, all biosynthetic products will require some utilisation of the energy producing pathways either for energy or building blocks or, more commonly, both.

Amino acids may be considered a simple biosynthetic product. However, when one examines the specific amino acid production rate and the amino acid yield as a function of the specific growth rate in continuous culture, a very complex interaction is seen to exist (Figures 5(a) and 5(b)).

It is evident from this complex profile that the control of the production of the biosynthetic product (i.e. the interrelationship between the catabolic and anabolic sections of the cell) must be considered.

From these examples, it can be seen that the relationships $Q_s = \alpha \mu + \beta$ and $Q_p = \gamma Q_s$ only <u>partially</u> describe microbial systems. In fact, there are significant "typical microbial growth processes" which are inadequately described by such a relationship.

Results and Discussion

From the preceding discussion concerning microbial systems, it is clear that the nature of the product (whether it is catabolic or anabolic in origin) and the nature of the cell (whether it has one or more catabolic pathways and whether the energetics of such pathways are fixed) have a significant effect on whether simplistic descriptions of substrate uptake and product formation are valid.

Figure 6 illustrates a typical hybridoma growth curve in continuous culture (schematic). It is clear from this figure that hybridoma growth resembles, in many major features, the aerobic growth of *S.Cerevisiae* (Figure 3). As such, from previous discussions, it does not easily fit into a simplistic growth and non-growth associated product formation model.

Similarly, when one examines typical antibody concentration profiles and specific antibody production profiles in continuous culture (Figures 7(a) and 7(b))(3-6), it becomes apparent that such profiles are similar to the amino acid profiles of Figures 5(a) and 5(b) and that they cannot be explained by simple growth and non-growth associated models.

Hybridoma cells may also exhibit growth features which further reduce the applicability of simple models. While this and other aspects of this work are described in more detail elsewhere, three aspects are considered briefly:

(a) Biomass in simple models is considered homogeneous. In our work and that of others, there is the possibility that individual cells do not necessarily produce antibody at equal rates. This may be from cell cycle considerations or from the "selection" of high rate secretors during cultivation.

(b) The use of substrates and nutrients may be associated with "pooling" after uptake. If this is shown to be of significance, then the assumption in simple models that all of the substrate is converted to growth or product would need to be modified (at least, with time delays), this would lead to a much more complex model.

(c) Futile cycles and maintenance of ionic gradients, the extent of which are affected by medium composition and the physiological state of the cell. These directly affect β and whether $\beta = f(\mu)$.

In previous work (8,9), we have described a detailed metabolic structure and computer simulation for monoclonal antibody production by hybridomas. In this structure, the following are described in detail:

(i)   Major Catabolic Pathways
(ii)   Major Anabolic Pathways including origin of Biosynthetic intermediates
(iii) ATP Provision and Utilisation
(iv)   NAD(H)/NADP(H) Provision and Utilisation
(v)   Biosynthetic Intermediate(s) Provision and Utilisation
(vi)   Interaction between catabolic and anabolic pathways

We have previously described detailed simulations for yeast (1,2) and have recently completed a simulation of amino acid production (7). These simulations are capable of reproducing the complex experimental profiles of Figures 3, 5(a), 5(b), 6, 7(a) and 7(b) and, hence, identifying the carbon and energy flow throughout the cell. In this way, we are able to both understand the regulation of antibody synthesis and begin to optimise its production. They replace simplistic growth and non-growth associated concepts for the analysis of biosynthetic products.

Conclusions

Simple kinetic expressions relating substrate uptake and product formation rates taken from microbial systems are unlikely to provide a rational basis for understanding antibody production and hence maximising antibody yield. A more fundamental approach is preferred. This will involve a detailed investigation of the interaction between catabolism and anabolism. Computer simulation will greatly assist this process.

References

1.   Barford, J.P. (1991) 'A General Model for Aerobic Yeast Growth: Batch Growth', Biotechnol.Bioeng., 35, 907-920.
2.   Barford, J.P. (1991) 'A General Model for Aerobic Yeast Growth: Continuous Culture',

Biotechnol.Bioeng., 35, 921-927.

3. Harbour, C., Marquis, C.P., Barford, J.P and Walker, K.Z. (1989) 'Stability of Antibody Secretion by Hybridomas' in R.E. Spier, J.B. Griffiths, J. Stephenne and P.J. Crooy, (eds), Advances in Animal Cell Biology and Technology for Bioprocesses, pp 192-194.

4. Low, K.S., Harbour, C., Barford, J.P., DeZwart, R. and Marquis, C.P. (1987) 'Hybridoma Cell Growth and AntibodyProduction Kinetics', Aust.J.Biotech.,1, No.2, 59-64.

5. Miller, W.M., Blanch, H.W.and Wilke, C.R. (1988) 'A Kinetic Analysis of Hybridoma Growth and Metabolism in Batch and Continuous Suspension Culture: Effect of Nutrient Concentration, Dilution Rate and pH', Biotechnol. Bioeng. 32, 947-965.

6. Ray, N.G., Karkare, S.B.and Stadler, P.J. (1989) 'Cultivation of Hybridoma Cells in Continuous Culture: Kinetics of Growth and Product Formation', Biotechnol.Bioeng. 33, 724-730.

7. Barford, J.P., Phillips, P.J., Marquis, C.P. and Harbour, C. (1993) 'Biosynthesis of Protein Products by Animal Cells - Are Growth and Non-Growth Associated Concepts Valid or Useful?', Biotechnol.Bioeng. (submitted for publication).

8. Barford, J.P., Phillips, P.J. and Harbour, C. (1993) 'Simulation of Animal Cell Metabolism', Cytotechnology (in press).

9. Barford, J.P., Phillips, P.J. and Harbour, C. (1992) 'Enhancement of Productivity By Yield Improvement Using Simulation Techniques' in H. Murukami, S. Shirahata and H. Tachibana (eds), Animal Cell technology: Basic and Applied Aspects, Vol4, Kluwer Academic Publishers, Dordrecht, pp. 397-403.

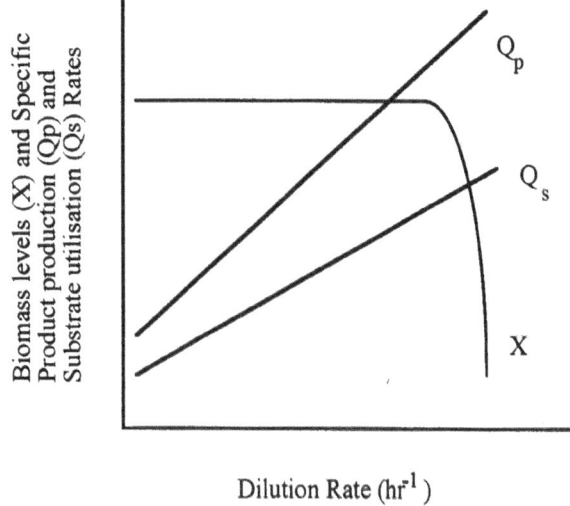

Dilution Rate (hr⁻¹)

Figure 1 Schematic diagram for growth and non-growth associated metabolism in continuous culture

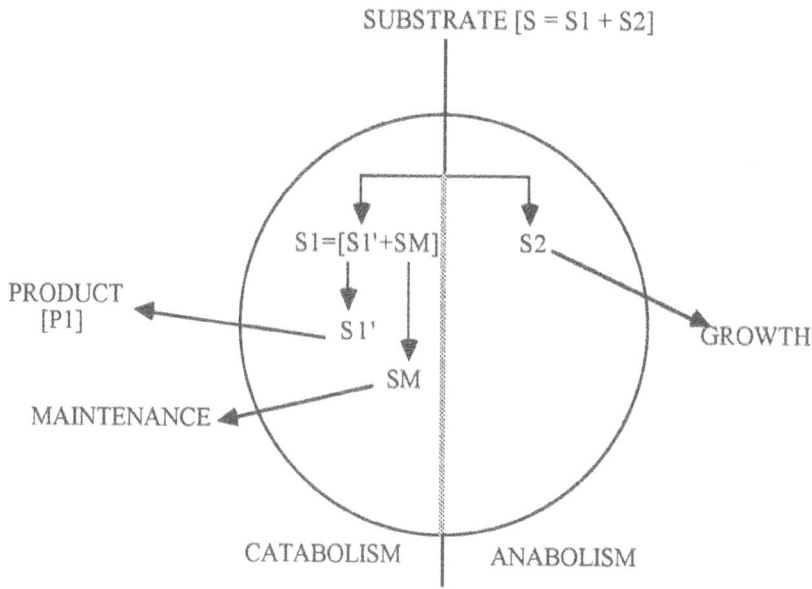

Figure 2(a) Catabolic product formation with growth and non-growth associated metabolism including the maintenance energy concept

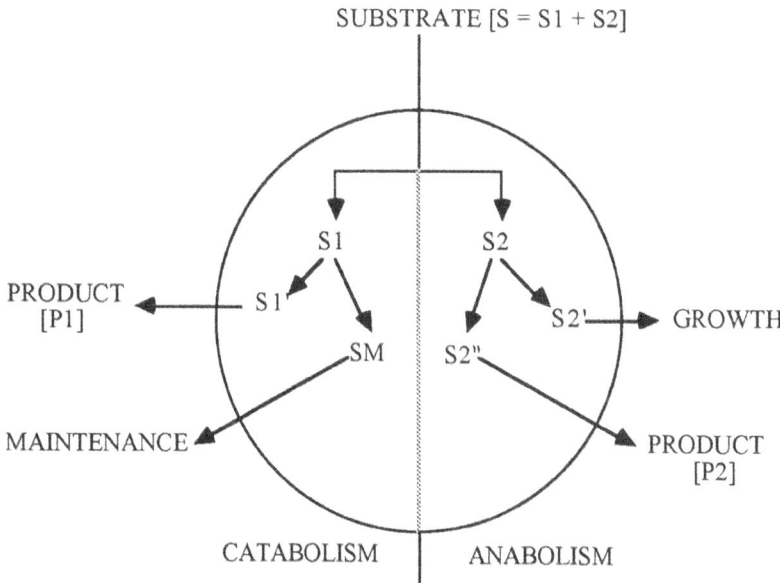

Figure 2(b) Anabolic product formation with growth and non-growth associated metabolism including the maintenance energy concept

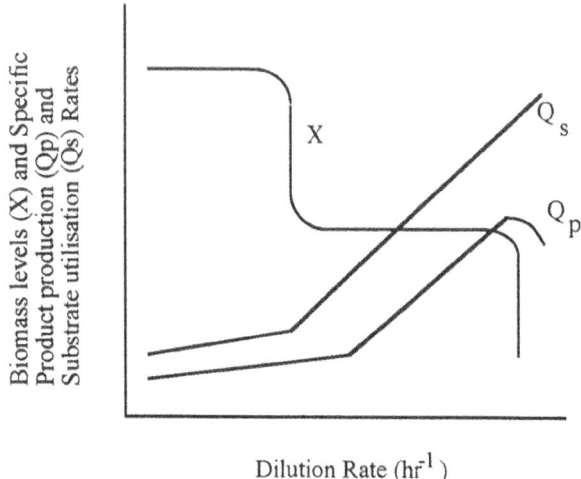

Figure 3 Aerobic growth of S.cerevisiae (schematic)

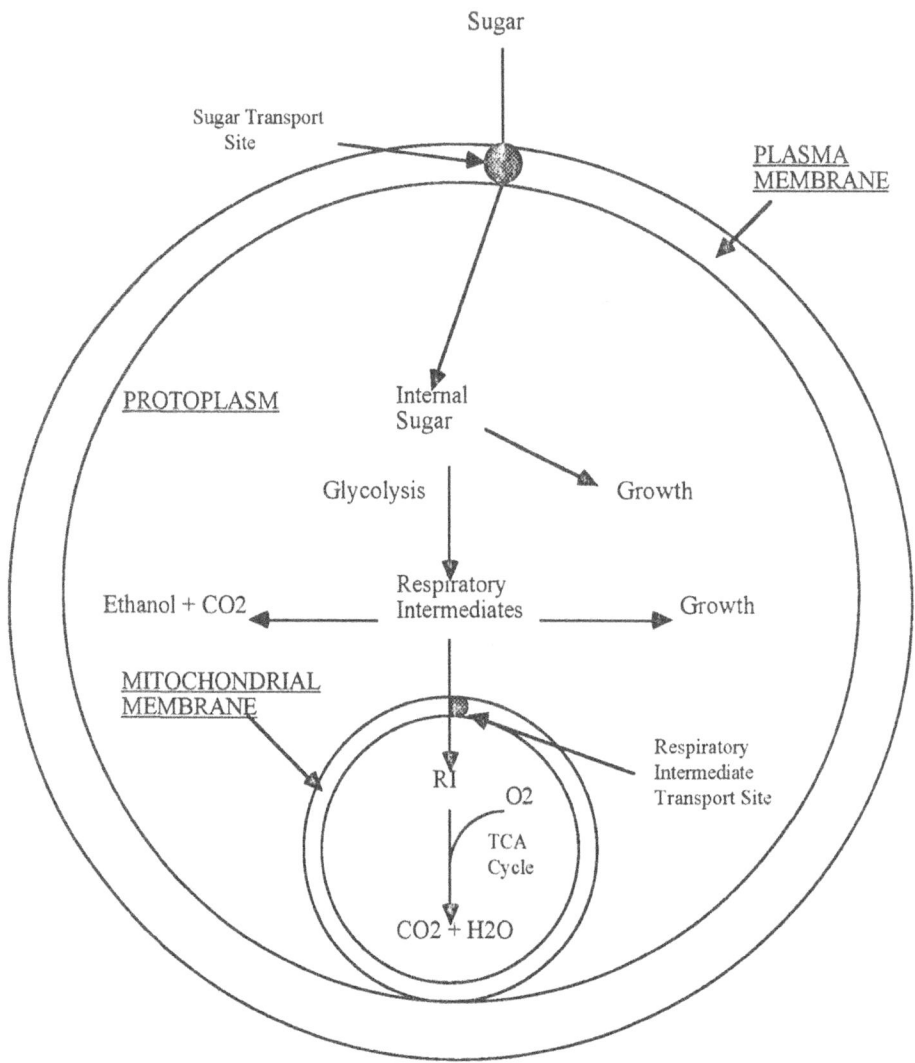

Figure 4 Structure of catabolic and anabolic pathways and their interaction in yeast.

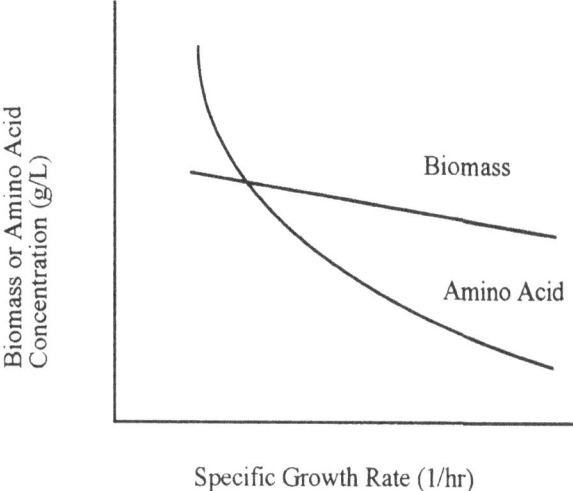

Figure 5(a) Amino acid concentration profile in continuous culture (schematic).

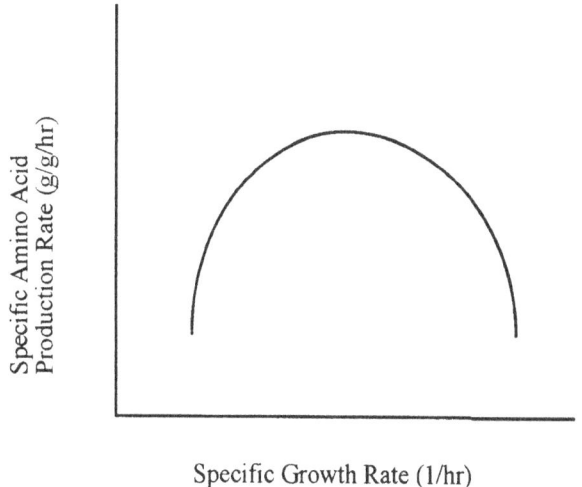

Figure 5(b) Specific amino acid production profile in continuous culture

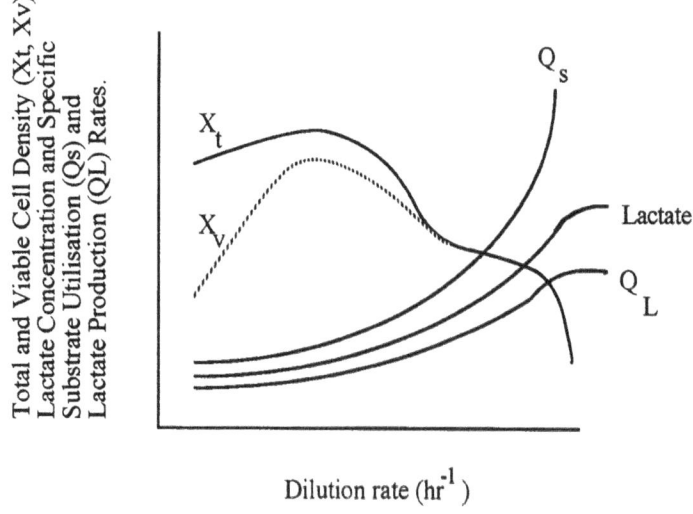

Figure 6 Schematic of hybridoma growth.

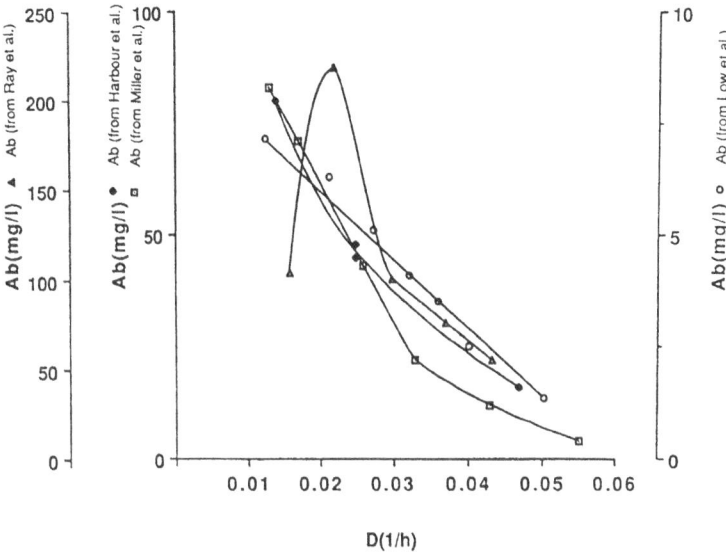

Figure 7(a) Antibody concentration profiles in continuous culture (compilation)

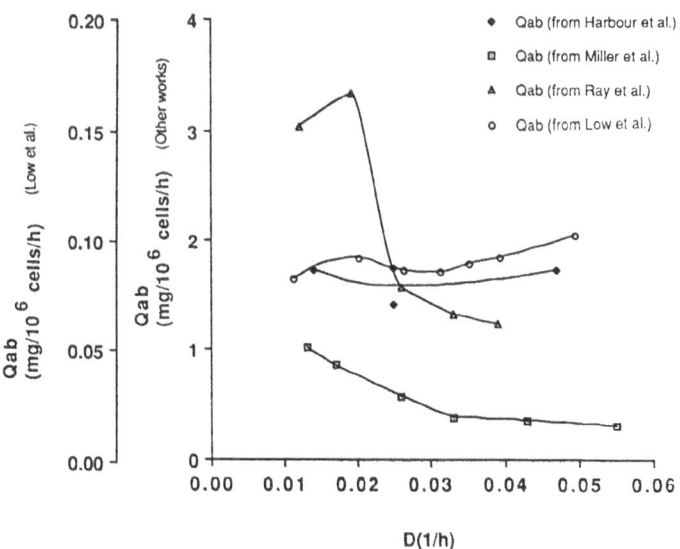

Figure 7(b) Specific antibody production profiles in continuous culture (compilation)

USE OF LIQUID MEDIUM CONCENTRATES TO ENHANCE BIOLOGICAL PRODUCTIVITY.

DAVID W. JAYME, RICHARD M. FIKE, JAMES M. KUBIAK, CHERYL R. NASH, and PAUL J. PRICE.
GIBCO-BRL/Life Technologies, Inc.,
2086 Grand Island Boulevard
Grand Island, NY 14072 USA

ABSTRACT. Liquid medium concentrates (LMC) were initially developed in response to industrial customer demand for improved efficiency and productivity of mammalian cell bioreactors. The resultant technology, which is undergoing its second generation of improvement, exploits biochemical properties intrinsic to the constituent nutrients to improve solubility, stability, and biological performance. Customer-perceived benefits, relative to liquid media produced by conventional technologies, include both enhanced biological productivity and improved manufacturing efficiency. The key concept which drives superior performance is that in concentrate technology all nutrient components of a complex biochemical formulation are fully pre-solubilized as a minimal number (generally three) of 50X LMC sub-groupings precedent to complete admixture. Under conventional procedures, critical nutrient potencies may be reduced relative to theoretical values due to poor solubilization and partial removal by precipitation and filtration. Use of LMC intermediates for the formulation of liquid media produces superior initial correlation with theoretical nutrient potency and may result in superior performance stability. Delivery of the full complement of nutrients to the biomass may be limiting to culture survival, attainment of maximal cell density, optimization of the specific cellular productivity, and prolongation of the bioreactor production cycle. The advantages of LMC become increasingly apparent with serum reduction or elimination, augmented cell densities, enhanced medium residence time and increasing development of balanced nutrient formulations. Complex nutrient media, formulated as LMC, are stable for up to one year in many cases, are compatible with various bulk containers, and exhibit superior biochemical potency and biological performance relative to conventionally-prepared culture media.

1.   Introduction

A decision common to biotechnology applications during the transition from product development to commercialization concerns the preparation and delivery of nutrient fluids to the bioreactor. Assuming that nutrient concentrations were optimized for the unique productive cell line and bioreactor system [1], process engineering must determine how, where, and with what starting materials the culture medium will be produced. The answers to these questions can have a profound impact on final product cost, design and renovation options for the manufacturing facility, and the product approval process [2].

1.1.   HISTORICAL ALTERNATIVES

Traditionally, there have existed only two culture medium source alternatives for large-scale

215

S. Kaminogawa et al. (eds.), Animal Cell Technology: Basic & Applied Aspects, Vol. 5, 215–222.
© 1993 *Kluwer Academic Publishers.*

mammalian cell culture production applications: purchase of sterile, bulk single-strength (1X) liquid medium from a qualified commercial vendor or internal manufacture by hydration of commercial powders or individually-weighed biochemical components. The advantages and disadvantages of these two medium formulation alternatives are summarized in Table 1.

Table 1. Comparison of Conventional Modes for Large-Scale Media Preparation.

LIQUID MEDIUM		POWDERED MEDIUM	
Advantages	Disadvantages	Advantages	Disadvantages
○Engineering convenience ○Reduced upfront capital expense ○Sterile product ○Reduced media kitchen labor cost ○Facilitated regulatory approval potential	○Cost per liter ○Refrigerated storage space ○Decreased nutrient stability ○Limited batch size ○Bulk container validation	○Cost per liter ○Greater nutrient stability ○Storage requirements ○Larger batch sizes	○Inconvenient scale-up process flows ○Sterile processing required ○Component insolubility ○Upfront capital investment ○Media kitchen training

1.2. LIQUID MEDIUM CONCENTRATES

Liquid medium concentrate (LMC) technology has emerged as a compromise alternative to conventional manufacturing options for large-scale preparation of liquid media for biological production applications. Derived from our biochemical analysis of spent medium to design supplemental nutrient concentrates for batch and perfusion bioreactor applications [1,3], LMC technology was rapidly expanded to complete media formulations [2,4].

Because many individuals have used various forms of concentrated media sub-groupings, it is useful to differentiate certain unique characteristics exclusive to liquid concentrates prepared by our proprietary methodology. The basic strategy for separation of a complex nutrient formulation into concentrated liquid medium sub-groups and their subsequent reconstitution has been described previously [4-5].

1.2.1. Minimal Number of Sub-groups. To minimize the possibility of confusion and error during reconstitution and to reduce costs associated with intermediate containerization and sterile processing, LMC's have been designed to provide the minimal number of sub-group components. For a complex nutrient formulation, all nutrient components (even elevated nutrient levels and relatively insoluble lipid constituents) may be incorporated into a standard three concentrated component configuration. Sodium bicarbonate is typically provided separately owing to technical and cost considerations.

1.2.2. Elevated Concentration. Each LMC sub-group component is prepared, using proprietary information regarding biochemical compatibility, solubility and order of addition, and nutrient co-stabilization, to produce the highest feasible concentrated solution. For simplicity and convenience in reconstitution and to facilitate automated continuous apportionate mixing devices, LMC sub-group components are typically prepared to equivalent fold concentrations. For example, a complex nutrient formulation prepared as three 50X LMC sub-groups would require 1 liter of each LMC component to produce 50 liters of fully-diluted culture medium, using either batch or continuous reconstitution procedures.

1.2.3. *Biochemical Stability.* Classical "concentrates" required either frozen storage or rehydration post-lyophilization and exhibited limited shelf life. LMC components demonstrated nutrient stability, as evidenced by HPLC and other quantitative biochemical and biological performance assays, for extended periods [4-5] under both refrigerated and ambient storage conditions. For example, concentrated liquid sub-group components of DMEM were stored refrigerated for over 12 months without any measurable nutrient deterioration.

2. Properties of Liquid Medium Concentrates

LMC technology does not add any extraneous components to the formulation to assist constituent solubility or to stabilize nutrient elements. It does not alter the qualitative or quantitative composition of the final product in any significant manner. Rather, LMC technology exploits the native biochemical interactive properties of medium constituents by appropriate methods of biochemical combination, segregation and order of addition to enhance intrinsic solubility and stability. Our internal validation procedures and collaborative studies with over a dozen leading biotechnology institutions on multiple continents have demonstrated the following advantageous properties of LMC's:

2.1. IMPROVED NUTRIENT POTENCY.

With conventional media preparation methods, molecules which tend to aggregate due to ionic and hydrophobic interactions and components not readily solubilized at neutral pH precipitate within the formulation tank or are differentially retained by the sterilizing filter, with a resultant nutrient sub-potency of the final filtered product which may vary from batch to batch. With media reconstituted from LMC's, all nutrient components are optimally pre-solubilized and biochemical analysis of media components reconstitution and post-filtration exhibits excellent correlation with theoretical potency and high batch-to-batch precision.

2.2. ENHANCED BIOCHEMICAL STABILITY.

LMC's have demonstrated unique biochemical stability against nutrient deterioration or precipitate formation under both refrigerated and room temperature storage. Of particular significance, glutamine-containing media prepared by conventional methods typically are shelf life limited and performance-constrained by spontaneous deamidation, liberating ammonia and resulting in glutamine deprivation and ammonia toxicity. Owing to glutamine stabilization by association with other nutrient components, glutamine exhibited minimal deterioration over the storage period and retarded spontaneous degradation upon reconstitution [4-5].

2.3. ACCELERATED MEDIA PREPARATION.

Reconstitution of LMC sub-groups is rapid and flexible, either in batch or continuous format. If desired to minimize bioburden and endotoxin in the prepared medium, reconstitution may be performed in a refrigerated vessel. Collaborating facilities have reported substantial (> 50%) reduction in medium preparation times, permitting more efficient use of media kitchen facilities and personnel. Homogeneous batches of LMC materials in the range of 25,000 to over 100,000 liter-equivalents have been demonstrated and validation is proceeding for further scale-up.

2.4. Consistent with Efficient Engineering and GMP Facility Design.

Engineering designs favor liquid over powder flows for large volume media preparation. Closed system manipulation of liquids is also superior for GMP and health containment considerations by eliminating concerns over powder aerosols. Although most current users of LMC's utilize batch reconstitution, continuous media preparation using an automated apportionate mixing device offers significant future opportunities for improved facilities design and manufacturing efficiency.

2.5. Competitive Overall Media Preparation Cost.

Although the purchase price for LMC's is intermediate between powdered and liquid media, collaborators report that the fully-burdened cost of producing a liter of medium at production scale is cost-effective. Contributory factors include more efficient use of media kitchen facilities and personnel, reduced batch-related quality control testing costs, reduced filter consumption, reduced media storage and container reuse/disposal expense, reduced media waste, reduced requirement for supplemental costly serum or growth factors, and incremental product yield.

2.6. Biological Performance.

Although diminished by serum supplementation and excessive nutrient replenishment, the superior nutrient delivery and stability afforded by LMC technology has resulted in enhanced bioreactor performance. In serum-free and reduced serum cultivation systems, the profound performance advantages of LMC-derived media were reported by collaborators in terms of increased cellular proliferative rate and reduced seed chain time, higher maximal cell density, extended productive bioreactor lifetime, and increased specific productivity.

2.7. Breadth of Use.

To demonstrate the universality of basic LMC principles and to prepare for future introduction of catalog LMC products, we prepared over fifteen different catalog formulations as 50X LMC sub-groups. Formulations ranged from traditional media intended for further supplementation by serum or growth factors to highly-complex serum-free media designed for specialty applications in biotechnology and biomedicine. In collaboration with industrial partners, we have also prepared in excess of forty different custom formulations for external evaluation, several of which are currently being used for pharmaceutical product or clinical trial material manufacture.

3. Batch and Continuous Reconstitution of LMC Intermediates.

Figure 1 illustrates a general scheme for LMC dilution, depicting A, B & C as LMC sub-groups delivered either sequentially (for batch processes) or simultaneously (for continuous apportionate mixing processes) into a mixing chamber, together with production water and any other desired additives. For batch processing, this chamber may represent a large formulation tank or a closed sterilize-in-place vessel containing sterile production water. For continuous processing, the

Figure 1. Generic Scheme for Reconstitution of Liquid Media from Concentrated Intermediates.

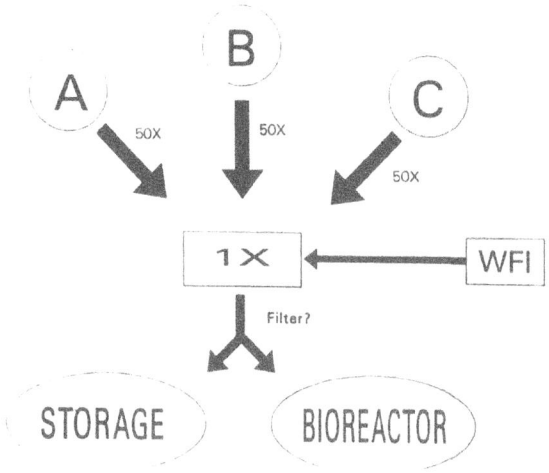

Figure 2. Campaigned Preparation of Similar Medium Formulations.

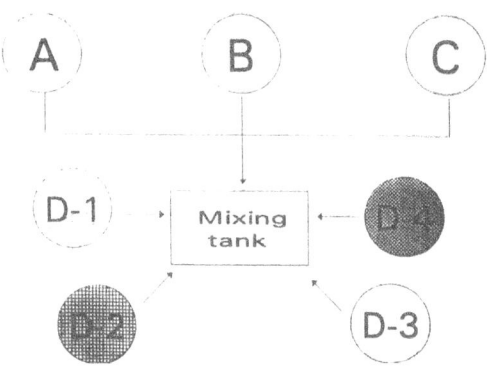

Figure 3. Nutrient Augmentation by Supplemental Concentrate Addition.

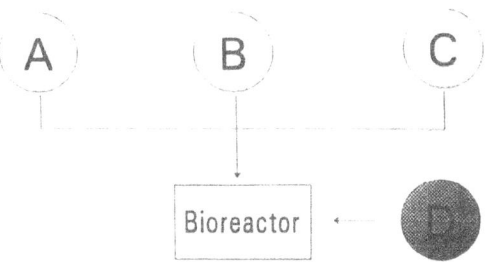

mixing chamber is substantially smaller, and designed for automated production of 100-1000 liters per hour of biochemically-precise, single-strength liquid media of pre-determined pH and osmolality as monitored by validated in-process probes. Since LMC intermediates are sterile-filtered (SAL $< 10^{-3}$), on-line filtration is optional. Fully-diluted medium may be stored in a large surge tank or manifolded to multiple stainless steel or flexible plastic containers, or it may be delivered directly to feed or perfuse the bioreactor(s).

3.1. FACILITATED PREPARATION OF MULTIPLE, SIMILAR MEDIUM FORMULATIONS.

Many institutions utilize multiple variants of a common parental cell line, genetically-engineered to produce different biological products, which require similar but non-identical nutrient media. LMC sub-groups (A, B & C) are designed to include basal nutrients common to all formulations. These generic LMC's may be augmented efficiently, either in batch or continuous media preparation modes, by a fourth, variable concentrated component (D-1, D-2, D-3 & D-4), corresponding to the unique nutrient requirements of a particular cell type (Figure 2).

3.2. ENHANCED EFFICIENCY WITH SUPPLEMENTAL NUTRIENT CONCENTRATES.

Rapid consumption of certain nutrient medium constituents in high density cultures has driven excessive perfusion or medium dilution rates to maintain culture viability and productivity. Analysis of spent perfusates and cell-free supernatants has revealed that only a few nutrients within most conventional formulations are truly growth or production rate limiting. Under these circumstances, increased media consumption in perfusion and fed-batch culture becomes a costly, inefficient process. It may also lead to dilution of paracrine stimulatory factors and complicate product harvest by requiring a concentration step. Rational design of a concentrated supplement (Figure 3) of rapidly-exhausted nutrients permits an overall reduction in medium consumption by lowering the primary medium feed rate and replenishing exhausted nutrients by pulsed or continuous administration of the supplemental nutrient concentrate. The net result is a reduction in the total volume of medium consumption, which may be accompanied by an increase in cell-specific productivity and biological product recovery. This mode of concentrated nutrient addition also avoids problems of secondary nutrient depletion and accumulation of toxic catabolites occasionally observed with medium recirculation techniques.

4. Collaborative Experiences with Liquid Medium Concentrates.

Table 2 summarizes non-proprietary information obtained from an international panel of thirteen biotechnology institutions over the past eighteen months. These field studies compared performance of media derived from LMC intermediates with their conventional method for obtaining the identical formulation, with the alternative methods approximately equally-divided among current liquid media and powdered media users. Evaluative phases ranged from early pilot scale studies to three institutions who are utilizing LMC's at the final biological product manufacturing scale. Eight of these institutions reported performance advantages with LMC-derived media, in terms of either improved manufacturing efficiency or elevated biological production. As noted previously, most of these institutions utilized serum-free or low serum cultivation systems which accentuate the nutrient delivery advantages of LMC-derived media. Four collaborators reported equivalent performance in serum-supplemented pilot scale studies, but are extending their investigations to larger-scale or reduced serum applications to test potential LMC opportunities. The remaining institution provided a technically-challenging custom formulation with novel ingredients which initially yielded inferior performance, but is currently examining a second generation LMC with more encouraging results.

Table 2. Field Studies of Liquid Media Concentrates by Collaborating Institutions.

β-site Location	Culture Condition	LMC-derived Media Performance vs. Medium Prepared Using Conventional Manufacturing Methods
U.S. (West Coast)	Serum-free (proprietary media)	Superior performance: internal manufacturing advantages (production scale)
U.S. (West Coast)	Serum-free (Hybridoma-SFM)	Superior performance: enhanced cell growth; enhanced specific MAb production (pilot scale)
U.S. (West Coast)	Serum-free (Hybridoma-SFM)	Superior performance: enhanced cell growth and MAb production (pilot scale)
U.S. (West Coast)	Serum-free (CHO-S-SFM II)	Superior performance: enhanced productivity; internal manufacturing advantages (production scale)
U.S. (West Coast)	Serum-free (CHO-S-SFM II)	Superior performance: enhanced cell growth; internal manufacturing advantages (pilot scale)
Japan	Serum-supplemented (RPMI 1640)	Superior performance: internal manufacturing advantages (pilot scale)
U.S. (Midwest)	Serum-supplemented (MEM) & low serum (OptiMEM)	Superior performance: internal manufacturing advantages (production scale)
U.S. (Midwest)	Serum-free (Hybridoma-SFM)	Superior performance: enhanced cell performance; internal manufacturing advantages (pilot scale)
Japan	Serum-supplemented (proprietary media)	Equivalent performance: internal manufacturing advantages (pilot scale)
U.S./Europe	Serum-supplemented (modified DMEM)	Equivalent performance: internal manufacturing advantages (pilot scale)
Japan	Serum-supplemented (modified eRDF)	Equivalent performance: internal manufacturing advantages (pilot scale)
Japan	Serum-supplemented (RPMI 1640)	Equivalent performance: internal manufacturing advantages; possible performance increase (pilot scale)
U.S. (East Coast)	Serum-free (proprietary media)	Inferior performance: novel components incorrectly sub-grouped; improved samples under evaluation

5.   Discussion.

The high interest and rapid acceptance LMC technology underscores its technical and practical advantages for the large-scale manufacture of complex nutrient media. LMC sub-grouping capabilities and manufacturing processes have progressed through several quantum level improvements since the original patent submission, based upon greater understanding of nutrient co-stabilization, biochemical interactions and techniques for constituent solubilization. Field reports of superior biological performance with LMC-derived media may become noteworthy with increased emphasis on serum-free cultivation and optimally-balanced nutrient formulations.

We encountered an interesting paradigm shift in understanding transitions among media forms with increasing volumetric media requirement. Our early assumption was that LMC's would be positioned intermediate between bulk liquid media and powdered media along the scale-up progression. With this assumption, potential interest in LMC's would come from current bulk liquid media users needing to scale-up media production volumes but desiring to forestall the transition to powdered raw materials. We also anticipated some recruitment of current powdered media users in response to greater convenience, cost-effectiveness and improved medium quality. These assumptions have both proved valid; however, we failed to anticipate the interest from current powdered media consumers who valued the LMC engineering systems approach as a superior alternative to high volume powdered media reconstitution and massive tank farms.

6.   Acknowledgments.

The pioneering work of Dennis A. DiSorbo is gratefully acknowledged. The authors also commend the commitment and flexibility of our Manufacturing Technology group under the direction of Ray Watkins for their assistance in manufacturing process design and validation. The invaluable support of Dennis Barger, Toshio Watanabe, David Cady and Tony Kazarian in interfacing with collaborators during field evaluations is recognized with appreciation. Finally, we wish to express our gratitude to those collaborating partners whose objective candor and willingness to evaluate our promising experimental techniques have helped to accelerate our understanding of the benefits of concentrate technology.

7.   References.

1.   Jayme DW, *Cytotechnology* (1991) *5*: 15-30, *Nutrient optimization for high density biological production applications.*
2.   Jayme DW and Greenwold DJ, *Bio/Technology* (1991) *9*: 716-721, *Media Selection and Design: Wise Choices and Common Mistakes.*
3.   Fike RM, Jayme DW and Weiss SA, *American Biotechnology Laboratory* (1991) June, pp. 40-42, *"Monoclonal antibody enhancement in protein-free and serum-supplemented hybridoma culture."*
4.   Jayme DW, DiSorbo DM, Kubiak JM and Fike RM In: Animal Cell Technology: Basic and Applied Aspects, Kluwer (1992) Dordrecht, *Use of nutrient medium concentrates to improve bioreactor productivity.*
5.   Jayme DW, DiSorbo DM, Kubiak JM, Fike RM, and Price PJ, In: *Proceedings of the International Biotechnology Exposition (IBEX '92),* (October, 1992, San Francisco), *Enhanced Biological Productivity Through Use of Liquid Medium Concentrates.*

OPTIMISATION OF α-INTERFERON EXPRESSION OF NAMALWA CELLS

S. C. MUSGRAVE, Y. DOUGLAS, G. LAYTON*, J. MERRETT*, M. F. SCOTT and C. A. CAULCOTT

*Biotechnology Development Laboratories and *Management Services Division, The Wellcome Foundation, Beckenham, Kent, UK.*

ABSTRACT.   Namalwa cells grown in culture can be readily induced to make natural α-interferon which, after extraction from the culture supernatant and subsequent purification, is marketed as Wellferon. The process by which Namalwa cells are induced to express interferon involves the treatment of the cells with sodium butyrate followed by addition of Sendai virus as the inducing agent. Work has been undertaken to determine whether the viral induction stage of the Wellferon production process could be improved upon, so as to obtain higher levels of interferon.

Experimental studies showed that the effect of two Sendai virus additions as opposed to the conventional single addition was to give a significant increase in the final interferon levels. These studies were expanded to investigate and optimise the effect of the time of addition of the two Sendai virus doses, the relative proportion of the initial (priming) dose and the main dose, the possibility of scaling the process up to 10L fermenters and the effect the process modification had on the final product composition.

It was found that a standardised process could be identified which gave a 1.5 - 2.5 fold increase in the level of interferon present in the culture supernatant. This process could be operated at a variety of scales, from laboratory shake flasks through to 10L (working volume) fermenters. No differences were observed between a batch of Interferon made by this method and Wellferon manufactured by the standard process.

1.   Introduction

α-interferon (Wellferon) is manufactured by adding Sendai virus to Namalwa cells which have been treated with sodium butyrate for 42 hours. The cells are then harvested 24 hours after Sendai induction and the α-interferon purified.

Investigations have been undertaken to determine whether the levels of interferon in the final harvest can be increased. It is known that treatment of some cell types *in vivo* with interferon prior to induction can cause a "priming" effect such that the cells, when stimulated by an inducer such as Sendai virus, produce greater quantities of interferon (1). Namalwa cells have previously

S. Kaminogawa et al. (eds.), Animal Cell Technology: Basic & Applied Aspects, Vol. 5, 223–230.
© 1993 *Kluwer Academic Publishers.*

been shown not to be susceptible to such induction (2,3). However, this may be associated with the precise mechanism of the 'priming' effect. It was therefore decided to investigate whether such a 'priming' effect could be achieved using the Sendai virus inducer rather than interferon.

2   Materials and Methods

2.1   CELL GROWTH

Namalwa cells were grown in a 3.5 litre fermenter on a fixed 2 day subculture cycle, diluting with growth medium to give a viable cell concentration of 1.5×10^6 cells/ml after subculture. The culture typically grew to between 4 and 6×10^6 cells/ml over the 2 day cycle. The fermenters were controlled at pH 7.0, 37°C and 15% dissolved oxygen tension. The growth medium was RPMI 1640 supplemented with 0.5% tryptone and 3.3% adult bovine serum.

2.2   INDUCTION OF α-INTERFERON PRODUCTION

Cells from the fermenter were diluted with maintenance medium (growth medium without serum) to give a cell count of 2×10^6 cells/ml. Sufficient sodium butyrate solution was added to give 1.5mM and the mixture was aliquoted into spinner bottles, Erlenmeyer flasks or fermenters. Spinners were stirred at 70 rpm, Erlenmyer flasks placed on an orbital shaker at 100 rpm. Sendai virus (1.5 ml of allantoic fluid per litre of culture) was added 42 hours after butyrate treatment. Samples for interferon assays were take at various time points, including the normal point at 66 hours after butyrate treatment.

2.3   SENDAI VIRUS

The Sendai virus was produced in eggs and the allantoic fluid harvested and stored frozen in 300ml bottles. A bottle of frozen allantoic fluid was thawed at 4°C, and the contents were aliquoted into 10ml quantities and stored frozen at -20°C. For each experiment a single aliquot was thawed and used for all of the additions within that experiment.
In all experiments where the Sendai virus was added as two additions,   the total Sendai virus dose received (prime plus main dose) was the same as that received by the control.

2.4   FERMENTERS

Six 2 litre SGi fermenters (1.0 litre working volume) and two 2 litre SGi fermenters (15L working volume) were used for the study. The fermenters were controlled at 36°C during butyrate treatment and 35°C during Sendai induction. Dissolved oxygen tension was controlled at 15% throughout the run by sparging with air. The pH was controlled at 6.9 by sparging with CO_2 or addition of 1M Na_2CO_3 solution. The cultures were continuously stirred at 100rpm. Measured values for temperature, pH, dissolved oxygen and agitator speed were automatically recorded on a Bio-i computer data logging system.

3   Results and Discussion

3.1   EFFECT OF 'PRIMING' NAMALWA CELLS WITH SENDAI VIRUS

An experiment was carried out in order to determine whether Sendai virus could be used to 'prime' Namalwa cells such that, when induced with further Sendai virus at the normal process time, increased levels of α-interferon were then synthesised by the cells. The results are shown in Table 1.

TABLE 1.   Harvest titres (IU/ml) from two groups of 4 spinners each containing Namalwa cells treated with butyrate.

	Split	Control
	6607	3388
	5495	4266
	6607	5495
	7244	5380
Mean	**6488**	**4630**

The priming dose of Sendai virus (50% of the total dose) was added 20 hours after the start of butyrate treatment, with the remaining 50% being added at the normal time of 42 hours after the start of the butyrate treatment. The control spinners received all the Sendai virus dose at 42 hours. For all cultures the total Sendai dose was 1.5ml per flask

This approach to the addition of Sendai virus gave an increase in yield of 40% over the control. Additional similar experiments provided further data supporting the effectiveness of this technique. Work was therefore undertaken to optimise the ratio between the initial and main Sendai doses and the time at which the priming dose was added.

3.2   RATIO OF PRIMING DOSE TO FINAL DOSE

The effect of priming Namalwa cells with different proportions of the total Sendai dose 18 hours before the addition of the remainder of the Sendai at the normal process time (42 hours after butyrate treatment) was investigated. It was found (Figure 1) that priming doses greater than 50% of the total dose gave final interferon levels equal to or less than the control. This was probably the result of there being insufficient Sendai remaining for the main dose. Further experiments showed that even a 50% priming dose was not always effective and that the optimum priming dose was between 10 and 20% of the total dose of Sendai (Figure 2).

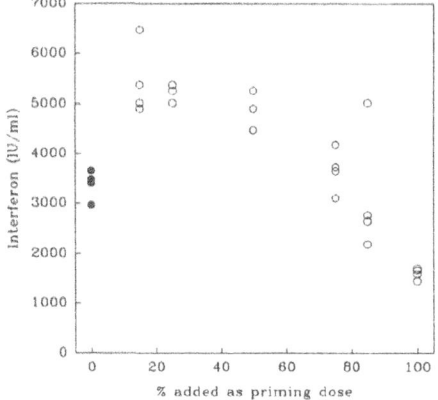

Figure1. The effect of the ratio of Sendai used as the priming dose on the harvest titre of α-interferon.

Total Sendai dose: 1.5ml/L.
Priming dose added 18 hours before main dose.

Control flasks (•) received only one dose of Sendai at 42 hours after the start of butyrate treatment. Each point represents the titre at the normal time of harvest (66 hours) of an individual flask

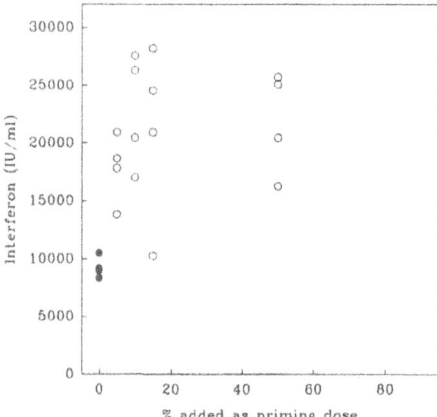

Figure 2. The effect of the ratio of Sendai used as the priming dose on the harvest titre of α-interferon.

Total Sendai dose: 1.0 ml/L.
Priming dose added 18 hours before the main dose.

Control flasks (•) received only one dose of Sendai at 42 hours. Each point represents the titre at the normal time of harvest (66 hours) of an individual flask.

From these results it was concluded that the optimal ratio for the priming dose to the final dose was 15% to 85%. This was known to always be effective, giving an increase over the control in the harvest levels of interferon.

3.3   TIME OF ADDITION OF PRIMING DOSE

The normal time for addition of the Sendai virus inducer is 42 hours after Namalwa cells have been treated with butyrate. The time of addition of the Sendai prime was initially investigated by adding 50% of the total dose at one of a variety of times after butyrate treatment, before addition of the main dose. The harvest interferon titres for the various times of addition of the Sendai prime are shown in Figure 3. From this it was concluded that the priming dose had to be added no earlier than 22 hours before the final dose (i.e. 20 hours after the initial butyrate treatment). Further experiments demonstrated that the optimal time of addition of the Sendai prime was between 18 and 10 hours before addition of the main dose of Sendai virus (Figure 4).

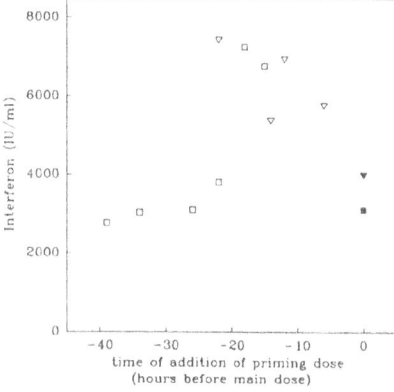

Figure 3. Two experiments examining the effect of the time of the priming dose on the harvest titre of α-interferon.

Total Sendai dose: 1.5ml/L, 50% of total dose added as the priming dose.

Closed symbols: controls. Each point represents the geometric mean of the harvest titre of 4 flasks.

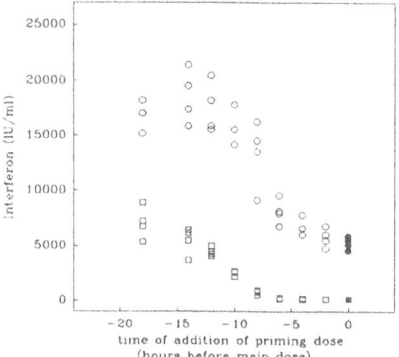

Figure 4. The effect of the time of the priming doe on the harvest titre of α-interferon.

Total Sendai dose: 1.5ml/L. 50% added as the priming dose. Each point represents the titre of an individual flask.

Titres at main dose ▢
Titres at harvest o
Closed symbols: controls

From these results it was concluded that the optimal time for addition of the Sendai priming dose was 12 hours before addition of the main dose. As with a 15% priming dose, this time of priming was shown to always give a significant increase in the interferon harvest titre as compared with the controls.

3.4   EFFECT OF PRIMING NAMALWA CELLS IN FERMENTERS

The experimental work described above was carried out in shake flasks or spinner culture. The Wellferon production process is operated under controlled conditions in 8000L fermenters. It was therefore appropriate to demonstrate that the technique of adding the total Sendai virus as a prime and main dose was also successful in small scale fermenters. Two 20L and four 2L fermenters were used, with the Namalwa cells being treated with either the standard, control process (all the Sendai virus added 42 hours after butyrate treatment) or in the optimised process (15% of the Sendai virus added 12 hours before the residual 85% was added). One 20L and two 2L fermenters were used for the controls, the remaining three fermenters were used for the optimised process. Samples were taken during the first seven hours after the addition of the main/control dose of Sendai from all four 2L fermenters in order to follow the kinetics of production of α-interferon. From Table 2 it will be seen that in both the small and large fermenters, the addition of Sendai in two doses gave a significant increase in harvest interferon levels.

TABLE 2. Harvest α-interferon titres in 2 and 20 litre fermenters

Working Volume (L)	Control	Sendai Split	Ratio Split/Control
1.0	10400 17900 (13600)[1]	21000 21600 (21300)[1]	1.57
14.00	34700	58600	1.69

[1]   Geometric mean

The levels of α-interferon in the cell cultures during the first 7 hours after addition of the main/control Sendai dose are shown in Figure 5. It can be seen that in the cultures treated with a priming dose of Sendai, the interferon levels increase rapidly after the addition of the main Sendai dose. Interestingly, it appears possible that the maximum levels of interferon are expressed by the primed cells within 10 hours of the addition of the main dose of Sendai.

α-interferon from the two 20L fermenters was extracted, purified and the material from the control and Sendai split processes compared. No difference between the two lots of α-interferon were seen. In addition, both control and Sendai split derived α-interferon gave similar reverse phase HPLC profiles to the control standard of Wellferon.

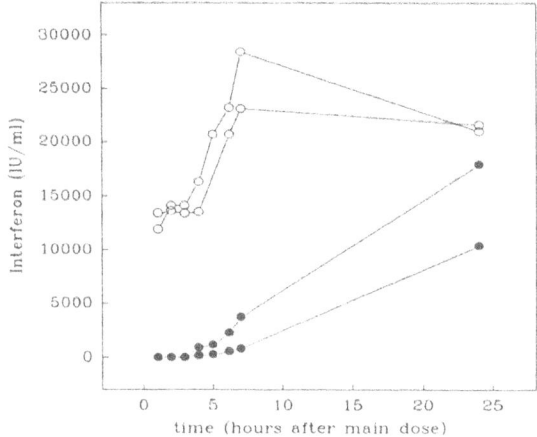

Figure 5. Kinetics of α-interferon synthesis by Namalwa cells induced with Sendai virus in four 1litre laboratory fermenters.

Total Sendai dose: 1.5ml/L. 15% of total added 12 hours before the main dose.   Sendai split o   Controls   •

4   Concluding Discussion

Although there had been previous reports that Namalwa cells were not susceptible to priming with α-interferon (2,3), it was found that they could be primed with the Sendai virus used as the principle induction agent. Statistical assessment of the results from a wide variety of experiments suggests that in general an increase of between 1.5 and 2.5 fold in the harvest α-interferon levels will be seen if Sendai virus is added in a split manner, rather than as one dose at 42 hours. It is likely that the Sendai split method is effective because in some manner the Namalwa cells are prepared (primed) by the initial dose of Sendai virus such that the α-interferon biosynthetic machinery is fully prepared prior to the introduction of the main Sendai dose. The Namalwa cells are therefore more effectively switched to α-interferon synthesis when the main dose of Sendai virus is added. The precise primary mechanism at the cellular and molecular level is not known, but there is some evidence that it is associated with an interferon intermediate.

The split addition of Sendai virus inducer to Namalwa cells offers a potential method for improving the productivity of the Wellferon manufacturing process. The full exploitation of this required production trials in order to establish the reproducibility of the method at large scale. However, it appears an attractive and logical process modification, leading to significant benefits in Production.

5   Acknowledgements

We would like to thank Mrs S Chapman and Miss C Beattie for ELISA assays, Dr H Griffiths and Ms J Oakes for the primary purification and Dr L Leadbetter, Mr R Raja and Dr W G Lewis for the HPLC analysis.

6   References

1.   De Mayer, E M and De Mayer-Guignard, J (1988)
      Interferons and other regulatory cytokines, John Wiley, New York

2.   Johnston, M D, Fantes, K H, Finter, N B and Chir, B (1978).
      In 'Human Interferon'. Stineberg W R and Chapple P J (Eds)
      Plenum Press, New York

3.   Morser, J, Flint, J, Graves, H E, Baker, P N, Coleman, A and Burke, D C.
      (1979). Journal of General Virology **44**, 231

CHANGES IN THE GLYCOSYLATION OF INTERFERON-γ DURING CULTURE

N. JENKINS, C. WINGROVE, P. STRANGE, A. BAINES,
E. CURLING, R. FREEDMAN AND P. PUCCI*

Biological Laboratory,   *Servizio di Spettrometria
University of Kent at Canterbury,   de Massa CNR,
Kent CT27NJ.   Universita de Napoli,
United Kingdom.   via Pangini 5, I-80131
                                          Napoli, Italy.

ABSTRACT. We have shown in previous studies that recombinant human interferon-γ (IFN-γ) expressed in CHO cells displays macroheterogeneity caused by variable N-glycosylation (Curling et al., 1990). In this paper, FAB-Mass Spectrometry of IFN-γ glycopeptides was used to assign the N-glycosylation preference of individual asparagine (Asn) sites within the IFN-γ molecule. Whilst both Asn_{28} and Asn_{100} N-glycosylation sites were used in doubly-glycosylated IFN-γ, a clear preference for the Asn_{28} site was apparent in the singly-glycosylated IFN-γ glycoform. The extent of glycosylation at the Asn_{28} site remained relatively stable during batch culture, whereas the percentage of Asn_{100} sites glycosylated decreased from 60% to 23% over the same period. These results demonstrate that marked differences exist in the capacity of CHO cells to glycosylate at individual Asn sites during cell culture.

1. Introduction

Many factors influence the degree of N-glycosylation of recombinant proteins by cells in culture. The glycosylation pattern tends to be cell type specific, and the position of the Asn site within the recombinant protein also influences its glycosylation status (Goochee & Monica, 1990). The presence of an Asn-X-Ser/Thr consensus sequence does not, by itself, guarantee its N-glycosylation. Indeed, we have shown in previous studies that the cell culture conditions greatly influence the degree of N-glycosylation (Curling et al., 1990, Hayter et al., 1991, 1992). We have found that recombinant CHO cells grown in batch culture secrete three glycoforms of the human IFN-γ product (doubly-glycosylated, singly-glycosylated and non-glycosylated), and the proportion of these glycoforms changes during culture (Fig.1). This macroheterogeneity arises from variable N-glycosylation on two possible sites (Asn_{28} & Asn_{100}).

Macroheterogeneity is also found in other recombinant molecules such as GM-CSF (Moonen et al., 1987) and tPA (Parekh et al., 1989). In this study we have used protease

231

S. Kaminogawa et al. (eds.), Animal Cell Technology: Basic & Applied Aspects, Vol. 5, 231–235.
© 1993 Kluwer Academic Publishers.

digests and mass spectrometry to measure the degree of N-glycosylation at each of the two Asn sites on the recombinant IFN-γ protein synthesised in CHO cells.

2. Materials and Methods

CHO cells expressing the gene for IFN-γ were donated by Wellcome Biotechnology. The gene was co-expressed with the enzyme DHFR, allowing gene amplification by methotrexate selection at 0.1μM. This resulted in an IFN-γ expression level of approximately 3,000 IU/ml. Cell culture media and general fermentation conditions have been described previously (Curling et al., 1990; Hayter et al., 1991). IFN-γ was purified from cell culture supernatant using an ion exchange chromatography column (S-Sepharose) followed by immunoaffinity column (Celltech antibody 20B8) on a Pharmacia FPLC system. The protein was concentrated by lyophilization, and found to be >98% pure by gel electrophoretic analysis.

Lyophilised IFN-γ was resuspended in 0.4% (w/v) ammonium bicarbonate to a final concentration of 1mg/ml. To isolate peptides containing individual glycosylation sites, Staphylococcal V8 protease was added at an enzyme to substrate ratio of 1:50 (w/w), and the digestion proceeded for 6 hours at 40°C. Peptide fragments (100μg) were resolved using reverse phase HPLC (Vydac 218TP54) with a C_{18} column. Separation was achieved using a linear gradient of 100% (v/v) buffer A (0.001% (v/v) TFA in HPLC grade H_2O) to 70% (v/v) buffer B (0.001% (v/v) TFA in 70% (v/v) acetonitrile in HPLC grade H_2O) over 40 minutes at a flow rate of 1 ml/min. The column was equilibrated in buffer A. Both buffers were degassed by continuous sparging with helium. Absorbance was measured for the peptide bond at 220nm. The two peptides containing glycosylation sites were identified by their shift in HPLC retention time following Endoglycosidase F digestion to remove sugars.

Samples for FAB-MS were desalted by either repeated freeze-drying or using Sep Pak HPLC followed by evaporation under vacuum. They were resolubilised in 2-5 μl of 5% (v/v) acetic acid, and 2 μl (0.1-1.0 nmol) was applied to the FAB probe. Mass spectra were recorded using a VG ZAB High Field FAB-MA with an M-SCAN Fast Ion Gun operating at 10 kV and 20μA Xenon beam energy.

3. Results & Discussion

The sequential cation exchange and immunoaffinity protocol yielded 2mg of >98% pure IFN-γ (as judged by gel electrophoresis) from 1.5l of cell culture supernatant generated in fermenter culture. Good resolution of Staphylococcal V8 protease-generated peptides was achieved by HPLC using a gradient of 0-70% (v/v) acetonitrile., and preparative rp-HPLC was used to prepare peptides for analysis by MS (sugars were removed using endoglycosidase F). A satisfactory mass spectrum for the proteolytic fragments was obtained, leading to the construction of peptide map of the protein. Almost the entire IFN-γ molecule (from Asn_{10} to Glu_{119}) was identified using FAB-MS, along with additional peptides Leu_{120} - Lys_{125} and Ser_{132} - Arg_{137} following subdigestion with trypsin.

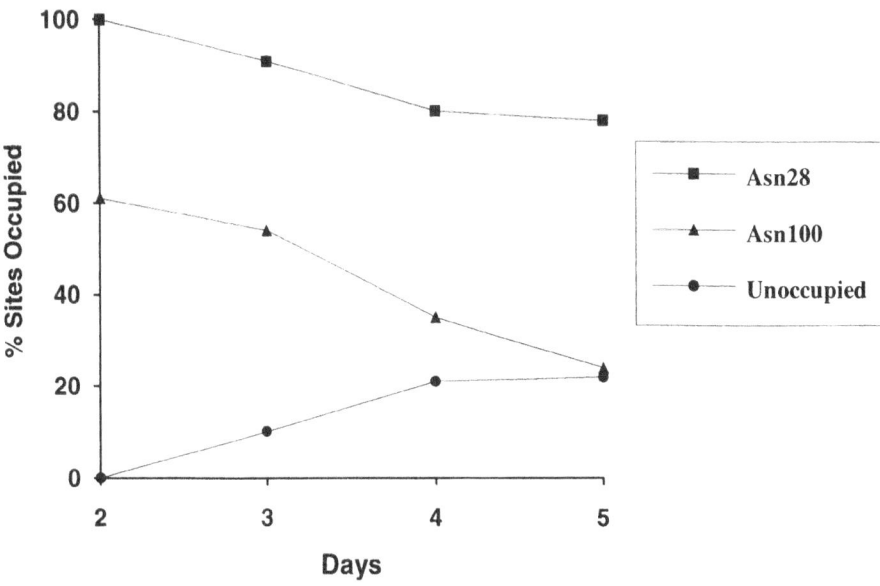

Figure 1: Percentage of Asn sites N-glycosylated within the recombinant IFN-γ molecule during batch culture of CHO cells. Asn assignments were based on FAB-MS analysis of glycopeptides (see text for details).

Endoglycosidase F cleaved the linkage between the polysaccharide and Asn in the two glycopeptides, yielding an aspartic acid residue. The mass of this amino acid is exactly 1 Da larger than asparagine, and this difference can be detected by FAB-MS. In the case of the Endoglycosidase F-digested peptide containing Asn_{28}, all molecular masses obtained were found to be 1 Da larger than predicted from their peptide sequence. In the case of peptides containing Asn_{100} a mixed population of masses was seen, i.e. the predicted mass for residues 95-108 is 1650 Da, but masses of both 1650 and 1651 Da were observed. This result indicates that Asn_{28} is exclusively used in singly-glycosylated IFN-γ, and is the preferred site for N-glycosylation within the IFN-γ molecule.

The information allows a re-interpretation of data gained from previous batch culture experiments in terms of Asn_{28} and Asn_{100} glycosylation profiles (Fig.1). N-glycosylation at Asn_{100} declines substantially towards the end of batch culture (coincident with the rise in non-glycosylated IFN-γ), whereas Asn_{28} glycosylation remains almost constant until the very late in culture. This preference for Asn_{28} glycosylation could be due to differences in the accessibility of N-glycosylation sites to oligosaccharide transferase during folding of the nascent peptide in the endoplasmic reticulum (Kellerher et al., 1992). Other groups have also shown that unoccupied sites are more likely to be found near the N-terminus of glycoproteins (Gavel & von Heijne, 1990).

Although the crystal structure of non-glycosylated human IFN-γ derived from *E.coli* has recently been published (Ealick *et al.*, 1991), any proposed differences in Asn accessibility must await further structural analysis of glycosylated IFN-γ (derived from mammalian cells). Alternatively, it is been shown that the oligosaccharide transferase shows a clear preference in transferase activity for peptide sequences containing Asn-X-Thr (as in the Asn_{28} of IFN-γ) compared to Asn-X-Ser (as in the Asn_{100} of IFN-γ), and this substrate discrimination may account for the differences observed (Bause, 1984). Variable N-glycosylation at a specific site is also shown by tPA expressed in CHO cells (Asn_{184}; Parekh *et al.*, 1989), and human GM-CSF expressed in the same cell type (Moonen *et al.*, 1987). Our strategy of defined proteolytic cleavage followed by FAB-MS analysis should prove widely applicable to studies on the macroheterogeneity of recombinant proteins.

4. Acknowledgements

This work was supported by the SERC Biotechnology Directorate and the following companies: Celltech, SK Beecham, Glaxo, British Biotechnology, Wellcome Foundation Ltd, ICI Pharmaceuticals. FAB-MS measurements were performed at the University of Naples.

5. References

Bause, E. (1984) Model studies on N-glycosylation of proteins. Biochem. Soc. Trans. 12, 514-517.

Curling, E.M.A., Hayter, P.M., Strange, P.G., Baines, A.J., Bull, A.T. and Jenkins, N. (1990) Recombinant interferon-γ: glycosylation and proteolytic processing leads to heterogeneity in batch culture. Biochem. J. 272, 333-337.

Ealick, S.E., Cook, W.J., Vijaykumar, S., Carson, M., Nagabhushan, T.L., Trotta, P.P. and Bugg, C.E. (1991) 3-Dimensional structure of recombinant human interferon-gamma. Science 252, 698-702.

Gavel, Y. and von Heijne,G. (1990) Sequence differences between glycosylated and nonglycosylated Asn-x-Thr/Ser acceptor sites - implications for protein engineering. Protein Engineering 3, 433-442.

Goochee, C.F. and Monica, T. (1990) Environmental effects on protein glycosylation. Bio/Technology 8, 421-427.

Hayter, P.M., Curling, E.M.A., Baines, A.J., Jenkins, N., Salmon, I., Strange,P.G. and Bull, A.T. (1991) Chinese hamster ovary cell growth and interferon production kinetics in stirred batch culture. Appl. Microbiol. Biotechnol. 34, 559-564.

Hayter, P.M., Curling, E.M.A., Baines, A.J., Jenkins, N., Salmon, I., Strange, P.G., Tong, J.M. and Bull, A.T. (1992) Glucose limited chemostat culture of CHO cells producing recombinant interferon-γ. Biotechnol. Bioeng. 39, 327-335.

Kellerher, D.J., Kreibich, G. and Gilmore, R. (1992) Oligosaccharyltransferase activity is associated with a protein complex composed of ribophorins I and II and a 48 kd protein. Cell 69, 55-65.

Moonen, P., Mermod, J.J., Ernst, J.F., Hirschi, M., and DeLamarter, J.F. (1987) Increased biological activity of deglycosylated recombinant human granulocyte macrophage colony-stimulating factor produced by yeast or animal cells. Proc. Nat. Acad. Sci. 84, 4428-4431.

Parekh, R.B., Dwek, R.A., Thomas, J.R. Opdenakker, G. and Rademacher, T.W. (1989) Cell-type-specific and site-specific N-glycosylation of type I and type II human tissue plasminogen activator. Biochem. 28, 7644-7662.

GROWTH CONTROL OF HYBRIDOMA CULTURES BY ADDING OR DEPRIVING INTERLEUKIN-6, CAFFEINE, AMINO ACIDS, GROWTH FACTORS, OR ANTISENSE C-myc

E. SUZUKI[1], K. TAKAHASHI[1], S.TERADA[1], AND F. MAKISHIMA[2]

1) Department of Chemical Engineering, Faculty of Engineering, University of Tokyo
7-3-1 Hongo, Bunkyo-ku, Tokyo 113, Japan
2) Institut für Genetik der Universität zu Köln, Weyertal 121, W-5000 Köln 41, Germany

ABSTRACT. mRNAs encoding antibodies accumulated in growth-suppressed hybridoma cells. The accumulation of stable mRNA was also predicted by the cell cycle model, which suggested that the growth suppression of hybridomas would enhance the antibody production rate per cell. Aiming to acheive this for the industrial production of antibodies, we layed down the criteria for desirable growth-suppression methods, including the cost and facility of application, and then examined several growth-suppression methods to find those that would satisfy the criteria. The addition of caffeine to the culture medium for suppressing growth satisfied the criteria best. The limitation of insulin and transferrin was the second best. With these growth-suppression methods, the antibody production rate per cell was enhanced threefold. Glutamine limitation in a glucose-deficient medium was also acceptable and enhanced the productivity by 70%.

1. Introduction

Increasing the amount of the mRNA coded for the antibody in hybridoma cells seems to be a rational strategy for enhancing antibody productivity (antibody production rate per cell). To achieve this, there may be several ways: 1) amplifying the antibody gene, 2) enhancing transcription of the gene, 3) stabilizing the mRNA, and 4) suppressing the growth (proliferation) of hybridoma cells. The first could be tried, for instance, by introducing antibody genes into myeloma cells by using vectors such as the papilloma virus vectors which can multiply as episomes in the cells (Karasuyama and Melchers, 1988). The second could be tried, for instance, by introducing vectors conveying a strong promoter such as the cytomegalovirus promoter upstream of the antibody genes into myeloma cells (Karasuyama et al.,1990). The third may not be possible, since mRNAs coded for antibodies are reportedly too stable to make any stabler (Storb, 1973). We report our work to achieve growth control for the last method.

This method is not straightforwardly connected to increasing the amount of mRNA in the cells. According to our cell cycle model (Suzuki and Ollis, 1990), stable mRNAs would accumulate in cells when the cells are growth-inhibited (Fig.1). Accordingly, the mRNAs coded for antibodies, which reportedly have twelve- to forty-hour half-lives would accumulate two- to four-fold in the cells during growth inhibition for four days after normal

237

S. Kaminogawa et al. (eds.), Animal Cell Technology: Basic & Applied Aspects, Vol. 5, 237–242.
© 1993 Kluwer Academic Publishers.

growth. We confirmed this prediction by measuring the mRNA concentration in the cells and the antibody productivity while the cells were growth-suppressed.

We then tried to establish an industrially applicable method for suppressing growth to enhance antibody productivity. First, we set criteria for selecting methods that fit best to this objective. They are 1) keeping high cell viability during growth suppression, 2) reversibility, 3) not suppressing protein synthesis and secretion, 4) being inexpensive, 5) being easy to apply, and 6) applicability to wide variety of hybridomas. Reversibility means here that the cells should be able to grow normally when the growth-suppressing conditions are reversed. The methods for growth-suppression we tried can be classified into two categories: class 1. addition of growth-suppressing compounds to the culture medium; and class 2, deprivation or depletion of some components required for growth from the culture (Table 1). We report here on caffeine, interleukin 6, and antisense c-myc oligonucleotides as the additives used for the methods in class 1, and then on glutamine, leucine, and insulin and transferrin as the essential components to be limited for the methods in class 2. The results for some other compounds or components for the methods in classes 1 and 2 have already been reported (Suzuki and Ollis, 1990; Suzuki *et al.*, 1992) some of them satisfying the criteria partly and others not at all.

2. Materials and Methods

2.1 CELL LINE, MEDIUM AND CULTIVATION METHDS

The cell line used was 2E3, which is a hybridoma between p3-X63- Ag.8U.1 mouse myeloma and spleen cells secreting IgG_1 against a trinitrophenyl-hapten (Mikami *et al.*, 1991).

RPMI 1640 culture medium (Life Technologies, Gaithersburg, Md.) deficient in glutamine or leucine, and supplemented with 10% CPSR-3 (Sigma, St. Louis, Mo.), a serum replacement, was used for glutamine or leucine limitation, respectively. ASF103 culture medium (Ajinomoto, Tokyo), a chemically defined serum-free medium, was used for all the other experiments.

The cells were cultured in 24-well plates at 37°C in a humid atmosphere of 5% CO_2. The initial cell concentration was about 10^4 cells/ml and the culture volume was 1.5ml/well. Each experiment was triplicated.

2.2 ANALYTICAL TECHNIQUES

The antibody was determined by sandwich ELISA, using anti-mouse IgG, peroxidase-conjugated anti-mouse IgG, and *o*-phenylenediamine dihydrochloride as a substrate for peroxidase. The optical density at 492 nm was measured, and calibration curves were prepared by using a known amount of purified IgG.

Cell numbers were counted with a hemocytometer, the viable cell population being distinguished from dead cells by the trypan blue exclusion method.

The amount of mRNA coded for the antibody was determined with Northern blotting. Total RNA was extracted by the guanidine isothiocyanate/hot phenol method from 5×10^6

cells, Northern-blotted and determined by using the ^{32}P-labelled fragment of the C region of the IgG heavy chain.

3. Results and Discussion

3.1 ACCUMULATION OF mRNA AND ENHANCED ANTIBODY PRODUCTIVITY

The concentration of γ1 mRNA in 2E3 cells that had been growth-suppressed by adding interleukin 6 at 20 units/ml was determined by Northern blotting as the relative amount to the mRNA of β-actin. Interleukin 6 suppressed the growth of the four cell lines of hybridomas or myelomas among the five we examined when the cells were adapted to the serum-free medium. The concentration of mRNA increased 3.5-fold (Fig. 2), and the antibody productivity was enhanced 2- to 3.5-fold during growth suppression (Terada *et al.*, 1992). These results supported our prediction. The addition of interleukin 6 to the culture medium satisfied the criteria for a desirable growth suppression method, except that interleukin is expensive, at least for the present.

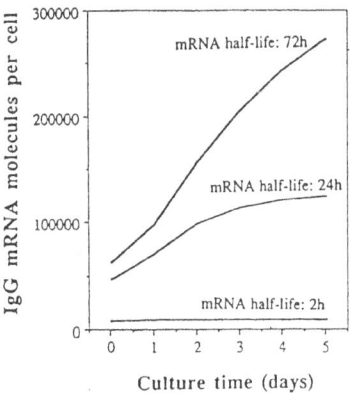

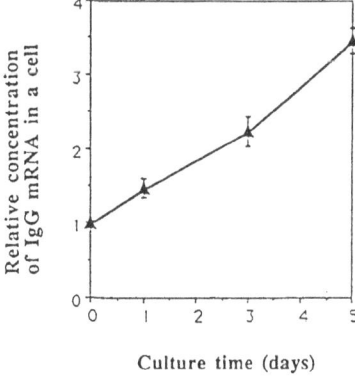

Fig. 1 Predicted accumulation of mRNA coded for the antibody in growth-inhibited hybridoma cells. The growth was assumed to have been completely inhibited on day 0.5 after exponential growth at a specific growth rate of 1.19day^{-1}. The rate of antibody mRNA synthesis was assumed as 3735 molecules/(h cell).

Fig. 2 Concentration of mRNA coded for the antibody in growth-suppressed 2E3 cells. The growth rate was slowed to 1/7 of its maximum by adding interleukin-6 to the culture medium at 20units/ml on day 0 after a two-day culture at the maximum growth rate. The relative concentration of IgG mRNA was determined as the radioactivity of the mRNA band on a filter prepared from the cells in each culture by Northern blotting divided by that for the culture on day 0.

3.2 ADDITION OF CAFFEINE

The cells were growth-suppressed, maintaining a viability of over 70%, when caffeine was added to the culture at a concentration of 1.92mM (Fig. 3). The antibody productivity increased four- to five-fold during growth suppression (Fig. 4). The addition of caffeine to the culture satisfied all the criteria for the industrially applicable growth-suppression method. Applicability of this method to a wide variety of cell lines was suggested by some reported results (Okuda and Kimura, 1988).

3.3 ADDITION OF ANTISENSE C-myc

Antisense c-myc, the oligonucleotide complimenting the first five codons at the 5' terminal of c-myc, was added to the culture at concentration from 0.64μM to 80μM. The growth was suppressed at the 80μM of the antisense level, although this result still needs to be confirmed by comparing with a negative control, using oligonucleotides of random sequence. However, the cell viability decreased rapidly, and we could not find any concentration of the antisense suitable for suppressing growth while maintaining a high cell viability.

3.4 DEPRIVATION OF GROWTH FACTORS

The cells were growth-suppressed by simultaneously depriving of growth factors, insulin and transferrin, from the serum-free medium. The cell viability was higher than 80% at all the growth rates above 20% of the normal rate. The antibody productivity increased three-fold as the growth rate decreased from the normal rate (Fig. 5). This growth-suppression method satisfies all the criteria, except for facile application. To maximize the total production of a batch culture of a hybridoma, we need to let the cells grow rapidly to a little below the over-growth concentration, then stop growth by replacing the culture medium with one deprived of the factors. To achieve this, we need some kind of perfusion culture system, which is more complicated than the batch culture system. This complication of application is more or less common to all the methods in class 2.

3.5 GLUTAMINE LIMITATION

The cells were adapted for a few weeks in a glucose-deficient medium at the normal glutamine concentration until they could grow at the normal rate. They were then cultured at a variety of glutamine concentrations in the glucose-deficient medium, and grew at a rate depending on the glutamine concentration, while maintaining a viability of over 80%. The antibody productivity increased by 70% as the growth rate decreased from the normal rate to 10% of it (data not shown). The glutamine limitation can be achieved without using a perfusion culture system by adding glutamine little by little into batch cultures started with a limited amount of glutamine to supply the amount consumed. Glutamine is a principal energy source for cell growth in a glucose-deficient culture. The appropriate concentration of glutamine for suppressing growth while enhancing antibody productivity is a little above that needed as a protein building block, but much below that required as the energy source for normal growth. The 70% enhancement of antibody productivity was less than that achieved by caffeine addition or growth-factor deprivation; this, we think, is because the enhancing effect of the increased amount of mRNA was partially cancelled by the suppression of protein synthesis due to the limited glutamine concentration. The glutamine limitation satisfies all the criteria, except for the minor suppression of protein synthesis.

3.6 LEUCINE LIMITATION

The cells were growth-suppressed by leucine limitation, while the antibody productivity was not enhanced (data not shown). Since this method is thought to suppress growth by means of suppressing protein synthesis, the lack of enhancement of antibody productivity is consistent with the prediction by our model (Suzuki *et al.*, 1992).

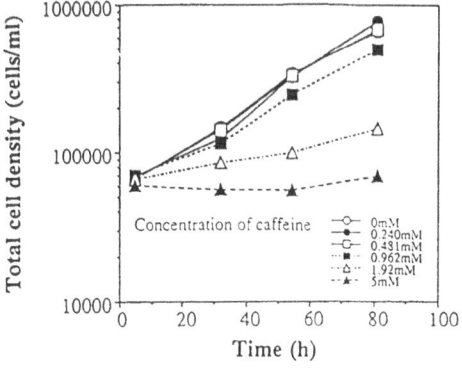

Fig. 3 Suppressive effect of caffeine on the growth of hybridoma 2E3.

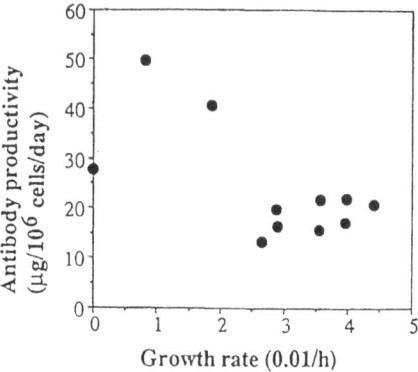

Fig. 4 Antibody production rate per cell *vs.* the growth rate of 2E3 cells. The growth rate was suppressed by caffeine.

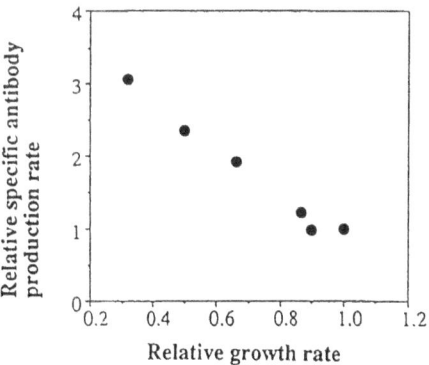

Fig. 5 Relative antibody production rate per cell *vs.* the growth rate of 2E3 cells. The growth rate was suppressed by depriving insulin and transferrin.

4. Conclusion

The prediction was experimentally confirmed that the mRNA coded for an antibody accumulates in hybridoma cells, and that antibody productivity increases concomitantly when the çells are growth-suppressed by a method that does not directly suppress the synthesis of mRNA and protein. Among the fifteen methods we tested for suppressing the growth, the addition of caffeine to the culture seems to be the best for enhancing the antibody productivity of the culture, considering the high productivity, low cost, and ease of application. This conclusion can probably be expanded to other hybridoma cell lines, judging from the parallelism between the prediction by the generalized model and the results from the various experiments.

Acknowledgements

This work was supported by Grant No. 03805088 for General Scientific Research from the Ministry of Education, Science and Culture, Japan, and partly by a grant for the `Biodesign Research Program' from RIKEN to F. Makishima.

References

Karasuyama, H. and Melchers, F. (1988) `Establishment of mouse cell lines which constitutively secrete large quantities of interleukin 2, 3, 4 or 5, using modified cDNA expression vectors', Eur. J. Immunol. 18, 97-104.

Karasuyama, H., Kudo, A., and Melchers, F. (1990) `The proteins encoded by the V_{preB} and λ_5 pre-B cell-specific genes can associate with each other and with μ heavy chain', J. Exp. Med. 172, 969-972.

Mikami, T., Makishima, F., and Suzuki, E. (1991) 'Enhancing effect of mouse peritoneal exudate cells and their products on antibody productivity of hybridoma cells: Application of in vivo factors to in vitro culture', Cytotechnol. 7, 93-101.

Okuda, A. and Kimura, G. (1988)`Elongation of G_1 phase by transient exposure of rat 3Y1 fibroblasts to caffeine during the previous and present generations', J. of Cell Sci. 89, 379-386.

Storb, U.(1973) `Turnover of myeloma messenger RNA', Biochem. Biophys. Res. Commun. 52, 1483-1491

Suzuki, E., Takahashi, K., and Ollis, D. F. (1992) `A simple structured model predicted positively-, negatively-, or non-growth associated antibody production rate depending on culture conditions', in H. Murakami, S. Shirahata and H. Tachibana (eds), Animal Cell Technology: Basic & Applied Aspects, 4, Kluwer Academic Publishers, Dordrecht, pp. 375-378.

Suzuki, E. and Ollis, D. F. (1990) 'Enhanced antibody production at slowed growth rates: experimental demonstration and a simple structured model', Biotechnol. Prog. 6, 231-236.

Terada, S., Makishima, F., and Takamatsu, H. (1992) `Enhancing effect of interleukin-6 on antibody productivity of a hybridoma cell line', in H. Murakami, S. Shirahata and H. Tachibana (eds), Animal Cell Technology: Basic & Applied Aspects, 4, Kluwer Academic Publishers, Dordrecht, pp. 413-418.

CHEMOSTAT CULTURE STUDIES ON CHO CELLS PRODUCING INTERFERON-γ

P.M. HAYTER, M.L. GOULD, I. SALMON, J.M. TONG, & A.T. BULL.
Biological Laboratory,
University of Kent at Canterbury,
Kent CT2 7NJ,
United Kingdom.

ABSTRACT. The physiology of a recombinant Chinese hamster ovary cell line in glucose-limited chemostat culture has been studied over a range of dilution rates (D=0.008-0.020h^{-1}). There was a deviation of the specific growth rate (μ) from D at low dilution rates due to an increased specific death rate. Extrapolation of these data suggested a minimum specific growth rate of 0.011h^{-1} (μ_{max}=0.025h^{-1}). The specific rate of glucose uptake increased linearly with μ and K_S was calculated to be 59.6μM. There was also a linear increase in the rate of lactate production with a higher yield of lactate from glucose at high growth rates. The specific rate of IFN-γ production increased with μ in a manner indicative of a growth related product. Despite changes in the IFN-γ production rate and cell physiology, the pattern of IFN-γ glycosylation was similar at all except the lowest growth rates where there was a decrease in the proportion of fully glycosylated IFN-γ.

1. Introduction

Mammalian cells are used extensively for the production of recombinant human proteins since it is considered that they possess the appropriate machinery for the correct processing of the product. However, since a number of proteins produced by mammalian cells show variable post-translational processing, the intensification of the production process requires consideration of not only the quantity but also the quality of the product.

Human interferon-γ, which has glycosylation sites at asparagine residues 28 and 100, is heterogeneous when produced by CHO cells due to variable occupation of

S. Kaminogawa et al. (eds.), Animal Cell Technology: Basic & Applied Aspects, Vol. 5, 243–250.
© 1993 Kluwer Academic Publishers.

these glycosylation sites. We have shown that there is a decline in the proportion of fully glycosylated interferon-γ produced by CHO cells in the later stages of a typical batch culture (Curling et al., 1990). We have also demonstrated that the pattern of IFN-γ glycosylation remains constant at steady state in chemostat culture but this pattern is influenced by changes to different steady state conditions and by nutrient additions during steady state (Hayter et al., 1992).

In this study we examine the effect of the dilution rate on the specific growth rate, cell physiology and the rate of production and glycosylation pattern of recombinant IFN-γ produced by CHO cells in glucose-limited chemostat culture. Some data from methionine-limited culture are compared.

2. Materials and Methods

The cell line used in this investigation (CHO320) was a Chinese hamster ovary cell producing human interferon-γ co-amplified with DHFR by means of methotrexate. All studies were made in a defined serum free medium based on RPMI1640 supplemented with BSA, insulin, transferrin, pyruvate, alanine, putrescine, selenium, $FeSO_4$, $ZnSO_4$ and $CuSO_4$ (Hayter et al., 1991).

Viable and non-viable cell numbers were determined using trypan blue dye exclusion. Glucose and lactate concentrations were determined by commercial assay kits (Sigma) and ammonia by the indophenol method of Fawcett and Scott (1960). Amino acid concentrations were determined by reverse phase HPLC of OPA derivatised amino acids. Interferon concentrations were determined by ELISA and glycosylation analysis by polyacrylamide gel electrophoresis of immunoprecipitated IFN-γ.

Chemostat culture studies were made in Bioengineering fermenters (Bioengineering, Wald, Switzerland) with a 1.7 litre working volume. The cultures were maintained at pH 7.2 and a dissolved oxygen tension of 40% air saturation. The feed glucose concentration was 2.75mM and it was confirmed that the cells were glucose-limited over the range of dilution rates since the cells responded to an increase in the glucose feed concentration.

3. Results

3.1. THE EFFECT OF THE DILUTION RATE ON CELL GROWTH AND VIABILITY

Figure 1 shows the effect of the dilution rate on the steady state cell number and viability. The decline in cell viability at low dilution rates can be attributed to an increase in the specific death rate (μ_d). The specific death rate is related to the

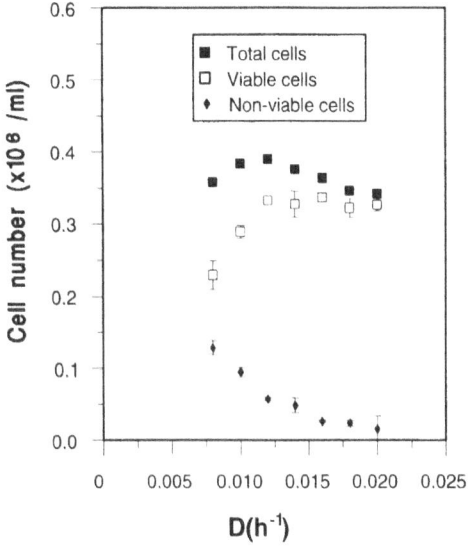

Figure 1. Steady state cell numbers

proportion of viable (X_v) and dead cells (X_d) and the dilution rate (D) by:

$$\mu_d = D\,(X_d/X_v)$$

Since only the viable cell population are able to divide, it follows that the specific growth rate (μ) deviates from D. The relationship between D, μ and μ_d is as follows:

$$\mu = D + \mu_d$$

This implies that $\mu=\mu_d$ at D=0, and indeed, extrapolation of the values for μ and μ_d calculated from the data in Fig 1. indicates that both μ and $\mu_d=0.011h^{-1}$ at D=0. This suggests a minimum specific growth rate (μ_{min}) of $0.011h^{-1}$ for CHO320 cells under glucose limitation.

3.2. THE EFFECT OF THE DILUTION RATE ON CELL METABOLISM

The residual glucose concentration (s) increased with dilution rate in accordance with the Monod model ($\mu=\mu_{max}.s/(K_s+s)$) with $K_s=59.6\mu M$ and $\mu_{max}=0.025h^{-1}$. Linear regression analysis of the plot of the specific rate of glucose uptake ($q_{Glucose}$) versus growth rate implies that there is no maintenance energy requirement for glucose since the intercept of q_s at $\mu=0$ is not significantly different from zero.

There was a linear increase in the specific rate of lactate production with increasing specific growth rate although the yield of lactate from glucose ($Y_{Lac,Glc}$) was increased threefold at the highest dilution rate (Table 1). Linear regression of the plot for $q_{Lactate}$ suggests that there would be no net production of lactate from glucose at $\mu=0.010h^{-1}$ which is close to the value postulated for μ_{min}. The production of serine, which is also derived from glucose metabolism, was relatively constant at high μ, but showed a steep decline as μ approached μ_{min} (Table 1). Again, regression of the curve for q_{serine} suggests that there would be zero production of serine at $\mu=0.011h^{-1}$.

Table 1. Glucose metabolism in glucose or methionine limited chemostat culture

Limiting nutrient	μ	$q_{Glucose}$	$q_{Lactate}$	q_{Serine}	$Y_{Lac.Glc}$
	(h^{-1})	$\mu mol/10^6$ cells/h		$nmol/10^6$ cells/h	mol/mol
Glucose	0.012	0.095	(0.047)	(3.96)	0.49
(*S_R=2.75mM)	0.013	0.093	(0.099)	(6.87)	1.06
	0.014	0.096	(0.078)	(8.18)	0.81
	0.016	0.115	(0.102)	(8.80)	0.89
	0.017	0.120	(0.162)	(8.70)	1.35
	0.019	0.146	(0.210)	(7.47)	1.44
	0.021	0.154	(0.243)	(8.71)	1.58
Methionine	0.012	0.344	(0.670)	(3.48)	1.95
(S_R=0.03mM)	0.023	0.464	(1.077)	(4.94)	2.32

*S_R=Limiting nutrient feed concentration. Numbers in parenthesis are specific production rates.

As might be expected, in methionine-limited culture with glucose excess, $q_{Glucose}$ and $q_{Lactate}$ were considerably higher than in glucose-limited culture with a higher lactate yield from glucose at both high and low dilution rates (Table 1). By contrast serine was produced at a lower rate in methionine-limited culture although the reasons for this are unclear.

Glutamine was consumed at a high rate and its specific rate of uptake ($q_{Glutamine}$) showed a linear increase with increasing μ (Table 2) whereas the specific rate of ammonia production ($q_{Ammonia}$) remained relatively constant. Thus the yield of ammonia from glutamine ($Y_{Amm,Gln}$) declined with increasing specific growth rate.

The net production of alanine by transamination probably accounts for a proportion of the nitrogen from glutamate and there was a linear increase in $q_{Alanine}$ with increasing specific growth rate. Regression of this line indicates that there would be no production of alanine at μ=0.010h^{-1} which is similar to the intercept for $q_{Lactate}$. This suggests that pyruvate, which is the the substrate for both lactate and alanine production, becomes limiting at low specific growth rates. Consequently, the increased ammonia yield at low μ may reflect a lack of substrates such as pyruvate for transamination reactions. This is supported by the results from methionine limited chemostat culture (glucose excess) where $q_{Glucose}$ and $q_{Alanine}$ remain high at D=0.009h^{-1} (Tables 1 and 2) and furthermore, $Y_{Amm, Gln}$ was similar at high and low D under methionine limitation.

Table 2. Glutamine metabolism in glucose or methionine limited chemostat culture

Limiting nutrient	μ	$q_{Glutamine}$	$q_{Ammonia}$	$q_{Alanine}$	$Y_{Amm.Gln}$
	(h^{-1})		$nmol/10^6$ cells/h		mol/mol
Glucose	0.012	64	(91)	(3)	1.42
($*S_R$=2.75mM)	0.013	69	(81)	(9)	1.17
	0.014	68	(84)	(11)	1.24
	0.016	80	(83)	(15)	1.04
	0.017	85	(78)	(15)	0.92
	0.019	92	(95)	(23)	1.03
	0.021	100	(89)	(25)	0.89
Methionine	0.012	74	(75)	(30)	1.01
(S_R=0.03mM)	0.023	86	(84)	(25)	0.98

$*S_R$=Limiting nutrient feed concentration. Numbers in parenthesis are specific production rates.

3.3. INTERFERON PRODUCTION AND GLYCOSYLATION.

The steady state concentration of IFN-γ was similar over the range of dilution rates (data not shown) and there was an increasing specific rate of interferon-γ production with increasing specific growth rate (Figure 2). Although this suggests that IFN-γ production is growth related, it should be noted that such growth related behaviour may not be apparent under different nutrient limitations.

The glycosylation pattern was comparable at specific growth rates between 0.014 and 0.021h⁻¹ (Figure 3) and was similar to that reported previously at D=0.015h⁻¹ with a comparable glucose feed concentration (Hayter et al., 1992). .However at specific growth rates of 0.012 and 0.013h⁻¹ there was a decline in the proportion of fully glycosylated IFN-γ and an increase in the proportion of non-glycosylated IFN-γ. Thus changes in the pattern of IFN-γ glycosylation only occur at growth rates approaching μ_{min}.

4. Discussion

The decrease in cell viability observed at low dilution rate suggests that there is a minimum specific growth rate for CHO320 cells as has been suggested for other mammalian cells (Tovey, 1980; Miller et al., 1988; Frame and Hu 1991). Although

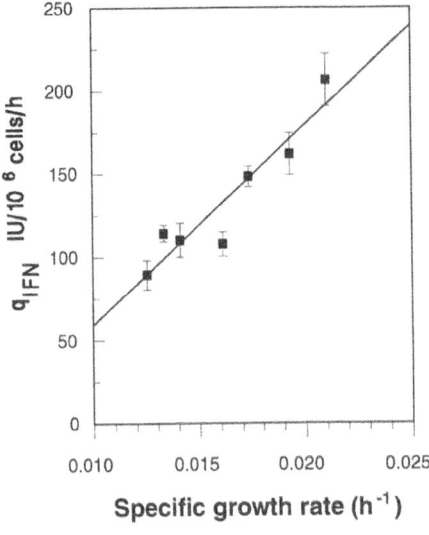

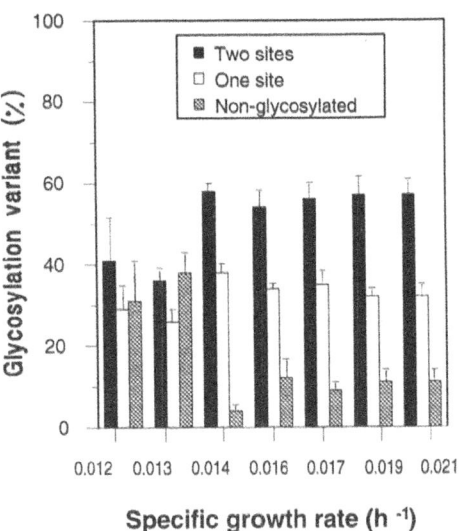

Figure 2. IFN-γ production   **Figure 3.** IFN-γ glycosylation

the reasons for the presence of μ_{min} are unclear, it might be postulated that below μ_{min} the glucose uptake rate is not sufficient to maintain the levels of intermediates necessary for cell survival. It is notable that products of glucose metabolism such as alanine, lactate and serine decline at low μ and regression implies that their production rate would be zero at $\mu=0.010$-$0.011h^{-1}$ which coincides with the calculated μ_{min}. This may result from a decline in the supply of precursors such as pyruvate and 3-phosphoglycerate which are derived from glucose. Studies on human fibroblasts have shown that at a glucose level of 40μM, cell growth is arrested due to the limitation of metabolites derived from glycolytic intermediates (Zielke *et al.*, 1976). In this study, the Monod model suggests that the steady state glucose concentration would be 47μM at μ_{min}.

The linear increase in the specific rate of interferon production with increasing μ is strongly indicative of growth related production kinetics. However since it is possible to uncouple interferon production from cell growth under certain conditions (Hayter *et al.*, 1991), the apparent growth relationship may be a function of the changes in cell metabolism under glucose limitation and not the specific growth rate *per se*. Ongoing studies on the effect of the dilution rate under methionine limitation may reveal whether this is the case.

The relative amounts of the glycosylated and non-glycosylated interferon variants were particularly stable at dilution rates of 0.012 h^{-1} and above. Thus over this range of dilution rates the efficiency of IFN-γ glycosylation did not appear to be influenced by its rate of production or changes in cell metabolism. However, the decline in the amount of fully glycosylated IFN-γ as μ approached μ_{min} may reflect the changes in glucose metabolism at low D. Since the production of several metabolites derived from glycolytic intermediates tend towards zero at μ_{min}, it might be suggested that the supply of sugar precursors for glycosylation might also become limiting at low μ. Therefore our present studies are concerned with the identifying which intermediates might be limiting the glycosylation pathway by the pulse addition of glycosylation precursors to steady state cultures.

5. Acknowledgements

We thank Lucy Gettins for excellent technical assistance, The Wellcome Foundation Ltd for the cell line, Celltech Ltd for the monoclonal antibodies and the SERC Biotechnology Directorate, British Biotechnology, Glaxo, ICI, SmithKline Beecham and The Wellcome Foundation for support of the research programme of which the work described here is a part.

6. References

Curling, E.M.A., Hayter, P.M., Baines, A.J., Bull, A.T., Gull, K., Strange, P.G. and Jenkins, N. (1990) Recombinant human interferon-γ: Differences in glycosylation and proteolytic processing lead to heterogeneity in batch culture. Biochem. J. 272, 333-337

Fawcett, J.K. and Scott, J.E. 1960. A rapid and precise method for the dtermination of urea. J. Clin. Pathol. 13, 156-159

Frame, K.K. and Hu, W.-S. 1991. Kinetic study of hybridoma cell growth in continuous culture. I. A model for non-producing cells. Biotechnol. Bioeng. 37, 55-64

Hayter, P.M., Curling, E.M.A., Baines, A.J., Jenkins, N., Salmon, I., Strange, P.G. and Bull, A.T. 1991. Chinese hamster ovary cell growth and interferon production kinetics in stirred batch culture. Appl. Microbiol. Biotechnol. 34, 559-564

Hayter, P.M., Curling E.M.A., Baines, A.J., Jenkins, N., Salmon, I., Strange, P.G., Tong, J.M. and Bull, A.T. 1992. Glucose-limited chemostat culture of Chinese hamster ovary cells producing recombinant human interferon-γ. Biotechnol. Bioeng. 39,327-335

Miller, W.M., Blanch, H.W. and Wilke, C.R. 1988. A kinetic analysis of hybridoma growth and metabolism in batch and continuous suspension culture: effect of nutrient concentration, dilution rate and pH. Biotechnol. Bioeng. 32, 947-965

Tovey, M. 1980. The cultivation of animal cells in the chemostat: application to the study of tumor cell multiplication. Adv. Cancer Res. 33, 1-37

Zielke, H.R., Ozand, P.T., Tildon, J.T., Sevdalian, D.A. and Cornblath, M. 1976. Growth of human diploid fibroblasts in the absence of glucose utilization. Proc. Natl. Acad. Sci. USA 73, 4110-4114

CHINESE HAMSTER OVARY (CHO) CELL GROWTH AND RECOMBINANT PROTEIN PRODUCTION IN SERUM-FREE MEDIA

PAUL J. BATTISTA, MARY LYNN TILKINS, DAVID A. JUDD,
STEPHEN F. GORFIEN and DAVID W. JAYME
GIBCO BRL/Life Technologies, Inc.,
2086 Grand Island Blvd.
Grand Island, New York 14072 USA

ABSTRACT. CHO cells have become increasingly important for recombinant gene expression, owing to their low rate of spontaneous transformation and biomanufacture of recombinant products that structurally and functionally resemble the native molecules. We recently developed a low protein (<100 $\mu g/ml$), low endotoxin (<0.25 EU/ml) serum-free medium (CHO-S-SFM II) formulated to support the growth of CHO cells and the production of recombinant proteins in suspension culture. Both wild type and recombinant CHO cells were adapted, maintained, cryopreserved and recovered in CHO-S-SFM II. Cells cultured in this serum-free medium out-perform parallel cultures in serum-supplemented medium, reaching peak densities of 3-4 X 10^6 viable cells/ml and producing over 1.0 $\mu g/ml$ of recombinant human chorionic gonadotropin. CHO-S-SFM II demonstrated superior growth performance compared to four commercial serum-free media for CHO cells. A prototype powdered form of CHO-S-SFM II exhibited performance equivalent to liquid medium. Serum-free medium eliminates problems associated with serum usage, such as lot-to-lot performance variability, presence of adventitious agents and fluctuations in price and availability. The low protein content of CHO-S-SFM II facilitates downstream processing of recombinant proteins and reduces final product cost. Additionally, the low endotoxin level of this medium reduces regulatory concerns for the production of therapeutic proteins.

1. Introduction

Chinese hamster ovary (CHO) cells are widely used for the production of recombinant gene products. Examples of recombinant proteins produced using CHO cells include: interferon (Scahill et al., 1983), tissue plasminogen activator (Kaufman et al., 1985), Factor VIII (Kaufman et al., 1988), CD4 cell surface antigen (Davis et al., 1990), prorenin (Portner et al., 1991) and antithrombin III (Yamauchi et al., 1992). CHO cells are well-suited for foreign gene expression owing to a relatively low rate of spontaneous transformation and to production of recombinant products that structurally and functionally resemble the native proteins (Utsumi et al., 1989). Serum-free culture of CHO cells is desirable since it facilitates downstream processing and recovery of products and minimizes problems associated with serum usage, such as lot-to-lot performance variability, presence of exogenous contaminants, and fluctuations in price and availability (Jayme et al. 1988).

Earlier work in the serum-free culture of CHO cells was done using adherent monolayer

251

S. Kaminogawa et al. (eds.), Animal Cell Technology: Basic & Applied Aspects, Vol. 5, 251–257.
© 1993 *Kluwer Academic Publishers.*

cultures. Nutrient Mixture F-12 was initially developed as a serum-free medium (SFM) for the culture of CHO cells (Ham, 1965). The inability of this formulation to support good clonal growth of CHO cells without protein supplementation led to the development of MCDB 301 (Hamilton and Ham, 1977). However, for large-scale industrial applications involving CHO cells, monolayer culture can be technically challenging. It is desirable to utilize some form of suspension culture to achieve higher cell densities and maximum levels of recombinant protein production.

Through comparison of earlier published formulations designed for adherent CHO cell growth, we developed a basal formulation to which several modifications were made to enhance cellular growth in suspension culture. High performance liquid chromatography analysis of nutrient consumption was employed to optimize medium components depleted during high density CHO cell culture. By adjusting the composition of the basal medium to compensate for nutrients depleted during growth, we were able to significantly improve viable cell density. We report here on the development of CHO-S-SFM II, a low protein, low endotoxin SFM formulated to support the growth and production of recombinant proteins by CHO cells in suspension culture.

2. Materials and Methods

2.1. CULTURE OF WILD-TYPE CHO CELLS

Wild-type CHO cells were obtained from Mr. Robert A. Tobey, Los Alamos National Laboratory, and maintained in DMEM/F12 supplemented with 5% FBS. For the evaluation of growth in SFM, CHO cells were directly adapted through three passages to CHO-S-SFM II and following the third passage were seeded in either control medium or SFM at a density of 3 x 10^5 cells/ml in 125 ml spinner flasks. Cultures were incubated without refeeding at 37°C in humidified air containing 8% CO_2 and the viable cell density determined at 24-hour intervals. For growth comparison studies, CHO-S-SFM II was evaluated against four other commercially available serum-free CHO media. Stock cultures maintained in EMEM supplemented with 5% FBS and 40 mg/L L-proline were sequentially adapted to growth in all SFM formulations. Following adaptation, spinner flasks were seeded at a density of 3 x 10^5 cells/ml in a final volume of 100 ml of SFM. Flasks were incubated at 37°C in a humidified atmosphere of 8% CO_2 in air and the viable cell density determined at 24-hour intervals.

2.2. CULTURE OF RECOMBINANT CHO CELLS

Recombinant CHO cells (rCHO) expressing the gene for human chorionic gonadotropin (rhCG) were obtained from Dr. Irving Boime, Washington University School of Medicine, St. Louis, MO. These cells were constructed as described by Matzuk and Boime (1988) using a pM^2 vector containing an ampicillin and neomycin resistant gene and the Harvey murine sarcoma virus long terminal repeat promoter. Selection was conducted in the presence of the neomycin analogue, G148. rCHO cells previously adapted to growth in either DMEM/F12 supplemented with 5% FBS or CHO-S-SFM II were seeded in spinner flasks at a density of 3 x 10^5 cells/ml in a final volume of 200 ml of media and cultured as described above. Viable cell densities were determined at 24 hour intervals and rhCG was quantitated in spent culture media using an immunosorbant assay system (Tilkins et al., 1992). For competitor evaluation, rCHO cell growth and rhCG production in CHO-S-SFM II were compared to Ex-Cell 301 (JRH Biosciences, Lenexa, KS) as described for wild-type CHO cells. For scaleup studies, a 5L Celligen™ bioreactor (New Brunswick Scientific, Edison, NJ) and a 30 L Chemap Airlift (Alfa Laval Inc., South Plainfield, NJ) were used. The 5 L bioreactor was seeded at a density of 3 x 10^5 cells/ml in a final volume of 3.75 L and samples removed at 24 hour intervals for the determination of viable cell density and rhCG production. Initial parameter settings were: dissolved oxygen concentration of 50% saturation, temperature 37°C, pH 7.2 and impeller speed 100 rpm. The Chemap Airlift was seeded and sampled as described for the 5 L bioreactor except that a 15 L volume was used. Settings for the airlift run were: dissolved oxygen concentration, 65% saturation, airflow rate 1.5 L/minute, temperature 37°C and pH 7.4.

3.   Results

Growth of wild-type CHO cells was six-fold higher in CHO-S-SFM II when compared to serum-supplemented medium (Figure 1). Peak cell densities of 3.9 x 10^6 cells/ml were achieved in SFM versus 0.6 x 10^6 cells/ml for control cultures. Both rCHO cell growth and rhCG production were higher in CHO-S-SFM II cultures than serum-supplemented control cultures (Figure 2). Specific productivity for days 3, 4 and 5 in SFM were 0.7, 0.6 and 0.9 pg of rhCG/cell, respectively. Specific productivity for control cultures were 0.1, 0.2 and 0.2 pg of rhCG/cell for days 3, 4 and 5, respectively. A powdered prototype version of CHO-S-SFM II formulation supported rCHO cell growth and rhCG production comparable to that observed with the liquid medium (Figure 3). Specific productivity on day 5 was 0.9 and 0.7 pg of rhCG/cell for liquid and powdered media, respectively. To demonstrate scalability, growth and rhCG production by rCHO cells were evaluated in a 5 L Celligen™ bioreactor and a 30 L Chemap airlift bioreactor. In the 5 L bioreactor, a peak cell density of 2.0 x 10^6 was obtained on day 5, which corresponded to a specific productivity of 1.8 pg of rhCG/cell. Specific productivity on day 6 was 2.6 pg of rhCG/cell (Figure 4). rCHO cells cultured in the airlift bioreactor achieved a peak viable cell density of 1.9 x 10^6 cells/ml on day 5, which corresponded to a specific productivity of 2.6 pg of rhCG/cell (Figure 5). Figure 6 shows the growth of wild-type CHO cells in CHO-S-SFM II versus other commercially available serum-free CHO media. CHO-S-SFM II demonstrated superior growth performance when compared to other commercially available media. A peak viable cell density of 3.0 x 10^6 cells/ml was achieved using CHO-S-SFM II while competitive products resulted in peak viable cell densities which were only 45% to 10% of that observed when cells were cultured in CHO-S-SFM II. Similarly, rCHO cell growth and rhCG production were greater for CHO cells cultured in CHO-S-SFM II than in JRH Ex-Cell 301 (Figure 7). We were unable to successfully adapt and maintain rCHO cells in serum-free CHO media supplied by Ventrex and BioWhittaker.

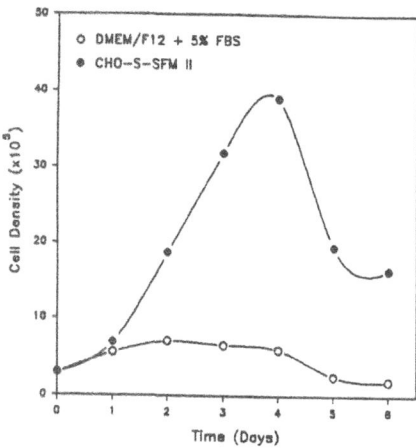

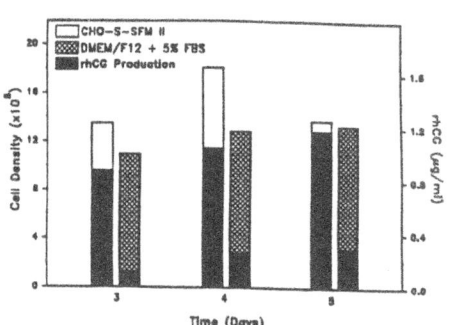

Figure 1: Daily growth kinetics of wild-type CHO cells cultured in CHO-S-SFM II versus DMEM/F12 supplemented with 5% FBS.

Figure 2: Growth and rhCG production by rCHO cells cultured in CHO-S-SFM II versus serum-supplemented DMEM/F12.

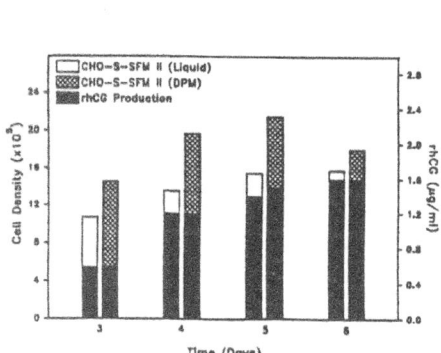

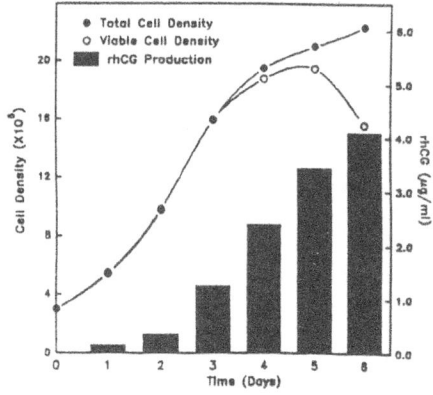

Figure 3: Daily growth kinetics of rCHO cells cultured in liquid versus dry powder CHO-S-SFM II.

Figure 4: Growth and rhCG production by rCHO cells cultured using reconstituted dry powdered CHO-S-SFM II in a 5L Celligen™ bioreactor.

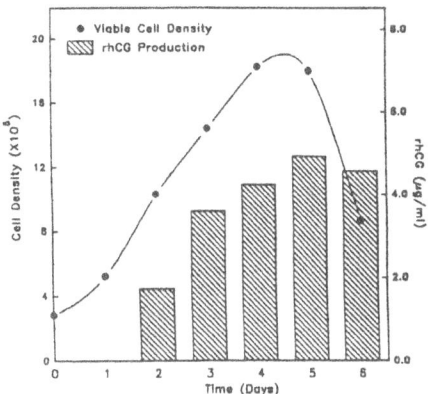

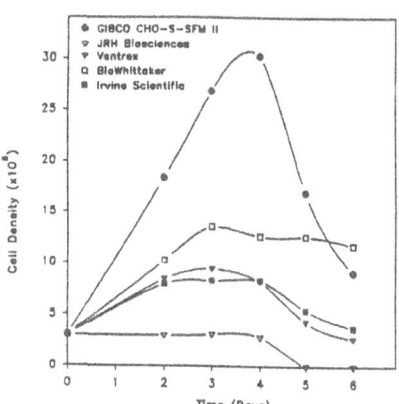

Figure 5: Growth and rhCG production by rCHO cells cultured in a 30L Chemap airlift bioreactor.

Figure 6: Comparison of wild-type CHO cell growth in CHO-S-SFM II versus other commercially available serum-free CHO media.

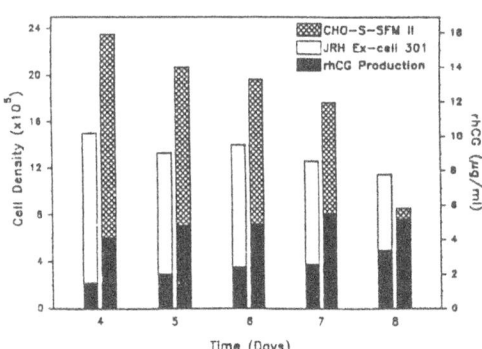

Figure 7: Comparison of growth and rhCG production for rCHO cells cultured in CHO-S-SFM II versus JRH Ex-Cell 301.

4. Discussion

CHO-S-SFM II is a low protein (< 100 µg/ml), low endotoxin (< 0.25 EU/ml), SFM formulated specifically to support the high density growth of wild-type and rCHO cells in suspension culture. Cells cultured in CHO-S-SFM II demonstrated biological performance superior to serum-supplemented media and have been shown to reach peak viable cell densities of 3-4 x 10^6 cells/ml and to produce in excess of 1.0 µg/ml of rhCG. When compared to other commercially available SFM, both wild-type and rCHO cells achieved higher viable cell densities when cultured in CHO-S-SFM II. Additionally, recombinant protein production was enhanced when rCHO cells were grown in this formulation. rCHO cells were successfully scaled-up using CHO-S-SFM II in a 5 L Celligen™ bioreactor and a 30 L Chemap airlift bioreactor. Trials of a powdered version of the CHO-S-SFM II formulation resulted in biological performance equivalent to the liquid medium.

Utilization of SFM eliminates many problems associated with using high supplemental concentrations of animal sera. The low protein content of CHO-S-SFM II facilitates downstream processing and recovery of recombinant proteins, thereby reducing overall production cost. Additionally, the low endotoxin level and reduced risk of adventitious contaminants and human immunogenic components diminish regulatory concerns for the bioproduction of therapeutic proteins.

5. References

Davis, S.J., Ward, H.A., Puklavec, M.J., Willis, A.C., Williams, A.F. and Barclay, A.N. (1990) "High level expression in Chinese hamster ovary cells of soluble forms of CD4 T lymphocyte glycoprotein including glycosylation variats", J. Biol. Chem. 265, 10410-10418.

Ham, R.G. (1965) "Clonal growth of mammalian cells in a chemically defined, synthetic medium", Proc. Natl. Acad. Sci. U.S.A. 53, 288-293.

Hamilton, W.G. and Ham, R.G. (1977) "Clonal growth of Chinese hamster cell lines in protein-free media", In Vitro 13, 537-547.

Jayme, D.W., Epstein, D.A. and Conrad, D.R. (1988) "Fetal bovine serum alternatives", Nature 334, 547-548.

Kaufman, R., Wasley, L. Spiliotes, A., Gossel, S., Latt, S., Larsen, G. and Kay, R. (1985) "Coamplification and coexpression of human tissue-type plasminogen activator and murine dihydrofolate reductase sequences in chinese hamster ovary cells", Mol. Cell. Biol. 5, 1750-1759.

Kaufman, R.J., Wasley, L.C. and Dorner, A.J. (1988) "Synthesis, processing, and secretion of recombinant human factor VIII expressed in mammalian cells", J. Biol. Chem. 263, 6352-6362.

Matzuk, M.M. and Boime, I. (1988) "The role of the asparagine-linked oligosaccharides of the α subunit in the selection and assembly of human chorionic gonadotropin", J. Cell. Biol. 106, 1049-1059.

Portner, R., Matsumura, M., Hatal, T., Murakami, K. and Kataoka, H. (1991) "Perfusion-microcarrier cultivation of rCHO cells in serum-free medium for production of human renin", Bioproc. Eng. 7, 63-69.

Scahill, S.J., Devos, R., Heyden, J.V. and Fiers, W. (1983) "Expression and characterization of the product of a human interferon cDNA gene in Chinese hamster ovary cells", Proc. Natl. Acad. Sci. U.S.A. 80, 4654-4658.

Tilkins, M.L., Judd, D.A., Weiss, S.A. and Gorfien, S.F. (1992) "Serum-free culture of Chinese hamster ovary cells", Focus 14, 95-98.

Yamauchi, T., Asakura, E., Amatsuji, Y., Uno, S., Furuta, R., Tujikawa, M. and Tanabe, T. (1992) "Production of human antithrombin III in a serum-free culture of CHO cells", Biosci. Biotech. Biochem. 54, 600-604.

Utsumi, J., Mizuno, Y., Hosoi, K., Okano, K., Sawada, R., Kajitani, M., Sakai, I., Naruto, M. and Shimizu, H. (1989) "Characterization of four different mammalian cell-derived recombinant human interferon - ß1s: Identical polypeptides and non-identical carbohydrate moieties compared to natural ones", Eur. J. Biochem. 181, 545-553.

6.   Acknowledgements

The authors would like to thank Glenn Godwin (GIBCO BRL, Grand Island, NY) and Mary Pixley (Washington University School of Medicine, St. Louis, MO) for technical assistance, and David DiRamio and Terrilyn M. Summers CPS (GIBCO BRL, Grand Island, NY) for assistance in manuscript preparation. The authors would also like to express their sincere appreciation to Stefan Weiss (Calypte Biomedical, Berkeley, CA) for his dedication to the development of serum-free cell culture technology.

MECHANISM OF IL-10 PRODUCTION BY A CD8[+] T CELL CLONE

Y. MINAI, T. HISATSUNE, K. NISHIJIMA,
A. ENOMOTO AND S. KAMINOGAWA
Department of Agricultural Chemistry
The University of Tokyo
Yayoi 1-1-1, Bunkyo, Tokyo 113, Japan

ABSTRACT   We have established CD8[+] suppressor T cell (Ts) clone 13G2 which produces a suppressive lymphokine, interleukin 10 (IL-10). In order to clarify signal transduction mechanisms of IL-10 production and proliferation of a CD8[+] suppressor T cell clone, we examined whether several T cell-specific stimulations could induce both production of IL-10 and proliferation of 13G2 cells. We detected IL-10 production from 13G2 cells stimulated with immobilized monoclonal antibody to CD3/T cell receptor complex molecules or with IL-2. These stimuli induced proliferation of 13G2 cells, while ^{3}H-thymidine uptake by 13G2 cells stimulated with immobilized anti-CD3 was less than that with IL-2. The result from the experiment using hydroxyurea indicates that 13G2 could produce IL-10 without those proliferation, suggesting that the production of IL-10 is unrelated to the proliferation of parental cells.

These results reveal that IL-2 is a good stimulator for both proliferation of Ts and production of IL-10 from Ts.

1. Introduction

It is conceivable that down-regulation of the immune system would be needed to maintain homeostasis *in vivo* (1). Many groups have descrubed that down-regulation of the immune response was mediated by suppressor T cells (Ts) or their producing factors (2, 30. In order to better understand the logical and clinical aspect of immune response, it is necessary to clarify the condition for stimulation that can induce both production of suppressive lymphokines from Ts and the multiplication of Ts. However, there is little information on lymphokine production and Td proliferation, because stable Ts clones which produce suppressive lymphokines have not been established.

We have recently established Ts clone 13G2 which has suppressive activity for Ag-induced proliferation of helper T cell clones, without exhibiting any cytotoxic activity (4). We also reported that 13G2 produced suppressive lymphokine IL-10, which inhibited type 1 of helper T cell proliferation to antigen plus antigen presenting cells (5).

S. Kaminogawa et al. (eds.), Animal Cell Technology: Basic & Applied Aspects, Vol. 5, 259–266.
© 1993 *Kluwer Academic Publishers.*

In this study, we examine which physiological activator could induce both the production of IL-10 from 13G2 and the proliferation of 13G2.

2. MATERIALS AND METHODS

2.1. MICE

Female C57BL/6 mice (6-8 weeks old) were purchased from Charles River Japan (Tokyo, Japan).

2.2. REAGENTS

Cycloheximide and hydroxyurea was obtained from Wako Pure Chemical Industries (Osaka, Japan). Recombinant human interleukin-2 (IL-2) was generously supplied by Takeda Chemical Industries (Osaka, Japan).

2.3. T CELL CLONES

$CD8^+$ suppressor T cell clone 13G2 was established from lymph node cells of C57BL/6 immunized with bovine αs1-casein, and was maintained by 50 U/ml of IL-2 in conditioned RPMI1640 medium, which was supplemented with 10% FCS, 5×10^{-5} M 2-mercaptoethanol, 100 U/ml of penicillin and 100 μg/ml of streptomycin. αs1-casein-specific $CD4^+$ T cell clone 3D20 was maintained as described previously (4).

2.4. ANTIBODY

Monoclonal antibody (mAb) 145-2C11 (6) recognizing the ε-chain of murine CD3, which is named anti-CD3 mAb, was produced by Dr. J. A. Bluestone of the University of Chicago (Chicago, IL) and kindly provided by T. Tada of the University of Tokyo (Tokyo, Japan). This mAb was purified from murine ascites by passing through protein A-agarose columns (MAPS-II, Bio-Rad, Richmond, CA). The purified antibodies were dialyzed against phosphate buffered saline (PBS), filter-sterilized, and then kept at -20°C until needed for use.

2.5. ASSAY FOR THE DETECTION OF IL-10

The supernatant of 13G2 cells suppressed the antigen-induced proliferation of type 1 helper T cell clone 3D20, and this suppression was completely blocked by the addition of anti-IL-10 antibody (5). Thus, this assay system was used for detecting the biological activity of IL-10. Briefly, 3D20 cells (2×10^4) were stimulated with irradiated spleen cells (2×10^5) plus as1-casein in the absence or presence of IL-10. The cells were cultured for 3 days, and were then pulsed for 20 hrs with 37 KBq of ^{3}H-thymidine (248 GBq/mmol, New England Nuclear, Boston, MA). The amount of radioactivity was measured by standard liquid scintillation counting. One U/ml of IL-10 activity is defined as the amount of IL-10 necessary for a 50 % suppression of 3D20 proliferation in a 200 μl culture. The data presented are the mean of triplicate determinations.

2.6. IL-10 PRODUCTION BY THE 13G2 CLONE

13G2 cells at a concentration of 5×10^5/ml were cultured with a various titers of IL-2 or with immobilized anti-CD3 mAb. Supernatants of the 13G2 cells were harvested after the indicated times, and then treated with Sephadex gel bound anti-IL-2 antibody (presented by Takeda) to remove IL-2. The samples were filtered and measured to detect IL-10.

2.7. TREATMENT OF THE 13G2 CLONE WITH METABOLIC INHIBITORS

13G2 cells at a concentration of 5×10^5/ml in RPMI 1640 conditioned medium were incubated for 2 hrs with cycloheximide (0 to 4 µg/ml) or with 2 mM hydroxyurea. The cells were then stimulated with the immobilized anti-CD3 antibody for 24 hrs. The supernatants were collected, dialyzed against PBS at 4 $^\circ$C, and sterilized by a 0.22 mm membrane filter. Each inhibitor was added to RPMI 1640 at 100 µg/ml and finally dialyzed against PBS at 4 $^\circ$C as a control.

2.8. CELL CYCLE ANALYSIS

13G2 cells were harvested at various times after IL-2 stimulation. The harvested cells were suspended in chilled PBS at 1×10^6 cells/ml, and were then fixed by slowly adding a 2-fold volume of cold ethanol. The cells were centrifuged and resuspended in PBS containing 40 mg/ml of RNase A (Sigma, St. Louis, MO) and 50 mg/ml of propidium iodide (Sigma), before being analyzed for cellular DNA content by a flowcytometry, using 488 nm excitation.

3. Results

3.1. CELL CYCLE ANLYSIS OF 13G2 CELLS

To determine the conditions under which all 13G2 cells would be arrested in the G_0/G_1 phase, the DNA distribution of the 13G2 cells was examined by flowcytometry, after staining with the DNA binding dye, propidium iodide. In this study, 13G2 cells were stimulated with 50 U/ml of IL-2. 37 % of the 13G2 cells being in the $S + M + G_2$ phase 1 day after stimulation (Fig. 1). However, the 13G2 population 6 days after stimulation were found to be arrested in the resting state (G_0/G_1) of the cell cycle. Hereafter, we used 13G2 cells that had been obtained from 6-day culture with IL-2 as resting cells for the subsequent experiments.

3.2. STIMULATION REQUIRED FOR IL-10 PRODUCTION BY 13G2

We assessed whether the production of IL-10 from 13G2 was constitutive or not. As shown in Table 1, IL-10 was not detected in the 13G2 culture without stimulation. We then investigated the effect of T cell-specific stimuli on the production of IL-10 from 13G2 cells. There are two main physiological activation pathways for T cells, one being via the T cell receptor (TCR) and the other via the IL-2 receptor (7). It has been reported that T cells could be stimulated with immobilized anti-CD3 antibody (8), because the binding the anti-CD3 antibody to CD3/TCR complex molecules might mimic antigenic stimulation (9). For 13G2 cells that did not respond to the antigen, anti-CD3 antibody was used as a substitute

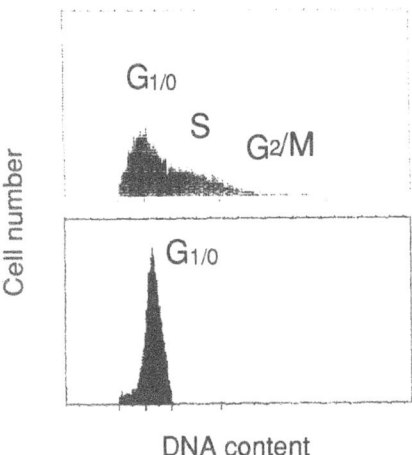

Fig. 1. Cell cycle analysis of 13G2 cells cultured in 50 U/ml of IL-2-containing medium.

for antigenic stimulation. 13G2 cells present in the resting phase were stimulated with IL-2 or immobilized mAb. As indicated in Table 1, IL-10 was detected in the culture of 13G2 stimulated with anti-CD3 antibody or with more than 10 U/ml of IL-2 for 24 hrs, 1 U/ml of IL-2 proving insufficient for inducing IL-10. These results shows that the 13G2 cells did not produce IL-10 constitutively, and that the production of IL-10 required stimulation by the specific antibody or the optimal amount of IL-2.

Table 1   Production of IL-10 from 13G2 cells

Stimulant	IL-10 (U/ml)
-	< 0.5
Anti-CD3	74 ±16
IL-2 1 U/ml	< 0.5
IL-2 10 U/ml	55 ±12
IL-2 40 U/ml	49 ±10

3.3. KINETICS OF IL-10 PRODUCTION BY 13G2

We next determined the minimal time that would be necessary for IL-10 production. This was accomplished by stimulating resting 13G2 cells with either 40 U/ml of IL-2 or with anti-CD3 antibody. As shown in Fig. 1, IL-10 activity was detectable as early as 3 hrs after stimulation with IL-2 or the antibody, indicating that a 3-hr pulse of IL-2 or immobilized mAb would be sufficient to initiate the production of IL-10 from 13G2. The two curves for IL-10 production from 13G2 cells stimulated with IL-2 and with mAb are similar, suggesting that the effect of either stimulant on IL-10 production was equal.

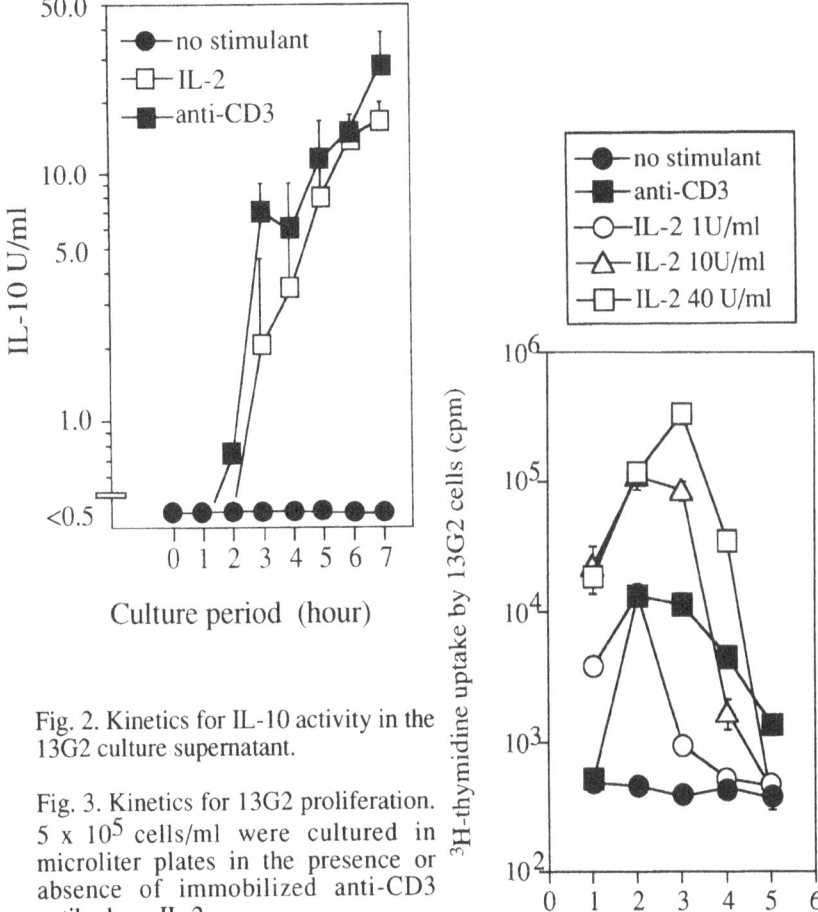

Fig. 2. Kinetics for IL-10 activity in the 13G2 culture supernatant.

Fig. 3. Kinetics for 13G2 proliferation. 5×10^5 cells/ml were cultured in microliter plates in the presence or absence of immobilized anti-CD3 antibody or IL-2.

3.4. STIMULATION REQUIRED FOR THE PROLIFERATION OF 13G2

For Ts to effectively exhibit suppressive activity, the population must be multiplied. We next examined the effect of IL-10-inducing stimuli on 13G2 proliferation. This was accomplished by stimulating resting 13G2 cells with either IL-2 or the immobilized antibody. A kinetic analysis of the ^{3}H-thymidine uptake by 13G2 cells is shown in Fig. 2. The peak for ^{3}H-thymidine incorporation into the 13G2 cells was on the 3rd day in 40 U/ml of the IL-2 culture. The proliferative response of the immobilized anti-CD3 antibody-stimulated 13G2 cells was maximal on day 2. The level of ^{3}H-thymidine in corporated by the anti-CD3 antibody was 1/15 less than that by 40 U/ml of IL-2 stimulation. No DNA synthesis was apparent in the 13G2 cells without stimulation.

3.5. EFFECT OF METABOLIC INHIBITORS ON IL-10 PRODUCTION

In order to determine the relationship between the proliferation of 13G2 cells and the production of IL-10, we arrested the 13G2 cells in the G_1 phase with hydroxyurea (10). Table 2 shows that hydroxyurea at 2 mM could completely block the ^{3}H-thymidine uptake of 13G2 cells stimulated with anti-CD3 antibody. On the other hand, the same dose of hydroxyurea did not reduce IL-10 production from 13G2 cells, suggesting that hydroxyurea had no direct effect on the production of IL-10. This result reveals that the proliferation of 13G2 cells was not necessary for the production of IL-10.

3.6. IL-10 IS *de novo* SYNTHESIZED BY STIMULATION

As described in Table 1, IL-10 was secreted upon appropriate stimulation, so we examined whether these stimuli would affect only the secretion of IL-10. The 13G2 clone was treated with various concentrations of cycloheximide, an inhibitor of protein synthesis (11), and the production of IL-10 from the treated cells was measured. As shown in Fig. 4, a dilution of cycloheximide less than 2 mg/ml did not affect the viability of 13G2 cells when assessed by trypan blue dye exclusion, while more than 0.5 mg/ml of cycloheximide could block IL-10 production, indicating that stimulation was necessary for the synthesis of IL-10. It is suggested that IL-10 was *de novo* synthesized by this clone.

Table 2   **Effect of hydroxyurea on the production of IL-10 from 13G2 cells**

Stimulant	Hydroxyurea	^{3}H-thymidine uptake (cpm)	IL-10 (U/ml)
-	-	735 ±440	<0.5
-	+	1078 ±386	<0.5
Anti-CD3	-	4525 ±578	64 ±25
Anti-CD3	+	784 ±159	77 ±6

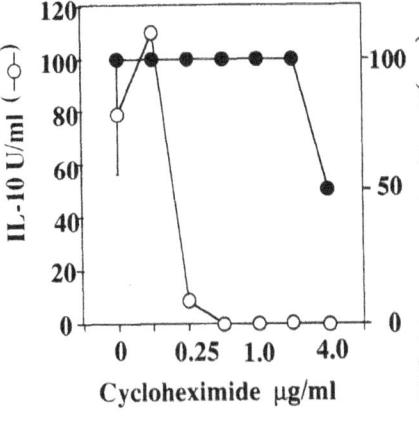

Fig. 4. Effect of cycloheximide on the production of IL-10 from 13G2 cells.

4. Discussion

13G2 cells responded in different ways for IL-10 production and for the proliferation of the cell when stimulated by the anti-CD3 antibody or IL-2 (Figs. 2 and 3). It has been established that there are two main pathways for activation of the T cell, one being through antigen-specific receptor CD3/TCR complex molecules and the other through the lymphokine receptor. It is general in T cells that both pathways work concertedly for their proliferation: stimuli through CD3/TCR transfers T cells from the G_0 to G_1 phase with the production of lymphokine, and this lymphokine transfers the cell from the G_1 to S phase through the lymphokine receptor (7, 12). Therefore, we confirm that stimulation through CD3/TCR can lead T cells to synthesize lymphokine and to this proliferation.

However, we could not find in the 13G2 cell the concerted reaction between the two pathways observed for the T cell. The anti-CD3 antibody did induce lymphokine IL-10, but not the high level of the proliferation compared with IL-2 stimulation. These results lead us to consider that activation by anti-CD3 works mainly for IL-10 production, and that activation by IL-2 works for both IL-10 production and proliferation. As shown in Table 1 and Fig 3, both anti-CD3 and IL-2 induced same level of IL-10 production, but the inductive ^{3}H-thymidine uptake by 13G2 was quite different between these stimuli. It is therefore considered that IL-10 production and proliferation of parental cells might be functionally separated. The result from Fig. 4, which shows that IL-10 production was induced without G_1/S transit, also supports IL-10 production being unrelated to the proliferation of parental cells.

However, it could seem plausible that the production of IL-10 via antigen stimulation is not a direct consequence of the T cell receptor complex being occupied by the anti-CD3 antibody, and that this is due to stimulation by IL-2 secreted through the action of the monoclonal antibody. This possibility can be excluded, because the ^{3}H-thymidine uptake of 13G2 stimulated by CD3/TCR occupancy was the same as that stimulated with 1 U/ml of IL-2, and this level of IL-2 cannot induce IL-10 production. Additionally, we did not detect IL-2 in the supernatant of 13G2 by a CTLL-2 assay (data not shown).

The distinct actions on antigen and lymphokine stimulation in the T cell clone have been reported by Harris et al. (13). They demonstrated that antigenic stimulation only induced the production of lymphokine (macrophage-activating factor), and not proliferation of the cytotoxic T cell clone, and that IL-2 induced proliferation of the cells without producing this macrophage-activating factor. Our findings are in contrast to the report that IL-2 did induce lymphokine production as well as the proliferation of parental cells. It is important to confirm whether the interesting behavior of the 13G2 cell described in this report is observed generally reproduced in IL-10 production or not. Further investigations on the induction of IL-10 from other CD8$^+$ T cell clones should be performed.

5. References

1. Cantor, H. and Gershon, R. K. (1979) 'Immunological circuits: Cellular composition', Fed. Proc. 38, 2058-2064.
2. Sercarz, E. and Krzych, U. (1991) 'The distinctive specificity of antigen-specific suppressor T cells', Immunol. Today 12, 111-117.
3. Hodes, R. J. (1989) 'T-cell-mediated regulation: Help and suppression', in W. E. Paul (ed.), Fundamental Immunology 2nd ed., Raven Press, New York, pp. 587-620.

4. Hisatsune, T,, Enomoto, A., Nishijima, K., Minai, Y., Asano, Y., Tada, T. and Kaminogawa, S. (1990) 'CD8$^+$ suppressor T cell clone capable of inhibiting the antigen- and anti-T cell receptor induced proliferation of Th clones without cytolytic activity', J. Immunol. 145, 2421-1378.
5. Hisatsune, T., Yuji, M., Nishijima, K., Enomoto, A., Moore, K. W., Yokota, T., Arai, K. and Kaminogawa, S. (1992) 'A suppressive lymphokine derived from Ts clone 13G2 is IL-10', Lymphokine Cytokine Res. 11, 87-93.
6. Leo, O., Michele, M., Sachs, D. H., Samelson, L. E. and Bluestone, J. A. (1987) 'Identification of a monoclonal antibody specific for a murine T3 polypeptide', Proc. Natl. Acad. Sci. USA 84, 1374-1378.
7. Weiss, A. (1989) 'T lymphocyte activation', in W. E. Paul (ed.), Fundamental Immunology 2nd ed., Raven Press, New York, pp. 359-384.
8. Geppert, T. D. and Lipsky, P. E. (1987) 'Accessory cell independent proliferation of human T4 cells stimulated by immobilized monoclonal antibodies to CD3', J. Immunol. 138, 1660-1666.
9. Chang, T. W., Kung, P.C,. Gingras, S.P. and Goldstein, G .(1981) 'Does OKT3 monoclonal antibody react with an antigen-recognition structure on human T cells ?' Proc. Nalt. Acad. Sci. USA 78. 1805-1808.
10. Young, C. W. and Hodas, S. (1964) 'Hydroxyurea: Inhibitory effect on DNA metabolism'. Science 146, 1172-1174.
11. Young, C. W., Robinsom, P. F. and Sacktor, B. (1963) 'Inhibition of protein synthesis in intact animals by acetoxycycloheximide (E-73) and a concomitant metabolic derangement', Fed. Proc. 22, 237.
12. Tepperman, K., Finer, R., Donovan, S., Elder, R. C., Doi, J., Ratliff, D. and Ng, K. (1984) 'Interleukin 2 regulates expression of its receptor and synthesis of gamma interferon by human T lymphocytes', Science 225, 429-430.
13. Harris, D. T., Kozumbo, W. J., Cerutti, P. and Cerottini, J. (1987) 'Molecular mechanisms involved in T cell activation 1. Evidence for independent signal transducing pathways in lymphokine production vs proliferation in cloned cytotoxic T lymphocytes', J. Immunol. 138, 600-605.

GENERATION OF CHO CELL MUTANTS FOR GROWTH CONTROL

N. JENKINS AND A. HOVEY

Biological Laboratory,
University of Kent,
Canterbury,
Kent CT2 7NJ.
United Kingdom

ABSTRACT. The use of a temperature switch to control the growth and productivity of temperature-sensitive (*ts*) mutants was investigated, in order to extend the productive life span of recombinant CHO cells in batch culture. Bromodeoxyuridine was used at 39°C to select mutagenized CHO-K1 cells, which resulted in the isolation of 31 temperature-sensitive mutants that were growth-inhibited at 39°C. Two of these mutants were successfully transfected with the gene for tissue inhibitor of metalloproteinases (TIMP) using glutamine synthetase amplification, and a permanent recombinant cell line established (5G1-B1) that maintains the *ts*-phenotype.

Continuous exposure to the non-permissive temperature (npt) of 39°C led to a rapid decline in cell viability, however a temperature regime using alternating incubations at 34°C and 39°C arrested the 5G1-B1 cells, whilst retaining a high cell viability for up to 170 hours in culture. The specific production rate of the growth-arrested cells was 3-4 times that of control cultures maintained at a constant 34°C over the crucial 72-130 hour period of culture, which resulted in a 35% increase in the maximum product yield.

1. Introduction

A major limitation on the batch culture productivity of animal cell lines such as CHO, hybridoma or NS/0 is the disturbing speed at which cells die after reaching maximum cell density (Bebbington *et al.*, 1992). Furthermore, recent chemostat (Miller *et al.*, 1988) and modelling studies (Suzuki and Ollis, 1989) have demonstrated an inverse correlation between hybridoma cell growth rate and the secretion of protein product, and have concluded that exploration of methods to arrest cells while retaining high viability is likely to improve specific productivity. Mutant cell lines (particularly temperature-sensitive mutants) have proved a fruitful source of reversibly arrested systems that have the potential to maintain high levels of viability in prolonged culture (Marcus *et al.*, 1985). However, the actual performance of such mutants as mammalian host lines for recombinant protein synthesis has never been tested.

267

S. Kaminogawa et al. (eds.), Animal Cell Technology: Basic & Applied Aspects, Vol. 5, 267–272.
© 1993 *Kluwer Academic Publishers.*

268

Here we report on the isolation of a temperature-sensitive (*ts*) mutant CHO line and its transfection with the recombinant gene for tissue inhibitor of metalloproteinases (TIMP). The growth, productivity and metabolism of this line were studied under different temperature regimes.

2. Materials and Methods

A mutant selection protocol based on that described by Dermody *et al* (1986) was employed, using ethyl-methane-sulphonate to mutagenize CHO-K1 cells and bromodeoxyuridine (BU) to select *ts*-mutants at the non-permissive temperature (npt) of 39°C. Putative mutants were validated in 24-well plates and replicate T25 flasks, before being cloned by limiting dilution. The plasmid DNA used for transfection (pH.TIMP2.GS) contained an hCMV-driven TIMP transcription unit combined with a Glutamine Synthetase (GS) selectable marker. Transfections and a single round of methionine sulphoximine (MSX, 25µM) selection were performed according to the method of Cockett *et al* (1990).

Recombinant TIMP production was assayed in cell supernatants by an enzyme linked immunosorbant assay (ELISA) adapted from that used by Cockett *et al* (1990). Culture supernatants were also assayed for glucose by the o-toluidine method (Sigma Assay 635), lactate by a lactate dehydrogenase procedure (Sigma Assay 836-uv) and ammonia by the indophenol method of Fawcett and Scott (1960). The DNA content of cells was measured using an Epics Profile II Flow Cytometer (Coulter Electronics Ltd.), and the proportion of cells in G_1/G_0, S, and G_2/M phases was calculated using Coulter Multicycle software.

3. Results

Many of the initial colonies surviving BU treatment were found to be false positive on replica testing, resulting in a final number of 31 *ts*-mutants (i.e. $1:10^6$ cells). The BU dose affected the number of *ts*- mutants isolated: 9% resulted from exposure to 200µM BU, 42% from 100µM BU, 18% from 50µM BU and 31% from 25µM BU. One mutant in particular (5G1, isolated using 100 µM BU) displayed good growth characteristics at 34°C together with a strong *ts*-phenotype that was stable over many generations. This line was used as the principal host to receive the recombinant gene for TIMP. The highest TIMP producer (5G1 subclone B1) was finally chosen as the most promising cell line.

Optimisation of temperature control was first explored using replicate cultures of 5G1-B1 cells exposed to intermittent pulses of either 8 hours or 12 hours at 39°C throughout the culture, followed by recovery periods at 34°C (Fig.1). Control cultures at either constant 39°C or constant 34°C were set up in parallel from the same seed stocks. Intermittent 8 hour exposures to 39°C did not significantly affect cell growth. However, recurrent 12 hour exposures to 39°C were able to effectively control cell growth over the first 48 hours of culture, as shown by a significantly lower peak cell density (p<0.05). Unfortunately, this temperature regime was not able to maintain a prolonged static culture and cell viability fell sharply after 72 hours.

In the next experiment cells were therefore exposed alternately for 12 hours at 39°C and 34°C only for the first 48 hours, after which they were cultured continuously at 34°C. Triplicate flasks were sampled daily and compared with control 5G1-B1 flasks run in parallel at 34°C throughout the experiment. This temperature regime was able to effectively control growth without compromising cell viability (Fig.2). Arrested cells produced greater quantities of TIMP at a faster rate than in controls during the critical mid phase of culture (Fig.3) resulting in a final TIMP concentration (290mg/l) 35% higher than that produced by control cells. At 72 hours of culture the specific production rate of the growth-arrested cells (3.4pg/cell/hr) was 3 fold greater than in control cells, and this value remained significantly above the control rate over 72-148 hours. Ammonia levels in the medium increased throughout the culture at a similar rate in both control and growth-arrested cells, resulting in a final ammonia concentration of 1.2-1.3mM in all cultures. Control cells utilised glucose steadily throughout the culture, whereas glucose utilisation declined from the point at which growth was arrested in npt-exposed cultures (data not shown). During growth arrest lactate levels decreased, coincident with the decline in glucose utilisation Arrested cells displayed a significant increase in the percentage of cells in G_1/G_0 and a significant decrease in the percentage of cells in S phase.

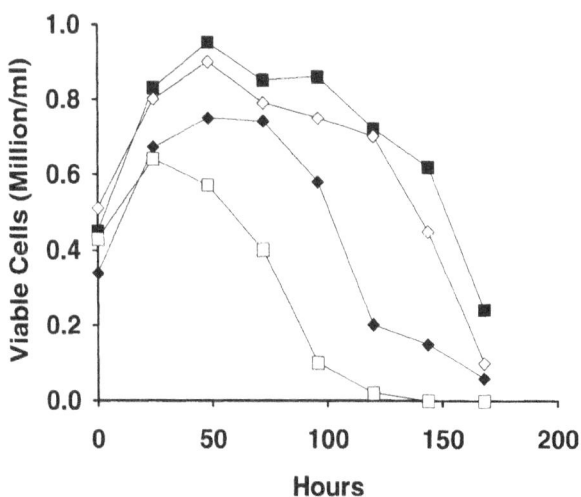

Fig. 1. Effect of four different temperature regimes on the growth of 5G1-B1 cells. Triplicate flasks were incubated at constant 34°C (■), constant 39°C (□), alternating 12hr periods at 34°C and 39°C (♦), or cycles of 8hr at 39°C followed by 16hr at 34°C (◊).

4. Discussion

This is the first reported incidence of a *ts* mutant being used to control growth and productivity in a recombinant mammalian cell culture. This system is clearly an improvement on the level of growth control achieved by temperature changes in wild-type cells such as rat mouse mouse triomas (Bloemkolk *et al.*, 1992) and mouse hybridomas (Sureshkumar and Mutharasan, 1991). Several reports, including this paper, have shown that slowing down growth increases the specific production rate. These studies include theoretical models (Suzuki & Ollis, 1989) and experiments on hybridomas (Miller *et al.*, 1988), where higher specific antibody production rates were obtained at lower specific growth rates (achieved by manipulating the dilution rate in chemostat culture).

Various explanations have been offered for this phenomenon, e.g. the high requirements for nucleotides and proteins during DNA synthesis and mitosis may limit the availability of substrates for recombinant RNA and protein synthesis. Interestingly, the 3-fold increased 5G1-B1

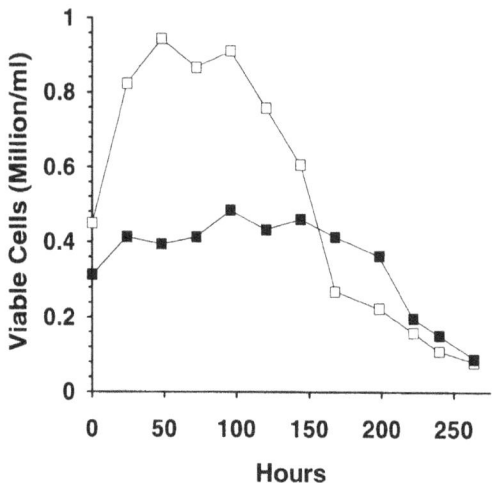

Figure 2. Mean viable cell counts of 5G1-B1 cells at constant 34°C (□, control) or with two 12hr exposures to 39°C (■, see text for details).

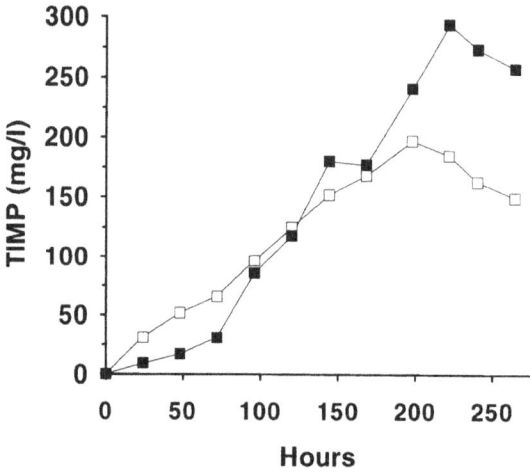

Figure 3. Concentration of TIMP (mg/l) produced by 5G1-B1 cells at constant 34°C (□, control) or with two 12hr exposures to 39°C (■, see text for details).

productivity following npt exposure was achieved despite a significant reduction in glucose uptake and lactate production, indicating a slower metabolic rate in arrested cells.

The Miller study (1988) and that of Ramirez & Mutharasan (1990) also suggest that a minimal growth rate is essential for preserving cell viability, although the precise rate required will obviously vary between cell lines. In contrast, the very low μ value obtained in arrested 5G1-B1 cells ($0.004h^{-1}$) did not compromise viability until late in the culture (>160 hours, Fig.2). Providing the extent of npt exposure is kept to the minimum possible for growth control (i.e. two 12 hour periods at 39°C), cell viability can be maintained at ≥80% up to 190 hours of culture. Furthermore, there was no ultrastructural evidence of major cell damage in 5G1-B1 cells arrested using the alternating temperature regime. This cell line accumulates in the G_1/G_0 phase of the cell cycle when transferred to npt.

In conclusion, this report has shown that it is possible to transfect a temperature sensitive CHO cell mutant with a recombinant gene and use the engineered cell line to improve productivity using temperature control. Our studies demonstrate that growth can be efficiently uncoupled from product synthesis in this *ts*-mutant line, and the system has potential for enhancing the productivity of mammalian cells grown in simple batch fermenters.

5. Acknowledgements

The authors are grateful for the support and advice of Celltech Biologics and Celltech Research Departments, and to the SERC Biotechnology Directorate and the Royal Society for financial support.

6. References

Bebbington, C.R., Renner, G., Thomson, S., King, D., Abrams, D. and Yarranton, G.T. (1992) High-level expression of a recombinant antibody from myeloma cells using a glutamine synthetase gene as an amplifiable selectable marker. Bio/Technology 10, 169-175.

Bloemkolk, J.W., Gray, M.R., Merchant, F. and Mosmann, T.R. (1992) Effect of temperature on hybridoma cell cycle and MAb production. Biotech. Bioeng. 40, 427-431.

Cockett, M.I., Bebbington, C.R. and Yarranton, G.T. (1990) High level expression of tissue inhibitor of metalloproteinases in chinese hamster ovary cells using glutamine synthetase gene amplification. Bio/technology 8, 662-667.

Dermody, J.J., Wojcik, B.E., Du, H. and Ozer, H.L. (1986) Identification of temperature-sensitive DNA⁻ mutants of chinese hamster cells affected in cellular and viral DNA synthesis. Mol. Cell. Biol. 6, 4594-4601.

Fawcett, J.K. and Scott, J.E. (1960) A rapid and precise method for the determination of urea. J. Clin. Path. 13, 156-159.

Marcus, M., Fainsod, A. and Diamond, G. (1985) The genetic analysis of mammalian cell-cycle mutants. Ann. Rev. Genetics 19, 389-421.

Miller, W.M., Blanch, H.W. and Wilke, C.R. (1988) A kinetic analysis of hybridoma growth and metabolism in batch and continuous suspension culture, effect of nutrient concentration, dilution rate, and pH. Biotech. Bioeng. 32, 947-965.

Miller, W.M., Blanch, H.W. and Wilke, C.R. (1988) A kinetic analysis of hybridoma growth and metabolism in batch and continuous suspension culture, effect of nutrient concentration, dilution rate, and pH. Biotech. Bioeng. 32, 947-965.

Ramirez, O.T. and Mutharasan, R. (1990) Cell cycle and growth phase-dependent variations in size distribution, antibody productivity, and oxygen demand in hybridoma cultures. Biotech. Bioeng. 36, 839-848.

Sureshkumar, G.K., and Mutharasan R. (1991) The influence of temperature on a mouse mouse hybridoma growth and monoclonal-antibody production. Biotech. Bioeng. 37, 292-295.

Suzuki, E. and Ollis, D.F. (1989) Cell cycle model for antibody production kinetics. Biotech. Bioeng. 34, 1398-1402.

REMOVABILITY AND PERMEABILITY OF DNA IN A SOLUTION BY CUPRAMMONIUM REGENERATED CELLULOSE HOLLOW FIBER (BMM™) FOR PROCESS VALIDATION OF PURIFICATION PROCESS OF BIO-DRUGS

T. HIRASAKI,[1]   K. YAMAGUCHI,[2]   A. KONO,[3]   T. NODA,[1]   S. UEMATSU,[1]
T. TSUBOI,[1]   K. IMADA,[4]   and N. YAMAMOTO[5]
[1]Asahi Chemical Ind. Co. Ltd., The Imperial Tower 18F, 1-1
Uchisaiwaicho 1-chome, Chiyoda-ku, Tokyo 100, Japan; [2]Yamaguchi
University School of Medicine, Kogushi 1144, Ube, Yamaguchi 755,
Japan; [3]National Kyushu Cancer Center, Notame 3-1-1, Minami-ku,
Fukuoka 815, Japan; [4]Miyazaki University, Gakuenkibanadainishi
1-1, Miyazaki, Miyazaki 889-21, Japan; [5]Tokyo Medical and Dental
University School of Medicine, 1-5-45 Yushima, Bunkyo-ku, Tokyo
113, Japan

ABSTRACT. The removability and/or permeability of DNA by the BMM virus
removal filter was investigated as a function of the dispersion state of
DNA in a bio-drug solution. Isolated DNA of various molecular weights
and packaged DNA in a virus were filtered by BMM. Dead-end filtration
was performed under a constant transmembrane pressure (TMP) of 200 mm Hg
. After filtration, the DNAs captured by BMM were observed by transmission electron microscopy (TEM). In the case of BMM with a mean pore size
of 15nm (BMM15), the permeability of the isolated DNA was more than 0.90
when the molecular weight was less than 10^7. Packaged DNAs in the virion
were removed at over 5 of the logarithmic rejection coefficient. Electron microscopic observation showed that DNAs were elongated by the flow
of the solution through BMM. It was concluded that the removal and reduction of contaminating DNA in bio-drugs needs to allow for the dispersion state of DNA in the solution.

1. INTRODUCTION

The strategy against the virus contamination of bio-drugs has been
given by various governments [1-3]. To ensure the safety of final products, inactivation and/or removal of viruses has to be used during the
purification process in manufacturing. In a previous paper, we have
shown that the BMM virus removal filter was applicable for process
validation in manufacturing of bio-drugs [4]. The mechanism for virus
removal by BMM depends on the size effect, regardless of the property of
a virus having DNA or RNA and whether of the envelope or non-envelope
type [4,5]. We could predict the virus removability with BMM by their
size [4,5].

The regulation of residual DNA in bio-drugs has been proposed in
governmental guidance [1-3]. Oncogenic DNA originating from host cells
and infectious DNA in viruses may cause adverse effects on the human
body. Therefore, a method for validating the contaminating DNA in bio-

273

S. Kaminogawa et al. (eds.), Animal Cell Technology: Basic & Applied Aspects, Vol. 5, 273–279.
© 1993 *Kluwer Academic Publishers.*

drugs is needed in the same way as that for viruses. We tried to apply the BMM virus removal filter to remove DNA.

In addition, we also wanted to clarify the permeation mechanism of such biopolymers as protein and DNA during filtration.

In this report, we will evaluate the removability and permeability of DNA with various dipersion states in a solution by BMM. Electron microscopic observation of DNA in BMM after filtration will be performed to identify the structure during filtration.

2. MATERIALS AND METHODS

(1) BMMs: The BMM with a mean pore size of 15nm (BMM15) and 35nm (BMM35) were from Asahi Chemical Ind. Co. Ltd. Japan. Each BMM was prepared from cellulose cuprammonium solution through the micro-phase separation method. The mean pore size, $2\overline{r}_f$, was obtained by the water flow method.

(2) Viruses: λ phage was suspended in pepton-yeast extract medium. Hepatitis B virus (HBV) and hepatitis C virus (HCV) were suspended in human plasma.

(3) Proteins: Commercially available proteins with various molecular weights (MW) were employed; thermolysin from bacillus thermoproteolytics (MW=2.8x10^4, Serva Feinbiochemica GmbH&Co.), bovine serum albumin (MW= 6.5x10^4, Sigma Chemical Co.), bovine serum γ -globulin (MW=1.5x10^5, Sigma Chemical Co.), urease (MW=4.8x10^5, P-L Biochemicals Inc.), human IgM (MW=9.0x10^5; Calbiochem-Behring Co.), bovine serum β -lipoprotein (MW=2.7x10^6, United States Biochemical Co.), and hemocyanin from keyhole limpet (MW=7.5x10^6, Calbiochem-Behring Co.). Blood coagulation factor VIII combined with von-Willebrand factor (F-VIII with vWF) was supplied by Chemo-Sero-Therapeutic Research Institute. The concentration of each commercial protein was determined by a protein-dye binding assay, using CBB-G250. The concentration of F-VIII with vWF was measured by the partial thromboplastin time method.

(4) DNAs

① Isolated DNAs: Phage vector λ gt-11 (MW=3x10^7, Takara Shuzo Co. Ltd.) and T4 phage DNA (MW=1x10^8, Wako Pure Chemical Ind. Ltd.) were commercially available. These were suspended in Tris/EDTA buffer (pH 7.5), the concentration being detected by spectrophotometry and ethidium bromide staining.

② Virus DNAs: DNAs of λ phage and HBV were detected by dot-blot hybridization. RNA of HCV was detected by a polymerase chain reaction (PCR), using a specific primer.

(5) Ultrafiltrations: The dead-end filtration was carried out under a constant transmembrane pressure (TMP) of 200 mm Hg at 20°C.

(6) Electron microscopy: After filtration, BMM was fixed with a glutaraldehyde/paraformaldehyde mixture, immersed in tannic acid, post-fixed with osmic acid, and then embedded in epoxy resin. Ultra-thin sections were sliced and stained with lead citrate and uranylacetate, and observed through transmission electron microscope (TEM).

3. RESULTS AND DISCUSSION

Figure 1 shows the dependence of permeability φ on the molecular weight (MW) of various proteins and DNAs for BMM15 and BMM35. When MW was less than 10^6, the permeability of a protein was more than 0.90 for BMM35. In the case of DNA, when MW was less than 10^8, φ was more than 0.90 for BMM35, and when MW was less than 3×10^7, φ was more than 0.90 for BMM15. Although the permeability of each DNA and protein was different, even with a common MW, both φ values depended on the MW value. Interestingly, the permeability of one kind of protein, blood coagulation factor Ⅷ combined with von Willebrand factor (F-Ⅷ with vWF) whose MW was about 3×10^7, was 0.95 for BMM 35 and was located nearer to the permeability line for DNA rather than that for protein. It is considered that the difference in structure between DNA and protein reflected their difference in permeability.

Table 1 shows the permeability of DNA and RNA with various dispersion states for BMM35. We used three viruses of λ phage, HBV and HCV as typical examples. In regard to the dispersion state of DNA, we classified this into "free" and "packaged". "Free" means molecularly dispersed, and "packaged" means enveloped in the virion. Free DNA passed through by over 0.70 of the original, but packaged DNA and RNA were removed at over 4 of the logarithmic rejection coefficient. Data for HCV and HBV are quoted from previous papers [6] and [7]. This result indicates that the removability of DNA depended on its dispersed state in the solution.

Figure 2 shows a transmission electron micrograph of the cross sectional view of BMM35 after filtration of isolated λ phage DNA. DNA in a solution is usually considered to exist in a random coiled state. DNAs observed in BMM were like hair, and apparently elongated by the flow of their solution through BMM.

Figure 3 shows a transmission electron micrograph of the cross sectional view of BMM15 after filtration of isolated λ phage DNA. The structure of BMM15 had more closely packed pores than that of BMM35, and similarly to Figure 2, DNAs in BMM were like hair.

We can conclude that necessary aspects for validating the removal of contaminating DNA are as follow:
1) Detection system of DNA under high sensitivity
2) Reduction and/or removal system of DNA with various dispersion states.

Table 2 summarizes the reduction and/or removal system of DNA according to its dispersion states. Packaged DNA and RNA in microbes could be effectively removed by BMM15. In the case of non-packaged DNA removal, for example isolated DNA or combined DNA with other molecules, the affinity method may be effective in addition to filtration by BMM15.

Concerning the filtration mechanism for biopolymers, the permeability of DNA and protein depends on their sizes in the same way as that for virus particles. We have classified biopolymers into two types, i.e. , "solid or rigid" and "flexible or liquid", according to the structural change under shear stress. Table 3 shows the structural change of microorganisms under shear stress, in which undeformable in flow is defined as "solid or rigid", and deformable as "flexible or liquid". For example , albumin, IgG, IgM and viruses are of the solid type, while DNA, RNA and F-VIII with vWF are of the flexible type. Mycoplasma is intermediate between solid and flexible. In addition to the size effect, such bio-rheological deformation would affect on the permeation of a biopolymer by BMM.

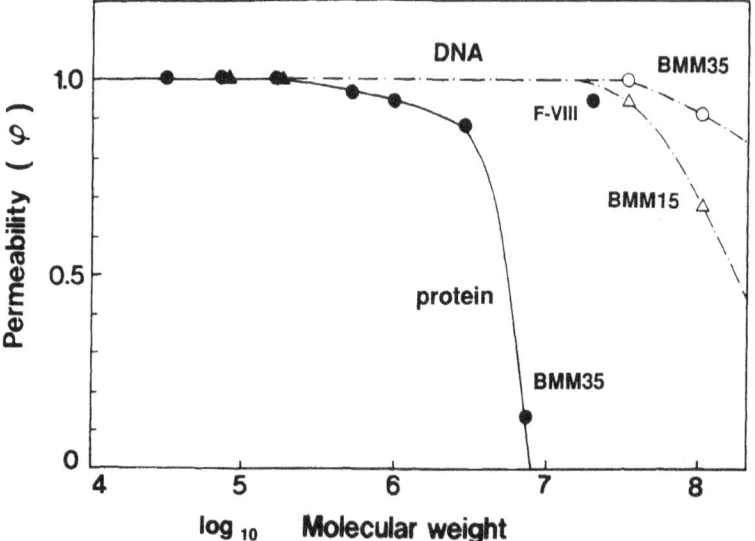

Figure 1. Dependence of φ on the molecular weight of various proteins and DNAs for BMM15 and BMM35.
The molecular weight scale is logarithmic, and the numbers attached to BMM indicate the mean pore size in nm. The chain line indicates data for DNA, and the solid line for protein.

Table 1. Permeability of DNA and RNA with various dispersion states for BMM35.

DNA and RNA		Permeability	Detection method
λ-phage	free DNA	1	ethidium bromide staining OD
	packaged DNA	$< 10^{-5}$	dot-blot hybridization
HBV	free DNA	0.7	dot-blot hybridization
	packaged DNA	10^{-5}	
HCV	free RNA	N. D.	PCR
	packaged RNA	10^{-4}	

The data for HCV are from a previous paper [6], and for HBV are taken from paper [7]. N.D. indicates "not detectable".

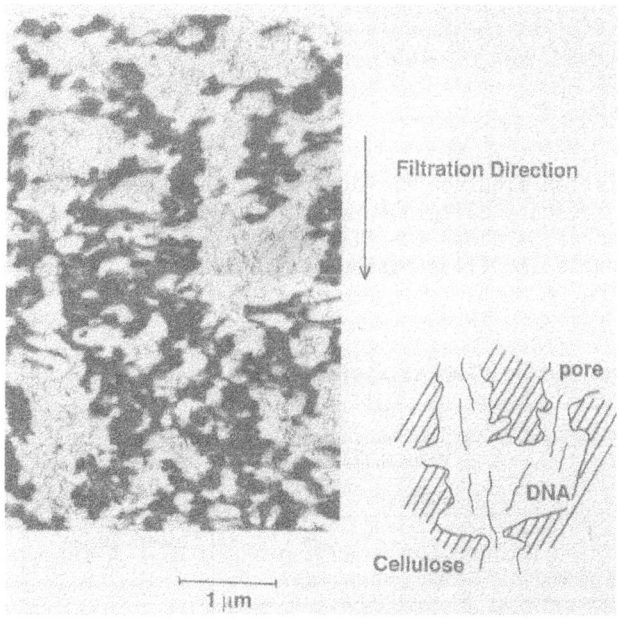

Figure 2. Transmission electron micrograph of DNA in BMM35.
The black area represents cellulose, white is pores, and the hair-like forms are DNA. The arrow shows the filtration direction, and the scale bar is 1μm in length. An illustration of the TEM is on the right-hand side.

278

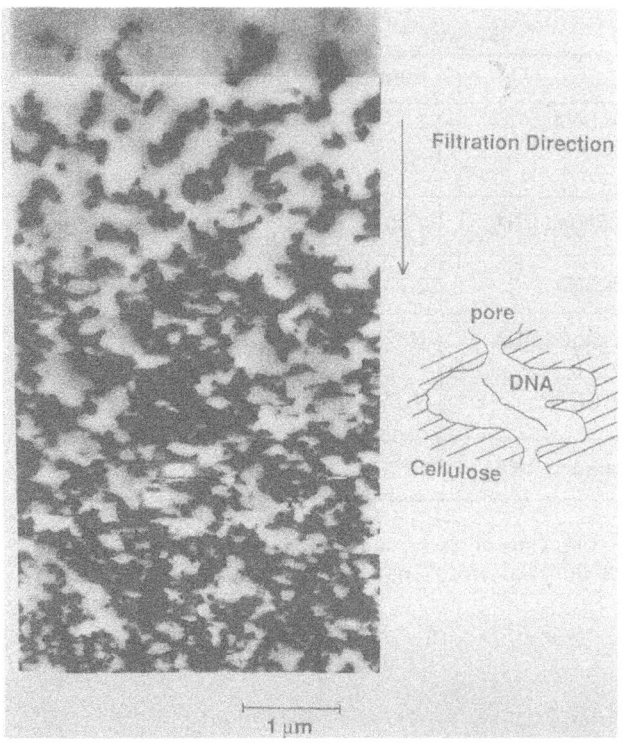

Figure 3. Transmission electron micrograph of DNA in BMM15.
The black area represents cellulose, white is pores, and the hair-like
forms are DNA. The arrow shows the filtration direction, and the scale
bar is $1\,\mu$m in length. An illustration of the TEM is on the right-hand
side.

Table 2. Reduction and/or removal system of DNA with various dispersion
states.

Dispersion state	Permeability by BMM		Removal system
	BMM15	BMM35	
Isolated DNA	1	1	BMM15+affinity
Combined DNA with protein	0.1	0.1	
Packaged DNA in a fragment of source material (cell) (200nm)	N. D.	N. D.	BMM15
Packaged DNA in a virus (50nm~)	$<10^{-10}$	$<10^{-8}$	
Packaged RNA in a virus (35nm~)	$<10^{-8}$	$<10^{-4}$	

N.D. indicates "not detectable".

Table 3. Deformation of microorganisms under shear stress.

Substance	Example	Deformation under shear stress
Protein	Albumin IgG IgM	rigid (solid)
Microbe	Virus	
Gene	DNA RNA	flexible (liquid)
Protein	F-VIII (with vWF)	
Microbe	Mycoplasma	

4. REFERENCES

[1] Commission of the European Community (1991) 'Validation of
    virus removal and inactivation procedure'.
[2] Office of Biological Research and Review, FDA (1987) 'Point to
    consider in manufacture and testing of monoclonal antibody
    products for human use'.
[3] Office of Biological Research and Review, FDA (1985) 'Point to
    consider in the production and testing of new drugs and biologi-
    cals produced by recombinant DNA technology'.
[4] T.Hirasaki, T.Tsuboi, T.Noda, S.Uematsu, G.Ishikawa, A.Kono,
    and N.Yamamoto (1992) 'Removability of virus particles and
    permeability of protein from cell culture medium using
    cuprammonium regenerated cellulose hollow fiber (BMM^{TM})',
    Animal Cell Technology: Basic & Applied Aspects, Kluwer Academic
    Publishers, Netherlands, 49-55.
[5] G.Ishikawa, T.Hirasaki, S.Manabe, S.Uematsu, and N.Yamamoto
    (1991) 'Novel determination method of size of virus in solution
    using cuprammonium regenerated cellulose membrane (BMM)',
    Membrane 16 (6), 376-386.
[6] T.Yuasa, G.Ishikawa, S.Manabe, S.Sekiguchi, K.Takeuchi, and
    T.Miyamura (1991) 'The particle size of hepatitis C virus
    estimated by filtration through microporous regenerated cellulose
    fiber', J. General Virology 72, 2021-2024.
[7] K.Sekiguchi, M.Ito, H.Kobayashi, T.Ikeda, T.Tsurumi, G.Ishikawa,
    S.Manabe, and T.Yamashiki (1989) 'Possibility of hepatitis B
    virus (HBV) removal from human plasma using regenerated cellulose
    hollow fiber (BMM)', Membrane 14 (4), 253-261.

PRINCIPLES AND METHODS OF PROCESS VALIDATION STUDIES FOR VIRUS REMOVAL AND INACTIVATION

ALLAN J. DARLING AND MALCOLM K. BRATTLE
QUALITY BIOTECH LTD., 6.04 KELVIN CAMPUS
WEST OF SCOTLAND SCIENCE PARK,
GLASGOW, G20 0SP, SCOTLAND.

ABSTRACT

A common feature in the manufacture of all pharmaceuticals whose production has involved the use of animal material is the risk of contamination of viruses. Viral contamination can occur through direct contamination of the starting material itself or by introduction of viruses through the use of animal material in the production process e.g. bovine serum in cell culture. Process validation studies for virus clearance and inactivation in association with rigorous testing of the starting material and product throughout the process provide an essential level of assurance that a particular product is safe and free from viral contamination.

1. INTRODUCTION

Three complimentary approaches have been adopted to try to control the problem of virus contamination of therapeutic antibody production:-

a) selection and testing of the start material for the presence of viruses and other contaminants.

b) testing of the product at various stages of production for the presence of viruses.

c) testing the capacity of the production process to remove or inactivate viruses.

Testing of the starting material or product through the production process can only assure a certain level of safety as it is not feasible to test for all possible contaminants in a cell line, or there may exist variants or as yet unidentified viruses which may not be picked up by existing techniques. All testing also suffers from the limitation that you can only ever sample a small percentage of the overall material and if virus contamination is present at low concentrations then there is a high probability that your sample will fail to reveal the presence of contamination in the overall material. This can be represented statistically by the Poisson distribution [Fig. 1].

S. Kaminogawa et al. (eds.), Animal Cell Technology: Basic & Applied Aspects, Vol. 5, 281–288.
© 1993 Kluwer Academic Publishers.

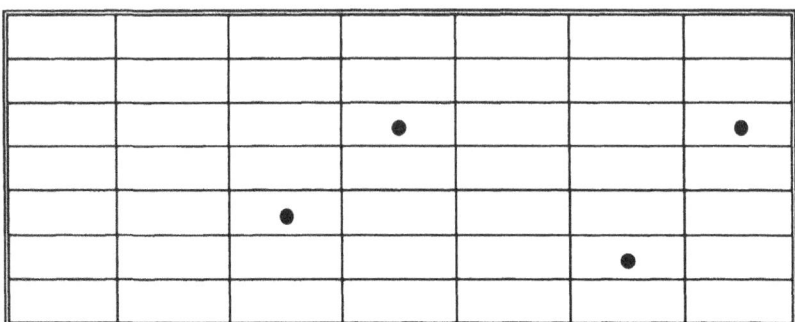

Figure 1. Probability of detecting viruses at low concentrations

The total sample volume is represented by the large rectangle. Aliquots taken for testing are represented by the small rectangles. Infectious virus is represented by the filled circles.

If only a small aliquot is removed for testing then there is a high probability that this aliquot will fail to contain any infectious virus even though infectious virus is present in the total sample.

The probability that an aliquot does not contain infectious virus can be represented by the Poisson distribution:-

$$P_- = e^{-cv}$$

Where c is the concentration of virus particles litre and v is the volume of the aliquot tested in ml. For a 1ml sample, the probability that it does not contain infectious virus at virus concentrations from 10 to 1000 particles per litre are:-

$\underline{c}$	$\underline{10}$	$\underline{100}$	$\underline{1000}$
P_-	0.99	0.90	0.37

Therefore direct testing can only give a certain level of assurance that a product is free from the presence of contaminating viruses. For this reason validation of the purification process to remove or inactivate viruses is an essential stage in ensuring that a product will be free of virus contamination but it is essential that a testing schedule be based on a balance between all three strategies. Performing process validation studies for virus removal or inactivation involves several basic steps.

a)   Scaling down of the process
b)   Consideration of virus inactivation
c)   Choosing the appropriate viruses for the study
d)   Spiking selected steps of the process with high titres of the appropriate viruses
e)   Titrating the output samples and calculating virus reduction factors
f)   Multiplying the virus reduction factors for each step to obtain an overall clearance factor for the whole process

In Europe, the CPMP have issued a "Notes for Guidance" covering virus validation (1) and the FDA are in the process of producing a "Points to Consider" document covering this same area.

2.   SCALING DOWN OF THE PURIFICATION PROCESS

Downscaling of the purification process is an essential part in performing process validation studies for virus removal/inactivation.   For several reasons including the scale of the production process, it is either impossible or impractical to perform these studies on the full manufacturing scale.   Therefore the steps to be studied are scaled down to laboratory scale which generally means a scale down factor of 1/100 - 1/200 although there are not set guidelines governing the size of a scale-downed process.   What is essential is that the downscaled steps of the process mimic as closely as possible the full manufacturing scale process.   Therefore, the downscale of the steps should be validated to ensure that the yield, purity, buffer compositions etc. are as close as possible to the full manufacturing process. This ensures that the virus clearance studies are performed on a process that is as close a representation of the full manufacturing process as possible.   In determining the stages to be validated cognise should be taken whether the clearance factors that will be generated by each of the individual stages can be multiplied together to give an overall clearance value for the process.   Where two stages remove or inactivate virus by independent mechanisms then the clearance factors for each stage can be multiplied together.   If the mechanism of virus removal or inactivation is similar e.g. where two column steps share a common buffer which is responsible for virus inactivation then the clearance factors for each step can only be considered additive not multiplicative.   This would lead to a lower overall virus clearance factor for the whole process and should therefore be avoided.

The scaled down study should also be designated to minimise the eluate volumes generated for each stage that will subsequently be assayed for infectious virus due to the considerations mentioned above concerning the ability to detect infectious virus at low concentrations.

3.   INCORPORATION OF A VIRUS INACTIVATION STEP

Whenever possible inactivation steps should be incorporated into the design of a validation. This can be done by monitoring the inactivation of virus in one of the process buffers e.g. the low pH buffers used in the elution of antibody from Protein-A columns, or may include the incorporation of a specific virus inactivation step into the process.   These studies are usually performed by looking at the kinetics of inactivation as virus inactivation is rarely linear over time and a persistent fraction is often present.   The overall constituents of the buffer can also affect virus inactivation.   For example temperature and pH inactivation steps are influenced by the divalent cation and protein concentration and these should be adjusted to reflect the levels in the full scale process.   Different viruses also have different responses to different inactivation steps and this should also be borne in mind.

Several specific methods for virus inactivation have been developed particularly for the blood and plasma fractionation industry and these are now finding a use in the monoclonal antibody production process.   Freeze drying and heat treatment have been used for several years to ensure inactivation of potential viral contaminants and several comprehensive studies have been carried out to study the efficacy of these processes [reviewed in 2,3]. The use of organic solvents such as ethanol and B-propiolactone have also been used for the inactivation of viruses [4,5].   Recent developments in the use of solvent inactivation has focused on the use of solvent/detergent combinations [6,7].   The solvent tri(n-butyl) phosphate (TBNP) in association with detergents such as Triton X-100 or Tween-80 can readily inactivate enveloped viruses.   Because the solvent and detergent act synergistically, relatively low concentrations of both reagents can be used and thus the treatment is relatively mild and is unlikely to be detrimental to the product.   The addition of a solvent/detergent step into a monoclonal antibody production process can thus generate excellent clearance data for retroviruses and for herpesviruses.   However the main shortcoming of this technique is that infectivity of viruses which do not possess a lipid envelope or which do not require their envelope for infectivity is not reduced.

Several studies have shown the effectiveness of U.V. irradiation particularly in the presence of chromophore dyes on virus inactivation [8-10]. This technique is not widely used in production processes however but remains a technique which has potential for the future. Similarly, several companies in Europe, USA and Japan are now manufacturing specific virus removal filters. These filters have the advantage that virus is physically removed from the solution so both enveloped and unenveloped viruses can be removed with efficiency of virus removal being a function of size and shape of virus. Although membrane filtration has a future in production processes its use will be limited to specific products and will not be the sole virus clearance step in a production process.

By far the most common method of virus inactivation involves the use of pH extremes. Retroviruses, which as we mentioned above are the viruses of most concern in hybridomas are extremely susceptible to inactivation at pH values below pH 5.0. Therefore the use of low pH buffers in the purification of monoclonal antibodies from Protein-A columns is an extremely effective step in ensuring that the product will not contain infectious retrovirus. Herpesviruses, another virus group of concern in hybridomas, are also readily inactivated at low pH values. This treatment is not effective for all viruses unfortunately and small unenveloped viruses such as parvoviruses and polioviruses are not readily inactivated at low pH values.

The incorporation of a specific virus inactivation step into a process can help generate the necessary virus clearance value especially where a purification process may be relatively "mild" and may have been designed to maximise yields and throughput without consideration being given to the potential virus problem.

4. SELECTION OF APPROPRIATE VIRUSES

Selection of the number and type of viruses to be used in a process validation study depends on a number of parameters. One of the most important considerations is that the viruses to be used must be available at high titre. The higher the titres of virus used to spike the process, the higher are the virus clearance factors that can theoretically be achieved. The viruses must also be easily assayed enabling accurate and reproducible titration results and should not pose an unnecessary hazard to the operators conducting the validation study. The number of viruses to be used in the study will depend on how much testing has been done on the master cell bank, working cell bank and final product and on the stage of development of the product. The virus validation data required for a product going into phase 1 clinical trials for example is less comprehensive than if the product was to be tested on healthy volunteers in phase 2 or 3 clinical trials.

In certain cases the viruses used should be the actual potential contaminants found in the starting material i.e. it is obligatory to use HIV when performing validations where blood or blood derivatives are involved in the production process. Similarly for murine hybridomas which contain endogenous C-type retroviruses and CHO cells which contain retroviral type-particles then murine leukaemia virus is used as one of the viruses for the validation studies.

It is often not possible to use the actual potential contaminant viruses in the process validation study for various reasons. The viruses may be too dangerous to work with without special containment facilities or they may not grow in cell culture or to a high enough titre to use in the studies. The actual number of potential contaminant viruses may also be too numerous to use every single potential contaminant in a validation study. To cover such eventualities model viruses are used in process validation studies. These viruses are chosen either to substitute for specific viruses or more commonly to cover a virus group such as herpesviruses which contains several different viruses which can be pathogenic to man. Model viruses should also be chosen to cover the different virus groups and to cover the range of resistance to physico-chemical inactivation which have not been covered in the selection of obligatory viruses.

In summary the viruses selected for the validation, whether known contaminants or models should cover both DNA and RNA viruses both enveloped and unenveloped, small and large. At least one of the viruses should be highly resistant to physico-chemical inactivation and this virus will provide a severe test of the clearance capacity of the purification system. A retrovirus should always be included in the validation study as the genomes of these viruses can be inherited through the germ line or in the case of infection can be integrated into the genome.

An example of viruses which may be chosen to validate a murine hybridoma are shown in Table 1.

TABLE 1

Viruses used in a virus validation for a murine monoclonal antibody

Virus	Size	Envelope	Nucleic Acid
MuLV	80-120nM	Yes	RNA
Poliovirus	20-30nM	No	RNA
Herpes simplex virus type 1	120-150nM	Yes	DNA
Adenovirus	70-90nM	No	DNA

As mentioned previously murine leukaemia virus is used as a representative of the C-type sub-group of retroviruses and this virus is medium sized enveloped RNA virus which has a low resistance to physico-chemical inactivation. A herpesvirus is also included as these viruses also have the capacity to establish latent infections within cells. Herpes simplex virus type 1 is an appropriate model to use for murine hybridomas and is a large DNA enveloped virus with a medium to low resistance to physico-chemical inactivation. Poliovirus may be included not because polioviruses are potential contaminant of murine hybridomas but because it is a small tough unenveloped RNA virus which is very difficult to inactivate. Thus poliovirus in this case is included as a model for other unidentified contaminants which may be highly resistant to inactivation. To complete the different range of viruses in the study a medium sized unenveloped DNA virus such as Adenovirus should also be used.

For a human hybridoma the choice of viruses is slightly different. It is necessary to show removal or inactivation of the lentivirus sub-group of retroviruses. This group, which includes HIV, are more resistant to inactivation than the C-type retroviruses or the oncovirus sub-group. In this case the ovine lentivirus Maedi-visna virus is used as a model lentivirus as it displays many of the physical properties of HIV but does not pose a hazard to the operators. Herpes simplex virus type 1 is also used as a model in this case where the hybridoma has been immortilised using the herpesvirus, Epstein-Barr virus. HSV-1 also is used as a model to cover for possible contamination with other herpesviruses such as cytomegalovirus, HHV-6 and HHV-7.

5.   PERFORMING THE PROCESS VALIDATION STUDY FOR VIRUS CLEARANCE AND INACTIVATION

One of the first manipulations to be performed in performing virus validation studies is to test whether the samples that will be generated by the steps examined will by cytotoxic to the indicator cells used to titrate the viruses. Cytotoxicity can be a function of the composition or pH of the buffer or can be due to the product itself having a deleterious effect on cells. Cytotoxicity can often be alleviated by diluting the sample in PBS or tissue culture medium but care must be taken not to generate large volumes which would exacerbate the problem of detecting virus present at low concentrations as mentioned previously.

The process validation experiments are performed by spiking the starting material for each step of the process with the relevant viruses. The virus spike is added in a volume that is <10% v/v of the total volume of the material to be spiked. This is to ensure that the nature of the starting material is not affected by adding virus in tissue culture medium. This material is then taken through the purification process and the appropriate fractions are collected for assay of infectious virus. Under conditions where virus clearance is generated by partitioning of the virus into a different fraction rather than inactivation of the virus it is appropriate to collect other fractions other than the eluate to examine where the virus has been distributed. Where the purification process involves the use of column purification steps and if these columns are re-used as is the norm in production, then the efficiency of the sanitization regime used must also be demonstrated as virus could build up on the chromatography matrix and desorb at any time. These experiments are performed as kinetic inactivation studies in the sanitization solutions. It is also recommended that for resins that are continuously regenerated and reused that the virus validation is performed on fresh resin and on resin at the end of life in the production process and the clearance factors compared. For certain resins this may be upwards of 100-200 cycles of purification and sanitization.

The fractions are then assayed for the presence of the relevant viruses. All assays for virus must be based on infectivity and not on other techniques such as Reverse transcriptase activity or PCR detection as these methods do not accurately quantify virus titre.

6.   CALCULATION OF VIRUS CLEARANCE FACTORS

The virus reduction factor for an individual purification or inactivation step is defined as the $\log_{10}$ of the ratio of the virus load in the pre-purification material divided by the virus load in the post purification material (Table 2).

TABLE 2

Calculation of Reduction Factors

The virus reduction factor of an individual purification or inactivation step is defined as the $\log_{10}$ of the ratio of the virus load in the pre-purification material divided by the virus load in the post-purification material.

Starting Material: vol v'; titre $10^{a'}$
                      virus load v' x $10^{a'}$

Final Material: vol v''; titre $10^{a''}$
                      virus load v'' x $10^{a''}$

$$10^{RI} = \frac{v' \times 10^{a'}}{v'' \times 10^{a''}}$$

The formula takes into account both the titres and volumes of the materials before and after the purification step. For example, if a Protein-A Sepharose step was spiked with 10ml of murine leukaemia virus at a titre of 5.0×10^7 pfu ml^{-1} and the eluate volume was 50ml which contained virus at a titre of 1.5×10^1. Therefore the reduction factor for this step would be:

$$10^{RI} = \frac{10 \times 5.0 \times 10^7}{50 \times 1.5 \times 10^1} = 6.7 \times 10^5 = 10^{5.8}$$

Therefore this step of the process can remove 5.8 logs of retrovirus under the conditions used in the purification step. Because of the inherent imprecision of some virus titrations, an individual reduction factor to be significant would have to be $> = 1$ log. A clearance factor of greater than 5 logs represents an effective virus clearance step.

7. CALCULATION OF THE CLEARANCE CAPACITY OF THE OVERALL PROCESS

Once all the individual reduction factors have been calculated then an overall clearance factor for the purification process can be calculated. As was mentioned above reduction factors for each individual step can be multiplied if the steps remove or inactivate virus by different mechanisms. Where the steps are similar or identical then the clearance factors can only be additive. Table 3 gives an example of a theoretical process for the purification of a monoclonal antibody. This purification regime incorporates three column purification steps, a specific detergent virus inactivation step and an ultrafiltration step. Summing the individual clearance factors for the individual steps of this process gives an overall clearance capacity of 18.5 logs. The clearance factor for the overall process gives an indication of the capacity of the purification process to inactivate viruses.

TABLE 3

Example of Typical Retrovirus Removal/Inactivation Study

	Purification Step	Log Reduction
1)	Affinity chromatography	2.8
2)	Detergent Inactivation	5.1
3)	Ion-exchange chromatography	3.9
4)	Size-exclusion chromatography	2.5
5)	Ultrafiltration	4.2
	Cumulative Reduction	**18.5**

It is important to relate this figure to the assessment of virus risk and the virus load either quantitated or estimated in the starting material. The risk assessment includes how the final product is administered, the dose and frequency of application and the patient population to be treated. For example testing may detect 10^7 particles ml^{-1} in a 5000-fold concentrate of pre-purified bulk material. This would give a retrovirus concentration of $10^{4.3}$ particles ml^{-1}. If one dose of the therapeutic was made from 1000ml of the pre-purified bulk then one dose could contain $10^{4.3} \times 1000 = 10^{7.3}$ virus particles. This gives an initial level of what virus clearance values should be obtained but still means that if your purification process could only clear 7.3 logs then there would still be one virus particle present per dose. To reduce the risk of virus contamination of the final product even further a safety factor of approximately 6 logs should be added to the initial clearance capacity that has to be obtained. For this example the clearance capacity that should be achieved is $10^{7.3} + 10^6 = 10^{13.3}$ logs. This means that on average there will be less than one virus particle per million doses of the therapeutic. If the purification process shown in Table 3 is used to purify the starting material which could remove or inactivate 18.5 logs of virus then it is obvious that this process fulfils and in fact exceeds the safety requirements for retrovirus clearance.

8.   LIMITATIONS OF VIRUS VALIDATION STUDIES

Validation studies are essential in contributing to the assurance that an acceptable level of safety in the final product is established but in itself does not categorically establish safety. A number of limitations in the design and execution of virus validation experiments may lead to an incorrect estimate of the ability of the process to remove virus infectivity. One of these limitations is in the use of model viruses or viruses that have been adapted for growth in tissue culture which may not behave the same as the viruses present as actual contaminants for example in their degree of purity and aggregation. This is important as aggregation of virus can protect from inactivation. There can exist resistant sub-populations of virus which are extremely difficult to remove or inactivate. Therefore a virus resisting a first inactivation step may be more resistant to subsequent steps and as a consequence the overall reduction factor is not necessarily the sum of the reduction factors calculated from individual steps each time with a fresh virus spike-suspension. Virus clearance may have also been over-estimated by summing the logarithmic clearance factors for steps where the virus removal was by a similar mechanism i.e. adsorption onto the column matrix. It is also almost impossible to completely miniaturise the full-scale manufacturing process and maintain exactly the equipment and conditions found in the full manufacturing scale. For these reasons it is generally recommended that at least one individual stage of the validation process can remove or inactivate at least 5 logs of retrovirus. This again gives an extra degree of assurance that the virus validation results are meaningful and contribute to the safety of the final product.

9.   REFERENCES

1)   Commission of the European Communities Committee For Proprietary Medicinal Products Ad Hoc Working Part on Biotechnology/Pharmacy.   Final Document 111/8115/89-EN (1991).

2)   Markus-Sekura, C.J., in: "Developments in Biological Standardisation.   Virological Aspects of the Safety of Biological Products", 75, 133-14, S. Karger, Basel (1991).

3)   Piszkiewicz D, Thomas W, Lieu MY et al. in: "Virus inactivation in plasma products. Curr. Stud. Hematol. Blood Trans., "Morgenthaler J-J, (ed): 56, 44-54, (1989).

4)   Morgenthaler J-J, Effect on ethanol on viruses in: "Virus inactivation in plasma products. Curr. STud. Hematol. Blood Trans." Morgenthaler J-J. (ed): 56, 109-121, (1989).

5)   Stephan W. in: "Virus inactivation in plasma products. Curr. Stud. Hematol. Blood Trans." Morgenthaler J-J (ed): 56, 122-127, (1989).

6)   Horowitz B, Wiebe ME, Lippin A et al,. Transfusion 25, 516-522, (1985).

7)   Horowitz B., in "Developments in Biological Standardisation.   Virological Aspects of the Safety of Biological Products", 75, 43-52. S. Karger, Basel, (1991).

8)   Wallis C, Melnick JL., Photochem Photobiol 4, 159-170, (1966).

9)   Chanh TC, Allan JS, Matthews JL et al, J Virol Methods 26, 125-132, (1989).

10)   Neundorff HC, Bartel DL, Tufaro F et al, Transfusion 30, 485-490, (1990).

COMBINATION OF VIRUS REMOVAL AND INACTIVATION FOR THE PROCESS VALIDATION OF BIO-DRUGS

T. Tsuboi,[1] T. Hirasaki,[1] T. Noda,[1] H. Nakano, [1] G. Ishikawa,[1] S. Uematsu, [1] and N. Yamamoto [2]
[1] BMM Development and Business Promotion Department, Fibers and Textiles Administration, Asahi Chemical Ind. Co.Ltd., The Imperial Tower 18F, 1-1 Uchisaiwaicho 1-chome, Chiyoda-ku, Tokyo 100, Japan
[2] Department of Microbiology, Tokyo Medical and Dental University, School of Medicine, 1-5-45 Yushima, Bunkyo-ku, Tokyo 113, Japan

ABSTRACT. The cumulative effects of viral removal and inactivation were investigated in view of the logarithmic reduction efficacy in virus infectivity, and an effective procedure is proposed to reduce the viral infectivity of bio-drug solutions. The two-step procedure involves filtration through a cuprammonium regenerated cellulose membrane (BMM) filter and other physical inactivation methods. Polio virus, Sindbis virus and Japanese encephalitis virus (JEV) were tested from the viewpoint of their sizes and existence of envelopes. A polio virus solution containing 10% FCS was prepared from FL cells, one for Sindbis virus was from BHK-21, and one for JEV was from Vero cells. Dead-end filtration was carried out at 200 mmHg with the BMM filter. The filtrate was thermo-inactivated at a given temperature and then passed through the BMM filter again. Viral infectivity was evaluated by the $TCID_{50}$ assay procedure. The log reduction efficacy of each treatment was cumulative or more than cumulative when these two procedures were combined in an appropriate order. Moreover, the virus inactivation effects was enhanced when a virus solution was prefiltrated by BMM, whose mean pore size is about twice that of virus. These results indicate that a part of viruses exists as aggregates which resist thermo-inactivation. These aggregates could be easily removed by BMM and the aggregate-free viruses then smoothly inactivated.

INTRODUCTION

The high quality and security of bio-drugs must be warranted prior to their marketing. Bio-drugs such as plasma-fractionated products and vaccines are faced with the hazard of viral contamination[1]. To prevent viruses from contaminating bio-drugs, three measures need to be considered, that is preventing viral contamination from the source materials, virus removal and/or inactivation in the purification processes (i.e., process validation for virus removal[2],[3]), and consistent checks on the products. Process validation is the only defence strategy against unknown viruses, the proposed degree of virus removal and/or inactivation being more than 12log10[4]. Although this degree of virus removal and/or inactivation is difficult to achieve by a single process, it can be attained by combining several processes. Under these circumstances, the "validatable membrane" for virus removal has been developed for application to

289

the purification processes of bio-drugs[5]. The cuprammonium regenerated cellose hollow fiber (BMM) is the "validatable membrane" having both reproducibility and predictability for virus removal[2],[3].

In this study, we will show that combined filtration by BMM with other physical inactivation methods, e.g. thermo-inactivation, is validatable for virus removal and/or inactivation, and propose a more effective combination to achieve this.

MATERIALS AND METHODS

(1) BMM filter: The BMM with a mean pore size of 18nm (BMM18), 35nm (BMM35) and 75nm (BMM75) were supplied by Asahi Chemical Ind. Co. BMM filters with an effective filtration area of $0.03m^2$ were employed.

(2) Prefiltration with BMM: The dead-end type of filtration was carried out under the constant pressure of 200 mmHg at ambient temperature. Polio virus with a diameter of about 25nm was prefiltrated with BMM35. Japanese encephalitis virus (JEV) whose diameter of about 43nm and Sindbis virus of about 60nm were both prefiltrated with BMM75. Each prefiltrate was then thermo-inactivated.

(3) Thermo-inactivation of the viruses: Each of the original virus solutions and its prefiltrate were separated 1ml into microtubes for thermo-inactivation. The temperature was set at 50 °C for polio virus, at 48 °C for JEV and at 56°C for Sindbis virus to achieve inactivation at the proper rate.

(4) Filtration with BMM: Each of the thermo-inactivated polio virus solutions was instantly ice-cooled and filtrated through a BMM18 single-fiber filter under the filtration conditions already mentioned.

RESULTS AND DISCUSSION

We attempted to reduce the infectivity of polio virus, Sindbis virus and JEV by combining filtration through BMM and thermo-inactivation. The strategy for this combination is summarized in Figure 1.

Figure 2 shows the dependence of the polio virus titer before and after BMM filtration and thermo-inactivation on the inactivation time. The original polio virus solution was thermo-inactivated at 50 °C for the indicated time (procedure I).In procedure C, the original polio virus solution was prefiltrated with BMM35, whose mean pore size is about twice of that of polio virus, and then thermo-inactivated at 50°C for the indicated time.

Sindbis virus (Figure 3) and JEV (Figure 4) were prefiltrated with BMM75 in the same way. The titer of these viruses before and after prefiltration was almost equal, indicating that prefiltration had a negligible effect (compare data points I_0 and C_0 in Figs. 2, 3 and 4). The infectivity of the prefiltrated viruses was decreased more sharply compared with the original virus solutions with increasing thermo-inactivation time. These results indicate either that part of viruses existing as aggregates, which may resist thermo-inactivation, can be removed by prefiltration, and that the residual virus in the filtrate may exist in as isolated state that can be effectively inactivated, or that particles larger than virus, which would disturb the effect of the inactivation treatment, can be removed by prefiltration. Each of the thermo-inactivated polio virus samples was then filtrated with the main BMM18 filter (Figure 2).

Logarithmic rejection coefficient Φ of the polio virus solutions both before
and after main filtration that had not been thermo-inactivated are ca. 2.7
(i.e., the difference between points I_0 and D_0, and C_0 and E_0). With each of
the samples which had been thermo-inactivated for the indicated time, the virus
removability expressed by the Φ value was same as that at time 0. These results
suggest that the effects of virus removal with BMM and inactivation were
retained, even when they were combined. Furthermore, since the prefiltration
effect was also retained, it was more effective to use BMM prior to other
physical inactivation methods.

We did not examine the combination of virus removal by filtration with the
main BMM filter first, and then inactivation. It is possible that filtration
with the main BMM filter would remove not only the aggregates and/or larger
particles but also single isolated viruses, so that the filtrate could then be
smoothly inactivated. Viruses in bio-drug solutions must be removed and/or
inactivated by a method that dose not reduce the drugs' activity. Physical
inactivation by the solvent detergent method or UV radiation can be applied to
many bio-drugs[6],[7], and we are now examining the combination of BMM
filtration and these other physical inactivation methods. Aggregates of a virus
which resists thermo-inactivation would tend to resist the other physical
inactivation methods. We thus conclude that the combination of virus removal by
BMM filtration and an inactivation method is suitable for the process validation
of bio-drugs, and that the most effective combination is initial filtration with
BMM and then, physical inactivation.

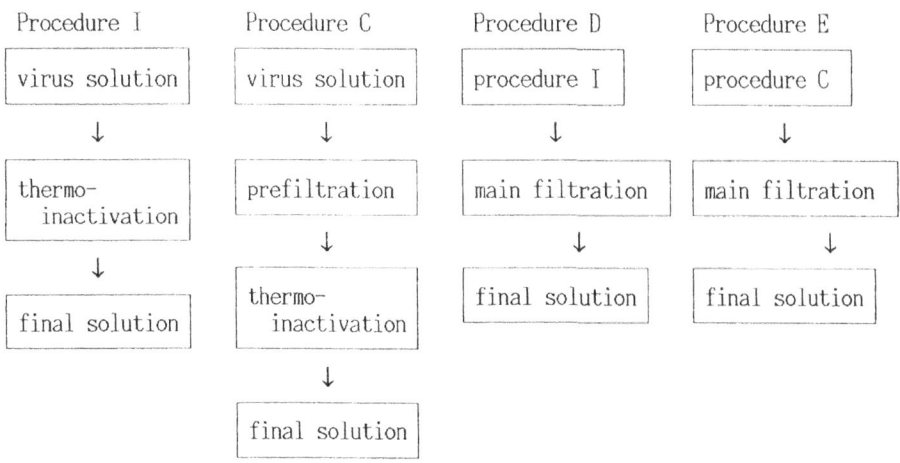

Figure 1. Procedures for combined virus removal and inactivation.

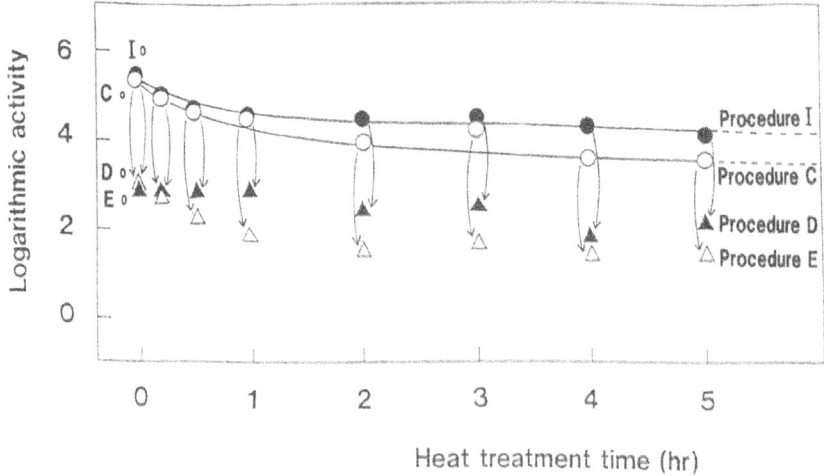

Figure 2. Effects of combined BMM filtration and thermo-inactivation on polio virus: The original polio virus solution (its titer is I_0) and its prefiltrate (its titer is C_0) were thermo-inactivated for the indicated time. Each of the thermo-inactivated samples was then filtrated through the main BMM18 filter (procedure D is derived from procedure I, and procedure E from procedure C).

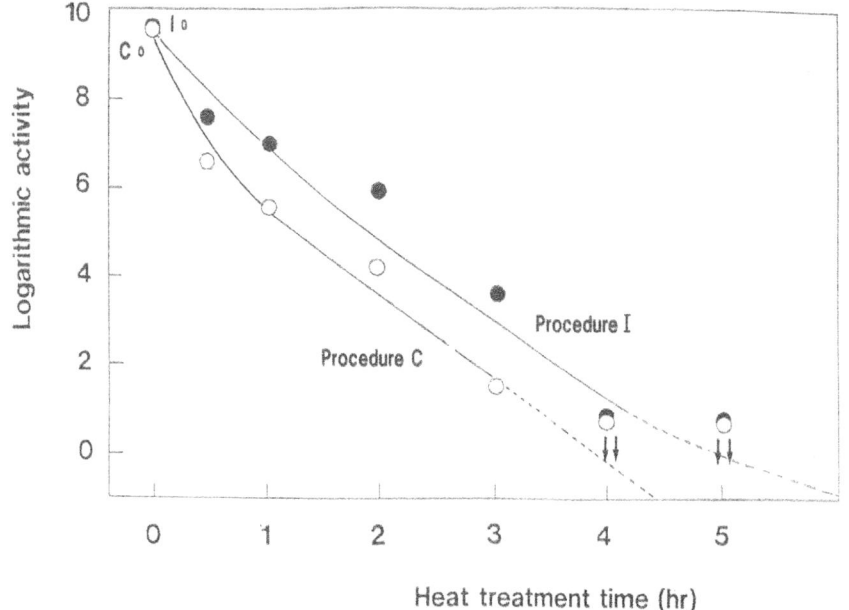

Figure 3. Effects of combined BMM filtration and thermo-inactivation on Sindbis virus: The original Sindbis virus solution (its titer is I_0) and its prefiltrate (its titer is C_0) were thermo-inactivated for the indicated time.

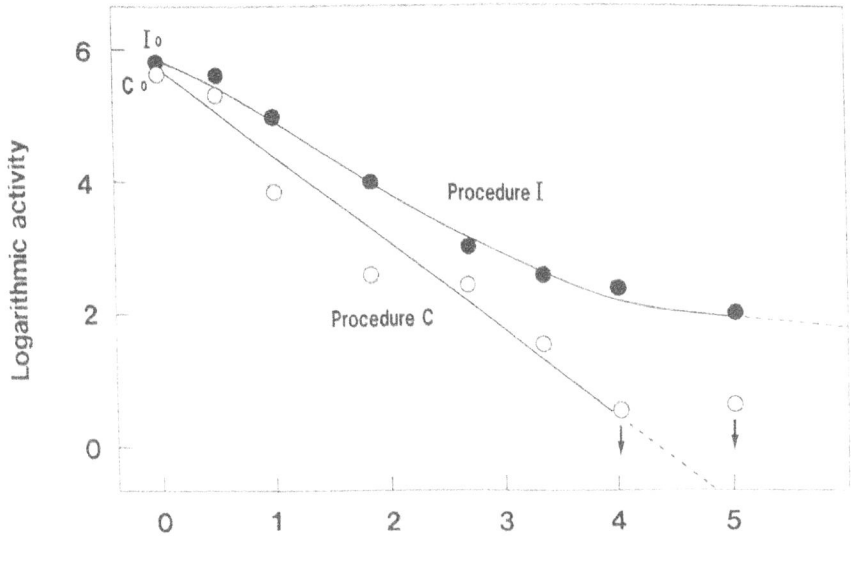

Heat treatment time (hr)

Figure 4. Effects of combined BMM filtration and thermo-inactivation on Japanese encephalitis virus (JEV): The original JEV solution (its titer is I_0) and its prefiltrate (its titer is C_0) were thermo-inactivated for the indicated time.

REFERENCES

[1] Sekiguchi, S., Ito, K., Kobayashi, M., Ikeda, H., Tsurumi, T., Ishikawa, G., Manabe, S., Satani, M., and Yamashiki, T. (1989) 'Possibility of hepatitis B virus (HBV) removal from human plasma using regenerated cellulose hollow fiber(BMM),' Membrane 14 (4), 253-261.

[2] Manabe, S. 'Virus removal and inactivation in process validation,' in Animal Cell Technology: Basic and Applied Aspects,Proceedings of the Fourth Annual Meeting of the Japanese Association for Animal Cell Technology 4, 15-30.

[3] Nakano, H., Manabe, S., Uematsu, S., Sato, T., Osawa, N., Hirasaki, T., Yamashiki, T., Sekiguchi, S., and Yamamoto, N. 'Novel validation method of virus removability for biological cell culture products using polymeric membrane filters,' in Animal Cell Technology: Basic and Applied Aspects, Proceedings of the Fourth Annual Meeting of the Japanese Association for Animal Cell Technology 4, 87-102.

[4] Commission of the European communities (1991) 'Validation of virus removal and inactivation procedure.'

[5] Manabe,S. et al., 'Elimination of microorganisms from cell culture medium using regenerated cellulose hollow fiber (BMM),' Animal Cell Technology: Basic and Applied Aspects, Proceedings of the Third Annual Meeting of the Japanese Association for Animal Cell Technology 3, 87-94.

[6] Prince, A.M., Stephan, W., and Brotman, B.(1983) ' β -propiolactone/

ultraviolet irradiation: a review of its effectiveness for inactivation of viruses in blood derivatives,' Rev. Infect. Dis. 5, 92-107.

[7] Horowitz, B., Wiebe, M.E., Lippin, A., and Stryker, M.H. (1985) 'Inactivation of viruses in labile blood derivatives: I. Disruption of lipid-enveloped viruses by tri(n-butyl) phosphate detergent combinations,' Transfusion 25, 516-522.

A NEW, VALIDATED METHOD FOR THE DESTRUCTION OF CONVENTIONAL VIRUSES AND UNCONVENTIONAL INFECTIOUS AGENTS (e.g. BSE AND SCRAPIE)[*] IN PROTEINS AND SERA

[1]Kosaku Iida and [2]Herwig E. Reichl.

[1]Veritas Corp., 1-1-9 Atago, Minato-ku, Tokyo 105, Japan and
[2]Haemosan GmbH., Neudorf 41, A-8262 Ilz, Austria

Abstract :

A process has been developed for the production of albumins which has been shown to inactivate all tested conventional viruses as well as unconventional infectious agents ("prions" causing diseases like scrapie and BSE (bovine spongiphorme encephalopathie)) (1, 2). BSA produced according to this process has been tested in tissue culture and is shown to be of at least equivalent quality as compared to other TC-grade BSA.
By variation of certain parameters, this process can be modified to produce other viral-free biologicals like proteins, sera and extracts.

Methods :

BSA was produced from bovine plasma following a newly developed method (3). Briefly, this method consists of a heating step in the presence of a detergent, acidification, filtration, the addition of a chaotropic agent, chromatography and diafiltration, followed by drying.

Tissue culture testing of BSA
Cell lines used were :
> IL-9 dependent T-cells C 1/9.4a2 in Iscove's ATL medium, supplemented with 200 pg/ml IL-9.
> Mouse myeloma cells Sp2 in DMEM

Virus validation studies

The full set of validation studies will be published elsewhere (4). The following bovine viruses were tested : Bovine Viral Diarrhoea (BVD), ParaInfluenza 3 (PI 3), Infectious Bovine Rhinotracheitis' (IBR), Maedi-Visna Virus (MVV), Foot-and Mouth-Disease (FMD) and ORF Virus.
The Scrapie-Mouse model was used to test for the destruction of BSE causing agents, as there is at least a close relationship between both diseases.

S. Kaminogawa et al. (eds.), Animal Cell Technology: Basic & Applied Aspects, Vol. 5, 295–300.
© 1993 Kluwer Academic Publishers.

Results :

The purity and other relevant data of BSA produced according to the new process can be seen in Table 1.

When applied in serum-free tissue culture, effects on cell growth and proliferation were found similar when compared to conventionally purified BSA for tissue culture (Table 2).

Supplementing serum-containing media (5 % FBS) with BSA (0.5 %) showed growth promotion of this newly-developed BSA comparable to 10 % FBS, chromatographically-purified and fatty acid-free BSA (Table 3).

When high concentrations of this BSA (up to 20 %) were added to tissue culture media, no adverse effects could be detected, showing complete removal of toxic chemicals added during processing (data not shown).

A summary of the results of the virus validation studies can be seen in Table 4.

A separate, stepwise validation study using Foot and Mouth Disease Virus showed a combined clearance rate of more than 20 logs of infectivity (data not shown).

Discussion :

From these data it can be concluded that this new process yields BSA of high purity, which is suitable not only for diagnostic applications, but also for tissue culture. In addition, it is free from viruses and even prions, as was proven by validation studies.

It should be possible to apply this process (with modifications, where necessary) to other proteins of in vitro and therapeutical interest. Preliminary results indicate that it can be modified to produce immunoglobulins, hemoglobin, human albumin, transferrin and even some enzymes. In addition, whole plasma and sera as well as tissue extracts can be subjected to this treatment. In all cases, careful validation studies will have to be conducted, using relevant viruses, to rule out possible matrix effects.

As a conclusion, we can offer a product which is suitable for sensitive applications in tissue culture as well as a process which holds promises to ban the threat of viral contaminations, which is imminent with all products of animal and human origin, but also with most of the new products produced by biotechnologal methods in tissue culture.

Table 1

HAEMOSAFE TC-grade BSA - **chemical analysis :**

Protein content	> 98 % of dry matter
Protein purity	> 98 %
Ash	< 2 %
pH (1 % in water)	6,5 - 7,5
Immunoglobulins	nil by Ouchterlony
Fatty acids (enzymatically)	< 0,1 %
Endotoxins (kinetic LAL - Test)	< 1 ng/mg
Total T3	< 10 ng/g
Total T4	< 50 ng/g

Table 2

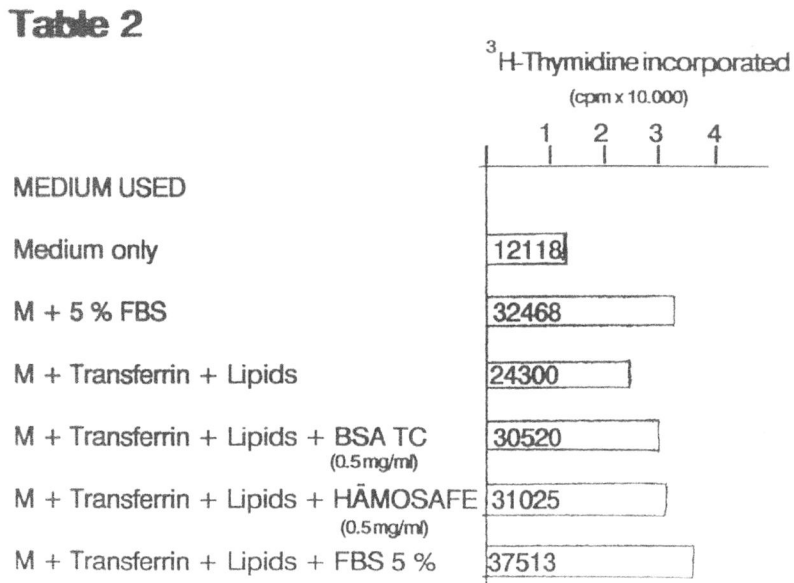

^{3}H-Thymidine incorporated

(cpm x 10.000)

MEDIUM USED	
Medium only	12118
M + 5 % FBS	32468
M + Transferrin + Lipids	24300
M + Transferrin + Lipids + BSA TC (0.5 mg/ml)	30520
M + Transferrin + Lipids + HÄMOSAFE (0.5 mg/ml)	31025
M + Transferrin + Lipids + FBS 5 %	37513

Table 2

Application of HÄMOSAFE TC Grade BSA in serum-free Tissue Culture

Cells used : Human T-cells C 1/9, da 2, IL-9 dependent

Medium :   ISCOVE'S ATC supplemented with 200 pg/ml IL-9

Table 3

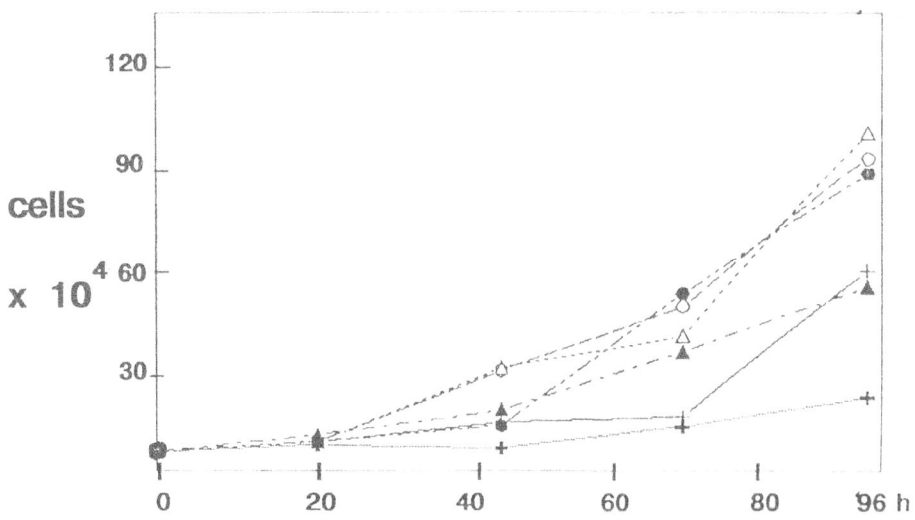

Cells :   SP-2, passage 26, were seeded at 400.000/ml

Medium :   DMEM + 5 % FBS; BSA at 0.5 %

Legend :   +————+   BSA Standard
           +- - - - - +   BSA Cohn F V
           ●- - - - - ●   FBS 10 %
           o- - - - - o   BSA HÄMOSAFE TC Grade
           ▼- - - - - ▼   FBS 5 %, no BSA
           ∇- - - - - ∇   BSA Chromat. purified

Table 4

Infectious agent	Starting Titer	Model (cell line)	detectable infectivity	CLEARANCE
Scrapie/BSE	2 x 10 exp.8	Mouse	n.d.	>10 exp.5
BVD virus Strain "Singer" pass.9	1 x 10 exp.8	Sec. bovine embryonic lung	n.d.	3 x 10 exp.6
IBR virus Strain "Ames" pass.18	1.8 x 10 exp.6	-	n.d.	7 x 10 exp.4
PI 3 virus Strain "Freistadt" pass.78	80 HTH units	-	n.d.	?
FMD virus O1 BFS 1860 pass 5	2 x 10 exp.7	BHK - 21 IB - RS - 2 suckling mice	n.d.	1 x 10 exp.6
MV virus (ATCC-VR-779)	1.2 x 10 exp.9	WSCP	n.d.	3 x 10 exp.6
ORF virus	2 x 10 exp.9	PAL-6	n.d.	3.4 x 10 exp.6

AMINO ACID METABOLISM OF A HIGH DENSITY PERFUSION CULTURE

Henrik Albøhn Hansen, Bo Damgaard,[§] & Claus Emborg
Department of Biotechnology, Block 223,
Technical University of Denmark, DK-2800 Lyngby, Denmark.

[§]Present address: Novo-Nordisk A/S, Novo Allé, DK-2880 Bagsvaerd, Denmark

Abstract

Using an external cell separator based on cell sedimentation and recycling, the density of viable cells reached a level around $10*10^6$ cells ml^{-1} with a perfusion rate of 1.0 day^{-1}. This is five times higher density than achievable in batch cultivations. The specific antibody productivity decreased by a factor two in the middle period of the cultivation from the value 0.4 μg $1*10^6$ cells^{-1} h^{-1} observed initially, when the culture was operated in batch mode. However, the productivity recovered in the final high density stage of the cultivation. In the late period of the cultivation when the viable cell concentration increased to $10*10^6$ cells ml^{-1} a shift in metabolism was observed. The two most important substrates for energy generation, glucose and glutamine, were reduced from feed concentrations of 25 mM respective 10 mM to 1.5 mM and 1 mM. At the same time the specific glycine production increased so the glycine concentration ended at 1.5 mM and aspartate was produced giving a concentration of 0.02 mM. It is suggested that this shift in metabolism leads to enhanced ATP production on glutamine consumption. Calculations of specific glucose consumption and lactate production also indicate that the glucose utilization when necessary, changes to be more energy economical. Anyhow this investigation shows a transition in metabolic state in high cell density perfusion culture, and that this transition may be correlated to a positive effect on monoclonal antibody production. High cell density cultures may have another amino acid metabolism than low cell density cultures, and such metabolic changes may have a beneficial influence on specific antibody productivity. The combination of these two factors caused a volumetric productivity of monoclonal antibody of 100 mg l^{-1} d^{-1} for the high cell density culture.

1 Introduction

High density perfusion cultures are gaining increased interest for production of monoclonal antibodies and complex recombinant DNA proteins, because of the possibility to increase the product concentration and the volumetric output. In order to increase cell concentrations above levels achievable in simple continuous systems different approaches have been used.[4,7,15] We developed a surprisingly simple external cell separator based on sedimentation for establishment of perfusion cultures. This system has been used successfully in production of monoclonal antibodies with hybridomas, as well as for yeast fermentations.[9] To elucidate parameters usable for control and optimization of perfusion cultures a thorough kinetic analysis of key metabolic components is necessary. In this paper the metabolism of a hybridoma cell line is discussed.

2 Materials and methods

2.1 CELL LINE

The investigation was carried out with a X63-Ag8.653-derived mouse/mouse hybridoma cell line, NUC 1-4, which produces monoclonal antibodies against nuclease (from *Serratia marcescence*).[1]

301

S. Kaminogawa et al. (eds.), Animal Cell Technology: Basic & Applied Aspects, Vol. 5, 301–307.
© 1993 *Kluwer Academic Publishers.*

The cells were maintained in T-flasks (NUNC, Roskilde, Denmark) in a thermostated incubator (Queue Systems, Parkersburg, WV, USA) with a humidified atmosphere (92% rel. humidity) of 5% CO_2 in atm. air. The bioreactor was inoculated with cells from T-flasks to an initial density of $0.11*10^6$ cells ml^{-1}.

2.2 MEDIUM

Standard DMEM powder medium (Gibco, Paisley, UK) containing 25 mM glucose and 4 mM L-glutamine was dissolved in Milli-Q water. Additionally 44 mM sodium bicarbonate, 6.2 mM L-glutamine, 1.1 mM sodium pyruvate, 70 μM ß-mercaptoethanol, 2.0 g l^{-1} PEG 20,000 all Merck (Darmstadt, Germany), 10 mM HEPES (Biochrom, Berlin, Germany), 5 mg l^{-1} transferrin, 2.5 nM ethanolamin, 2.4 nM sodium selenite all Sigma (St. Louis, MO, USA), 2.00 g l^{-1} dextran (Pharmacia, Uppsala, Sweden) were added to the medium, and the solution was sterilized by filtration through a 0.22 μm filter and stored at 4 °C. Before use 3 % fetal calf serum (Batch 3S06, Biochrom), 100,000 units l^{-1} penicillin and 100 mg l^{-1} streptomycin (Biological Industries, Kibbutz Beth Haemek, Israel) were added to the medium.

2.3 CULTIVATION

A 2 l total volume Biostat® MC reactor (B. Braun Diessel Biotech, Melsungen, Germany) equipped with a variable-pitch paddle impeller was head space aerated. DOT at 45 % and pH at 7.2 were maintained with O_2, N_2 and CO_2 and the temperature was kept at 37°C. The cultivation was started with gentle mixing 35 rpm and later the stirring was increased to 100 rpm for proper oxygen transfer. The perfusion culture was established by recycling cells from a modified Erlenmeyer flask separator unit (volume 300 ml, made in-house[9]) which was connected to the product stream from the reactor. The Erlenmeyer flask was mounted on top of the reactor with an angle of 45° for the underside of the flask so the cells could slide out of the separator back into the reactor. Before use the flask was siliconized with dichloromethylsilan. The separator was isolated to

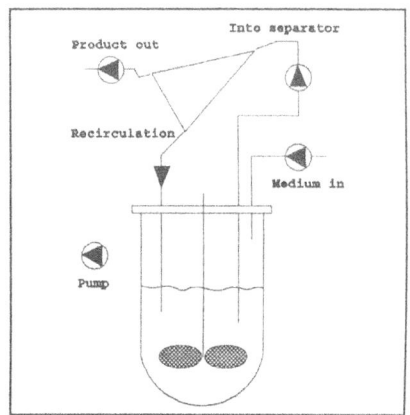

Figure 1 Cultivation set-up

maintain the temperature at 37 °C. Reactor, cell separator, medium reservoir, and product reservoir were connected with silicone tubes as depicted in Figure 1. A peristaltic pump model MHRE 7 (Watson-Marlow, Falmouth, UK) pumped culture broth from the reactor to the cell separator with a constant flow of 250 ml h^{-1}. A peristaltic pump model MHRE 2 (Watson-Marlow) pumped spent medium (product) out of the cell separator to the product reservoir with a flow that determined the perfusion rate. The product reservoir was shifted daily for the determination of the cell numbers. A constant working volume of 1.5 l in the reactor was maintained by a level sensor in the reactor coupled to a peristaltic medium pump (model FE 411, B. Braun Diessel Biotech), i.e. the total working volume was 1.8 l in perfusion mode, when the 0.3 l separator was filled with culture liquid. Both medium reservoir and product reservoir were kept at 4°C.

2.4 ASSAYS

Cell numbers for both the reactor culture and for cultivation broth collected in the product reservoir were determined daily with a haemocytometer using the Trypan blue dye exclusion method. The viable and dead cell numbers were mean values from two individual determinations. Glucose and lactate were determined daily using a YSI 2000 semi-automatic D-glucose/L-lactate analyzer (Yellow Springs Instruments, Yellow Springs, Ohio, USA). The product concentration was determined by a three-layer sandwich ELISA (catching ELISA) method.[11] Analysis of ammonium ions were based on an enzymatic kit (Cat. no. 1112732, Boehringer Mannheim, Mannheim, Germany). Free amino acids were determined with the *ortho*-phthaldialdehyde/2-mercaptoethanol (OPA/2-ME) precolumn derivatization procedure, reversed phase HPLC, and fluorescence detection as described earlier.[6]

3 Results and discussion

3.1 THE CELL CULTURE AND ANTIBODY PRODUCTION

Cell concentrations and perfusion rates for the hybridoma culture are seen in Figure 2a. After 72 h in batch mode where the total cell concentration increased from $0.11*10^6$ cells ml^{-1} to $1.09*10^6$ cells ml^{-1}, 650 ml of fresh medium was pumped into the reactor system while the separation chamber was filled. After the filling of the separation chamber the perfusion is started.

Until 215 h culture time the total cell concentration continued to increase to $4.76*10^6$ cells ml^{-1}. At this time the perfusion rate was adjusted to 0.045 h^{-1}, which, however, caused a slight decrease in cell number. The increasing trend of the cell number was reestablished by reduction of the perfusion rate to about 0.037 h^{-1}. After 402 h at a total cell concentration of $11.0*10^6$ cells ml^{-1}, wash out once again occurred when the perfusion rate was adjusted to 0.047 h^{-1}. Again a stable culture was established by decreasing the perfusion rate.

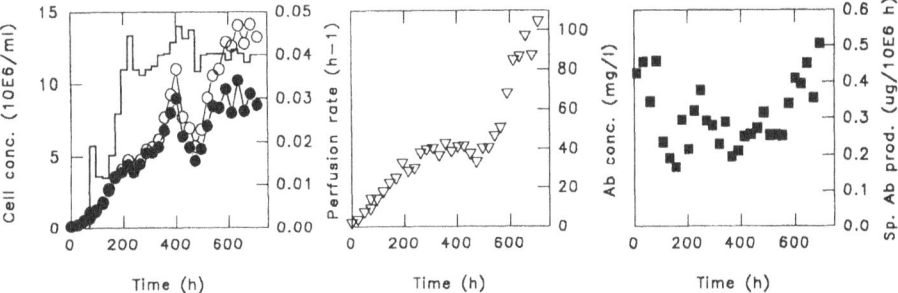

Figure 2 Time course for a: Viable (●) and total (○) cell concentrations and perfusion rate (—) b: Antibody concentration (▽) c: Specific antibody production (■).

Until termination of the culture viable cell concentrations oscillated between $8*10^6$ cell ml^{-1} and $10*10^6$ cell ml^{-1}. The highest possible perfusion rate before wash out occur seems to be 0.043 h^{-1}. The development in cell densities display a trend similar to what is often seen in perfusion cultures independent of how the cell retention is established.[8,12] At the same time as the cell density increases the viability decreases from nearly 100 % to 65 % due to accumulation of dead cells. It is important to note, that the drop in viability only is a matter of accumulation of dead cells - the viable cell density is unaffected in our case.

In Figure 2b the antibody concentration is presented. During the initial 200 h the antibody concentration increased to 40 mg l^{-1}. This level, kept until 550 h, is in the high range of what is seen in comparable batch cultivations in the same reactor where antibody concentrations around 20 mg l^{-1} were obtained.[11] However, the antibody concentration increased significantly in the late part of the cultivation to a level of 100 mg l⁻¹. The specific antibody production seen in Figure 2c explains the increase in antibody concentration. Although a decrease in productivity occurred in the middle of the cultivation compared to the batch part of the cultivation, the specific antibody production recovered in the late part of the cultivation to the initial level. The high product concentration in the final part of the cultivation is a good example of a perfusion culture, where the idea behind the method is demonstrated in a positive fashion - the cell density is increased and the specific production rate is retained. With the established viable cell concentration and the specific productivity 0.4 μg $(1*10^6$ cells h)$^{-1}$ an antibody concentration of 100 mg l⁻¹ and a volumetric productivity of 100 mg l⁻¹ d⁻¹ were obtained. In the literature very different trends for specific productivity in high cell density perfusion systems are reported. In some investigations the specific productivity increased, in other investigations it decreased in the stage when high cell densities occured.[2,8,12,14]

3.2 THE METABOLISM

From Figure 3a it is seen that glucose concentrations of 1.5 mM and lactate concentrations about 30 mM were reached in the final stage. The profile for lactate is nearly the opposite of the one for glucose. In Figure 4a the specific consumption and production rates for glucose respective lactate are shown. The first days in batch mode exhibited high rates indicating intense energy formation by glycolysis. After the perfusion started the rates were more constant. However it can be seen that the yield of lactate on glucose decreased to nearly 1 mole mole⁻¹ in the periods where the cell density was high. This means a better utilization of glucose by energy generation through oxidative metabolism.

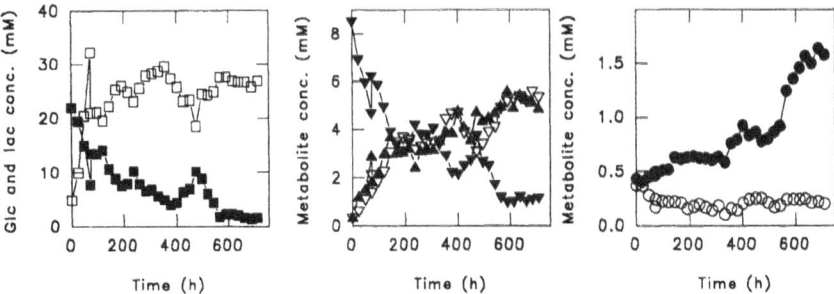

Figure 3 Metabolite concentrations vs. culture time. a: (■) glucose and (□) lactate. b: (▼) glutamine, (▽) alanine and (▲) ammonia, c: (○) serine and (●) glycine,

Examples of amino acid profiles and specific consumption/production rates during the cultivation are seen in Figure 3 and 4. Glutamine, histidine, arginine, tyrosine, valine, isoleucine, leucine, and lysine were consumed by the cells. In the final part of the cultivation with the highest cell density all these components were at the lowest level obtained. However, none of them were totally depleted. The glutamine concentration from the inlet stream was reduced ten-fold to a level about 1 mM which was the largest decline. Compared to the inlet concentration lysine was reduced to 50 %, isoleucine, valine, and histidine to 40 %, while arginine, tyrosine, and leucine were reduced to about 30 % in the final part of the cultivation.

Five of the amino acids were produced during the cultivation. Of these, alanine was the most prominent ending at a level of 5 mM. In Figure 3c it is seen that the glycine concentration changed dramatically during the cultivation. After 350 h the concentration increased from a level

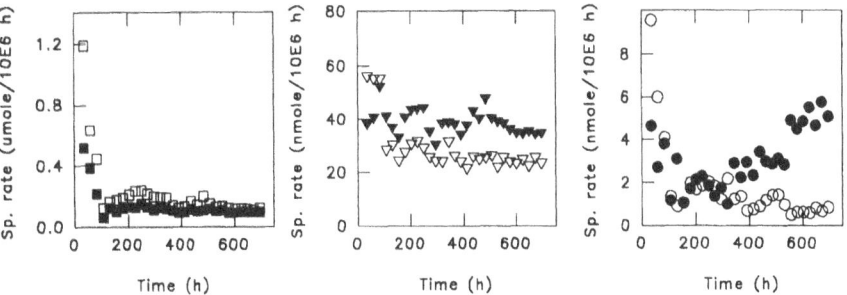

Figure 4 Specific metabolite consumption and production rates vs. culture time. a: (■) glucose consumption and (□) lactate production b: (▼) glutamine consumption and (▽) alanine production c: (●) glycine production and (○) serine consumption.

around 0.6 mM to 0.8 mM and after 550 h to a level around 1.5 mM. Low amounts (0.05-0.15 mM) of glutamate and asparagine were released through out the cultivation while aspartate occurred at a very low level (0.02 mM) only in the final part of the cultivation where the cell density was highest. Except for the batch part of the cultivation the threonine concentration was fairly constant during the cultivation. The serine concentration decreased in the first half of the cultivation to a level around 0.15 mM. Then after 400 h the concentration increased to a level over 0.2 mM. It is also in this period that the high glycine concentration occurred. During the first 72 h in batch mode ammonium was formed giving a concentration of 3 mM. Throughout the rest of the cultivation this concentration increased at a lower rate to a level of 5 mM (Figure 3b). The specific ammonia production was near constant during the perfusion cultivation (not shown).

For most of the amino acids specific consumption and production rates were fairly constant through the perfusion cultivation while large fluctuations were seen for some of the components in the batch part of the cultivation illustrated in Figure 4b for glutamine and alanine. In Figure 4c the specific glycine production and serine consumption rates are shown. From this graph it is seen that the increase in glycine concentration during the cultivation was due to an increase in the specific glycine production. At the same times as the specific glycine production changed the specific serine consumption also changed, however, to a relative lesser extent.

It is often reported that cells in culture at the same time as they consume some amino acids (notably glutamine) produce other amino acids (e.g. alanine and aspartate), but specie and extent vary between cell lines.[3,13,16] Some of these differences are explained by the interplay between different enzymes, e.g. glutamate dehydrogenase, and alanine and aspartate amino transferase in the two step conversion of glutamine through glutamate to 2-oxoglutarate.[10] The extensive metabolism of glutamine as respiratory fuel and for anabolic purposes usually referred to as glutaminolysis is also seen in the present work. Here especially alanine is produced which is explained by alanine amino transferase being the dominant enzyme in the conversion of glutamate to 2-oxoglutarate. For the metabolism of glutamine ending with alanine the energy generation was calculated as up to 9 mole ATP per mole glutamine (dependent on which pathway is used from the TCA cycle to pyruvate).[5]

The production of glycine has also been seen by others, but is probably more seldom[13,16] and may be dependant of cell line and physiologic conditions. As will be discussed the shift in metabolism seen by the change in glycine production during the cultivation may be necessary for proper generation of energy and anabolic precursors under certain physiologic conditions.

By examination of Figure 5 it is seen that a key point in the metabolism is pyruvate. It is

306

generated from glucose by glycolysis, it can enter the tricarbocylic acid (TCA) cycle through acetyl CoA, but it can also be generated from TCA cycle components, either directly from malate or from oxaloacetate through P-enolpyruvate. Therefore glutamine can be converted to pyruvate after 2-oxoglutarate derived from glutamine has passed through a part of the TCA cycle. It is also seen that the pyruvate carbon molecules can end in alanine or lactate. Different textbooks and review articles often describe different pathways and the reason becomes clear exploring the original literature - different biological systems make use of different metabolic enzymes. This contribute to uncertainty in the explanations and conclusions made in the present kind of investigations. Having this in mind we turn to a calculation of energy generation under production of glycine.

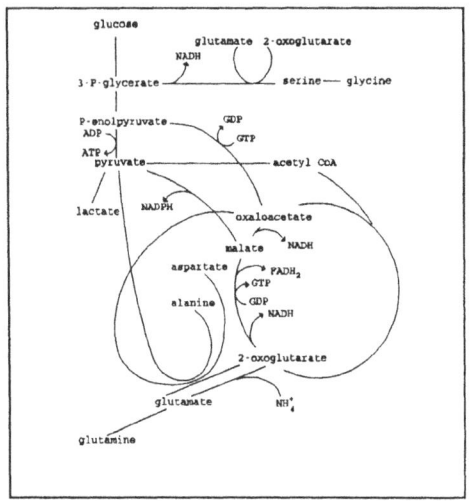

Figure 5 Simplified diagram of the pathways in the energy metabolism

If there for energy formation in the cells is sufficient pyruvate generated from glutamine (through part of the TCA cycle) the glycolytic intermediate 3-P-glycerate can be saved for other purposes, i.e. the pathway ending with glycine. There is a continuous glycolytic flux of glucose carbon to pyruvate and the conversion of P-enolpyruvate to pyruvate is unidirectional - although gluconeogenetic reactions can convert pyruvate to P-enolpyruvate, probably no glutamine carbon can be found in glycine; nevertheless a theoretical calculation of ATP generated from glutamine with glycine as end product can be made.

Glycine is formed from 3-P-glycerate through 3-P-hydroxypyruvate, 3-P-serine, and serine. One mole NADH is gained under the formation of 3-P-hydroxypyruvate from 3-P-glycerate, and 3-P-hydroxypyruvate can be transaminated to 3-P-serine concomitantly with the conversion of glutamate to 2-oxoglutarate making the connection with the glutamine metabolism. From 3-P-serine serine is formed and glycine is produced from serine, when the ß-carbon atom of serine is transferred to tetrahydrofolate, a carrier of one carbon units. The tetrahydrofolate derivatives serve as donors of one carbon units in a variety of biosynthesis. Using 3-P-glycerate for the pathway ending in glycine means that one mole ATP is lost in the conversion of glucose carbon from P-enolpyruvate to pyruvate. The energy output for the theoretical conversion of glutamine to glycine is 11-12 mole ATP per mole glutamine consumed dependent on which pathways that are involved. This under assumption that one mole NADH (NADPH) gives three mole ATP. This energy yield should be compared with the 9 mole ATP which can be generated per mole glutamine under alanine formation. In Fig. 4c the specific glycine production and serine consumption are compared. Part of the increase in glycine production can originate from the consumption of serine, but some must originate from other sources e.g glutamine. Comparison of the specific glycine and antibody production support the idea that more energy and tetrahydrofolate derivatives are generated for antibody synthesis.

Our experiment indicate that changes in the glucose and amino acid metabolism are dependant on physiologic factors, like metabolite concentrations and cell densities. These physiologic factors appear to be important in the design of cultivation procedures.

4 Acknowledgement

The authors would like to acknowledge Miss A. Jensen and Miss M. Bjerre for excellent technical assistance. This work was supported by a grant from the Nordic Programme on Bioprocess Engineering under the auspices of the Nordic Fund for Technology and Industrial Development and by the Center for Process Biotechnology, Technical University of Denmark.

References

1. Andresen, L.O., Koch, C., Sørensen, B.B., Emborg, C. 1989. Production of monoclonal antibodies against the enzyme nuclease from Serratia marcescens and evaluation of the individual antibodies for their use in EIA and in affinity chromatography. Biotechnology Techniques 3: 407-411.
2. Batt, B.C., Davis, R.H., Kompala, D.S. 1990. Inclined sedimentation for selective retention of viable hybridomas in a continuous suspension bioreactor. Biotechnol. Prog. 6: 458-464.
3. Duval, D., Demangel, C., Munier-Jolain, K., Miossec, S., Geahel, I. 1991. Factors controlling cell proliferation and antibody production in mouse hybridoma cells: I. Influence of amino acid supply. Biotechnol. Bioeng. 38: 561-570.
4. Feder, J., Tolbert, W.R. 1983. The large-scale cultivation of mammalian cells. Sci. Am. 248: 24-31.
5. Glacken, M.W. 1988. Catabolic control of mammalian cell culture. Bio/Technology 6: 1041-1050.
6. Hansen, H.A., Emborg, C. 1992. Experimental design in the development and characterization of a high-performance liquid chromatographic method for amino acids. J. Chromatogr. 626: 171-180.
7. Himmelfarb, P., Thayer, P.S., Martin, H.E. 1969. Spin filter culture: The propagation of mammalian cells in suspension. Science 164: 555-557.
8. Kitano, K., Shintani, Y., Ichimori, Y., Tsukamoto, K., Sasai, S., Kida, M. 1986. Production of human monoclonal antibodies by heterohybridomas. Appl. Microbiol. Biotechnol. 24: 282-286.
9. Lassen, K.M.K., de Caldeano, P.R.E.C., Emborg, C. 1992. Methods for cell separation based on sedimentation. Biotechnology Techniques 6: 121-126.
10. McKeehan, W.L. 1986. Glutaminolysis in animal cells. p. 111-150 In: M.J. Morgan. (ed.), Carbohydrate metabolism in cultured cells, Plenum Press, New York.
11. Persson, B., Emborg, C. 1992. A comparison of three different mammalian cell bioreactors for the production of monoclonal antibodies. Bioprocess Eng. 8: 157-163.
12. Reuveny, S., Velez, D., Miller, L., Macmillan, J.D. 1986. Comparison of cell propagation methods for their effect on monoclonal antibody yield in fermentors. J. Immunol. Meth. 86: 61-69.
13. Schmid, G., Johannsen, R. 1990. Metabolic quotients for recombinant CHO and BHK cell lines producing human antithrombin III. Biotech. Lett. 12: 317-322.
14. Seamans, T.C., Hu, W.-S. 1990. Kinetics of growth and antibody production by a hybridoma cell line in a perfusion culture. J. Ferment. Bioeng. 70: 241-245.
15. Tokashiki, M., Arai, T., Hamanoto, K., Ishimaru, K. 1990. High density culture of hybridoma cells using a perfusion culture vessel with an external centrifuge. Cytotechnology 3: 239-244.
16. Wagner, R., Ryll, T., Krafft, H., Lehmann, J. 1988. Variation of amino acid concentrations in the medium of HU β-IFN and HU IL-2 producing cell lines. Cytotechnology 1: 145-150.

KINETICS OF CELL METABOLISM AND ANTIBODY PRODUCTION IN HIGH CONCENTRATION PERFUSION CULTURES

S. Zhang, A. Handa-Corrigan and R.E. Spier
School of Biological Sciences
University of Surrey
Guildford
Surrey GU2 5XH, UK

ABSTRACT. High concentration cell cultures were conducted in a perfusion system based on external microfiltration with direct micron-sized bubble oxygenation. Kinetics of cell metabolism with respect to the consumption of glucose, glutamine and the production of lactate, ammonia and monoclonal antibody were examined and compared in different culture media. Cell-specific antibody production rate increased gradually as the cell concentration increased, whilst, cell-specific glucose, glutamine uptake rates and lactate, ammonia production rates decreased. Antibody production rate increased linearly with both the nutrient uptake rates and metabolite production rates in cultures without either nutrient limitation or metabolite inhibition. A stepwise decrease in cell concentration during the course of cultivation reversed the direction of the trends of changes in cell metabolic activities. Close relationship between the cell-specific metabolic rates and cell-specific growth rates was observed. The implications of the cellular metabolic information on the optimization of high concentration animal cell perfusion culture systems were discussed.

1.   Introduction

A variety of medium perfusion techniques have been developed to increase the process intensities of stirred tank bioreactors for animal cell cultures. These include external and internal membrane filtration (Smith, et al., 1991; Blasey and Jäger, 1991), rotating wire cage (Varecka and Scheirer, 1987), external centrifuge (Tokashiki, et al., 1990; Hamamoto et al., 1989), and gravitational sedimentation (Tokashiki and Arai, 1989; Kitano, et al., 1986). Recently, we have developed a perfusion culture system based on external membrane filtration to examine different methods of oxygenation in high density animal cell cultures (Zhang et al., 1992c). It was found that direct micron-sized bubble oxygenation, which had an oxygen transfer capability in the order of 100 h^{-1} (Zhang, et al., 1992b), could be used successfully in high concentration animal cell cultures. In addition, it was found that the accumulation of ammonia in the culture medium rather than oxygen limitation was one of the likely problems that inhibited cell growth. Three possible strategies can be exercised to reduce the accumulation of ammonia in the culture medium. Firstly, one can increase the medium perfusion rate so as to dilute the ammonia concentration in the culture medium. This strategy is not economical and is likely to be limited by the filtration efficiency of the hollow fibre cartridge. Secondly, one can remove ammonia specifically from the culture medium using ammonia specific zeolite cartridge (Nayve, et al., 1991). However, this method increases substantially the complexity and contamination risk of the culture system. Moreover, the ammonia removal efficiency is likely to be limited by the saturation concentration of the zeolite cartridge. Finally, one can control ammonia formation by manipulating cellular metabolism in culture. This is based on the understanding that ammonia is the waste product of glutamine metabolism and that glucose and glutamine, which are the main nutrient and energy sources for animal cells in culture, are metabolically closely inter-related (Glacken,

S. Kaminogawa et al. (eds.), Animal Cell Technology: Basic & Applied Aspects, Vol. 5, 309–324.
© 1993 Kluwer Academic Publishers.

1988). This method is expected to be the most efficient in controlling ammonia formation without increasing the cost and complexity of the culture process. However, the manipulation of cellular metabolism should also be assessed from a product formation point of view. This requires the knowledge of metabolic and product formation rates of cells in high concentration culture system and the corresponding changes in the cells. We think that this knowledge is crucial to the optimization of high concentration animal cell cultures which are operated without oxygen limitation effects.

In this paper, metabolic and antibody production rates of cells were determined in high concentration perfusion cultures with culture media having different glucose and glutamine concentrations. Based on the metabolic information obtained, strategies for optimizing homogeneous high concentration animal cell cultures were proposed.

2. Materials and Methods

2.1 CELLS

A mouse-mouse hybridoma cell line designated AFP-27, which produced anti-α foetal protein IgG, was used in the present study.

2.2 CULTURE MEDIA

It was reported previously that supplementation with Pluronic F-68 in culture medium was essential for the cultivation of hybridoma cells under direct bubble oxygenation (Zhang, et al., 1992a). Consequently, in the present study, RPMI-1640 basal media (Flow Labs.) supplemented with 5% New Born Calf Serum (Gibco), 0.2% (w/v) Pluronic F-68 (BASF, Ludwigshafen, Germany) and various concentrations of glucose and glutamine were used as the culture media.

2.3 METABOLIC ASSAYS

Samples of the homogeneously suspended cells and culture medium were taken daily from the bioreactor for metabolic assays. Each sample was spun down at 3000 rpm for 6 min. Glucose and lactate concentrations in the supernatant were measured by a glucose/lactate analyzer (Analox Instrument Ltd, England). The ammonia and glutamine concentrations in the same sample were determined with an ammonia electrode (Kent Industrial, Kent, England).

2.4 MONOCLONAL ANTIBODY ASSAY

The monoclonal antibody (MAb) concentration in the same sample taken from the bioreactor was measured by the ELISA technique.

2.5 MEASUREMENT OF CELL CONCENTRATION AND VIABILITY

Cell concentration and viability were determined by trypan blue dye exclusion in a haemocytometer. The stained cells were treated as dead cells.

2.6 PERFUSION CULTURE SYSTEM

Details of the perfusion culture system have been reported previously (Zhang, et al., 1992c). Briefly, a 2-L stirred tank bioreactor (SGi, Toulouse, France) was connected to an externally mounted hollow fibre microfiltration cartridge (Minikros, Microgon, USA) for continuous medium perfusion. Dissolved oxygen tension (DOT) was controlled at $35\pm 5\%$ of air saturation with intermittent, direct micron-sized bubble oxygenation. pH was controlled at 7.20 ± 0.02 by either adding CO_2 into the headspace or delivery of 0.06M NaOH solution into the culture medium. Temperature was controlled at $36.5 \pm 0.2°C$ by a hot air chamber around the bioreactor, whilst agitation speed was controlled at 60 rpm.

2.7   CALCULATION OF CELL-SPECIFIC METABOLIC RATES

The cell-specific nutrient uptake rates, product formation rates, and the apparent cell-specific growth rate were calculated with respect to the cell numbers in the bioreactor. This was based on the observation that negligible changes in the cell sizes were found during the course of the perfusion cultures. For cultures which involve significant changes in cell size, however, cell-specific metabolic rates based on biomass need to be considered (Modha, *et al.*, 1992). Cell-specific nutrient uptake rates, U_n

$$U_n = \frac{R_p(2C_{no}-(C_{nt_2}+C_{nt_1}))}{X_{t_2}+X_{t_1}} \qquad (1)$$

Cell-specific antibody production rate, U_a

$$U_a = \frac{R_p(C_{at_2}+C_{at_1})}{X_{t_2}+X_{t_1}} \qquad (2)$$

Apparent cell-specific growth rate, μ_{app}

$$\mu_{app} = \frac{2(X_{t_2}-X_{t_1})}{(X_{t_2}+X_{t_1})(t_2-t_1)} \qquad (3)$$

where R_p is medium perfusion rate (ml/h), C_{no} is the initial nutrient concentration in the medium, C_{nt1} and C_{nt2} are the nutrient concentrations assayed at culture time of t_1 and t_2, respectively, X_{t1} and X_{t2} are the cell concentrations at culture time of t_1 and t_2, respectively, C_{at1} and C_{at2} are the antibody concentrations determined at culture time of t_1 and t_2, respectively.

3.   Results

3.1   CULTURE MEDIUM WITH HIGH GLUTAMINE AND GLUCOSE CONCENTRATIONS

The culture medium was formulated with glutamine and glucose concentrations of 5 mM and 450 mg/dl, respectively. The medium was supplemented with 5% NBCS and 0.2% Pluronic F-68 prior to use for medium perfusion. Figures 1 (a) and (b) show the perfusion rate, cell growth, MAb concentration and metabolic assay results, respectively. Similar to the results reported previously (Zhang *et al.*, 1992c), consistently high ammonia concentrations (> 2.5 mM) were found at the later stages of the cultivation. This is expected to contribute to the inability to attain higher cell concentrations in the bioreactor. The result suggests that simply increasing the hollow fibre filtration surface area and the medium perfusion rate are not effective methods to control ammonia accumulation.

Figures 2 (a) and (b) show the cell-specific growth, antibody production, and metabolic

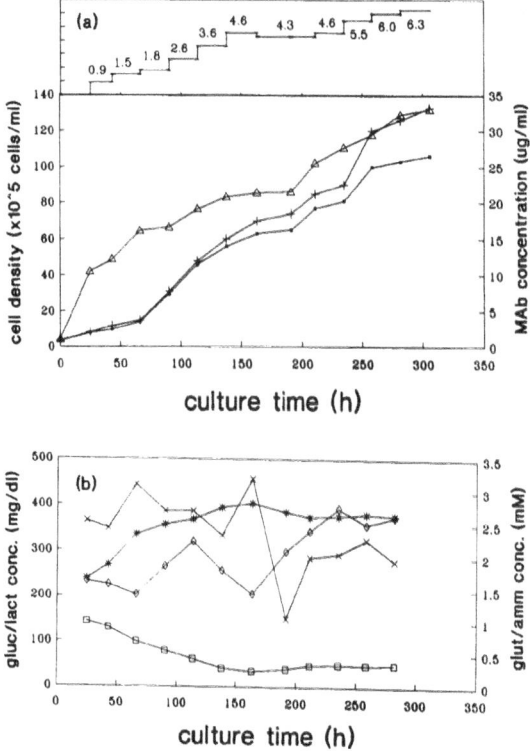

Figure 1. Perfusion culture with high glutamine concentration medium: (a) perfusion rate, cell and MAb concentration profiles; (b) nutrient and metabolite concentration profiles. Symbols show viable cell (—), total cell (+), MAb (△), glucose (✳), glutamine (✳), lactate (⊟), and ammonia concentrations (◇). Numbers in the graph denote the perfusion rate (d⁻¹).

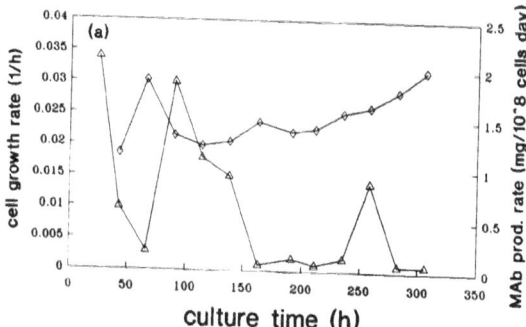

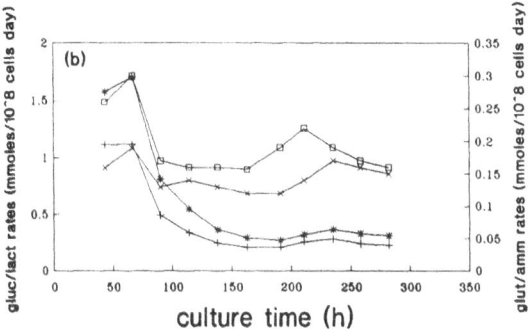

Figure 2. Cell-specific growth and metabolic rates in perfusion culture with high glutamine concentration medium: (a) cell-specific growth and MAb production rates; (b) cell-specific metabolic rates. Symbols show cell-specific growth rate (-△-), MAb production rate (-◇-), glucose uptake rate (-+-), glutamine uptake rate (-□-), lactate production rate (-*-), and ammonia production rate (-×-).

rates, respectively, in the above perfusion culture. After an initial rise and subsequent decline, the cell-specific antibody production rate increased gradually as the cultivation progressed. The decrease in the cell-specific growth rate observed between culture times of 25 - 50 h was attributed to the high shearing medium recirculation rate of 730 ml/min, which was used during this period of cultivation. An inverse relationship appeared to exist between the cell-specific growth and antibody production rates during the initial stage of the culture. Unlike the cell-specific antibody production rate, cell-specific metabolic rates showed opposite trends as the cultivation progressed. Initially, increases in cell metabolic rates were observed although cell growth rates decreased. This could be attributed to the cell-damaging effects at high medium recirculation rates (730 ml/min), suggesting that cells are metabolically more active under conditions of mechanical stress. This is in agreement with the metabolic results reported by Smith and Greenfield (1992) and Abu-Reesh and Kargi (1991). Following these initial increases, cell-specific metabolic rates decreased rapidly before reaching relatively constant values. These results suggest that as the cell concentration increased, cells became more efficient in antibody production although the metabolic activities of the cells decreased. This effect is likely to be the result of reduced cell growth rate, and the channelling of metabolic energy and intermediates to the synthesis of antibody. Similar changes in cell metabolic activities have been reported by Runstadler and Young (1991) and Hayter, *et al.* (1992).

3.2 CULTURE MEDIUM WITH LOW GLUTAMINE CONCENTRATION

Since ammonia is the main waste product of glutamine metabolism, ammonia production is expected to be reduced in culture medium having a lower glutamine concentration. Consequently, culture medium with a glutamine concentration of 2 mM was used in this experiment. Figures 3 (a) and (b) show the perfusion rate, cell growth, MAb concentration, and metabolic assay results, respectively. Ammonia concentrations were successfully controlled below 1.5 mM at the later stage of the cultivation without causing glutamine limitation in the culture. Unfortunately, this perfusion culture run was not long enough to show the effects of reduced ammonia concentration on cell growth and MAb production, because the hollow fibre cartridge was damaged at the culture time of 209 h. However, it is

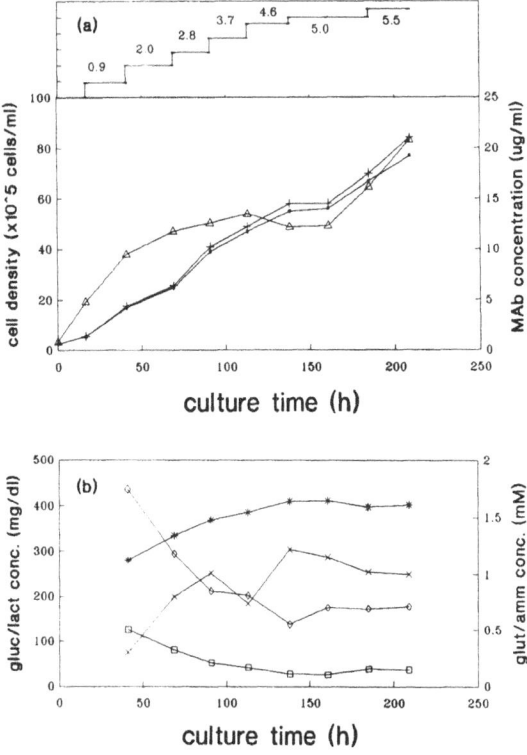

Figure 3. Perfusion culture with low glutamine concentration medium: (a) perfusion rate, cell and MAb concentration profiles; (b) nutrient and metabolite concentration profiles. Symbols show viable cell (-•-), total cell (-+-), MAb (-△-), glucose (-*-), glutamine (-×-), lactate (-☐-), and ammonia concentrations (-◇-). Numbers in the graph denote the perfusion rate (d⁻¹).

However, it is expected that these effects will become apparent in long term cultivations. (In the next section, we present data showing increased cell growth and MAb production in an extended perfusion culture).

Figures 4 (a) and (b) show the cell-specific growth, antibody production, and metabolic rates, respectively, in medium having an initial glutamine concentration of 2 mM. The cell-specific growth rate remained relatively stable at a value of 0.043 h⁻¹ for approximately 24 h, before decreasing rapidly to a value of about 0.014 h⁻¹ in the next 30 h cultivation. This rapid decrease in the cell-specific growth rate was accompanied with a large increase in the cell-specific antibody production rate. This again indicates the existence of an inverse relationship between the cell-specific growth rate and the cell-specific antibody production rate. As the cultivation progressed the cell-specific growth rate decreased relatively gradually, except at the culture time of 100 h when increased cell growth rate was observed. On the other hand, the cell-specific antibody production rate increased slowly during the course of the culture.

The cell-specific metabolic rates decreased gradually between 50 -150 h of cultivation, and

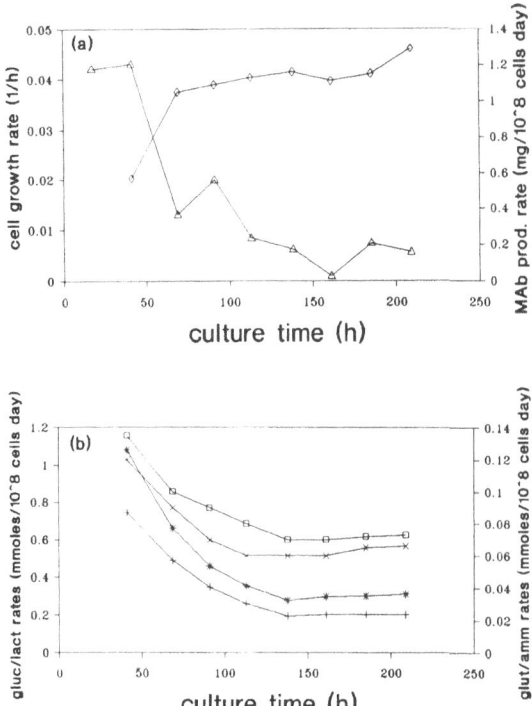

Figure 4. Cell-specific growth and metabolic rates in perfusion culture with low glutamine concentration medium: (a) cell-specific growth and MAb production rates; (b) cell-specific metabolic rates. Symbols show cell-specific growth rate ($-\triangle-$), MAb production rate ($-\diamond-$), glucose uptake rate ($-+-$), glutamine uptake rate ($-\square-$), lactate production rate ($-*-$), and ammonia production rate ($-\times-$).

then remained relatively constant during the remaining 70 h of cultivation. Figure 5 (a) and (b) show the comparison of cell-specific glutamine uptake rates, ammonia production rates, and antibody production rates in culture media having initial glutamine concentrations of 5 mM and 2 mM, respectively. Cell-specific glutamine uptake rate and ammonia production rate in the low glutamine medium were approximately half of those in the high glutamine medium. Except in the early stages of the cultivation, significant changes in the cell-specific antibody production rate were not observed in the two different media. The high cell-specific antibody production rate observed in the high glutamine medium at culture times between 50 -100 h is resulted from the high shearing medium recirculation rate used during this period of cultivation(730 ml/min). Although the glutamine assay cannot differentiate changes in glutamine concentration caused by either cell utilization or spontaneous degradation (Tritsch and Moore, 1962), the latter is expected to be negligible because of the short residence time of the medium in the culture system ($<$ 12 h), and the relatively low temperature at which the medium was kept ($<$ 20°C). The reduced ammonia concentration in the culture medium is, therefore, likely to be the result of reduced glutamine utilization.

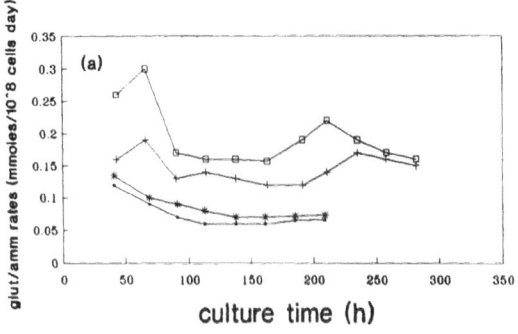

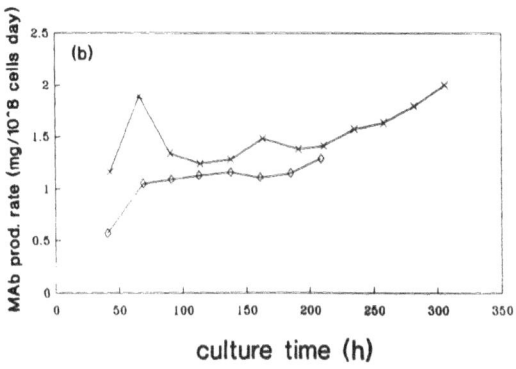

Figure 5. Comparison of cell-specific glutamine uptake rates, ammonia and antibody production rates in culture media having both low and high glutamine concentrations: (a) glutamine and ammonia rates; (b) antibody production rates. Symbols show glutamine uptake rate (-□-), ammonia production rate (-+-), and antibody production rate (-*-) in high glutamine concentration medium, and glutamine uptake rate (-*-), ammonia production rate (-•-), and antibody production rate (-◇-) in low glutamine concentration medium.

Despite the changes in the cell-specific glutamine uptake rate and ammonia production rate, significant changes in the MAb production, glucose uptake, and lactate production rates were not observed between the high and low glutamine culture media. These results suggest that change in the medium glutamine concentration does not affect other cell metabolic parameters and productivity profiles.

Although both the cell-specific glucose and glutamine uptake rates decreased as the cell concentration increased, the cell-specific glucose uptake rates at the later stages of cultivation were approximately 4-fold lower than those at the early stages, while those for the cell-specific glutamine uptake rates were only 1.5 - 2.0-fold lower. Furthermore, this significant drop in the cell-specific glucose uptake rate was associated with a substantial drop in the cell-specific lactate production rate. This suggests that cells used in the present study became more glutamine dependent as the cell concentration increased. This may account for the consistently high glucose and low lactate concentrations observed at the later stages of the cultivation (shown in Figs. 1 (a) and 3 (a)).

3.3   CULTURE MEDIUM WITH BOTH LOW GLUTAMINE AND GLUCOSE CONCENTRATIONS

As shown previously, very high glucose concentrations were present in the culture medium. To test their effects on cell growth and MAb production, the medium glucose concentration was reduced to 200 mg/dl, while keeping the glutamine concentration at 2 mM. Furthermore, in order to verify the inverse relationship observed between the cell-specific antibody production rate and the apparent cell-specific growth rate in the previous cultures, cells were bled out of the system during the course of cultivation in order to create a stepwise change in the cell concentration. Figures 6 (a) and (b) show the perfusion rate, cell growth, MAb concentration, and metabolic assay results, respectively. Cells were bled out of the system at the cultivation time of 240 h when the cell concentration was 1.0×10^7 cells/ml. After this bleeding-out, cells continued to grow from an initial cell density of 3.5×10^5 cells/ml, and reached a final cell concentration of 2.5×10^7 cells/ml (viable). On the other hand, MAb concentration increased from 10 μg/ml to 170 μg/ml after the cell bleeding. Ammonia concentrations were successfully controlled below 1.5 mM in the whole culture process. Compared to the cell density of 8×10^6 cells/ml and MAb concentration of 20 μg/ml obtained in the previous perfusion cultures, which were also devoid of ammonia inhibition (shown in Fig. 3 (a)), the increased cell and MAb concentrations in the present study are attributed to the low ammonia concentrations during this extended culture run.

The cell bleeding procedure caused significant changes in the nutrient and metabolite concentrations in the culture medium. Substantial decreases in both the glucose and glutamine concentrations were observed one day post the cell bleeding. These decreases were accompanied by significant increases in both the lactate and ammonia concentrations in the culture medium. These variations are attributed to the sharp increase in the cell growth rate observed after cell bleeding (shown in Fig. 7 (a)). However, as the cultivation progressed further, most of the nutrient and metabolite concentrations resumed to the previous levels observed before cell bleeding.

Figures 7 (a) and (b) show the cell-specific growth, MAb production, and metabolic rates, respectively, for the above perfusion culture. Before cell bleeding, the cell-specific antibody production rate increased gradually, whilst the cell-specific growth rate decreased as the cultivation progressed. After the cell-bleeding, the cell-specific growth rate increased sharply to a value of about 0.03 h^{-1}, before started to decrease again in a manner similar to that observed before the cell bleeding. On the other hand, after the cell bleeding, the cell-specific antibody production rate decreased to a value of 0.8 mg/10^8 cells day, which is close to the value observed at the beginning of the culture. As the culture progressed further, the cell-specific antibody production rate increased again as the cell-specific growth rate decreased. These results are in agreement with the previous findings that cell-specific growth rate changed inversely with the cell-specific antibody production rate. Therefore, it is suggested that maximization in antibody production can be achieved under conditions where cell growth rate is kept to minimal values without compromising too much on the viability of the cells. From the results presented previously, continuous high density perfusion animal cell culture appears to be an optimal culture system for antibody production, because of the high cell density, minimal cell growth rate and relatively high cell viability.

Large increases in the cell-specific metabolic rates were observed only after the cell-bleeding, and these are likely to be related with the associated increases in the cell-specific growth rates. At other times, however, similar trends to that observed previously were seen, i.e. the cell-specific metabolic rates decreased as the cell concentration increased. Unlike the reduction of the medium glutamine concentration (shown in Fig.4 (b)), a significant change in the cell-specific metabolic rates was not induced by changing the medium glucose concentration from 450 mg/dl to 200 mg/dl. This is in contrast to the cell-specific glutamine

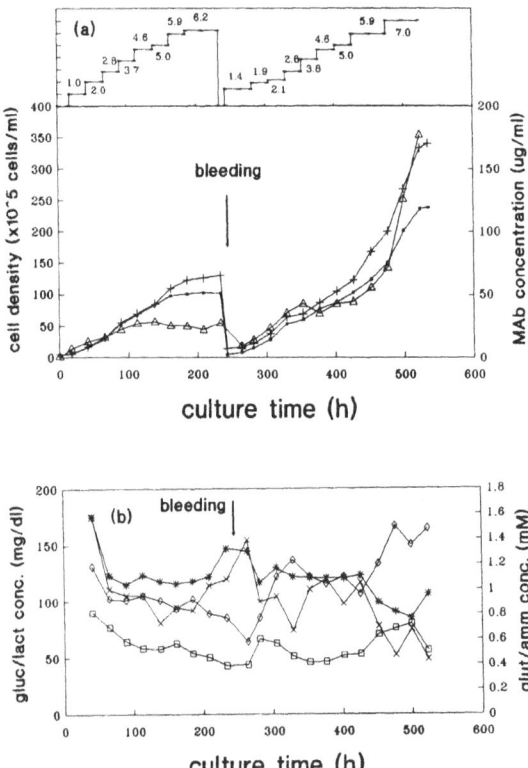

Figure 6. Perfusion culture with medium having both low glutamine and glucose concentrations: (a) perfusion rate, cell and MAb concentration profiles; (b) nutrient and metabolite concentration profiles. Symbols show viable cell (⊶), total cell (+), MAb (-△-), glucose (✳), glutamine (⨯), lactate (-□-), and ammonia concentrations (-◇-). Numbers in the graph denote perfusion rate (d^{-1}).

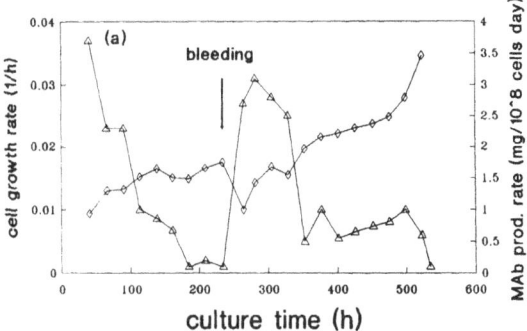

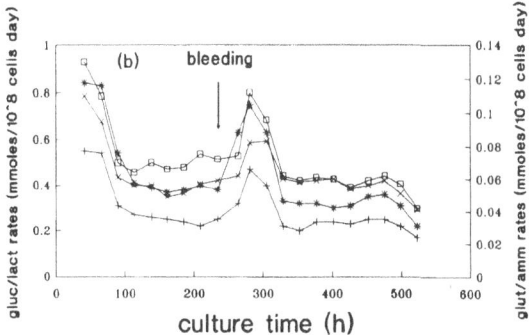

culture time (h)

Figure 7. Cell-specific growth and metabolic rates in perfusion culture with medium having both low glucose and glutamine concentrations: (a) cell-specific growth and MAb production rates; (b) cell-specific metabolic rates. Symbols show cell-specific growth rate ($-\triangle-$), MAb production rate ($-\diamond-$), glucose uptake rate ($-+-$), glutamine uptake rate ($-\boxminus-$), lactate production rate ($-*-$), and ammonia production rate ($-\times-$).

uptake and ammonia production rates, which were decreased by nearly 50% when a reduction in the medium glutamine concentration was made from 5 mM to 2 mM. It appears that neither glutamine nor glucose metabolism is dependent on glucose at a concentration range between 200 - 450 mg/dl. Even with the low glucose medium, consistently high glucose concentrations (> 100 mg/dl) were still present in the culture medium. These results suggest that glucose concentrations of 450 mg/dl and 200 mg/dl in the basal RPMI media are both in excess, with respect to the requirement of the cells. Since there has been a report that high glucose concentration has inhibitory effect on recombinant protein production in animal cell cultures (Sigiura, 1992), further reduction in glucose concentration is recommended in order to examine its effects on cell growth and MAb production in high density hybridoma cell cultures.

3.4 RELATIONSHIP BETWEEN ANTIBODY PRODUCTION, NUTRIENT UPTAKE AND METABOLITE PRODUCTION RATES

Although cell-specific metabolic and antibody production rates are important parameters in determining the efficiency of a perfusion culture system, they do not directly relate to the control and prediction of antibody production in this perfusion culture system. For the purpose of control, metabolic and antibody production rates based on bioreactor volume rather than cell numbers are more relevant. Consequently, metabolic results of the perfusion culture carried out with medium having an initial glucose and glutamine concentration of 200 mg/dl and 2 mM, respectively, were rearranged based on the bioreactor volume. Figures 8 (a) and (b) show the relationships between antibody production rate and glucose, glutamine uptake rates, respectively. Antibody production rate was directly proportional to the uptake rates of both glucose and glutamine. For each milligram of antibody produced, approximately 0.11 mmole glucose and 0.028 mmole glutamine were utilized. The intercepts at zero rates of MAb production observed in the graphs are suggested to be the minimum nutrient uptakes required for cell maintenance. Similar linear relationships were also observed between antibody production rate and metabolite production rates, as shown in Figures 9 (a) and (b). For each milligram of antibody produced, approximately 0.15 mmole lactate and 0.027 mmole

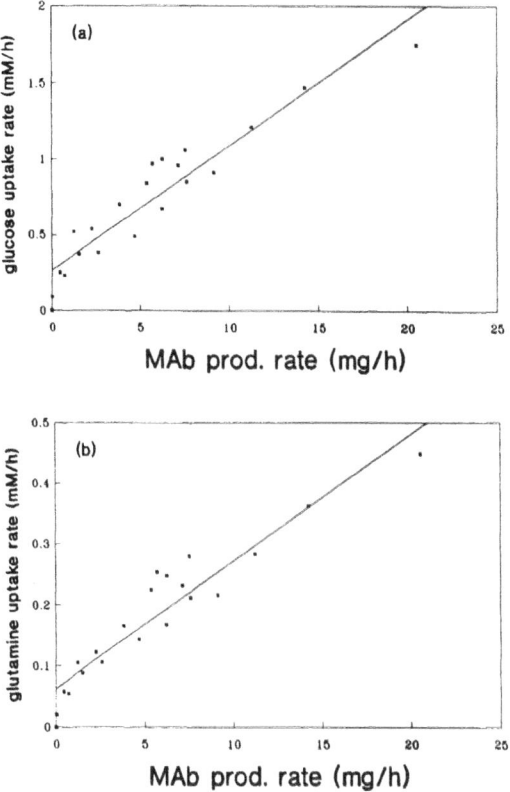

Figure 8. Relationships between nutrient uptake rates and MAb production rate: (a) glucose uptake rate and MAb production rate; (b) glutamine uptake rate and MAb production rate.

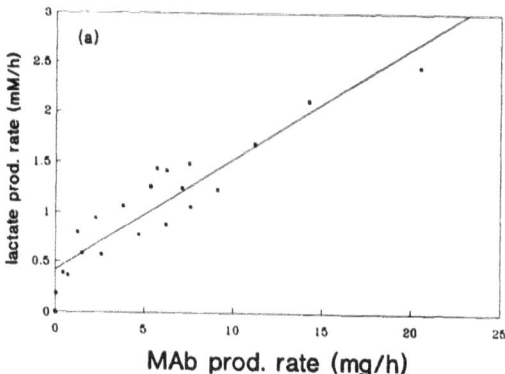

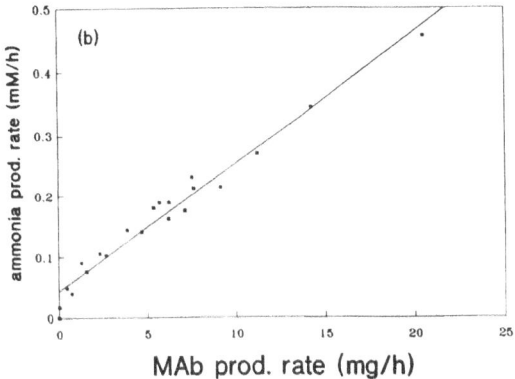

MAb prod. rate (mg/h)

Figure 9. Relationships between metabolite production rates and MAb production rate: (a) lactate and MAb production rates; (b) ammonia and MAb production rates.

ammonia were produced. Comparable linear relationships between antibody production rate, nutrient uptake rates and metabolite production rates were reported by Handa-Corrigan *et al.* (1992a) in hollow fibre bioreactors, which were optimized using similar strategies employed in the present study. Therefore, it seems that under optimized growth conditions, cells perform rather similarly in different configurations of high cell concentration culture systems.

4. Discussion

The accumulation of high concentrations of ammonia was again shown to be one of the problems that inhibited cell growth and MAb production in medium having a high glutamine concentration, even under a relatively high medium perfusion rate. This suggests that simply increasing the medium perfusion rate is not an effective method for controlling ammonia accumulation in a perfusion culture process. By making use of the linear relationship between ammonia production and glutamine consumption rates, ammonia concentrations were successfully controlled at a non-inhibiting level by reducing the glutamine concentration from 5 mM to 2 mM in the perfusing medium. The spontaneous degradation of glutamine is negligible in these experiments and therefore, the reduction in ammonia production is expected to be mainly the result of reduced glutamine utilization in low glutamine medium. Comparable cell growth and MAb production were found in both the low and high glutamine concentration media. Therefore, the reduced glutamine utilization in low glutamine media may be attributed to the reduced energy and intermediate requirements for cell maintenance under such conditions. In the other words, when cells are under either mechanical or metabolic stresses (such as ammonia inhibition), the metabolic activities of the cells are enhanced to meet the increased demand for energy and intermediate required for cell maintenance. This is in agreement with the results reported by Smith and Greenfield (1992) and Abu-Reesh and Kargi (1991).

The reduction in the medium glutamine concentration did not cause any significant change in the consumption of glucose and the production of lactate. Compared with the cell-specific glucose uptake rate, relatively small reductions in the cell-specific glutamine uptake rates were observed as the cell concentration increased. This suggests that glutamine is preferentially used for cell maintenance and product formation, while glucose consumption is required for rapid cell growth at low cell densities in the early stages of perfusion cultivation. Similar results have been reported by Miller *et al.* (1988; 1989) in continuous hybridoma cell cultures. This raises some questions about the role of glucose on cell growth and product formation in high density perfusion cultures. The results obtained in the present study show that both the glucose and glutamine metabolism are not dependent on the glucose concentrations in the range between 450 - 200 mg/dl, which are in great excess relative to that which is actually required for cell growth. In addition, the relatively low lactate concentrations observed in the culture media also suggest a low consumption of glucose, since lactate has been recognized as one of the major end products of glucose metabolism through glycolysis. In addition, some of the lactate may have accumulated from glutamine consumption through glutaminolysis (McKeehan, 1982). Therefore, it is suggested that further reduction in medium glucose concentration be carried out in order to elucidate the possible glucose effects on cell growth and product formation in high density perfusion cultures.

The cell-specific glucose and glutamine uptake rate decreased as the cell concentration increased in media with different glucose and glutamine concentrations. This is attributed to the decreased cell growth rate observed at high cell concentrations. Also associated with the decreased cell growth rate, was increased cell-specific antibody production rate. Since significant changes in cell sizes were not observed during the course of the perfusion cultures, the results suggest that, at high cell concentrations, more energy and intermediates derived from nutrient metabolism are directed to the synthesis of MAb. This may suggest that high concentration, perfusion animal cell culture systems are likely to be the optimal culture systems for continuous MAb production, because of the increased efficiency in MAb production and the high cell concentrations. Moreover, the observed linear relationship between nutrient uptake rates and MAb production rate makes the control and prediction of MAb production feasible in an extended high density perfusion culture system, provided that cells are not under either nutrient or oxygen limitation, or metabolite inhibition.

It is suggested that the development of an efficient high concentration perfusion culture system will eventually be based on the cell metabolic information, provided that enough oxygen can be supplied to the cell mass without affecting the cells' normal metabolic activities. To this end, micron-sized bubbles are highly recommended for high density animal cell cultures. In addition to the study of glutamine and glucose metabolism, the study of amino acid metabolism could provide further information necessary for the optimization of high density perfusion culture system.

5.   Conclusion

It was observed that as the cell concentration increased in perfusion cultures, the cell-specific antibody production rate increased, whilst the cell-specific nutrient uptake rates and metabolite production rates decreased. These observations were found to be related to the apparent cell-specific growth rate. At low cell-specific growth rate, cells became more efficient in antibody production in spite of the reduced cellular metabolic activities. When nutrient limitation or metabolite inhibition were not apparent in a perfusion culture, linear relationships between antibody production rate (based on bioreactor volume) and both the nutrient uptake rates and metabolite production rates were observed. This relationship is expected to be very useful in the control and prediction of antibody production in extended high concentration perfusion cultures, and has been shown to work effectively for hollow fibre cultivations (Handa-Corrigan et al., 1992a).

In high glutamine concentration media, high ammonia concentrations were observed during the perfusion cultivations, and was probably a major factor contributing to decreased cell growth. The utilization of high medium perfusion rates as a strategy for flushing out the ammonia was not successful in reducing the ammonia concentration.   The ammonia concentration in perfusion cultures could only be successfully maintained at a non-inhibiting level by reducing the medium glutamine concentration to a level of about 2 mM. Compared to glutamine, the glucose utilization of the cells was relatively lower. Even after the glucose concentration was reduced from 450 mg/dl to 200 mg/dl, excessive glucose was still present in the culture medium. It is suggested that further work be carried out to examine the glucose effects in high density perfusion cultures.

6. Acknowledgment:

The technical assistance of Marion Chadd is greatly appreciated.

7. References:

Abu-Reesh and Kargi (1991) 'Biological responses of hybridoma cells to hydrodynamic shear in an agitated bioreactor', Enzyme Microb. Technol. 13, 913-919.

Blasey, H.D. and Jäger, V. (1991) 'Strategies to increase the efficiency of membrane aerated and perfused animal cell bioreactors by an improved medium perfusion', in: R. Sasaki and K. Ikura (eds.), Animal Cell Culture and Production of Biologicals, Kluwer Academic Publishers, Dordrecht, pp. 61-73.

Glacken (1988) 'Catabolic control of mammalian cell culture', Bio/Technol. 6, 1041-1050.

Hamamoto, K., Ishimaru, K., Tokashiki, M. (1989) 'Perfusion culture of hybridoma cells using a centrifuge to separate cells from culture mixture', J. Ferment. Bioeng. 67, 190-194.

Handa-Corrigan, A., Nikolay, S., Jeffery, D., Heffernan, B., and Young, A. (1992) 'Controlling and predicting monoclonal antibody production in hollow-fibre bioreactors', Enzyme Microb. Technol. 14, 58-63.

Hayter, P.M., Kirkby, N.F., and Spier, R.E. (1992) 'Relationship between hybridoma growth and monoclonal antibody production', Enzyme Microb. Technol. 14, 454-461.

Kitano, K., Shintani, Y., Ichimori, Y., Tsukamota, K., Sasai, S., and Kida, M. (1986) 'Production of human monoclonal antibodies by heterohybridomas', Appl. Microbiol. Biotechnol. 24, 282-286.

McKeehan, W.L. (1982) 'Glycolysis, glutaminolysis and cell proliferation', Cell Biology Int. Rep. 6, 635-650.

Miller, W.M., Blanch, H.W., and Wilke, C.R. (1988) 'A kinetic analysis of hybridoma growth and metabolism in batch and continuous suspension culture', Biotechnol. Bioeng. 32, 947-965.

Miller, W.M, Wilke, C.R., and Blanch, H.W. (1989) 'Transient responses of hybridoma cells to nutrient addition in continuous culture: I. glucose pulse and step changes', Biotechnol. Bioeng. 33, 477-486.

Modha, K., Whiteside, J.P., and Spier, R.E. (1992) 'Dissociation of MAB production from cell division using DNA biosynthesis inhibitors', in: R.E. Spier, J.B. Griffiths, and C. MacDonald (eds.), Animal Cell Technology: Developments, Processes & Products, Butterworth-Heinemann, Oxford, pp. 81-87.

324

Nayve, Jr. F.R.P., Motoki, M., Matsumura, M., and Kataoka, H. (1991) 'Selective removal of ammonia from animal cell culture broth', Cytotechnol. 6, 121-130.

Runstadler, P.W. and Young, M.W. (1991) 'GMP production of biopharmaceuticals using high density, fluidized-bed cell culture technology', in: R. Sasakai and K. Ikura (eds.), Animal Cell Culture and Production of Biologicals, Kluwer Academic Publishers, Dordrecht, pp. 103-119.

Smith, C.G., Guillaume, J.m., Greenfield, P.F., and Randerson, D.H. (1991) 'Experience in scale-up of homogeneous perfusion culture for hybridomas', Bioprocess Eng. 6, 213-219.

Smith, C.G. and Greenfield, P.F. (1992) 'Mechanical agitation of hybridoma suspension cultures: metabolic effects of serum, Pluronic F-68, and albumin supplements', Biotechnol. Bioeng. 40, 1045-1055.

Sugiura, T. (1992) 'Effects of glucose on the production of recombinant protein C in mammalian cell culture', Biotechnol. Bioeng. 39, 953-959.

Tokashiki, M. and Arai, T. (1989) 'High density culture of mouse-human hybridoma cells using a perfusion culture apparatus with multi-settling zones to separate cells from the culture medium', Cytotechnol. 2, 5-8.

Tokashiki, M., Arai, T., Hamamoto, K., and Ishimaru, K. (1990) 'High density culture of hybridoma cells using a perfusion culture vessel with an external centrifuge', Cytotechnol. 3, 239-244.

Tritsch, G. L. and Moore, G. E. (1962) 'Spontaneous decomposition of glutamine in cell culture media', Exp. Cell Res. 28, 360-364.

Varecka, R. and Scheirer, W. (1987) 'Use of a rotating wire cage for retention of animal cells in a perfusion fermenter', Develop. Biol. Standard. 66, 269-272.

Zhang, S., Handa-Corrigan, A., and Spier, R. E. (1992a) 'Foaming and media surfactant effects on the cultivation of animal cells in stirred and sparged bioreactors', J. Biotechnol. 25, 268-283.

Zhang, S., Handa-Corrigan, A., and Spier, R. E. (1992b) 'Oxygen transfer properties of bubbles in animal cell culture media', Biotechnol. Bioeng. 40, 252-259.

Zhang, S., Handa-Corrigan, A., and Spier R. E. (1992c) 'A comparison of oxygenation methods for high density perfusion cultures of animal cells', Biotechnol. Bioeng. (in press).

GROWTH YIELDS OF HYBRIDOMA CELLS AND MONOCLONAL ANTIBODY PRODUCTION IN HIGH DENSITY CULTURE

Y. SHIRAI[*], K. Hashimoto, T. Aoki, and
T. Yoshimi
Department of Chemical Engineering,
Faculty of Engineering, Kyoto University,
Sakyo-ku, Kyoto 606, JAPAN
[*]Department of Biochemical Engineering & Science,
Faculty of Computer Science & Systems Engineering,
Kyushu Institute of Technology, Iizuka, Fukuoka 820,
JAPAN

ABSTRACT The growth yields and monoclonal antibody production kinetics of hybridoma cells were investigated for a high
density perfusion culture, and compared with those obtained
for an ordinary batch culture. We have previously found that
a large difference in the growth yields of a 4C10B6 hybridoma
cell for glucose and glutamine in a high density culture when
compared to a normal culture[1]. In this work, a similar
trend was observed with other hybridoma cells.

1. Introduction

A perfusion culture system of hybridoma cells is a powerful
tool for producing monoclonal antibodies with high efficiency. This system, involving simultaneous feeding of nutrients
and removing waste while keeping the cells in a fermentor,
makes the concentration level of toxic waste produced by the
cells low enough for their safety, resulting cells cultivated
at a high density of over 1×10^7 cells/ml, and achieving
high productivity of monoclonal antibodies.

The key point of a successful perfusion system is efficient separation of the cells from the waste medium. A poor
device for separation makes long-term monoclonal antibody
production difficult because of the frequent trouble caused
by such a device, although many suitable separation devices
for perfusion systems have already been developed. Saving the
medium supplied can also contribute to stabilizing the operation of a perfusion system because it makes the life of the
device longer.

We have previously shown that the specific uptake rates
of glucose and glutamine of mouse-mouse 4C10B6 hybridoma

325

cells was decreased significantly in high density culture. This suggests that less medium is required for a high density perfusion culture than for an ordinary batch culture to produce the same amount of products. This paper deals with the changes in growth and monoclonal antibody yields for a high density perfusion culture of another cell line, mouse-mouse A3 hybridoma cells.

2.Materials and Methods

2.1.Experimental

Mouse-mouse A3 hybridoma cells, which were established in our laboratory by fusing NS1 mouse myeloma and splenocytes from a mouse immunized by glutaraldehyde to produce an IgG[2], were used in this work.

A serum-free RDF medium was used, only ITES components being used as serum free elements[3].

A perfusion system with 100 ml of working volume was used for the high density culture, a schematic diagram of the fermentor used being shown in **Fig. 1**. The medium and cells were separated by a membrane filter with a pore diameter of 1.8 μm. The dissolved oxygen concentration in the medium was adjusted to half the level saturated by adjusting the air pressure with an on-off controller connected to a DO electrode. Perfusion culture was performed at constant perfusion rates (the ratio of the feeding rate to the working volume of the fermentor) by synchronizing two pumps for feeding and withdrawing the medium.

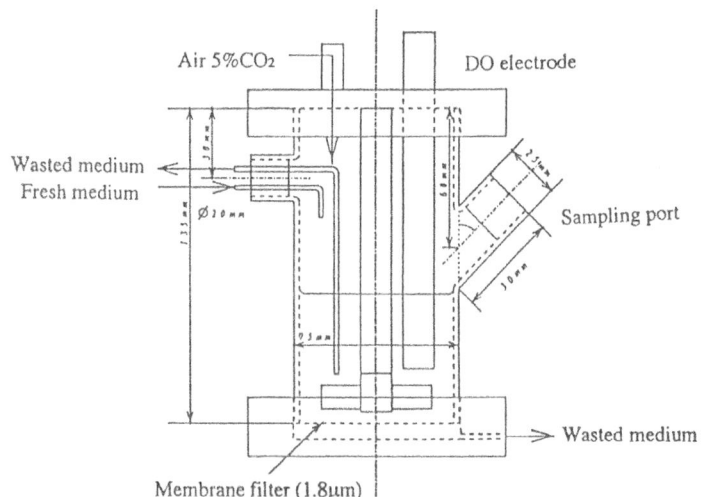

Fig. 1. Experimental apparatus

The concentration of each component in the medium was assayed by using an enzymatic reaction. The concentration of IgG was determined by an enzyme-linked immuno-sorbent assay method.

2.2.Mathematical Equations

For cell proliferation, Eq. (1) can be used, where μ is the specific growth rate and μ_d is the specific death rate. Equation (2) is for cell death in a perfusion system.

$$\begin{cases} \dfrac{dX}{dt}=(\mu-\mu_d)X & \cdots(1) \\[2mm] \dfrac{dX_d}{dt}=\mu_d X & \cdots(2) \end{cases}$$

Each specific rate can be determined from a cell growth curve experimentally obtained.

The growth yield for a substrate, $Y_{x/s}$, is the ratio of the total amount of cells proliferated to the total amount of the substrate consumed by the cells. The total amount of the substrate consumed can be obtained from the material balance of a perfusion system, which is given by Eq. (3):

$$V(\frac{dS}{dt})_v=V(\frac{dS}{dt})_x+F(S_0-S) \qquad \cdots(3)$$

where V is the working volume of the fermentor, F is the feeding rate of fresh medium, S_0 is the initial substrate concentration, $(dS/dt)_v$ is the change in substrate concentration in the fermentor, and $(dS/dt)_x$ is the substrate consumption rate by the cells. Equation (3) can be rearranged into Eq. (3)'.

$$\int_0^t (\frac{dS}{dt})_x dt=\int_0^t (\frac{dS}{dt})_v dt- \frac{F}{V}\int_0^t (S_0-S)dt \quad \cdots(3)'$$

The values for each term on the right-hand side of Eq. (3)' can be determined by experiments over a fixed cultivation period, thus yielding the total amount of the substrate consumed.

The production rate of a monoclonal antibody by the A3 hybridoma cells is given by the following equation[2], because the production of a monoclonal antibody by this cell line can be described by a growth-associated model.

$$\frac{dP}{dt}=\alpha\frac{dX}{dt} \qquad \cdots(4)$$

The MAb (monoclonal antibody) yield is the ratio of the total amount of MAb produced to the total amount of substrate consumed. The total amount of MAb produced by the cells can also be obtained from a similar treatment to that for the growth yield with the material balance in a fermentor given by Eq.

328

(5).

$$V(\frac{dP}{dt})_v = V(\frac{dP}{dt})_x - FP \qquad \cdots (5)$$

The specific substrate uptake rate and specific product production rate are given by the following equations:

$$q_s = \mu/Y_{x/s} \qquad \cdots (6)$$

$$q_p = Y_{p/s} \cdot q_s \qquad \cdots (7)$$

where $Y_{p/s}$ is the product yield from the substrate.

3. Results and Discussion

3.1. Growth of A3 Hybridoma Cells in High Density Culture

Figure 2 shows a growth curve of the A3 hybridoma cells in high density culture. The numerical values shown at the bottom of **Fig. 2** indicate the perfusion rate (F/V), and the supply of fresh medium was suspended for one week. The total cell numbers were not changed during this week, indicating that a negligibly small amount of cells were miscounted due to fragments originating from dead cells. This insures sufficient accuracy for the specific growth and death rates shown in **Fig. 2**.

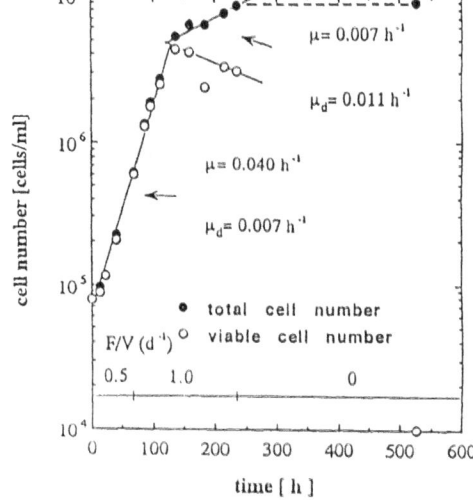

Fig. 2. Growth Curve of A3 Cells in High Density Culture

A stepwise change in the specific growth and death rates can be seen, drastically decreasing and increasing, respectively. In **Fig. 3**, changes in the substrates, waste

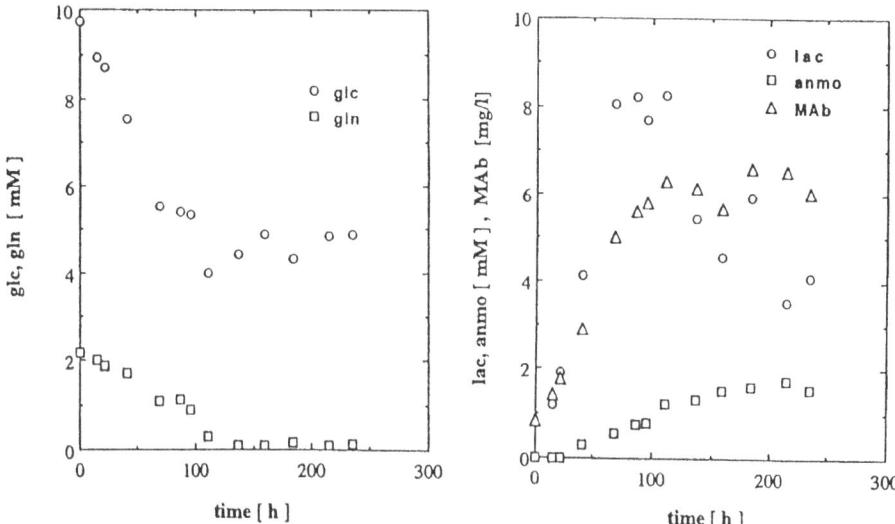

Fig. 3. Concentration Changes in Substrates & Products in
High Density Culture

products, and monoclonal antibodies are shown. **Figures 2 and
3** indicate that the specific growth and death rates changed
after the glutamine concentration became close to zero.

The total amounts of substrates consumed and products
produced by the cells at any point measured were calculated
from the slope between that point and the previous adjacent
point, $(dS/dt)_V$ and $(dP/dt)_V$, and from the substrate and
product concentrations at that point, S or P, using Eqs (3)'
and (5)'.

3.2. Change in the Growth Yield for glucose in High Density Culture

Figure 4 shows the relationship between total amount of
glucose consumed and the cells that proliferated. The slope
of each straight line in the figure indicates the growth
yield for glucose. The growth yield increased more than three
times in the exponential growth phase, although the specific
growth rate is identical (0.040 h^{-1}). The early stage in
which the smaller value of the growth yield was obtained is
called Phase I, and the late stage in which the larger value
was obtained is Phase II. The same trend was also found in a
batch culture of A3 hybridoma cells[2]. The value of the
growth yield in each Phase was also almost the same in the
high density culture and batch culture[2]. After the
specific growth rate had drastically changed, the growth
yield for glucose further increased, and we call this stage
Phase III.

Figure 5 shows the relationship between the total amount
of glucose consumed and the total amount of lactate produced
by the cells. The slope indicates the yield coefficient of

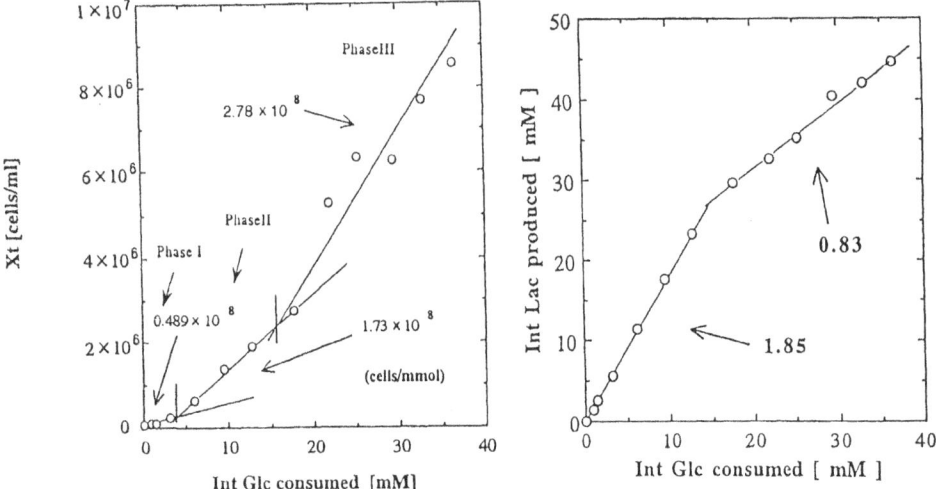

Fig. 4. Growth Yield for Glucose in High Density Culture

Fig. 5. Yield Coefficient of Lactate from Glucose in the Culture

lactate from glucose and how much glucose was converted to lactate. One mole of glucose is changed to two moles of lactate under anaerobic conditions. Therefore, the slopes of the two suggest that all the glucose was oxidized only via the glycolytic pathway for any lactate flux, i.e. that from glutamine via the TCA cycle was negligible. The yield coefficient of lactate was much reduced in Phase III, indicating that part of the glucose was oxidized via the TCA cycle.

3.3. Change in Growth Yield for Glutamine in High Density Culture

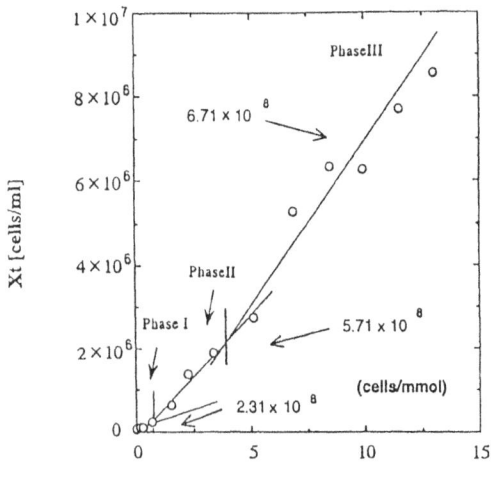

Fig. 6. Growth Yield for Glutamine in High Density Culture

Changes in the growth yield for glutamine were also observed and are shown in **Fig. 6.** The growth yield for glutamine also increased by more than tow in Phase II, and increased further after the exponential growth phase in which glutamine was almost depleted.

3.4. Change in the Specific Substrate Uptake and Product Production Rate in High Density Culture

The specific glucose and glutamine uptake rates were determined by using Eq. (6).

The specific growth rate in each phase was used for the calculation. The specific oxygen uptake rate was determined separately by measuring the change in oxygen concentration of a cell suspension in a small bottle with no air space. The specific oxygen uptake rates in Phases I and II were 1.83 x 10^{-10} mmol/cell h and 1.88 x 10^{-10} mmol/cell h, respectively. That in Phase III was not measured.

The specific lactate, ammonium and alanine production rates were determined by using Eq. (7). The yield coefficient of lactate from glucose was determined from **Fig. 5.** **Figs. 7 and 8** show relationships between total amount of glutamine consumed and the total amount of ammonium or alanine produced, respectively. The ammonium and alanine yield coefficients were found from the slopes in the figures.

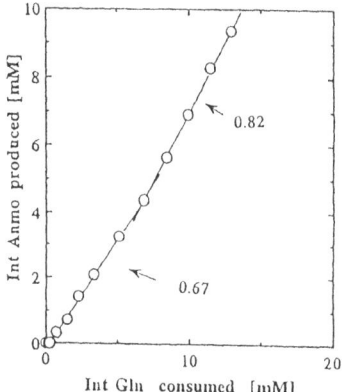
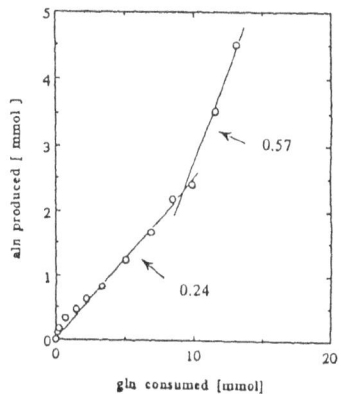

Fig. 7. Yield Coefficient of Ammonium from Glutamine in High Density Culture

Fig. 8. Yield Coefficient of Alanine from Glutamine in High Density Culture

The specific rates in each phase are summarized in **TABLE 1,** each specific rate decreasing during the later stage of cultivation. Especially in Phase III, the value was reduced drastically probably due to the lower specific growth rate. Two papers presented in this 5th JAACT meeting have suggested the same trends[4,5], and indicate that less medium is needed

TABLE 1. Specific Uptake & Production Rates of Components in High Density Culture

[mmol/h cell]	Phase I	Phase II	Phase III
$q_{glc.}$	8.2×10^{-10}	2.3×10^{-10}	2.5×10^{-11}
q_{lac}	1.5×10^{-9}	4.3×10^{-10}	2.1×10^{-11}
$q_{gln.}$	1.7×10^{-10}	7.0×10^{-11}	1.1×10^{-11}
$q_{ammo.}$	1.2×10^{-10}	4.7×10^{-11}	0.8×10^{-11}
$q_{aln.}$	4.1×10^{-10}	1.7×10^{-11}	0.6×10^{-11}

in the late stage of high density culture of hybridoma cells.

3.5. Monoclonal Antibody Production in High Density Culture

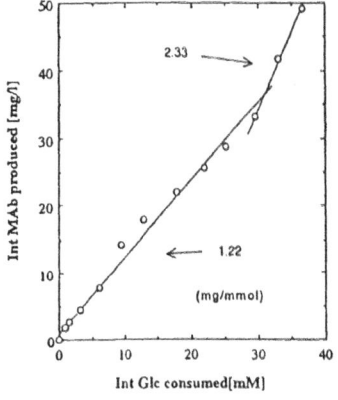
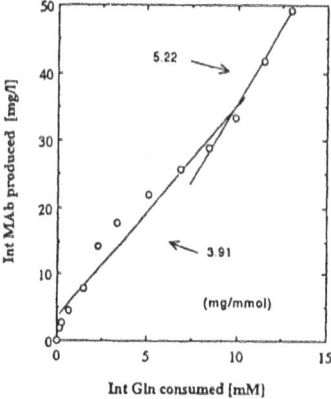

Fig. 9. MAb Yield from Glucose in High Density Culture

Fig. 10. Mab Yield from Glutamine in high Density Culture

The MAb yields from glucose and glutamine were examined, **Figs. 9 and 10** showing the relationships between the total amount of glucose or glutamine and monoclonal antibody produced. The slope of each indicates the MAb yield. The MAb yield increases at the late stage of culture, indicating that less medium was needed for the production of monoclonal antibodies in high density culture.

4. Conclusions

1. The growth kinetics of hybridoma cells at the early stage of high density perfusion culture was examined.
2. Stepwise changes and increases in the growth and MAb yields for glucose and glutamine were found.
3. The specific uptake and production rates of components decreased with the culture time.
4. It is suggested that energy sources are utilized more efficiently by hybridoma cells in a high density culture.

5. References

1. Shirai, Y. Hashimoto, K. and Takamatsu, H. (1992) 'Growth Kinetics of Hybridoma Cells in High Density Culture', J. Ferment. Bioeng., **73**, 159-165.
2. Shirai, Y. Yoshimi, T. and Hashimoto, K. (1992) 'Effects of Conditioned Medium on the Growth Kinetics and the Monoclonal Antibody Productivity of Hybridoma Cells', in H. Murakami et al. (eds.) Animal Cell Technology: Basic and Applied Aspects, **Vol. 4**, 279-285.
3. Murakami, H. Masui, H. Sato, G. H. Sueoka, N. Chow, T. P. and Sueoka, T. P., (1982) 'Growth of Hybridoma Cells in serum free medium: ethanolamine is a essential component', Proc. Natl. Acad. Sci. USA, **79**, 1158-1162.
4. Hansen, H. A. Damgaard, B. and Emborg, C (1992) 'Amino Acid Metabolism of a High Density Perfusion Culture', Abstracts of the Fifth Annual Meeting of JAACT, Omiya, 63.
5. Zhang, S. Handa-Corrigan, A. and Spier, R. E. (1992) 'Kinetics of Cell Metabolism and Antibody Production in High Density Perfusion Cultures', Abstracts of the Fifth Annual Meeting of JAACT, Omiya, 64.

TOWARDS A STRATEGY FOR HIGH DENSITY CULTURES OF VERO CELLS IN STIRRED TANK REACTORS

R.X. VAN DER MEER[1], M.C. PHILIPPI[1], B. ROMEIN[2], C.A.M. VAN DER VELDEN-DE GROOT[1] AND E.C. BEUVERY[1]

[1] National Institute for Public Health and Environmental Protection (RIVM), PO Box 1, 3720 BA Bilthoven, The Netherlands
[2] Department of Biochemical Engineering, Delft University of Technology, project group control, Julianalaan 67, 2628 BC Delft, The Netherlands

ABSTRACT. Results from batch and continuous perfusion cultivations of Vero-cells are presented. Using standard medium, highest cell densities were reached by continuous perfusion, whereas cell yield per unit of medium was highest in batch cultivation. Amino acid consumption- and production patterns were in both systems comparable, except for alanine. Alanine was produced during continuous perfusion and slightly consumed during batch cultivation. Results from measurement and control equipment can be used to calculate cell density on-line. Image analysis can be employed to analyse cell distribution over microcarriers.

1. Introduction

In our laboratory, Vero cells are cultivated on a scale ranging from 3 l up to 350 l working volume. The cells are utilized as a substratum for the cultivation of polioviruses. Results from small scale studies indicated that D-antigen production is dependent of cell density at the point of time of virus inoculation, but independent of the mode of cell cultivation: batch, fed-batch, recirculation and continuous perfusion (Fig. 1). Therefore, the objective of the cell cultivation must be to reach the maximal cell density.

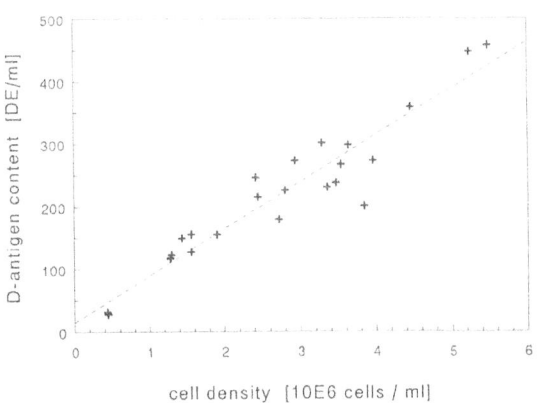

Figure 1.   D-antigen content as a function of cell density for type I poliovirus.

S. Kaminogawa et al. (eds.), Animal Cell Technology: Basic & Applied Aspects, Vol. 5, 335–340.
© 1993 Kluwer Academic Publishers.

It is well known that the highest cell densities can be obtained by means of continuous perfusion mode cultivation. However, compared to batch mode cultivation with the same medium composition, cell yield per unit of medium is generally lower (Table 1). Therefore, study of metabolism of the cell line cultivated is essential to increase this yield.

Table 1. Cell yield for different cultivation modes. The cell yield was calculated by dividing the maximum cell density reached by the total volume of medium used

mode of cultivation	cell yield [million / l]
batch	1.4
fed-batch	1.2
recirculation	1.6
continuous perfusion	0.5 *

* cell yield was much lower than for the other cultivation modes, because the perfusion scheme was not optimized.

The role of glucose and glutamine, the nutrients with the highest specific consumption rates, has been examined extensively by many authors for many different cell lines [1, 2]. Feeding strategies can be employed to control production of ammonia and lactate.

The metabolism of amino acids, although studied for more than three decades [3], is still less well elucidated. Amino acid metabolism seems to be not only strongly dependent of cell line, but also of stage of cultivation, mode of cultivation, reactor type, etc. This makes it difficult to design an optimal medium composition or feeding strategy.

Besides aspects of medium utilization, other aspects need to be considered when maximizing cell density. One of these is the distribution of cells over microcarriers. An ill-balanced distribution at the start of the cultivation may lead to submaximal growth rates because a minimal number of cells per microcarrier is needed for growth to occur. Especially at large scale, ill-balanced cell distribution may become prominent.

2. Materials and methods

Vero cells (passage 143) were cultivated in a 3 l (working volume) bioreactor, coupled to a CF500 measurement and control system (Applikon BV, Schiedam, The Netherlands) [4]. Setpoints were temperature: 37 °C, pH: 7.2 and DO: 50% air saturation. The culture medium was Eagle's MEM, supplied with 7.5% bovine serum and 2.5% fetal calf serum (5% bovine serum during perfusion).

Glucose, lactate and ammonia were analyzed using Boehringer Mannheim enzyme kits. Amino acids were analyzed on a Waters HPLC-system with automated pre-column derivatization (Waters Millipore, Milford, MA), using OPA (ortho phtaldialdehyde) with ß-mercaptoethanol as derivatization reagent.

Culture amples, consisting of suspensions of microcarriers with adhering cells, were incubated with haemotoxylin and washed with PBS. Images of these preparations were recorded with a CCD black and white camera (Model MX, High Technology Holland, Eindhoven, The Netherlands), coupled to a PC equipped with a frame grabber (PCvision plus, Imaging Technology Inc., Woburn, MA). After digitalization, the images were analyzed with the software package TIM (Difa Measuring Systems, Breda, The Netherlands). For each microcarrier, an average grey value was determined. Grey values were divided in classes, and for each class the relative frequency of microcarriers with corresponding grey values was counted.

3. Results

Continuous perfusion resulted in a much higher cell density than batch cultivation (Fig 2.). As can be expected, ammonia and lactate concentration were lower during perfusion (results not shown), but specific production rates in both systems were comparable. Also the yields for lactate from glucose and ammonia from glutamine were similar, which points to comparable medium utilization and metabolism in both systems.

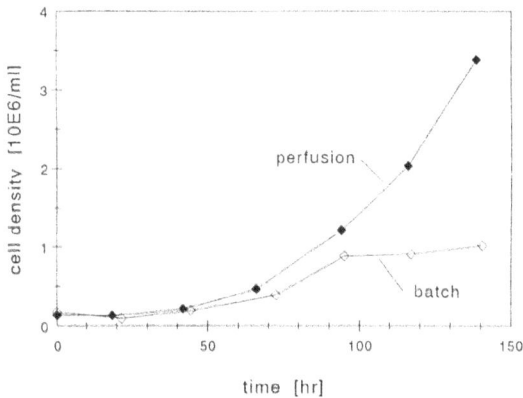

Figure 2.   Growth of Vero cells during batch and continuous perfusion cultivation

3.1 MEASUREMENT AND CONTROL EQUIPMENT

During a DO-controlled cultivation, the course of the DO itself does not provide much additional information. However, the course of the DOCO (controller output of the DO-controller) can provide information about the course of the oxygen uptake rate (Fig. 3). The DOCO is the value that is calculated by the PID-controller after comparing the actual value of the DO with the setpoint value. It can be seen that the course of the DOCO corresponds with the course of the cell density. Using an algorithm that takes into account stirrer speed and k_la, the OUR can be calculated. When the acid production rate is monitored, the cell density can be estimated [1].

3.2 AMINO ACID METABOLISM

Alanine was the only amino acid that showed a marked difference in behaviour between batch and continuous perfusion cultivation. During continuous perfusion cultivation, alanine was produced, whereas during batch cultivation a slight consumption occurred (Fig. 4). The other amino acids showed comparable consumption and production patterns, e.g. serine (Fig 5). Glutamic acid was the only other amino acid that was produced (results not shown). Other amino acids were consumed at a high (e.g. gln) or low (e.g. asp, gly and val) specific consumption rate or neither consumed nor produced.

338

The average grey values measured can be considered as a measure of the number of cells per microcarrier (results not shown). From the peak shape at day 1 (narrow and tall) it can be concluded that there exists a uniform distribution of the cells over the microcarriers (Fig 6). During the course of the cultivation, the peak shifts towards higher grey values, which agrees with the increase of the number of cells per microcarrier. The widening of the peak shape points to an ill-balanced growth on the different microcarriers. The narrowing during the last three days and the increase of the steepness of the right flank indicates surface limitation during the later stages of the cultivation.

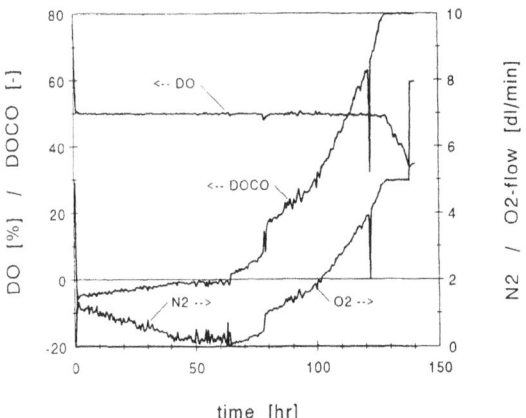

time [hr]

Figure 3.   Time course of dissolved oxygen concentration (DO) and controller output (DOCO), nitrogen flow (N2) and oxygen flow (O2) during continuous perfusion cultivation

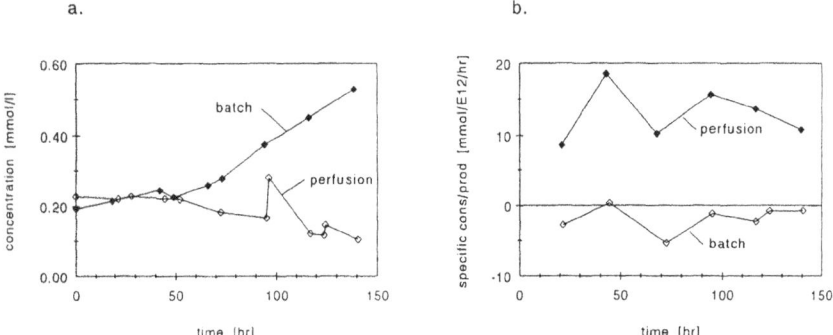

Figure 4.   Alanine concentration (a) and consumption (negative values) / production (positive values) (b) during batch and continuous perfusion cultivation

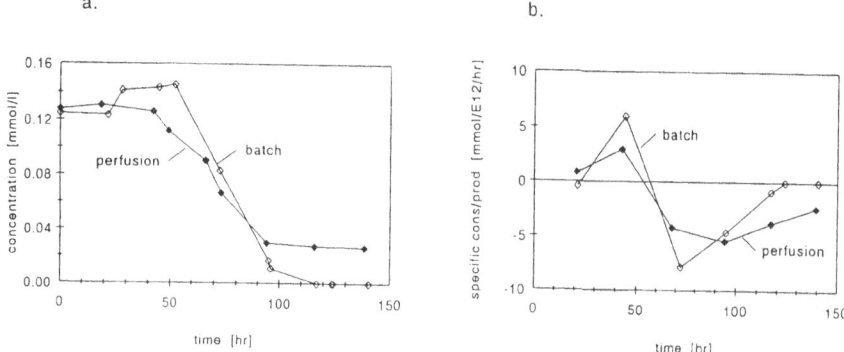

Figure 5.   Serine concentration (a) and consumption (negative values) / production (positive values) (b) during batch and continuous perfusion cultivation

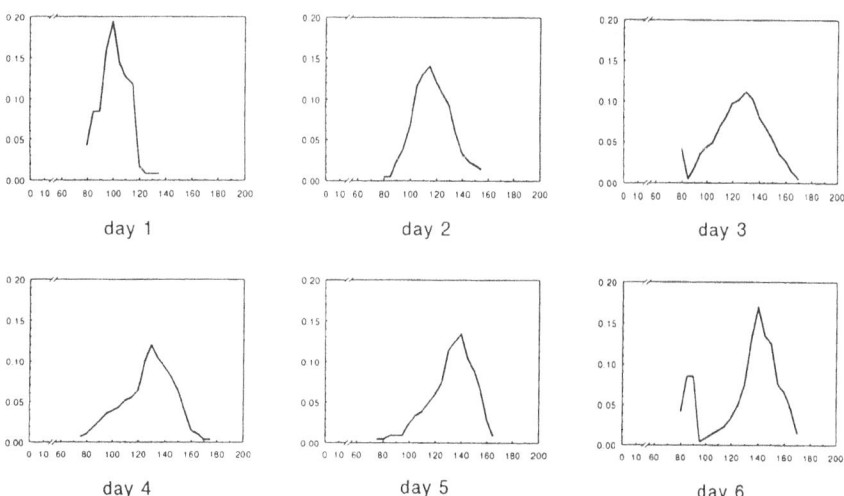

Figure 6.   Distribution of average grey values during consecutive days of a batch cultivation

4.   Discussion

The ultimate objective of optimizing a production process will be to obtain large amounts of the desired product with high quality at low cost prize. To reach this objective, the complete process must be considered, starting from the precultivation of the cells up to the purification of the product.

Since medium and serum costs contribute largely to the total cost prize, knowledge about medium and serum utilization wil be of the highest importance. Therefore, metabolism is studied to be able to design the optimal medium composition and feeding strategy. When consumption and production kinetics are elucidated, low cost components such as glucose, glutamine or certain amino acids may be added or toxic wastes such as lactate or ammonia removed [5]. Since medium utilization may be scale dependent, results from small scale experiments may not be translated directly to large scale.

The results show that the measurement and control equipment can be used not only to register variables such as temperature, pH and DO but also, from the control actions to maintain these variables at setpoint, changes in the activity or behaviour of the cells. Growth and cell density can be determined from the DO-controller output, whereas glucose limitation and lactate accumulation may be calculated from the pH-controller output (results not shown).

Results from image analysis have shown that the distribution of cells over microcarriers can be analyzed in a relatively simple way. Results from small scale and large scale have to be compared to determine whether attachment at a large scale is different from the attachment process on small scale.

5. References

[1] Glacken M.W., Fleischaker R.J. and Sinskey A.J. (1986) 'Reduction of waste product excretion via nutrient control: possible strategies for maximizing product and cell yields on serum in cultures of mammalian cells', Biotechnology & Bioengineering XXVIII, 1376-1389

[2] Miller W.M., Wilke C.R. and Blanch H.W. (1989) 'Transient responses of hybridoma cells to nutrient additions in continuous culture: I. glucose pulse and step changes' Biotechnology & Bioengineering 33, 477-486

[3] Eagle H. (1959) 'Amino acid metabolism in mammalian cell cultures', Science 159, 432-437

[4] Wieten et al. G. (1989) 'The CF 1000: second generation instrumentation for monitoring and control of bioreactors' in R.E. Spier, J.B. Griffiths, J. Stephenne and P.J. Crooy (eds.), Advances in animal cell biology and technology for bioprocesses, Butterworth & Co. Ltd., Sevenoaks, 408-411

[5] Kempken R., Büntemeyer H. and Lehmann J. (1991) 'The medium cycle bioreactor (MCB): monoclonal antibody production in a new economic production system' Cytotechnology 7, 63-74

50L SCALE PERFUSION CULTURE OF HYBRIDOMA CELLS BY GRAVITATIONAL SETTLING FOR CELL SEPARATION

Takami Arai[1], Seiichi Yokoyama[1] and Michiyuki Tokashiki[2]

1) Biotechnology Research Laboratories, Teijin Limited, 4-3-2 Asahigaoka, Hino-shi, Tokyo 191, Japan
2) Research Planning Department, Teijin Limited, 2-1-1 Uchisaiwaicho, Chiyoda-ku, Tokyo 100, Japan

Summary

A mouse - human hybridoma, X32 was cultured in a perfusion system with a gravitational settling zone for cell separation. The viable cell density and IgG concentration reached 10^7 cells/mL and 30 mg/L respectively in 50L scale, which were almost equivalent to those in the smaller culture.
In the case of scaling-up a gravitational settling type of perfusion culture vessel, the larger settling area is required for cell separation in proportion to the culture volume. As the culture volume increases, the cell separation becomes more difficult because it is susceptible to be influenced by any liquid turbulence in the settling zone.
Then, the clear separation of cells from the culture medium was achieved by improving agitation and controlling liquid turbulence in the settling zone. In consequence, we succeeded in 4L, 20L and 50L perfusion culture as in 120 mL one.
The system reported here satisfied the requirement in practical application.

Introduction

Mammalian cells have been conventionally cultured in batch systems for commercial use. However, the productivity of target materials is low in batch culture because of slow proliferation of cells, a low maximum cell density and the death of cells in several days after the maximum cell density is reached. Thus, it has been recognized that culture systems should be developed for high cell density and long term culture, and a variety of proposals have been made.[1]-[5] Since mammalian cells have generally a low ability to produce target materials, perfusion culture which can continuously maintain high cell density for a long period of time is considered advantageous for economical mass production. The authors developed a gravitational settling type of perfusion culture vessel free from clogging and obtained good results.[6] In scale up, the larger settling area is required for cell separation in

S. Kaminogawa et al. (eds.), Animal Cell Technology: Basic & Applied Aspects, Vol. 5, 341–346.
© 1993 Kluwer Academic Publishers.

proportion to the culture volume, but, as the culture volume increases, the cell separation becomes more difficult because it is susceptible to any liquid turbulence in the settling zone. In this report, the large scale perfusion culture including the maximum volume of 50L, in which the sufficient separation of cells was achieved by improving agitation and controlling liquid turbulence in the settling zone, is described as the high density culture of mouse-human hybridoma cells.

Materials and Methods

Experimental apparatus

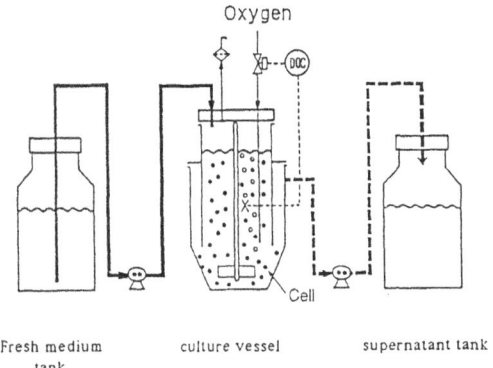

Fresh medium tank   culture vessel   supernatant tank

Fig.1   Perfusion culture apparatus separating cells from culture medium by gravity

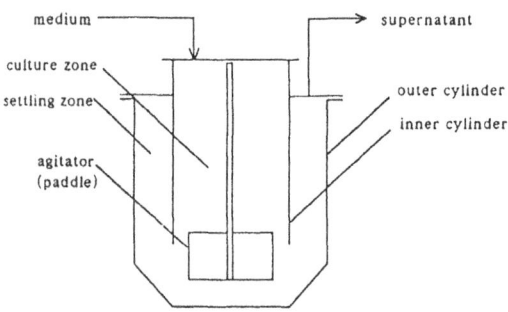

Fig. 2   Culture vessel with a gravitational settling zone

The apparatus comprised a fresh medium tank, a culture vessel, a supernatant tank and peristaltic pumps (Fig.1). The culture vessel was composed of an inner cylinder and an outer cylinder (Fig.2). The inside of the inner cylinder and the lower part of the culture vessel (outer cylinder) worked as the cell culture zone and the annulus between the inner and the outer cylinders served as the settling zone. Four culture vessels of net volume, 120mL, 4L, 20L, and 50L were used. The culture vessel was equipped with an agitator (a paddle type), an oxygen gas sparger and a dissolved oxygen (DO) sensor (Oriental Electric Co. Ltd.). Each of culture vessels (except for that of 120mL net volume) was equipped with another concentric cylinder between the inner and the outer to prevent the influence of liquid turbulence in the settling zone. Each settling area of these culture vessels is shown in Table 1.

Cells

Mouse-human hybridoma X32 and X 87 lines were used. These are fusion products between mouse myeloma P3-X63-Ag8-U1 cells and in vitro immunized human spleen cells, and produce human monoclonal antibodies (IgG_1).

Table 1   Settling area and liquid velocity in the settling area

Vessel volume (L)	Settling area (cm^2)	Liquid velocity in the settling zone (cm/hr) *
0.12	14.5	0.7
4	269	1.2
20	1050	1.6
50	1630	2.6

* at the specific perfusion rate of 2 vol/vol/day

Culture medium

Serum-free eRDF (Kyokuto Pharmaceuticals) containing 9 mg/L of insulin (Novo), 10 mg/L of human transferrin (Sigma), 10 μM of ethanolamine (Wako Pure Chem.) and 20 nM of sodium selenite (Wako Pure Chem.) was used as the medium.

Assay

IgG was determined by the single radial immunodiffsion with LC-Partigen IgG⊛ (Behring werke). The viable cell density was determined by a haemacyto meter with trypan blue.

Procedure

The fresh medium was fed into the culture vessel which was immersed in a water bath controlled at 37°C. Then, the cells prepared from stationary culture were inoculated and agitation was started. So as to sustain DO concentration at about 3ppm, air containing 5% CO_2 or pure oxygen gas was fed over the free surface of the culture medium and then oxygen gas was directly sparged into the culture medium when DO level could not be kept at 3ppm due to the increase of the cell density. The cells were separated from the culture medium in the settling zone by gravitation. The supernatant was taken out of the upper side of the settling zone, and the fresh medium was added to the culture vessel through a peristaltic pump to keep the level of the culture medium constant.

Result

120 mL culture

344

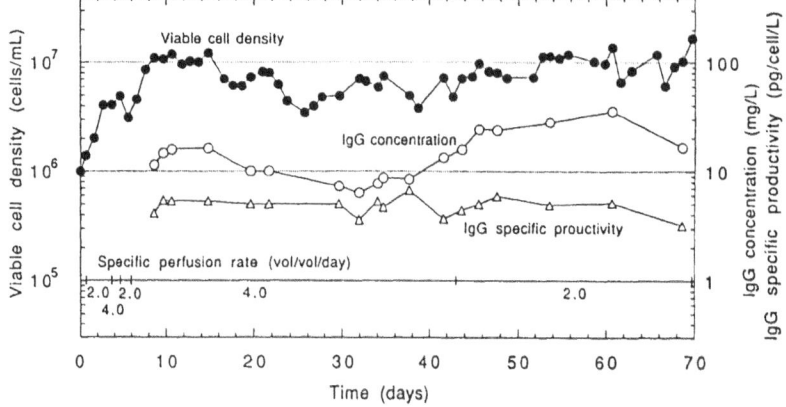

Fig. 3 120 mL Perfusion culture
(Cells: X32 line)

The culture was started with mouse-human hybridoma X32 line at the viable
cell density of 1.0×10^6 cells/mL (Fig. 3).The viable cell density reached 4.0×10^6 cells/mL at the specific perfusion rate of 2 vol/vol/day. The specific
perfusion rate was increased to 4 vol/vol/day on the 7th day. The viable cell
density reached more than 10^7 cells/mL, and thereafter the viable cell density
was kept at about $6 - 8 \times 10^6$ cells/mL while IgG concentration was about 7 - 15
mg/L. Then, the specific perfusion rate was decreased to 2 vol/vol/day on the
44th day. In consequence, the viable cell density increased to about 10^7
cells/mL and IgG concentration reached 20 - 35 mg/L. The culture was
maintained stable for 70 days without any substantial trouble, and IgG specific
productivity was constantly about 5 pg/cell/day.

4L culture

The result is shown in Fig. 4, which was already reported by Hamamoto
et al.[7]. The culture was started with mouse-human hybridoma X87 line from
the initial cell density of 3.9×10^5 cells/mL. The viable cell density reached 2.5
$\times 10^6$ cells/mL on the 3rd day at the specific perfusion rate of 0.5 vol/vol/day.
On the 3rd - 7th day, the specific perfusion rate was kept at 1 vol/vol/day, and
after the 7th day it was raised to 2 vol/vol/day, at which the maximum viable
cell density was 1.5×10^7 cells/mL. IgG concentration was 40 - 70 mg/L, thus
IgG specific productivity was about 10 pg/cell/day.

20L culture

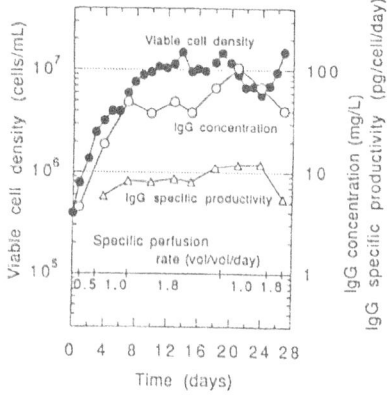

Fig. 4   4L Perfusion culture[7)]
(Cells : X87 line)

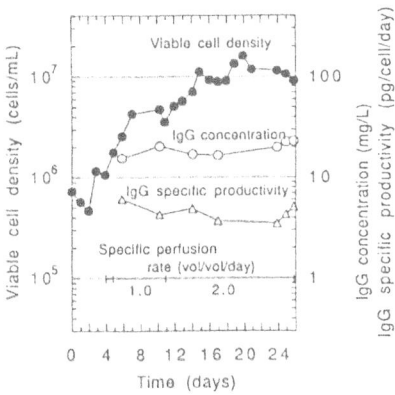

Fig. 5   20L Perfusion culture
(Cells : X32 line)

The result is shown in Fig. 5. The culture was started with mouse-human hybridoma X32 line from the initial cell density of 7.2 x 10^5 cells/mL and the initial culture volume of 5.5L, then the culture volume was increased stepwise to 22L until the 3rd day. After that, the perfusion was started and continued at the specific perfusion rate of 1 vol/vol/day. Then, it was increased to 2 vol/vol/day on the 11th day and the perfusion was continued until the 26th day. The viable cell density reached 10^7 cells/mL on the 15th day, and the maximum was 1.6 x 10^7 cells/mL. IgG concentration was 20 mg/L and IgG specific productivity was 4 to 5 pg/cell/day.

50L culture

The result is shown in Fig.6. Mouse-human hybridoma X32 line was cultured. The initial culture volume was 10L, then increased to 20L on the 2nd day, 40L on the 3rd day and 50L on the 8th day. The viable cell density was initially 1.6 x 10^6 cells/mL and the cells grew

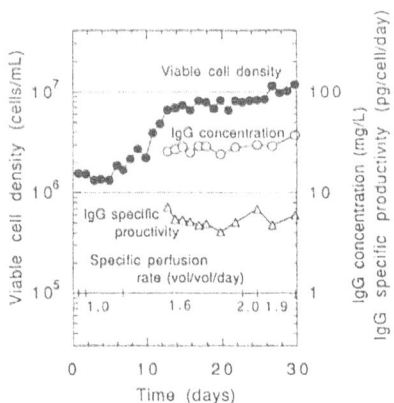

Fig.6   50L Perfusion culture
(Cells : X32 line)

up very well. The specific perfusion rate was started at 1.0 vol/vol/day on the 2nd day and increased to 1.6 vol/vol/day on the 7th day, so the viable cell density was kept at about 8×10^6 cells/mL. After the 23rd day, the specific perfusion rate was maintained at 2.0 - 1.9 vol/vol/day and in consequence the viable cell density reached 1.2×10^7 cells/mL. IgG concentration was 24 - 34 mg/L and IgG specific productivity was 4 -5 pg/cell/day as much as in smaller scale culture.

Discussion

We developed a gravitational settling type of perfusion culture vessel of maximum 50L net volume which can be used for practical use. Before scaling-up this perfusion system, the settling velocity of mouse-human hybridoma X32 cells was measured in the perfusion culture of 1.2L net volume and found to be about 3 cm/hr (data not shown), and each culture vessel was designed on the basis of this data. Table1 shows the settling area of each culture vessel in addition to the liquid velocity in the settling zone at the specific perfusion rate of 2 vol/vol/day. The results of large scale perfusion culture with these vessels (max. 50L of net culture volume) showed the sufficient separation of cells in spite of substantially low cell separation efficiency because of very small difference in density between the culture medium and the cell of a small size and were satisfactory as mentioned above. Thus, it has been proved that the high density culture of hybridoma cells and the long term continuous production of monoclonal antibodies are possible in a large volume by a gravitational settling type of perfusion culture.

References

1. Duff RG (1985) Microencapsulation technology: a novel method for monoclonal antibody production. Trends in biotechnology 3: 167-171.
2. Himmelfarb P, Thayer PS and Martin HE (1969) Spin filter culture: The propagation of mammalian cell in suspension. Science 164: 555-557.
3. Lydersen K, Pugh G, Paris S, Sharma P and Noll A (1985) Ceramic matrix for large scale animal cell culture. Bio/Technology 3: 63-67.
4. Murakami H,Shimomura T,Nakamura T, Ohashi H, Shinohara K and Omura H (1984) Development of a basal medium for serum-free cultivation of hybridoma cells in high density.J.Agric.Chem.Soc.Japan, 58:575-583.
5. Tolbert W and Feder J (1983) Large-scale cell culture technology. Annual Reports on Fermentation Process 6: 35-74.
6. Tokashiki M, Hamamoto K, Takazawa Y and Ichikawa Y (1988) High density culture of mouse-human hybridoma cells using a new perfusion culture vessel. KAGAKU KOUGAKU RONBUNSHU 14:337-341.
7. Hamamoto K,Ishimaru K and Tokashiki M (1989) Perfusion culture of hybridoma cells using a centrifuge to separate cells from culture mixture. J.Ferment.Bioeng. 67:190-194.

PRODUCTION OF MONOCLONAL ANTIBODIES WITH A RADIAL-FLOW BIOREACTOR

H. YOSHIDA, S. MIZUTANI,and H. IKENAGA
Central Laboratories for Key Technology
Kirin Brewery Co.,Ltd.
1-13-5 Fukuura, Kanazawa-ku, Yokohama-shi, Kanazawa 236,
Japan

ABSTRACT. Mouse-mouse hybridoma cells producing IgG against α-amylase were cultivated with a radial-flow bioreactor (RA-400), and continuous production of monoclonal antibodies was tested. After inoculation, the cells grew exponentially, and both the fresh medium feed rate and glucose consumption rate also increased exponentially. After 7 days from the start of cultivation, the fresh medium feed rate attained almost a steady state, the final feed rate being 18 l/day. The IgG concentration obtained in this time was about 80 mg/l, the same as that in batch culture with a flask and spinner flask. We finally obtain 1.5 g of monoclonal antibody per day, and totally 10 g after 14 days of continuous cultivation. In this way, we could easily obtain a large quantity of antibody on a production scale (about 400 g per year).

1. Introduction

Monoclonal antibodies (mAbs) are applied for therapy, diagnosis, as a tool for the purification of biologically active substances, and as analysis antigens. Monoclonal antibodies are generally produced by cultivating hybridoma cells, and many mAbs are still produced *in-vivo*. However, *in-vivo* techniques have such problem as lack of reproducibility, high cost, and difficulty of purification. Many kinds of serum-free medium can also be used for cultivationing hybridoma cells, and it is easly to produce mAbs *in vitro* and to purify these mAbs. There are several systems to achieve a high cell density *in vitro*. We have developed a radial-flow packed bed bioreactor for cultivationing un-collage dependant mammalian cells. Using recombinant CHO and BHK cells, the performance of the bioreactor was previously confirmed. Since medium flows radially across the bed, a supply of nutrients and oxygen with low shear stress is possible, and the matrix entraps cells at a high density. In this present study, we applied the bioreactor to the cultivation of suspension cells. Although it is an advantage of hybridoma cells that no matrix is required to be attached, it is otherwise difficult to immobilize hybridoma cells and to continuously attain high-cell density cultivation. These problems were overcome by using an ion-charged matrix (Cellsnow EX™) or porous glass spheres. The advantages of fixed bed bioreactor are that the system can be made compact, simple, and easy to operate. The productivity of mAbs is higher than in batch culture systems[1,2]. Since this type of matrix could entrap the hybridoma cells well, high-cell density cultivation can be achieved.

347

S. Kaminogawa et al. (eds.), Animal Cell Technology: Basic & Applied Aspects, Vol. 5, 347–353.
© 1993 *Kluwer Academic Publishers.*

2. Materials and Methods

2.1. CELLS AND MEDIUM

Mouse-mouse hybridoma cells producing IgG against α-amylase were kindly provided by Dr. T. Kobayashi. [3] All culturing were done in DF-ITES medium[2](DF medium: 5.24 g of Dulbecco's MEM (Gibco), 5.58 g of Ham's F12 (Gibco), 3.58 g of Hepes buffer (Dojin), $NaHCO_3$, 100,000 U potassium penicillin G and 90 mg streptomycine sulfate per liter; supplemented with ITES components (Sigma): 5 mg insulin, 35 mg of transferrin, 1.2 μl of ethanolamine, and 4.3 μg of sodium selenite per liter).

2.2. BATCH CULTIVATION

The cells were cultured in a 175-cm^2 T-flask under 95% air and 5% CO_2 at 37 ºC. The concentration of inoculated cells was 2 x 10^5 cells/ml.

2.3. MATRIX

The 0.6-mm diam. Siran porous glass beads (Schott Glaswerke) were packed into the bioreactor, small diameter beads bing used because a large quantity of medium could flow radially with a low shear stress.

2.4. RADIAL-FLOW BIOREACTOR

The consumption type of radial-flow bioreactor is shown in Fig. 1. The 950-ml reactor contained a central tube that was wrapped with a stainless steel mesh. The 400-ml matrix was packed into the central tube, there were 6 syringes around the central tube. The medium was supplied to the bioreactor through the syringes, flowing radially across the bed, and was recovered through a metal tube made located in the middle of the bed. The cells could pass through this metal tube. When the circulation rate was 530 l/day, the linear flow velocity was 2.0-9.7 cm/min. This linear velocity increased with flow from the outer layer of beads towards the center.

2.5. CONTINUOUS CULTIVATION

Figure 2 shows the experimental apparatus, which consisted of a radial flow bioreactor, control vessel (fermentor), and the instruments to monitor and control the system. Cells used to inoculate continuous cultures were from spinnerflask (2 l) cultures. To the system start the radial flow bioreactor, about 7x10^8 cells were inoculated to the fermentor, and the broth was pumped into the bioreactor. Until the cells become entrapped in the matrix, the circulation rate of the medium was slow. After 1 hour, the circulation rate was elevated. Dissolved oxygen (DO) concentrations in the fermentor and at the exit from the bioreactor were measured by DO electrodes (Ingold), and kept at appropriate levels so as not to become too low at the exit from the bioreactor. The pH in the fermentor was measured by a pH electrode (Ingold) and kept at pH 7.3. At the start of a culture, 95% air and 5% CO_2 was supplied into the head space of the fermentor. As the cells grew, the CO_2 and air flow rate was decreased, and the O_2 flow rate was increased. After the partial pressure of CO_2 had reached zero, 3 N NaOH was added to maintain the pH level. The feed rate of fresh medium was then gradually increased.

2.6. ASSAYS

Glucose was analyzed by an enzymatic assay, and the antibody production by 16-3F cells was measured by using sandwich ELISA.

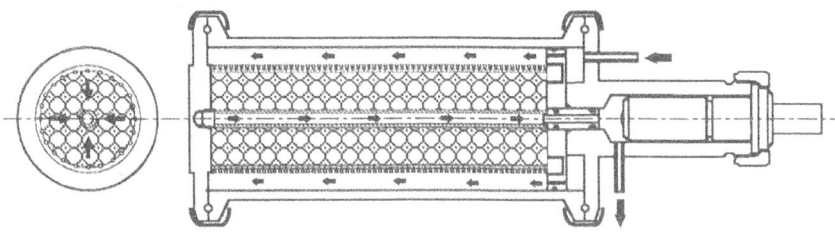

Fig. 1 Radial Flow Bioreactor

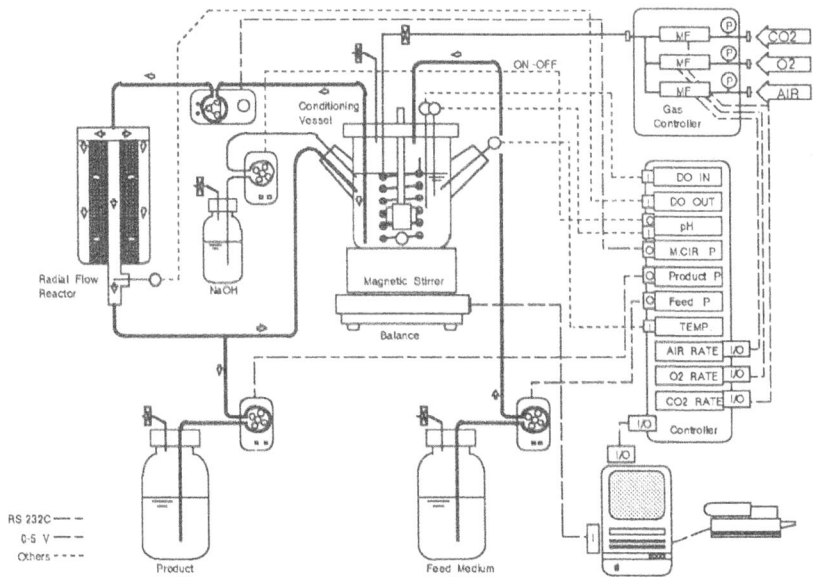

Fig. 2 Schematic diagram of the experimental apparatus. After cell inoculation, the bioreactor system is automatically controlled by a microcomputer.

3. Results and Discussion

Figure 3 shows the results from a batch cultivation of 16-3F cells in the T-flask. The antibody concentration amounted to 80 mg/l after 5 days. The concentration of inoculated cells was 1×10^5 cells/ml and increased to 1.5×10^6 cells/ml. After cell growth had stopped, the mAb concentration hardly increased further. It was necessary for the hybridoma cells to be kept in the growth phase to produce mAbs efficiency, although other hybridomas produced more mAbs in the stationary phase and not in the exponential growth phase.

The results from continuous antibody production in the adial flow bioreactor are shown in Fig. 4. The medium feed rate was controlled so that the glucose was not depleted, and the circulation rate was controlled to keep DO at the exit from the reactor at above 1.0 ppm. 16-3F cells can hardly produce mAbs in oxygen-limited conditions[3], so it was important to maintain DO above 1.0 ppm. The medium feed rate reached 18 l/d after 13 days, and circulation rate reached 530 l/d. In spite of this high circulation rate, most of the cells remained in the reactor, the cell concentration in the spent medium being about $3 - 8 \times 10^5$ cells/ml after 5 days. This level scarcely increased as the circulation rate and medium feed rate were increased, so the medium feed rate could be increased without fear of washing out the cells. The cell concentration in the bed, which was calculated by the oxygen consumption rate, was 8×10^7 cells/ml-bed, and 1.3×10^8 cells/ml-matrix. It is thought that the matrix filled up with cells, and the cells outside the matrix overflowed to the bed.

The antibody concentration is shown in Fig. 5, and reached a high level similar to that with batch cultivation. By supplementing with some other nutrients, the mAb concentration can be made a little higher, and it will save medium. Before the produced mAbs are purified or used, it is necessary to have as high an mAb concentration as possible *in vivo*, because a serum-free medium is used and it will easily be condensed by ultrafiltration. The daily glucose consumption rate and antibody production rate are shown in Fig. 6. The glucose consumption rate increased to 50 g/day, and the antibody production rate increased at the same rate to reach 1.5 g/day. Antibody productivity was heavily affected by low DO conditions in the serum-free medium, in spite of the fact there was no influence on the cell growth and glucose consumption rate [4,5]. On the other hand, cell growth was also affected under high DO conditions [5]. Consequently DO in the conditioning vessel and the circulation rate were gradually increased in order to maintain DO at the exit from the bioreactor at more than 1.0 ppm.

The total antibody production is shown in Fig. 6, and attained 11 g after 14 days of cultivation. If the cultivation was continued for a month, over 30 g of mAbs could be obtained, and if continued for 100 days, about 140 g of mAbs could be obtained.

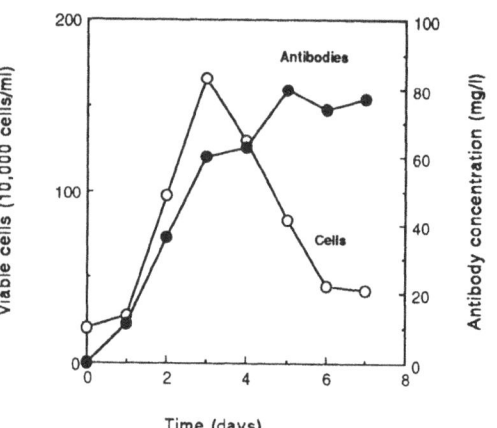

Fig. 3 Batch culture in the T-flask

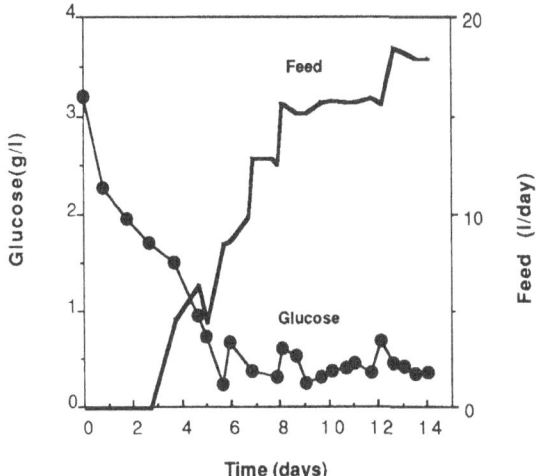

Fig. 4 Glucose and feed rates with the radial-flow bioreactor

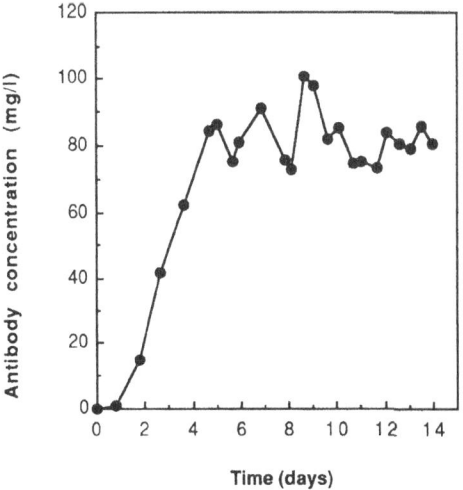

Fig. 5 Continuous production of mAbs by the radial-flow
bioreactor

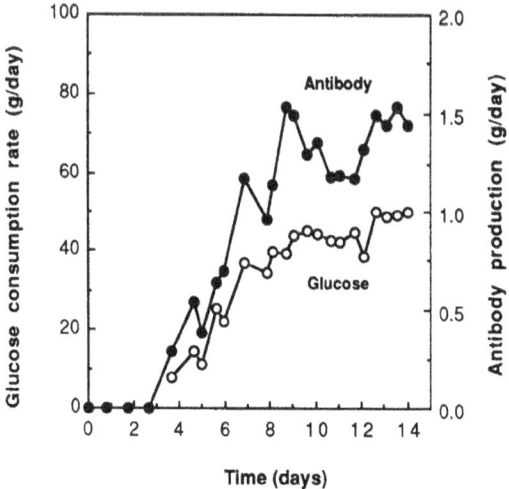

Fig. 6 Daily mAb production and daily glucose consumption

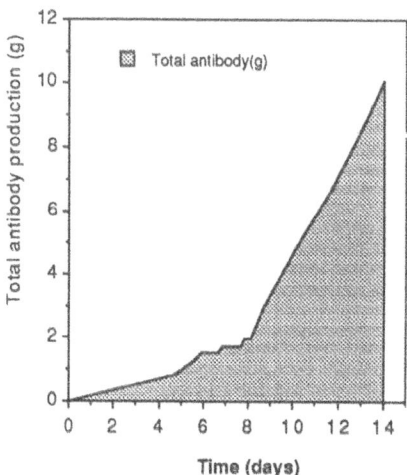

Fig. 7 Total antibody production

6. Conclusion

The radial-flow bioreactor was easily used to achieve continuous production of monoclonal antibodies. The calculated cell concentration in the bed was 1.3×10^8 cells/ml-matrix. The system can be used for not only anchorage-dependent cells, as well as for high-density cultivation of suspension cells. The total mAb amounted to 11 gafter 14 days cultivation. If same amount of mAbs were produced in 100 mm diam. petri dishes, 1800 dishes would be wasted every day.

7. References

[1] Griffiths, J. B. *et al.* (1992) 'Maximization of perfusion systems and process comparison with batch-type cultures, ' Cytotechnology vol. 9, 3-9.

[2] Bohmann, A. *et al.* (1992) 'The membrane dialysis bioreactor with integrated radial-flow fixed bed - a new approach for continuous cultivation of animal cells,' Cytotechnology vol. 9, 51-57.

[3] Kamihira, M. *et al.* (1988) 'Production and characterization of monoclonal antibody to recombinant α-amylase, ' J. Chem. Eng. Jpn. vol. 21, 357-362.

[4] Kamihira, M. *et al.* (1990) 'Effect of Oxygen Aeration on Monoclonal Antibody in Immobilized Hybridoma-Cell Bioreactor' Journal of Fermentation and Bioengineering vol. 69, no.5, 311-312.

[5] Ogawa, T. *et al.* (1992) 'Effect of Dissolved Oxygen Concentration on Monoclonal Antibody Production in Hybridoma Cell Cultures' Journal of Fermentation and Bioengineering vol. 74, no. 6, 372-378.

Factors Affecting the Yield of a Baculovirus Insect Cell System by Hollow-fiber Culture

Hideki Shimoizu, Yuko Naka and Nobuyuki Etou

Tabai Espec Corp. Development Group, Business Development Dept.
3-5-6, Tenjinbashi, Kita-ku, Osaka 530, Japan

ABSTRACT. Directly measurement of the concentration of dissolved oxygen in the medium of a T-flask demonstrated that the oxygen concentration in the medium had a great effect on the virus production. While a high concentration was necessary for the infection process, this was harmful for production yield. A low concentration of oxygen in the production process was required for a high yield of protein.

1. Introduction

A method for utilizing a recombinant baculovirus and an insect cell has recently been developed for the production of protein. It should be possible to produce high amounts of physiologically active proteins by modifying glycochains, phosphorylating and so on by this system. [1]

We have studied several factors related to the yield of the baculovirus-insect cell system in order to develop an efficient process for hollow-fiber culture, which is now generally used for producing many kinds of recombinant proteins and monoclonal antibodies. [2, 3]

There are several aims for using a hollow-fiber bioreactor in this system :

1) Maintaining a safe laboratory environment

   Although the virus is said not to infect mammalian and mammalian cells, it is important to protect the experimental environment, production environment and sub-culturing cells. An ultrafiltration membrane of hollow-fiber allows no baculovirus to be transferred from the extracapillary space (ECS) to outside.

2) Easily changed culture conditions

   The specific growth factors in insect cells have not been identified. This

355

S. Kaminogawa et al. (eds.), Animal Cell Technology: Basic & Applied Aspects, Vol. 5, 355–360.
© 1993 *Kluwer Academic Publishers.*

allows serum to be used in a culture, although a serum-free medium has been
developed for culturing Sf-9 cells. Even if serum is being used, the phase
between growth and production can be separated. The hollow-fiber culture method
enables easy exchange of fresh serum-free medium without the troublesome process
of centrifugaion and medium filtration.

We previously found that little beta-galactosidase was produced in a hollow-fiber
culture in spite of a large quantity of infected cells, in which numbered at least
2×10^8 cells per milliliter. This was thought to have been due to two factors,
oxygen and serum. In a hollow-fiber culture, the serum is passed through an
ultrafiltration membrane(UF membrane), only the molecules larger than the pore size
of the UF membrane remaining in the extracapillary space. Fetal bovine serum ultrafiltrated under the same ECS conditions resulted in reduced cell growth and cell
production in comparison with intact FBS.[3] Another reason is related to the concentration of oxygen in the medium, because, in a hollow-fiber culture, the concentration of dissolved oxygen is maintained at a high value under 100mm Hg at
28°C. By directly measuring the concentration of dissolved oxygen in the T-flask
medium, we studied the relationship between cell production and oxygen.

2. Materials and Methods

Culture cell and baculovirus

An insect cell, Spodoptera frugiperda (Sf-21 AE Ⅱ),[4] was cultured in IPL-41
medium supplemented with 10 % fetal bovine serum in a 25-cm^2 T-flask (polystyrene)
AcNPV with a cDNA encoding for E. Coli beta-galactosidase production was used
to infect Sf-21 cell by using transfer vector pAc360.

Measuring the concentration of dissolved oxygen in the medium

Cells were cultured in the normal way, and the dissolved oxygen concentration
in the T-flask medium was measured shown in Fig. 1. Using the same sterilized T-flask
as that for culture, the concentration was measured with a dissolved oxygen electrode
(Ingold Inc.) in the medium supplemented with serum. It was measured in each
condition: with no cells cultured as control, with T-flask's cap fastened, with cap
loosened, with cells cultured and with cells infected baculovirus (cap fastened).
A rubber seal was used between the electrode and T-flask to protect against
contamination.

Cell culture and cell production

1.62×10^6 cells were inoculated into the T-flask for culture, the concentration of
dissolved oxygen was measured at 28 °C and 95 % relative humidity in a using CO_2
incubator (BNR-110, Tabai Espec Corp.). After changing the medium, AcNPVs with a
cDNA encoding for E. Coli beta-galactosidase were used to infect the cells.
The activity of beta-galactosidase was measured by Miller's method.[5]

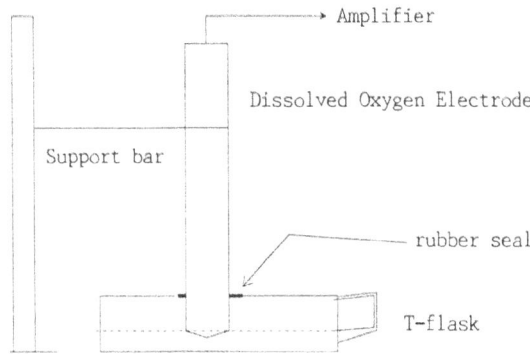

Fig. 1. Method for measuring the concentration of dissolved oxygen
in the T-flask medium

3. Results and Discussion

We first checked if the concentration of dissolved oxygen in the meidum could be
correctly measured by this method. As shown Fig. 2, equilibrium was clealy reached
in the T-flask when no cells were cultured under 142mm Hg at 28 °C with the cap
fastened. This indicates that the dissolving rate of oxygen into the medium
and the consumption rate of the dissolved oxygen electrode were equivalent at about
142mm Hg and 28 °C, proving the validity of this method.
The concentration of dissolved oxygen gradually increased in the 3 days after
infection, which agrees with the time when the cells began to die (data are not
shown) (Fig. 3).

With the T-flask's cap fastened, the concentration of dissolved oxygen was about
30 to 40mm Hg after inoculating 1.62×10^6 cells by the 3rd day (Fig. 4). 30-40mm Hg
is the actual value by calculated from the consumption of the dissolved oxygen
electrode.
With the T-flask's cap loosened, it was about 100mm Hg under the same conditions
after one day(Fig. 5), this level continuing the 6th days.
With the T-flask's cap fastened, 3.9×10^5 units/1.0×10^6 cells of beta-galactosidase was produced by Sf-21 cells infected with recombinant baculovirus. However,
only 3.3×10^4 units/ 1.0×10^6 cells was produced with the cap loosened (Table 1).

Table 1. Beta-galactosidase activity by Miller's method

cap fastened	3.9×10^5 units/ 1.0×10^6 cells
cap loosened	3.3×10^4 units/ 1.0×10^6 cells

358

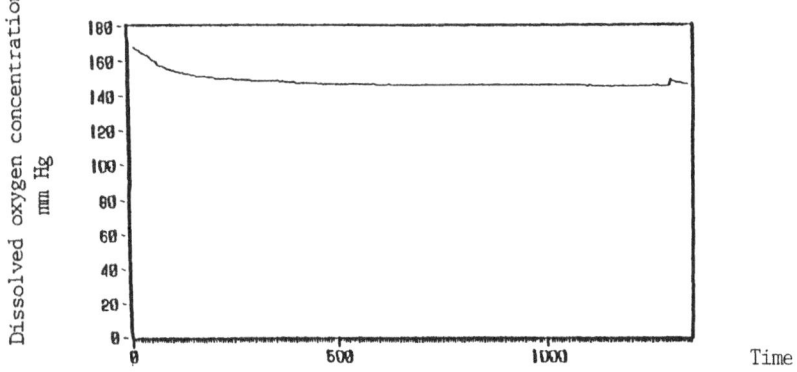

Fig. 2 Equilibrium value for the concentration of dissolved oxygen
about 142mm Hg with no cells being cultured in T-flask at 28°C

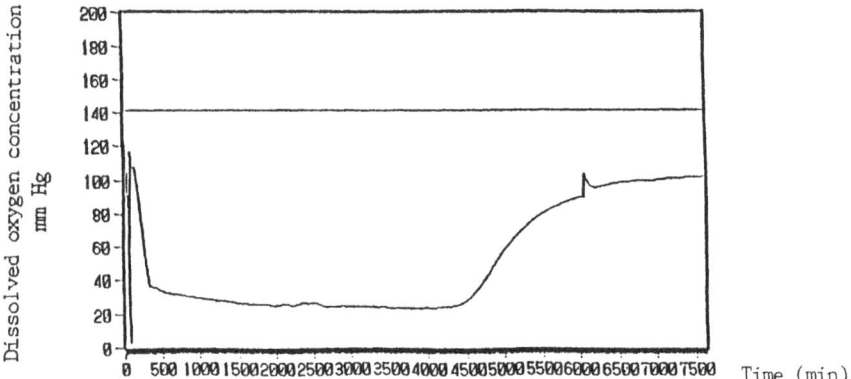

Fig. 3 Change in the concentration of dissolved oxygen in the T-flask
with the cap fastened due to infected Sf-21 cells

These results imply that the concentration of oxygen in the medium is related to
the production process.

With the cap loosened, the concentration of dissolved oxygen remained at about 100
mm Hg, and when fastened, it was about 30 to 40mm Hg. This difference was due to
the ingress of air into the T-flask by agitating the atmosphere with a fan in the CO_2
incubator. The former case promoted oxygen dissolution into medium, but in the latter
case, the rate of dissolution into the medium would be much slower.

The value with the cap loosened is almost the same as that for hollow-fiber culture.
We have already found little production of beta-galactosidase by hollow-fiber culture
in spite of the large quantity of infected cells. This was examined by an immunochemical study and the reaction of ONPG reagent (o-nitro-phenyl-beta-D-galactopyrano-
side)

We conclude that a high concentration of oxygen would reduce cell production due to
some physiological mechanisms and metabolic functions. Although data were not shown,

the concentration of dissolved oxygen was more effective in the cell production than serum ultrafiltrated.

In contrast, although not shown in the data, a high concentration of oxygen was required for the infection process. Infection without changing the medium resulted in very low cell production because of the low concentration of dissolved oxygen. If the medium is not changed, a longer duration of culture resulted in a reduced concentration of oxygen in the medium of T-flask or spinner flask. A low oxygen concentration results in baculoviruses not infecting insect cells due to some physiological mechanisms and metabolic functions.

Although other factors, probably inhibitors, promoters, pH value, MOIs, cell phases and serum, will be involved, the concentration of dissolved oxygen seems to have the greatest effects on infection and production from these results.

We are now investigating the optimal value for the concentration of dissolved oxygen, and it seems to be good to control the level at 20-50mm Hg as in the case of the T-flask with its cap fastened.

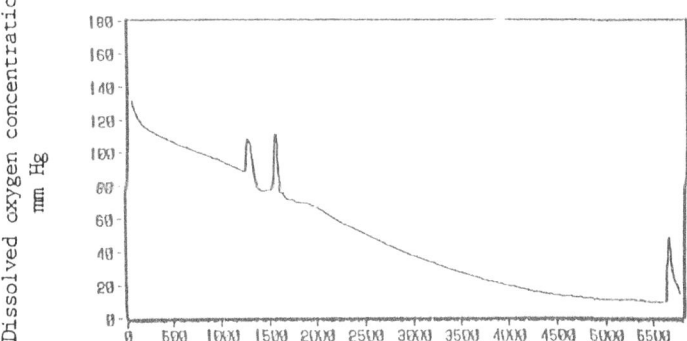

Time (min)

Fig. 4 Declining curve indicates the concentartion of dissolved oxygen
with the T-flask's cap fastened

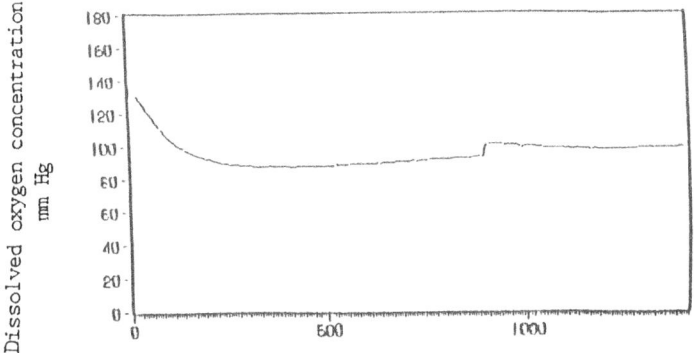

Time (min)

Fig. 5 Declining curve indicates the concentartion of dissolved oxygen
with the T-flask's cap loosened

We also have to solve some problems, for instance, if a baculovirus can not intrude into a cell in a low concentration of dissolved oxygen, and/or about the reason why a baculovirus can not multiply in a cell in a high concentration of dissolved oxygen and so on.

In conclusion, when producing recombinant proteins, it is necessary for the infection process to have a high level of dissolved oxygen concentration, but this high level is harmful to the production process.

A low concentration of dissolved oxygen is required to produce high yield of protein.

4. References

1. Smith, G.E., Summers, M.D. and Fraser, M.J. (1983) 'Production of human beta inter-feron in insect cells infected with a baculovirus expression vector,' J. Mol. Cell. Biol. 3, 2156-2165
2. Knazek, R.A., Gullino, P.M.,Kohler, P.O. and Dedrick, R.L.,(1972) 'Cell culture on artificial capillaries:an approach to tissue growth in vitro,'Science 176, 65-67
3. Shimoizu, H.,Naka, Y. and Etou, N., manuscript in preparation
4. Vaughn, J.L., Goodwin, R.H., Tompkins,G.J. and Mccawley,P. (1977) 'The establishment two cell lines from the insect Spodoptera frugiperda (Lepidoptera; Noctuidae),' In Vitro 13, 213-217
5. Miller, J.H. (1972) 'Assay of beta-galactosidase,'in ' Experiments in Molecular Biology,' 352-355, Cold Spring Harbor Laboratory, Cold Spring Harbor, N.Y.

High-Density Animal Cell Culture by Gas Sparging

Sei Murakami, Yoshihisa Yamaguchi
Kasado Works, Hitachi Ltd.
794 Higashitoyoi, Kudamatsu
Yamaguchi, Japan

Ryoichi Haga, Hikari Murakami
Hitachi Research Laboratory, Hitachi Ltd.
7-1-1 Omika-cho, Hitachi
Ibaraki, Japan

ABSTRACT. In order to enhance the oxygen supply for an animal cell culture, several foam-suppressing methods which enable direct gas sparging into culture medium were investigated. When an anti-foaming reagent was added to the medium to suppress foaming, the amount of reagent needed to be controlled in accordance with the aeration rate. With sparger optimization, increasing the sparger pore size reduced the remaining foam time. However, with the sparger pore size larger than certain value, the remaining foam time became constant, and this pore size can be regarded as the optimum. A defoaming device with a hydrophobic net has also been developed, and a culture system equipped with this defoaming net was constructed. A series of culture experiments was performed to confirm the efficacy of the defoaming net, and enhanced cell density and productivity were confirmed.

1. Introduction

Animal cell culture has already been applied for producing proteins on an industrial scale. Many applications use stirred tank reactors that contain suspended cells or microcarriers. Their operation is often by a batch process with a relatively low cell density. This is because the stirred tank batch culture technique is well established for bacterial culture and stable production can be expected. However, in order to improve the culture productivity by this conventional method, the animal cell characteristics need to be carefully considered.

Of the possible productivity improvements, enhancing the oxygen supply is one of the most important methods. Conventional surface aeration cannot maintain a high cell density, especially on a large scale. On the other hand, gas sparging is known to be the simplest and most efficient method for enhancing oxygen supply. However, sparging often causes over-foaming, and this prevents gas sparging from being applied to animal cell cultures. We have studied the foaming mechanism and several foam-suppressing methods. To achieve high oxygen transfer without a complicated defoaming mechanism, a device with a hydrophobic net has been developed[1], and a culture system utilizing this defoaming net was constructed. In some cases, cell damage by gas sparging has been reported[2,3]. Therefore, a series of culture experiments was performed to confirm the efficacy of the defoaming net.

S. Kaminogawa et al. (eds.), Animal Cell Technology: Basic & Applied Aspects, Vol. 5, 361–367.
© 1993 *Kluwer Academic Publishers.*

2. Defoaming Methods

2.1 Anti-foaming Reagent

The most common foam-suppressing method is adding an anti-foaming reagent to the medium. However, some anti-foaming reagents may be toxic to the cells, and these reagents often require additional downstream processing.

2.2 Sparger Optimization

Another foam-minimizing method is sparger optimization. The oxygen-transfer capacity coefficient, kla, can be increased by reducing the sparger pore size. However, smaller pore sizes increase foam stability, which makes it more difficult to disperse, so that the optimum sparger pore size needs to be selected.

2.3 Defoaming Net

To enhance the oxygen supply without a complicated defoaming mechanism, Ishida *et al.*[1] have developed a defoaming device with a hydrophobic net, which enables vigorous gas sparging without adding any anti-foaming reagent to the medium.

3. Materials and Methods

3.1 Anti-foaming Reagent

Simethicone-based anti-foam is known to be non-toxic to many cells[4], and the cell toxicity of a simethicone-based anti-foaming reagent containing methyl cellulose (Dow-Corning, Medical Antifoam C Emulsion) was investigated by applying the reagent to a CHO cell culture. In addition, a 4-liter airlift loop reactor was used for measuring the amount of anti-foaming reagent necessary to suppress foaming with various gas sparging rates. The medium used was Dulbecco's Modified Eagle's Medium / F12 nutrient mixture (1:1) supplemented with 1.2% fetal bovine serum.

3.2 Sparger Optimization

A 30-liter vessel equipped with various diameter spargers was prepared to measure the foam accumulation with several gas sparging rates. Five liters of DM-160AU medium (Kyokuto Seiyaku) supplemented with 15% newborn bovine serum was fed into the vessel, and, $27Ncm^3/min$ of N_2 gas was then sparged from various sized nozzles. All experiments were performed at under 37 °C.

3.3 Defoaming Net Characteristics

The oxygen-transfer capacity coefficient, kla, was measured with a 5-liter stirred tank reactor containing distilled water. The time for effective defoaming was measured with the same reactor containing Eagle's essential medium and 10% fetal calf serum.

3.4 Cell Culture Experiments

We have developed several culture systems equipped with the defoaming net as follows:

3.4.1 *Stirred Tank Reactor.* The membrane cell separator, sparger, and defoaming net were installed in a 5-liter stirred tank as shown in Fig. 1.

3.4.2 *Airlift Suspension Reactor.* The stirrer was replaced with an airlift system to minimize cell damage as shown in Fig. 2.

3.4.3 *Airlift Packed-bed Reactor.* In order to apply the foregoing airlift suspension culture system to anchorage-dependent cells, the liquid circulating direction was reversed, and a packed bed was inserted at the downcomer of the airlift where no bubbles exist. Porous glass beads (Siran, Schott Glaswerke) were used as the packing material. The airlift packed-bed reactor is well known[5], and its mass transfer characteristics have been well examined[6]. To provide the optimum medium flow across the packed bed, the cross section of the medium path needs to be kept constant. If this is not achieved, the smallest cross section of the medium path will significantly restrict the total medium flow rate and the achievable cell density. Therefore, we selected a conventional cylindrical packed bed with the medium flowing parallel to its axis, a schematic diagram of the reactor being depicted in Fig. 3.

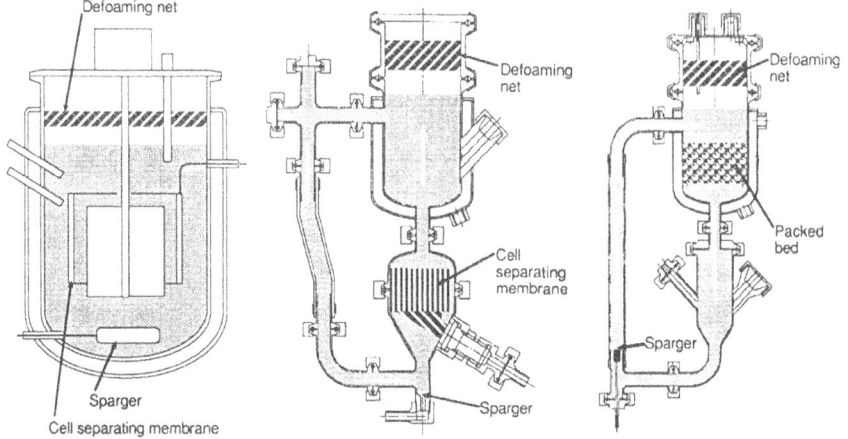

Fig. 1. Stirred tank reactor   Fig. 2. Airlift suspension reactor   Fig. 3. Airlift packed-bed reactor

By using the aforementioned culture systems, the series of culture experiments shown in Table 1 was performed to confirm the efficacy of the defoaming net.

Table 1. Cell culture experiment

Culture system	Cell line	Product	Medium
Stirred tank reactor	Rat ascites hepatoma JTC-1		MEM+10% FCS
Airlift suspension culture system	Mouse-mouse hybridoma STK 1	Anti-DNA polymerase α IgG	DME+10% FCS
Stirred tank reactor	Mouse-human hybridoma JHB-6	Anti-HBs IgG	RPMI1640-F12-DME (1:2:2)+ITES+ PEG50000
Airlift packed-bed reactor	CHO-K1		DME-F12 (1:1) +10% FCS

4. Results and Discussion

4.1 Anti-foaming Reagent

CHO cells were successfully cultured by using the simethicone-based anti-foaming reagent containing methyl cellulose. To suppress foaming, the concentration of the anti-foaming reagent needed to be increased in accordance with the aeration flow rate as shown in Fig. 4.

4.2 Sparger Optimization

At the start of sparging, the foam volume above the medium increased almost linearly, and after thirty minutes, the foam volume approached a plateau as shown in Fig. 5. By increasing the sparger pore size, the accumulated foam volume went down. However, with a sparger pore size larger than 1 mm, the accumulated foam volume was not reduced much more as shown in Fig. 5. This sparger pore size can be regarded as the optimum under these experimental conditions. Different optimum sparger pore sizes are possible with different gas flow rate and medium conditions.

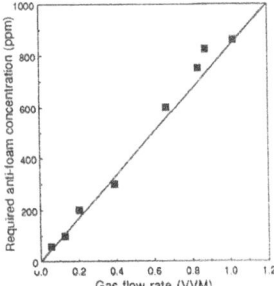

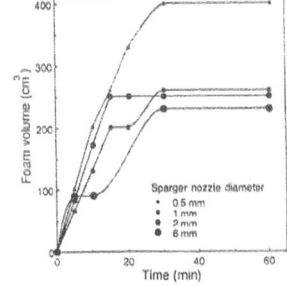

Fig. 4. Required amount of anti-foaming reagent   Fig. 5. Foam accumulation

4.3   Defoaming Net Characteristics

With the defoaming net, a higher kla, value when compared with surface aeration, was obtained as shown in Fig. 6. The time spans of defoaming ability depended on the surface reagent and aeration rate (see Fig. 7).

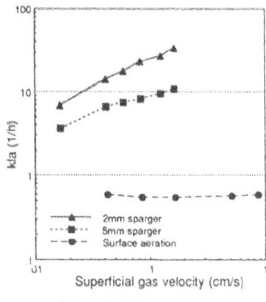

Fig. 6. kla enhancement

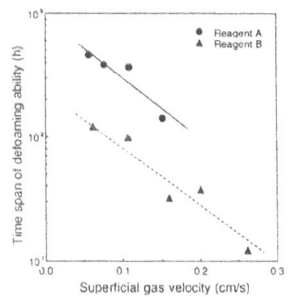

Fig. 7. Time span of defoaming ability

4.4   Cell Culture Experiment

4.4.1 *Rat Ascites Hepatoma (JTC-1)*.   With the stirred tank reactor and a defoaming net, a 31-day cultivation was performed. With an inoculum cell density of 1×10^6 cells/ml, a cell density of 2.7×10^7 cells/ml was reached by the 27th day as shown in Fig. 8.

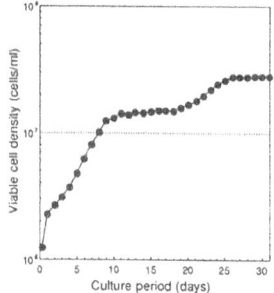

Fig. 8. JTC-1 cell growth

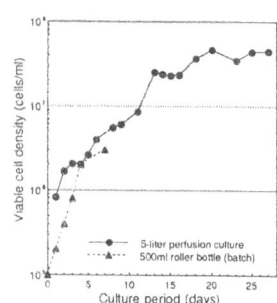

Fig. 9. STK1 cell growth

4.4.2 *Mouse-Mouse Hybridoma (STK 1)*.   In order to assess the culture productivity, STK 1 was cultured, and anti-DNA polymerase α monoclonal antibody was measured. The cell density was 8×10^5 cells/ml at the beginning, and finally reached 4×10^7 cells/ml as shown in Fig. 9. The volumetric monoclonal antibody productivity is shown in Fig. 10, the productivity increasing in proportion to the cell density. Both the cell density and productivity were much higher than those from batch culture with a 500-ml roller bottle as shown in Figs. 9 and 10. During this culture, the glucose and lactate concentrations were maintained at an adequate level as shown in Fig. 11 by adjusting the perfusion rate shown in Fig. 12.

366

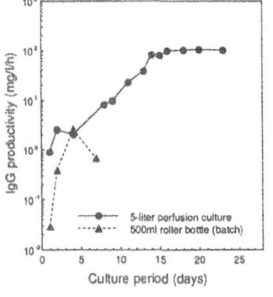

Fig. 10. STK1 productivity

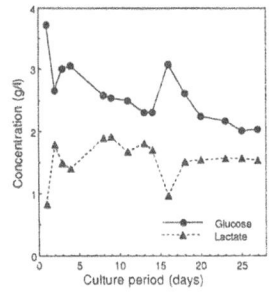

Fig. 11. Medium conditions

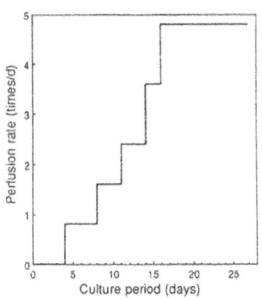

Fig. 12. Perfusion rate

4.4.3 *Mouse-Human Hybridoma (JHB-6).* A serum-free culture was also tested with the defoaming net. Even with this serum free culture, the protein produced by the cells caused foaming, especially at the high cell density stage. With the defoaming net, however, a high cell density and productivity were obtained as shown in Figs. 13 and 14, respectively.

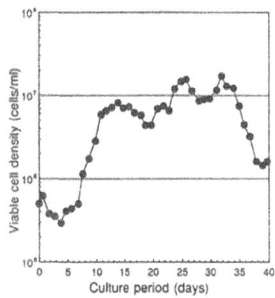

Fig. 13. JHB-6 cell growth

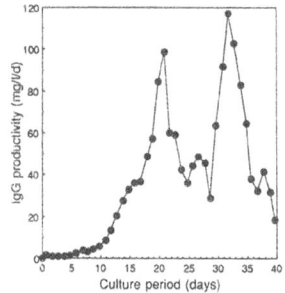

Fig. 14. JHB-6 productivity

4.4.4 *Chinese Hamster Ovary (CHO-K1).* By adding a packed bed to the system, the anchorage-dependent cell, CHO-K1, could also be cultured with the airlift and defoaming net system. In this system, cell separation is not necessary, because the cells are immobilized in the packed bed. The cell density was counted at the end of culture, and during the culture, the cell density was estimated from the glucose consumption and lactate production rates. The estimated cell density was lower than the counted cell number at the end of culture as shown in Fig. 15, which indicates lower cell activity due to the confluent cell density. The culture medium was kept almost constant, except in the initial stage and the no-perfusion period around the 50th day, as shown in Figs. 16 and 17.

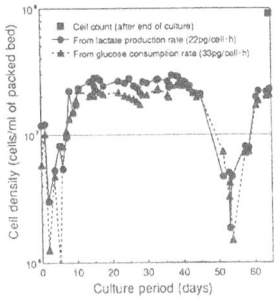

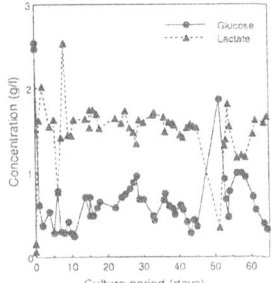
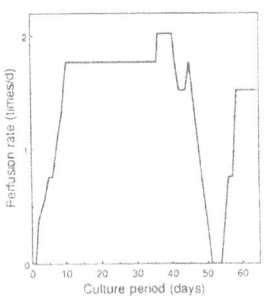

Fig. 15. CHO-K1 cell growth   Fig. 16. Medium conditions   Fig. 17. Perfusion rate

5. Conclusion

Several foam suppressing methods which enable direct gas sparging into the culture medium were investigated to enhance the oxygen supply for animal cell culture. When an anti-foaming reagent was added, the amount needed to be controlled in accordance with the aeration rate. With a sparger pore size larger than a certain value, the foam accumulation remained constant, and this sparger pore size can be regarded as the optimum. Defoaming device with a hydrophobic net enabled culture to be successfully performed, including suspension culture, packed-bed culture for anchorage-dependent cells, and serum-free culture.

6. Acknowledgement

The authors thank Yuji Ishikawa of the Chemo-sero-therapeutic Research Institute for supplying the JHB-6 cell line,for giving advice on the culture medium, and for the product assay.

7. References

[1] Ishida, M., Haga, R., Nishimura, N., Matsuzaki, H. and Nakano, R. (1990) 'High Cell Density Suspension Culture of Mammalian Anchorage Independent Cells: Oxygen Transfer by Gas Sparging and Defoaming with a Hydrophobic Net,' Cytotechnology, 4, 215-225.

[2] Handa-Corrigan, A., Emery, A. N. and Spier, R. E. (1988) 'Effect of Gas-Liquid Interfaces on the Growth of Suspended Mammalian Cells: Mechanisms of Cell Damage by Bubbles,' Enzyme Microb. Technol., 11, 230-235.

[3] Spier, R. E. and Griffiths, J. B. (1988) Animal Cell Biotechnology, Academic Press, London.

[4] Aunins, J. G., Croughan, M. S., Wang, D. I. C. and Goldstein, J. M. (1986) 'Engineering Developments in Homogeneous Culture of Animal Cells: Oxygenation of Reactors and Scaleup,' Biotechnol. Bioeng. Symp., 17, 699-723.

[5] Whiteside, J. P. and Spier, R. E. (1981) 'The Scale-up from 0.1 to 100 Liter of a Unit Process System Based on 3-mm-diameter Glass Spheres for the Production of Four Strains of FMDV from BHK Monolayer Cells,' Biotechnol. Bioeng., 23, 551-565.

[6] Murakami, S., Chiou, T.-W. and Wang, D. I. C. (1991) 'A Fiber-Bed Bioreactor for Anchorage-dependent Animal Cell Cultures: Part II. Scaleup Potential,' Biotechnol. Bioeng, 37, 762-769.

NON-WOVEN FABRICS FOR IMMOBILIZING MAMMALIAN CELLS

YOSHIKI KAGEYAMA[1], FUMINORI KIMURA[1], YOSHIAKI MATSUDA[2]
and SHINJIRO MITSUDA[2]
1)Tokyo Research and Development Center, Japan Vilene Co.,LTD.
  7 Kitatone, Sowa-machi, Sashima-gun, Ibaraki 306-02,Japan
2)Research Institute of Life Sciences, Snow Brand Milk
  Products Co., LTD.
  519 Shimo-Ishibashi, Ishibashi-machi, Shimotsuga-gun,Tochigi
  329-05,Japan

Abstract.   We have developed two types of Non-woven fabrics (NWF)
with urethane binder for tissue culture use.   Most of cell lines
tested, L-929, BHK-21, IMR-90 and hybridoma cells, could grow well
on the NWF coated with Gelatin + Fibroin + Chitosan (M type).
A long-term continuous production of Erythropoietin by recombinant
BHK-21 cells was performed by using a fixed-bed type of reactor
packed with the M type NWF.

1.Introduction

A high density cultivation is an important means for an efficient
production of biologically active substances with mammalian cells.
Some reports[1)-4)] have presented the effect of the material with a
three-dimensional structure on tissue cultures of mammalian cells.
The NWF has such a three-dimensional structure with vacant space.
    In this report, we describe some characteristics and availability
of our NWF for the tissue culture use.

2.Method

2.1.Preparation of the NWF

The NWF was prepared as follows.
    A web was prepared by opening a rayon staple as a principal
fiber, and subjected to a needle punch process. And the sheet was
strengthened by a Poly-urethane binder at the cross-point of the
staples.   In the case of applying cell adherence factors, the NWF
sheet was treated by a dipping process and an insolubilization
process with an alkaline solution.

2.2.Cultivation and cell counting

The NWF was used after an autoclave sterilization, and DMEM
containing 10 % CS was used as cultivation medium. The cultivation
was carried out according to following procedures.
    The cell suspension (cell concentration :$10^4 - 10^6$ cells/ml) was
inoculated onto the NWF (sized 15*15 mm, thickness approx.3mm)
putting in the 12-well plate. After 3 — 4 hours, the NWF was
transferred into the other plate with fresh medium, and cultivated
for fixed number of days. The cell count was done by a tryptic
treatment and a hematocytometer method at the fixed number of days.

369

3.Results

3.1.Examination of cell adherence factors

TABLE 1. shows the result of cell growth on NWF prepared using different cell adherence factors. This examination was performed with 10^5 (cells/cm³) of BHK-21 cell suspension and cell counting was done after 4 days cultivation.

In the case of Gelatin + Fibroin (Silk) + Chitosan, the growth was the best.

TABLE 1. Influence for cell growth on the cell adherence factors

P V A	Gelatin	Fibroin	Chitosan	Cell Density (cells/cm³)
	○	○	○	$2.3*10^6$
		○	○	$1.8*10^6$
			○	$1.8*10^6$
			○	$1.6*10^6$
○	○	○		$1.3*10^6$
○	○			$1.2*10^6$
○				$0.5*10^6$

We examined the effect of the combination of adherence factors on growth of L-929 cells by a microscopic observation. The result is shown as Photo. 1. This experiment was carried out on a glass dish which was coated with a thin layer of each factor. In the case of Gelatin + Fibroin + Chitosan, good adherence and extension were observed.

3.2.Effects of the cell inoculum size on growth

We prepared two types of NWF which were named as U type and M type. U type was consisted of a staple and a binder, and M type was prepared by adding three factors to U type. BHK-21 cells were used in this study.

Fig.1 shows the time course of cell growth using two kinds of NWF. There was not a significant difference between M type and U type. But the cells adhered to M type were not easily recovered by a tryptic treatment. A typical photograph of M type by SEM (omitted) was obviously showing the three-dimensional structure .

In order to investigate the adherence (or growth) ability of M type, we examined the influence of the cell inoculum size. The result is shown in Fig.2. The BHK-21 cell showed good adherence and growth at the lower initial cell density such as $2*10^4$ (cells/cm³). In this experiment, the density of BHK cell reached $1*10^7$ (cells/cm³) on the 9th day.

By the LEM observation of slice section, BHK cells were well growing on the adherence factor's layer covering staples.

3.3.Cultivation of several cell lines on the NWF

Most of cell lines tested, L929, CHO, IMR-90 and Hybridoma cells (secreting anti-Prolactin antibody) could grow well on the M type.

In particular, fibrous cell lines gave good adherence and growth, and formed some clusters in the NWF.

3.4.Erythropoietin production by recombinant BHK-21 cells in the reactor packed with the M type NWF

A diagram of the reactor is shown in Fig.3. The M type NWF was cut to a disk shape of 25mm diameter (totally 30cm^3), and packed into the reactor of 50cm^3 capacity.
　　The perfusion cultivation was operated in following conditions.
　Cell line : Recombinant BHK-21 producing Erythropoietin
　Inoculum size:2*10^5 cells/cm^3-NWF
　Cultivation medium :～4th day, DMEM + 10% CS for batch cultivation
　　　　　　　　　　4th day～ , DMEM + 2% CS for perfusion cultivation
　Concentration of dissolved Oxygen : controlled at 40% air saturation
　　Fig.4 means that the NWF realized stable cultivation of the BHK-21 cells and high ability of Erythropoietin production during 30 days.

4.Discussion
We have demonstrated the efficiency of NWF on mammalian tissue cultures. In the results, the difference of cell growth rate between M type and U type was a little in appearance (Fig.1). However, we think M type should be regarded as more effective one on the experiments. Because, M type is superior to U type on the point of cell adherent ability, and it has a tendency to decline of cell recovery rate with a trypsin treatment consequently.
　　The cell adherence factors, Gelatin + Fibroin + Chitosan, could work to change the property of U type's surface, that is, an optimum hydrophilicity for the cells.

　　A microscopic observation suggested that the NWF could give the comfortable places for cultivation, mainly at a staple's wide surface and the cross-point regions. Further, it is possible that the three-dimensional structure of NWF contributes to create a micro-environment that is essential to cell growth or extension.

　　In the experiment on the reactor, the NWF applying BHK cells contributed to good erythropoietin production. It suggested that the BHK cells formed a network, like nervous system, in three-dimensional structure of NWF, and NWF could maintain the nutritious conditions at enough vacant space of its for the cells.
　　We expect that the NWF could be applicable to organ culture, a primary culture and a hybridoma culture.

5.References
1)D.J.GIARD, D.H.LOEB et al.(1979) 'Human Interferon Production with Diploid Fibroblast Cells Grown on Microcarriers',Biotechnology and Bioengineering,XXI,pp.433-442.
2)A.KADOURI and Z.BOHAK (1989) 'NON-WOVEN FABRIC MICROCARRIERS FOR MAMMALIAN TISSURE CULTURE',INDA-TEC-1989,17th Annual International Nonwovens Technological Conference,pp.477-487.
3)S.MITSUDA, Y.MATSUDA et al.(1990) 'Non-Woven Fabrics as New Cell Matrices for IMR-90 Human Embryonic Lung Diploid Fibroblast Cells', J.FERMENT.BIOENG., Vol.70, No.4, pp.289-291.
4)T.OGAWA, M.KAMIHIRA et al.(1992) 'High Cell Density Cultivation of Anchorage-Dependent Cells Using a Novel Macroporous Cellulosic Support', J.FERMENT.BIOENG., Vol.74, No.1, pp.27-31.

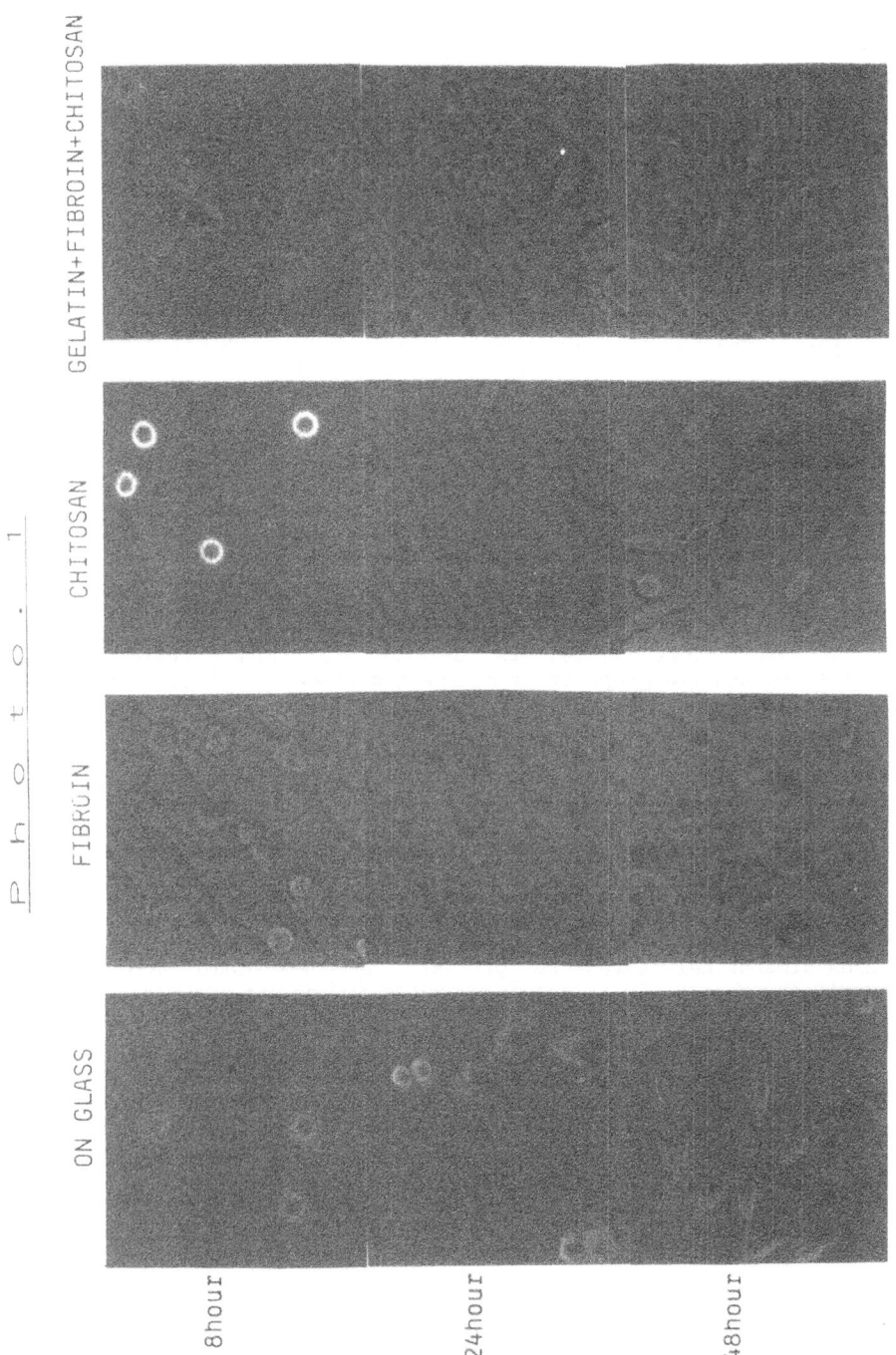

Photo. 1

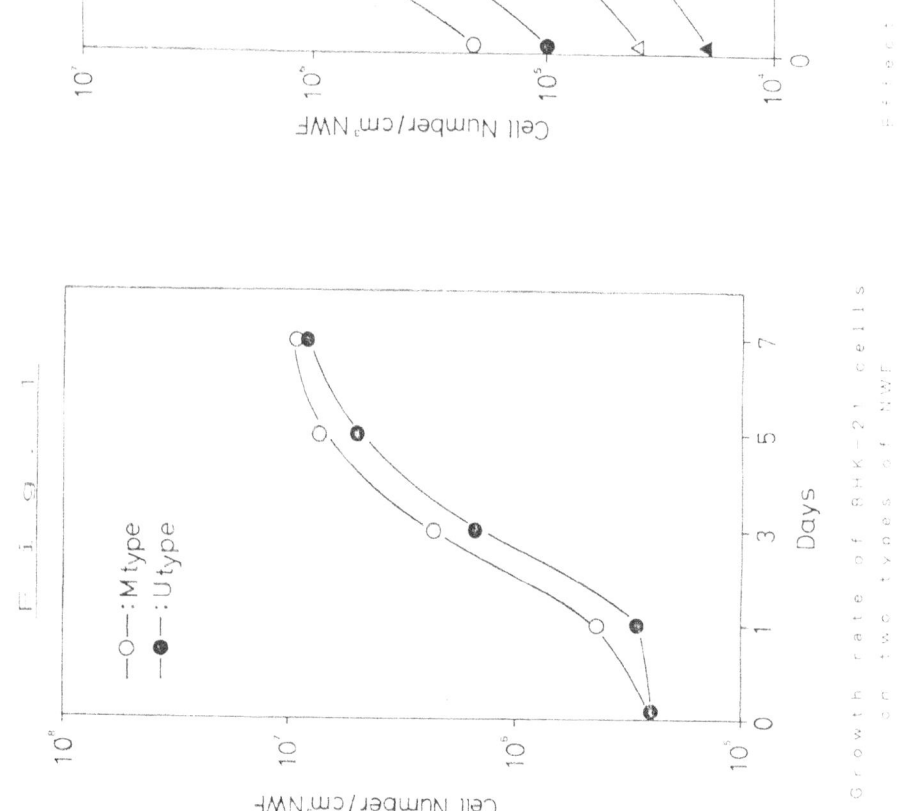

Fig. 1

Growth rate of BHK-21 cells
on two types of NWF

Fig. 2

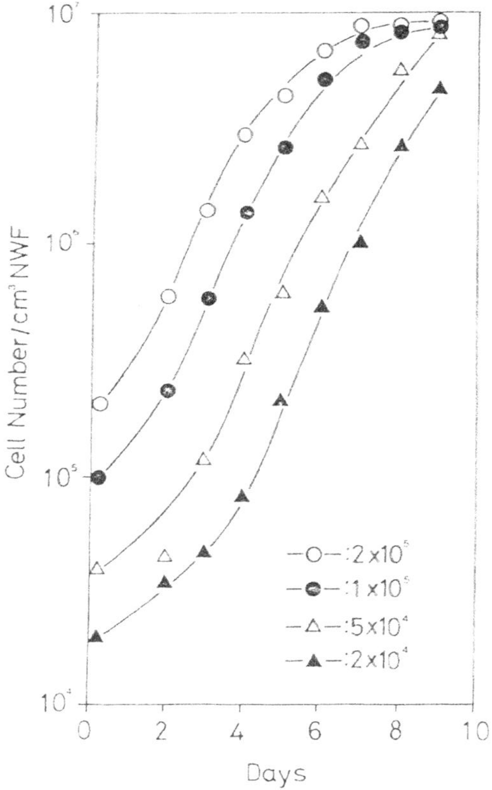

Effect of cell inoculum size
on growth of BHK-21 cells

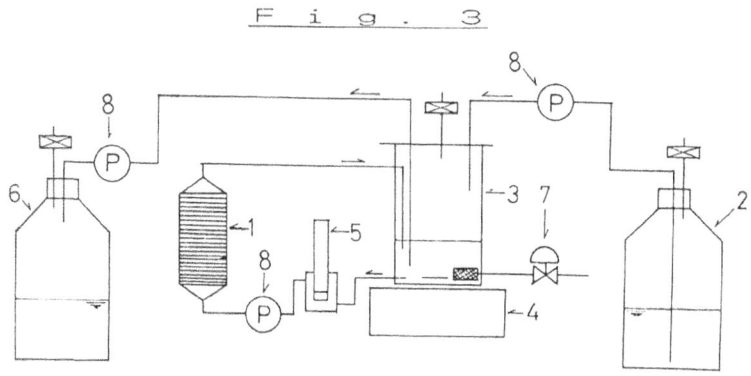

Fig. 3

Schematic diagram of the perfusion
culture system with NWF

1. reactor   5. DO electrode
2. medium reservoir   6. harvest reservoir
3. recirculating vessel   7. electromagnetic valve
4. magnetic stirrer   8. pump

Fig. 4

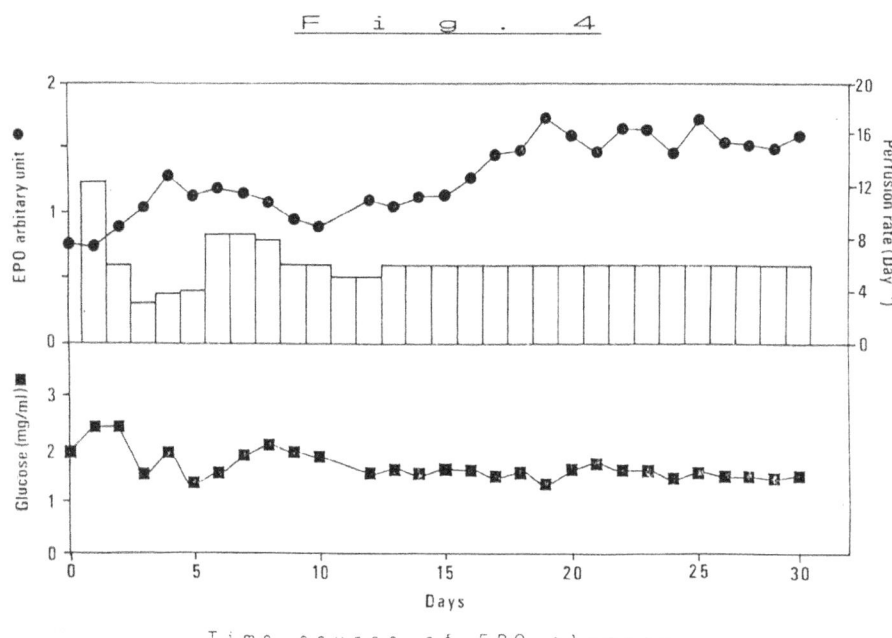

Time course of EPO, glucose
and perfusion rate

STATISTICAL DESIGN OF CULTURE MEDIUM FOR RECOMBINANT INTERFERON-γ PRODUCTION

P.M.L. CASTRO, P.M. HAYTER, A.P. ISON† and A.T. BULL
Biological Laboratory
University of Kent
Canterbury CT2 7NJ
UK

†*Advanced Center for Biochemical Engineering*
Department of Chemical and Biochemical Engineering
Torrington Place, London WC1E 7JE
UK

ABSTRACT. A statistical optimization approach based on a Plackett-Burman design was used to identify important nutrient components for both CHO cell growth and IFN-γ production. Nutrient components of the serum-free culture medium were first subjected to this analysis. BSA was shown to be one of the major medium components; preliminary studies indicated that its stimulating growth effect was a nutritional one. Cell growth in BSA-reduced cultures supplemented with a combination of a commercial lipid mixture plus Pluronic F68 was similar to control cultures, but IFN-γ production was adversely affected. A second statistical analysis was then carried out in order to investigate the importance of individual lipid components on those cultures. Linoleic acid was identified as a positive variable for cell growth while cholesterol was identified as a negative variable for both cell growth and IFN-γ production. When a combination of linoleic acid plus Pluronic F68 was included in the formulation of BSA-reduced medium, cell growth was similar to cultures with high BSA concentration. However, IFN-γ concentration was depressed to a large extent (ca. 45%). No accumulation of IFN-γ was found within the cell.

1. Introduction

Animal cells in general have the ability to secrete proteins in a correctly folded functional state and with the correct post-translational modifications, such as glycosylation (Berman and Lasky, 1985), unlike bacterial or yeast systems. Human IFN-γ is a glycosylated protein and it has been expressed in CHO cells (Mutsaers et al. 1986). However, product yields obtained in animal cell cultures tend to be very low. Manipulation of the nutritional environment of the cell may significantly alter this typical pattern.

At the present, the trend in industry is to move towards more well-defined culture medium. BSA is a component that introduces great variability to the culture media and it is usually included at high concentrations. Albumin may serve as a carrier for metals, lipids and hormones (Barnes and Sato 1980) and may also provide the cells with protection from

S. Kaminogawa et al. (eds.), Animal Cell Technology: Basic & Applied Aspects, Vol. 5, 375–381.
© 1993 *Kluwer Academic Publishers.*

shear damage (Lambert and Birch 1985). Lipids have major functions in animal cells, both as structural components of the membranes and as an energy source (King and Spector 1981). The growth promoting effects of lipids have been sometimes referred to and they are included in most serum-free medium formulations (Barnes and Sato 1980, Glassy et al.1987). However, lipid requirements in cell culture seem to vary much between cell lines. Pluronic F68 has been shown to protect animal cells from shear damage (Handa-Corrigan et al. 1989). Mizrahi (1975) also suggested that low concentrations of pluronic could enhance nutrient uptake into cultured human lymphocytes. Recently, commercial grade Pluronic F68 has been reported to have either growth stimulating (Bentley et al., 1989) or growth detrimental and production stimulating (Al-Rubeai et al., 1992) effects.

The use of statistical experimental designs that allow the screening of a large number of variables becomes particullarly useful for process optimization of animal cell cultures (Bull et al., 1990). In this study, a statistical methodology based on a Plackett-Burman design (Plackett and Burman, 1946) was used. The investigation was carried out in two sequential steps. First, the components of the serum-free medium were screened for their effects on both cell growth and IFN-γ production. Subsequently the BSA concentration of the medim was reduced and the importance of individual lipids on the same parameters was investigated.

2. Materials and Methods

2.1. CULTURE CONDITIONS

The cell line used was a dihydrofolate reductase deficient (DHFR-) mutant of CHO-K1 which has been cotransfected with the DHFR gene and the human gamma-IFN gene by methrotrexate selection. The culture medium was a serum-free medium described previously (Hayter et al. ,1989) and based on RPMI 1640 supplemented with bovine serum albumin (BSA), human transferrin, bovine insulin, sodium pyruvate, alanine, putrescine, $FeSO_4$, $ZnSO_4$, Na_2SeO_3 and $CuSO_4$. Free fatty-acids and cholesterol were complexed with fatty acid free BSA at a 1:100 ratio (w/w).

Cell counts were made in a Neubauer counting chamber and cell viability determined by trypan blue dye exclusion. IFN-γ titres were determined using an ELISA assay.

2.2. STATISTICAL METHODOLOGY

The experimental design was based on a Plackett-Burman statistical approach, as previously described (Castro et al., 1991). It allows the study of up to N-1 variables with N experiments. The experiments were carried out according to a working matrix, where the concentration of the variables is systematically changed between two levels. After completion, a statistical data analysis allows the ranking of the variables that significantly affect the system, either positively or negatively. Only significance levels greater than 70% were accepted.

The variables under study were oleic and linoleic acids, cholesterol and choline/ethanolamine. The low level of each component was zero, except for choline, which is already present in the medium at a concentration of 3 μg/ml. The high levels were 5 μg/ml for oleic and linoleic acids and cholesterol and 40 and 10 μg/ml for choline and ethanolamine respectively.

3. Results

On the first stage of the study the importance of low molecular weight nutrients, including glucose and amino acids and also protein supplements used in serum-free medium, was investigated. 20 different variables in total were tested (Castro et al., 1991). When high concentrations of the positive variables were included in the culture medium, specific growth rate, cell and IFN-γ concentrations were increased by 43, 41 and 45% above the original formulation. BSA was among those positive variables. Subsequent experiments done with different types and concentrations of BSA indicated that the positive effect found for this medium component is probably due to differences in the source of BSA rather than to the actual amount present in the culture medium.

Taking into account the advantags of reducing the BSA level of the medium, at the second stage of the study the search for BSA substitutes was made. Preliminary work carried out with BSA-reduced cultures (1 mg/ml) showed that supplementation of the culture with a combination of a commercial lipid mixture plus Pluronic F68 had a growth stimulating effect. Growth patterns similar to control cultures (5 mg/ml BSA) could be obtained, i.e only about 10% reduction in specific growth rate and viable cell titre. However, a detrimental effect on IFN-γ production was observed, with approximately 35% reduction in the titre obtained. Non supplemented BSA-reduced cultures showed about 40% reduction in both cell growth and IFN-γ production.

It became important to search for lipid components responsible for such behaviour. The effect of individual lipids on cell growth and IFN-γ production was investigated using the Plackett-Burman statistical procedure, using a N=8 matrix (Table 1). BSA-reduced medium (1 mg/ml) suplemented with 0.1% Pluronic F68 was used as the basis of the matrix.

TABLE 1. Plackett -Burman matrix for the study of 7 variables with 8 experiments

Experiments	Variables						
	A	B	C	D	E	F	G
1	+	+	+	-	+	-	-
2	-	+	+	+	-	+	-
3	-	-	+	+	+	-	+
4	+	-	-	+	+	+	-
5	-	+	-	-	+	+	+
6	+	-	+	-	-	+	+
7	+	+	-	+	-	-	+
8	-	-	-	-	-	-	-

+, high level of a particular variable; -, low level of the same variable

The results are presented in Table 2. The statistical analysis showed a significantly positive effect of linoleic acid on both cell growth and maximum viable cell number; on the other hand, cholesterol had a negative effect on both parameters. Product titre at different days was determined and cholesterol again showed significantly negative effects. No indication of positive effects on product formation by any of the components was found. A positive control (BSA 5 mg/ml) was included in the study. Some of the experiments belonging to the working matrix showed growth patterns similar to the control culture but IFN-γ titres were always lower.

TABLE 2. Culture variables and their effects on cell growth and IFN-γ production

Variable		Oleic acid	Linoleic acid	Cholesterol	Choline/Ethanolamine
Cell growth rate	a	0.002175	0.004625	-0.002625	0.001425
	b	1.44	3.06	-1.74	0.94
	c	70%	90%	80%	
Viable cell	a	0.375	1.575	-1.425	-0.025
production	b	0.48	2.01	-1.82	-0.03
	c		80%	80%	
IFN-γ titre	a	-0.1225	-0.0075	-0.3275	0.525
(day 3)	b	-1.42	-0.09	-3.78	-0.61
	c	70%		99%	

a- effect; b- t-value; c- significance level. Positive variables are shown in bold and negative variables in italics

Based on the results obtained, the effect of a combination of linoleic acid plus Pluronic F68 on cell growth and IFN-γ production was tested (Table 3). Again, growth patterns similar to control cultures were obtained, but IFN-γ production was much lower. The possibility of accumlation of IFN-γ within the cell by inhibition of the secretion mechanism was investigated by measuring the intracellular IFN-γ over time. No differences were found against control cultures. It is interesting to note that the values obtained are of the same order of magnitude as the specific production rates (data not shown).

4. Discussion

Early in this study, BSA was identified as a major component for both cell growth and IFN-γ production. It is likely that its stimulating properties are due to the effects of unidentified molecules bound to it. Fatty acids and other molecules have been found to be tightly bound to BSA (Iscove et al. 1980). Its quality varies with the lot and source, introducing much variability into the culture medium. The fact that our studies indicated that the source of material used had implications on culture performance further supports this view.

In an attempt to identify important nutrients that BSA might provide to the culture medium, the effect of individual lipids on cell growth and IFN-γ production was investigated. Two of the components under this study must be pointed out, linoleic acid and cholesterol, the former for its growth positive effect and the latter for its negative effect on both cell growth and IFN-γ production. Fatty acids and cholesterol constitute the basic lipid building blocks for the cells. The growth promoting effects of these two components have been reported (Sato et al., 1987, Kovar and Franek 1986, Darfler and Insel 1982, Minamoto et al., 1991). Concentrations used tipicaly vary from 1 to 10 μg/ml for oleic and linoleic acid and from 1 to 20 μg/ml for cholesterol. Low density lipoprotein, which contains chlolesterol, has been shown to have either growth promoting or inhibiting effects, depending on the cell line (Kawamoto et al., 1986). Schmid et al. (1991) found that low serum medium supplementation with different combinations of fatty acids, namely oleic and linoleic acids, did not have any effect on rBHK cell growth but that product titre was a positive function of those supplements. Of the lipids used in the present study, there was no indication of any positive effect on IFN-γ production.

TABLE 3. Cell growth and IFN-γ production for cultures with (A) 5 mg/ml BSA , (B) 1 mg/ml BSA +0.1% Pluronic, (C) 1mg/ml BSA + 0.1% Pluronic + 8 μg/ml linoleic acid

	A	B	C
Specific growth rate (h^{-1})	0.024 ± 0.002	0.024 ± 0.001	0.026 ± 0.001
Viable cell number (x 10^{-5})	9.7 ± 0.2	8.1 ± 0.1	9.7 ± 0.3
IFN-γ production (IU/ml)	8540 ± 900	6888 ± 675	4871 ± 404

The fatty acid composition of membrane phospholipids is similar to that of the lipids present in the culture medium. Modification of the groups present in the membrane will be reflected in cell function, most likely affecting cell growth rates. The maintenance of the bilayer fluidity seems to be essential for normal cell growth. Doi et al. (1978) correlated the unsaturated fatty acid content of membrane phospholipids with cell growth and found severe growth inhibition if reducing the content of unsaturated fatty acids in the membrane to less than 50%. Cholesterol is also involved in the regulation of the membrane fluidity and permeability (Phillips et al., 1987) and an increase in cholesterol content tends to reduce the fluidity of the membrane. The fact that this study indicated cholesterol as having negative effects on cell growth and production may be related to this function.

Under the conditions of this study oleic acid, choline and ethanolamine did not show any significant effects. An increase in the medium level of choline and ethanolamine, to 75 and 20 μg/ml respectively, has been recently reported to substitute the requirement of lipoproteins by hybridoma cells (Maiorella 1992).

Although cell growth in BSA reduced cultures supplemented with 0.1% Pluronic and linoleic acid was similar to control cultures, IFN-γ production was much lower. The reduction in product titre seems to be due to the reduction of BSA level but the addition of linoleic acid itself seems to be detrimental for IFN-γ production, which is in accordance with our previous findings. This work further reinforces the importance of BSA on culture performance and the value of experimental statistical techniques applied to process optimization.

5. Acknowledgments

We thank Wellcome Biotechnology for the CHO cell line and Celltech for the monoclonal antibodies. PML Castro would like to thank JNICT, Portugal, and SERC for financial support of the project.

6. References

Al-Rubeai, M., Emery, A.N. and Chalder, S. (1992) The effect of Pluronic F68 on hybridoma cells in continuous culture, Appl Microbiol Biotechnol 37, 44-45.

Bentley, P.K., Gates, R.M.C., Lowe, K.C., de Pomerai, D.E. and Lucy Walker, J.A. (1989) *In vitro* cellular responses to a non-ionic surfactant, Pluronic F-68, Biotech Letts Vol 11, 111-114.

Barnes, D. and Sato, G. (1980) Methods for growth of cultured cells in serum-free medium, Anal Biochem 102, 25-270.

Berman, P.W., Lasky, L.A. (1985) Engineering glycoproteins for use as pharmaceuticals, Tibtech 3:51-53.

Bull, A.T., Huck, T.A. and Bushell, M.E. (1990) Optimization strategies in microbial process development and operation, in R.K.Poole, M.J.Bazin and C.W.Keevil (eds), Microbial Growth Dynamics, IRL Press, Oxford, pp145-168.

Castro, P.M.L., Hayter, P.M., Ison, A. P. and Bull, A.T. (1991) Application of a statistical design to the optimization of culture medium for recombinant interferon-γ production by CHO cells, Appl Microbiol Biotechnol 38, 84-90.

Darfler, F.J. and Insel, P.A. (1982) Serum-free culture of resting, PHA-stimulated, and transformed lymphoid cells, including hybridomas, Exp. Cell Res. 138, 287-295.

Doi, O., Doi, F., Schroeder, F., Alberts, A.W. and Vagelos, P.R. (1978) Manipulation of fatty acid composition of membrane phospholipid and its effects on cell growth in mouse LM cells, Biochim et Biophys Acta 509, 239-250.

Glassy, M., Tharakan, J.P., Chan, P.C. (1988) Serum-free media in hybridoma culture and monoclonal antibody production, Biotech Bioeng 32, 1015-1028.

Handa-Corrigan, A, Emery, A.N. and Spier, R.E. (1989) Effect of gas-liquid interfaces on the growth of suspended mammalian cells: mechanisms of cell damage by bubbles, Enzyme Microb Technol 11, 230-235.

Hayter, P.M., Furlotte, D., Wilcox, M., Curling, E.M.A. and Jenkins, N. (1989) Recombinant gamma-interferon production by CHO cells in serum-free medium, in R.E.Spier, J.B.Griffiths, J.Stephenne and P.J.Crooy (eds), Advances in Animal Cell Biology and Technology for Bioprocesses, Butterworths, Sevenoaks, pp 280-282.

Iscove, N.N., Guilbert, L.J. and Weyman, C. (1980) Complete replacement of serum in primary cultures of erythropoietin-dependent red cell precursors (CFU-E) by albumin, transferrin, iron, unsaturated fatty acid, lecithin and cholesterol, Exp. Cell Res. 126, 121-126.

Kawamoto, T., Sato, J.D., McClure, D.B., and Sato, G.H. (1986) Serum-free medium for the growth of NS-1 mouse myeloma cells and the isolation of NS-1 hybridomas, Methods in Enzymology 121, 267-277.

King, M.E. and Spector, A.A. (1981) Lipid metabolism in cultured cells, in C. Waymouth, R.G. Ham and P.J. Chapple (eds.), The growth requirements of vertebrate cells in vitro, Cambridge Univ. Press, New York, pp. 293-312.

Kovar, J. and Franek, F. (1986) Serum-free medium for hybridoma and parental myeloma cell cultivation, Methods in Enzymology 121, 277-292.

Lambert, K.J. and Birch, J.R. (1985) Cell growth media, in R.E. Spier and J.B. Griffitts (eds), Animal Cell Biotechnology, Academic Press, New York, vol 1, pp. 85-112.

Maiorella, B. (1992) "Hybridoma culture- optimization, characterization, cost", in M.R. Ladisch and A. Bose (eds), Harnessing Biotechnology for the 21st Century, pp. 25-29.

Minamoto, Y., Ogawa, K., Abe, H., Iochi, Y. and Mitsugi, K. (1991) Development of a serum-free and heat-sterilizable medium and continuous high-density cell culture, Cytotechnology 5, 35-51.

Mizrahi, A. (1975) Pluronic polyols in human lymphocyte cell line cultures, J. Clin Microbiol 2, 11-13.

Mutsaers, J.H.G.M., Kamerling, J.P., Devos, R., Guisez, Y., Fiers, W. and Vliegenthart, J.F.G. (1986) Structural studies of the carbohydrate chains of human γ-interferon, Eur J Biochem 156, 651-654.

Philips. M.C., Johnson, W.S. and Rothblat, G.H. (1987) Mechanisns and consequences of cellular cholesterol exchange and transfer, Biochim et Biophys Acta 906, 223-276.

Plackctt, R.L., Burman, J.P. (1946) The design of of optimum multifactorial experiments, Biometrica 33:305-325.

Sato, J.D, Kawamoto, T. and Okamoto, T. (1987) Cholesterol requirements of P3-X63-Ag8 and X63-Ag8.653 mouse myeloma cells for growth in vitro, J. Exp. Med. 165, 1761-1766.

Schmid, G., Zilg, H., Eberhard, U. and Johannsen, K. (1991) Effect of free fatty acids and phospholipids on growth of and product formation by recombinant baby hamster kidney (rBHK) and Chinese hamster ovary cells (rCHO) in culture, J Biotechn 17, 155-167.

Takazawa, Y., Tokashik, M., Murakami, H., Yamada, K. and Omura, H. (1988) High-density culture of mouse-human hybridoma in serum-free defined medium, Biotech Bioeng. 31, 168-172.

MEDIUM OPTIMIZATION FOR PERFUSED CULTURE TO PROVIDE HIGH PRODUCT TITER

N. G. Ray, S. S. Ozturk, A. S. Tung, R. H. L. Pang, M. W. Young and P. W. Runstadler
Verax Corporation
6 Etna Road
Lebanon, New Hampshire 03766 U.S.A.

ABSTRACT. Culture productivity and product titer can significantly affect the economics of large-scale production of biotherapeutics. Bioreactor space-time productivity is determined by cell density and cell-specific productivity, while product titer is determined by medium feed rate per cell and cell-specific productivity. Therefore, high space-time productivity and product titer from a bioreactor system can be obtained with increased cell density, increased cell-specific productivity and reduced specific medium feed rate.

Perfused, fluidized-bed, porous collagen microsphere systems have consistently demonstrated high cell density and high space-time productivity with a large number of hybridoma and recombinant mammalian cell lines. Cell-specific productivity in fluidized, collagen microsphere culture has been observed to be significantly higher compared to cells grown in suspension. It has also been observed that requirements for serum and external growth factors for cell growth are substantially reduced in high density collagen microsphere cultures leading to the development of low cost culture medium.

We have optimized culture media to further enhance specific productivity and product titer in fluidized-bed culture. Enrichment of medium components allowed a reduction in specific medium feed rate while maintaining culture nutrient levels. Product titer is expected to go up to the extent specific medium feed rate is reduced at a constant cell-specific productivity. However, we have observed in the fluidized-bed culture that with an optimized, enriched medium and an optimized perfusion strategy, the increase in product titer was greater than what was expected from the reduction in specific medium feed rate alone -- indicating an increase in cell-specific productivity.

1. Introduction

Culture productivity and product titer can significantly affect the economics of large-scale production of biotherapeutics. In animal cell culture processes, the cost of medium is usually one of the major contributing factors to the production cost of biotherapeutics. This is particularly important when media contain serum or expensive components (e.g. growth factors, hormones, etc.).

Production cost can be significantly lowered by reducing the cost of culture medium. Reduction in relative cost of medium in a production campaign can be achieved by (1) reducing the unit cost of medium (i.e. $/liter) and (2) increasing product titer (i.e. gm product/unit volume of

383

S. Kaminogawa et al. (eds.), Animal Cell Technology: Basic & Applied Aspects, Vol. 5, 383–390.
© 1993 *Kluwer Academic Publishers.*

medium) without a significant increase in unit cost of medium. In addition, increase in product titer also reduces costs associated with medium preparation, handling and harvest processing.

Use of enriched medium to enhance product titer and culture productivity is discussed in this paper. In a perfusion culture system, enrichment of medium nutrients allows one to reduce the medium perfusion rate (MPR) per cell while maintaining the levels of nutrients in the culture. If cell specific productivity (CSP) remains unchanged, then an increase in product titer in proportion to the reduction in MPR is realized. However, along with the desired product, the concentrations of culture metabolites and other cell secreted components will also increase as a result of the decrease in cell specific MFR. Therefore, a proper balance among levels of nutrients, inhibitory metabolites, and potentially stimulating proteins secreted by the cells in the culture medium is important to achieve an optimum steady-state environment for enhancing product titer and, in some instances, culture productivity.

It has been observed that cell growth, metabolic characteristics and CSP can be profoundly affected by culture physical and biochemical environment in both continuous suspension and high cell density perfusion bioreactors (1-7). For example, in chemostat hybridoma cultures, the dilution rate is shown to have significant influence on cell metabolism and antibody productivity (1, 5-7). A change in dilution rate alters the concentration of nutrients and cell secreted components in the culture liquid, thereby influencing culture metabolism and growth. As the dilution rate is decreased, cell specific growth and metabolic rates decrease (1, 5). These changes in growth and metabolic characteristics are caused by lower concentration of rate limiting nutrient(s), higher concentration of inhibitory metabolites or both of these factors. To what extent the CSP will be influenced by the levels of nutrients or secreted components will depend on a particular cell line. A number of cell lines were observed to exhibit different product secretion characteristics in response to variations in culture pH, osmolarity, dissolved oxygen, dilution rate, etc. (1-4).

If CSP is maintained in a culture, a decrease in medium perfusion rate would result in a proportional increase in product concentration. Therefore, for a given medium formulation, the steady-state product titer is maximized at the lowest perfusion rate which can maintain a steady-state culture population and culture productivity in the bioreactor. Any further decline in perfusion rate will cause a decline in culture performance. If such a decline is solely associated with decreased levels of limiting nutrient(s) instead of an increase in inhibitory metabolites, the product titer can be further increased in proportion to the net increase in limiting nutrient(s) by using enriched culture medium. For example, a 2x enrichment of the nutrients (assuming physiological osmolarity is maintained) can potentially increase product titer by two-fold with a decrease in perfusion rate by half while maintaining culture nutrient levels similar to levels observed with a medium having substrate concentration equal to x. Furthermore, since this strategy, which adjusts perfusion rate and, therefore, the medium nutrient concentrations, maintains a constant substrate environment for the cells in the reactor, the cells will be in a constant physiological status relative to specific substrate uptake rates, product expression, etc.

In this paper we first present a model describing the relationships among perfusion rate, concentration of nutrients in the perfusion medium, nutrient levels in the bioreactor culture and product titer. This model assumes that the levels of metabolites do not cause toxic inhibition to culture performance. Based on this approach, experimental results with two hybridoma cell lines are presented demonstrating significant improvements in both product titer and culture productivity.

2. Model

Figure 1 illustrates that medium perfusion rate (MPR) and medium substrate levels can be manipulated together to control culture nutrient concentrations (e.g. glucose, glutamine, amino acids, etc.) in the bioreactor.

As indicated in this figure, titers of cell secreted metabolites and proteins, including the desired product, increase as a result of reduced MPR while incoming medium nutrients will decrease unless incoming nutrient levels are purposely increased to maintain nutrient levels unchanged.

Reactor production rate:

$$\text{Reactor production rate} = \text{CSP} \times \text{total cell number in bioreactor} \qquad (1)$$

where,

$$\text{CSP} = \text{cell specific productivity}$$

Product titer:

$$\text{Titer} = \text{Reactor production rate/MPR} \qquad (2)$$
$$= \frac{\text{CSP} \times \text{Total Cell Number}}{\text{MPR}}$$

where,

$$\text{MPR} = \text{medium perfusion rate}$$

Glucose consumption rate (GCR):

$$\text{GCR} = \text{MPR} \times [\text{Glu}_i - \text{Glu}_o] \qquad (3)$$

where,

$$\text{Glu}_i = \text{Glucose concentration in incoming medium}$$
$$\text{Glu}_o = \text{Residual glucose concentration in the bioreactor culture}$$

Now, combining equations (2) and (3),

$$\text{Titer} = \left[\frac{\text{CSP} \times \text{Total Cell Number}}{\text{GCR}} \right] \times [\text{Glu}_i - \text{Glu}_o] \qquad (4)$$

The above model can be significantly simplified with the following assumptions:

(1) CSP is constant,
(2) constant viable cells in the bioreactor,
(3) glucose is the limiting substrate, and
(4) cell specific glucose consumption rate is constant. This assumption is valid for constant Glu_o.

With the above assumptions, Equation (4) indicates that an increase in product titer can be achieved by increasing glucose concentration in the feed medium while maintaining the same glucose level in the bioreactor culture. In a glucose limiting perfusion culture, this is achieved by reducing the perfusion rate of higher glucose feed medium. An increase in product titer in proportion to the increase in $[Glu_i - Glu_o]$ is realized. However, it must be recognized that in a continuous or perfused cell culture system, more than one substrate can become rate limiting at a given time. For simplicity, we have designated glucose as an indicator for the limiting substrate(s).

3. Experimental Study

Two hybridoma cell lines were investigated to study the effects of media on product titer and culture productivity. One of the cell lines, designated as VX-12, secretes IgG_{2a} and a second cell line, designated as VX-944, secretes IgG_1. The cell lines were cultured in Verax fluidized-bed perfused bioreactors.

3.1 MATERIALS AND METHODS

VX-12 hybridoma: This is a mouse-mouse hybridoma cell line. Serum-free medium was used during the entire study. The base medium consists of a mixture of Dulbecco's Modified Eagle (DME) medium (4.5 gm/L glucose) and Ham's F-12 (1.8 gm/L glucose). The final glucose concentration in the base medium was 3.5 gm/L. In addition to the base medium, two concentrated media formulations, 1.9x and 2.9x of the base medium nutrients, with osmolarity adjusted to the base level by adjusting sodium chloride and potassium chloride concentrations, were also used in the study. The bioreactor had a fluidized-bed volume of 10 mL with a total fluid system volume of 90 mL.

VX-944 hybridoma: This is a human-mouse cell line which produces IgG_1. The experimental study was carried out entirely in serum-free medium. The base medium consists of a mixture of DMEM (4.5 gm/L glucose) and Ham's F-12 (1.8 gm/L glucose). Several enriched media formulations were used. Primarily the base medium was enriched with (i) glutamine alone, (ii) glucose and glutamine and (iii) glucose, all the amino acids, choline chloride and aminoethanol. The bioreactor was a Verax System 10 perfusion bioreactor, having a maximum fluidized-bed volume of 150 mL and a total fluidized system volume of 750 mL.

3.2 ANALYSES

Liquid cell number was determined using a Coulter Counter (Model: 2M, Coulter Electronics). Cell viability was determined with a hemocytometer by erythrosin B dye exclusion technique. Glucose and lactate were assayed by an enzyme electrode analyzer (Model: 2000, Yellow Springs Instrument Co.). Glutamine and ammonia levels were measured using enzymatic kits. Antibodies were assayed by enzyme-linked immunosorbent assay (ELISA).

4. Results

Figure 2 shows medium perfusion rate and product titer. The base (1x) medium was used during the first 15 days of culture after inoculation. Between days 15 - 25 the 1.9x concentrated feed medium was used. The nutrient concentration was further increased to 2.9x base formulation on day 25 until the end of the run. As the figure indicates, the medium perfusion rate was greatly reduced as the feed media were changed to concentrated formulations. The culture product titer was enhanced by more than eight-fold using 2.9x concentrated medium. As indicated in Table 1, the perfusion rate was reduced a maximum of approximately ten-fold with 2.9x concentrated medium. This resulted in an increase of 8.5-fold in product titer. Bioreactor productivity and calculated cell specific productivity remained relatively unchanged during the course of this study. Lactate concentration in the bioreactor culture reached 7.8 gm/L. The glucose consumption rate remained relatively stable up to a culture lactate level equal to approximately 5.0 gm/L.

TABLE 1: Culture Results
Titer Optimization
(VX-12 Hybridoma Culture)

Medium Conc. Factor	$(PR)_0/PR$	T/T_0	Productivity (mg/day)	CSP (pg/cell/hr.)	Medium Conc. Required
1.00	1.00	1.00	4.03	0.20	1.00
1.95	1.88	1.41	3.45	0.13	1.29
1.95	2.15	2.23	4.80	0.19	1.93
2.87	3.18	3.14	4.63	0.22	2.59
2.87	5.15	4.86	4.36	0.27	4.07
2.87	10.10	8.48	4.38	0.33	6.87

TABLE 1: Culture Results (continued)
Titer Optimization
(VX-12 Hybridoma Culture)

Medium Conc. Factor	Reactor Glucose Conc. (g/L)	Reactor Lactate Conc. (g/L)	Glucose Consump. Rate (g/day)
1.00	1.10	2.360	1.24
1.95	2.36	3.74	1.27
1.95	1.64	4.02	1.26
2.87	3.15	5.08	1.12
2.87	1.72	5.63	0.85
2.87	0.58	7.80	0.56

VX-944 HYBRIDOMA

Figure 3 presents medium perfusion rate and product titer of the VX-944 culture in a System 10 fluidized-bed system. The figure shows data between days 134 - 194 of the bioreactor run. Initially, between days 136 - 140, product titer was enhanced in response to a decrease in perfusion rate. Between days 140 - 167 there appears to be only a slight increase in product titer at a constant perfusion rate. During that period, the feed medium was further enriched with glutamine and other amino acids as indicated in the figure. Between days 168 - 174, the perfusion rate was decreased by one-half and the product titer was further increased by approximately two-fold. Near the end of the run, the product titer dropped returning to the original level when the original culture conditions relative to feed medium nutrients and perfusion rate were imposed. Overall, the product titer was increased by greater than ten-fold, during the conditions imposed between days 174 - 180, over the product titer prior to day 134.

5.   Conclusion

We have demonstrated the feasibility of enhancing product titer in high cell density perfusion culture by medium optimization. Indeed, in these studies using two hybridoma cell lines, secreting IgG_2 and IgG_1 respectively, an increase in product titer with the corresponding decrease in MFR has been observed. Accordingly, a very favorable economics for manufacture can be achieved. In addition, medium optimization to enhance CSP will further improve production economics.

6.   References

1   Miller, W.M., Blanch, H.W., and Wilke, C.R., (1988) 'A Kinetic Analysis of Hybridoma
        Growth and Metabolism in Batch and Continuous Suspension Culture: Effect of Nutrient
        Concentration, Dilution Rate, and pH', Biotech Bioeng., 32, pp. 947-965.

2   Ozturk, S.S. and Palsson, B.O., (1991) 'Metabolic, Growth and Antibody Production
        Kinetics of Hybridoma Cell Culture: I. Analysis of Data from Controlled Batch
        Reactors', Biotechnology Progress, 7, pp. 741-480.

3   Ozturk, S.S. and Palsson, B.O., (1991) 'Metabolic, Growth and Antibody Production
        Kinetics of Hybridoma Cell Culture: II. Effect of Serum Concentration, Dissolved
        Oxygen and pH in Batch Reactor', Biotechnology Progress, 7, pp. 481-494.

4   Ozturk, S.S. and Palsson, B.O., (1991) 'Effect of Media Osmolarity on Hybridoma Cell
        Growth, Metabolism, and Antibody Production', Biotech. Bioeng., 37, pp. 989-993.

5   Ray, N.G., Kalkare, S.B., and Runstadler, P.W., (1989) 'Cultivation of Hybridoma Cells
        in Continuous Cultures: Kinetics of Growth and Product Formation', Biotech. Bioeng.,
        33, pp. 724.

6   Robinson, D.K., Widmer, J., and Memmert, K., (1992) 'Effect of Specific Growth Rates on Productivity in Continuous Open and Partial Cell Retention Animal Cell Bioreactors', Journal of Biotechnology, 22, pp. 41-50.

7   Schmid, G., Wilke, C.R., and Blanch, H.W., (1992) 'Continuous Hybridoma Suspension Cultures With and Without Cell Retention: Kinetics of Growth, Metabolism and Production Formation', Journal of Biotechnology, 22, pp. 31-40.

Figure 1: Perfusion Culture

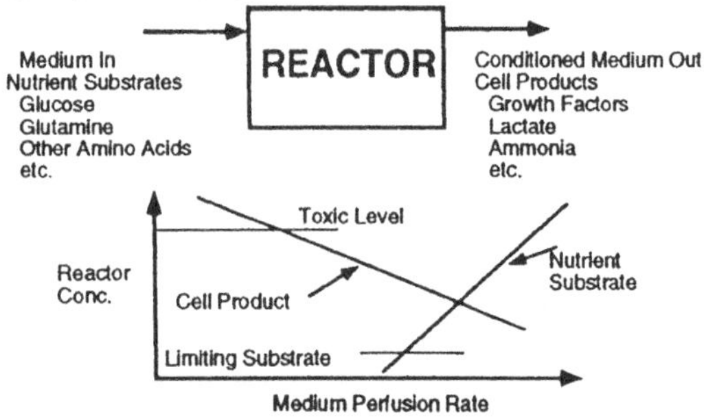

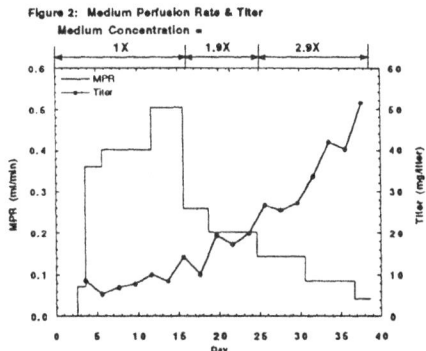

REACTOR

Medium In
Nutrient Substrates
  Glucose
  Glutamine
  Other Amino Acids
  etc.

Conditioned Medium Out
Cell Products
  Growth Factors
  Lactate
  Ammonia
  etc.

Toxic Level

Reactor Conc.

Cell Product

Nutrient Substrate

Limiting Substrate

Medium Perfusion Rate

Figure 2: Medium Perfusion Rate & Titer

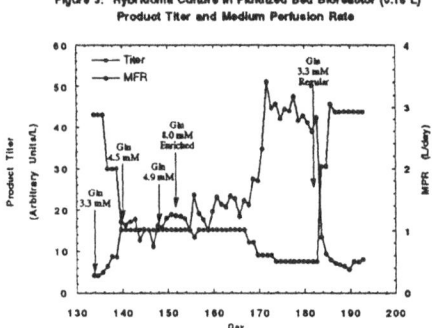

Figure 3: Hybridoma Culture in Fluidized Bed Bioreactor (0.16 L)
Product Titer and Medium Perfusion Rate

eRDF Medium Enhances Antibody Production Of IgG And IgM Secreting Hybridomas.

CHUA Florence[1], OH Steve Kah Weng[2], YAP Miranda [1,2]and TEO Wah Koon[1]

Chemical Engineering Dept[1] and Bioprocessing Technology Unit [2],
National University of Singapore,
10, Kent Ridge Crescent, SINGAPORE 0511

ABSTRACT

In an attempt to determine the best basal medium type for hybridoma culture we systematically studied the performance of 2 cell lines, an IgG producer, 2HG11 and an IgM producer, B10 in 3 types of basal media (eRDF, DMEM/F12 and RPMI). Both DMEM/F12 and RPMI are commonly available world wide but eRDF is only available in Japan.

The growth performance and antibody (Mab) yields of both hybridomas are briefly summarised below.

Media type with 10%FCS	IgG producer (2HG11)		IgM producer (B10)	
	Cell Density $(\times 10^6/ml)$	Ig Levels (mg/l)	Cell Density $(\times 10^6/ml)$	Ig Levels (mg/l)
eRDF	2.9	150	2.5	45
DMEM/F12	1.4	50	1.2	22
RPMI	1.0	40	0.8	7

Maximum cell numbers in eRDF medium were up to 3 times higher than the other 2 media whilst maximum Mab levels were 2-6 times greater. These results show that eRDF, highly enriched in amino acids and vitamin supplements is the best media for both cell types and may be the nutrient of choice for Mab production.

When 2HG11 cells were adapted to serum free conditions, eRDF enabled the fastest adaptation within 10 days, followed by DMEM/F12 in 15 days and RPMI required 30 days to stabilise at ¯90% viability. Results also showed that cells perform best in eRDF under serum free conditions.

S. Kaminogawa et al. (eds.), Animal Cell Technology: Basic & Applied Aspects, Vol. 5, 391–396.
© 1993 *Kluwer Academic Publishers.*

Introduction

A wide range of basal media have been formulated for the growth of hybridomas and choosing the right basal medium is very important for monoclonal antibody (Mab) production. As is evident in many reports, different basal media support cell growth differentially, which in turn affect the productivity as Mab production is often a reflection of cell growth (Murakami, 1989). It has further been observed that the ease of adapting a hybridoma cell line to serum free conditions depends on the basal media used (Kovar, 1989). In this investigation we examined the performance of both IgG and IgM producing hybridomas in 2 commonly available media, RPMI and DMEM/F12 (DF) compared to a newly formulated medium, eRDF. We also examined the adaptability of the IgG producing hybridoma to serum free conditions in these 3 media. Finally, its performance in serum free and serum supplemented conditions are reported for all 3 basal media types.

Materials and Methods

Basal media comparisons
Hybridoma cells 2HG11, an IgG producer and B10, an IgM producer were obtained from the World Health Organisation and the Institute of Molecular and Cell Biology in Singapore respectively. The basal media used in this study were RPMI, DMEM/F12 and eRDF. The first 2 were obtained from Gibco, and the last one from Kyokuto Pharmaceutical. In the serum supplemented experiments, 10% foetal calf serum (FCS) was used. All batch cultures were performed in shake flask cultures at a shaker speed of 150 rpm.

Serum free studies
In the serum free experiments, the following compounds insulin (5 mg/l), transferrin (10 mg/l), ethanolamine (1.5 mg/l), sodium selenite (4.3 µg/l) and BSA (1 g/l) were added. Adaptation to serum free conditions was carried out in 10 ml tissue culture flasks. Cell counts and viability were checked daily and the cultures diluted to 2×10^5/ml. This procedure was followed for a period until the cells maintained a viability of 90% or above. Batch culture experiments were conducted for the serum supplemented and serum free conditions for all 3 basal media types in shake flasks set at 150 rpm.

Results

The growth of 2HG11, an IgG producing hybridoma and B10, an IgM producing hybridoma in shake flask cultures are depicted in Figs. 1a) and b) respectively. It is evident that RPMI is the least effective media for cell growth, followed by DF whilst eRDF provides a dramatic 2-fold increase in cell yields for both cell types. More importantly, Fig. 2a) and b) also shows that IgG and IgM expressions respond in a similar manner, with eRDF media allowing 2 to 6 fold higher yields than in the other 2 media. Table 1 summarises the data on maximum viable cell numbers, and maximum Ig levels achieved in the 3 types of serum supplemented media for 2HG11 and B10 cells.

Table 1. Comparison of the Performance of IgG and IgM producers in 3 Types of Basal Media.

Media type with 10%FCS	IgG producer (2HG11)		IgM producer (B10)	
	Cell Density ($\times 10^6$/ml)	Ig Levels (mg/l)	Cell Density ($\times 10^6$/ml)	Ig Levels (mg/l)
eRDF	2.9	150	2.5	45
DMEM/F12	1.4	50	1.2	22
RPMI	1.0	40	0.8	7

In the following experiments, 2HG11 hybridomas were adapted to serum free (SF) conditions in the 3 media types. As shown in Fig. 3a), hybridomas readily adapted to eRDF/SF medium within 10 days. Daily passaging caused cell to double or triple with ease. The viability of the culture dropped gradually around 80% on day 5 during adaptation. Thereafter viability rose and reached 90% by day 11. In DF/SF medium, the cells required 15 days to adapt and there was a greater drop in viability to 70% during adaptation but viability stabilised at 90% eventually as depicted in Fig. 1b). Cell densities generally doubled with daily feeding. Viability of 2HG11 cells dropped to as low as 60% while the culture was adapting to RPMI/SF medium as seen in Fig. 3c). Cell numbers did not always double on feeding and sometimes the culture required two days to duplicate in numbers. Viability of the adapted culture fluctuated around 85% after 30 days. Thus RPMI/SF was the least favourable medium as adaptation required the longest time and the viable cell fraction was the lowest.

Table 2 briefly summarises the results of the batch experiments with 2HG11 cells in serum supplemented and serum free media. Once again, eRDF basal media appears to be the medium of choice for IgG production. Although maximum cell density and specific growth rates were much reduced in eRDF/SF medium compared to eRDF with serum, the maximum antibody level achieved was fractionally higher for the SF culture. In terms of cell specific productivity, cells in serum free medium seem to be more productive. In contrast, hybridomas in DF and RPMI media with serum achieved better growth performance compared to their serum free counter-parts and Ig levels were also 10-20 % higher than the SF conditions. Thus it appears that the ingredients of eRDF are much superior to either RPMI or DF (which are commonly used worldwide) in serum free as well as in serum supplemented cultures for antibody production.

Table 2. Comparison of Cell Performance in Serum Supplemented and Serum Free Media.

Medium	Max. Cell Density (x 10⁶/ml)	Growth Rate (hr⁻¹)	Max. Ig Levels (mg/l)
eRDF + FCS	2.9	0.050	150
eRDF/SF	1.8	0.038	170
DF + FCS	1.4	0.038	50
DF/SF	1.0	0.028	40
RPMI + FCS	1.0	0.037	40
RPMI/SF	0.7	0.025	35

Conclusions

1. IgG and IgM producing hybridomas were shown to produce the highest cell numbers and Mab yields in eRDF medium compared to either DF or RPMI, 2 more commonly available commercial media.

2. Adaptation to serum free conditions was much easier and faster in eRDF compared to the other two media types.

3. Cells adapted to eRDF/SF conditions were much more productive than the serum supplemented cells, whereas cells in DF/SF and RPMI/SF conditions were less productive.

References

1. Murakami, H. (1989) Serum-free media used for cultivation of hybridomas. In: Mizrahi, A. (Ed.), Advances in Biotechnological Processes. Vol. 11, Monoclonal Antibodies: Production and Application. Alan R. Liss, New York, pp. 107-141.

2. Kovar J. (1989) Various cell lines grow in protein-free hybridoma medium. In Vitro Cellular and Development Biology 25, 395-396

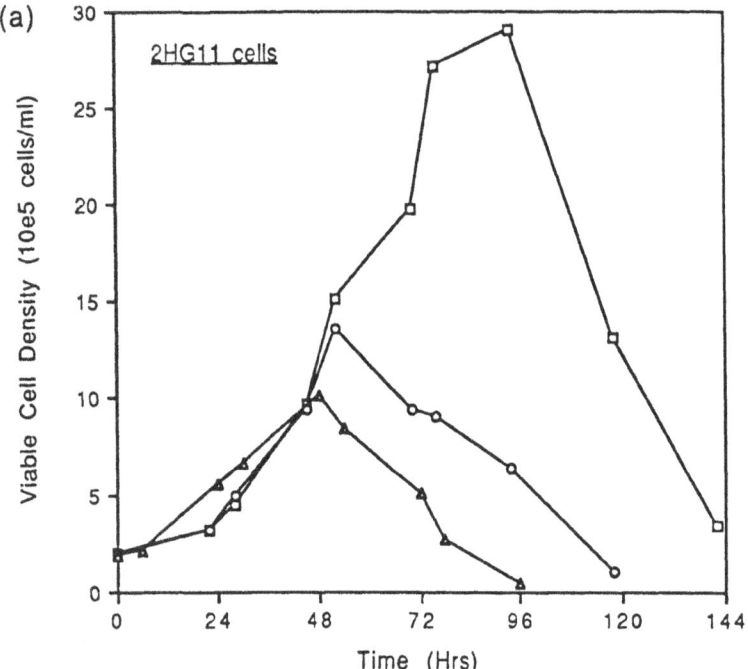

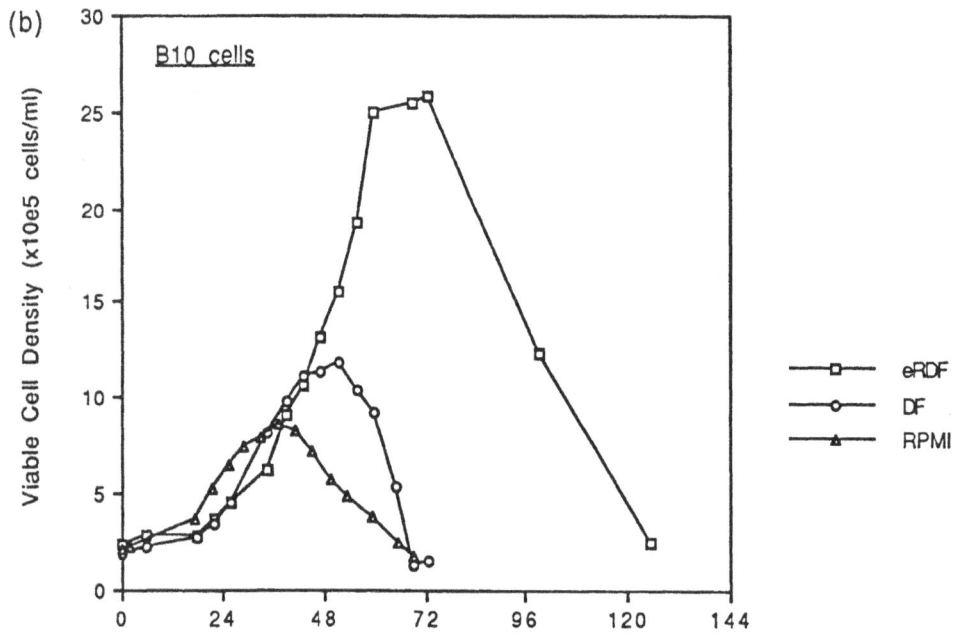

Fig. 1 Growth of (a) 2HG11 and (b) B10 cells in eRDF, DF and RPMI media

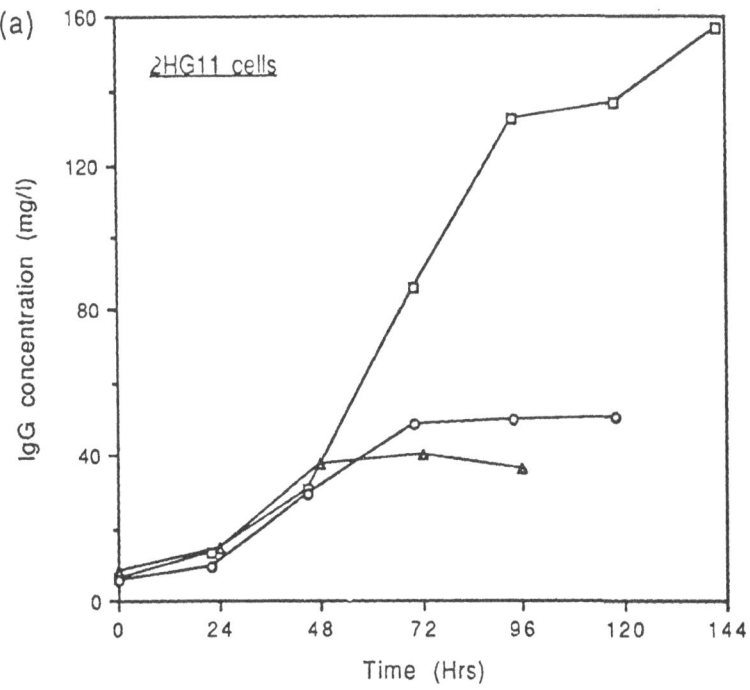

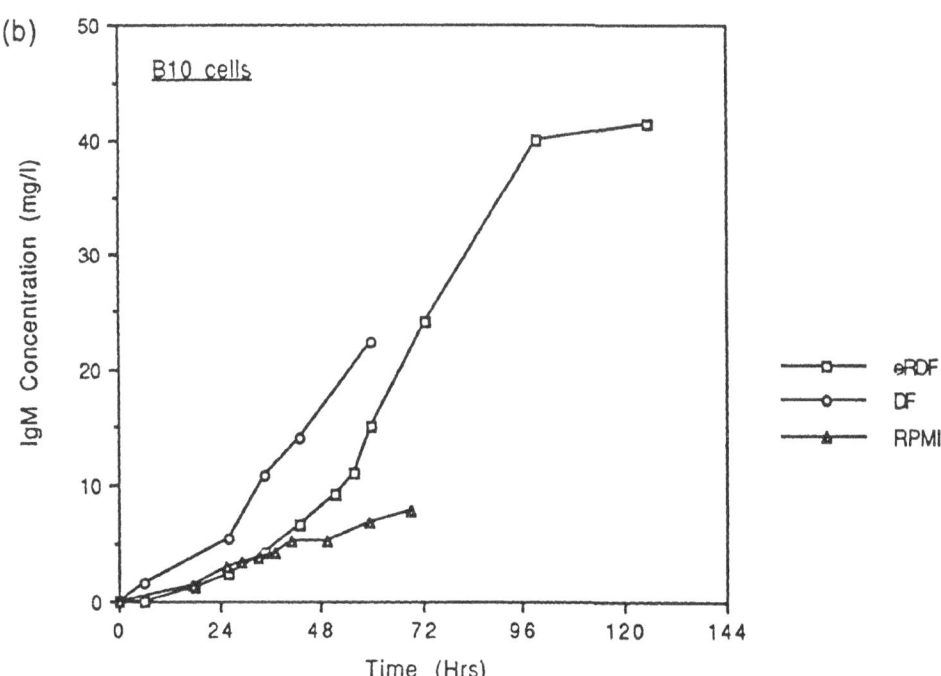

Fig. 2   Mab concentration of (a) 2HG11 and (b) B10 cells in eRDF, DF and RPMI media

(a) <u>eRDF/SF Medium</u>

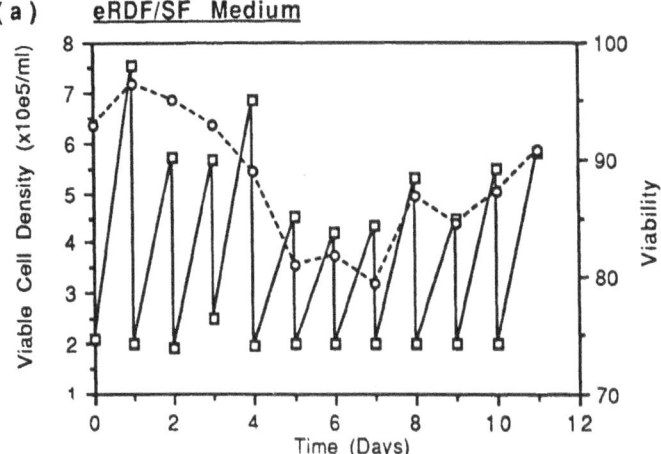

(b) <u>DF/SF Medium</u>

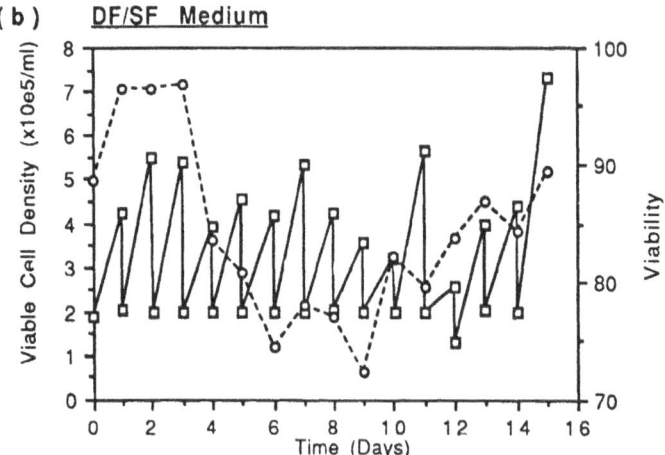

(c) <u>RPMI/SF Medium</u>

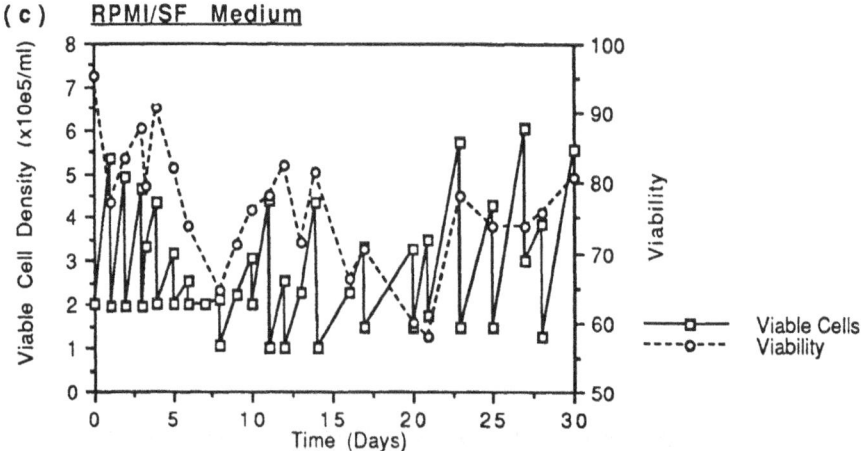

<u>**Fig. 3**</u> <u>**Adaptation of 2HG11 Cells to Serum-free Media**</u>

SERUM-FREE MEDIA FOR LARGE-SCALE SUSPENSION CULTURES

Yuka Tawara, Masa-aki Kitajima, and Masatoshi Togami
Tsukuba Research Laboratories
Japan Synthetic Rubber Co. Ltd.
25 Miyukigaoka, Tsukuba, Ibaraki 305, Japan

Abstract.

Two types of powdered serum-free media suitable for the large-scale culture of animal cells were developed, each consisting of a powdered basal medium and a small amount of a growth factor solution. The JSR-2 medium can be sterilized by filtration, and the JSR-4 medium, by autoclaving. Cell growth and monoclonal antibody production in the JSR-2 and JSR-4 media were investigated, using a myeloma, two hybridomas and two types of CHO cell lines without any weaning procedure. In both media, anchorage-independent cell lines such as hybridomas, and the anchorage-independent CHO cell line grew as well as in a serum-supplemented medium, the productivity of monoclonal antibody being equivalant to 10% FBS.

Introduction

Serum-containing media have been used for animal cell cultivation. However, serum contains various proteins in a large amount, and has prevented purification of the cellular products and limited an understanding of cellular physiology. Many investigators have attempted to develop serum-free media.[1] They have substituted serum by hormones, growth factors, trace elements, attachment factors, and other substances, and have developed a number of serum-free media for various cell lines, which are specific to each cell line. To use these media for producing biological materials and for investigating the cells themselves, we often have to wean the cells to the media over a long period of time.

Weaning cells for a prolonged period is time consuming and also has the possibility of transforming the cells. We focussed our efforts on the development of serum-free media that would not need a weaning treatment for the cells. For large-scale culture, such a medium should be provided in a concentrated or powdered form to make transportation easy and to save stock space.

We developed two types of powdered serum-free media for the large-scale culture of anchorage-independent cells. JSR-2 and JSR-4 both contain a powdered basal medium which requires different sterilization methods and utilize the same supplement of growth factors in a liquid form.

S. Kaminogawa et al. (eds.), Animal Cell Technology: Basic & Applied Aspects, Vol. 5, 397–401.
© 1993 Kluwer Academic Publishers.

We will compare the JSR-2 and JSR-4 media with serum-free media on the market in terms of the cell growth and productivity.

Materials and Methods

Cell Lines and Media

Mouse-mouse hybridoma 1 was used which secreted a monoclonal antibody (IgG) against tissue plasminogen activator and originated from the fusion of mouse myeloma cell lines NS-1 and Balb/c spleen cells from immunized mice; this hybridoma was kindly provided by Dr. Usuda (Institute of Immunis, Tochigi, Japan). NS-1 was obtaind from ATCC. CHO-K1 and CHO-K1(SC) (anchorage-independent CHO-K1) were obtaind from RIKEN Cell Bank (Tsukuba, Japan), and was maintaind in a static culture and in a suspension culture respectively.

These cells were subcultured every 3 or 4 days in RPMI1640 or Ham's F12 supplemented with 10% FBS under a humidified 5% CO_2 atmosphere.

Serum-free media (designated as A-I) were purchased from chemical and pharmaceutical companies.

Media Preparation

(1) JSR-2 : Powdered basal medium, JSR-2, HEPES and $NaHCO_3$ were dissolved in deionized distilled water and sterilized by filtration. A sterilized growth factor solution was added to the medium.

(2) JSR-4 : Powdered basal medium, JSR-4 and HEPES were dissolved in water and sterilized by autoclaving (121 C, 30min). L-Gln, $NaHCO_3$ and a growth factor solution sterilized by filtration were added to the medium.

(3) Serum-free media : All were prepared according to the appropriate manual.

Cell Counting and Monoclonal Antibody Determination

Cells were inoculated at 1 or 5×10^4 cells/ml in 35- or 60-mm dishes. The cell number was counted by a hemocytometer with or without trypan blue dye, all experiments being carried out in duplicate or triplicate. Mouse IgG in the culture supernatant was determined by the ELISA method.

Results

Culture of the Hybridoma Cell Lines

Both hybridoma cell lines grew well in the JSR-2 medium with no lag phase, the cell growth rates being more than 80% of those in RPMI1640 with 10% FBS. The cell growth behavior in the JSR-4 medium was different for each cell line. Hybridoma1 had a lag phase for the first 20 hrs after plating and then grew at the same rate as that in a serum-supplemented medium. Hybridoma2 grew slowly but steadily, and the cell number reached 5×10^5 cells/ml at 120 hrs, 70% of that in the H serum-free control medium with BriClone. (Figs.1-A and 2-A) The cell growth activity of the JSR-2 medium was better than that of the JSR-4 medium. The amount of antibody production was different according to the cell line and medium, that for hybridoma 1 in the JSR-4

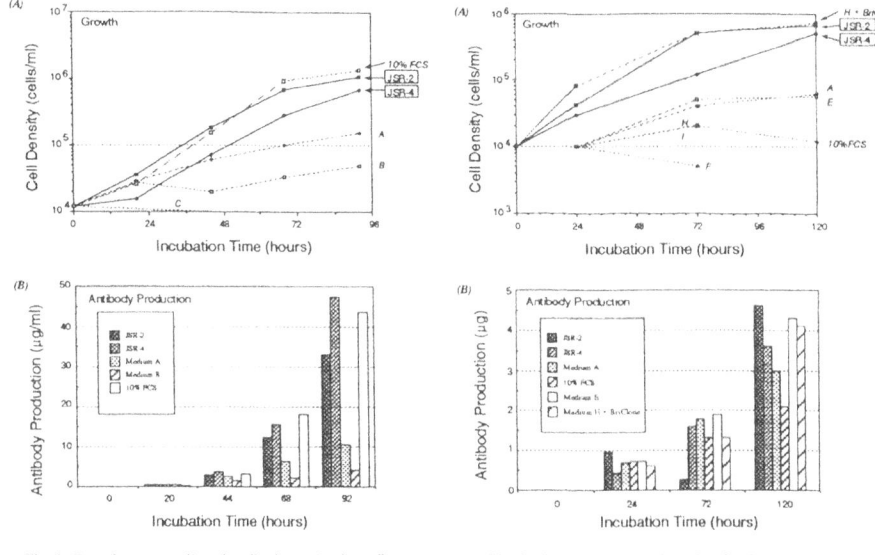

Fig. 1 Growth curves (A) and antibody production (B) for hybridoma cell line 1 in various serum-free media.

The antibody production in medium C was neglected.

Fig. 2 Growth curves (A) and antibody production (B) for hybridoma cell line 2 in various serum-free media.

The antibody production in media F, H and I was negligible.

Assayed by Dainippon Pharmaceutical Co. Ltd.
* "BriClone" : IL-6-containing culture supplement (Dainippon Pharmaceutical Co. Ltd)

medium being more than that in the serum-supplemented medium at 92 hrs. The antibody production of hybridoma2 in the JSR-2 medium was more than that in the control medium (H+BriClone) at 120 hrs.

The cells did not grow well in the serum-free media on the market, and the growth rates being less than 10% of that in the serum-supplemented medium. The viability of hybridoma1 in these media at 68 hrs was about 50%. On the other hand, the viability of hybridoma 1 in the JSR-2 and JSR-4 media at 68 hrs was more than 90%.

Culture of the NS-1 Cell Line

The growth rates of NS-1 in the JSR-2 and JSR-4 media were equivalent to that in RPMI1640 supplemented with 10% FBS (Fig.3), the viability being more than 95% at any stage.

In the serum-free media on the market, the cells did not grow well. The viability in media A and B decreased slowly and was about 50% at 68 hrs. In media C and D, the cells did not grow at all and gradually died.

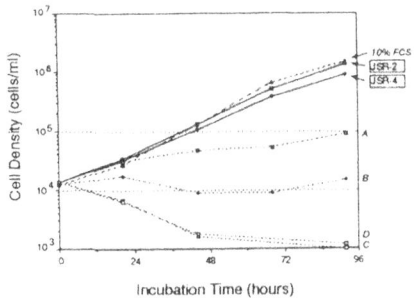

Fig. 3 Growth curves for the NS-1 cell line in various serum-free media.

400

Culture of the CHO Cell Lines

Two types of Chinese hamster ovary (CHO) cell lines, one being anchorage-dependent (CHO-K1) and the other anchorage-independent (CHO-K1(SC)), were tested.

CHO-K1(SC) grew in the JSR-2 medium as well as in serum-supplemented Ham's F12, the growth rate being 105%. In the JSR-4 medium, the growth rate was 70% of that in the serum-supplemented medium(Fig. 4). The growth rate of CHO-K1 in JSR-2 was about 30 % of that in Ham's F12 supplemeted with 10% FBS throughout the experiment (Fig. 5). No CHO-K1 attached to the dishes and grew in an aggregated form in JSR-2.

The growth of CHO-K1(SC) in the JSR-2 medium was separately examined in a spinner flask (100ml). CHO-K1(SC) grew as well as in the serum-supplemented Ham's F12 medium (data not shown).

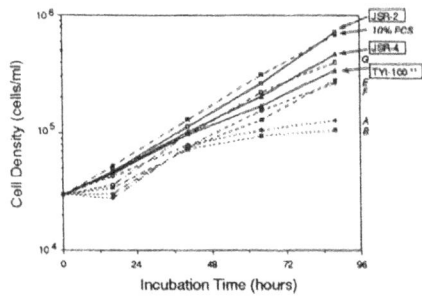

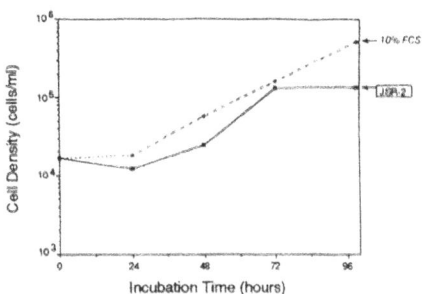

Fig. 4 Growth curves for theCHO-K1 (SC) cell line in various serum-free media.

Fig. 5 Growth curves for theCHO-K1 cell line

* CHO-K1 (SC) cells : RIKEN Cell Bank Cell No. RCB403
** TYI-100 : Liquid serum-free medium (Japan Synthetic Rubber Co. Ltd.)

Discussion

We developed two types of powdered serum-free media named JSR-2 and JSR-4. These media supported the cell growth of five cell lines without any weaning procedure.

The sensitivity of both the JSR-2 and JSR-4 media to cell growth appeared to be different with the two hybridomas. We consider that this difference resulted from the different medium requirements of each cell line.

The relationship between growth and productivity was also different between cell lines. With hybridoma 1, the cell growth in JSR-2 was better than that in JSR-4, but the productivity in JSR-4 was better than that in JSR-2. With hybridoma2, both the growth and productivity in JSR-2 were better than those in JSR-4. We think that the necessary components for growth and productivity were different between these cell lines.

In the serum-free media on the market, three cell lines (the two hybridomas and NS-1) did not grow well, the viability being very low. It is suggested that these media were each specific for a certain cell line and that weaning was needed.

The cell growth of the two CHO cell lines was quite different, the derived strains from CHO cells having a wide variety of medium requirements. JSR-2 and JSR-4 optimized for anchorage-independent cells successfully supported the growth of CHO-K1(SC), but needed the addition of an attachment factor for CHO-K1.

In the JSR-2 and JSR-4 media, various anchorage-independent cell lines, including the hybridomas, and anchorage-independent CHO cells (CHO-K1 (SC)) grew better than in any serum-free medium on the market without any weaning procedure.

Both these media are supplied in a powdered basal form and a small volume of sterilized growth factor solution is added. They are therefore convenient for delivery and storage, and can also be provided in a large quantity. One of the two basal media can be sterilized by filtration, and the other by antoclaving, so that they can be used for a wide variety of cell-culture methods. These powdered serum-free media are suitable for the production of pharmaceutical substances using animal cells.

Acknowledgements

The authors thank Dainippon Pharmaceutical Co. Ltd for evaluating the media and for useful discussions.

References

1) Mark C. Glassy, John P. Tharakan, Pao C. Chau (1988), in 'Serum-Free Media in Hybridoma Culture and Monoclonal Antibody Production,' Biotechnology and Bioengineering, Vol. 32, pp. 1015-1028.

SUSPENSION CULTURE OF MOUSE HYBRIDOMA HB8852 CELLS WITH A PROTEIN-FREE MEDIUM

Hiroshi Shinmoto and Shun'ichi Dosako
Technical Research Institute, Snow Brand Milk Products Co. Ltd.,
1-1-2 Minamidai, Kawagoe 350, Japan

A protein-free, ERDF-based medium was developed containing ethanolamine, sodium selenite, and inorganic iron as growth factors. Mouse hybridoma HB8852 cells were cultured in a spinner flask with the protein-free medium to evaluate the cell growth and monoclonal antibody production, which were comparable to those with a chemically defined, serum-free medium (ITES-ERDF) and a 10% FCS-ERDF medium. However, the cell viability was slightly lower. SDS-PAGE analysis of the spent, protein-free medium revealed the recovery of monoclonal antibodies in the medium with high purity. It is concluded that the medium is highly beneficial to produce and purify monoclonal antibodies in a large-scale mammalian cell culture.

Introduction

A serum-free, chemically defined medium is highly useful to culture mammalian cells and to produce biological materials such as monoclonal antibodies, lymphokines, and hormones[1-3]. We have recently developed a protein-free medium based on ERDF medium and containing ethanolamine (2-aminoethanol), sodium selenite, and an inorganic iron compound as growth factors, and then cultured mouse hybridoma HB8852 cells for 7 months in the medium [4]. The advantages of the protein-free medium are its low material and purification costs, making it potentially useful for a large-scale cell culture, and we evaluated the cell growth the monoclonal antibody production from a suspension culture.

Materials and methods
1. Mediaum and cell line
The chemically defined, protein-free medium was prepared according to the previous report [4], and contained 20 μM ethanolamine, 25 nM sodium selenite, and 80 μM

403

iron (II) sulfate as growth factors. Mouse hybridoma HB8852 cells secreting the anti-bovine lactoferrin antibody (IgG1 class) [5] were cultured with the protein-free medium for 3 months in 10-cm culture dishes (Falcon 3003), before the cells were transferred to a spinner flask for suspension culture.

2. *Suspension culture of HB8852 cells*
HB8852 cells were seeded into 200 ml of the protein-free medium, and for comparison into ITES-ERDF medium (ERDF medium [6] containing 5 µg/ml to insulin, 10 µg/ml of human transferrin, 20 µM ethanolamine, and 25 nM sodium selenite), and into 10% FCS-ERDF medium (ERDF medium containing 10% FCS; Flow Lab., USA) in a spinner flask (200 ml, Shibata, Japan) at a density of 2×10^5 cells/ml. Each was then cultured in a CO_2 incubator (5% CO_2) at a rotation speed of 100 rpm for 4 days. The viable cell density was counted by the tripan blue dye exclusion method, and the concentration of the monoclonal antibody (mouse IgG) was measured by an enzyme-linked immunosorbent assay (ELISA) [7].

3. *Polyacrylamide gel electrophoresis (SDS-PAGE) of the spent medium*
The spent medium harvested from the culture was concentrated ten times by ultrafiltration through Mol-Cut (Millipore, USA, molecular weight cut-off of 20 kDa). The concentrated medium was subjected to SDS-PAGE [8] under non-reduced and reduced conditions, using a 10-20% gradient of acrylamide gel (Daiichi Kagaku, Japan). The proteins on the gel were stained with Quick-CBB dye (Wako Pure Chemicals, Japan).

Results and discussion

1. *Long-term culture of HB8852 cells in the protein-free medium*
Figure 1 shows growth curves for mouse hybridoma HB8852 cells cultured in the protein-free medium and ITES-ERDF medium in a 25 cm^2 tissue culture flask (Corning 25102, USA). The cell density in the protein-free medium reached the same level as that in the ITES-ERDF medium.

The HB8852 cells were then serially cultured with the protein-free medium for three months in a 10-cm culture dish. As shown in Figure 2, the cells grew well in the medium without any significant change in the cell density during the three months of culture.

2. *Suspension culture of HB8852 cells*
After the three-month culture, the cells were transferred into a 200-ml spinner flask in order to evaluate the cell growth and antibody production. Figure 3 shows cell growth curves and viability levels in the protein-free medium (A), ITES-ERDF medium (B), and 10% FCS-ERDF medium (C). Although the cell density at day 4 in the protein-free medium was the highest, the cell viability was lower than that in the ITES-ERDF medium or 10% FCS-ERDF medium. The lower viability with the protein-free medium was due to the share stress from rotation, and it might be better to add a share stress protectant such as Pluronic F-68 [9].

The IgG concentration in the spent protein-free medium was higher than that in the ITES-ERDF medium or 10% FCS-ERDF medium (Table I). This is compatible to the

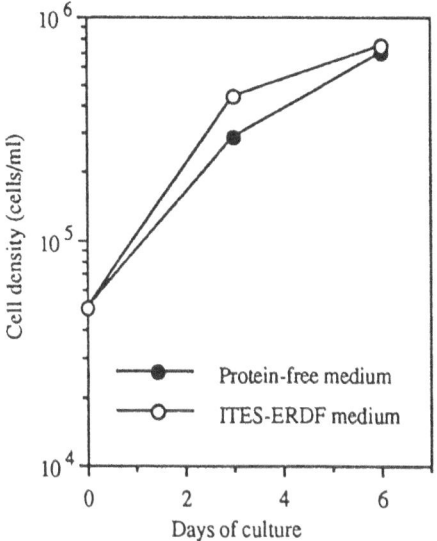

Figure 1. Growth curves for HB8852 Cells in the protein-free medium and ITES-ERDF medium.
Mouse hybridoma HB8852 cells were cultured with 10 ml of the protein-free medium or ITES-ERDF medium in 25 cm²-tissue culture flasks for 6 days.

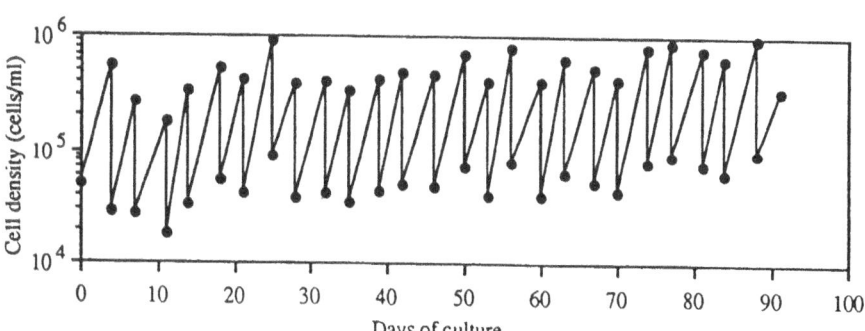

Figure 2. Long-term culture of HB8852 cells with the protein-free medium.
HB8852 cells were serially cultured in 10-cm culture dishes for three months.

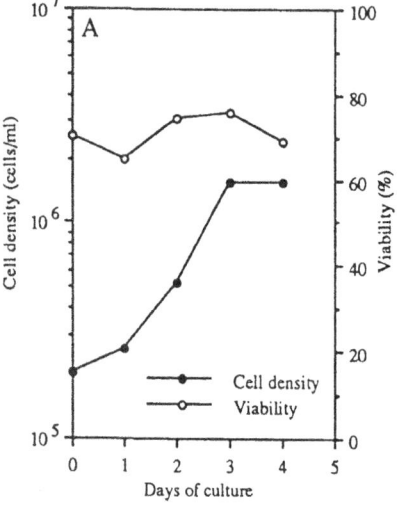

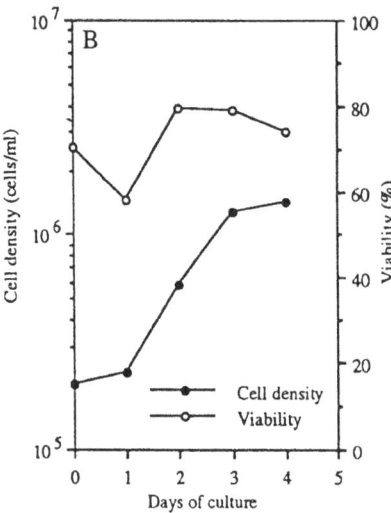

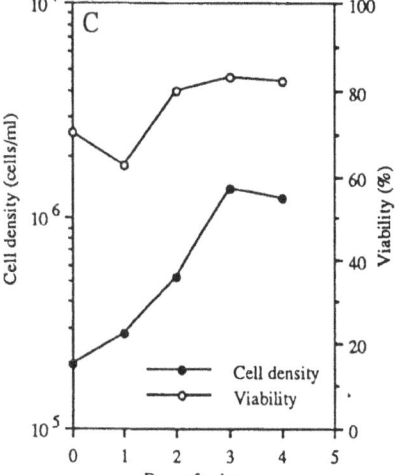

Figure 3.   Suspension culture of HB8852 cells in spinner flask.
HB8852 cells were seeded into 200 ml of the protein-free medium (A), ITES-ERDF medium (B), and 10% FCS-ERDF medium (C), and cultured in 200-ml spinner flasks for 4 days.

previous report and suggests the enhanced production of IgG in the protein-free medium.

The high IgG productivity in the protein-free medium is attributable to the lower cell viability in this medium. Hybridomas increase the specific antibody production rate when subjected to environmental stress [10], and HB8852 cells in the death phase account for the production of IgG. Although the viability of the HB8852 cells in the protein-free medium was approximately 70%, the HB8852 cells were continuously maintained by batch-fed culture in a spinner flask for two weeks (data not shown).

Table I. IgG concentrations of the spent media from the spinner culture at day 4.

Medium	IgG (µg/ml)
Protein-free	25.6
ITES-ERDF	13.4
10% FCS-ERDF	15.0

Figure 4. Polyacrylamide gel electrophoresis of the spent protein-free medium.
The spent medium from spinner culture of HB8852 cells (Fig. 3-A, day 4) was concentrated andsubjected to SDS- PAGE under non-reduced (lane 2) and reduced (lane 3) conditions.

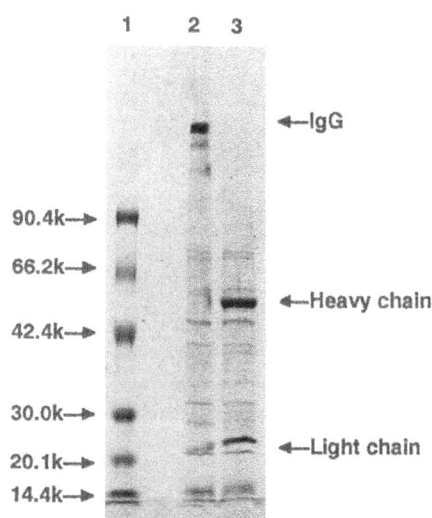

3. *SDS-PAGE analysis of the culture medium*
The spent protein-free medium from the suspension culture of HB8852 cells was subjected to SDS-PAGE analysis, Fig. 4 showing the SDS-PAGE patterns of the spent medium. The majority of the spent medium was found to be IgG (150 kDa), along with several minor bands. The purity of the IgG was approximately 80%,

based on an analysis with a densitometer. Thus, it was easy to purify monoclonal antibodies of interest from the protein-free medium.

From these results, we conclude that the protein-free medium we prepared would be useful for producing and purifying monoclonal antibodies from mouse hybridoma cells.

References
[1] H. Murakami, in Methods for Serum-Free Culture of Neuronal and Lymphoid Cells, Alan R. Liss, New York, pp.197-205 (1974).
[2] J. Kovar and F. Franek, Biotechnol. Lett., **9**, 259-264 (1987).
[3] N. Eto, K. Yamada, T. Shito, S. Shirahata, and H. Murakami, Agric. Biol. Chem., **55**, 863-865 (1991).
[4] H. Shinmoto, S. Dosako, and S. Taneya, Biotechnol. Lett., **13**, 683-386 (1991).
[5] H. Kawakami, H. Shinmoto, S. Dosako, and Y. Sogo, J. Dairy Sci., **7 0**, 752-759 (1987).
[6] H. Murakami, T. Shimomura, T. Nakamura, H. Ohashi, K. Shinohara, and H. Omura, Nippon Nogeikagaku Kaishi., **58**, 575-583 (1984).
[7] E. Engvall and P. Perlmann, Immunochemistry, **8**, 255-270 (1980).
[8] U.K. Leammli, Nature, **227**, 680-685 (1970).
[9] O.-W. Merten and J. Litwin, Cytotechnology, **5**, 69-82 (1991).
[10] S. Reddy, K.D. Bauer, and W.M. Miller, Biotechnol. Bioeng., **40**, 947-964 (1992).

IMPROVED MONOCLONAL ANTIBODY PRODUCTION VIA CONTROLLED FEEDING STRATEGIES DURING FEDBATCH CULTURES OF A HYBRIDOMA CELL LINE UTILIZING PROTEIN FREE MEDIA

W. Niloperbowo, L.K. Nielsen, S. Reid, P.F. Greenfield
Department of Chemical Engineering
The University of Queensland
St. Lucia - Qld 4072
Australia

ABSTRACT. A quantitative study on the effect of feed concentration and feed rate during a fedbatch cultivation system on the performance of the SPO1 hybridoma cell line using a protein free media is presented. The objective of this experiment was to study the kinetics of cell growth, antibody production and metabolism of this cell line. With an appropriate feeding strategy a long stationary condition of viable cell concentration was obtained and cultures fed with a high concentration of feed medium and fed at a low rate maintained a higher concentration of viable cells. Up to a four fold improvement in final antibody concentration was observed in fedbatch cultivation compared to the normal batch cultivation. However, a significant reduction in specific antibody productivity was observed during fedbatch cultivation and it was linearly correlated with the reduction in cell growth rate. During the cultivations, glutamine was found to be the limiting substrate, the consumption rate of this substrate was determined by its concentration and followed monod kinetics and the glucose consumption rate was influenced by the growth rate. It was observed that the yield coefficient of biomass on glutamine and glucose was constant throughout the cultivation period and the yield coefficient of ammonium on glutamine and lactate on glucose decreased as the concentration of ammonium and lactate increased leading to only a slow increase in the concentration of these metabolites during the stationary phase.

1. Introduction

For a commercial scale of monoclonal antibody production, the cost of production is crucial. One of the parameters that influences the cost is the concentration of antibody obtained at the end of the cultivation. A high concentration of product not only increases the yield per cultivation but also reduces the downstream processing costs (Hacking, 1986) which commonly dominate the production cost of a monoclonal antibody.

S. Kaminogawa et al. (eds.), Animal Cell Technology: Basic & Applied Aspects, Vol. 5, 409–415.
© 1993 Kluwer Academic Publishers.

Fedbatch cultivation. The fermenter used and physical parameters applied during fedbatch cultivations were the same as that in batch. The initial culture volume was 700 ml and feeding was initiated 50 hours after inoculation with half of the strength of the desired feed medium for 24 hours before shifting to the full strength of the feed. If necessary, every time the culture volume reached 1000 ml, 300 ml of the culture was withdrawn from the fermenter.

Analysis. About 1 ml of culture was taken every 12 - 24 hours. Cell enumeration was carried out with a dye exclusion method using a hemocytometer. The sample was then centrifuged at 2000 rpm and the supernatant was frozen at -20 °C for further analysis. Glucose and lactate were determined using HPLC, glutamine was determined using a glutamic acid assay kit (Boehringer Mannheim, Germany) and ammonium was determined using an ammonia probe (Orion, USA). Antibody concentration was determined using an ELISA.

3. Results

The growth of SPO1 cells in batch and fedbatch cultures are presented in Figure 1A. In batch culture the viable cell concentration peaked at about 1.6×10^6 cells per ml, 60 hours after inoculation, followed by a sharp decrease in cell concentration. By employing a feeding strategy, as described in section 2, the culture was successfully brought to a long stationary phase with a viable cell concentration ranging from 1.2×10^6 to 2.8×10^6 cells per ml depending on the feed rate and feed concentration employed.

A significant increase in the final antibody concentration was observed in fedbatch cultures compared to that seen in batch culture (Figure 1B). This finding is consistent with other reports (Luan et.al., 1987b; Glacken et.al., 1989; Trembley et.al., 1992).

As indicated in Figure 1C, the glutamine concentration in batch culture decreased from about 3.4 mM and was exhausted at about 60 hours after inoculation. In all fedbatch cultures the glutamine decreased from about 3.3 mM to a steady concentration (below 1 mM) during the stationary phase. The ammonium concentration, increased steeply during the early stage of all cultivations followed by a slow increase through to the end of the cultivations (Figure 1D). The glucose concentration decreased quite fast for all cultures from about 15.4 mM and reached its lowest concentration (0.5 mM) at about 70 hours after inoculation. Unlike glutamine, the glucose concentration gradually increased through to the end of the fedbatch cultivations (Figure 2A). Lactate accumulation in the culture medium showed a similar trend to that observed for ammonium (Figure 2B). The results of this study are summarized in Table 1.

For the cell line employed in these studies, a normal batch cultivation yields only about 35 g/ml of antibody by the end of the run. A number of methods can be applied to improve the production of monoclonal antibody in batch cultivation. One of the methods is enriching the culture medium by supplementing a single nutrient or a mixture of nutrients (Low and Harbour, 1985; Luan et.al., 1987a; Geaugey et.al., 1989; Jo et.al., 1990). However this method is limited by substrate inhibition. Previously it was found in our laboratory that the growth of SPO1 cells and their antibody production rate were inhibited in medium containing a five fold increase of nutrients compared to those found in DMEM : Ham's F12 (1:1) (Data to be published).

Such a limitation for cultivating hybridomas in a batch mode makes fedbatch systems more attractive. This Paper presents the results of fedbatch cultivations of SPO1 cells in protein free medium employing various feed concentrations and feed rates.

2. Materials and Methods

Cell line and culture conditions. The cell line used in this study was referred to as SPO1 cells. It is a fusion product of the SP2/O-AG-14 myeloma with mouse spleen cells, producing an IgG1 against paraquat. This cell line was a gift from Dr. S. Pond of Princess Alexandra Hospital, Brisbane - Australia. Culture medium used was DMEM : Ham's F12 (1:1) (Gibco BRL, USA) supplemented with glutamine to a final concentration of 4.5 mM, 0.1 % (v/v) component A of serum synthetic replacement (SSR A) (Medicult, Denmark), ethanolamine to a final concentration of 15 M and 1.2 g sodium bicarbonate for every liter of medium. pH of the medium was adjusted to 7.2 with sodium hydroxide followed by filter sterilization. For fedbatch cultivation three feed media were formulated ie : two (2x feed), four (4x feed) and eight (8x feed) times of the strength of DMEM : Ham's F12 (1:1), using 100x stock solutions of amino acids (without glutamine), Vitamins and salt (without sodium chloride). The mixture was then supplemented with glutamine, glucose, ethanolamine and SSR A according to the strength of the desired feed medium. These feed media were then supplemented with 1.2 g sodium bicarbonate per liter of feed medium and an appropriate amount of sodium chloride to adjust the osmolarity to about 300 mOsmol.

Batch cultivation. A normal batch culture was conducted as a control for the fedbatch cultures. This batch culture was carried out in a 2 L LH fermenter (LH Fermentation, England) with a one liter working volume. pH of the culture was maintained at 7.1 - 7.2 with 0.5 M sodium hydroxide, DOT was controlled at 30 % air saturation with air sparging, temperature was maintained at 37 °C and the culture was agitated at 60 rpm using a marine impeller.

412

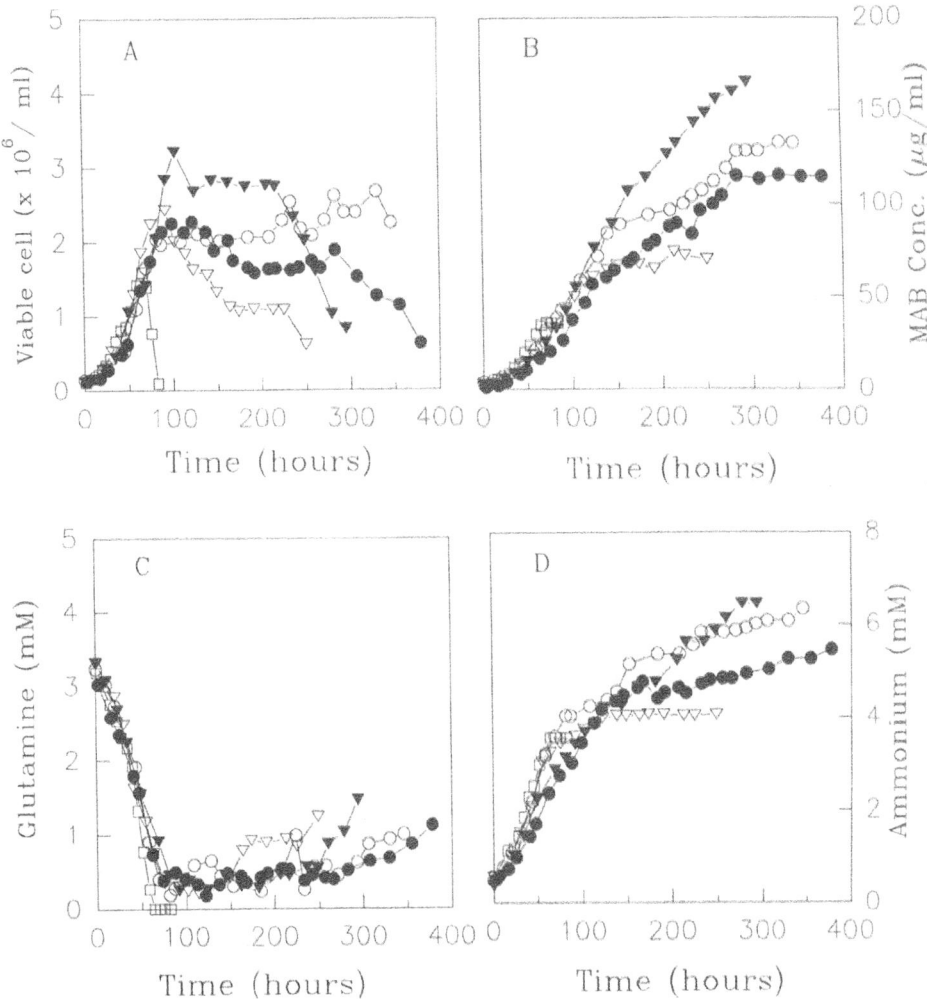

Figure 1. Concentration of viable cells (A), antibody (B), glutamine (C) and ammonium (D) observed in Batch (□), F1 (4x feed, 4 ml/hr) (○), F2 (4x feed, 2 ml/hr) (●), F3 (2x feed, 4 ml/hr) (▽) and F4 (8x feed, 2ml/hr) (▼).

Table 1. Growth, antibody production and metabolic parameters observed in batch and fedbatch cultivations of SPO1 cells

Parameter	Batch	2x feed (4 ml/h)	4x feed (2 ml/h)	4x feed (4 ml/h)	8x feed (2 ml/h)
Run time (h⁻¹)	104	249	378	346	294
[a]Viable cell conc. (x 10^6 per ml)	1.6	1.2	1.8	2.2	2.8
[a]Growth rate (h⁻¹)	0.040	0.013	0.012	0.013	0.014
Final MAB conc. (g/ml)	35.42	70.16	114.37	132.85	165.35
Vol of Culture (ml)	1000	1300	1000	1600	1000
Total MAB obtained (mg)	35.42	94.15	114.37	212.56	165.35
Final NH^+_4 conc. (mM)	3.78	4.06	5.46	6.34	6.45
Final Lactate conc. (mM)	28.35	22.54	38.12	37.20	35.74

a) The value is the average during the growth phase (batch) or stationary phase (fedbatch)

4. Discussion

As indicated in Figure 1A, the growth of SPO1 cells in batch culture ceased about 65 hours after inoculation due to glutamine exhaustion. Feeding the culture with an appropriate feeding strategy successfully extended the growth of SPO1 cells in culture. The concentration of viable cells during the stationary phase was influenced by the concentration of feed and the feed rate. Where the culture was fed with a high concentration of feed medium and fed at a low rate it exhibited a high viable cell concentration. It was observed that the rate of cell growth decreased from about 0.055 per hour at the early stage of the cultivation to about 0.013 per hour during the stationary phase and this drop was determined by the concentration of glutamine in the medium. It was observed in all of the fedbatch cultures that the stationary phase was achieved following a sudden jump in the cell death rate at about 100 hours after inoculation (40 hours after feeding was initiated). The cell death rate jumped from about 0.005 per hour to about 0.030 per hour at this time and then fell back to about 0.013 per hour through to the end of the cultivation. It is suggested that this event occurred as a result of an insufficient supply of nutrients to support a further increase of cell growth, thus the culture corresponded by matching the viable cell concentration to the rate of nutrient supply.

414

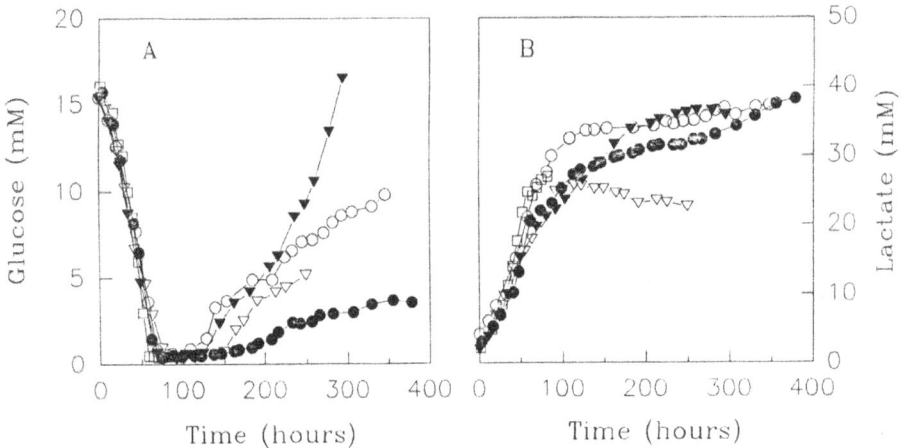

Figure 2. Concentration of glucose (A) and lactate (B) observed in F1 (4x feed, 4ml) (○), F2 (4x feed, 2 ml/hr) (●), F3 (2x feed, 4 ml/hr) (▽), F4 (8x feed, 2 ml/hr) and Batch (□) culture.

Despite a significant improvement in the final antibody concentration obtained in the fedbatch culture, the result was not as expected. It was found that the specific antibody production in all of the fedbatch cultures decreased from about 0.7 $g/10^6$ viable cells per hour at the beginning of the cultivation to about 0.3 $g/10^6$ viable cells per hour during the stationary phase. A number of workers have suggested that such a reduction is associated with genetic drift (Ozturk and Palsson, 1990; Lee et.al., 1991). However, when a sample was taken from the fedbatch culture, centrifuged, resuspended in a fresh DMEM : Ham' F12 (1:1) medium and grown in batch mode, it exhibited a similar result to that of a normal batch cultivation which suggests that the reduction in antibody productivity was not associated with genetic drift. Furthermore, it is unlikely that the reduction was caused by a limitation of nutrients other than glutamine since it was observed that other essential amino acids and other nutrients remained present in a reasonable ratio compared to glutamine (data not shown). It was observed that the reduction of antibody productivity coincided with the reduction of cell growth and it is suggested that the antibody production kinetics of SPO1 cells is positively growth associated.

It was observed that the rate of glutamine consumption decreased as the concentration of this substrate in the medium decreased and it followed monod kinetics. Furthermore it was found that the yield coefficient of ammonium on glutamine decreased as the concentration of ammonium in the medium increased.

This resulted in a slow accumulation of ammonium during the stationary phase. The rate of glucose consumption appeared to be determined by the cell growth rate. Despite an increase in the glucose concentration during the stationary phase, the glucose consumption rate remained low. Also it was found that the yield coefficient of lactate on glucose decreased as the concentration of this metabolite increased which lead to a slow accumulation of lactate during the stationary phase. It was found that the yield coefficient of biomass on glutamine and glucose was constant throughout the fedbatch cultivation period.

5. References

Geaugey, V., Duval d., Gahael I., Marc a. and Engasser J.M. (1989) 'Influence of amino acids on hybridoma cell viability and antibody secretion'. Cytotechnology, 2, 119-129.

Glacken. M.W., Huang C. and Sinskey A.J. (1989) ' Mathematical description of hybridoma cell culture kinetics : III Simulation of fedbatch bioreactors', Journal of Biotechnology, 10, 36-66.

Hacking J.A. (1986) Economic Aspect of Biotechnology, Cambridge University Press, Cambridge.

Jo E., Park H. and Kim K. (1990) 'Balance nutrient fortification enables high density hybridoma cell culture in batch culture', Biotechnology and Bioengineering, 36,717-722.

Lee G.M., Varma A. and Palsson B.A. (1991) 'Application of population balance model to the loss of hybridoma antibody productivity', Biotechnology Progress, 7, 72-75.

Low K. and Harbour C. (1985) 'Growth kinetics of hybridoma cells : (2) The effect of varying energy source concentration', Developmental Biology Standardization, 60, 73-79.

Luan Y.T., Mutharasan R. and Magee W. (1987a) 'Effect of various glutamine and glucose ratios on hybridoma growth, viability and monoclonal antibody formation', Biotechnology Letter, 9, 535-538.

Luan Y.T., Mutharasan R. and Magee W. (1987b) 'Strategies to extend longevity of hybridomas in culture and promote yield of monoclonal antibodies, Biotechnology Letter, 9, 691-696.

Ozturk S.S. and Palsson B. (1990) 'Loss of antibody productivity during long term cultivation of a hybridoma cell line in low serum and serum free media', Hybridoma 9/2,167-175.

Trembley M., Perrier M., Chavarie C. and Archambault J. (1992) 'Optimization of fedbatch culture of hybridoma cells using dynamic programming : Single and multi feed cases', Bioprocess Engineering, 7, 229 - 234.

PREDICTION OF DISRUPTION OF ANIMAL CELLS BY LAMINAR AND TURBULENT FLOWS

Z. Zhang, C. Born, M. Al-Rubeai and C. R. Thomas
SERC Centre for Biochemical Engineering
School of Chemical Engineering
University of Birmingham
Edgbaston, Birmingham B15 2TT, UK

ABSTRACT. Models which describe cell-hydrodynamic interactions have been developed for estimation of cell disruption in laminar and turbulent flows. Mechanical properties of animal cells were measured by micromanipulation for use in these models. Cells were disrupted in laminar flow in a cone and plate viscometer, and in turbulent flows in capillaries. Predictions of the models have been shown to be in good agreement with measured losses of viable cell numbers in these flow devices.

Introduction

The fragility of animal cells often limits the design and operation of suspension cultures at high cell densities. Many studies on cell damage in bioreactors and various simple flow devices such as laminar cone and plate viscometers and turbulent capillary tubes have been reported and have been reviewed recently by Papoutsakis (1991). Although these studies suggest some factors which may determine cell damage, it is difficult to use this information to estimate whether disruption can be expected for a particular culture in a given flow field. Ideally, a mechanistic model is needed to describe cell-hydrodynamic interactions so that cell disruption can be predicted using information about the hydrodynamics and about cell mechanical properties, such as cell bursting membrane tension, elastic area compressibility modulus and cell size. These mechanical properties can be determined by a recently developed micromanipulation technique and cell model (Zhang et al, 1991, 1992). Cell-hydrodynamic models have been developed to predict disruption of cells in laminar flows in a cone and plate viscometer and in turbulent flows in capillaries. To validate the models, cells in suspension were disrupted in these devices, and predicted and measured damage were compared. Validation of the model would demonstrate the potential usefulness of micromanipulation in prediction of cell damage in bioreactors.

S. Kaminogawa et al. (eds.), Animal Cell Technology: Basic & Applied Aspects, Vol. 5, 417–423.
© 1993 *Kluwer Academic Publishers.*

The Model of Cell Disruption in Laminar Flow

In laminar flow, a cell is modelled as if it behaves like a drop in an analogous two liquid phase emulsion (Born et al, 1992). It is assumed that the original cell radius equals the undeformed drop radius, the membrane tension of the cell corresponds to the surface tension of the drop, and the equations proposed by Taylor (1934) for describing small deformation of drops under shear can also be applied to calculate the cell distortion D,

$$D = \xi \tau \frac{R}{T} \tag{1}$$

where τ is the shear stress, R is the original cell radius, T is the membrane tension, and ξ is a very weak function of the ratio of the internal viscosity of cell to the viscosity of the suspension.

For a given applied shear stress, a cell will burst if

$$T \geq T_B \tag{2}$$

where T_B is the cell bursting membrane tension. Using equations (1) and (2), and by considering the Gaussian distributions of cell bursting membrane tension and radius, the percentage of cell disruption in laminar flow can be predicted (Born et al, 1992).

The Model of Cell Disruption in Turbulent Capillary Flow

Suppose a cell suspension passes through a pipe in turbulent flow. The flow field can be considered to consist of eddies of different sizes. Cells can interact with eddies of similar or smaller sizes, resulting in velocity differences across a given cell, which will be deformed locally. These deformations will cause an increase in cell surface area and consequently an increase in membrane tension. The cell disruption mechanism is presumed to be that a cell will be disrupted if its bursting membrane tension is exceeded. Suppose a cell behaves like a liquid drop in that both of them accept energy from surrounding eddies, but the former stores the energy in the form of its surface energy, whilst the latter manifests it in its kinetic energy. Suppose therefore that the surface energy of a cell is equal to the kinetic energy of a liquid drop of the same size. The cell disruption condition is then that a cell will burst if its surface energy, i.e. the kinetic energy of a liquid drop of the same size, E_k, exceeds the cell bursting surface energy E_{sb}

$$E_k \geq E_{sb} \tag{3}$$

The bursting surface energy of a cell can be expressed as

$$E_{sb} = T_B A \tag{4}$$

The velocity difference between any two points within a drop might be described by equation (5), for the inertial convention subrange (Hinze, 1955):

$$\overline{u^2}(r) = 2.0(\varepsilon r)^{2/3} \tag{5}$$

where ε is energy dissipation rate and r is the distance between the two point. The kinetic energy of the liquid drop can be estimated by integrating across all scales up to the radius of the cell:

$$E_k = \frac{1}{2}\rho\int_0^R \overline{u^2}(r)4\pi r^2 dr = \frac{12\pi}{11}\rho\varepsilon^{2/3}R^{11/3} \tag{6}$$

where ρ is the specific density of fluid. The cell bursting condition in turbulent capillary flows can therefore be written as:

$$\frac{3\rho\varepsilon^{2/3}R^{5/3}}{11T_B(1+T_B/K)} \geq 1 \tag{7}$$

The percentage of cells in a suspension, which can be disrupted in a single pass through a capillary with a given energy dissipation rate ε, i.e. the specific lysis rate (McQueen et al, 1987) can therefore be calculated by considering the probability distribution of cell mechanical properties.

In capillary turbulent flow, there exists a laminar sublayer close to the pipe wall. Normally the thickness of the laminar sublayer is less than the cell diameter. Here it is supposed that no cells can be entrained into this region, but that a significant proportion of the energy dissipation can occur there. The energy dissipation rate in equation (7) can be determined from pipe-flow theory after allowing for the laminar sub-layer.

Materials and Methods

CELL DISRUPTION IN LAMINAR FLOWS

TB/C3 murine hybridomas producing antibodies against human IgG were used for cell disruption studies in laminar flows. For all shear experiments a Contraves Rheomat 30 rotaviscometer with a cone and plate shear device was used. To achieve a laminar stress sufficient to disrupt the cells dextran was added to the cell suspensions. The mechanical properties of these resuspended cells were then examined by micromanipulation. Several samples of 500 μL were then sheared at shear stresses up to 600 N/m^2 for times up to 180 s.

CELL DISRUPTION IN TURBULENT CAPILLARY FLOWS

NS1 myeloma and TB/C3 hybridoma cells were used. The flow apparatus consisted of a pump, a holding flask and a capillary test section connected into the loop wide bore tubing. Flow rates through the capillary tubes were set to ensure that the Reynolds number was always more than 3500. The cell suspension was circulated around the flow loop for less than one and half hours, the duration of the experiments. The mechanical properties of the cells in the holding flask were measured by micromanipulation at room temperature before and after the pumping.

The actual cell losses in both systems were determined at several times by counting cells in samples, using Trypan Blue stain and a hemocytometer. The actual specific lysis rate in

420

capillaries was calculated based from the relationship between the remaining viable cell concentration and the number of passes (McQueen et al, 1987).

RESULTS AND DISCUSSION

Fig. 1 shows the dependence of total and viable cell concentration of the hybridomas against time of laminar shearing, for a shear stress capable of causing some cells to burst. Conceptually, cell disruption under shear in the cone and plate viscometer should occur as soon as the vulnerable cells are sufficiently distorted. However, it takes a few seconds for the viscometer to accelerate to its eventual steady rotational speed after it is switched on. Cell deformation to equilibrium or to disruption will also take some time. It was not expected therefore that the disruption would be instantaneous, and in practice a first measurement of damage at 15 s was considered to be reasonable. Fig. 1 clearly shows a very rapid and large initial loss of the cell viability and number. The predictions of the model were tested in a series of experiments using different shear stresses. Cells from semicontinuous cultures were subjected to shear stresses between 124 and 577 N/m^2 for 180 s. Fig. 2 presents the actual and the predicted cell losses for these experiments. The linear regression line, forced through the origin, has a slope of 1.06 and correlation coefficient of 0.97. Clearly this is an excellent correlation between actual and predicted damage, although the latter is some 6% on average higher than the former. In any case the agreement is remarkable, and validates the micromanipulation technique, the cell model, and the mode of cell-laminar shear flow interactions.

Figure 3 shows the fraction of viable hybridoma cell remaining in the holding flask versus number of passes through a turbulent capillaries. The predicted cell disruption and actual cell disruption are compared in Fig. 4. The energy dissipation rate ranged from 800 to 20000 m^2s^3. It can be seen that the predicted cell disruption is about 15% less than the actual disruption. Nevertheless, this result demonstrates that the present approach to modelling cell-hydrodynamic interactions in turbulent flows is reasonable.

Conclusion

Models for prediction of cell disruption in laminar flows and turbulent capillary flows have been proposed. The model for the former can predict cell losses within a maximum error of 30% and the model for the latter underestimated the cell disruption on average by about 15%. The success of the models implies that they might be extended to predict cell disruption in bioreactors, once adequate knowledge about the spatial distribution of energy dissipation rates and circulation times in stirred vessels, and about the hydrodynamics of bubble bursting, is available.

Acknowledgements

This research was supported by the Science and Engineering Research Council, UK.

421

References

Born, C., Zhang, Z., Al-Rubeai, M. and Thomas, C. R. (1992) `Estimation of disruption of animal cells by laminar shear stress', Biotechnol. Bioeng. 40, 1004-1010.

Hinze, J. O. (1955) `Fundamentals of the hydrodynamic mechanism of splitting in dispersion processes', AIChEJ 1, 289-295.

McQueen, A., Meilhoc, E. and Bailey, J. E., (1987) `Flow effects on the viability and lysis of suspended mammalian cells', Biotechnol. Letts. 9, 831-836.

Papoutsakis, E. T. (1991) `Fluid-mechanical damage of animal cells in bioreactors', Trends Biotechnol. 9, 427-437.

Taylor, G. I. (1934) `The formation of emulsions in definable fields of flow', Proc. Roy. Soc. 146, 501-525.

Zhang, Z., Ferenczi, M. A., Lush, A. C., Thomas, C. R. (1991) `A novel micromanipulation technique for measuring the bursting strength of single mammalian cells', Appl. Microbiol. Biotechnol. 36, 208-210.

Zhang, Z., Ferenczi, M. A., Thomas, C. R., (1992) `Micromanipulation technique with theoretical cell model for determining mechanical properties of single mammalian cells', Chem. Eng. Sci. 47, 1347-1354.

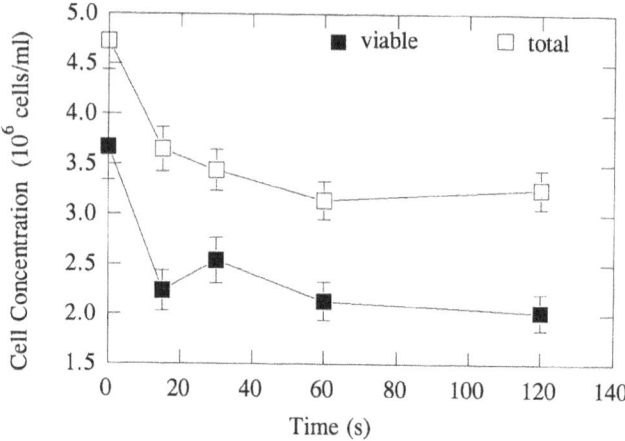

Fig. 1 Total and viable cell concentrations of TB/C3 hybridomas versus duration of shear in a cone and plate viscometer (shear stress 208 N/m^2). The error bars represent the 95% confidence intervals.

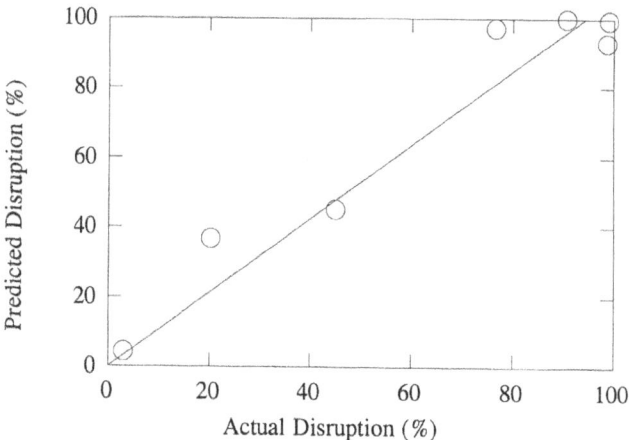

Fig. 2 Comparison between predicted and actual disruption after 180 s exposure to shear stresses from 124 to 577 N/m^2 in a cone and plate viscometer.

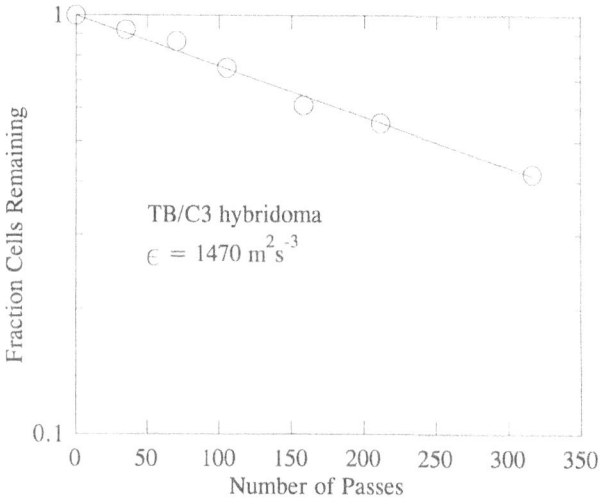

Fig. 3 Concentration of viable hybridoma cells versus number of passes in a turbulent capillary.

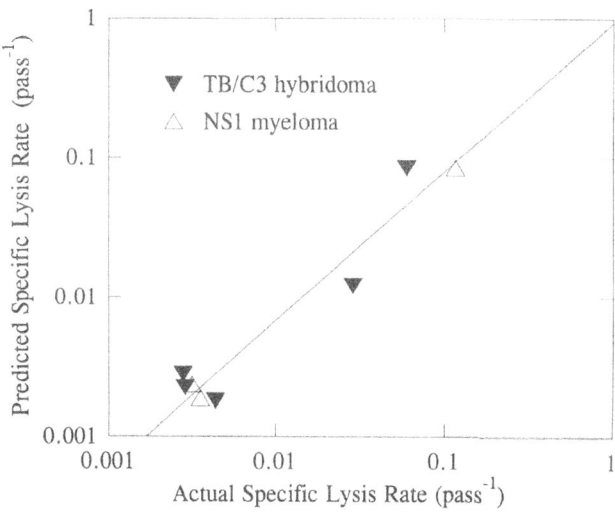

Fig. 4 Comparion between predicted and actual disruption in turbulent capillaries; mean energy dissipation rate ranged from 800 to 20000 m^2s^{-3}.

FED-BATCH PROPAGATION OF MOUSE HYBRIDOMA AND HUMAN LYMPHOBLASTOID CELL LINES (LCLs)

Marquis C.P.[1], Barford, J.P.[1] & Harbour, C.[2]

Dept. Chemical Engineering[1] & Dept. Infectious Diseases[2]
University of Sydney, NSW 2006 AUSTRALIA

ABSTRACT

Data is presented for fed-batch cultures of two different mammalian cell lines, producing IgG monoclonal antibodies. Supplementation of the hybridoma cultures by controlled pulse feeding of glucose and glutamine alone did not enhance antibody production. Use of more complex pulse feeding elevated production by up to 6-fold. Also described is a constant dilution rate fed-batch of the mouse hybridoma. Similar pulse feeding experiments of the human LCL did not result in elevated antibody production.

INTRODUCTION

Efficient methods for large scale cultivation of animal cells are required to meet current and projected demands for mammalian derived biologicals. Recently reactor design and operation research has focused largely on small volume-high cell density systems to achieve this. Developments have included hollow fibre and ceramic supports as well as modifications to fermenters with cell retention systems added such as spin filters to enable cell concentration. Disadvantages of these systems include difficulties in process control, heterogeneity in cell environment (and thus potentially in product) and the start-up expense of some commercially available systems. Comparatively low productivities in low-density large-volume bioreac·o·ᵤ may in some cases be improved by adoption of appropriate feeding strategies. Homogenei⁺y of cell environment is also achievable, enabling fundamental studies of cell behaviour under process conditions and facilitating scale-up.

An appropriate feeding strategy should ideally fulfill the following objectives:

1. Lower the production of metabolites that compromise cell viability and/or specific productivities via some sort of catabolite repression

2. Maintain nutrients at appropriate levels

3. Increase specific productivity of the desired product and yield with respect to expensive medium components

4. Minimise product dilution (per total volume and per unit of undesired protein)

5. Be applicable to large-scale commercial application.

425

S. Kaminogawa et al. (eds.), Animal Cell Technology: Basic & Applied Aspects, Vol. 5, 425–431.
© 1993 Kluwer Academic Publishers.

Simple feeding of glucose and glutamine was previously reported to be beneficial (Luan et al., 1987) and more recently pulsing of large quantities of glutamine have proved stimulatory (Flickinger et al., 1991, Omasa et al., 1992). As an alternative to pulse feeds, continuous feeds may be employed using programmed or controlled feeds. Improved process productivity was achieved by developing a feeding strategy based on optimal control theory (Glacken et al., 1989).

A survey of continuous culture data (Phillips et al., 1991) indicates that many mammalian cell lines may exhibit a saturated or even negatively growth associated profile of specific antibody production rate with respect to growth rate (Suzuki and Ollis, 1989). Strategies could be developed to exert physiological control and impose sub-maximal growth rates to exploit this property, via nutrient limitation.

In this work, stirred flask experiments were performed using both cell lines and a variety of feeds with the aim of improving antibody yield. A controlled fermentation is described where a constant dilution rate was used in an attempt to maintain the cells in a "quasi" steady-state at sub-maximal growth rates, to achieve higher productivities.

MATERIALS & METHODS

Cells

Experiments were performed using a mouse hybridoma, AFP-27 (a kind gift from Prof. W-S Hu, University of Minnesota), producing an IgG antibody to human a-fetoprotein and a human lymphoblastoid cell line (a kind gift from Dr. Ann Fletcher, Red Cross Blood Transfusion Service, Sydney), producing an IgG monoclonal antibody to human Rh^+ red blood cells.

Media

The mouse hybridoma was routinely cultivated in a 1:1 mixture of glucose- and glutamine free DMEM and RPMI (Cytosystems, Sydney) with 2% foetal calf serum (CSL, Parkville). Glutamine (Sigma) and Glucose (Ajax, Sydney) were supplemented as indicated. Feed supplements also included MEM amino acids (Sigma), MEM vitamins (Gibco) and foetal calf serum. The composition of the feeds used in the various experiments is summarised in Table 1. The amount fed on each occasion was based on the result of glucose measurements after sampling. The LCL was routinely cultivated in the same basal medium, supplemented with insulin and transferrin (10 mg/L, Sigma), a linoleic acid-bovine serum albumin complex (1:100, 0.5 g/l) and trace elements including selenium.

Bioreactors

Experiments described here were performed in magnetically stirred, 250 ml flasks (Corning, NY)and a 5L stirred tank reactor equipped with 2 top driven marine impellers and a pH and dissolve oxygen probe (Ingold, Switz.). The stirred flasks were pulse fed and the stirred tank was fed using a peristaltic pump (Watson-Marlow, UK). The stirred tank was monitored and controlled using a computer-controlled FC-4 data logger (Real Time Engineering, Sydney).

Table 1   Feed compositions

Expt	Feed	Gluc. (mM)	GLN (mM)	AA	Vit	I,T,S	FCS
AFP flask 1	1	50	0	0	0	0	0
	2	0	20	0	0	0	0
	3	50	20	0	0	0	0
AFP flask 2	1	50	20	4X*	2X	0	2%
	2	50	20	4X	2X	2X	0
	3	200	80	16X	8X	5X	0
AFP reactor	initial	0	60	0	0	0	0
	main	50	20	2X	1X	0	2%
LCL flask	1	50	20	4X	2X		
	2	50	20	4X	2X	2X	
	3	50	20	4X	2X		2%

* indicates extent to which normal concentration was exceeded

Assays

Glucose and lactate were measured using commercially available enzymatic kits (Roche and Behring). Ammonia was measured using an enzymatic method (Kun & Kearney, 1985). Total immunoglobulin was measured using an ELISA technique. Amino acids were quantitated by HPLC. Briefly, dried deproteinated samples were reacted with N,N-diethyl-2,4-dinitro-5-fluoroaniline (FDNDEA) according to Fermo et al. (1988). The derivitised amino acids were detected at 360 nm after gradient elution through a C-18 column (Beckman, Ultrasphere).

RESULTS

Figures 1 and 2 compare growth and antibody production of the mouse hybridoma in batch (25 mM glucose, 4 mM glutamine) to flasks where pulse feeding was performed approximately every 12 hours, according to measured glucose levels. The strategy was to return glucose levels to 5 mM and to feed glutamine based upon yield coefficients obtained previously (Frame & Hu, 1990). Growth profiles are seen to be almost identical, with the terminus in viable growth corresponding to both glutamine and tryptophan limitation. In the flask where glucose only was fed, antibody production ceased at 67 hours.

Figures 3 and 4 indicate the result for the same cell line when more complex feeds were utilised. In all three, maximum viable cell levels were higher, the cells plateaued for longer and declined in viability more slowly than batch cultures. This is despite ammonia levels rising to levels significantly higher than maximal batch concentrations of approximately 3.5 mM (figure 9).Total antibody production (volume X concentration) reached levels above 30 mg (for a 100 ml culture), which is six times greater than for a batch culture of the same volume.

Figures 6 and 7 indicate the performance of the mouse hybridoma in a controlled reactor. Feeding was commenced at 60 h and ramped up every 2 hours to impose a constant dilution rate of 0.017 +/- 0.001 h^{-1}. As indicated in figure 7, viable cell levels plateaued for approximately 45 hours. Conditions at this "quasi-steady state" are summarised below in table 2.

Table 2 Quasi steady-state conditions in reactor experiment

Variable	Level	Variable	Level
Viable cells, Xv (* 10^9/L)	1.64 +/- 0.20	Glucose (mM)	25.7 +/- 2.8
Total cells, Xt (*10^9/L)	1.92 +/- 0.21	Lactate (mM)	23.1 +/- 2.7
Antibody (mg/L)	44.5 +/- 3.8		

Glutamine levels increased during this period, indicating that it was not limiting and ammonia levels steadily rose from a normal batch level of 3.7 mM to 4.9 mM, presumably due largely from an increased contribution from spontaneous glutamine degradation.

Figures 7 and 8 compare batch and fed-batch cultures of the human LCL. No improvement in cell levels or antibody production was observed by the use of this pulse feeding technique. Feeding of glucose and glutamine alone resulted in significantly lower antibody and cell levels, due possibly to dilution of other components.

DISCUSSION

The significance of waste product accumulation versus nutrient limitation in batch cultures of mammalian cells is not fully understood. The flask culture experiments here are largely supportive of findings by Luan et al., (1987), that pulse feeding of nutrients extends cell viability and protein productivity, despite accumulation of ammonia to 2-3 times normal batch levels. Some productivity improvement is observed with incorporation of amino acids and vitamins alone (results not shown), but the best result is obtained when serum or serum-substitutes are also included. The most likely reason is the replenishment of heat or pH labile proteins required for growth stimulation. The results suggest that in any case, the most expensive medium components such as serum, insulin or transferrin would be most efficiently used by gradual or periodic addition.

Continuous culture studies indicate that many hybridoma cell lines exhibit either a saturated or negatively associated specific antibody productivity with respect to dilution rate. An explanation for higher specific productivities at low growth rate has been that under this condition, significantly higher proportions of the cell population are held in G1 phase, normally associated with protein synthesis (Suzuki & Ollis, 1989). It was felt that by maintaining growth at lower than maximum growth rates, by implementation of an exponential feed may result in higher specific productivities. However, in this experiment, specific productivity almost halved during exponential feeding. Antibody levels rose from 50 to 78 mg during the second batch phase, after feeding had ceased.

Finally, it was felt during development of feeding strategies it was important that they should be generally applicable. The human LCL has quite different growth characteristics to mouse hybridoma. As evidenced here, feeding could at best only emulate normal batch production levels. It is unlikely that nutrient limitation is the dominant effect for this cell line and perhaps some metabolite or protein is preventing viability and productivity enhancement.

CONCLUSIONS

The productivity of a normal batch culture of a mouse hybridoma could be greatly improved by implementation of pulse feeding. This productivity gain was not observed for serum-free cultures of a human lymphoblastoid cell line. Controlled exponential feeding did not enhance production to the extent of pulse feeding. These findings suggest that fed-batch culture may in some cases provide an economic industrial process for mammalian cell cultures, depending on cell line characteristics. Care however should be made in applying strategies to new or uncharacterised cell lines. Economic utilisation of the most expensive medium components is best achieved with feeding over the whole culture period.

430

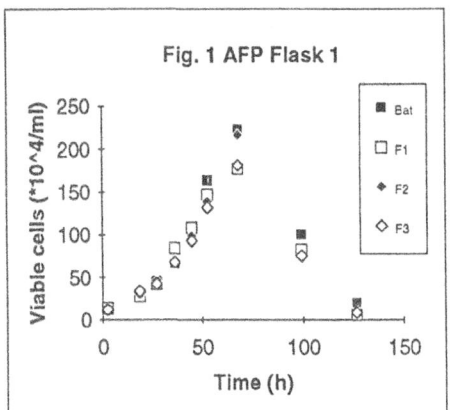

Fig. 1 AFP Flask 1

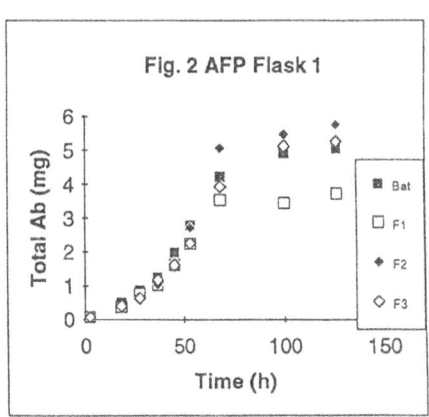

Fig. 2 AFP Flask 1

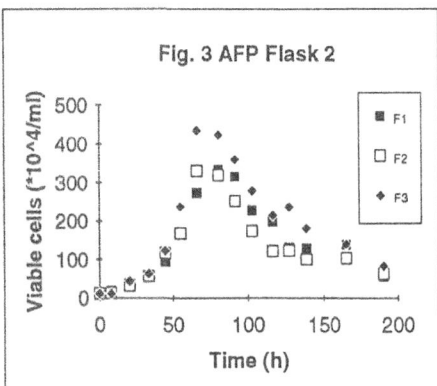

Fig. 3 AFP Flask 2

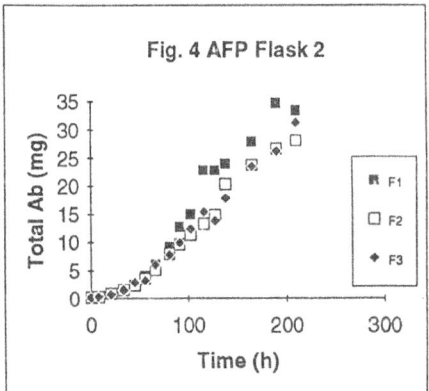

Fig. 4 AFP Flask 2

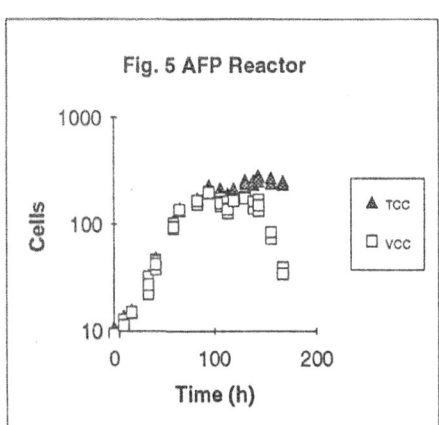

Fig. 5 AFP Reactor

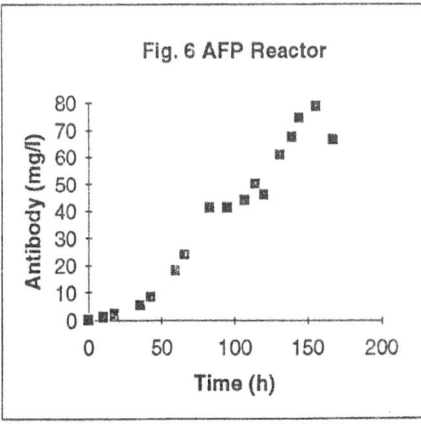

Fig. 6 AFP Reactor

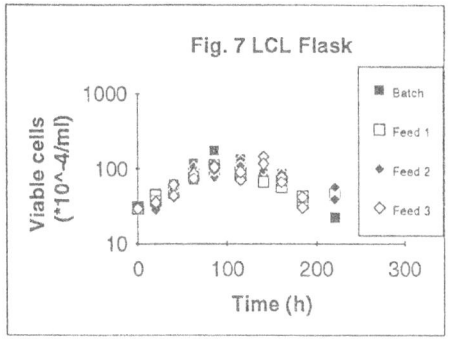

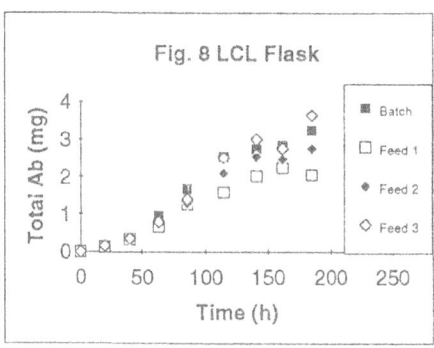

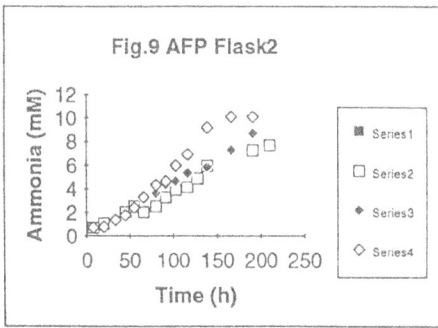

REFERENCES

Fermo,I., Rubino,F.M., Nolzacchini,E.,Arcelloni,C. & Bonini,P.A. (1988)
J. Chromatogr. Biomed. Appl. 433:53-62

Flickinger, M.C., Goebel, N.K., Bibilia, T. & Boyce-Jacino, S. (1992) J. Biotechnol. 22:201-226

Frame,K.K. & Hu,W-S. (1991) Enz. Microb. Technol. 13:690-696

Glacken,M.W., Huang,C. & Sinskey,A.J. (1989) J. Biotechnol. 10:39-66

Kun,E. & Kearney,E.A. (1985) in Methods of Enzymatic Analysis 3rd Edn. Vol 3.: 1803-1806

Luan,Y.T., Mutharasan,R. & Magee,W.E. (1987) Biotechnol. Letts 9(10): 691-696

Omasa,T., Ishimoto,M.,Higashiyama,K-i.,Shioya,S. & Suga,K-i. (1992)
Cytotechnology 8(1):75-84

Phillips,P.J., Marquis,C.P., Barford,J.P. & Harbour,C. (1991) Cytotechnology 6:189-195

Suzuki,E. & Ollis,D. (1989) Biotechnol. Bioeng. 34:1398-1402

Bioreactor with a Radial-flow Non-woven Fabric Bed for Animal Cell Culture

M. Matsumura,[1] M. Motobu,[1] S. Matsuo,[2] Y. Yamazaki,[3] and H. Kataoka[1]

1 Institute of Applied Biochemistry, University of Tsukuba, 1-1-1 Tennodai, Tsukuba, Ibaraki 305, Japan
2 Fuji Photo Film Co. Ltd., Minato-ku, Tokyo 106, Japan
3 Kirin Brewery Co. Ltd., Tsurumi-ku, Yokohama 230, Japan

ABSTRACT. A non-woven fabric packed-bed reactor with radial-flow was developed, and the stimulation of human renin production induced by shear stress was observed in the reactor. It was possible to expose cells to a moderate controlled shear stress with sufficient oxygen and nutrient supply with relatively low growth medium flow.

1. Introduction

Several effective support materials for cells like porous carriers have been developed for establishment of dense cell cultivation. By using these porous carriers, a high cell density of 10^8 cells/ml of carrier volume is now attainable in a fluidized-bed reactor with a relatively high circulating flow of the growth medium. One of the advantages of porous carriers is their high protective effect from shear stress. However, shear stress is not always detrimental, and a low level of shear stress may have beneficial effects on cellular metabolism. In order to study the effects of shear stress, cells need to be exposed to a liquid stream. A fixed-bed reactor with solid carriers is one of the most suitable devices for this purpose. However, in the conventional axial-flow packed-bed reactor, the problem of concentration gradients for the nutrients and metabolic waste becomes more serious when compared to a fluidized-bed reactor.

In our previous work on hybridoma cultivation with a packed-bed reactor of porous glass spheres, radial flow was found to be effective for limiting concentration gradients [1]. This advantage of a radial-flow packed-bed reactor is due to its large flow area and shallow path for circulating the culture broth. In this work, a non-woven fabric mat with a low pressure drop was employed as the packing material, which enabled a wide range of shear stress to be applied to the cells without depleting the nutrients. A perfusion culture of recombinant CHO cells was carried out for renin production, and its performance will be compared with our previous results

433

S. Kaminogawa et al. (eds.), Animal Cell Technology: Basic & Applied Aspects, Vol. 5, 433–442.
© 1993 Kluwer Academic Publishers.

obtained from a perfusion culture with the solid microcarrier, Cytodex 1 [2].

2. Materials and Methods

2.1. CELL LINE

All experiments in this study were carried out with a recombinant CHO cell line provided by Prof. Kazuo Murakami of the University of Tsukuba [3]. This cell line, designated as Ar-2 is a dihydrofolate reductase-negative CHO transfected with an expression vector containing human renin cDNA. Human renin-producing cells were established with 800 nM MTX (methotrexate). The renin-angiotensin system plays an important role in hypertension, renin cleaving angiotensinogen and converting it to angiotensin I. This reaction is known to be a rate-limiting step in this system.

2.2. MEDIUM USED FOR RENIN PRODUCTION

DMEM or SF-02 (Sanko Junyaku, Japan) was used as a basal medium, and the following supplements were employed for a perfusion culture of the rCHO cells: 2-5% dialyzed fetal calf serum (dFCS) or 2-10 mg/l of fibronectin, 2 mmol/l of L-glutamine, 0.1 mmol/l of MEM nonessential amino acids, 100 units/ml of penicillin, and 100 μg/ml of streptomycin.

2.3. NON-WOVEN FABRIC MAT

The Non-woven fabric mat consisted of polyester fibers of 55 μm in diameter with 130°C melting point. Thinner polyester fibers (39 μm dia.) with a melting point of 110 °C were used to bind the main fibers, the content of the binding fibers being 10% by weight. The non-woven fabric mat was coated with 0.03% or 0.1% type I atelocollagen, and the collagen surface was cured by crosslinking with hexamethylene diisocyanate or by ultraviolet radiation. The apparent fiber density of the non-woven fabric mat was thus changed from 29.3 to 74.6 kg/m^3 of mat volume. The ratio of the cell attachment surface to carrier volume (S/V) was also changed from 15.8 to 40.3 with increasing apparent fiber density.

2.4. BIOREACTOR WITH A RADIAL-FLOW NON-WOVEN FABRIC BED

The schematic construction of the radial-flow packed-bed reactor is shown in Fig. 1. A non-woven fabric mat of 6.7 mm in thickness is spirally wound together with a spacer around a central core with holes, and set in a cylinder made of sintered stainless steel to form a cartridge. This cartridge is placed in a glass or stainless steel housing. The spacer between the non-woven fabric mat layers allows for the homogeneous inoculation of cells, and the sintered stainless steel cylinder with a pore size of 30 μm around the mat acts as the distributor for the circulated culture broth.

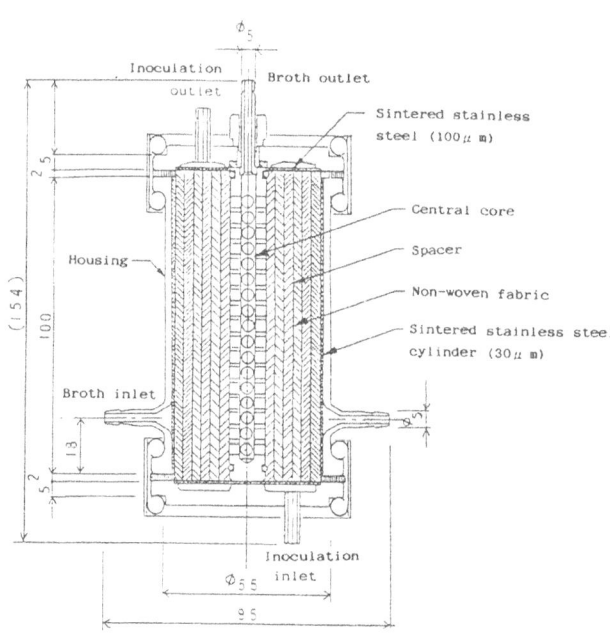

Fig. 1. Construction of the radial-flow non-woven fabric packed-bed reactor.

The radial-flow packed-bed reactor was connected to an external medium reservoir consisting of stirred 500-ml stirred glass vessel equipped with control devices for pH, DO, temperature and liquid level, as shown in Fig. 2. A UF-membrane with a molecular cut-off of 10 kD is fixed to the bottom of the reservoir to accumulate high-molecular materials such as growth factors and renin in the culture broth. Oxygen is supplied by both surface and bubble-free aeration via a porous PTFE tube, and DO is controlled by supplying oxygen-enriched air into the PTFE tube. Fresh medium was continuously supplied to the reservoir during the perfusion culture.

After conditioning the non-woven fabric mat with DMEM containing 10% FCS, inoculation was achieved by circulating a seed cell suspension

436

through the packed-bed reactor until the cells attached themselves to the fibers. Flow of the culture medium from the reservoir into the packed-bed reactor was then started in the radial direction. The linear velocity of the medium was gradually increased from 40 to 100 cm/h to match the cell growth in the reactor.

2.5. FLAT-BED TYPE OF REACTOR FOR EVALUATING THE NON-WOVEN FABRIC MAT

The non-woven fabric mat was evaluated by using a flat-bed module containing a mat of 5x5x0.5 cm in size. This flat-bed module connected to a medium reservoir was placed in a CO_2 incubator, and a repeated-batch culture was performed. The culture medium in the reservoir was replaced with a fresh one before the glucose became depleted.

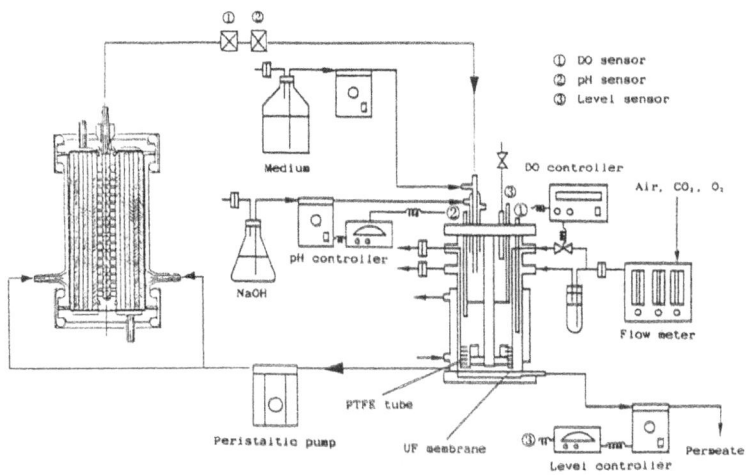

Fig. 2. Apparatus for perfusion culture with the radial-flow packed-bed reactor.

2.6. ANALYTICAL METHODS

2.6.1. *Cell Counting.* To achieve complete dissociation of the cells attached to fibers, the non-woven fabric mat was treated by trypsinization and subsequent centrifigation. All the detached cells were collected, and then diluted with a 0.1% trypan blue solution for cell counting with a hemocytometer. Cell growth on the fibers was observed by an optical microscope after Giemsa staining or by a scanning electron microscope.

2.6.2. *Substrates and Metabolic Waste.* Glucose and ammonia were analyzed with commercially available assay kits, and lactate by HPLC.

Amino acids were measured by an amino acid analyzer.

2. 6. 3. *Product.* The renin concentration was determined from the enzyme activity for converting angiotensinogen to angiotensin I, and is expressed by the unit of ng-AI/ml/h.

3. Results and Discussion

3. 1. OPTIMIZATION OF THE NON-WOVEN FABRIC MAT

3. 1. 1. *Crosslinking Method.* There are several reagents for crosslinking collagens like glutaraldehyde, formaldehyde, polyepoxy compound and hexamethylene diisocyanate. Ultraviolet radiation is also employed. Considering the toxicity of these reagents to the cells, we tried to crosslink type I atelocollagen with hexamethylene diisocyanate and by ultraviolet radiation. The collagen crosslinked by ultraviolet radiation was found to be unstable and easily detached itself from the fibers during autoclaving. Therefore, in the following experiments, we employed hexamethylene diisocyanate as the crossliking reagent.

3. 1. 2. *Apparent Density and Collagen Concentration of the Non-woven Fabric Mat.* Using mats coated with 0. 03% and 0. 1% collagen, repeated-batch cultures were performed with a flat-bed type of reactor. The medium was continuously circulated by a peristaltic pump at a linear velocity of 2. 2 cm/h in the initial cultivation stage, and this was then increased to 11 cm/h. The change in glucose concentration during the repeated-batch culture was measured. The maximum glucose uptake rate during the repeated-batch culture and the final concentration of cells attached to the fibers are summarized in Table 1. Owing to the high S/V ratio, the non-woven fabric mat with an apparent density of 74. 6 g/l gave the highest cell concentration of $3. 7 \times 10^6$ cells/ml of mat volume. However, we could not identify any large difference in glucose consumption rate between the mats with different apparent density. This might have been due to the low circulating flow rate of the culture medium, and glucose may have become depleted in the flat-bed module.

For the concentration of collagen used for coating the non-woven fabric mat, no remarked difference was found between 0. 03% and 0. 1%. Therefore, we fixed the collagen concentration at 0. 03%.

3. 2. PERFUSION CULTURE WITH THE RADIAL-FLOW NON-WOVEN FABRIC BED REACTOR

Two experiments with the radial-flow packed-bed reactor were carried out. In both experiments, the non-woven fabric mat with an S/V ratio

Table 1. Effects of the apparent fiber density and collagen
concentration on rCHO cell cultivation.

	S/V ratio of mat (cm^2/cm^3)	Collagen concentration (%)	Glucose uptake rate (g/l/day)	Attached cell concentration (cells/ml)
NW-1	15.8	0.03	0.91	5.6×10^5
NW-2	21.3	0.03	0.87	4.7×10^5
NW-3	40.3	0.03	1.14	31.0×10^5
NW-4	40.3	0.10	1.07	27.0×10^5

of 40 was installed to a cartridge of 85 ml in volume (total surface area was 3400 cm^2). In the first experiment DMEM supplemented with 5 or 2% dFCS was used as the feed medium. The feed rate of fresh medium was fixed at 350 ml/day (dilution rate of 1 day^{-1}), but the glucose content was increased from 4 mmol/l to 9.4 mmol/l to match the glucose consumption rate. The inoculum cell concentration was 2×10^5 cells/ml of mat volume (5000 cells/cm^2), and the circulating flow of medium through the packed bed was held at a constant linear velocity of 44.3 cm/h. In the second experiment, the circulating flow was gradually increased from 52.5 cm/h to 106 cm/h to investigate the effects of the flow rate on cell growth and renin production.

The results of the first experiment are shown in Fig. 3. Using DMEM supplemented with 5% dFCS, the perfusion culture started 73 hours after inoculation. To meet the increase in glucose consumption rate, the glucose concentration in the feed medium was increased 1.7 times. In the late stage from 260 hours, the content of dFCS was reduced to 2%. When the cultivation was terminated, the culture broth and mat inside the module were taken out to measure the suspended and attached cells in the packed-bed reactor, the results of the cell count being shown in Table 2. 6×10^4 units/ml of renin had accumulated in the culture broth after 384 fours. The specific glucose consumption and renin production rates were obtained by dividing the average consumption and production rate during the perfusion culture by the final cell density, these results also being summarized in Table 2.

In the second experiment (Fig. 4), using DMEM supplemented with 5% dFCS, perfusion was started 65 hours after inoculation at a feed rate of 300 ml/day. The initial circulating flow rate was gradually increased from 52.5 cm/h to 70, 91.1 and finally up to 106.3 cm/h. The feed medium was also changed to SF-02 supplemented with 2% dFCS, whose glucose content was 1.7 times higher than that of DMEM, and the feed rate was also increased to 350 ml/day. After changing the circulating flow rate to 70 cm/h, the production of renin was greatly enhanced, the final concentration attained being 9×10^5 unit/ml.

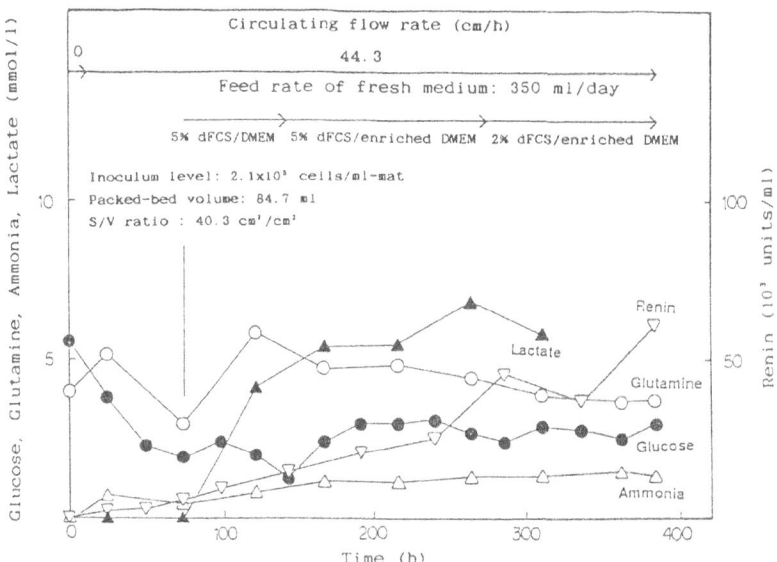

Fig. 3. Perfusion culture of rCHO cells in the non-woven fabric packed-bed reactor under a circulating linear velocity of 44.3 cm/h.

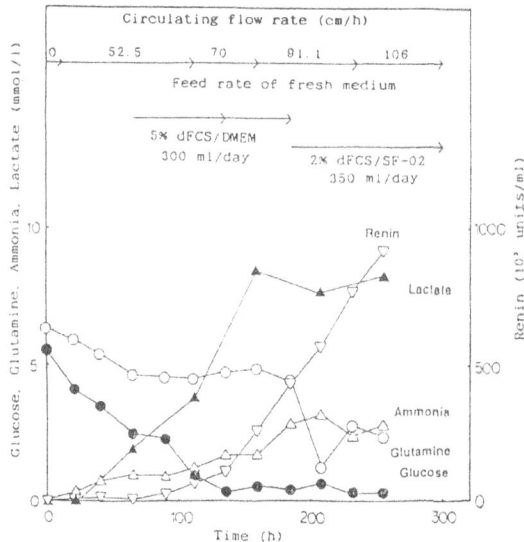

Fig. 4. Perfusion culture in the non-woven fabrics packed-bed reactor under increasing linear velocity from 52.5 to 106 cm/h.

3. 3. COMPARATIVE PERFUSION CULTURE WITH A MICROCARRIER

A perfusion culture with 3 g/l of Cytodex 1 microcarrier, whose S/V ratio was 18 cm^2/cm^3, was carried out, and the renin productivity was compared with the results obtained with the radial-flow packed-bed reactor. In the early stage of the perfusion culture with the microcarrier, DMEM supplemented with 10% dFCS was fed into the reactor at a dilution rate of 1.8 h^{-1}. The feed medium was then changed to SF-02 with 3.3 μg/ml of fibronectin, and the dilution rate was also reduced to 1.2 h^{-1}. When the culture was terminated, the cells on the microcarrier had formed multilayers, the renin concentration attained being 2x10^4 units/ml. The specific glucose consumption and renin production rates were calculated, and these are shown in Table 2 together with the results obtained for the packed-bed reactor.

As can be seen in this table, the effect of the circulating flow rate on the attached number of cells and on the specific glucose consumption rate are not marked, 80 to 90% of the total cells in the packed-bed reactor being attached, and 45 to 48% of the surface of the fibers being covered with the cells. The surface area fraction covered with cells was estimated from the maximum number of cells attaining confluence in a T flask (200,000 cells/cm^2 for CHO-K1). In contrast, the specific renin production rate was greatly influenced by the circulating flow rate, the difference between the two experiments being more than 20 times. This high stimulation of renin production in the fixed-bed reactor was beyond expectation when compared with the production using microcarriers, being more than 100 times greater.

Previous works have also reported the stimulation of protein synthesis induced by shear stress. Giard and co-workers observed that human fibroblasts secreted up to 30-fold greater amounts of interferon when the cells were exposed to a high shear stress [4]. From the experimental results on prostacyclin production with endothelial cells, Frangos and co-workers demonstrated that shear stress in certain ranges stimulated the synthesis of proteins [5].

Using a glass-fiber packed-bed reactor with axial flow, interferon production with recombinant CHO cells was carried out by Perry and co-workers [6]. The S/V ratio of their packed-bed reactor was 30 cm^2/cm^3, and the linear velocity of the circulating culture broth was 134 cm/h. These operating conditions are similar to ours. They also observed an unexpected productivity increase in their packed-bed reactor (19 times as much as that in a culture with microcarriers). The reasons for this higher specific rate was attributed to stress or adverse conditions for cell growth. The stress in the fiber packed-bed reactor may have been the surface curvature and shear stress induced by the liquid stream.

4. Conclusion

The critical shear stress level which is detrimental to animal cells is around 1 N/cm² [7]. Cells grown in a fiber packed-bed reactor with a 90% void fraction and perfused with medium at 3600 cm/h would experience a shear stress of 0.36 N/m² [6], which is lower than the critical shear stress causes cell damage. The configuration of the fiber packed-bed reactor is appropriate to expose the cells to moderate and controlled shear stress. Especially, in a fiber packed-bed reactor with radial flow, a sufficient oxygen and nutrient supply to support cell growth was attainable at much lower medium circulating flow rate.

Table 2. Comparison with perfusion-microcarrier cultivation.

	Run 1	Run 2	Microcarrier
Attacted cell conc. (cells/ml-mat)	3.60x10⁶	2.98x10⁶	4.0x10⁶
Suspended cell conc. (cells/ml-mat)	0.43x10⁶	1.02x10⁶	—
Total cell conc. (cells/ml-mat)	4.03x10⁶	4.00x10⁶	4.00x10⁶
Cell number per unit contact area (cells/cm²)	89,300	73,900	222,200
Fraction of area covered with cells (%)	44.7	37.0	111
Specific glucose uptake rate (mmol/cell/h)	3.01x10⁻¹⁰	3.93x10⁻¹⁰	(1.0~4.4)x10⁻¹⁰
Yield of lactate from glucose (mmol/mmol)	0.94	0.85	0.75
Specific renin production rate (units/cell/h)	1.81x10⁻⁴	4.86x10⁻³	3.48x10⁻⁵

5. References

1. Kurosawa, H., Markl, H., Niebuhr-Redder, C. and Matsumura, M. (1991) "Dialysis Bioreactor with Radial-Flow Fixed Bed for Animal Cell Culture", J. Ferment. Bioeng. 72, 41-45.
2. Portner, R., Matsumura, M., Hatae, T., Murakami, K. and Kataoka, H. (1991) "Perfusion-Microcarrier Cultivation of rCHO-Cells in Serum-Free Medium for Production of Human Renin", Bioprocess Eng., 7, 63-69.
3. Poormam, R.A., Palermo, D.P., Post, L.E., Murakami, K., Kinner, J.H., Smith, C.W., Reardon, I. and Heinrikson, R.L. (1986) "Isolation and Characterization of Native Human Renin Derived from Chinese Hamster Ovary Cells", Proteins: Structure, Function, and Genetics 1, 139-145.
4. Giard, D.J., Loeb, D.H., Thilly, W.G., Wang, D.I.C. and Levine, D.W. (1979) "Human interferon production with diploid fibroblast cells grown on microcarriers", Biotechnol. Bioeng., 21, 433-442.
5. Frangos, J.A., McIntire, L.V. and Eskin, S.G. (1987) "Shear Stress Induced Stimulation of Mammalian Cell Metabolism", Biotechnol. Bioeng., 32, 1053-1060.
6. Perry, S.D. and Wang, D.I.C. (1988) "Fiber Bed Reactor Design for Animal Cell Culture", Biotechnol. Bioeng., 34, 1-9.
7. Stathopoulos, N.A. and Hellums, J.D. (1985) "Shear Stress Effects on Human Embryonic Kidney Cells in Vitro", Biotechnol. Bioeng., 27, 1021-1026.

Partial Cell Culture Medium Recirculation
Experience in Pilot Scale

U. Riese, R. Heidemann, D. Lütkemeyer, J. Lehmann

Institute for Cell Culture Technology, University of Bielefeld
P. O. Box 10 01 31, 4800 Bielefeld 1, FRG

Keywords: **Medium Recirculation, CHO cells, Antithrombin III, Recycling Rate, Reduction of Waste Water**

Summary

Based on bench scale datas a continuous process was developed in which a partial recycling of the spent medium was employed for the cultivation of a human active Antithrombin III producing CHO cell line in 100 l scale. Saving of 44 % of the medium costs were achieved.

Introduction

The first attempt to recycle spent medium in mammalian cell cultivation has been described by Misrahi et al. (1977). Kempken (1992) created a medium cycle bioreactor (MCB) to re-use medium for production of monoclonal antibodies. Büntemeyer et al. (1992) described a modified version of the MCB for the investigation of effects of accumulated inhibitors. It was shown that cell death was caused by toxic effects of inhibitors when medium was totally recycled. Therefore, a detoxification was necessary and could have been easily performed by dilution of spent medium with a fresh medium concentrate. This technic was used in a repeated batch experiment in large scale as reported by Riese et al. (1992).

This paper is dealing with a **continuous** large scale cultivation of a recombinant CHO cell line producing human active Antithrombin III (AT III) in 110 l working volume. The bioreactor was equipped with several components to perform the medium recirculation (Fig. 1). The cell bleed stream (F_B), the product harvest stream (F_P) and the waste stream of spent medium (F_W) was kept constant and compensated by the stream of the fresh medium concentrate (F_C). Cell retention was performed by microfiltration. After separation of the product by ultrafiltration the UF-filtrate was partially fed back as recycle stream (F_R).

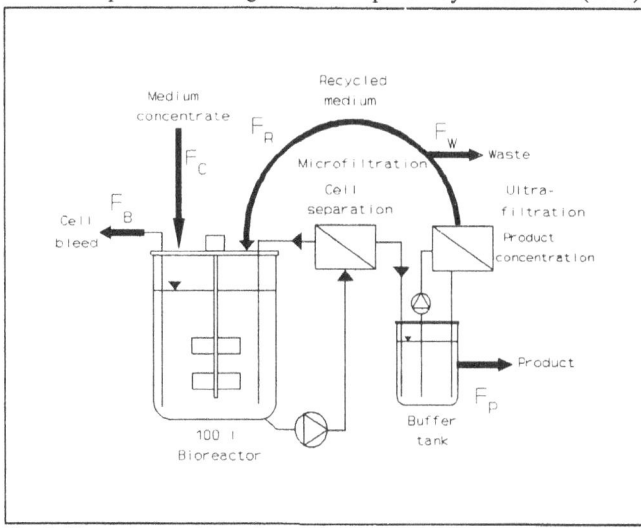

Fig. 1: Flow scheme with medium streams of continuous large scale medium recycling.

443

S. Kaminogawa et al. (eds.), Animal Cell Technology: Basic & Applied Aspects, Vol. 5, 443–446.

444

Materials and Methods

The cultivated recombinant CHO-AT III SS3 A2 cell line producing human active AT III (Zettlmeissl et al., 1987) was kindly provided by the Behringwerke AG, Marburg. The cultivation temperature was 37 °C, pH = 7.1, pO_2 = 40% air saturation and impeller speed 30 rpm.

The cells were cultivated in a serum-free medium based on a 1:1 mixture of DMEM/HAM'S F12 (Boehringer, Mannheim), supplemented with 8.1 mg/l phenol-red, 100 μM ethanolamine, 10 mg/l human transferrin (Fe saturated), 10 mg/l bovine insulin and 1 ml/l Pentex Excyte I (lipoprotein, Bayer AG). During recycling a nutrient concentrate was added (Tab. 1). These concentrate was composed to avoid limitations of nutrients up to cell densities of $5 \cdot 10^6$ living cells per ml at a recycling rate of 50 %. For a recycling rate of 70 % the nutrient concentrations have been changed as needed (datas not given). The medium was free of antibiotics.

A 100 l bioreactor was used (BIOSTAT 100, B. Braun Biotech International, Melsungen, FRG). This bioreactor was equipped with a fixed silicone aeration tube and a marine-type impeller.

Substrate	DMEM + F12 Basal medium [μM]	Nutrient concentrate [μM]
Glucose	18149	32023
ASN	50	275
SER	250	635
HIS	150	300
GLN	2500	8320
THR	450	660
ARG	700	800
TYR	215	430
ETA	-	150
TRP	45	135
MET	116	290
VAL	450	835
PHE	215	409
ILE	415	915
LEU	450	1080
LYS	500	1350
CYS	260	520
PRO	150	375

Tab. 1: Nutrient concentration in medium and medium concentrate.

Cell separation was performed with a hollow-fibre module (Enka, Wuppertal, FRG) with 0.9 m² filtration area, 0.2 μm pore size. The module was in-line steam sterilisable.

For product separation a Sartocon II cross-flow system (Sartorius, Göttingen, FRG) was used. It was equipped with one Ultrasart-module (0.7 m² filtration area, 10 kD cut-off). From the UF-concentrate stream the product was subsequently purified by chromatographic means. The UF-system was in-line steam sterilised.

From the cell bleed stream the cells were harvested by continuous centrifugation with a 300 MD System (Biofuge 17 RS, titan-rotor 8575; Heraeus, Osterode/Harz, FRG). The rotor could be autoclaved. The cells were separated at 6058 • g.

Results

Experiments in 1 l scale were used to estimate the steady state conditions for cell concentration, nutrient concentration and medium supply according to the rates and equations given in Tab. 2. The scale up process was performed in two periods: first batch and second perfusion with medium recycling. The process started with a volume of 35 l and was filled up to the final working volume of 110 l. After 8 days of cultivation time the perfusion mode was started (Fig. 2). From the start of the perfusion

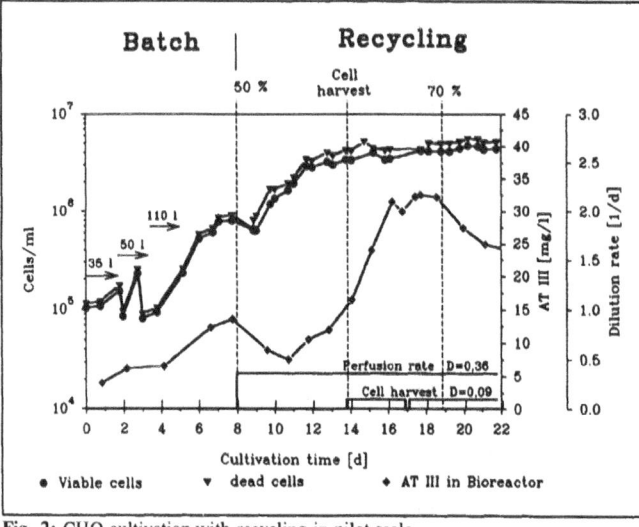

Fig. 2: CHO cultivation with recycling in pilot scale.

spent medium was recycled with a recycling rate R = 50%. After 18 days cultivation time the recycling rate was increased to R = 70 %. The cell bleed started after 13.7 days. At 17 days cultivation time it was necessary to clean the hollow-fibre module. It was disconnected, cleaned, linked back to the bioreactor and finally sterilised. This procedure caused a lack of cell mass. Therefore, the cell bleed was stopped at this day. The cultivation was finished after 21.8 days process time.

After the start of the perfusion the product concentration increased with the growth of cells. After starting cell bleed the cell density reached a stable value of $4.5 \cdot 10^6$ living cells per ml. During the final phase of the cultivation at 70 % recycling rate the cell concentration was constant.

Based on a constant working volume of 110 l and an overall dilution rate of D = 0.45 [d⁻¹] (49.5 l/d; including a cell bleed F_B of 10 l/d) the recycling rate was set to R = 50%. So the streams of medium concentrate and recycled medium became equal to $F_C = F_R = 24.75$ l/d. The spent medium was returned into the bioreactor directly without sterile filtration. The cell bleed rate was set to $D_B = 0.091[d⁻¹]$ (F_B = 10 l/d). Under this conditions a stable

Concentrate Stream	Recycling Rate
$F_C = F_B + F_P + F_W$	
Cell Bleed Rate	$R = \dfrac{100 \cdot F_R}{F_c + F_R}$
$D_B = \dfrac{F_B}{V_R} = 0.091$	Overall Dilution Rate
	$D = \dfrac{F_C}{V_R} = \dfrac{F_P + F_B + F_W}{V_R}$
Reactor Volume:	
V_R (l)	Indices:
	B = Bleed
Flow Rate:	C = Concentrate
F (l/d)	P = Product
	R = Reactor
Recycling Rate:	W = Waste
R (%)	

Tab.2: Definition of flow parameters and rates.

cell concentration was achieved. At day 14 the product concentration reached a steady state value of 32 mg AT III/l.

There was a small decrease in product concentration at the beginning of the perfusion mode and after the cleaning of the hollow-fibre module. This is due to coating of the cleaned microfiltration membrane by proteins. After the change of the recycling rate R to 70 % the product concentration was less than that at 50 % recycling rate.

During the whole process a total amount of 17 g AT III was produced in 22 days.

Medium cost reduction

As shown in Fig. 4 medium costs for the cultivation dropped after starting the recycling process. The savings were calculated according to the procedure, described by Kempken et al. (1991).

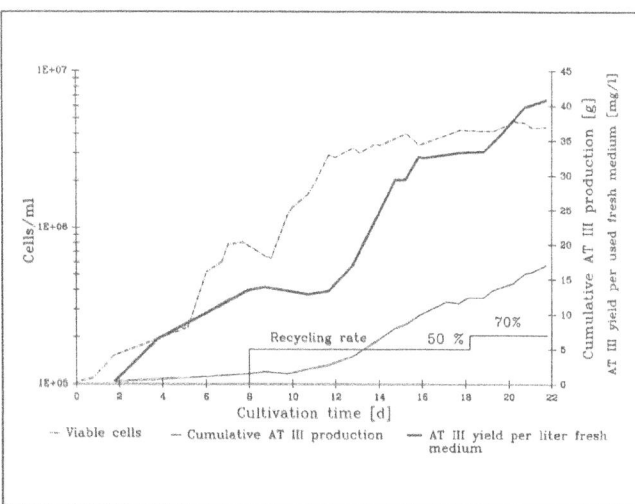

Fig. 3: Productivity and product yield per litre used fresh medium.

446

This results in a better economy of the whole process, because instead of 720 l basal medium for a common perfusion process only 420 l medium concentrate were necessary for the whole cultivation. By using a serum free medium at costs of 11 DM/litre the medium costs dropped to 4718 DM instead of 7995 DM calculated for a conventional perfusion process.

Conclusion

The partial medium recirculation could be employed for the AT III production. In perfusion stage

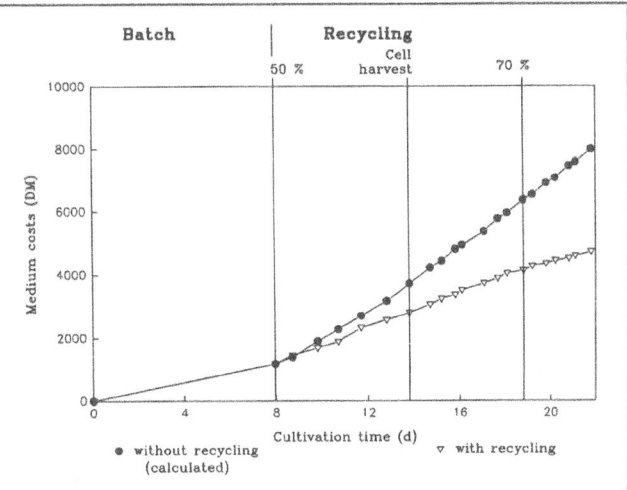

Fig. 4: Saving of costs by medium recycling.

of the process the AT III concentration was four times higher than in the batch process.
At R = 50 % recycling rate the volumetric yield of AT III reached 34 mg AT III per litre used medium concentrate, at R = 70 % the volumetric yield increased to 40.5 mg AT III per litre used medium concentrate. In batch mode a volumetric yield of only 8.25 mg AT III per litre used basal medium was achieved (Fig. 3). Additionally instead of 750 litre fresh medium only 420 litre of medium were necessary. This leads also to a reduction of the volume of waste water.

Acknowledgements

This study was supported by the project 'Development of a procedure and a plant for the recirculation of nutrient media for animal cell culture' (BMFT project reference no. 031 93 46A) of the German Federal Ministry of Science and Technology.
We like to thank the Behringwerke AG Marburg, FRG, for the kindly supply of the CHO cell line.

References

Büntemeyer H, Wallerius C, Lütkemeyer D, Lehmann J (1992) Optimal medium use for continuous high density perfusion processes. Cytotechnology 9: 59-67

Kempken R, Büntemeyer H, Lehmann J (1991) The medium cycle bioreactor (MCB): Monoclonal antibody production in a new economic system. Cytotechnology 7: 63-74.

Kempken R (1992) PhD Thesis, University of Bielefeld, FRG.

Misrahi A, Avihoo A (1977) Growth medium utilization and its re-use for animal cell cultures. Journal of Biological Standardization 5: 31-37.

Riese U, Heidemann R, Lütkemeyer D, Büntemeyer H, Lehmann J (1992) Re-use of spent cell culture medium. Engineering Foundation Conference Cell Culture III, Palm Coast Florida

Zettlmeissl G, Ragg H, Karges HE (1987) Expression of biologically active human antithrombin III in chinese hamster ovary cells. Biotechnology 5: 720-725.

Adaptation To Stirred Culture - A Study Of A Moderately Shear Sensitive Cell Line (PQXB1/2).

OH Steve Kah Weng[1], *AL-RUBEAI Mohammed,[2]* *EMERY Nick [2]and NIENOW Alvin[2].*

[1]Bioprocessing Technology Unit ,National University of Singapore, 10, Kent Ridge Crescent, SINGAPORE 0511 and [2]School of Chemical Engineering, University of Birmingham, ENGLAND.

ABSTRACT

A hybridoma was shown to be sensitive to low levels of agitation. Hybridoma PQXB1/2 (from ICI, Pharmaceuticals Division) normally grew in RPMI with 5% FCS in static culture, however it was shown that even at speeds as low as 15 or 40 rpm, growth performances were severely reduced. Both growth rate and viability index were reduced at these speeds but these effects were ameliorated by increasing the serum levels.

Therefore to develop a "resistance" to higher shear levels, the following protocol was implemented which successfully allowed PQXB1/2 to be cultured at speeds up to 500 rpm and eventually these cells were able to adapt to suspension cultures in 3 types of bioreactors agitated by radial flow turbine impellers.

Protocol:-
1] Place cells in stirred culture at 25 rpm containing RPMI + 10% FCS and passage every 2 days.
2] When growth rate was identical to the control (static) culture the speed was increased to 100 rpm.
3] When cells become more robust, the speed was increased to 200 rpm.
4] When good growth performance was achieved the speed was increased to 500 rpm
5] Healthy cells were then placed in RPMI + 5% FCS at the lower speed of 200 rpm.
6] Cells were passaged for another 2 weeks and finally healthy cells were used for bioreactor studies.

Healthy cells, which achieved a similar growth rate to the control cells after adaptation, were able to grow well at a higher agitation speed of 100 rpm in 3 commercially available stirred bioreactors (LH, SGi and LSL) fitted with turbine impellers after adaptation of 40 passages.

INTRODUCTION

It has previously been shown with 3 hybridoma cell lines[1] that high agitation in the absence of air bubbles did not adversely affect cell growth. However, sparging in combination with agitation was detrimental to cells.[2] In this paper a hybridoma was discovered that was susceptible to even very gentle stirring. Hybridoma PQXB1/2 (obtained from ICI plc) which was partially sensitive to a stirred environment was studied for its response to agitation at different serum levels. Subsequently, due to the poor performance in stirred conditions, a process of gradual adaptation was employed to develop resistance to high agitation levels. Finally, we demonstrated that cells were able to grow at higher speeds in 3 different bioreactors equipped with different impeller types after a long period of adaptation.

MATERIALS AND METHODS

Serum studies
Cells from medium supplemented with 5% serum were inoculated into 8 Duran bottles with magnetic bars. The first 4 bottles contained medium supplemented with 1, 5 10 and 15% FCS and the stirring speed for suspension was set at 15 rpm. The second 4 bottles contained these same medium

447

compositions but all were stirred at 40 rpm. Cell number, growth rate, μ and viability index, V_i were determined for all cases in batch experiments.

Adaptation to increasing agitation
Cells were thawed from a frozen stock and adapted in medium supplemented with 10% FCS as follows:-
1] Place cells in stirred culture at 25 rpm containing RPMI + 10% FCS and passage every 2 days.
2] When growth rate was identical to the control (static) culture the speed was increased to 100 rpm.
3] When cells become more robust, the speed was increased to 200 rpm.
4] When good growth performance was achieved the speed was increased to 500 rpm
5] Healthy cells were then placed in RPMI + 5% FCS at the lower speed of 200 rpm.
6] Cells were passaged for another 2 weeks and finally healthy cells were used for bioreactor studies.

Bioreactor studies
Inocula for bioreactors were prepared from adapted hybridomas. Cells were cultured in Duran bottles with medium supplemented with 5% FCS and stirring speed was maintained at 200 rpm. In the first experiment, cells were inoculated into 2 types of bioreactors with axial flow impellers at a density of 10^5/ml. Growth performances were compared to the control (Duran bottle) condition. In the second experiment, cells were tested in 3 types of bioreactors with radial flow impellers and compared to the control.

RESULTS

Fig. 1a) shows that cells cultured at 15 rpm performed less well at all serum levels below 15% compared to the static control with 5% serum. Cells in 1% serum, at this low speed, remained at the inoculum density and stayed in the stationary phase. At the slightly higher agitation level of 40 rpm, Fig. 1b) shows that maximum cell numbers were depressed at all serum levels including the one containing 15% FCS. Performance was particularly poor at 1% serum where cells died rapidly. The results are represented in terms of growth rates, μ and viability indices, V_i in Figs. 2a) and b) respectively. The former shows that μ fell gradually at serum levels below 10% when cells were stirred at 15 rpm, whilst μ were reduced sharply at all serum levels when agitated at 40 rpm. Viability indices however, which are a combined measure of culture longevity and maximum cell density were severely reduced at all speeds and serum levels except for the culture with 15% serum at the speed of 15 rpm. Therefore high serum levels provided negligible protection to cells, even at such low shear levels, prior to adaptation.

To determine whether shear sensitivity was time dependent, cells were taken from static cultures at the earlier passages of 2, 3 and 7, and inoculated into stirred and static conditions. Fig. 3 shows the cells, prior to adaptation, always performed worse at 100 rpm compared to the static condition. Growth rates and maximum cell densities were always lower than the control. Thus it was decided to examine whether cells could be gradually adapted to enable them to grow at 100 rpm in stirred bioreactors by following the protocol described in the method section.

Table 1 shows the passage numbers, or time points, where agitation speed and serum levels were changed during the adaptation protocol. The criteria for increasing the speed or decreasing the serum level was based on the growth rate measured every 2 days. When μ had stabilised at a similar or better value than the control, the speed was increased. Once cells had adapted to 500 rpm, serum level was lowered to further select for hybridomas requiring low serum levels. The specific growth rate of the cells as shown in Fig. 4 gradually increased with adaptation to higher speeds and increasing passage number. In comparison, control cells in static culture which were cultured in parallel, exhibited a more stable specific growth rate.

After cells had been successfully grown in stirred bottles, it was still necessary to determine their performance in stirred bioreactors. Thus, cell were first grown in 2 bioreactor types fitted with "low shear", axial flow impellers at 100 rpm and compared to a Duran bottle control culture with a magnetic bar stirring at the same speed. At a later passage, cells were also grown in 3 types of bioreactors fitted with "high shear", turbine impellers and compared to a control. Fig. 5 shows that growth performances were similar to the control in all 5 bioreactor types. Intracellular Mab content as determined by flowcytometry, was found to be identical both before and after adaptation to high agitation (data not shown). The biological implications of adaptation to agitation is not yet known, however it is possible that strengthening of the cytoskeletal and cytofilamental structures could enhance the ability of the cells to resist high shear rates.

Table 1. Comparison of growth rates between static and stirred cultures at various passages.

Passage Number	μ of static culture (hr^{-1})	μ of stirred culture (hr^{-1})	Speed (rpm)	Serum level
1	Cells thawed from frozen stock and placed in medium with 10% FCS			
2	0.028	0.006	25	(10% FCS)
7	0.039	0.007	25	(")
10	0.037	0.020	25	(")
14	0.022	0.023	100	(")
17	0.024	0.029	100	(")
19	0.035	0.035	200	(")
22	0.027	0.048	500	(")
25	0.029	0.037	200	(5% FCS)
28	0.050	0.037	100	(")
33	0.041	0.040	100	(")
37	0.034	0.033	100	(")

CONCLUSIONS

1. PQXB1/2 hybridomas were initially "shear sensitive" to very low agitation levels even in the presence of high serum levels.

2. Shear resistivity did not improve with passage time.

3. These cells were gradually adapted to higher stirring conditions by regular passaging every 2 days with step increases in stirring speed followed by a serum reduction.

4. After ~40 passages, hybridomas were able to grow in bioreactors fitted with either "low shear" or "high shear" impellers. Growth performances were identical to the controls.

5. Thus it was demonstrated that hybridomas can be adapted to a high shear environment by gradual step increases in agitation speed.

REFERENCES

1. Oh S. K. W., Nienow A. W., Al-Rubeai M. and Emery A. N. (1989) The effects of agitation intensity with and without continuous sparging on the growth and antibody production of hybridoma cells. J. Biotechnol. 12, 45-62.

2. Oh S. K. W., Nienow A. W., Al-Rubeai M. and Emery A. N. (1992) Further studies of the culture of mouse hybridomas in an agitated bioreactor with and without continuous sparging. J. Biotechnol. 22, 245-270.

450

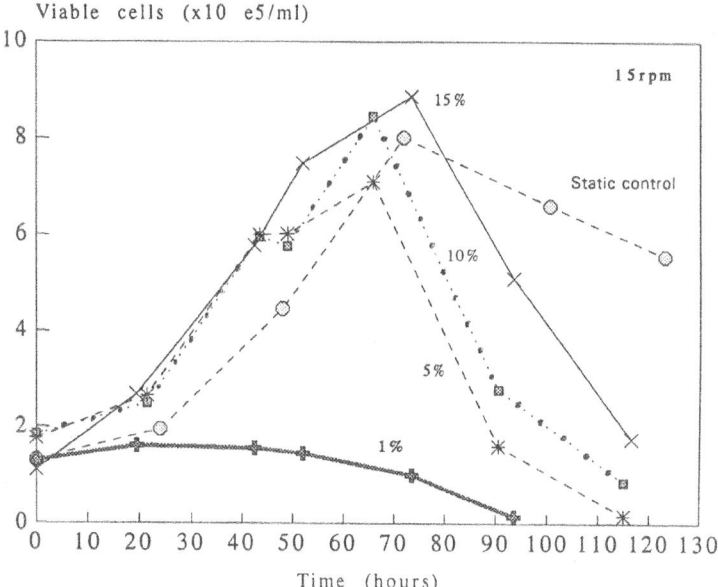

Fig. 1a) Effect of increasing serum levels on
the growth of PQXB1/2 cells at 15 rpm

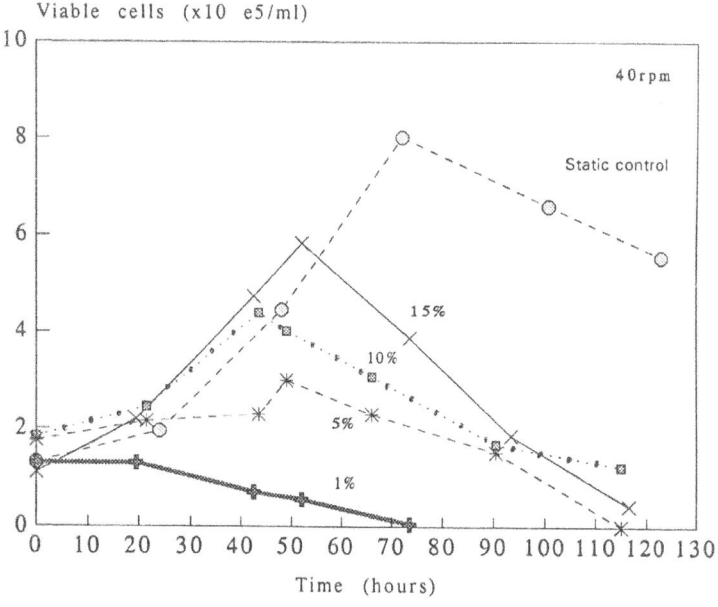

Fig. 1b) Effect of increasing serum levels on the
growth of PQXB1/2 cells at 40 rpm.

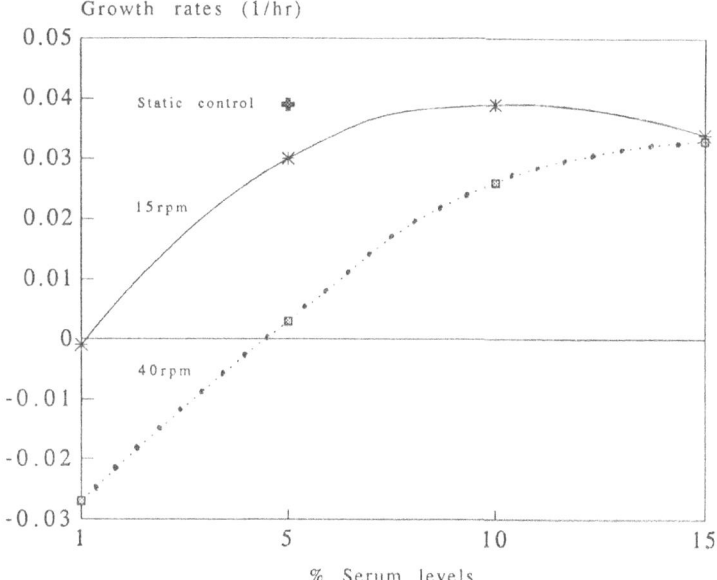

Fig. 2a) Effect of serum on the growth rates of
PQXB1/2 at 15 & 40rpm compared to a control.

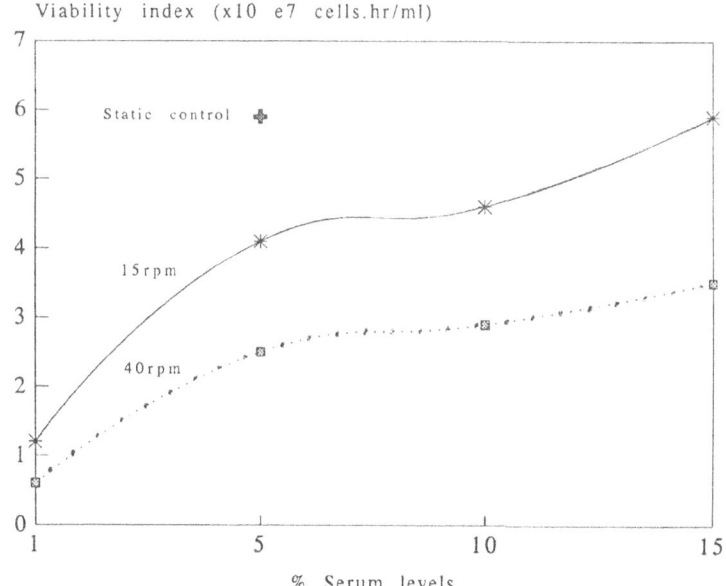

Fig. 2b) Effect of serum on the viability index of
PQXB1/2 at 15 & 40rpm compared to a control.

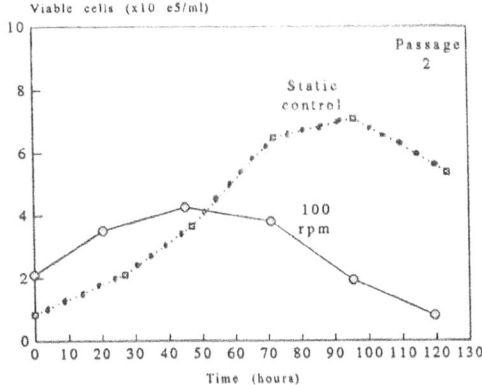

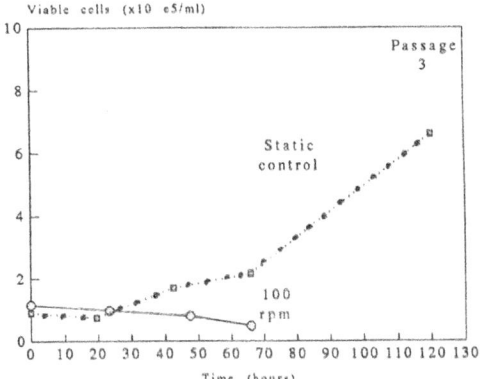

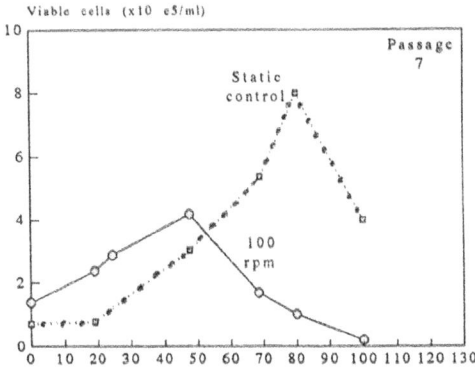

Fig. 3 Comparison of growth performance between
static and 100 rpm cultures from various passages.

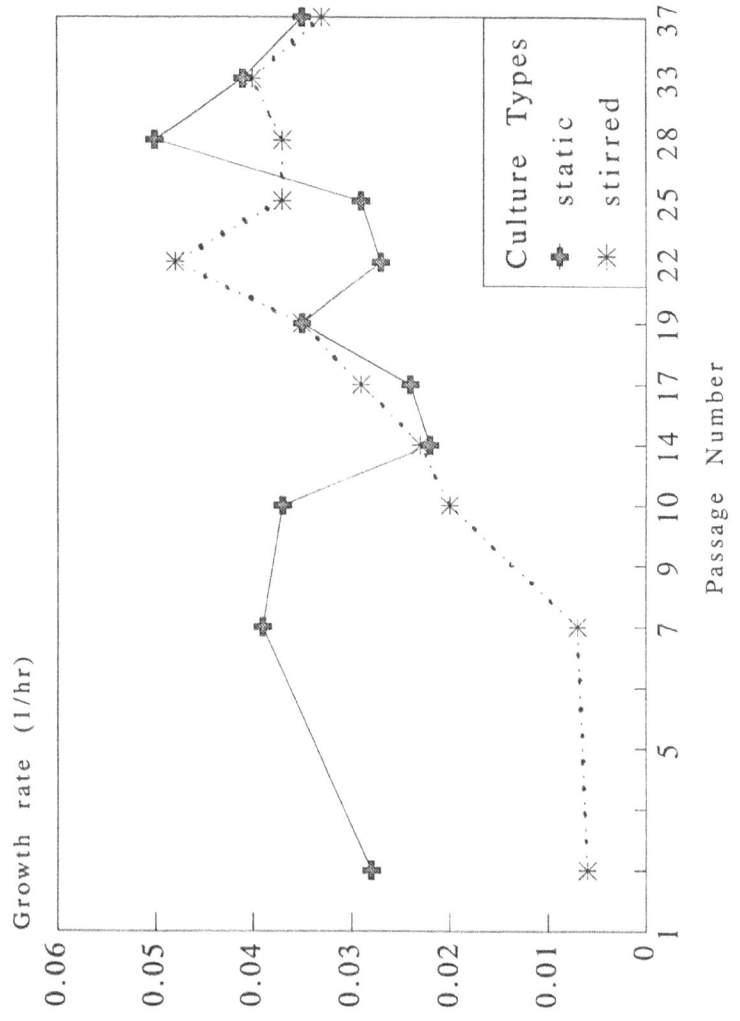

Fig. 4 Performance of PQXB1/2 cells adapting to
increased agitation compared to static controls.

454

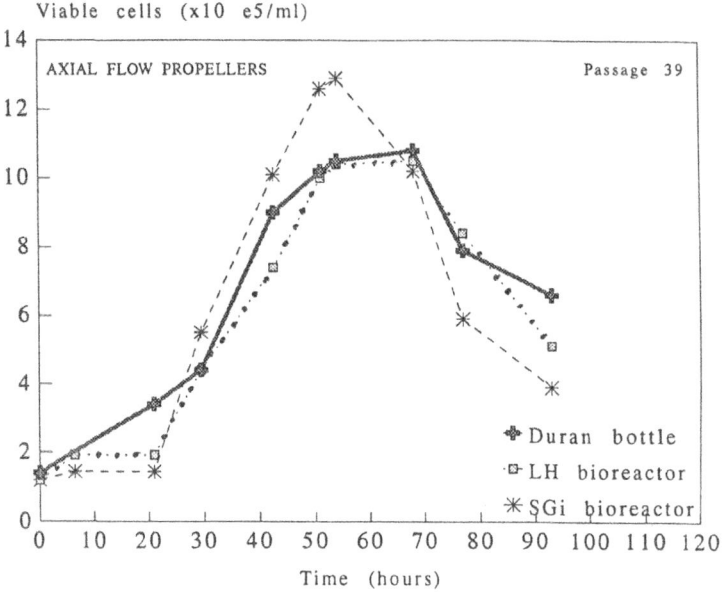

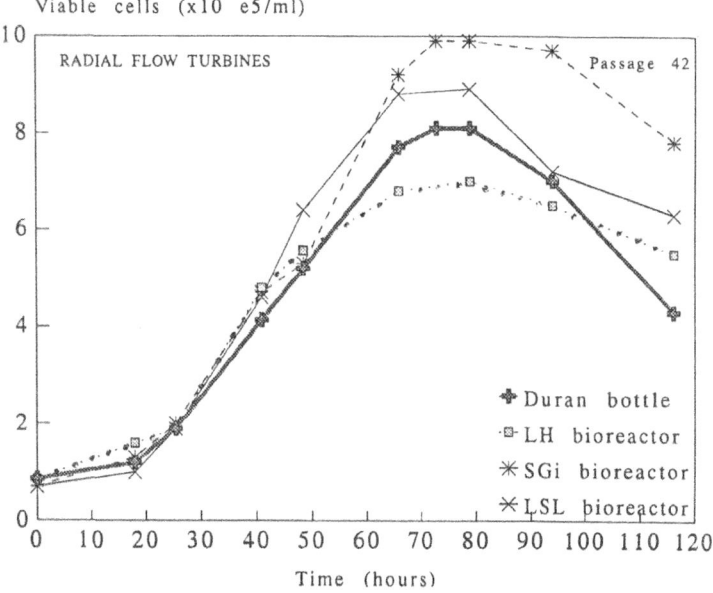

Fig. 5 Growth of PQXB1/2 cells in bioreactors, using axial or radial flow stirrers, agitated at 100rpm after adaptation.

Evaluation of optimal process parameters for the membrane dialysis bioreactor with integrated radial-flow fixed bed

Bohmann, A.; Pörtner, R.; Märkl, H.;

Department of Bioprocess and Biochemical Engineering,
Technische Universität Hamburg-Harburg, Denickestr. 15,
2100 Hamburg 90, Germany

Abstract

It could be shown that continuous cultivation of animal cells, e.g. hybridomas producing monoclonal antibodies (MAb's) against penicillin-G-amidase, is possible in a membrane dialysis bioreactor with integrated radial-flow fixed bed over a period of 76 days. Because of the advantageous design of the reactor, the feed of serum has been proven to be necessary to the inner chamber only and therefore operation costs are reduced significantly compared to conventional reactor systems. Membrane fouling due to high protein titers was not observed. An operation mode of the reactor aiming at a high MAb productivity and concentration in the harvest stream was realized under the prerequisite of a constant glucose level. A mean maximum MAb concentration of 368 mg l^{-1} was reached in continuous cultivation, ten times the value of stationary culture flasks after 4 days of incubation.

Introduction

Within the last years several reactor concepts were tested that referred especially to the requirements of animal cell cultivation. Among the reactor systems proposed fixed bed reactors [1] and reactors with integrated dialysis membranes [2] are of growing interest. The features of the reactor presented here are the following: Medium supply and removal of waste products via a dialysis membrane, immobilization and protection of the cells from mechanical stress using a radial-flow fixed bed and enrichment of the high molecular product, here a monoclonal antibody. The intention is to validate the reactor

455

S. Kaminogawa et al. (eds.), Animal Cell Technology: Basic & Applied Aspects, Vol. 5, 455–462.
© 1993 *Kluwer Academic Publishers.*

components and to develop an optimal operation mode. The elements to consider are the fixed bed, the dialysis membrane and the availabilty of two chambers. Basic data on the immobilization of hybridoma cells in fixed beds consisting of porous glass beads were published recently [3]. The work presented here focussed on the mass transfer characteristics of the dialysis membrane during a long-term high density cultivation of animal cells and on the dilution rates in the two chambers with regard to volume specific glucose uptake rate, volume specific MAb production rate, MAb concentration and the efficiency of the dialysis membrane to remove toxic metabolites like ammonia and lactate.

Materials and Methods

1. Cell line and culture medium

A mouse-mouse hybridoma cell line [4] producing monoclonal antibodies against penicillin-G-amidase was cultivated in a 1:1 mixture of Iscove's MEM and Ham's F12 supplemented with 3% FCS, 2 mmol l^{-1} glutamine and 2 g l^{-1} NaHCO$_3$.

2. Analyses

Glucose, lactic acid and ammonia were determined enzymatically, monoclonal antibody (MAb) by a mouse-IgG-ELISA (all kits purchased from Boehringer, Germany). Amino acids were determined by o-phthaldialdehyde precolumn derivatization.

3. Membrane dialysis bioreactor with integrated radial-flow fixed bed

The membrane dialysis bioreactor [5] consists of two chambers (outer chamber 5.0 l, inner chamber 1.2 l) which are separated by a dialysis membrane (Cuprophan, ENKA, Germany) with a cut-off of 10,000 Dalton (Fig.1).

The radial-flow fixed bed (volume 600 ml, inner radius 0.5 cm, outer radius 4 cm) containing Siran™ (Schott, Germany) carriers (diameter 3-5 mm) is arranged in the inner chamber. Low molecular nutrients and oxygen are supplied to the cells mainly via the membrane. Toxic metabolites such as lactic acid and ammonia are dialysed through the membrane, while antibodies are retained and accumulated in the inner chamber.

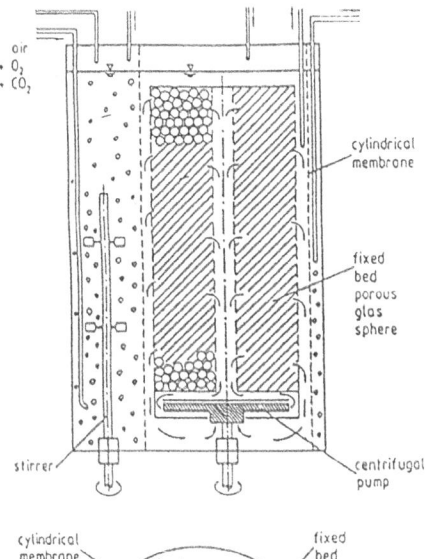

In continuous mode serum-free medium is fed to the outer chamber and serum containing medium to the inner chamber.

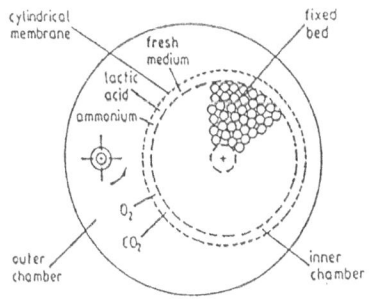

Fig. 1: Membrane diaysis bioreactor with radial-flow fixed bed

Results

1. The influence of high protein titers on mass transfer across the dialysis membrane

For modelling a fermentation using the membrane dialysis reactor the permeability of the membrane for each single component has to be determined. Cultivating animal cells in fixed bed reactors always implies high protein concentrations. As this titer varies according to the operation modes, its influence on mass transfer across the membrane was studied because of possible membrane fouling. Although the membrane was exposed to fermentation broths containing 0, 1.2 and 5 g protein l^{-1} no significant changes in the permeability coefficient could be detected neither for glucose nor for lactic acid, ammonia or amino acids tested (data not shown).

2. Continuous cultivation in the membrane dialysis bioreactor with radial-flow fixed bed

The intention of this work was to prove that a serum containing feed is necessary only to the inner chamber going along with a minimized serum input and to study the reactor's behaviour on varying the inner chamber's dilution rate. This is of great interest because of two reasons. First, a high product concentration should be reached making down-stream processing easier. Second, a high volume specific MAb production rate should be obtained.

Concentration of glucose and of MAb

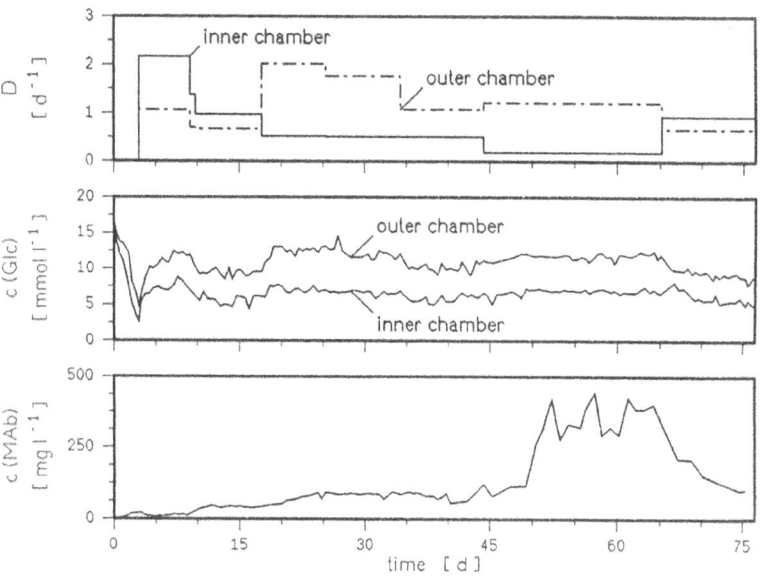

Fig. 2: Concentration of glucose c(Glc) in the inner and outer chamber, concentration of MAb c(MAb) and dilution rate D in the inner and outer chamber.

The dilution rate of the inner chamber was varied between 2.2 d^{-1} and 0.2 d^{-1}. From fig. 2 it can be seen, that it is possible to run the reactor stable with different dilution rates. Therefore the intention to operate the reactor in a way producing high product titers can successfully be realized. This is also true for the glucose level aspired to be constant by adjusting the dilution rate in the outer chamber. A mean maximum MAb concentration of about 368 mg l^{-1} was achieved at a dilution rate of 0.2 d^{-1} in the inner chamber.

To prove reproducibility and to make sure that possible effects are not of an origin independent of the operation mode, meening an irreversible change in the fixed bed, one of the first combinations of inner and outer chamber's dilution rates was set again at the end. The system behaved the same way as at the beginning of the fermentation when the same dilution rates for inner and outer chamber were used.

Concentration of lactate and of ammonia

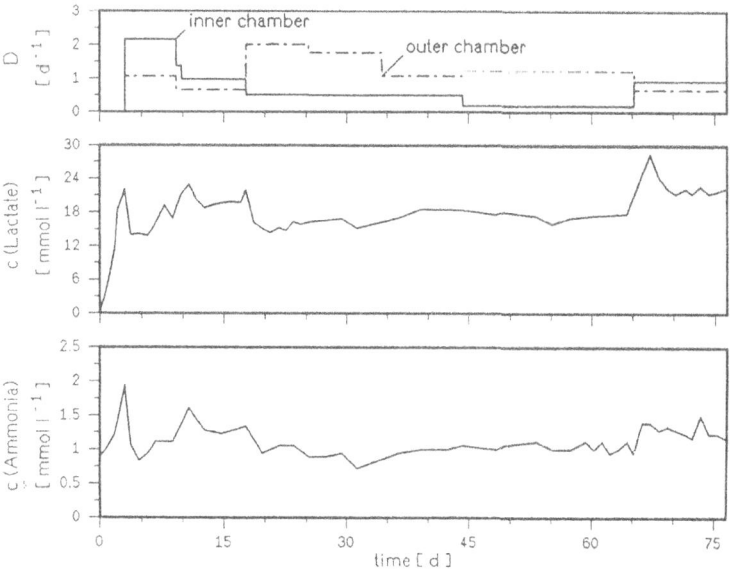

Fig. 3: Concentration of lactate and ammonia in the inner chamber of the membrane dialysis reactor and dilution rate D in the inner and the outer chamber.

Fig. 3 shows that both ammonia as well as lactic acid are kept at a very low level in the inner chamber. The highest concentrations reached for both compounds at the end of the batch phase were lower than the concentrations regarded as critical for hybridomas (lactate: 40 mmol l^{-1}, ammonia: 5 mmol l^{-1} [6]). Ammonia never exceeded a concentration higher than 2 mmol l^{-1} and lactic acid 28 mmol l^{-1}.

Specific glucose uptake rate and specific MAb production rate

An answer to the question whether variation of the inner chamber's dilution rate can lead to a further improvement of the reactor's performance is given in fig. 4

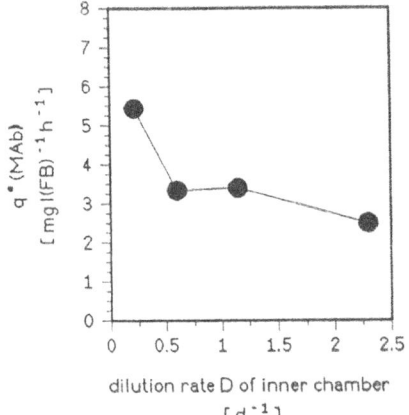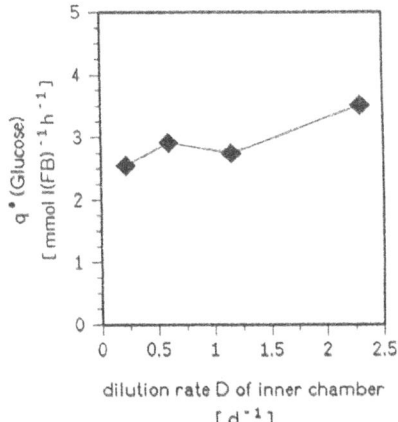

Fig. 4: Volume specific glucose uptake rate and volume specific MAb production rate as a function of the dilution rate in the membrane dialysis reactor's inner chamber

The volume specific glucose uptake rate seems to be more or less constant regardless of the dilution rate of the inner chamber. This is contradictory to the volume specific MAb production rate that nearly doubled when the dilution rate was lowered from 0.5 d^{-1} to 0.2 d^{-1}.

Discussion

The successful continuous run proved the suitability of the reactor concept compared to other perfusion systems.

It is a well known phenomenon that porous membranes tend to block if they are perfused. This membrane fouling leads to a change in the separation characteristics and flux [7]. As our results show this does not occur in non perfused membranes. The mass transfer characteristics across the dialysis membrane were constant regardless of time and fermentation mode.

In most of the reactor systems dilution of waste products such as ammonia and lactate goes along with dilution of the product of desire, e.g. MAb's. Compartimization of the reactor into an outer cell-free chamber and a cell containing inner chamber via a membrane leads to a decoupling of the mass streams of low molecular and high molecular substances in the reactor system. Concomitant different dilution rates can be realized in both compartments. A high dilution rate in the outer chamber for removal of low molecular waste products and a small dilution rate in the inner chamber where the high molecular components are enriched, results in a low waste product level and a high MAb level. By adjusting the dilution rate in the outer chamber a constant glucose level can be maintained even if the inner chamber's dilution rate is varied.

The fact that in our experiments the volume specific MAb production rate increased with a decreasing dilution rate in the inner chamber is of great importance. This means that more MAb's are produced per amount of glucose and this even if a fermentation mode is set leading to high product titers. Like other effects of the technical system on the cells reported in literature as shear stress, oxygen availabilty and others, the inner chamber's dilution rate must be taken into consideration as an important process parameter for comparable reactor systems.

462

Acknowledgement

This project is integrated in the "Forschungsverbund Bioverfahrenstechnik tierischer Zellen" and is sponsored by the BMFT (grant-nr.: 0319347A).
The authors would like to thank Prof. Dr.-Ing. J. Lehmann and his colleagues for amino acid analysis at the Department of Cell Culture Technology, University of Bielefeld, Germany and for helpful discussion.

References

[1]   Racher, A.J.; Looby, D.; Griffiths, J.B. (1990)
       J. Biotechnol. 15, 129-146.
[2]   Linardos, Th.,I.; Kalogerakis, N.; Behie, L.A.; Lamontagne, L.R. (1992)
       Biotechnol. Bioeng. 39, 504-510.
[3]   Bohmann, A.; Pörtner, R.; Schmieding, J.; Kasche, V.; Märkl, H. (1992)
       Cytotechnology 9: 551 - 557.
[4]   Niebuhr-Redder, Ch.; Kasche, V. (1990)
       Proc. German-Japanese Workshop on Animal Cell Culture Technology, Dec. 4 - 5, 1990
[5]   Kurosawa, H.; Märkl, H.; Niebuhr-Redder, Ch.; Matsumura, M. (1991)
       J. Ferm. Bioeng. 72, 1, 41-45.
[6]   Miller, W.M.; Wilke, C.R.; Blanch, H.W (1988)
       Bioprocess Eng. 3, 113-122.
[7]   McDonogh, R.M.; Bauser, H.; Stroh, N.; Chmiel, H. (1992)
       Chem. Engng. Sci. 47 (1), 271-279.

HIGH CELL DENSITY PERFUSION CULTURE OF HYBRIDOMA CELLS FOR PRODUCTION OF MONOCLONAL ANTIBODIES IN THE CELLIGEN PACKED BED REACTOR

GUOZHENG WANG, WENYING ZHANG, COREY JACKLIN,
LEE EPPSTEIN, DAVID FREEDMAN
NEW BRUNSWICK SCIENTIFIC CO., INC.
44 TALMADGE ROAD
EDISON, NEW JERSEY 08818-4005
USA

ABSTRACT. A packed-bed system employing a polyester fiber disk support matrix packed in a stationary basket has been developed for hybridoma cell culture to enhance antibody productivity. The system used a cell lift impeller inside a CelliGen Vessel. At the end of 35 days of continuous perfusion, a final cell density of 9.4×10^7 cells per cm^3 of bed volume was achieved, and a total of 11.6 grams of antibody was collected. The productivity of antibody during the production phase was 277 mg per liter per day, which represents more than a 12-fold increase when compared to static and stirred suspension culture systems. Variations in glucose consumption, lactate, and ammonia production between these systems were observed.

INTRODUCTION

The increased use of animal cell cultures for the production of biologicals has focused its attention on the economics inherent in high cell density systems. To maintain high cell concentrations, it is necessary to continually feed the cells by perfusion of medium. The enhanced productivity resulting from immobilization of hybridomas makes the methodology attractive for the development of commercial scale production systems. In animal cell biotechnology, several systems for cell immobilization have been described. These include hollow fiber reactor (Knazek et al (1972)), micro encapsulation (Lim and Moss (1981)), gel entrapment (Nilssen et al (1983)), and packed bed reactors (PBR), (Spier et al (1976), Kadouri, et al (1989)). We previously described perfusion systems for cultivation of animal cells in the CelliGen polyester fiber matrix column reactor and packed bed basket filled with a polyester fiber disk matrix (Wang et al (1992)). In comparison to other culturing systems, the new bioreactor offered several significant advantages. The polyester disk bed provides a high surface to volume ratio ($120 \ cm^2/cm^3$ of bed) for cell growth and results in high cell densities. Both anchorage dependent and suspension cells are well entrapped in the polyester disk bed. The bubble-free medium flowing through the packed bed provides the cells with optimum levels of oxygen, pH, and nutrients. The harvested product is free of cells and is produced in a low-protein medium which makes subsequent purification easier. This unique design allows more efficient and easier scale up because of the low pressure drop and low mass transfer resistance across the fiber bed. This paper reports comparisons of antibody productivity, glucose consumption, and lactate and ammonia production by hybridoma 123A cell line grown in a PBR and in comparable static and stirred culture systems.

S. Kaminogawa et al. (eds.), Animal Cell Technology: Basic & Applied Aspects, Vol. 5, 463–469.
© *1993 Kluwer Academic Publishers.*

464

MATERIALS AND METHODS

CELL LINE AND MEDIUM

The hybridoma 123A cell line produced an IgG type antibody against human gamma interferon and was kindly supplied by Dr. Sidney Pestka of the Department of Molecular Genetics & Microbiology, University of Medicine & Dentistry of NJ. DMEM was used for maintenance and growth of cells in T-flasks and bioreactors. Additional supplements were added, such as oxalacetic acid, mercaptoethanol, insulin, and antibiotics. Bovine calf serum was added in a concentration of 5% by volume in this experiment. In the packed bed basket perfusion culture the glucose concentration in fresh medium was 4.5 g/L - 8.0 g/L.

CELLIGEN PACKED BED BASKET PERFUSION CULTURE

Figure 1 shows a CelliGen basket-type packed bed and a cell lift impeller (New Brunswick Scientific). This bioreactor has a total volume of 2.5 L, working volume of 1.8 L, and basket volume of 1.0 L which was filled with 100 grams of New Brunswick Scientific's Fibracel™ polyester disks. The bioreactor was sterilized with PBS, the sterile solution was removed and replaced with fresh medium containing 1.0×10^9 total cells. Following inoculation the impeller speed was controlled at 60 rpm, then increased from 80 rpm to 130 rpm during the cultivation process (Figure 2). The bioreactor was operated in batch mode for the first two days and perfusion mode for the remainder of the run. The medium perfusion rate was increased from 1.2 L/day to 2.5 L/day in order to meet the nutrient requirements of the culture as indicated by the increase in glucose consumption rate. Dissolved oxygen at the top of the basket was maintained at 50% air saturation by controlling the flow rate of four gases. Increasing the impeller speed increased liquid flow rate which allowed for dissolved oxygen at the bottom of the basket to be maintained at 25% air saturation or greater. The pH was initially controlled at 7.2 by the four gases for the first week and then by the addition of a bicarbonate solution (8.0% W/V) $NaHCO_3$ in distilled water during the production phase.

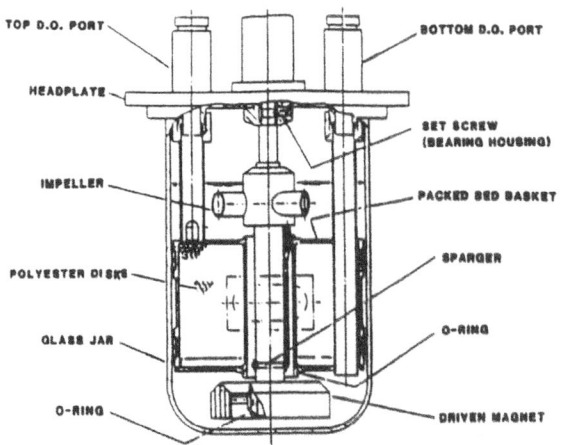

Fig 1. CelliGen Packed-Bed Basket Cell Culture System. The polyester non-woven fabric disks were packed into the basket of the reactor. Cells are immobilized on the bed. Air is introduced from the sparger, and oxygenates the medium, disengaging the bubbles at the upper fluid surface. The CelliGen impeller causes bubble-free medium to circulate through the packed-bed.

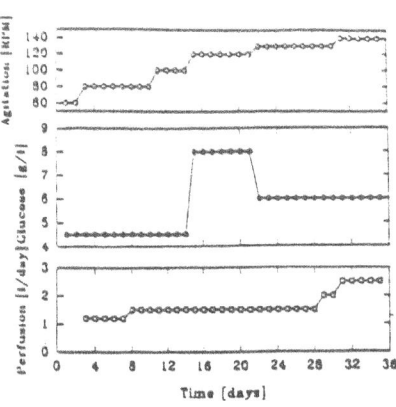

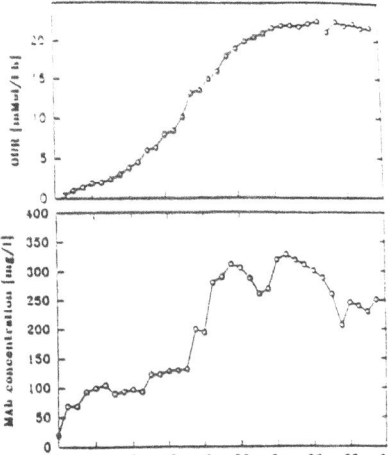

Fig 2. Culture conditions in the CelliGen packed bed basket reactor. (o) agitation [rpm]; (•) glucose concentration in perfused medium [g/l]; (□) perfusion rate of medium [l/day].

Fig. 3. Oxygen uptake rate (OUR) and MAb concentration during the cultivation process in the packed bed reactor.

Cultures were also grown in 75 cm^2 static T-flasks (50 ml of medium) and in the 2.5 L stirred tank reactor (1.9 L of medium). The pH and DO in the reactor were controlled at 7.2 and 50% air saturation respectively through direct sparging of four gases. The flask and stirred systems were run as repeated batch cultures, sampled daily over a period of 28 days and 21 days, respectively.

ASSAYS

The cell concentration was determined by manual cell counting with a hemocytometer with cell viability determined by the Trypan blue dye exclusion method. Glucose concentration was determined spectrophotometrically with a glucose hexokinase reagent from ALPKEM. Concentration of lactate and ammonia were determined using assay kits from Sigma Diagnostics: No. 16-UV and No. 170-UV respectively. Monoclonal antibody was determined by radial immunodiffusion (RID) assay utilizing anti-mouse IgG.

RESULTS AND DISCUSSION

CELL DENSITY AND VIABILITY

The viable cell density in the static and stirred systems is shown in Figures 4 and 5. During the 28 days of static culture, the cell density was increased from 1.3×10^6 cells/ml at day 5 to 2.8×10^6 cells/ml at day 26, and the viability was stable at about 68%. For the suspension culture in a stirred tank system, the viable cell density was 3.2×10^6 cells/ml after the first 3 days growth, and then decreased to 1.4×10^6 cells/ml at the end of the run. The viability of cells was also decreased from 76% at the end of the first batch to 25% at the end of the last batch.

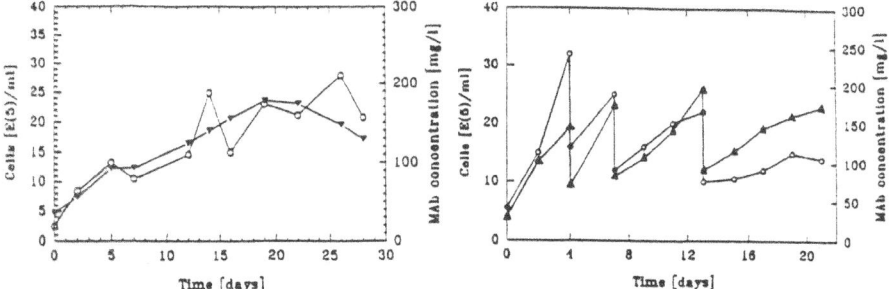

Fig 4. Cell growth and monoclonal antibody production in a static flask (30 ml). Hybridomas (123A) were grown in repeated batch culture and fed fresh medium of 25 mls every 2-3 days. (o) viable cells [10⁵ cells/ml]; (▲) Mab concentration [mg/l].

Fig 5. Cell growth and monoclonal antibody production in a 2.5 L CelliGen stirred tank. Cells (123A) were grown in repeated batch culture and fed fresh medium of 950 mls periodically. (o) viable cells [10⁵ cells/ml]; (▲) MAb concentration [mg/l].

The culture in the CelliGen packed bed basket was started with a cell concentration of 5×10^5 cells/ml. The rate of cell entrapment in the polyester matrix was very fast. Within one hour after cell inoculation, more than 95% of the cells were immobilized just by physical entrapment. There was no direct way to sample cells in the polyester disks to monitor cell growth. However, propagation of cells could be evaluated by oxygen uptake rate (OUR) values during the run. The final cell density of 9.7×10^7 cells per cm^3 of packed bed volume and the cell viability of 60% were determined by Trypan blue cell counts at the end of the 35-day run when the packed basket was dismantled. From the initial and final cell numbers, a calculated multiplication ratio of 94 was achieved.

Suspension cells in the harvested media were very low, with average daily cell concentrations of 3.9×10^4 cells/ml. This represented less than 0.2% of the cells in the polyester bed. The cell viability from daily medium samples was 67% which was almost the same as in the static flasks and much higher than in the stirred tank system. These results show that the polyester disk bed can provide a very high surface to volume ratio and can achieve high cell densities for long periods of propagation. The basket bioreactor eliminated the exposure of cells to gas-liquid interfacial forces as well as impeller shear forces and minimized hydrodynamic effects while maintaining adequate nutrient and oxygen transfer.

MAB CONCENTRATION AND PRODUCTIVITY

A comparison of the MAb concentration and productivity of the various systems of cell propagation tested is shown in Table 1. For the perfusion culture in the CelliGen packed bed, a total of 54 liters of culture medium containing 11.6 grams of antibody was collected. The average MAb concentration of harvested media in the packed bed basket system was 46% higher than in the static flasks, and 34% higher than in the stirred tank system. These results showed that the total media consumption per unit of product in perfusion culture is considerably lower than in batch culture. The productivity during the steady-state phase in the packed bed system was 277 mg/liter/day, representing 12 and 14-fold increases in production compared to the static and stirred reactor.

Table 1, MAb concentration and productivity in different cell propagation systems

Bioreactor Type and Mode	MAb Concentration (mg/L)	Productivity (mg/L/day)
Static T-flask with repeated batch culture	117	23
Stirred tank with repeated batch culture	142	20
Packed bed basket with perfusion culture	215	277

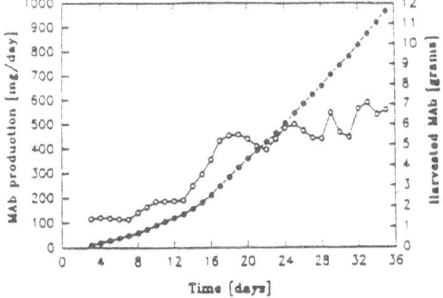

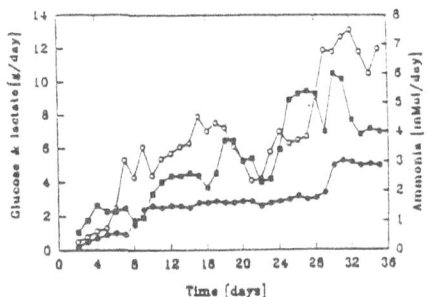

Fig. 6. Mab production from hybridoma 123A cultured in the CelliGen packed bed reactor. Total yield from this run was 11.6 grams.

Fig 7. Glucose consuption and lactate and ammonia production during the cultivation process in the packed bed reactor. (o) glucose [g/day]; (•) lactate [g/day]; (■) ammonia [mmol/day].

CELL METABOLISM

Table 2 shows the overall glucose consumption rates, and lactate and ammonia production rates of hybridoma 123A cell line grown in various reactors. It can be seen that glucose consumption rates generally mirrored the lactate production rates. Ammonia production rates varied between the culture systems, approximately 140% greater in the stirred tank than in packed bed basket cultures. The ammonia concentration in the stirred tank culture reactor was 1.5 mmoles at the end of the first batch and increased to 3.8 mmoles at the end of the last batch. With the 123A cell line, we have observed a decrease in cell viability and increase in waste products during later periods of cell growth in the suspension culture system. This indicates that the 123A cells used in these studies are very sensitive to shear forces and difficult to grow in stirred tank under direct sparging. Due to continuous dilution with fresh media, the concentration of lactate (1.0 - 2.5 g/L) and ammonia (1.0 - 3.0 mmol) remained at a lower level than that detected in batch or repeated batch stirred tank cultures, even though cell concentrations are more than 10-fold higher in the perfusion system.

Table 2 Cell metabolism in different systems of cell propagation.

Bioreactor Type and Mode	Glucose Consumption Rate $(mg/10^6$ cells/day$)$	Lactate Production Rate $(mg/10^6$ cells/day$)$	Ammonia Production Rate $(\mu mol/10^6$ cells/day$)$
Static T-flask with repeated batch	0.27	0.17	0.15
Stirred tank with repeated batch	0.21	0.09	0.22
Packed bed basket with perfusion	0.19	0.08	0.09

CONCLUSION

In comparison to other systems, the CelliGen packed bed basket reactor provides several desirable advantages for animal cell cultivation. Experiments have shown that hybridoma cells can be grown in a polyester fiber matrix achieving high multiplication ratios and high cell densities. In addition, adequate conditions for cell growth and product synthesis in the bioreactor can be maintained for a long period of time.

Several other advantages of the packed bed basket increase its potential as a useful technology for commercial production of secreted antibodies. The potential savings in downstream processing of antibodies are very high due to the high amount of antibody concentration and lack of cells in the harvested medium. Obviating the need for any cell separation device also reduces the overall cost of the packed bed basket.

Shear forces are kept at a minimum in the packed bed basket. This allows higher agitation rates which increases mass transfer of oxygen and nutrients. As a result, high cell densities are easily maintained during the steady state phase along with productivities that are 12 times higher than traditional stirred tank and static cultures.

Currently, benchtop reactors from 2 to 7.5 Liters have been successfully tried with the packed bed basket impeller. Studies are being conducted for the potential scale-up of this technology in the 30 to 150 Liter range.

REFERENCES

Kadouri, A. and Zipori, D. (1989) "Production of Anti-leukemic Factor From Stroma Cells in a Stationary Bed Reactor on a New Cell Support". In: Spier, R.E., Griffiths, J.B., Stephenne, J. and Crooy, P.J. (eds.) *Advances in animal cell biology and technology for bioprocesses.* pp 327-330, Courier International Ltd., Tiptree, Essex.

Knazek, R.A., Gullino, P.M., Kohler, P.O., and Dedrick, R.L. (1972) "Cell Culture on Artificial Capillaries: An Approach to Tissue Growth in Vitro." Science 178: 65-67

Lim, F., and Moss, R.D. (1981) "Microencapsulation of Living Cells and Tissues." J. Pharm Sci. 70: 351-354

Nilsson, K., Scheirer, W., Merten, O.W., Ostberg, L., Liehl, E., Katinger, H.W.D., and Mosbach, K. (1983) "Entrapment of Animal Cells For Production of Monoclonal Antibodies and Other Biomolecules." Nature 302: 629-630

Spier, R.E., and Whiteside, J.P., (1976) "The Production of Foot and Mouth Disease Virus From BHK 21 C 13 Cells Grown On The Surface of Glass Spheres." Biotechnol. Bioeng. 18: 649-657

Wang, G., Zhang, W., Jacklin, C., Freedman, D., Eppstein, L., and Kadouri, A. (1992) "Modified CelliGen Packed Bed Bioreactors For Hybridoma Cell Cultures". Cytotechnology 9: 41-49

Metabolic parameters of a hybridoma cell line in batch and continuous cultivation

Ralf Pörtner, Armin Bohmann, Ines Lüdemann,
Herbert Märkl
Technische Universität Hamburg-Harburg
Bioprozeß- und Bioverfahrenstechnik
Denickestr. 15,   2100 Hamburg 90, Germany

ABSTRACT

For a hybridoma cell line a decrease of the glucose uptake rate was observed during the course of batch cultures, even when the growth rate
remained constant. These results are valid for cell densities between $2*10^5$
and $2*10^6$ cells/ml. On the other hand these parameters remained constant
at cell densities between $2*10^6$ and $1.5*10^7$ cells/ml during continuous
cultivation. A relationship between the specific glucose uptake rate and the
cell number could be found.

1. Introduction

Specific substrate uptake rates (glucose, glutamine, amino acids) or
production rates for metabolites (ammonia, lactate) are important parameters for the performance and control of cell culture reactors. For
hybridoma cells literature data on these parameters are in some cases
inconsistent. Miller et. al. /1/ described an increase of the specific glucose
and glutamine uptake rate with increasing growth rate. Similar to this
Seaman and Hu reported an increase of the specific glucose uptake rate
with increasing growth rate and increasing cell number during the first stage
of a perfusion culture /2/. But at high cell densities in the same experiment
the specific glucose uptake rate remained constant, even when the specific
growth rate decreased further. Finally Shirai et. al. found a higher growth
yield for glucose in conditioned medium compared to fresh medium and
decreasing glucose and glutamine uptake rates at identical specific growth
rates in the course of a high density culture /4/.
Until now a satisfying theory for the effects of growth rate, substrate or
metabolite concentrations or other culture conditions on metabolic parameters is not available. Here metabolic parameters of a hybridoma cell line
determined in batch and continuous suspension cultures will be discussed.

471

S. Kaminogawa et al. (eds.), Animal Cell Technology: Basic & Applied Aspects, Vol. 5, 471–478.
© 1993 *Kluwer Academic Publishers.*

2. Materials and Methods

A mouse-mouse hybridoma cell line /4/ producing monoclonal antibodies (MAb) against penicillin-G-amidase was cultivated in a 1:1 mixture of Iscove's MEM and Ham's F12 supplemented with 3 % FCS, 2 mmol/l L-glutamine and 2 g/l $NaHCO_3$.

Glucose, lactate, ammonia and L-glutamine were determined enzymatically, monoclonal antibody by a mouse-IgG-ELISA (all kits from Boehringer, Germany). Cells were counted using a haemocytometer. Cell vialbility was determined by exclusion of trypan blue.

The cells were cultivated under static conditions in T-25 culture flask. A 1 l-glass reactor (working volume 600 to 800 ml) with DO and pH control, agitated with a flat-blade impeller, was used for repeated batch and continuous chemostat cultivations.

For high density cultivation a membrane dialysis reactor was used (Fig. 1). In this reactor (outer chamber 4.2 l, inner chamber 1.35 l) fresh medium and oxygen are supplied to the cells mainly via a dialysis membrane (Cuprophan, Enka) with a cut off of 10,000 Dalton /5/. Toxic metabolites such as lactic acid and ammonia are dialysed through the membrane, while MAb's are retained and accumulated in the inner chamber. In continuous mode the reactor is fed with serum-free medium in the outer and a serum containing medium in the inner chamber. Here the reactor was started in a repeated batch mode and finally run continuously in both chambers.

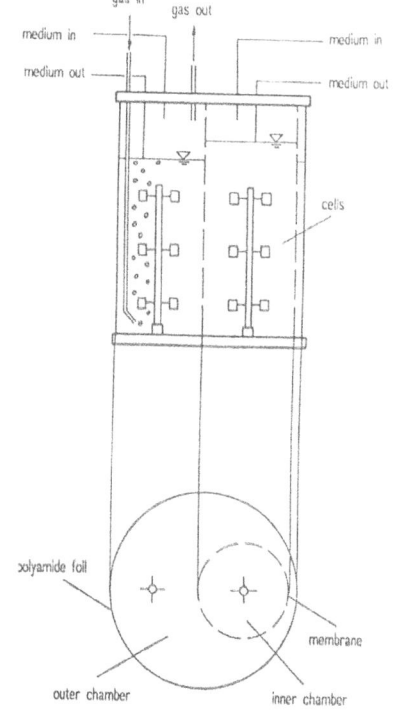

Fig. 1:   Membrane dialysis reactor

3. Results and discussion

A typical course of a repeated batch culture experiment performed in the
1 l-reactor is shown in Fig. 2. During the first batch phase the viable cell
number increased from $1.1*10^5$ to $6*10^5$ cells/ml. The growth rate μ
remained more or less constant, while the specific glucose uptake rate
q(Glc) decreased drastically.
50 % of the medium were exchanged after 73 h. The cells continued to
grow with the same growth rate. But with the medium exchange the glucose
uptake rate jumped to a higher value, inversely to the cell number, and
decreased during the progress of the culture with increasing cell concentration. The same occured after the second medium exchange. Similar
effects were obtained for the specific glutamine uptake rate and for
production rates of lactate and ammonia, respectively (data not shown).

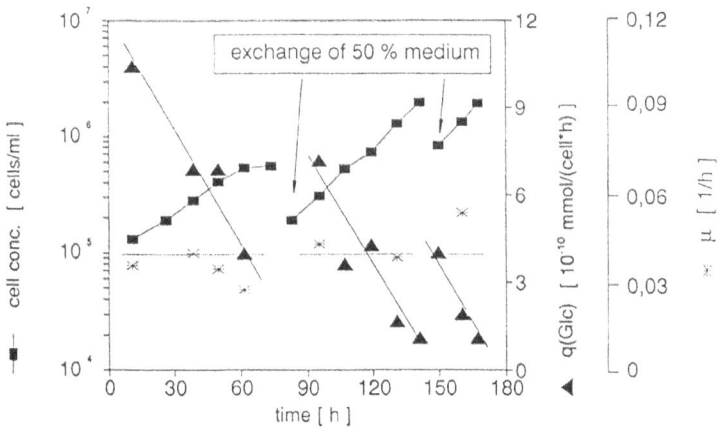

Fig. 2:   Cell number, specific growth rate μ and specific glucose uptake
rate q(Glc) of hybridoma cells during repeated batch cultivation in
a stirred 1 l-reactor

Several continuous cultivations were performed in the stirred 1 l-reactor, which was run as a chemostat. Steady state values obtained at different cell densities are summerized in Tab. 1. From the data for the chemostat a decreasing glucose uptake rate can be detected with increasing cell number and decreasing glucose concentration.

	cell conc. [cells/ml]	q(Glc) [mmol/(cell*h)]	c(Glc) [mmol/l]	μ [1/h]
chemostat	$6 * 10^5$	$3.3 * 10^{-10}$	7.7	0.026
	$1.53 * 10^6$	$2.5 * 10^{-10}$	6.8	0.034
	$2.5 * 10^6$	$1.8 * 10^{-10}$	3.5	0.028
membrane dialysis reactor	$1.4 * 10^7$	$1.5 * 10^{-10}$	1.7	0.023

Tab. 1: Steady state values for continuous cultivations in a 1 l-reactor run as a chemostat and the membrane dialysis reactor

A fermentation in the membrane dialysis reactor is shown in Fig. 3. The experiment was started as a repeated batch, where part of the medium in inner and outer chamber was changed twice (30 % after 5 days, 55 % after 6 days). After 8 days the reactor was run continuously in the outer and inner chamber. The cell density finally reached $1.4*10^7$ cells/ml at a dilution rate of 0.2 1/d in the inner and 1.8 1/d in the outer chamber. The specific growth rate dropped slightly during the cultivation. But the specific glucose uptake rate decreased during the first 5 days and remained constant at cell densities higher than appr. $2*10^6$ cells/ml.

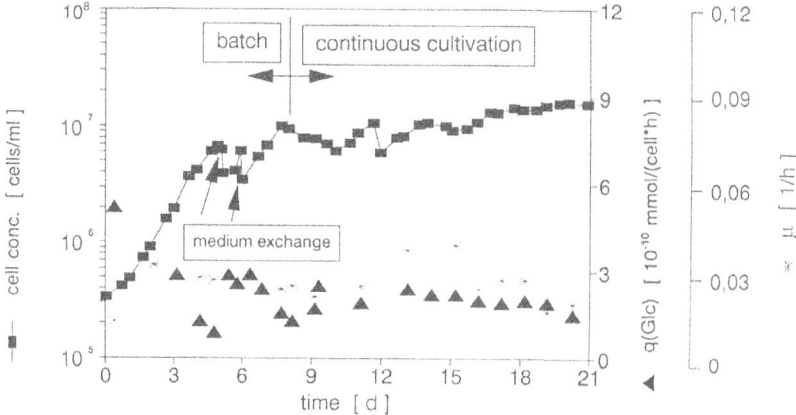

Fig. 3: Cell number, specific growth rate μ and specific glucose uptake
rate q(Glc) of hybridoma cells cultivated in the membrane dialysis
reactor

Several possible reasons for changing metabolic parameters are discussed
in literature. Miller et al. /1/ proposed a relationship between the substrate
uptake rate and the specific growth rate (maintenance energy model). But
the above results can not be related to the changing specific growth rate.

Another explanation might be seen in the changing substrate concentration
during a batch fermentation or at different dilution rates in continuous
cultivations. At high substrate concentrations the cells might have a higher
substrate uptake not leading to an increased growth rate but possibly to
larger cells. The method used for determination of the cell number doesn't
give any information on the cell size. Fig. 4 shows a plot of the specific
glucose uptake rate q(Glc) vs. the glucose concentration c(Glc). Additionally
to the data presented above results from batch experiments in T-25 flasks
are included (data not discussed here). At high glucose concentrations a
significantly higher glucose uptake rate can be observed, underlining the
theory discussed above. But the plot doesn't show a satisfying fit in one
curve.

Appart from the medium components discussed above, which are well known and analyzable, autocrine components are regarded as contributing to animal cell growth and metabolism. Shirai et al. /3/ postulated, that autocrine components might be responsable for changes in the metabolic activity of hybridoma cells. They observed a higher growth yield of glucose corresponding to a lower glucose uptake rate, when cells were grown in conditioned medium. In perfusion systems the growth yield was higher at lower perfusion rates, were the concentration of autocrine components can be expected as higher as at higher perfusion rates.

Until now only little is known, what kind of autocrine components might be responsible for the changes in the metabolic activity. But in a first approach it can be postulated, that these components are produced by the cells with a constant production rate. A relationship to the cell concentration might be considered. Fig. 4 shows a plot of the specific glucose uptake rate q(Glc) vs. the cell concentration. The data for all experiments fit quite well in one curve.

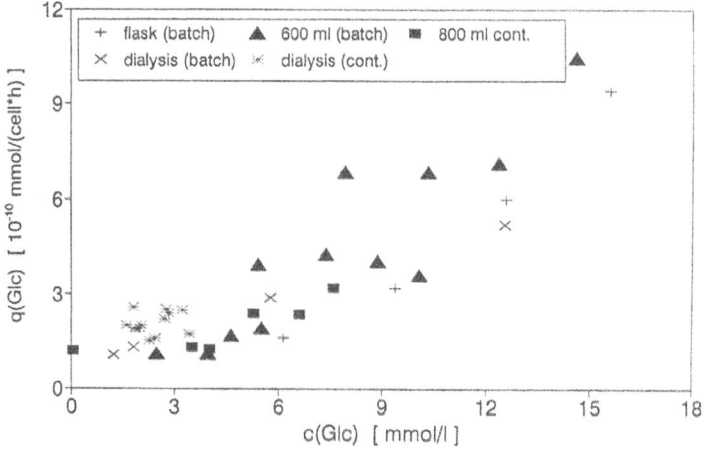

Fig. 4:   Relationship between the specific glucose uptake rate and the glucose concentration for batch and continuous cultivations Summary of different batch and continuous cultivations

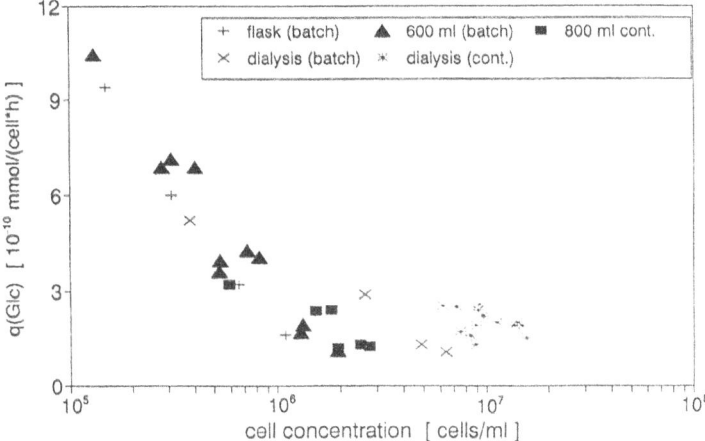

Fig. 5:   Relationship between the specific glucose uptake rate and the
           cell concentration for batch and continuous cultivations
           Summary of different batch and continuous cultivations

The specific glucose uptake rate decreases between 10^5 and $2*10^6$ cells/ml
and remains constant up to 10^7 cells/ml. Similar results were obtained for
the specific glutamine uptake rate and production rates for ammonia and
lactate, respectively (data not shown).

The two possible effects on the metabolic activity of the hybridoma cell line
investigated here can not be seen separatly, because usually high glucose
concentrations occure at low cell densities whereas the glucose concentration is low at high cell densities. At least in Fig. 5 the data fit better in one
curve than in Fig. 4, indicating, that a relationship between the specific
glucose uptake rate and the cell concentration gives a better correlation
usebal for the control of cell cultures.

Further studies have to scope on the mechanisms responsible for effects on
the specific metabolic parameters and on kinetic models for their description
to enable the development of control strategies for animal cell culture
reactors.

5. Acknowledgement

This project is integrated in the "Forschungsverbund Bioverfahrenstechnik tierischer Zellen" and is sponsored by the BMFT (grant-nr.: 0319347A).

6. References

/1/   Miller, W.M.; Blanch, H.W.; Wilke, C.R.: A kinetic analysis of
        hybridoma growth and metabolism in batch and continuous
        suspension culture: Effect of nutrient concentration, dilution rate and
        pH; Biotechnol. Bioeng. 73, 2, 159-165, 1992
/2/   Seamans, T.C.; Hu, W.-S.: Kinetics of growth and antibody
        production by a hybridoma cell line in a perfusion culture; J. Ferment.
        Bioeng. 70, 4, 241-245, 1990
/3/   Shirai, Y.; Hashimoto, K.; Takamatsu, H.: Growth kinetics of
        hybridoma cells in high density culture; J. Ferment. Bioeng. 73, 2,
        159-165, 1992
/4/   Niebuhr-Redder, Ch.; Kasche, V.: Proc. German-Japanese Workshop
        on Animal Cell Culture Technology, Dec. 4-5, 1990
/5/   Märkl, H.; Matsumura, M.; Kurosawa, H.: Dialysis bioreactor with
        radial-flow fixed bed for animal cell culture; J. Ferment. Bioeng. 72,
        1, 41-45

DEVELOPMENT OF A NEW DISPOSABLE CULTURE SYSTEM FOR HUMAN LYMPHOKINE-ACTIVATED KILLER (LAK) CELLS

M. MIYAMOTO,[1] M. MURATA,[2] I. YOSHINO,[3] M. TOGAMI,[2] T. YANO,[4]
K. YASUMOTO,[5] K. SUGIMACHI,[6] G. KIMURA,[3] AND K. NOMOTO[1]

[1] Department of Immunology, Medical Institute of Bioregulation, Kyushu University, Maidashi 3-1-1, Higashi-ku, Fukuoka 812, Japan;

[2] Japan Synthetic Rubber Co. Ltd., Tokyo, Japan; [3] Department of Virology, Medical Institute of Bioregulation, Kyushu University, Fukuoka Japan;

[4] Department of Chest Surgery, National Kyushu Cancer Center, Fukuoka, Japan;

[5] Department of Chest Surgery, Kitakyushu Municipal Medical Center, Kitakyushu, Japan; [6] Department of Surgery II, Faculty of Medicine, Kyushu University, Fukuoka, Japan

ABSTRACT. Lymphocytes of the regional lymph nodes obtained from 4 patients with lung cancer by surgical operations were cultured with rIL-2-containing medium for expansion and induction of the lymphokine-activated killer (LAK) cells. Using a new disposable culture system, the maximal cell density reached approximately ten times higher than that in a conventional dish culture system, and the LAK cells induced had high cytotoxicity. This disposable system should improve the routine performance of adoptive immunotherapy with LAK cells with respect to greater efficiency, easier manipulation, and lower cost.

INTRODUCTION. Adoptive immunotherapy with lymphokine-activated killer (LAK) cells and recombinant interleukin-2 (rIL-2) has been demonstrated to mediate tumor regression in some patients with advanced cancer (Rosenberg et al., 1987). The major obstacle to this therapy has been the complexity and expense of the ex vivo cultivation of a large number of lymphocytes by a conventional tissue culture system.

We have recently developed a unique culture device of 400 ml in volume (LIFROC, Liquid Filled Rotary Culture; formerly designated JCC device) for cultivating LAK cells at a high cell density that is essentially based on a dialysis-perfusion culture (Murata et al., 1990 and 1991). With LIFROC, we were able to culture LAK cells at a cell density approximately 10 times higher (2.7×10^7 cells/ml) than that with the conventional tissue culture system, and to reduce the cost and manpower required.

In this study, we developed a disposable version of LIFROC for clinical use. Using the disposable LIFROC system and a culture medium supplemented with rIL2, we cultured lymphocytes that were obtained from the regional lymph nodes removed by a surgical operation for lung cancer, and examined the cellular expansion efficiency, phenotype and function in comparison with the data obtained with the conventional cell culture system.

MATERIALS AND METHODS

rIL-2. Human rIL-2 (Kato et al., 1985) was kindly provided by Takeda Chemical Industries, Osaka, Japan. The specific activity was 1.4×10^7 JRU (Japan reference units)/mg of protein. Usually, 2,000 JRU/ml (1714 Biological Response Modifier Program units/ml) of this rIL-2 is able to induce optimum LAK activity *in vitro.*

Lymph node lymphocytes (LNL). Regional lymph nodes were obtained from lung cancer patients in a surgical operation. The lymph nodes were squeezed between a pair of slide glasses in Hanks' balanced salt solution (HBSS) and then passed through a gauze filter. The cells were then washed twice with HBSS and resuspended in RPMI1640 (Nissui Seiyaku Co. Ltd.,Tokyo, Japan) containing 20 mM HEPES, 100 U/ml of penicillin G, and 100 μg/ml of streptomycin sulfate (refered to as the basal medium, BM).

Cell culture with the disposable LIFROC system for LAK cells. The prototype of the LIFROC system, which is essentially based on a dialysis-perfusion culture, has been described previously (Murata et al., 1990). In this study, we developed a disposable version of LIFROC(400ml in culture volume). As shown in Fig. 1, the system is

S. Kaminogawa et al. (eds.), Animal Cell Technology: Basic & Applied Aspects, Vol. 5, 479–486.
© 1993 *Kluwer Academic Publishers.*

composed of 3 sections: culture vessels, a rotary device and medium reservoirs. Disposable parts were the culture vessels (made of polymethylmetacrylate), and the medium reservoirs (made of vinylchloride). If desired, 4 culture vessels (400ml each) are applicable to the rotary device at a time. The culture vessel comprises a culture unit and a reflux unit separated by a dialysis membrane (SPECTRA/POA 2, Spectrum Medical Institute, Los Angeles, CA) which is supported by a plastic mesh (Fig. 2). Cells were cultured in the culture unit, and beads of 1 mm in diameter (specific gravity of 1.04 g/cm^3, made of polystyrene, Sumitomo Bakelite Co., Tokyo, Japan) are added to the culture unit as agitators of the culture medium at a concentration of 2 g/l. These beads can be completely removed from LAK cell preparations at the time of harvesting as described later. The culture vessel, the medium reservoir, and the agitation beads are sterilized with ethylene oxide. A silicon tube is pre-set helically inside the culture unit (Fig. 2), and CO_2-mixed air is supplied to the culture unit from the surface of this silicon tube. Fresh medium is pumped into the reflux unit through a silicon tube from the medium reservoir, and the medium components and culture wastes are exchanged through the dialysis membrane, and before being returned to the medium reservoir through another silicon tube. The culture vessels and the rotary device are operated in a 37℃ incubator with a sterile laminar flow hood (MK-1 type, Ikemoto Scientific Technology Co., Tokyo, Japan). The culture vessel is rotated on its horizontal axis at approximately 6 rpm.

LNL were suspended at 2×10^6 cells/ml in BM supplemented with 5 % heat-inactivated (56℃ for 30min) pooled human AB serum, 2,000 JRU/ml of rIL-2 and 10 ng/ml of anti-CD3 monoclonal antibody (NU-T3, Nichirei Co., Tokyo, Japan), and incubated in two 175-cm^2 tissue culture flasks (200ml of medium, Falcon) at 37℃ in 5 % CO_2 for static culture (pre-incubation). After a period of 4-8 days of this pre-incubation, the whole culture content was transferred (without medium refreshment) to the culture unit of the culture vessel in the disposable LIFROC system. During the cultivation by the LIFROC system, 8×10^5 JRU of rIL-2 was added to the culture unit every 4 days. BM in the reservoir was replaced with fresh BM at appropriate intervals, the volume of BM in the reservoir being 2-4 l, and the frequency of the medium reservoir replacement was increased depending on the cell density.

At the time of cell harvesting, medium perfusion was stopped, the cell suspension was drained by gravity and passed through a nylon mesh filter (DIN 110-50, Nytal, Switzerland) to remove the agitation beads, and the cells were collected in centrifuge tubes. One-hundred ml of HBSS was added to the culture vessel before rotating for 1 min to harvest the remaining cells. This procedure was repeated 2 times.

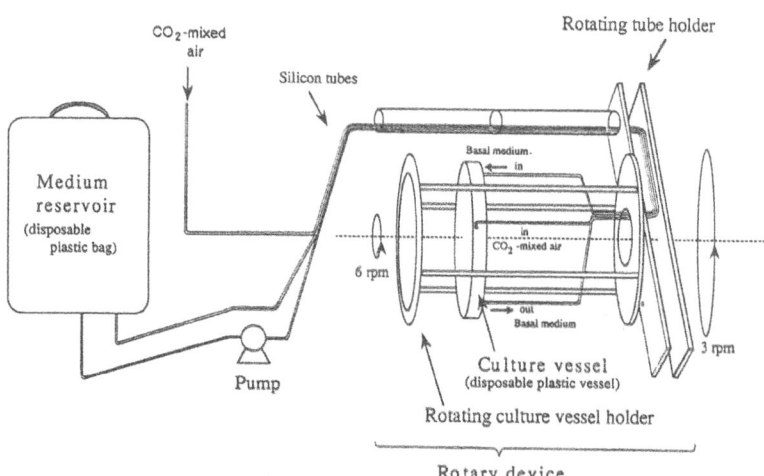

Fig. 1. Simplified representation of the disposable LIFROC system

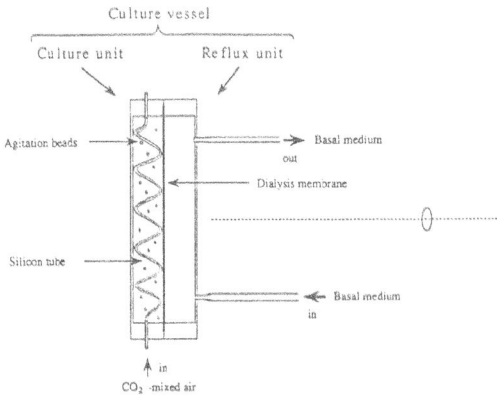

Fig. 2 Simplified representation of the culture vessel of the disposable LIFROC system

Conventional static culture of LAK cells. LNL was suspended at 1×10^6 cells / ml in BM containing 10 % heat-inactivated pooled human AB serum (complete medium, CM) and 2,000 JRU/ml of rIL-2 . Five ml each of the cell suspension was placed in 60 mm tissue culture dishes (Falcon 3002, Lincoln Park, NJ) and incubated in 5 % CO_2 at 37℃. Before the cell density had reached more than 2×10^6 cells/ml at each cell passage, the cells were subcultured at a density of 1×10^6 cells/ml in fresh CM with rIL-2. Subcultures were also performed two days before cytolysis assays, allowing the LAK cells to exhibit their optimum cytotoxicity (Yano *et al.*, 1989).

Cytolytic assay. The cytolytic activity of LNL was assayed by a 4-h ^{51}Cr-release assay against two human tumor cell lines: Raji, a natural killer (NK)-resistant B-cell lymphoma cell line (Epstein *et al.*, 1966),and K-562, an NK-sensitive erythroleukemia cell line (Lozzio *et al.*, 1973). Detailed procedures have been described elsewhere (Murata *et al.*, 1990). All assays were made in triplicate, and the percentage specific cytolytic activity was calculated according to the following formula:

$$\% \text{ cytolysis} = \frac{\text{experimental (cpm)} - \text{spontaneous (cpm)}}{\text{maximal (cpm)} - \text{spontaneous (cpm)}} \times 100$$

Data are reported as the mean of triplicates, the standard deviation always being less than 10 %.

Phenotypic analysis. Fluorescein isothiocianate (FITC)-conjugated monoclonal antibodies of anti-CD3 (NU-T3), anti-CD4 (NU-TH/I), and anti-CD20 (NU-B2) were purchased from Nichirei Corporation, Tokyo, Japan. FITC-conjugated anti-CD16 (Leu16) and phycoerythrin (PE)-conjugated anti-CD8 (Leu2a) were purchased from Becton Dickinson, Moutain View, CA. Cells were stained for 30min at 4℃ with each antibody, and after washing three times with PBS, the cells were analysed with a FACscan (Becton Dickinson).

RESULTS
Expansion of LNL by using the disposable LIFROC system. The new disposable LIFROC culture system for inducing and expanding LAK cells from LNL for the adoptive immunotherapy of cancer is shown in simplified form in Figs. 1 and 2.

Fig. 3 shows the results of one of the experiments for the growth curve of cultured LNL, and the concentration levels of glucose (as a carbon source for the medium) and lactate (as a waste from the culture medium) in the culture unit compared with that in the medium reservoir. As shown in Fig. 3A, final cell number reached 1.1×10^{10} cells at 12 days of culture, and the cell density at this time reached 2.75×10^7 cells/ml.

As shown in Fig. 3B, the concentration of glucose in the culture unit on day 8 (at the end of pre-incubation) was 40 % of the initial level, but after a 2-day cultivation with dialysis perfusion by the disposable LIFROC system, it had recovered to 80 % of the fresh medium level. During cultivation by this system, the glucose level recovered at every replacement of the medium reservoir being kept high throughout the cultivation.

As shown in Fig. 3C, the level of lactate in the culture unit at day 8 (at the end of pre-incubation) was very high, but after 2 days cultivation by the disposable LIFROC system, it had decreased to one third of the level on day 8. The concentration of lactate was kept at a low level by the disposable LIFROC system.

The expansion of LNL (LAK cells) by the disposable LIFROC system is shown in Table 1. Using the system, we cultured LNL separated from lymph nodes obtained from 4 patients. As a control, we cultured LNL by a conventional dish culture system in parallel. The maximal final cell density reached 2.75×10^7 cells/ml, and the average final cell density was 2.45×10^7 cells/ml. This value for the average final cell density was approximately ten times higher than that by the conventional dish culture system (data not shown).

The maximal cell number harvested was 1.1×10^{10} cells (average of 9.8×10^9 cells), which is thought sufficient for adoptive immunotherapy with LAK cells. The degree of expansion was from 12.1- to 24.3-fold (average 17.1-fold), but these values are not significantly different from those by the conventional dish culture.

The culture period required to reach the stationary phase with

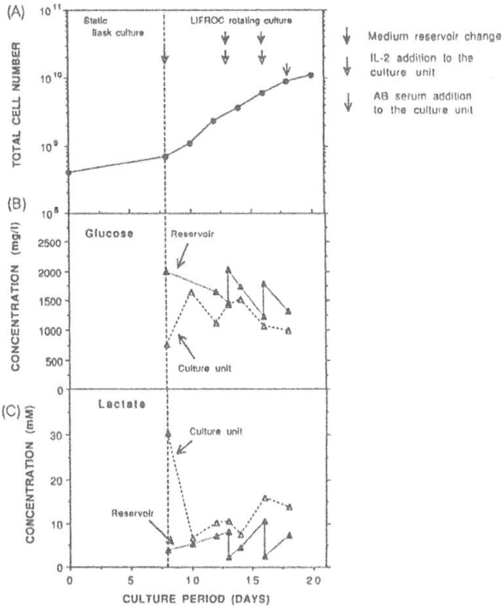

Fig. 3. Growth of LNL and concentration levels of glucose and lactate during the cultivation with the disposable LIFROC system.
(A) Growth curve for LNL, (B) glucose concentration in the culture unit and medium reservoir, (C) lactate concentration in the culture unit and medium reservoir.

the disposable LIFROC system depended on the number of cells inoculated, which itself differed among the patients, and ranged from 2 to 3 weeks.

Table 1. Expansion of LNL (LAK cells) by the disposable LIFROC system

Patient	LIFROC system					Dish culture
	No. of cells inoculated	period of culture (days)	No. of cells harvested	Final cell density (cells/ml)	Degree of expansion	Degree of expansion
1	7.5×10^8	22	9.0×10^9	2.25×10^7	12.1	ND
2	4.1×10^8	14	1.0×10^{10}	2.50×10^7	24.3	25.5
3	6.8×10^8	12	1.1×10^{10}	2.75×10^7	16.2	16.3
4	5.9×10^8	14	9.2×10^9	2.30×10^7	15.6	27.5

ND ; not determined

Phenotypic analysis of LAK cells induced by the disposable LIFROC system. A phenotypic analysis of the LAK cells induced by the disposable LIFROC system, compared with conventional dish culture, is shown in Table 2. Phenotypes were analysed by using specific antibodies against the cell surface markers of lymphocytes. The phenotypes of the LAK cells induced by the disposable LIFROC system were not significantly different from those induced by the conventional dish culture nor from previous data (Yano *et al.*, 1989).

Table 2. Phenotypes of LNL (LAK cells) cultured by the disposable LIFROC system

Patient	Culture system	Period of culture (days)	Positive cells (%)				
			CD3	CD4	CD8	CD16	CD20
1	LIFROC	22	94.3	26.6	64.3	0.45	1.4
	Dish		ND	ND	ND	ND	ND
2	LIFROC	14	90.5	16.9	67.7	1.3	3.4
	Dish		92.3	17.3	68.1	1.4	3.4
3	LIFROC	12	73.1	12.7	73.8	0.15	1.1
	Dish		74.3	12.6	75.3	0.15	0.4
4	LIFROC	14	87.1	16.1	74.9	0.04	6.3
	Dish		74.8	13.5	75.4	0.12	4.6

Phenotypes were analyzed by FACScan.
ND : not determined
CD3 : T cell marker,   CD4 : helper T cell marker,   CD8 : cytotoxic T cell marker
CD16 : NK cell marker,   CD20 : B cell marker

Cytotoxicity of LAK cells induced in the disposable LIFROC system. The cytotoxicity of the LAK cells induced by the disposable LIFROC system is shown in Table 3, being compared with that by the conventional dish culture. Cytotoxicity was assayed when the LAK cells were finally harvested by a 4-h ^{51}Cr-release assay against two human tumor cell lines (K562 and Raji) with the E/T ratio indicated in the table. The LAK cells induced by the disposable LIFROC system had sufficiently high cytotoxicity for adoptive immunotherapy with LAK cells, and there was no significant difference between these two culture systems.

Table 3. Cytotoxicity of LNL (LAK cells) cultured by the disposable LIFROC system

Patient	Period of culture (days)	Cytotoxicity (E/T=20/1) (%)			
		K562		Raji	
		LIFROC	Dish	LIFROC	Dish
1	22	ND	ND	ND	ND
2	14	65.1	68.2	60.3	62.1
3	12	70.8	83.4	82.1	82.1
4	14	51.2	44.1	61.2	47.8

ND : not determined

Estimated costs of the medium ingredients for expanding of lymphocytes. The estimated costs of medium ingredients for expanding and inducing LAK cells are shown in Table 4. These were calculated for the case of expansion from 4×10^8 to 1×10^{10} cells by the disposable LIFROC system in comparison with the conventional flask culture (175 cm^2 tissue culture flask). In the case of the disposable LIFROC system, we used 2 flasks and a disposable culture vessel (400 ml) . The total amount of medium used in the disposable LIFROC system was about 10 l, 400 ml in the culture vessel and 10 l of reservoir medium. The serum volume was calculated as 10% of the medium in the culture unit of the culture vessel, and there was no serum addition to the medium reservoirs.
The quantity of rIL-2 in the disposable LIFROC system was calculated as 8×10^5 JRU for static culture, and 32×10^5 JRU (3 times addition of 8×10^5 JRU each) for the rotating culture. It is clear from Table 4 that the serum volume, quantity of rIL-2, and number of vessels used with the disposabl LIFROC system were less than those used for conventional flask culture. Therefore, there is a large difference in the cost for inducing and expanding LAK cells between the two culture systems.

Table 4. Estimated costs of medium ingredients for expanding lymphocytes from
4×10^8 to 10^{10} cells : comparison of disposable LIFROC and conventional flask culture

	Culture system	
	LIFROC	Flask [1]
Medium (l)	10	15
Serum (ml)	40	1,500
rIL-2 ($\times 10^5$ U)	40	300
Number of vessels	3 [2]	75

[1] : 175 cm² tissue culture flask.

[2] : 2 flasks [1] and 1 culture vessel for disposable LIFROC

DISCUSSION. We recently developed a culture system for LAK cells based on dialysis perfusion (LIFROC ; Murata et al., 1990 and 1991). With this system, the final cell density reached approximately ten times more than that with conventional dish culture, and the cytotoxicity level was high enough for adoptive immunotherapy. When we used this system, however, we spent a lot of time putting together the parts of the culture vessel, because of its complex structure. Furthermore, the system could only accomodate one culture vessel in its rotary device. Although the system had already reduced the consumption of the culture medium, rIL-2, serum and manpower, in order to make manipulation easier, to further reduce manpower, and to prevent patients from the risk of contagious diseases, we decided to develop a disposable version of the system with a simpler structure for clinical use. We modified the original system in several ways to make the disposable system. First, the culture vessel and the medium reservoir, which were sterilized with ethylene oxide, were made disposable for easy manipulation and to prevent the risk of contagious disease being transferred from patient to patient. Second, to culture the lymphocytes of several patients and to provide more cells needed for therapy at one time, four culture vessels could be set up on the rotary device.

The average final cell density of LNL (LAK cells) with the disposable LIFROC system was 2.45×10^7 cells/ml, this cell density being approximately ten times higher than that by conventional dish culture. When a conventional dish culture method is used for the expansion of LNL (LAK cells) with high viability and cytotoxicity, it has been neccessary to keep the cell density under 1×10^6 cells/ml by periodical passages (Hoyer et al., 1986; Ochoa et al., 1987). Such a batchwise culture system forced us to consume large amounts of culture medium, rIL-2, and human serum, and also required extensive manpower.

Several other systems have recently been tried for culturing LAK cells at a high cell density, one being hollow-fiber system (Alter et al., 1987; Tanji et al., 1988). This system was originally developed in order to obtain cell products, rather than cells themselves. When such a system is applied to culture LAK cells, it can be difficult to harvest the cultured single cells, and a suspension culture is regarded as more suitable.

The conventional suspension culture system, which has impellers for agitation, may damage the cells by shear force (Tyo et al., 1981). The LIFROC system suspends the cells by rotating the culture vessel itself using agitation beads which do not touch the cells, and does not use impellers (Murata et al., 1990).

The LAK cells induced by the disposable LIFROC system had a high cytotoxicity of about 50-80% at an E/T ratio of 20/1 against K562 and Raji target cell lines at an average LAK cell density of 2.5×10^7 cells/ml. This maintenance of a high cytotoxicity level is thought to depend on the concentration of dissolved oxygen in the culture medium, and on supplying serum to the culture (Murata et al., 1990 and 1991). Supplying serum once to the culture a few days before harvesting the cells was enough to recover high cytotoxicity.

Culture bags made of gas-permeable plastics have been introduced (Colborn et al., 1989). This type of culture bag is usually used for the induction of LAK activity but not for cellular expansion. Because of the absence of active agitation of the medium in a culture bag, it may be difficult to expand LAK cells over a cell density of 10^7 cells/ml. More recently, a culture bag based on dialysis perfusion with an inner layer made of a dialysis membrane, had been introduced, and used for the cultivation and activation of lymphocytes (Nakamura *et al.*, 1992). With this system, the final cell density was estimated to be 1×10^7 cells/ml, but this study and our study were done under different conditions, because they cultured separated CD4$^+$ T cells.

We estimated the costs of medium ingredients for the expansion of lymphocytes from 4×10^8 cells to 1×10^{10} cells in a comparison between the disposable LIFROC system and conventional flask culture. As for the amount of basal medium, conventional flask culture needs 15 l of basal medium in the culture vessels, while the disposable LIFROC system needs 10 l of medium in the medium reservoir. rIL-2 consumed by the conventional flask culture is 7.5-fold more than that by the disposable LIFROC system, and serum consumed by conventional flask culture is 37.5-fold more than that by the disposable LIFROC system. There are large differences in cost and manpower between these two culture systems. The disposable LIFROC system should improve the routine performance of adoptive immunotherapy with LAK cells with respect to the efficiency, ease of manipulation, and low cost.

REFERENCES

Alter, B.J., Ochoa, A.C., Bach, F.B., *et al.* (1987) 'The growth of cells with LAK activity in an automated tissue culture system (Acusyst P)', In R.L. Truitt, R.P. Gale and M.M. Bortin (Eds), Progress in Clinical and Biological Research, Vol. 244, Cellular Immunotherapy of Cancer, Alan R. Liss, New York, p.289.

Colborn, D., Rinehalt, J., Joseph, I., *et al.* (1989) 'Expression of lymphokine-activated killer cells for clinical use utilizing a novel culture device', J. Immunol. Methods 119, 247.

Epstein, M.A., Achong, B.G., Barr, Y.M., Zajac, B., Henle, G., and Henle, W. (1966) 'Morphological and virological investigation on cultured Burkitt tumor lymphoblasts (strain Raji)', J. Natl. Cancer Inst. 37, 549.

Hoyer, M., Meineke, T., Lewis, W., Zwilling, B., and Reinhart, J. (1986) 'Characterization and modulation of human lymphokine(interleukine 2) activated killer cells induction', Cancer Res. 46, 2834.

Lozzio, C.B., and Lozzio, B.B. (1973) 'Cytotoxicity of a factor isolated from human spleen', J. Natl. Cancer Inst. 50, 535.

Murata, M., Yano, T., Togami, M., Yasumoto, K., Sugimachi, K., Kimura, G., and Nomoto, K. (1990) 'Development of a new culture system for lymphokine activated killer cells', J. Immunol. Methods 129, 89.

Murat, M., Yano, T., Yoshino, I., Togami, M., Sogabe, M., Yasumoto, K., Sugimachi, K., Kimura, G., and Nomoto, K. (1991) 'Development of a new culture system for human lymphokine-activated killer cells : Comparison with a conventional static culture method', Cytotechnology 7, 75.

Nakamura, Y., Tokuda, Y., Iwasawa, M., Tsukamoto, H., Kidokoro, M., Kobayashi, N., Kato, S., Mitomi, T., Habu, S. and Nishimura, T. (1992) 'Large-scale culture sysytem of human CD4$^+$ helper / killer T cells for the application to adoptive tumor immunotherapy', Br. J. Cancer 66.20.

Ochoa, A.C., Gromo, G., Alter, B.J., Sondel, P.M., and Bach, F.H. (1987) 'Long-term growth of lymphokine-activated killer (LAK) cells: role of anti-CD3, β-IL-1, Interferon-γ, and β', J. Immunol. 138, 2728.

Rosenberg, S.A., Lotze, M.T., Muul, L.M., Chang, A.E., Avis, F.P., Leitman, S., Linehan, W.M., Robertson, C.N., Roberta, E.L., Rubin, J.T., Seipp, C.A., Simpson, C.G., and White, D.E. (1987) 'A progress report on the treatment of 157 patients with advanced cancer using lymphokine-activated killer cells and interleukin-2 or high dose interleukine-2 alone. N. Engl. J. Med 316, 889.

Tanji, M., Tanaka, T., and Taguchi, T. (1988) 'The culture of LAK cells with hollow-fiber bioreactor', Biotherapy (Tokyo) 2, 361 (in Japanese).

Tyo, M., and Wang, D.I.C. (1981) 'Engineering characterization of animal cell and virus production using controlled charge microcarriers', In, M.M. Young, C.W. Robinson and C. Vezina (Eds.), Advances in Tiotechnology, Vol. 1, Scientific and Engineering Principles. Pergamon Press, New York, p.141.

Yano,T., Murata,M., Ishida,T., Mitsudomi,T., Kimura,G., Sugimachi,K.,and Nomoto,K.
(1989) 'Phenotypic characterization of lymphokine-activated killer cells from human
lymph node lymphocytes', Cell. Immunol. 122,281.

NEW METHOD FOR OBSERVING CELLS INSIDE A MACROPOROUS MICROCARRIER

T. MORI, K. KONOMOTO, J. SHIROKAZE and K. SAGARA
Cellulose Fibers Development Department
Asahi Chemical Industry Co. Ltd.
4-3400-1 Asahi-machi, Nobeoka-shi, Miyazaki 882, Japan

ABSTRACT.   A simple new method was established for observing cells inside a macroporous Asahi microcarrier, because such a microcarrier is not transparent in a culture medium due to diffused reflections which make it difficult to observe the cells inside. After microcarrier immobilizing CHO-K1 had been stained, it was soaked in several kinds of liquid having a similar refractive index to that of the microcarrier and then observed with an optical and confocal fluorescent microscope. Transparency of the microcarrier was successfully achieved with methyl salicylate, which made it easy to observe cells inside.

1. Introduction

Microcarrier cell culture is widely used for the propagation of anchorage-dependent cells in suspension.   DEAE-Sephadex beads with a controlled charge density have been well characterized (Levine *et al.*, 1979, Hu *et al.*, 1985) and are now among the most commonly used carriers in the industry. A number of macroporous microcarriers have also recently been developed which allow cell growth on internal surfaces as well as on the external surface (Nilsson *et al.*, 1986, Reiter *et al.*, 1990), making it possible to achieve a higher cell density culture.

One disadvantage of the macroporous microcarrier is its lack of light transmittance, which makes microscopic examination of cell behavior difficult. Routine microscopic examination is important in microcarrier culture to observe the cell distribution, morphology and growth throughout the cultivation period. The macroporous beads are amber colored when observed by ordinary-light microscopy; discerning the cells from the microcarrier matrix is difficult on the external surface and impossible inside the beads (Nikolai *et al.*, 1991).

Cells inside the microcarrier can be observed with an optical microscope if the microcarrier is embedded in resin, sliced carefully to make a cross section with a thickness of 1 μm, and then dyed for easy observation.   However, this method is too complicated for routine microscopic examination.   Certain fluorescent dyes such as fluorescein diacetate (FDA) and ethidium bromide (EB), which stain only cells and do not stain the supporting matrix, have been used as a tool for the microscopic examination of a macroporous microcarrier cell culture (Nikolai *et al.*, 1991).

S. Kaminogawa et al. (eds.), Animal Cell Technology: Basic & Applied Aspects, Vol. 5, 487–491.
© 1993 *Kluwer Academic Publishers.*

Since the macroporous microcarrier is opaque, visualizing the attachment and spread of cells by conventional microscopic techniques is difficult (Kim *et al.*, 1992). Kim *et al.* have demonstrated the use of optical sectioning by confocal microscopy for visualizing the distribution of cells within macroporous microcarriers.

In a culture medium, the macroporous microcarrier is not transparent, because diffused reflections make it difficult to observe cells inside the microcarrier. The refractive index of the microcarrier is different from that of the medium, so that diffused reflection occurs at the interface between the medium and the irregular surface of the microcarrier.

The purpose of our study was to soak the microcarrier in liquids which have nearly the same refractive index as that of the microcarrier in order to observe more easily cells stained with dye inside the macroporous microcarrier.

2. Materials and Methods

2.1. MICROCARRIER CULTURE

CHO-K1 was cultivated for 1 week in a 30-ml spinner flask with Asahi microcarrier (Asahi Chemical Ind.). This microcarrier is a macroporous type made from cellulose, the diameter of a bead being 200 μm, and the pore diameter being about 30 μm. Cells were grown in culture dishes and inoculated at a density of 2×10^5 cells/ml. All cells attached themselves to the surface of the microcarrier, both inside and out.

2.2. PREPARATION FOR STAINING

Five ml of the culture medium containing the microcarrier and immobilized cells was extracted by pipet and put into a 15-ml centrifuge tube. When the microcarrier had settled to the bottom of the centrifuge tube, the supernatant was removed by a pasteur pipet. The microcarrier was cleaned twice with PBS and then treated with a 2% glutaraldehyde solution for 6 hours at room temperature. After the treated microcarrier had been cleaned with PBS, the correct quantity of a dye solution was added and mixed for staining, before the microcarrier was cleaned twice again with PBS.

Before soaking the microcarrier in an organic solvent, PBS was substituted twice for ethyl alcohol.

2.3. MICROSCOPE

The microcarrier was examined with an IMT-2 optical microscope (Olympus) and an LSMGB-200(H) confocal laser scanning fluorescent microscope (Olympus) after the cells and the microcarrier had been stained.

The confocal microscope was used to make "optical sections" of the microcarriers and to visualize the distribution of cells within them.

3. Results and Discussion

Several kinds of liquids were tested for soaking the microcarrier. Finally, methyl salicylate was selected, because the refractive index of methyl salicylate is 1.537, which is very close to the 1.540 value for cellulose (Table 1), and because it's not harmful to the human body.

TABLE 1. Refractive index of the soaking liquid

Soaking liquid	Refractive index
PBS	1.334
Saturated aqueous solution of BaI_2	1.528
Chlorobenzene	1.524
Methyl salicylate	1.537

Many dyes were experimented with for staining the cells. However, they would be exposed to the methyl salicylate soaking liquid, and would subsequently be extracted and turn pale. Hematoxylin and quinacrine were finally selected because their staining stability was not affected by methyl salicylate (Table 2), hematoxylin being selected for optical microscopy, and quinacrine for fluorescent microscopy.

TABLE 2. Selection of the optimal dye

	Staining stability	
	in PBS	in methyl salicylate
Dye		
Giemsa	Good	Not good
Hematoxylin	Good	Good
Fluorescent dye		
Acridine orange	Good	Not good
Nile blue	Good	Not good
Auramine	Good	Not good
Quinacrine	Good	Good

The cells inside the microcarrier were dyed violet after the microcarrier had been soaked in the hematoxylin solution for 1 minute at room temperature. Because the microcarrier turned completely transparent in methyl salicylate (Photo 1), the cells inside the microcarrier could be observed very clearly with an optical microscope.

Cells soaked in a quinacrine solution for 30 minutes at room temperature were dyed to be fluorescent (Photo 2a). The microcarrier was also dyed slightly, well-defined images of an optical section could be taken. It was quite easy to observe the cell distribution, morphology and growth throughout the cultivation period (Photo 2b).

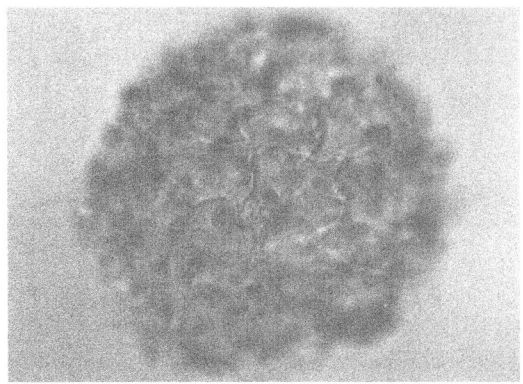

PHOTO 1. Image of cells cultivated on the surface of and inside the macroporous microcarrier. The cells were dyed violet with hematoxylin.

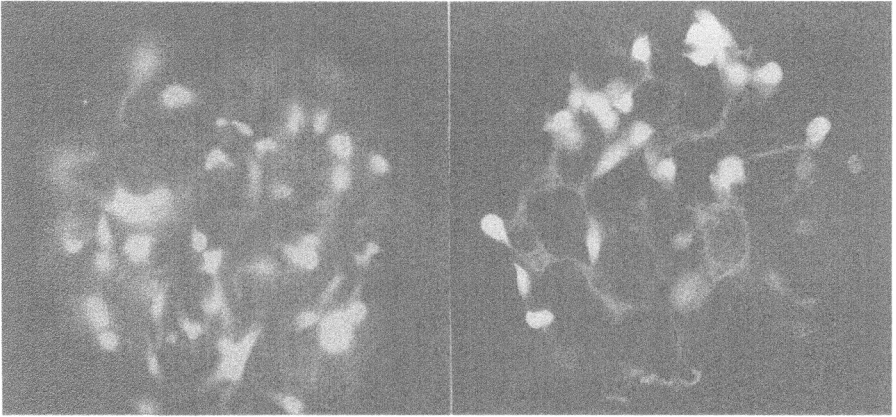

PHOTO 2. Fluorescent images of cells dyed with quinacrine.
(2a) Non-confocal image   (2b) Optical section taken with a confocal microscope

4. Conclusion

A cellulose-based macroporous microcarrier was used for mammalian cell culture. This microcarrier was opaque in a culture medium because diffused reflections occurred at the interface between the medium and the irregular surface of the microcarrier, which had a different refractive index from that of medium. It was proved when the microcarrier was soaked in methyl salicylate, which has a refractive index very similar to that of the microcarrier, that the microcarrier turned transparent. Furthermore, two dyes, hematoxylin and quinacrine, were selected for staining cells, because their staining stability was not affected by methyl salicylate. The combination of methyl salicylate and the two dyes made it very easy to observe cells inside the microcarrier with an optical microscope and a confocal microscope.

5. References

1. Levine, D.W., Wang, D.I.C., and Thilly, W.G. (1979) 'Optimization of growth surface parameters in microcarrier cell culture,' Biotechnol. Bioeng. 21, 821-845.

2. Hu, W.S., Giard, D.J., and Wang, D.I.C. (1985) 'Serial propagation of mammalian cells on microcarriers,' Biotechnol. Bioeng. 27, 1466-1476.

3. Nilsson, K., Buzsaky, F., and Mosbach, K. (1986) 'Growth of anchorage-dependent cells on macroporous microcarriers,' Bio/Technology 4, 989-990.

4. Reiter, M., Hohenwarter, O., Gaida, T., Zach, N., Schmatz, C., Bluml, G., Weigan, F., Nilsson, K., and Katinger, H. (1990) 'The use of macroporous gelatin carriers for cultivation of mammalian cells in fluidized bed reactors,' Cytotechnology 3, 271-277.

5. Nikolai, T.J., Peshwa, M.V., Goetghebeur, S., and Hu, W.S. (1991) 'Improved microscopic observation of mammalian cells on microcarriers by fluorescent staining,' Cytotechnology 5, 141-146.

6. Kim, J.H., Lim, H.S., Han, B.K., Peshwa, M.V., and Hu, W.S. (1992) 'Characterization of cell growth and improvement of attachment kinetics on macroporous microcarriers,' Animal Cell Technology: Basic & Applied Aspects, Kluwer Academic Publishers, Dordrecht 77-80.

PRODUCTION OF AN α-AMIDATING ENZYME (α-AE) IN RECOMBINANT CHO CELLS

K. FURUKAWA[1], K. OKUNO[1], S. ONAI[1], K. SUGIMURA[1], Y. YOKO-O[1],
Y. ISHIBASHI[1], T. OSHIMA[1], N. TSURUOKA[2], K. MAGOTA[2], S. TANAKA[2],
and K. OHSUYE[1]

[1]Suntory Bio-Pharma Tech Center, 2716-1 Kurakake, Akaiwa, Chiyoda, Ohra, Gunma
370-05, Japan ; [2]Suntory Biomedical Research Center, 1-1-1 Wakayamadai, Simamoto,
Mishima, Osaka 618, Japan

ABSTRACT. C-terminal α-amidating enzyme (α-AE) is a post-translational modification enzyme that catalyzes the conversion of glycine-extended prohormone substrates to biologically active α-amidated peptide hormones. In this report, we describe the production of an α-AE derived from *Xenopus laevis* skin by a recombinant CHO cell line in a serum-free suspension culture. Expression was achieved by using a truncated 799*Bg/II*α-AE cDNA to secrete the enzyme to the culture medium. A 3μ-1S cell line, which secreted 799*Bg/II*α-AE to the medium, was cultured on a 50 L scale by multiple cultivation for about 6 weeks. A stable cell growth rate and 799*Bg/II*α-AE production were maintained, and 3.8×10^8 units of the enzyme were produced during the culture. The structure of the integrated 799*Bg/II*α-AE gene was also stable in the absence of methotrexate during the culture. The secreted 799*Bg/II*α-AE was purified to homogeneity by anion-exchange and affinity chromatography. Furthermore, the purified 799*Bg/II*α-AE could convert the glycine-extended human calcitonin (hCT[Gly]) to the mature hCT.

1. INTRODUCTION

A large number of biologically active peptide hormones have amidated carboxyl termini. The C-terminal α-amide forms appear to be important both for biological activity and for protection against proteolytic degradation. Bradbury et al. have identified enzymatic activity in the porcine pituitary gland for the α-amidation of glycine-extended peptides [1]. α-AE catalyzes this reaction in the presence of copper, ascorbate and molecular oxygen [2]. We have isolated cDNAs encoding α-AE from *X. laevis* skin [3,4], and others have also isolated cDNAs from bovine pituitary, rat medullary thyroid carcinoma and rat atrial tissue. Kato and others have recently demonstrated that the α-AE reaction consists of two enzymatic reactions: the conversion of peptidylglycine into an α-hydroxyglycine intermediate, and the conversion of this intermediate into the α-amidated product, which are catalyzed by the N-terminal and C-terminal domains of α-AE, respectively [5].

Many bioactive peptide hormones are expressed as fusion proteins in *E. coli* and can be recovered from the fusion proteins by enzymatic or chemical methods, but we did not directly obtain a mature α-amidated peptide in an *E. coli* expression system. Therefore, it was important to establish a large-scale production method for α-AE, in order to produce biologically active α-amidated peptide hormones such as hCT. In this report, we describe an efficient production method for the 799*Bg/II*α-AE, which was derived from *X. laevis* skin, with recombinant CHO cells.

2. MATERIALS AND METHODS

2.1. Materials

Restriction enzymes were purchased from Takara-shuzo (Japan) and Toyobo (Japan). A DNA

493

preparation kit was obtained from BIO 101 (USA) and [^{32}P]dCTP from Amersham. Polyvinylidene difluoride (PVDF) membrane for Western blotting was obtained from Millipore, and nitrocellulose filter from Schleicher & Schuell (Germany). Fetal bovine serum (FBS) was obtained from Gibco, and F12 medium from Ajinomoto (Japan). Bovine insulin and transferrin were purchased from Intergen (USA), and methotrexate (MTX) was from Sigma. Horseradish peroxidase avidin D was obtained from Vector (USA). DEAE-Sepharose and ECH-Sepharose were products of Pharmacia.

2.2. Cell Line

We cloned and sequenced two types of cDNAs (XAE457 and XAE799) for α-AE from a *X. laevis* skin cDNA library. The second cDNA (XAE799) encoded a membrane-bound type of α-AE, consisting of 875 amino acid residues. To produce a secretory α-AE with a eukaryotic expression system, we constructed a pKD799*Bgl*II plasmid. The plasmid, pKD799*Bgl*II, contains a truncated 799*Bgl*IIα-AE gene, corresponding to amino acid sequences 1 to 731 of the XAE799 cDNA encoding protein, in which XAE799 cDNA was modified to terminate translation at a site preceding the transmembrane and cytoplasmic domains. In the plasmid, the 799*Bgl*IIα-AE gene and DHFR gene were transcribed under the controls of the SV40 early promoters, respectively. DHFR-deficient CHO cells were transfected with the pKD799*Bgl*II by a calcium phosphate method. A high 799*Bgl*IIα-AE producing cell line, named 3μ-1, was obtained by gene amplification with increasing concentrations of MTX (10 nM to 3 μM) [6].

2.3. Establishment of a Cell Line for 799*Bgl*IIα-AE Production

The 3μ-1 cells obtained were converted to a suspension culture by shaking in siliconized 300 mL flasks with a modified F12 medium supplemented with 10 % FBS and 1.0 μM of MTX. Furthermore, the cells were adapted to a serum-free medium, which was the modified F12 medium containing 1.0 μM of MTX, and 5 μg/mL each of insulin and transferrin. In this way, a 3μ-1S cell line, which could grow in suspension with the serum-free medium, was established from the 3μ-1 cell line.

2.4. Seed Culture

The 3μ-1S cells were cultured by shaking with the serum-free medium. This seed culture was then scaled up by using spinner flasks (up to 8 L) with the serum-free medium without MTX.

2.5. Cell Culture for 799*Bgl*IIα-AE Production

After the seed culture in spinner flasks, the 3μ-1S cells were inoculated to a 50 L reactor at 2×10^5 cells/mL. The cells were then cultured with the serum-free medium without MTX. When the cell density had attained 1×10^6 cells/mL, four-fifths of the cell suspension (40 L) was harvested, and 40 L of fresh medium was added to the residual cell suspension. In this way, the next culture was restarted at 2×10^5 cells/mL and repeated. During the culture, the temperature was controlled to 37 ℃ , DO to 60% air saturation by sparging with air, and pH to 7.2 by CO_2 or 7% $NaHCO_3$ addition.

2.6. Assay of α-AE Activity

The assay of α-AE activity was carried out as previously reported, using [^{125}I]-Ac-Tyr-Phe-Gly as a substrate [7]. One unit of activity is defined as the amount of enzyme that gives 50% conversion of Ac-Tyr-Phe-Gly to Ac-Tyr-Phe-NH$_2$ under the standard assay conditions.

2.7. Western Blot Analysis

Aliquots of each culture supernatant corresponding to 15 units of α-AE activity were electrophoresed in 10% polyacrylamide gel containing SDS, and then transferred to a PVDF membrane for Western blotting. After transfer, the membrane was incubated with rabbit anti-α-AE polyclonal antibody conjugated biotin and was allowed to react with horseradish peroxidase avidin D. Immunoreactive bands were visualized by using a POD immunostain set (Wako Pure Chemical, Japan).

2.8. Genomic DNA Analysis

Genomic DNA was isolated with a DNA preparation kit, and 6 μg was digested to completion with EcoRI and PvuII restriction enzymes. The digested DNA (3.5 μg) was used for 1% agarose gel electrophoresis for Southern blot transfer. The filter was hybridized to a [^{32}P]-labeled probe prepared by primer extension of the 799BglIIα-AE gene, and then autoradiographed.

2.9. Purification of 799BglIIα-AE

The culture supernatant was concentrated by ultrafiltration and precipitated by adding ammonium sulfate to a 50% saturation. The precipitate was dissolved in a 10 mM ammonium acetate buffer (pH 6.3), and then dialyzed against the same buffer. The dialyzate was applied to a column of DEAE-Sepharose, which had been equilibrated with a 10 mM ammonium acetate buffer (pH 6.3). The column was washed with the same buffer, and the adsorbed protein was then eluted with a linear gradient of 0-300 mM NaCl. The enzyme-active fractions were pooled and applied to a substrate affinity column (ECH-Sepharose conjugated Tyr-Phe-Gly) equilibrated with a 10 mM ammonium acetate buffer (pH 6.3) containing 300 mM NaCl. The bound protein was then eluted with a linear gradient of 100 mM Tris-HCl (pH 9.3) containing 300 mM NaCl. The purity of the protein obtained from each purification step was analyzed by SDS-PAGE.

3. RESULTS AND DISCUSSION

3.1. Cell Culture for 799BglIIα-AE Production

The seed culture of the 3μ-1S cells for 799BglIIα-AE production was performed as described in the materials and methods section, and the cell growth in this seed culture is shown in Fig. 1. Although initial lags were found just after the beginning of the shaking and spinner cultures, the 3μ-1S cells proliferated exponentially in the subsequent culture. The doubling times of 3μ-1S cells in the shaking and spinner cultures were 31 and 45 hours, respectively. 8 L of the cell suspension from the spinner culture was inoculated to the 50 L reactor at 2x10^5 cells/mL, and a large-scale cultivation was performed as described in the materials and methods section. As shown in Fig. 2, the 3μ-1S cells had an initial lag after inoculation, however, they retained a logarithmic phase in the following culture with a viability of approximately 90%. The doubling time of the cells was 31 hours. Furthermore, from the culture of the 3rd batch (6 days post-

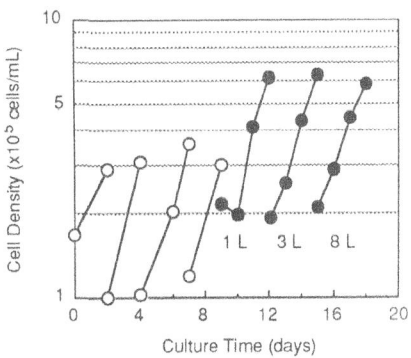

Fig. 1. Seed culture of 3μ-1S cells. Cell densities of the shaking culture (○), and the spinner culture (●) were measured using a hemacytometer.

inoculation), the consumptions of glucose and 7% NaHCO$_3$ were at constant levels of 2.2 g/L/batch and 20 g/L/batch, respectively. These results indicate that the culture remained stable for about 6 weeks.

The α-AE activity of the harvested culture supernatant was measured with the synthetic substrate, [^{125}I]-Ac-Tyr-Phe-Gly. As shown in Fig. 3, the 3μ-1S cell line constantly secreted from 600 to 800 units/mL (average of 730 units/mL) of 799BglIIα-AE in the absence of MTX for about 6 weeks. To analyze the stability of the 799BglIIα-AE protein secreted to the culture medium, immunoblotting with rabbit anti-799BglIIα-AE antibody was carried out. The results of the Western blotting are shown in Fig. 4. Immunoblotting revealed two major immunoreactive 75 and 80 KDa bands, as can be seen in Fig. 4, from lanes 2 to 5. The difference in the molecular weight may have been caused by different modifications of sugar chains. The results indicate that the 3μ-1S cells produced 799BglIIα-AE consistently, and that the secreted protein was not decomposed remarkably during the culture period.

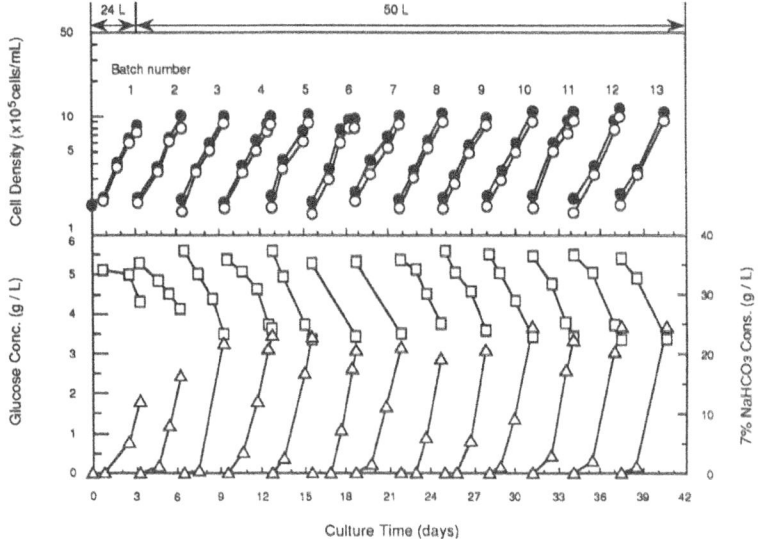

Fig. 2. Suspension culture of 3μ-1S cells in a 50 L reactor for 799BglIIα-AE production. The total cell density (●) was determined using a hemacytometer. The viable cell density (○) was calculated from the cell viability which was determined by the dye exclusion test with 0.1% trypan blue. The glucose concentration (□) and 7% NaHCO₃ consumption (△) were also measured at the same time.

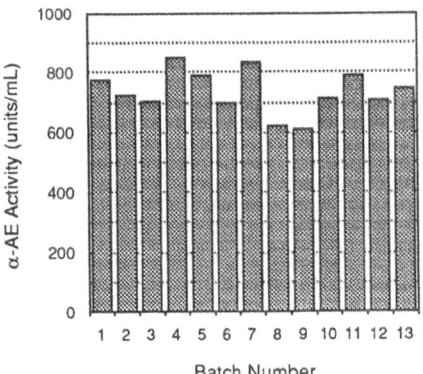

Fig. 3. Production of 799BglIIα-AE. The α-AE activity of each harvested culture supernatant was assayed as described in the materials and methods section. The batch numbers are the same as those shown in Fig. 2.

3.2. Analysis of Genetic Stability

The 3μ-1S cell line could consistently produce the enzyme as shown in Fig. 3 and Fig. 4. On the other hand, it is well known that many CHO cell lines which have been gene amplified with MTX showed instability of the amplified genes during a long culture period in the absence of MTX. Therefore, we attempted to analyze the stability of the amplified 799BglIIα-AE gene during the culture period. High-molecular-weight chromosomal DNA was isolated and digested with EcoRI and PvuII restriction enzymes to analyze the structure of the integrated 799BglIIα-AE gene. After agarose gel electrophoresis, Southern blot hybridization was performed as described in the materials and methods section. As shown in Fig. 5 from lanes 1 to 4, no alterations in the molecular size and the intensity of the digested bands are apparent.

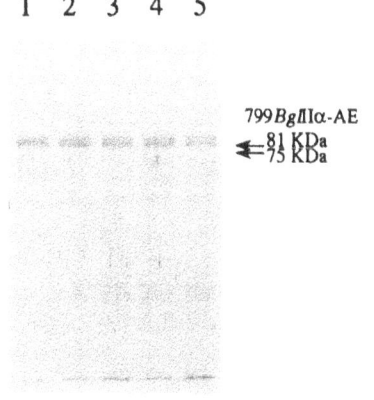

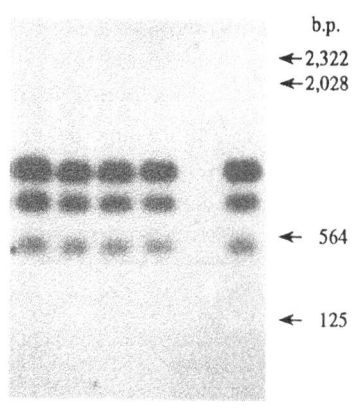

Fig. 4. Immunoblot analysis of secreted 799*Bg*III-α-AE. Immunoblotting was carried out by using a polyclonal antibody against the 799*Bg*IIα-AE protein as described in the materials and methods section. Aliquots of each culture supernatant corresponding to 15 units/mL of α-AE activity were used. Lanes: (1) standard 799*Bg*IIα-AE; (2) 1st batch; (3) 7th batch; (4) 11th batch; (5) 13th batch.

Fig. 5. Stability of amplified 799*Bg*IIα-AE gene. Genomic DNA was isolated and digested with *Eco*RI and *Pvu*II, and 3.5 μg of the digested DNA was used for Southern blot hybridization. Lanes: (1) 1st batch; (2) 7th batch; (3) 11th batch; (4) 13th batch; (5) master working cell bank. The numbers on the right indicate the *Hind*III digested λ molecular weight markers.

Therefore, the integrated gene was stably maintained during the culture period, and the 3μ-1S cell line was validated up to about 30 generations on the 50 L scale.

3.3. Purification of 799*Bg*IIα-AE

799*Bg*IIα-AE secreted to the medium was pooled and purified by DEAE-Sepharose and substrate-affinity chromatography as described in the materials and methods section. A summary of this purification is shown in Table 1. The specific activity increased about 10-fold and the recovery of the 799*Bg*IIα-AE protein was 48%. Interestingly, the recovery of the enzyme increased 1.3-fold after precipitation by ammonium sulfate. It is possible that inhibitor(s) which existed in the culture supernatant might have been removed by this step. The SDS-polyacrylamide gel electrophoresis of the purified 799*Bg*IIα-AE revealed only two bands of 75 and 81 KDa (Fig. 6).

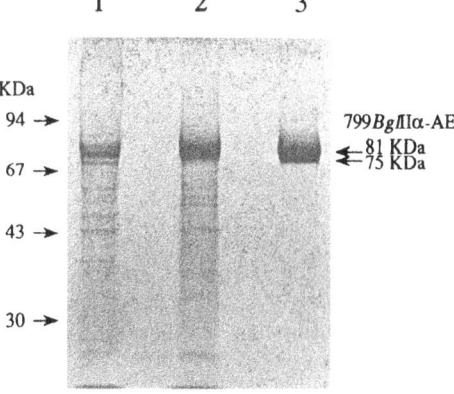

Fig. 6. SDS-PAGE for each purification step. The protein was stained by Coomassie brilliant blue. Lanes: (1) culture supernatant; (2) active fraction from DEAE-Sepharose chromatography; (3) active fraction from substrate-affinity chromatography.

Table 1. Purification of 799*Bg*IIα-AE

Step	Specific Activity (units/mg of protein)	Recovery (%)
1. Culture supernatant	1.3×10^4	100
2. 50% sat. (NH4)2SO4 ppt.	2.0×10^4	136
3. DEAE chromatography	4.3×10^4	51.4
4. Affinity chromatography	12.0×10^4	48.0

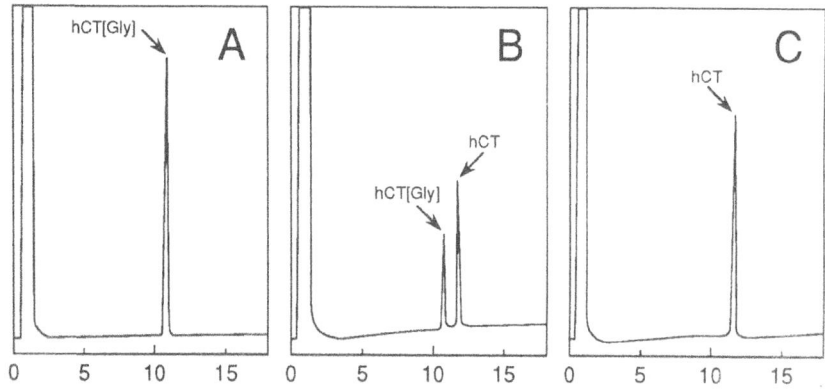

Retention Time (min.)

Fig. 7. HPLC analysis of amidating reaction. A portion of the amidating reaction mixture was subjected to reversed-phase HPLC (ODS A-302, 4.6x150 mm; YMC Shimakyu Co., Japan) equilibrated with 10 mM AcONH₄ and analyzed with a linear gradient from 30 to 57% of CH₃CN in 10 mM AcONH₄ over 18 minutes. The elution patterns are for reaction times of (A) 0 min, (B) 30 min and (C) 120 min.

3.4. Conversion of hCT[Gly] to hCT with 799*Bg*IIα-AE

We analyzed the conversion of the precursor hCT[Gly] to the mature hCT by the purified 799*Bg*IIα-AE with reversed-phase HPLC. 2.5 mg/mL of the hCT[Gly] was incubated with 1000 units/mL of the enzyme under standard conditions. Fig. 7 depicts the HPLC pattern of the amidation. After 30 minutes, about 40% of the hCT[Gly] had been converted to hCT (Fig. 7, B) and was completely converted to hCT in 120 minutes (Fig. 7, C). Therefore, we conclude that the purified 799*Bg*IIα-AE could produce the mature hCT.

4. CONCLUSION

In this study, we have reported an efficient production method with the recombinant CHO cell line for 799*Bg*IIα-AE derived from *X. laevis* skin. The 3μ-1S cell line, which could grow in suspension with the serum-free medium, was cultured on a 50 L scale for about 6 weeks. The 799*Bg*IIα-AE secreted was purified to homogeneity by two chromatography steps. Moreover, purified 799*Bg/*IIα-AE catalyzed the conversion of the precursor hCT[Gly] to the mature hCT. The production method described in this paper provides the possibility to produce biologically active C-terminal α-amidated peptide hormones on an industrial scale. The availability of purified 799*Bg/*IIα-AE is also of interest for other structural and functional research.

5. REFERENCES

[1] Bradbury, A. F., Finnie, M. D. A. and Smyth, G. G. (1982) 'Mechanism of C-terminal amide formation by pituitary enzymes', Nature 305, 686-688.

[2] Eiper, B. A., Mains, R. E. and Glembotski, C. C. (1883) 'Identification in pituitary tissue of a peptide α-amidation activity that acts on glycine-extended peptides and requires molecular oxygen, copper, and ascorbic acid', Proc. Natl. Acad. Sci. USA 80, 5144-5148.

[3] Mizuno, K., Ohsuye, K., Wada, Y., Fuchimura, K., Tanaka, S. and Matsuo, H. (1987) 'Cloning and sequence of cDNA encoding a peptide C-terminal α-amidating enzyme from *Xenopus laevis*', Biochem. Biophys. Res. Commun. 148, 546-552.

[4] Ohsuye, K., Kitano, K., Wada, Y., Fuchimura, K., Tanaka, S., Mizuno, K. and Matsuo, H. (1988) 'Cloning of cDNA encoding a new peptide C-terminal α-amidating enzyme having a putative membrane-spanning domain from *Xenopus laevis* skin', Biochem. Biophys. Res. Commun. 150, 1275-1281.

[5] Kato, I., Yonekura, H., Tajima, M., Yanagi, M., Yamamoto, H. and Okamoto, H. (1990) 'Two enzymes concerned in peptide hormone-amidation are synthesized from a single mRNA', Biochem. Biophys. Res. Commun. 172, 197-203.

[6] Tsuruoka, N., Magota, K., Ohsuye, K., Kitano, K. and Tanaka, S. (in preparation)

[7] Mizuno, K., Sakata, J., Kojima, M., Kangawa, K. and Matsuo, H. (1986) 'Peptide C-terminal α-amidating enzyme purified to homogeneity from *Xenopus laevis* skin', Biochem. Biophys. Res. Commun. 137, 984-991.

DEVELOPMENT OF INSTRUMENTATION AND CONTROL STRATEGY IN BIOREACTOR CULTURE OF HYBRIDOMA CELLS

R. M. MATANGUIHAN, R. PAMBAYUN, T. IZUISHI
K. B. KONSTANTINOV and T. YOSHIDA
International Center of Cooperative Research in Biotechnology
Faculty of Engineering, Osaka University
2-1 Yamada-oka, Suita-shi, Osaka 565 Japan

D. M. GRYTE and W. S. HU
Department of Chemical Engineering and Materials Science
University of Minnesota
421 Washington Ave.S.E., Minneapolis, MN 55455 USA

ABSTRACT. Batch cultures of hybridoma AFP-27 cells were used to elucidate the stoichiometric relationships among glucose, lactate, glutamine and ammonia. Good linear correlation between glucose consumption and lactate production, as well as between glucose and glutamine consumption was observed during the exponential growth stage. These stoichiometric relationships were used for the dynamic feeding of nutrients in fed-batch cultures. However, a feeding strategy based on stoichiometric relations is difficult to apply during unbalanced growth. Various on-line instruments were thus examined for their effectiveness in providing an accurate estimation of cell concentration or metabolic activities. A laser turbidity probe was found to provide a good estimation of cell concentration. Oxygen uptake rate correlated well to other indicators or cell metabolism. A capacitance probe was also examined, however, its use was limited by the high noise in the cell concentration range employed. These on-line instruments are currently being integrated with the aim of establishing an intelligent system for fed-batch culture of mammalian cells.

1. Introduction

Fed-batch and continuous perfusion cultures of hybridoma cells have been employed to increase cell and antibody concentrations. In some fed-batch cultures amino acids which are identified as deficient are fed continuously and resulted in increased productivity (Jain et al. (1991)). In some other cases, glucose is fed continuously to reduce the lactic acid production (Hu et al. (1987)). Dynamic feeding of the medium or selected nutrients is thus important in order to attain the optimum growth and productivity of the culture. The feeding strategy can be determined and implemented effectively if the required inputs of process parameters such as the concentration of glucose and glutamine are obtainable. Yet, these medium components can not be readily measured on-line; thus, other techniques based on available on-line variables are used to determine their concentrations indirectly.

S. Kaminogawa et al. (eds.), Animal Cell Technology: Basic & Applied Aspects, Vol. 5, 501–507.
© 1993 *Kluwer Academic Publishers.*

Also lacking is a reliable on-line estimation of cell concentration and other variables pertinent to the growth or physiological state of cells. Some new methods have been introduced recently to improve the on-line measurement in bioreactors. For example, a laser turbidity sensor was found to give an accurate estimate of hybridoma cell concentration (Konstantinov et al. (1992)). A capacitance sensor was also reported to be a potential tool for detecting viable cells in animal cell culture (Davey et al. (1988)). Another variable closely related to the physiological state of cells is oxygen uptake rate (OUR). However, with the relatively small OUR observed in hybridoma cell culture, material balances based on off-gas analysis often do not give a reliable estimation. Dynamic measurement based on the dissolved oxygen can be employed to estimate the OUR of hybridoma cell periodically. In addition, on line determination of the cell concentration can lead to the calculation of other physiological variables such as specific rates which are important to determine the cell physiological states and enhance the process control strategy.

In this study we examined the stoichiometric correlations among glucose consumption, lactate production, oxygen consumption and alkali addition and tested a procedure for the dynamic feeding of nutrients based on these correlations.

2. Materials and Methods

2.1. CELL LINES AND CULTURE MEDIA

The cell line used was mouse-mouse hybridoma, AFP-27, which produces IgG antibody against α-fetoprotein. The cells were grown in a 1:1 mixture of WRC-935 (Amicon, USA) and DMEM (Gibco Laboratories, USA) and supplemented with 5% fetal bovine serum (FBS, Sigma, USA). Penicillin G (100 units/mL) and streptomycin (100 ug/mL) were also added to the mixed media. The cells were maintained in T-flasks and kept in an incubator set at 37 $^{\circ}$C and 5% CO_2.

2.2. CELL CULTIVATION AND BIOREACTOR SYSTEM

The inoculum was prepared by transferring the cells from T-flasks to 100 mL spinner flask followed by a subsequent transfer to a 300 ml spinner flask. The inoculum was added to the bioreactor to attain an initial cell concentration of about 1 x 10^5 cells/mL.

The experiments on batch cultivation were performed in a stirred-type bioreactor (Chemap, FZ-2000) coupled with a microcomputer system for bioprocess monitoring and control. The working volume in the reactor was 1.5 L and the culture was maintained at 30 rpm agitation speed and 37 $^{\circ}$C. pH was controlled at 7.2 by the addition of 1.0 N sodium hydroxide. Air or oxygen were added to the headspace to maintain the dissolved oxygen level. The bioreactor was equipped with a laser turbidity sensor (ASR, Japan) for the on-line monitoring of total cell concentration. The capacitance was measured using a capacitance meter (Kobe Steel Co., Japan) with a sterilizable sensor (Mishima et al. (1991)).

Total and viable cells were counted in the hemacytometer after staining with trypan blue in phosphate buffer saline PBS. A glucose and lactate analyzer (YSI-2000, Yellow Springs Inc., USA), glutamine-kit (Boehringer-Mannheim) and ammonia-kit (Boehringer-Mannheim) were used to determine the residual concentrations of nutrients and metabolites.

2.3. ON-LINE DETERMINATION OF OXYGEN UPTAKE RATE

In a batch culture the dissolved oxygen concentration was controlled at 40% of saturation level. To measure OUR, the dissolved oxygen in the culture was raised to approximately 70-80% of saturation by injection of air/oxygen mixture. Afterwards air or air/nitrogen mixture was injected to allow the dissolved oxygen to decrease. The rate of decrease was caused by both cell consumption and the transfer of oxygen from the liquid to the gas phase. On-line integration was performed and the contribution of oxygen transfer was subtracted from the measurement using predetermined value of Kla (volumetric oxygen transfer coefficient) (Hu and Oberg (1990)). Since Kla was relatively constant during the cultivation period, an accurate estimation of OUR could be obtained by using this method.

3. Results and Discussion

3.1. MONITORING OF OPTICAL DENSITY AND CAPACITANCE

The output of laser OD sensor correlated very well with the total cell concentration in a batch culture (Fig. 1). This allows the accurate measurement of the cell concentration in real-time. The on-line signal exhibited minimal noise thus making the reliable estimation of the specific rates possible. However, deviation in the measurement was noted in early stationary phase and this could be due to a change in cell size distribution or in cell viability which could affect the light scattering property of the cells (Sen et al. (1989)).

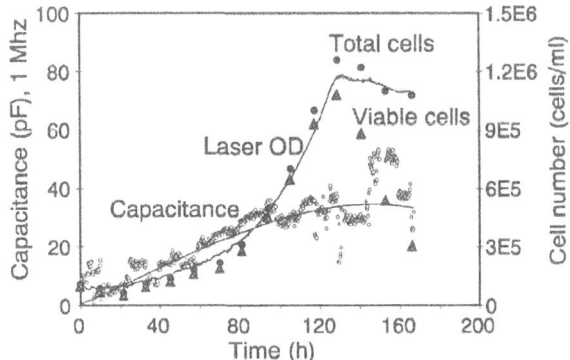

Fig. 1. On-line measurement of optical density and capacitance in a batch culture.

The relationship between capacitance and the viable cell concentration was examined. Capacitance measurement could compliment the turbidity measurement of the laser OD sensor to give an on-line estimation of cell viability. The on-line signal of the capacitance sensor was relatively noisy, although an increasing trend was observed during the early stage of cultivation. Marked fluctuations in the capacitance measurements were observed in the stationary phase. The relatively low cell concentration and high medium conductance could have an effect on the fluctuation in the measured value. Further study is necessary to establish experimentally the relationship between capacitance and viable cell concentration.

To test the feasibility of using the turbidity sensor for on-line control, a fed-batch culture was performed in a batch mode until 70g of NaOH was added. Subsequently fresh medium was fed continuously to maintain the on-line turbidity measurement at a constant level.

504

Shown in Figure 2 are the off-line measurement of cell concentration. The cell concentration was maintained relatively constant during that time period. The results thus indicate that the on-line turbidity measurement is sufficient for feedback control.

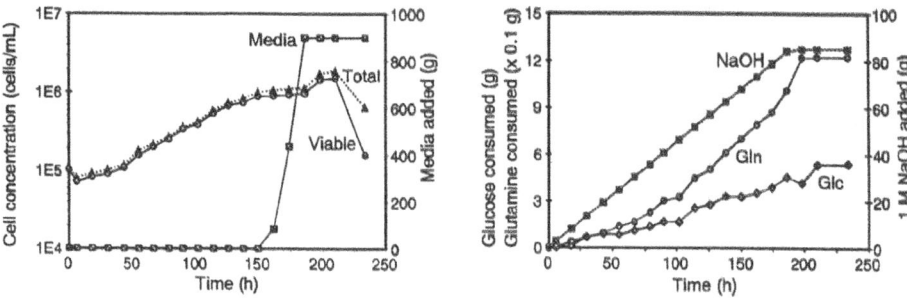

Fig. 2. Fed-batch culture of AFP-27 cells. From 150 to 180h, medium was fed to maintain a constant turbidity.

3.2. STOICHIOMETRIC RELATIONSHIP BETWEEN NUTRIENTS AND METABOLITES.

The cumulative consumption of glucose, glutamine and the cumulative production of lactate and ammonia during the cultivation are correlated stoichiometrically. One such set of data is shown in Figure 3a. A highly linear relationship was obtained between glucose consumption and lactate production. This relationship corresponded to about 0.86 g of lactate produced per gram of glucose consumed. The stoichiometric ratio between glutamine consumption and ammonia production was subject to slightly higher variation which could be related to the interactive nature of the metabolism of glutamine and other amino acids.

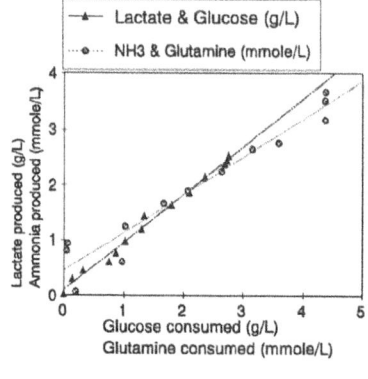

Regression
Glc, Lac: $y = 0.1 + 0.86x$ $r = 1$
Gln, NH$_3$: $y = 0.45 + 0.68x$ $r = 0.97$

Fig. 3a. Stoichiometric relationship between nutrients and metabolites.

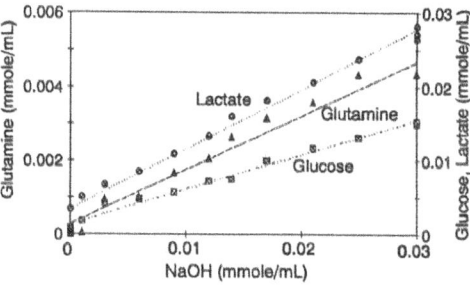

Regression
Lactate: $y = 0 + 0.81x$ $r = 0.99$
Glutamine: $y = 0 + 0.15x$ $r = 0.99$
Glucose: $y = 0 + 0.47x$ $r = 0.99$

Fig. 3b. Stoichiometric relationships between the NaOH added and metabolites.

The amount of NaOH added for maintaining a constant pH was directly proportional with the amount of lactate produced (Fig. 3b). Similarly, NaOH addition was indicative of the glucose and glutamine consumption. The high correlation between the NaOH added and the concentration of metabolites and residual nutrients showed the possibility of using the on-line measurement of NaOH addition to estimate the amount of lactate produced or the concentration of residual nutrients.

NaOH addition was used to estimate the amount of glucose consumed and the corresponding residual glucose concentration. AFP-27 cells in a 1.3 L reactor was grown initially in batch culture and subsequently fed with concentrated (20x) glucose solution (data not shown). When glucose concentration decreased to below the set point, the computer initiated the feeding of glucose at a stoichiometric ratio to NaOH addition. This feedback control of glucose feeding allowed a relatively constant glucose concentration to be maintained. The lactic acid concentration continually increased which also marked glucose consumption. However, the eventual insulin depletion in this experiment may have caused the loss of cell viability in the latter part of culture.

One source of error which should be accounted for in this estimation method is the presence of sodium bicarbonate buffer in the medium. The amount of bicarbonic acid is affected by the CO_2 partial pressure and the latter should be maintained constant.

3.3. ON-LINE OUR MEASUREMENT IN FED-BATCH CULTURE

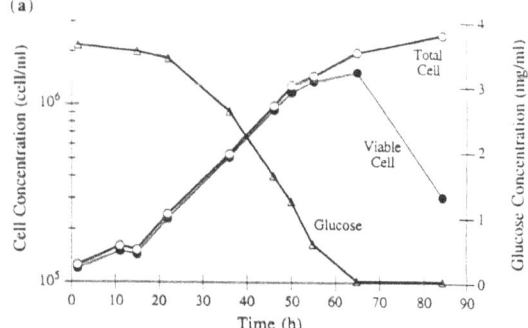

In a batch culture of AFP-27 cells, OUR was measured on-line (Fig. 4b) while total and viable cell concentrations and glucose concentration were measured off-line (Fig. 4a). OUR increased with increasing cell concentration and glucose consumption until the glucose in the medium was depleted ca. 65h. During the transition from exponential phase to the decline phase, OUR decreased rapidly as the viability decreased. The results indicated that OUR could be a sensitive indicator of the change in the viability. The specific OUR was calculated off-line basing on viable cells. This parameter showed an increasing trend during the early experimental growth

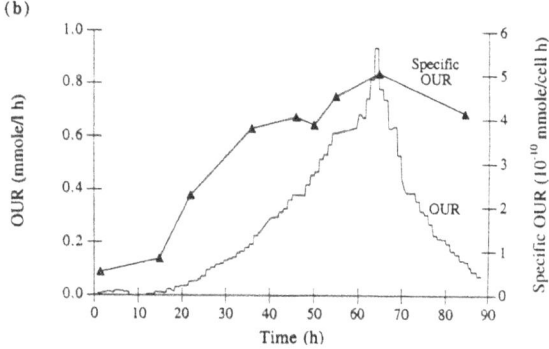

Fig. 4. On-line OUR measurement in a batch culture of AFP-27 cells.

506

phase indicating an increasing oxidative requirement by the cell, but only a small decrease in specific OUR was noted in the decline phase. In another experiment, the specific OUR remained nearly constant from the exponential to the decline phase which indicated that "dead" cells were not consuming oxygen while "viable" cells maintained a relatively constant specific oxygen consumption. The ratio of OUR and turbidity measurements could possibly give an on-line estimate of the specific oxygen consumption rate which would be another useful physiological variable.

OUR measurement could also be related with the consumption or oxidation of major nutrient components such as glucose. OUR was integrated over the entire cultivation period and plotted against the cumulative glucose consumption (Fig.5). A relatively linear stoichiometric relationship existed between oxygen consumption and glucose consumption equivalent to about 200.16 mg glucose per mole of oxygen. This showed that the time integral of on-line OUR measurement could provide an alternative means to estimate the amount of glucose consumed.

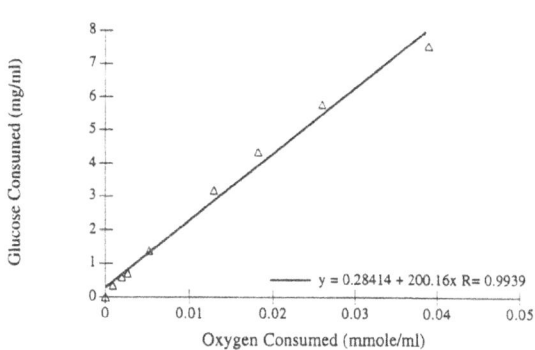

Fig. 5. Stoichiometric relationship between oxygen consumption and glucose consumption.

In fed-batch culture with a desired level or set point of residual glucose, it would be possible to use the time-integral of OUR to estimate the glucose consumed, hence the residual glucose, and the feeding rate could be controlled to maintain the appropriate glucose level in the medium.

4. Summary and Conclusion

In our batch culture studies we observed that good stoichiometric correlations exist among glucose consumption, lactate production, oxygen consumption and NaOH addition. The on-line measurement of alkali addition for pH control could be employed as an estimation of lactate production which was stoichiometrically related with glucose metabolism.

In order to increase our capability of on-line measurement, we investigated an on-line laser optical density sensor and a capacitance sensor. The laser optical sensor, with proper adjustment of sensitivity, gave an excellent estimation of total cell concentration. The physiological variable specific growth rate could then be calculated from the on-line cell concentration. On the other hand, substantial improvement and further study on the capacitance probe will be needed for the sensor to accurately give an estimate of the viable cell concentration. OUR was calculated by dynamic measurement and the time integral of OUR correlated well to the amount of glucose consumed. Furthermore, the profile of OUR in batch culture suggested that this variable could be a valuable indicator of the physiological state of the cells.

Work is in progress in our laboratories to integrate these on-line measurements to further advance our capability of monitoring and manipulating the physiological state of hybridoma cells in fed-batch and perfusion bioreactors.

5. References

Davey, C. L., Kell, D. B., Kemp, R., Meredith, R. W. (1988) 'On the audio- and radio-frequency dielectric behavior of anchorage-independent mouse L929-derived LS fibroblasts', Bioelectrochem. Bioenergetics, 20, 83-98.

Hu, W-S., Dodge, T.C., Frame, K.K. and Himes, V.B. (1987) 'Effect of glucose and oxygen on the cultivation of mammalian cells', Dev. Biol. Standard, 66, 279-290

Hu, W-S. and Oberg, M.G. (1990) 'Monitoring and control of animal cell bioreactors:Biochemical engineering considerations'. In: Large Scale Mammalian Cell Culture Technology, Ed. A.S. Lubineicky, pp. 451-481, Marcel Dekker, Inc., New York, NY.

Jain D, Gold S, Distefano D, Cuca G, Benincasa D, Ramasubramanyan K, Lenny A, Mark G, E, Silberklang M: (1991) 'Application of nutrient utilization analysis to obtain improved production of recombinant antibody', Abstract, General Meeting of European Society of Animal Cell Technology, September 2-6, 1991, Brighton, UK.

Konstantinov, K. B., Pambayun, R., Matanguihan, R. M., Yoshida, T., Perusich, C. M. and Hu, W. S. (1992) 'On-line monitoring of hybridoma cell growth using a laser turbidity sensor', Biotechnol Bioengineer, 40, 1337-1342.

Mishima, K., Mimura, A. and Takahara, Y. (1991) 'On-line monitoring of cell concentrations during yeast cultivation by dielectric measurements', J. Ferment. Bioengineer, 72, 296-299.

Sen, S., Srienc, F. and Hu W. S. (1989) 'Distinct volume distribution of viable and nonviable hybridoma cells: A flow cytometric study', Cytotechnology, 2, 85-94.

EFFECT OF DISSOLVED OXYGEN CONCENTRATION ON GROWTH OF MOUSE-MOUSE HYBRIDOMA

Takuo YANO and Yoshinori NISHIZAWA
Department of Fermentation Technology,
Faculty of Engineering, Hiroshima University,
Kagamiyama 1 chome, Higashihiroshima 724, Japan

ABSTRACT.   The effects of the dissolved oxygen concentration (DO) in the culture broth on the cell growth, consumption rates of nutritional elements such as glucose and amino acids, and production of anti-IgE antibody, lactic acid and ammonia were studied. Four batch cultures were carried out under the different DO conditions, 75, 50, 20 (corresponded to air saturation, ca.7 ppm) and 6%. With the increase of DO, maximum values of viable cell number and antibody produced decreased, but the amount of ammonia produced was almost constant. The specific growth rate was almost constant below 50% of DO, but at 75% of DO decreased remarkably. The specific production rates of inhibitory compounds, ammonia and lactic acid were minimum at 6% of DO. Cell yield from glucose or glutamine was maximum at 6% of DO, and decreased with the increase of DO. In conclusion, DO level at 6% seemed to be favorable for the effective antibody production by the hybridoma cells.

1. Introduction

In an animal cell culture, oxygen supply is one of the important factors for supporting good growth of cells, because oxygen is one of the nutritional elements for cell growth. In this study, the effects of the dissolved oxygen concentration (DO) in the culture broth on the cell growth, consumption rates of nutritional elements such as glucose and amino acids, and production rates of anti-IgE antibody, lactic acid and ammonia were studied.

2. Materials and Methods

2.1 Organism

The organism used in this study was a mouse-mouse hybridoma cell, which was produced by the fusion of the mouse myeloma P3-NS1/1-Ag4-1 with spleen cells from BALB/c mice. The hybridoma cell supplied from Sanyo Chemical Industries Co. (Kyoto, Japan) produces a monoclonal antibody, anti-IgE.

509

S. Kaminogawa et al. (eds.), Animal Cell Technology: Basic & Applied Aspects, Vol. 5, 509–514.
© 1993 *Kluwer Academic Publishers.*

2.2 Culture Conditions

RPMI-1640 medium containing 10% calf serum was used as a medium. Seed culture was prepared with 50ml medium in a 250ml T-flask. The culture was kept at 37℃ in a CO_2-incuvator (5% CO_2 content) until cell concentration reached 1×10^6 cells/ml broth. Then cell suspension was prepared by re-suspension with the medium after centrifigation of the seed culture broth.

The main culture was started after the injection of 10ml cell suspension to 490ml medium in a 1L jar-fermentor (Tokyo Rikakikai, MBF-100A), *i.e.*, working volume was 0.5 liter and initial viable cell concentration was ca. 1×10^5 cells/ml.

2.3 Control of pH and DO

To control pH and DO, N_2, CO_2 and O_2 gases were supplied to the head space in the jar-fermentor by the ON-OFF mode operation of a solenoid valve fitted to each gas flow line with a hand made control system [1]. Before the operation, each gas flow rate was adjusted by a needle-valve with a flow meter.

To control pH, carbonic acid concentration in the broth was regulated by the supply of CO_2 or N_2 gases, *i.e.*, when the pH value was above the upper set value, CO_2 gas was supplied, and when the pH was below the lower set value, N_2 gas was supplied.

For DO control, when DO was above the upper set value, N_2 gas was supplied in order to discharge DO from the broth. While O_2 gas was supplied when the DO value was below the lower set value.

2.4 Analysis

Cell number was counted by a haemocytometer after addition of trypan blue solution to the culture broth drawn out from the fermentor. Glucose and lactic acid concentrations in the supernatant of broth were measured by analytical kit applied by the enzymatic method, Diacolor·GC (Toyobo Co., Japan) and F-kit (Boehringer Co., Germany), respectively. Ammonia concentration was measured with an ion electrode (model 95-12, Orion Research, USA). The antibody was detected with ELISA method. Amino acids were measured with an automatic amino acid acid analyzer (655A, Hitachi, Japan).

3. Results and Discussion

To make clear the effect of DO on the growth of hybridoma cells, four batch cultures were carried out under different DO level (6, 20, 50 and 75% of DO). 20% of DO corresponds to air saturation, about 7 ppm. As an example, time course of the culture at 6% of DO is shown in Fig. 1. After about 30 h lagtime, cell number increased with the consumptions of glucose and glutamine, and stopped by the lack of glutamine. Glutamine may be first growth limiting factor in batch culture with RPMI-1640 medium. Antibody was produced with cell growth. Growth-inhibiting products, lactic acid and ammonia were also produced.

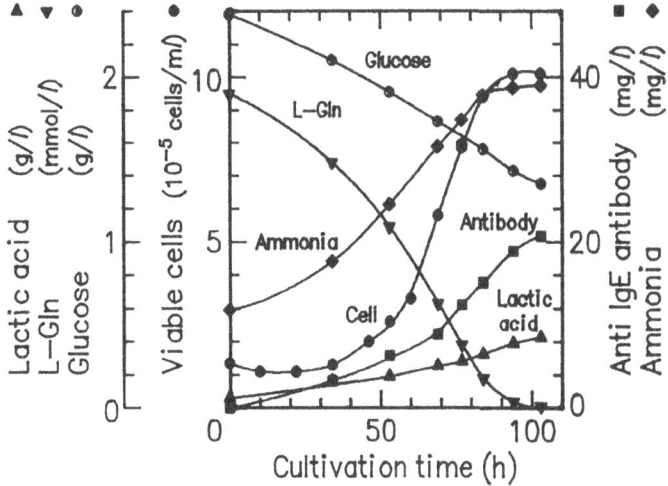

Figure 1 . Time courses of growth (●), glucose (◐), glutamine (▼), antibody (■), ammonia (◆) and lactic acid (▲) in the culture at 6% of DO.

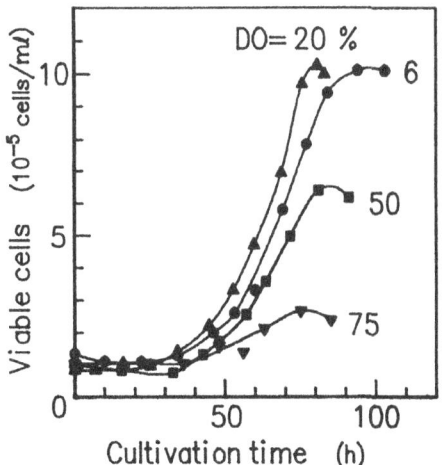

Figure 2   Effect of DO on cell growth. Cultivation was carried out at 6 (●), 20 (▲), 50 (■) and 75% (▼) of DO.

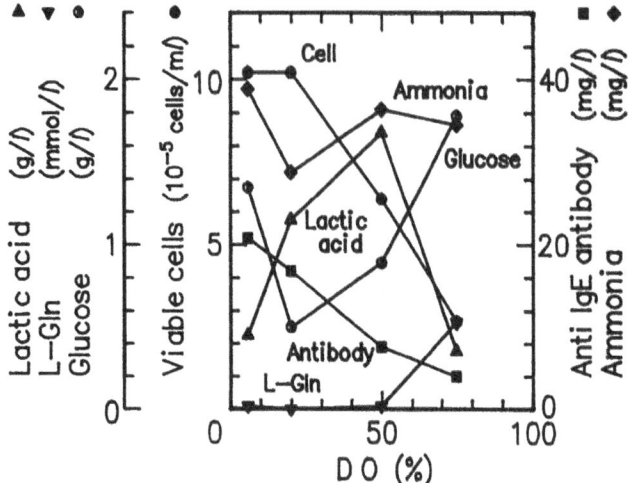

Figure 3　Effects of DO on final cell number (●), residual glucose (◐) and glutamine (▼) concentrations, final concentrations of products, antibody (■), ammonia (◆) and lactic acid (▲).

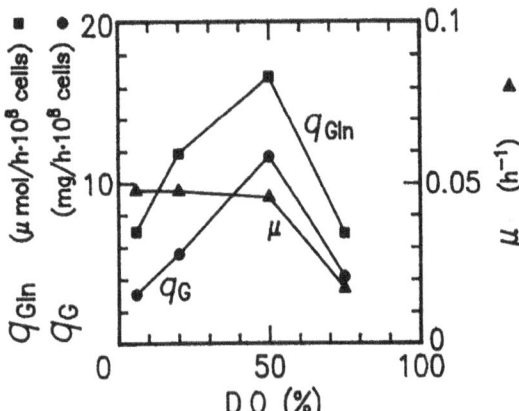

Figure 4　Effects of DO on specific growth rate (▲) and specific consumption rates of glucose (●) and glutamine (■).

In Fig. 2, only the growth curve in each run is summarized to simplify the effect of DO on the growth. At 20% of DO, growth was similar to that at 6%. But both growth rate and maximum viable cell number decreased with increasing DO value.

Concentrations of nutrients and products at the end of each run are plotted in Fig. 3. While glucose remained in all cases, glutamine was completely consumed except for the run at 75% of DO. Larger amount of lactic acid was produced at 20 and 50% of DO. The amount of ammonia was almost constant. The amount of the antibody as well as the viable cells decreased with the increase of DO.

Specific growth rate and specific consumption rates of glucose and glutamine are shown in Fig. 4. The values of specific growth rate (μ) were almost constant below 50% of DO, but at 75% of DO decreased remarkably. Specific consumption rates of glucose (q_G) and glutamine (q_{Gln}) increased with the increase of DO until 50%, but at 75% of DO both of them decreased also remarkably as well as μ. High concentration of DO such as 75% seems to give an extreamly strong effect, i.e., cell growth stopped inspite of the presense of glutamine and glucose (see Figs. 2 and 3).

Figure 5 shows the effect of DO on the specific production rates of antibody (q_{Ab}), ammonia (q_{NH3}) and lactic acid (q_L). For ammonia and lactic acid production, smaller value is better. The minimum value was observed at 6% of DO. On the specific production rate of antibody, the effect of DO seemed not to be strong.

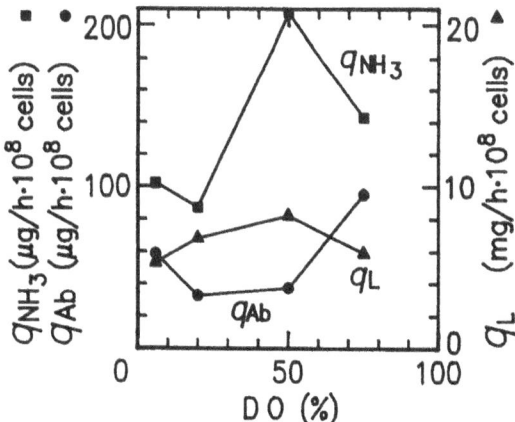

Figure 5   Effects of DO on specific production rates of antibody (●), ammonia (■) and lactic acid (▲).

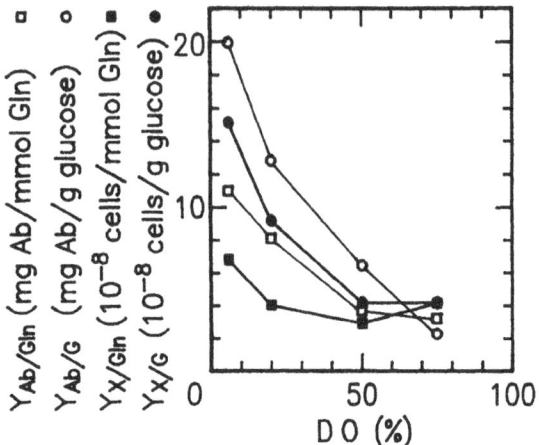

Figure 6   Effects of DO on cell yields (● and ■) and antibody yields
(○ and □) from glucose and glutamine, respectively.

Finally, the effect of DO on the yields of cells, $Y_{x/G}$ and $Y_{x/Gln}$ and the
yields of antibody, $Y_{Ab/G}$ and $Y_{Ab/Gln}$ from glucose and glutamin consumed,
respectively are summarized in Fig. 6. All of them gave maximum at 6% of DO,
and decreased with increasing DO value.

At lower value of DO, the cell could grow efficiently and produce the
antibody with high efficiency. The reason of this phenomenon was not yet
clear. Whereas oxygen was one of the essencial growth factors, it might have a
negative effect on the metabolisms of glucose and glutamine for cell growth
and antibody production. In conclusion, for good growth and effective
production of the antibody by the hybridoma cells, 6% of low DO level might be
favorable.

4. Acknowledgement

The authors are very grateful to Sanyo Chemical Industries Co. for the kind
gift of the mouse-mouse hybridoma cell.

5. References

1. Yano, T. and Nishizawa, Y. (1992) 'A simple automatic control system of pH
    and DO', in H. Murakami *et al.* (eds.), Animal cell technology: basic &
    applied aspects, Kluwer Academic Publishers, the Netherlands, pp. 295-302.

ISOLATION OF AUTOCRINE GROWTH FACTORS PRODUCING MOUSE HYBRIDOMA CELL LINES

Y.K.LEE* & P.K.YAP
Department of Microbiology
Faculty of Medicine
National University of Singapore
Lower Kent Ridge Road
Singapore 0511

*corresponding author

ABSTRACT. Isolation of low serum requiring mouse-mouse hybridoma populations was done by continuous dilution of the cultures with media containing lower serum concentrations. The isolated cell lines are stable in antibody production and growth even through multiple passages. Analysis of the selected clones from two cell lines revealed contrasting cellular physiology in terms of the affinity for serum components and the maximum specific growth rate. Our studies showed that the ability of isolated cell lines to grow in chemical defined media without serum supplements was attributed to the production of autocrine growth factors.

1. Introduction

Mammalian cell growth is principally regulated by highly specialized peptides known as growth factors, which interact with great specificity and high affinity with cell surface receptors[1,2,3]. It is these growth factors, that largely responsible for the well-known serum requirement for mammalian cell growth in culture.

Serum is traditionally included in animal cell culture media to support normal cell growth. It is however costly, difficult to sterilize and its inclusion in culture media interferes with subsequent product purification. Its presence in culture media also renders the cell free supernatant unsuitable for direct *in vivo* applications.

The phenomenon termed "inoculum density effects" whereby a decrease in the inoculum density causes lowering in the initial growth rate of mammalian cultures[4] and the increase in growth yield when cells are grown in conditioned medium[5], signals the involvement of factors produced by cells, which facilitate its growth.

We believe that all cells are capable of producing their own growth-promoting factors by stimulating them to express dormant genes or selecting cells which express these growth factor producing genes.

It is the aim of this work to isolate and characterize hybridoma cell lines which produce sufficient quantity of autocrine growth factors to support their growth, thus makes it possible to exclude serum from the culture media used.

515

S. Kaminogawa et al. (eds.), Animal Cell Technology: Basic & Applied Aspects, Vol. 5, 515–520.
© *1993 Kluwer Academic Publishers.*

2. Materials and Methods

2.1. CELL LINE AND MEDIUM

Hybridoma cell lines 43B3 and 6BB (provided by Professor SH Chan of this department) are mouse hybridoma that produces IgG and IgM antibody to human blood group B antigen respectively. The cells were grown in medium RPMI 1640 (Gibco) and supplemented with fetal bovine serum (Gibco).

2.2. ISOLATION PROCEDURE

Culture media with decreasing serum concentrations (from 10% to 8%, 6%, 4%, 2%, 1%, 0.5% and 0.125%) were fed continuously into suspension cultures of the desirable cell lines in a thermoregulated (37°C) stirred vessel, over a period of one month. The low serum requiring cell lines isolated were further cultured in RPMI medium in static culture flasks.

2.3. SAMPLE ANALYSIS

2.3.1. *Cell number and viability.* Determination of cell number and viability was done by inclusion of 0.4% (w/v) trypan blue and counted on a hemacytometer.

2.3.2. *Monoclonal Antibodies.* IgG and IgM monoclonal antibodies were determined by sandwich ELISA (enzyme-linked immunorsorbant assay) method.

2.3.3. *Protein separation.* Analysis on culture supernatant was performed using standard Laemmli SDS-PAGE, 3% stacking gel, 7% running gel.

2.4. DIFFUSION EXPERIMENT

Transwell™ plates (Costar) were used. A polycarbonate membrane with 3.0 um pores separated the culture well into upper and lower chambers. Selected clones were placed in the upper chamber while the unselected clones were placed in the lower chamber.

3. Results and Discussion

3.1. ISOLATION OF LOW SERUM REQUIRING SUB-CLONES

Using the system that we have designed, clones of cell lines which are able to grow in chemical defined medium with no serum were obtained. Two low serum requiring clones designated 43B3LS and 6BBLS, derived from cell line 43B3 and 6BB respectively were studied in detail.

Clones 43B3LS and 6BBLS were stable throughout multiple passages without apparent drop in antibody titre. In fact, antibody productivity increased in the low serum requiring clones. This may be due to the effect of " serum inhibition " as suggested by some workers[6].

3.2. ANALYSIS OF SELECTED CELL LINES

The effects of serum on the growth rate can be expressed mathematically by a Monod-type model which considers serum to be the limiting factor[7],

$$u = u_{max}S/(Ks + S)$$

where u is the growth rate, S is the serum concentration, u_{max} is the maximum specific growth rate, and Ks is the Monod constant (which indicates the affinity for serum components).

Theoretically, it is possible to select low serum or serum non-requiring cells of 3 types, as shown in Figure 1.

Besides having cells with altered u_{max} and Ks, there are possibilities that selected cells have increased their ability of producing cell-derived growth factors (autocrine growth factors), or there are changes in receptor concentrations on the cell membrane[8].

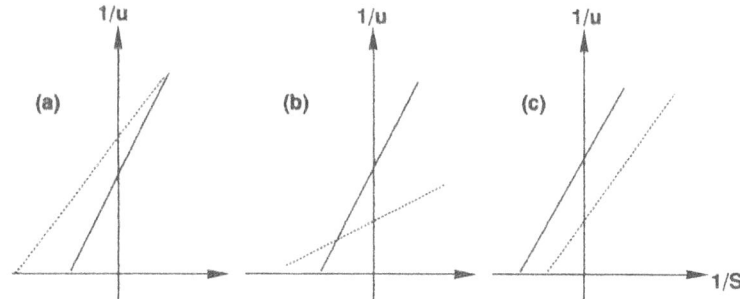

Figure 1. Double reciprocal plot of Monod relation: (a) cells with increase in affinity for serum components and decrease in maximum specific growth rate (b) cells with increase in affinity for serum components and increase in maximum specific growth rate (c) cells with decrease in affinity for serum components and increase in maximum specific growth rate compare to original clone (—).

518

In the case of 43B3 (Figure 2), it has a Ks value of 5.25% serum (v/v) and u_{max} of 0.05 hour^{-1} for the unselected clone and Ks value of 0.19% serum (v/v) and u_{max} of 0.04 hour^{-1} for the selected clone 43B3LS.

Under low serum condition, subclone 43B3LS was isolated from the parental clone due to its high affinity for serum components. The high affinity for serum components explains partly why the selected clone is able to grow at very low serum supplemented medium, but cannot account for the fact that it grows well in non-serum supplemented basal medium.

By the use of permeable cell chambers (Transwell^{TM}), where 43B3 and 43B3LS were separated by a 3 um-pore polycarbonate membrane, the unselected cell clone could maintain its viability in non-serum supplemented basal medium for at least 24 hours.

Whereas, the unselected cells died rapidly in non-serum supplemented basal medium (Figure 3). This gives indication that the selected cells produce factors for maintenance of viability or regulating cell growth.

Analysis of the culture supernatant of selected 43B3LS cultures by SDS-PAGE showed a protein of > 66 kDa (figure 4) which may in this case, be involved in conferring the ability of such selected cell line to grow in low serum or basal RPMI medium.

However, in the case of 6BB (Figure5), the Ks value of 6.68% serum (v/v) and u_{max} of 0.14 hour^{-1} for the selected cells, are much higher than the unselected cells having Ks value of 2.94% serum (v/v) and u_{max} of 0.05 hour^{-1}

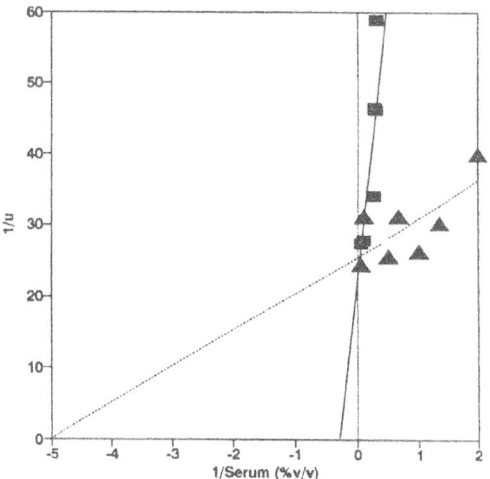

Figure 2. Double reciprocal plot of Monod relation for 43B3

(■)Unselected clone   (▲) Selected clone

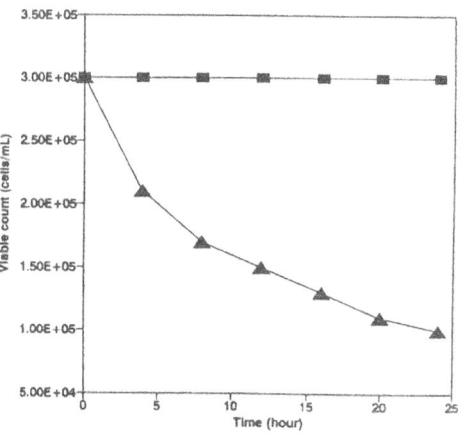

Figure 3. Stability of unselected clone with (■) and without (▲) the presence of selected clone grown in Transwell plates.

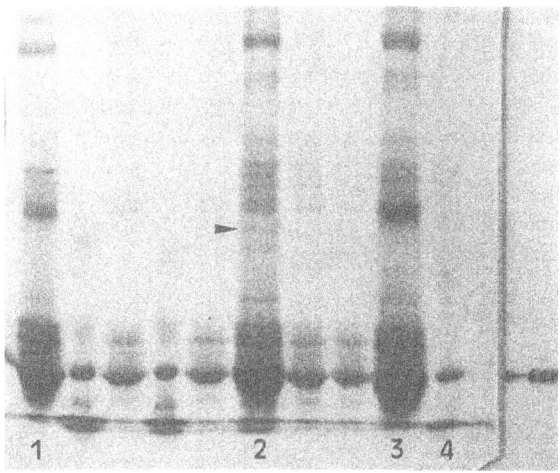

Figure 4. Gel Electrophoresis of : (1) Fetal bovine serum; (2) Culture supernatant from 43B3LS with an extra protein band indicated; (3) Culture supernatant from 43B3; (4) M.W. standard

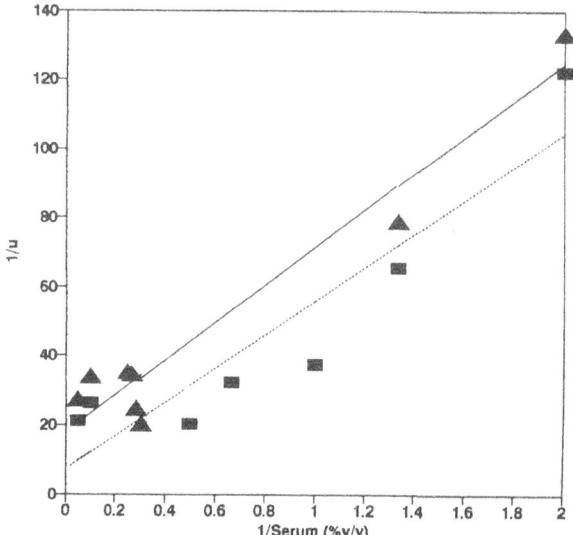

Figure 5. Double reciprocal plot of Monod relation for 6BB

(▲) Unselected clone   (■) Selected clone

We were unable to detect apparent differences in the protein electrophoretic profile of culture supernatant from 6BB and 6BBLS. There are possibilities that the growth factors produced is in very small quantity and are membrane-bound or the selected cells have an increase receptor concentrations on the cell membrane for the factors mentioned.

Thus, having an increase in Ks value for serum may not be all that detrimental, for it indicates the possibility that such clone produces its own growth factors, making itself insensitive to serum factors.

The present study has demonstrated that it is possible to select for low serum or non-serum requiring hybridoma cell lines, which either posses high uptake affinity for serum factors, or produce sufficient quantity of autocrine growth factors.

4. Acknowledgement

We would like to thank AP TEOH who have given her kind assistance to this work.

5. References

1. Burgess, A. and Nicola, N. (1983) Growth Factors and Stem Cells, Academic Press, New York.

2. Boynton, A.L. and Leffert, H.L., eds, (1985) Control of Animal Cell Proliferation, vol. 1, Academic Press, New York.

3. Guroff, G. ed., (1987) Oncogenes, Genes, and Growth Factors, John Wiley & sons, New York.

4. Hu, W.S., Meier, J. and Wang, D.I.C. (1985) 'A Mechanistic Analysis of the Inoculum Requirement for the cultivation of Mammalian Cells on Microcarriers', Biotech. Bioeng. 27, 585-595.

5. Shirai, Y., Yoshimi, T. and Hashimoto, K. (1991) 'Effects of Conditioned Medium on the Growth Kinetics and the Monoclonal Antibody Productivity of Hybridoma Cells', The Fourth Annual Meeting of JAACT, pp 96.

6. Glassy, M.C., Tharakan, J.P. and Chau, P.C. (1988) 'Serum-Free Media in Hybridoma Culture and Monoclonal Antibody Production', Biotech. Bioeng. 32, 1015-1028.

7. Pirt, S.J. (1975) Principles of Microbe and Cell Cultivation, Blackwell Scientific Publication, Cambridge.

8. Ozturk, S.S. and Palsson, B.O. (1991) 'Physiological changes during the adaptation of hybridoma cells to low serum and serum-free media', Biotech. Bioeng. 37, 35-46.

PRODUCTION OF MULTISUBUNIT PARTICLES FOR USE AS VACCINES USING THE BACULOVIRUS EXPRESSION VECTOR SYSTEM (BEVS)

KYM BAKER, YUAN-ZHI ZHENG, STEVEN REID AND
PAUL F. GREENFIELD
Department of Chemical Engineering
The University of Queensland   Qld   4072   Australia

ABSTRACT. A major attraction of the baculovirus expression vector system (BEVS) over other expression systems is the relative ease with which it can be used to express simultaneously two or more proteins to form multisubunit particles. Professor Polly Roy and her group at Oxford in England have developed recombinant baculoviruses capable of producing bluetongue virus (BTV) like particles which have potential for use as a vaccine against BTV which causes disease in sheep and cattle. Through collaboration with Professor Roy's group we have obtained a recombinant baculovirus capable of producing BTV core like particles (CLPs). In this work we show we can achieve high yields (50 mgm/litre) of such CLPs in suspension culture using serum free media. Multiplicity of infection (MOI), time of infection (TOI) and time of harvest (TOH) parameters are discussed in terms of maximising the yield of such multisubunit particles.

Introduction

A number of unique features distinguish BEVS from other expression systems including the ability for two or more proteins to be expressed simultaneously, (Luckow, 1991). BEVS is now often the first system tested when abundant expression of soluble biologically active eukaryotic proteins are desired, particularly given the recent improvements in methods for producing and selecting pure recombinant viruses (Bishop et al, 1992).

Serum free media developed by Gibco (New York) allows 2-3 fold improvements in production yields using BEVS and insect cells utilising simple batch cultures (Radford et al, 1992, Weiss et al 1992). Such increases in yield, and further improvements in our ability to model and optimise production processes with this system (Power et al, 1992), allows one to view this expression system as being economical for the low cost production of complex vaccines for the animal industry.

Professor Polly Roy and her group at Oxford in England have developed recombinant baculoviruses capable of producing bluetongue virus (BTV) like particles (French et al, 1990, French and Roy 1990). These virus like particles

521

S. Kaminogawa et al. (eds.), Animal Cell Technology: Basic & Applied Aspects, Vol. 5, 521–528.
© 1993 Kluwer Academic Publishers.

(VLPs) have great potential for use as an effective vaccine against BTV which causes disease in sheep and cattle. These BTV particles are currently being tested in South Africa with encouraging results (Roy *et al*, 1992). Bluetongue virus (BTV) exists as a double-shelled virus particle composed predominantly of four proteins VP3, VP7, VP5 and VP2. VP3 and VP7 form an inner core particle. VP2 and VP5 form an outer shell surrounding the VP7/VP3 CLP. VP7, VP3 CLP's alone may provide some protection against subsequent BTV infection.

An essential early step in commercialising such a vaccine will be the demonstration that such multigene particles can be produced with high yield using serum free media in simple (relatively inexpensive) batch culture. To this end, our laboratory has obtained the dual recombinant baculovirus capable of forming VP3, VP7 CLP's in order to conduct such basic production studies.

Materials and Methods

Recombinant baculovirus's (rAcNPV) used, expressed either VP7 alone or VP7, VP3 (CLPs) simultaneously. Both viruses were provided by Professor Roy, Institute of Virology and Environmental Microbiology, Oxford, U.K., (French and Roy, 1990, Oldfield *et al*, 1990). Sf9 (*Spodoptera Frugiperda*) cells were obtained from the ATCC (CRL 1711) or from Gibco (New York) preadapted to SF900II media. The serum free media, SF900II, was obtained from Gibco, New York. Suspension cultures were performed in shakers or in a bioreactor. Shaker cultures were conducted in 250 ml erlenmeyer flasks with a 100 ml working volume and were placed on a orbital shaker operating at 100 rpm within a 27°C refrigerated incubator. Bioreactor cultures were conducted in a 2 litre stirred tank (Setric Genie Industrial, France) with a 1 litre working volume and operated at 100 rpm, 30% DOT and 27°C. pH was monitored but not controlled (range observed 6.0-6.2). Samples were analysed by SDS polyacrylamide gel evectrophoresis and quantification of VP7, VP3 bands carried out by densitometry following coomassie blue staining.

Results

Initially some experiments were conducted using a virus expressing VP7 only. These studies were done in order to assess the quantity of VP7 expressed when it was the only foreign protein being produced and to get some data on the kinetics of VP7 production.

Figure 1 and Table 1 show data from a shaker culture where VP7 alone was expressed. The culture was infected at a cell density of 2.4×10^6 cells/ml and at a multiplicity of infection (MOI) of 1.0 plaque forming unit (PFU) per cell. Note that at this MOI, considerable cell growth post infection took place with the culture peaking at around 5×10^6 cells/ml.

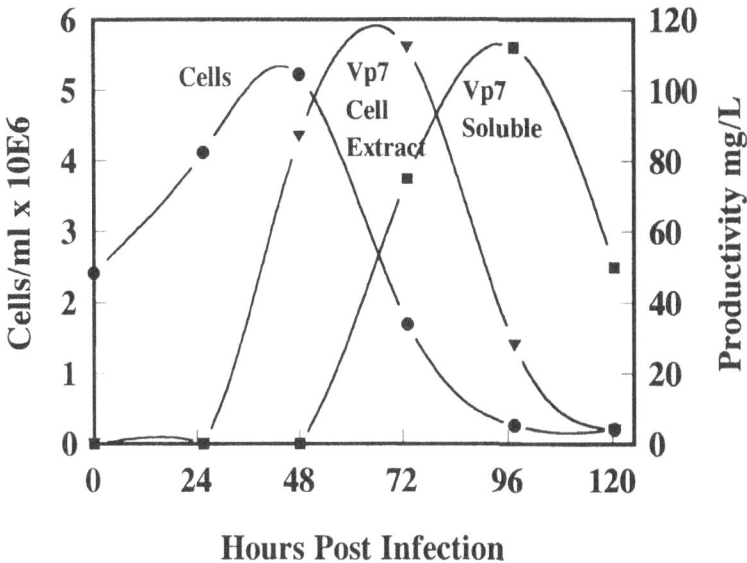

Figure 1.Production of Vp7 via BEVS

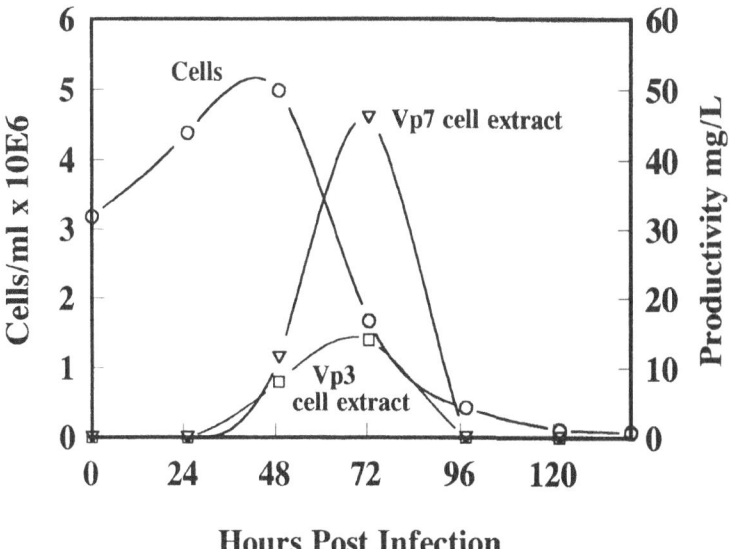

Figure 2.Production of CLPs,MOI of 1 PFU/cell

A peak level of VP7 production occurred at 3 days post infection (dpi), with 187 mg/l of VP7 being produced, 112 mg/l being cell associated at this time (Table 1, 73 hpi). The VP7 yield decreased by 4dpi presumably due to proteolysis following further drops in cell viability. To avoid potential product instability following cell lysis an appropriate time to harvest would be between 2-3 dpi. Note that cell viabilities drop rapidly between 2-3 dpi (Table 1, 48-73 hpi).

Table 1. VP7 Production

Time Post Infection (h)	Viable Cell Density (cells/ml x $10^{6)}$	Cell Viability (%)	+VP7 Cell Extract (mg/l)	VP7 Supernatant (mg/l)	Total VP7 (mg/l)
26	4.1	98	0	0	0
48	5.2	94	87	0	87
73	1.7	30	112	75	187
98	0.3	4	28	112	140
121	0.2	3	4	50	54

+VP7 cell extract refers to VP7 released from cells following low speed centrifugation and subsequent exposure to hypoosmotic shock following resuspension in water.

Figure 2 and Table 2 show data from shaker cultures where CLPs (VP7/VP3) were expressed. Note again that at a MOI of 1 PFU/cell considerable cell growth post infection takes place (Fig 2, Table 2a) but at a MOI of 10 PFU/cell, limited cell growth post infection is observed, Table 2(b).

For the CLP data, only intracellular (cell extract) yields are shown, as these particles are unstable following cell lysis and so will be harvested while cell viabilities remain relatively high. For the culture conducted at an MOI of 1 PFU/cell a peak yield of 60 mg/l is obtained at 3 dpi (73 dpi). When an MOI of 10 PFU/cell was used 53 mg/l was observed at only 1.5 dpi (37 hpi). Viabilities are already starting to drop by 1.5 dpi when the higher MOI is used (Table 2b). For an MOI of 1.0 PFU/cell viabilities again were observed to drop rapidly between 2-3 dpi, while the CLP yield increased 3 fold over this period. For the high MOI case between 37-60 hpi the cell associated yield of CLPs drops from 53 to 44 mg/l. The VP7 yield in particular drops (35 to 27 mg/l) suggesting a rapid loss from lysing cells of VP7 that perhaps has not had time to associate in a stable manner with CLP particles.

The molar ratios of VP7:VP3 are indicated (Table 2). The ratios determined, vary dramatically, in some cases, over the culture period (Table 2a). The kinetics of production of VP7 and VP3 and the kinetics of their interaction need to be studied in more detail, in relation to the kinetics of cell death if a time of harvest (TOH) is to be determined which will allow a CLP product to be obtained with a 'native like' VP7:VP3 ratio of 13:2.

Table 2 Production of Bluetongue Core Like Particles (CLPS)

(a) TOI: 3.2×10^6 cells/ml
  MOI: 1.0 PFU/cell

Time Post Infection (h)	Viable Cell Density (cells/mlx10^6)	Cell Viability (%)	+VP7 (mg/l)	+VP3 (mg/l)	+Total CLP Yield (mg/l)	*VP7:VP3 Molar Ratio
49	5.0	91	12	8	20	8:2
73	1.7	27	46	14	60	18:2

(b) TOI: 5.8×10^6 cells/ml
  MOI: 10.0 PFU/cell

Time Post Infection (h)	Viable Cell Density (cells/mlx10^6)	Cell Viability (%)	+VP7 (mg/l)	+VP3 (mg/l)	+Total CLP Yield (mg/l)	*VP7:VP3 Molar Ratio
37	4.8	82	35	18	53	10:2
60	1.8	23	27	17	44	9:2

+ The yield indicated refers to the cell associated (intracellular) yield.

* The VP7:VP3 molar ratios were calculated assuming a Mwt for VP7 of 38, 548 and for VP3 of 103, 416. Native CLPs have a molar ratio of 13:2 (French and Roy, 1990).

Figure 3 shows a bioreactor culture of Sf9 cells using SF900 II media. The cells were infected at 5×10^6 cells/ml using a MOI of 20 PFU/cell (CLP rAcNPV). For the bioreactor culture, cell death was delayed compared to the shaker (60 hpi, still 84% viable) and the intracellular production peaked at only 23 mg/l of CLPs.

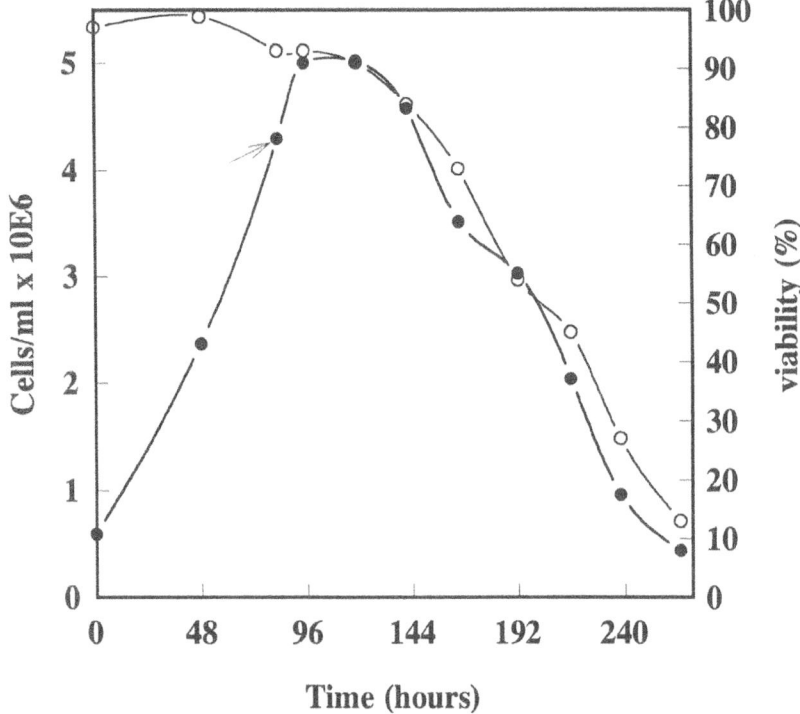

Figure 3.Bioreactor Production of CLPs

The arrow indicates the point of infection.
MOI: 20 PFU/cell
TOI: 4.3 x 10^6 cells/ml
CLP yield: 23mg/ L

Discussion

Our laboratory has carried out extensive studies with the baculovirus expression system using ß-Galactosidase as a model protein due to its ease of assay, (Power *et al*, 1992, Radford *et al*, 1992). This work indicates that when using shaker cultures, and SF900 II media, infected cells will display a peak cell specific productivity until the cell density exceeds 5-7 x 10^6 cells/ml, in batch culture using the original growth media (Radford *et al*, 1992). Data from this work (Fig. 3) and recent related work with ß-Gal indicates that the peak specific productivity at a bioreactor level decreases earlier, between 4-5 x 10^6 cells/ml. Further work is required to clarify this. SF900 II is a nutrient-rich media which allows peak specific productivities at cell densities much higher than can be observed using less optimised media such as IPL 41 plus foetal bovine serum (Radford *et al*, 1992). Less optimised media will not support high specific productivities above 2 x 10^6 cells/ml at either a shaker or bioreactor level. Whether further improvements in the upper cell density at which batch cultures will support peak specific productivities is only a matter of further nutrient supplementation is unclear. It is also unclear why bioreactors fail to maintain maximal specific productivities at 5-7 x 10^6 cells/ml as is observed with shakers.

The data indicates that production of VP7 alone leads to a higher yield of this protein 112 mg/l, (cell extract) at 73 hpi, compared to when it is co-expressed with VP3, 46 mg/l (cell extract) at 73 hpi, (Tables 1 and 2a).

It is shown that a yield of CLPs of 50-60 mg/l is possible (shaker cultures) but the quality of these CLPs with respect to their VP7:VP3 molar ratio, requires further work. The TOH parameter is critical when producing multisubunit particles, as it becomes a fine balance between maximising the yield of the individual proteins involved and the time for their interaction, against avoiding excessive cell lysis and subsequent particle damage.

References

Bishop, D.H.L., Hill-Perkins, M., Jones, I.M., Kitts, P.A., Lopez-Ferber, M., Clarke, A.T., Possee, R.D., Pullen, J. and Weyer, U. (1992). "Construction of Baculovirus Expression Vectors". In: Baculovirus and Recombinant Protein Production Processes, Eds. J.M. Vlak, E.J. Schlaeger and A.R. Bernard, Editiones Roche Basel, Switzerland, 27-49.

French, T.J., and Roy, P. (1990). "Synthesis of Bluetongue Virus (BTV) Corelike Particles by a Recombinant Baculovirus Expressing the Two Major Structural Core Proteins of BTV", J. Virology, 64, 1530-1536.

French, T.J., Marshall, J.J.A., and Roy, P., (1990). "Assembly of Double-Shelled, Viruslike Particles of Bluetongue Virus by the Simultaneous Expression of Four Structural Proteins", J. Virology, 64, 5695-5700.

528

Luckow, V.A. (1991). "Cloning and Expression of Heterologous Genes in Insect Cells with Baculovirus Vectors". In: Recombinant DNA Technology and Applications. Eds. A. Prokop, R.K. Bajpai and C.S. Ho, McGraw-Hill, Inc. New York, 97-152.

Oldfield, S. Abachi, A., Urakawa, T., Hirasawa, T. and Roy, P., (1990). "Purification and Characterization of the Major Group-specific Core Antigen VP7 of Bluetongue Virus Synthesized by a Recombinant Baculovirus". J. Gen. Virol. 71, 2649-2656.

Power, J., Greenfield, P.F., Nielsen, L. and Reid, S. (1993). "Modelling the Growth and Protein Production by Insect Cells following Infection by a Recombinant Baculovirus in Suspension Culture". Cytotechnology (in press).

Radford, K.M., Reid, S. and Greenfield, P.F. (1992), "Improved Production of Recombinant Proteins by the Baculovirus Expression System using Nutrient Enriched Serum Free Media". In: Baculovirus and Recombinant Protein Production Processes, Eds. J.M. Vlak, E.J. Schlaeger and A.R. Bernard, Editiones Roche, Basel, Switzerland, 297-303.

Roy, P., French, T. and Erasmus, B.J., (1992). "Protective Efficacy of Virus-like Particles for Bluetongue Disease". Vaccine, 10, 28-32.

Weiss, S.A., Whitford, W.G., Godwin, G.P. and Reid, S., (1992). Media Design: Optimizing of Recombinant Proteins in Serum-Free Culture. In: Baculovirus and Recombinant Protein Production Processes, Eds. J.M. Vlak, E.J. Schlaeger and A.R. Bernard, Editiones Roche, Basel, Switzerland, 306-315.

SPONTANEOUS APOPTOSIS IN MOUSE HYBRIDOMA CULTURE

F. FRANĚK and T. VOMASTEK
Department of Fundamental Cytotechnology
Institute of Molecular Genetics
Vídeňská 1083
CS-14220 Praha 4
Czechoslovakia

ABSTRACT. Previous investigations have shown that a substantial fraction of mouse hybridoma cells in suspension culture die by apoptosis. This work represents a first attempt at timing the onset of spontaneous apoptosis in hybridoma batch culture. Markers of apoptosis were followed in batch cultures carried out in protein-free medium. A steep increase of the apoptotic index, i.e. the relative number of bodies insoluble in 6M guanidinium hydrochloride, was observed on day. A notable increase of the degree of DNA fragmentation, that represents an early event of the death program, was apparent already on day 4. Thus, the critical period of the spontaneous onset of apoptosis is the exponential phase of the culture. This finding indicates that spontaneous apoptosis in mouse hybridomas is a process associated with cell proliferation and full metabolic activity.

1. Introduction

Several recent findings support a novel view on the modes of death of cultured animal cells. According to this view necrosis, i.e. a general collapse of all cell functions caused by a severe injury, represents the less frequent mode of death observed in bioreactor cultures. Events like total exhaustion of nutrients, sudden accumulation of toxic metabolites or sudden shift of pH out of the physiological range, that may cause the necrotic type of death, occur in optimized processes seldom if at all. Mild noxious stimuli, such as gradual depletion of growth factors or trophic factors or gradual accumulation of toxic metabolites are likely to induce apoptosis, i.e. a programmed way of cell death characterized by distinct morphological and biochemical features. The physiological triggering stimuli include loss of function of a repressor of apoptosis or gain of function of a specific inducer. Spontaneous triggering by an unknown stimulus is also possible in cultured cells (Fesus et al. 1991, Gerschenson and Rotello 1992, Cohen 1991) (Table 1).

Apoptosis is a broadly operating mechanism serving to maintain homeostasis in the animal organism. Besides embryogenesis, this mode of cell and organ elimination plays a key role in systems of proliferating cells, such as cells of the immune system (Cohen 1991). The present evidence for the occurrence of apoptosis in mouse hybridoma culture is based on the following findings: (1) presence of mono- and oligonucleosomes in the culture fluid (Franěk and Dolníková 1991), (2) presence of apoptotic bodies insoluble in 6 M guanidinium hydrochloride (Franěk and Dolníková 1991a), and (3) existence of fragmented nuclear DNA in hybridoma cells of the stationary phase (Franěk et al. 1992, 1992a). This work represents a first attempt at timing the onset of spontaneous apoptosis in hybridoma batch culture.

S. Kaminogawa et al. (eds.), Animal Cell Technology: Basic & Applied Aspects, Vol. 5, 529–534.
© 1993 *Kluwer Academic Publishers.*

TABLE 1. Alternatives for a viable cell

	Phenomenon (Cell physiology)	Significance (Cell technology)
Physiological:	Mitosis	Desirable in growth phase
	Growth arrest	Desirable in production phase
	Death by apoptosis (caused by inducer-repressor disbalance)	Undesirable
Pathological:	Death by apoptosis (caused by mild injury)	Undesirable
	Death by necrosis (caused by severe injury)	Undesirable

2. Materials and Methods

2.1 CELL CULTURE

A mouse hybridoma producing an IgG1 kappa antibody was used as a model cell line. The basal medium RPMI 1640 was enriched with MEM essential amino acids and MEM vitamins. The basal medium was combined with the iron-rich supplement (Franěk et al. 1992, 1992a). The cells were grown in T-flasks, seeded in the spinner flask (1 L volume) and cultured batch-wise at 35°C.

2.2 SAMPLE ANALYSIS

Cell growth was measured by counting cell number in a haemocytometer. Viability was determined by the trypan blue exclusion test. The apoptotic index, i.e. the percentage of apoptotic bodies, was determined by counting the particles insoluble in 6 M guanidinium hydrochloride as described in detail before (Franěk et al. 1992a). The antibody concentration was determined using turbidimetric immunoassay.

DNA isolation and analysis was carried out as described before (Franěk et al. 1992a). Briefly, the cell pellet was lysed by a standard method (SDS+proteinase K), DNA extracted by phenol-chloroform, and precipitated by ethanol. Electrophoretic patterns of ethidium bromide-visualized DNA served for qualitative assessment of the state of fragmentation. For the quantitative evaluation of the degree of DNA fragmentation the DNA samples were resolved on a column of Sepharose CL-2B, the contents of individual tubes were lyophilized and DNA in the tubes determined using the Dische diphenylamine reaction.

3. Results and Discussion

The phenomenon of apoptosis was found to be more pronounced in cultures conducted in basal media relatively poor in nutrients (Franěk and Dolníková 1991a). The present experiments were therefore carried out in a moderately fortified RPMI 1640 basal medium (Fig.1). The electrophoretic analysis of DNA isolated from the hybridoma cells revealed patterns typical of apoptotic cells, namely the "ladder" of oligonucleosomal fragments (Fig.2).

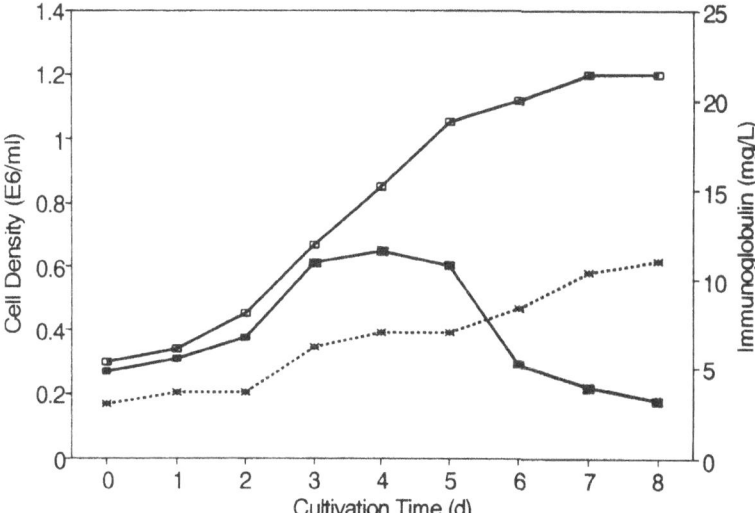

Figure 1. Growth and production characteristics of TSH-5.07 hybridoma in batch culture. Empty squares - total cells; filled squares - viable cells; asterisks - immunoglobulin concentration.

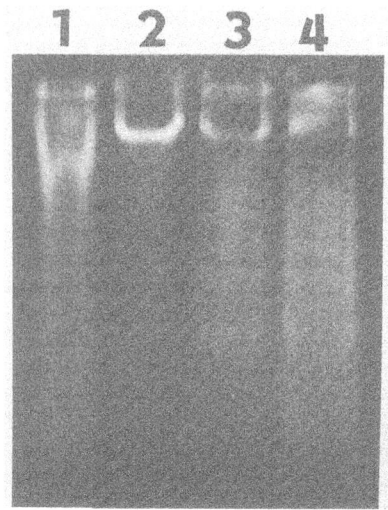

Figure 2. Agarose gel electrophoresis of DNA isolated from TSH-5.07 hybridoma cells collected at various phases of the batch culture. Staining with ethidium bromide. Lane 1 - molecular size markers (142 bp "ladder"); lane 2 - early exponential phase; lane 3 - stationary phase; lane 4 - decline of the culture.

532

The quantitative balance of various DNA forms in the sample collected on day 8 of the batch culture revealed that the integrity of DNA was seriously impaired (Table 2). Only about one half of DNA remained apparently intact. The rest was split into mono- and oligonucleosomes that partly remained inside the cell, partly were released to the culture fluid.

TABLE 2. Growth and DNA distribution characteristics of TSH-5.07 hybridoma in batch culture

Parameter	Day 0	Day 8
Total cells $(x10^{-6})(ml^{-1})$	0.30	1.20
Viability (%)	91	13
Apoptotic index (%)	3.1	26
Intracellular DNA (pg/cell)	8.0	6.7
Intracellular DNA (mg/1L culture)	2.80	8.50
Extracellular DNA (mg/1L culture)	0.15	1.45
Fragmented DNA (% intracellular)	17.3	44.5

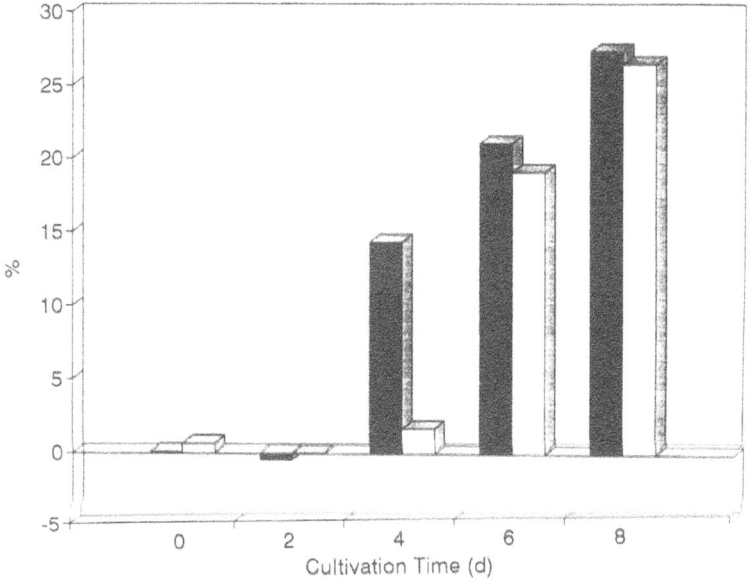

Figure 3. Apoptosis markers in TSH-5.07 hybridoma culture samples collected at various days of the batch culture. Filled columns - degree of cellular DNA fragmentation (per cent fragmented DNA fraction); empty columns - apoptotic index. The values of both markers were corrected for mean background values determined earlier (Franěk et al. 1992a).

The quantitative analysis of the degree of DNA fragmentation and of the apoptotic index, carried out in samples collected every 48 hours, allowed a more detailed timing of the onset of the apoptotic process in the time course of the batch culture. The increase of both marker values was found to be non-linear (Fig.3). While a steep increase of the apoptotic index could be observed on day 6, a notable increase of the degree of DNA fragmentation was apparent already on day 4.

The time-shift between the two apoptosis markers is not surprising, because the sequence of events of the apoptotic program is known to take many hours. The internucleosomal DNA fragmentation belongs to the early events, while the formation of apoptotic bodies represents the very end of the process (Gerschenson and Rotello 1992). The endogeneous signal triggering the program must necessarily precede the activation of the endonuclease and the fragmentation of DNA. Thus, the most likely timing of the spontaneous onset of apoptosis is the exponential phase of the culture. Consequently, the spontaneous apoptosis in mouse hybridomas seems to be a process associated with cell proliferation and full metabolic activity.

4. Conclusion

At the present state of the research it is not yet possible to conclude whether the triggering of spontaneous apoptosis in mouse hybridoma culture is of physiological or of pathological character (cf Table 1). The relatively early onset of the apoptotic process in the batch culture is in favor of the view that the tendency to undergo apoptosis is an inherent feature of lymphocyte hybridomas, similarly as it is an inherent feature of other cells of haemopoietic origin, and, to a certain degree, of all animal cells.

This view has been so far almost neglected, and animal cell culture has been treated , both in theory and in practice, as analogy to the culture of unicellular microorganisms, such as yeasts. In contrast to unicellular organisms animal cells are prone to undergo disciplined suicide, whenever apoptotic stimuli signalize or mimic a disbalance between cell generation and cell deletion. Future investigation will show whether some cultured cell lines exist that would be virtually resistant to apoptotic stimuli. Mouse hybridomas apparently belong to cell lines responding to apoptotic stimuli with significant sensitivity. So far a single way of suppressing or postponing apoptosis in hybridoma culture has been demonstrated: the employment of a highly fortified basal medium (Franěk and Dolníková 1991a).

As a consequence of the above considerations a novel task emerges for the engineering of animal cell lines. The goal would be the generation of cell lines devoid of responsiveness to apoptotic stimuli. This might be possibly achieved, e.g., by overexpression of a gene (bcl-2) acting as a repressor of lymphocyte death (Korsmeyer 1992). A significant progress toward maintaining long-term viability in animal cell technology may possibly result from this line of research.

534

5. Acknowledgements

The authors are indebted to Miss Jana Bimková for her skilful technical assistance. The work was partly supported by the Research grant No. 75267 of the Czechoslovak Academy of Science.

6. References

Cohen, J.J. (1991) 'Programmed cell death in the immune system', Adv. Immunol. 50, 55-85.

Fesus, L., Davies, P.J.A. and Piacentini, M. (1991) 'Apoptosis: Molecular mechanisms in programmed cell death', Eur. J. Cell. Biol. 56, 170-177.

Franěk, F. and Dolníková, J. (1991) 'Nucleosomes occurring in protein-free hybridoma cell culture: Evidence for programmed cell death', FEBS Letters 284, 285-287.

Franěk, F. and Dolníková, J. (1991a) 'Hybridoma growth and monoclonal antibody production in iron-rich protein-free medium: Effect of nutrient concentration', Cytotechnology 7, 33-38.

Franěk, F., Dolníková, J. and Vomastek, T. (1992) 'Hybridoma culture in iron-rich protein-free medium: Cell growth, apoptosis, and nature of macromolecules recovered from the culture fluid', in H. Murakami, S. Shirahata and H. Tachibana (eds.) Animal Cell Technology: Basic and Applied Aspects, Vol. 4, Kluwer Academic Publishers, Dordrecht 1992, pp. 303-309.

Franěk, F., Vomastek, T. and Dolníková, J. (1992a) 'Fragmented DNA and apoptotic bodies document the programmed way of cell death in hybridoma cultures', Cytotechnology 9, 117-123.

Gerschenson, L.E. and Rotello, R.J. (1992) 'Apoptosis: A different type of cell death', FASEB J. 6, 2450-2455.

Korsmeyer, S.J. (1992) 'Bcl-2: A repressor of lymphocyte death', Immunol. Today 13, 285-288.

CELL CYCLE DYNAMICS OF ANTIBODY FORMATION IN HYBRIDOMAS

S.J. KROMENAKER AND F. SRIENC
Department of Chemical Engineering & Materials Science and Institute for Advanced Studies in Biological Process Technology
University of Minnesota
1479 Gortner Avenue
St. Paul, Minnesota, 55108
U.S.A.

ABSTRACT. The effect of lactate on cell growth, cell cycle distributions, and intracellular antibody content distributions during batch cultivation of a mouse hybridoma was investigated. The specific growth rate was reduced by 30% relative to a control culture when the initial lactic acid concentration was 18 mM and by 40% when the initial lactic acid concentration was 33 mM. Elevated lactic acid concentrations caused no significant reduction of viability during the exponential phase of growth. All cultures achieved balanced growth during the exponential phase. Intracellular antibody content frequency distributions and DNA content frequency distributions were the same for all cultures during balanced growth indicating that although lactate reduces the specific growth rate of a culture, it effects all cells throughout the cell cycle in a similar manner.

1. Introduction

A rational approach towards the optimization of the synthesis and secretion of antibodies by hybridoma cell cultures is difficult because these processes are not well understood at the cellular level. We have developed and applied flow cytometry analysis procedures to obtain data on growing hybridoma cell populations which, in combination with population balance equations, permit a rigorous assessment of the single-cell dynamics of antibody formation during the cell cycle. In this approach, we are able to determine the antibody accumulation rate pattern during the cell cycle, cellular transition rates between individual cell cycle phases, and the partitioning function at cell division. These functions are characteristic physiological quantities which completely describe the growth dynamics of a culture. Results obtained for several different cell lines have shown that the rate of antibody accumulation is greatest at an early point in the cell cycle and that the $G_2 + M$ phase cells are characterized by a negative rate of accumulation of antibody. This latter result indicates that the rate of antibody secretion exceeds the rate of antibody synthesis in this cell cycle phase (Kromenaker and Srienc, 1991).

We have used this methodology to study culture conditions which altered growth rates and which might lead to enhanced productivity. Modulation of the specific growth rate was achieved by addition of lactate to hybridoma batch cultures. Lactate is a normal by-product of glucose metabolism (Butler and Jenkins, 1989) and has been shown to inhibit growth of hybridoma cells (Reuveny et al., 1986). The effect of lactate on specific antibody production

535

S. Kaminogawa et al. (eds.), Animal Cell Technology: Basic & Applied Aspects, Vol. 5, 535–541.
© *1993 Kluwer Academic Publishers.*

rates has not been thoroughly characterized and conflicting results have been reported in the literature. Glacken et al. (1988) found that the specific rate of antibody production decreased at higher lactate concentration. In contrast, Ozturk et al. (1992) found that the specific production rate increased at elevated lactate concentration. In all of these earlier studies, however, only population average data were obtained. In this report, the effects of lactate concentration and decreased specific growth rate on the distribution of intracellular antibody content and on the cell cycle distribution were determined. This work forms the foundation for detailed investigation into the effect of growth rate modulators on the pattern of the rate of accumulation of antibody during the cell cycle.

2. Materials and Methods

2.1. CELL LINES AND CELL MAINTENANCE

The cell line used in this study and its maintenance have been described previously (Kromenaker and Srienc, 1991). AFP-27E is a mouse-mouse hybridoma which produces IgG_1 (κ) monoclonal antibodies against human α-fetoprotein. The medium used was Dulbecco's Modified Eagle's Medium containing 4.5 mg/mL glucose (GIBCO Laboratories, Grand Island, NY) supplemented with 10% (v/v) heat inactivated horse serum (GIBCO), 100 units/mL sodium penicillin G (Sigma Chemical Company, St. Louis, MO) and 100 μg/mL streptomycin sulfate (Sigma).

2.2. BATCH GROWTH IN THE PRESENCE OF ADDITIONAL LACTIC ACID

Cells were grown in 250 mL spinner vessels (Corning, Inc., Corning, NY) in a 7% CO_2 humidified incubator at 37°C. Cultures were inoculated at cell densities of approximately 1.6×10^5 cells/mL. The lactate concentration during normal batch culture of this cell line varies from approximately 4 mM at inoculation to 15 mM at the end of the exponential phase of growth (Frame and Hu, 1991). Therefore, three separate spinner cultures were conducted simultaneously: one with an initial lactic acid concentration of 18 mM, one with an initial lactic acid concentration of 33 mM, and one with no additional lactic acid (control). It was necessary to use small amounts of 1 N HCl to reduce the pH of the medium in each flask to approximately 6.8 to completely dissolve the lactic acid. Cell counts were performed with a hemocytometer periodically during the experiment. Viability was assessed with light microscopy and the trypan blue dye exclusion method. Each time cell concentrations were determined, samples of 1×10^6 cells were harvested from each culture and centrifuged (700g, 4°C, 7-8 min). Supernatants were removed and stored at -20°C. Cells were washed once in 1 mL of cold PBS, centrifuged as above, and fixed in 1 mL of a 90% methanol-in-PBS solution. This fixation was accomplished by resuspending cells in 0.1 mL of PBS and then adding dropwise 0.9 mL of pure methanol during vigorous vortexing. Fixed cells were stored in plug-seal polystyrene centrifuge tubes (Corning) at -20°C for between 2 days and 2 weeks before further analysis.

2.3. INTRACELLULAR STAINING

Methanol fixed samples were transferred to 16 X 100 borosilicate glass tubes because cells tend to stick to the walls of polystyrene tubes during staining. Approximately 250,000 cells were recovered by centrifugation (1100g, 4°C, 15 min), washed once in 1 mL PBS and resuspended in 2 mL of PBS containing 1% (w/v) bovine albumin (Sigma), 0.02% (w/v) sodium azide (Sigma), and 2 mg/mL ribonuclease A (Sigma). This BSA-containing buffer is referred to as PBA. The samples were mixed by vortexing and incubated at room temperature (≈ 25°C) for 10 min. Next 2 mL of intracellular staining solution were added and the samples were vortexed and incubated at 37°C in the dark for 60 min. The intracellular staining solution consisted of fluorescein isothiocyanate (FITC)-conjugated goat anti-mouse IgG F(ab')$_2$ antibody fragments specific for both heavy and light chains (Tago, Inc., Burlingame, CA) diluted 1:250 in PBA. After staining was complete, cells were recovered by centrifugation as above, resuspended in 0.6 mL PBA, vortexed, and stored at 4°C in the dark. Samples were stable for up to 12 h. Just prior to flow cytometric analysis, 0.4 mL of PBS containing 50 μg/mL of propidium iodide (PI, Sigma) were added to each sample (final PI concentration: 20 μg/mL). The samples were kept on ice for approximately 20 min and analyzed with the flow cytometer without further dilution.

2.4. FLOW CYTOMETRY

The flow cytometer instrument used was an Ortho Cytofluorograf IIs (Ortho Diagnostics Systems, Westwood, MA) equipped with a Coherent Innova 90-5 argon ion laser (Coherent, Palo Alto, CA). The laser was operated at a wavelength of 488 nm and a light-stabilized beam power of 100 mW. An FITC filter (Ortho Diagnostics Systems) was used to determine green fluorescence intensity and a 590 long-pass filter (Corning) was used to determine red fluorescence intensity. Approximately 50,000 cells were analyzed for each sample. Forward-angle light scattering (FALS), green fluorescence and red fluorescence were measured simultaneously. The green fluorescence represented the amount of FITC-conjugated F(ab')$_2$ fragment bound to intracellular immunoglobulin proteins, and the red fluorescence represented the amount of PI bound to DNA. Cellular debris and cell doublets were excluded from analysis by examination of the FALS versus red fluorescence cytogram. In order to obtain accurate, quantitative data, the flow cytometer instrument was standardized each day prior to use. This calibration was performed with uniform green fluorescent beads (2.062 $\pm$ 0.025 μm, Duke Scientific Corp., Palo Alto, CA). After acquisition, list mode data were analyzed with either the Ortho 2151 data acquisition system or Acqlist data acquisition software (Phoenix Flow Systems, San Diego, CA). Cytogram and histogram data were analyzed with Multi2D and Multicycle software, respectively (both Phoenix Flow Systems).

3. Results and Discussion

3.1. EFFECT OF ADDITIONAL LACTIC ACID ON SPECIFIC GROWTH RATE

The AFP-27E hybridoma cells were grown in batch spinner cultures with an initial lactic acid concentration of either 0 mM, 18 mM, or 33 mM. Growth data for all three are shown in Figure 1. A lag phase was observed for all three cultures. This period lasted approximately 15 h when the initial concentration was 0 mM and approximately 20 h when the initial concentration was 33 mM. All three cultures grew exponentially to maximum viable cell concentrations of approximately 1.3-2.0X10^6 cells/mL. Viability during the exponential phase was at least 95% for all three cultures. The specific growth rate was reduced 30% relative to

538

the control culture when the initial lactic acid concentration was 18 mM and 40% relative to the control culture when the initial lactic acid concentration was 33 mM (Figure 2).

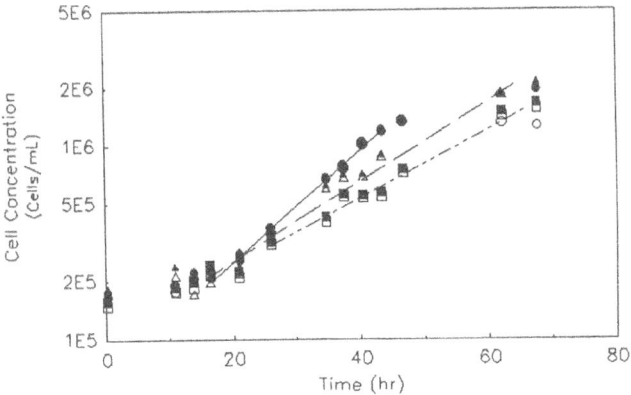

Figure 1. Batch growth of AFP-27E hybridomas in the presence of lactic acid. Open symbols represent viable cell concentrations and closed symbols represent total cell concentrations. The initial lactic acid concentrations were 0 mM (O), 18 mM (Δ), or 33 mM (□).

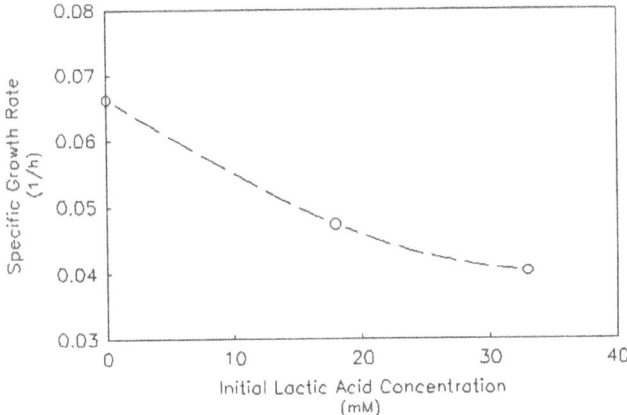

Figure 2. Effect of lactic acid concentration on specific growth rate.

3.2. BALANCED GROWTH CONDITIONS

A population of cells in balanced growth increases in number at an exponential rate and has property frequency distributions which are constant over time (Campbell, 1957). In order to determine the interval during which the spinner cultures were in balanced growth, samples were removed from the vessels periodically over the course of the batch experiment. The cells in these samples were double-stained for intracellular antibody content and DNA content as described. The flow cytometer was used to measure the frequency distributions of these properties. The variation of these distributions over time are shown in Figure 3 for the culture which had an initial lactic acid concentration of 0 mM. The shapes of these distributions are constant between 26.75 and 45.0 hours after inoculation indicating the population was in balanced growth for at least this period of time. Similar data were obtained for the other two cultures (data not shown).

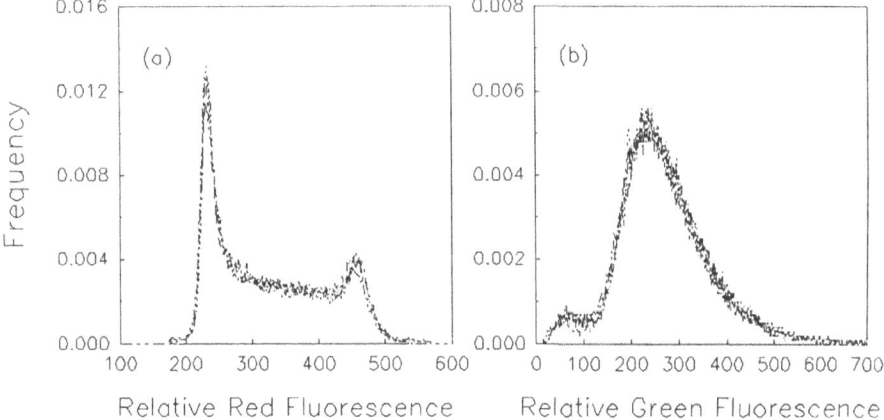

Figure 3. Frequency distributions of (a) DNA content and (b) intracellular antibody content during balanced exponential growth with an initial lactic acid concentration of 0 mM. Samples were removed and stained for intracellular antibody and DNA content as described at 26.75 h (– – –), 36.0 h (········), 38.25 h (– · –), and 45.0 h (– ·· –) after inoculation.

3.3. EFFECT OF ADDITIONAL LACTIC ACID ON INTRACELLULAR IgG AND DNA DISTRIBUTIONS

The stable frequency distributions of DNA content and intracellular antibody content characteristic of each of the three growing hybridoma populations during the period of balanced growth are compared in Figure 4. Each set is comprised of three distributions that are nearly identical. In fact, the differences in frequencies seen at any given DNA or antibody content are similar in magnitude to differences typically seen for duplicate samples. We conclude, therefore, that there are no significant differences between these samples.

The DNA content distributions were deconvolved in order to determine the fraction of the population in each of three cell cycle phases: G_1, S, and G_2+M. If one assumes that these populations grow according to an exponential age distribution, that cell division is symmetric, and that there is no dispersion in cell cycle times, then the cell cycle fractions and the specific growth rate can be used to estimate the time spent in each of these three cell cycle phases

(Slater et al., 1977). The results of this analysis are shown in Figure 5. One can see that the fraction of the population in each of the three cell cycle phases is not affected by elevated initial lactic acid concentration. The time spent in each cell cycle phase therefore increases by the same factor that the doubling time of the culture increases.

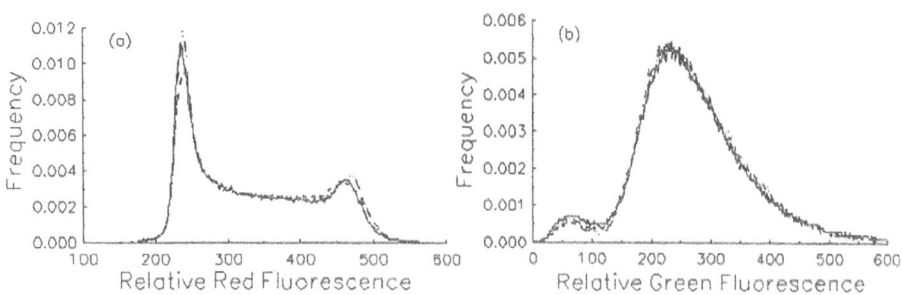

Figure 4. Frequency distributions of (a) DNA content and (b) intracellular antibody content during balanced growth of hybridoma cell populations grown in at three different initial lactic acid concentrations: 0 mM (———), 18 mM (– – –), and 33 mM(– ⋯ –).

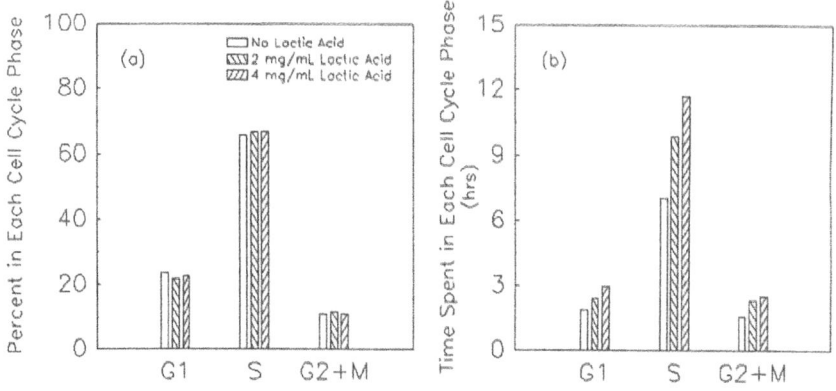

Figure 5. Analysis of cell cycle distributions for balanced growth of hybridoma cell populations grown at three different lactic acid concentrations (see key on figure). Both (a) cell cycle fractions and (b) time spent in each cell cycle phase were determined as described.

Other researchers have observed a relatively greater increase in the amount of time spent in the G_1 phase when growth is slowed (Smith and Martin, 1973 and Pardee et al., 1978). Thus our data indicate that during balanced exponential growth in batch culture, an elevated initial lactic acid concentration reduces the specific growth rate yet it affects the growth of cells in all stages of the cell cycle in a similar manner.

4. Acknowledgements

The authors wish to acknowledge the Graduate School of the University of Minnesota and the National Science Foundation for partial support of this work (grant BCS-9100385). SJK is the recipient of a National Science Foundation Graduate Fellowship.

5. References

Butler, M. and Jenkins, H. (1989) 'Nutritional aspects of the growth of animal cells in culture', Journal of Biotechnology, 12, 97-110.

Campbell, A. (1957) 'Synchronization of cell division', Bacteriological Review, 21, 263-272.

Frame, K.K. and Hu, W.-S. (1991) 'Comparison of growth kinetics of producing and nonproducing hybridoma cells in batch culture', Enzyme and Microbial Technology, 13, 690-696.

Glacken, M.W., Adema, E., and Sinskey, A.J. (1988) 'Mathematical descriptions of hybridoma culture kinetics: I. initial metabolic rates', Biotechnology & Bioengineering, 32, 491-506.

Kromenaker, S.J. and Srienc, F. (1991) 'Cell-cycle-dependent protein accumulation by producer and nonproducer murine hybridoma cell lines: a population analysis', Biotechnology & Bioengineering, 38, 665-677.

Ozturk, S.S., Riley, M.R., and Palsson, B.O. (1992) 'Effects of ammonia and lactate on hybridoma growth, metabolism, and antibody production', Biotechnology & Bioengineering, 39, 418-431.

Pardee, A.B., Dubrow, R., Hamlin, J.L., and Kletzien, R.F. (1978) 'Animal cell cycle', Annual Review of Biochemistry, 47, 715-750.

Reuveny, S., Velez, D., Macmillan, J.D., and Miller, L. (1986) 'Factors affecting cell growth and monoclonal antibody production in stirred reactors', Journal of Immunological Methods, 86, 53-59.

Slater, M.L., Sharrow, S.O., and Gart, J.J. (1977) 'Cell cycle of *Saccharomyces cerevisiae* in populations growing at different rates', Proceedings of the National Academy of Sciences, 74, 3850-3854.

Smith, J.A. and Martin, L. (1973) 'Do cells cycle?', Proceedings of the National Academy of Sciences, 70, 1263-1267.

EFFECT OF THE CELL–CYCLE PHASE ON HYBRIDOMA CULTURE

TAKESHI OMASA, MASAKI KOBAYASHI, SUTEAKI SHIOYA, and
KEN–ICHI SUGA
*Department of Biotechnology, Faculty of Engineering, Osaka University,
Suita, Osaka 565, Japan*

ABSTRACT. The effects of cell–cycle phase on the antibody production of 3A21 hybridoma were investigated, using synchronous and glucose–limited continuous cultures. From the synchronous culture experiment, it was found that the specific antibody production rate increased between the late G_1 and early S phases. Moreover, a comparison between the antibody productivity of a normal cell and a cell with the cell–cycle blocked by an excess thymidine dose indicated that the specific antibody production rate of the blocked cell was higher than that of the normal cell. The effects of specific growth rate on the cell–cycle phase and antibody production were investigated in a glucose–limited continuous culture. With decreasing specific growth rate, the proportion of the G_1–phase cells increased, and hence, the specific antibody production rate increased.

1. Introduction

According to observations of DNA duplication in a eukaryotic cell, the cell cycle is divided into four discrete phases, G_1 (gap period I), S (DNA synthesis), G_2 (gap period II), and M (Mitotic). In a eukaryotic cell, most of the protein synthesis is dependent on the cell–cycle phase. The antibody production is also dependent on the cell cycle from the results of synchronous culture experiments on lymphocyte [1] and myeloma cells [2]. Therefore, in a hybridoma cell, the antibody production is also considered to be cell–cycle dependent. The recent development of flow cytometry has enabled a cell–cycle phase analysis to be rather easily performed without using synchronous culture nor a radio isotope. In this report, we investigate the effect of cell–cycle phase on the antibody production in a synchronous hybridoma culture, and also examine the relationship between the cell–cycle phase distribution and the specific antibody production in a glucose–limited continuous culture experiment by using flow cytometry.

S. Kaminogawa et al. (eds.), Animal Cell Technology: Basic & Applied Aspects, Vol. 5, 543–549.
© 1993 *Kluwer Academic Publishers.*

2. Materials and Methods

The cell line employed in the experiments was the mouse–mouse hybridoma, 3A21 [3,4],and the culture medium was serum–free RDF–ITES [3,4]. The synchronous culture was performed by using the double thymidine block method, while the glucose–limited continuous culture experiments were performed as previously described [5]. Each sample was analyzed as previously described [3,4], and the specific growth, glucose consumption, lactate production, and antibody production rates were calculated as shown in the previous report [3,4]. The cell–cycle distribution analysis was carried out by using an argon–laser flow cytometer (Becton–Dickinson FACScan).

3. Results and discussion

3.1 SYNCHRONOUS CULTURE EXPERIMENT

In order to investigate the relationship between cell–cycle phase and antibody production in the 3A21 hybridoma cell line, a synchronous culture experiment was performed by using double thymidine block method [1,2]. The specific antibody production rate was calculated to evaluate the effect of cell–cycle phase on antibody production. Figure 1 shows the time–course of the specific antibody production rate and the cell–cycle phase fraction in the synchronous culture. The specific antibody production rate at each sampling time indicates the moving average of three sets of data, this rate changing according to the cell–cycle progress. Both peaks for the specific antibody production rate in the figure are between the late G_1 phase and early S phase.

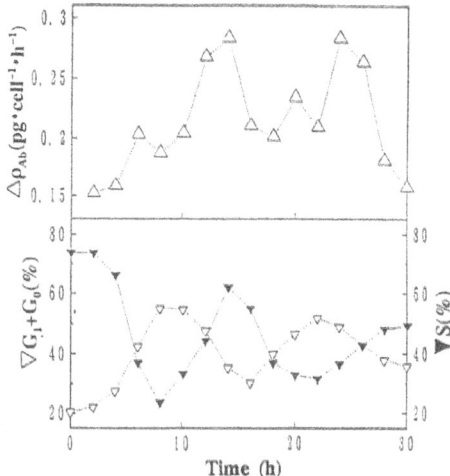

Figure 1. Time–courses for the specific antibody production rate and cell–cycle phase fraction

3.2 KINETIC ANALYSIS OF THE SYNCHRONOUS CULTURE

Based on the results from the synchronous culture, we calculated the specific antibody production rate in the G_1 (including G_0–phase cells) and S phases. The mass balance equation for antibody in a batch culture becomes

$$\frac{dAb}{dt} = \rho_{Ab}^{G1}X_{G1} + \rho_{Ab}^{S}X_{S} + \rho_{Ab}^{G2+M}X_{G2+M} \qquad (1)$$

where ρ_{Ab}^{G1}, ρ_{Ab}^{S}, ρ_{Ab}^{G2+M}, X_{G1}, X_S and X_{G2+M} are the specific antibody production rate in the G_1, S and G_2+M phases, and viable cell concentration in the G_1, S and G_2+M phases, respectively. It has often been reported that, in myeloma or lymphocyte cell lines, the antibody production was limited to between the G_1 and S phases [1,2]. We also assumed that no antibody was produced during the G_2 and M phases, and calculated the specific antibody production rate in the G_1 and S phases by integrating equation 1, using the data for each cell-cycle phase. The results are

$$\rho_{Ab}^{G1} = 0.490 \quad (pg \cdot cell \cdot h^{-1})$$
$$\rho_{Ab}^{S} = 0.00975 \quad (pg \cdot cell \cdot h^{-1})$$

In the G_1 phase, the specific antibody production rate increased about 50 times compared with that in the S phase. Using these values, we calculated the antibody concentration in synchronous culture. The calculated values agree well with the experimental results. Therefore, these estimated specific antibody production rates are considered to be reasonable values.

In the synchronous culture shown in Fig. 1, the intracellular antibody was also measured by using fluorescein isothiocyanate (FITC)–conjugated rabbit anti–mouse IgG (H+L) F(ab)'. The intracellular antibody was at a low level at first, but became constant during cultivation. In other words, the antibody production was dependent on the cell–cycle phase, while the intracellular antibody concentration was not affected by the cell–cycle phase. Therefore, it is considered that an increasing G_1–phase cell fraction is important for increasing the specific antibody production rate.

3.3 THYMIDINE BLOCK CULTURE

The cells in the S phase were arrested by excess thymidine, while the cells in the other cell-cycle phases were allowed to proceed to just before the S phase. Thus, the fraction of G_1-phase cells increased in the medium containing excess thymidine. In order to confirm the results from the synchronous culture, the cell cycle was blocked just before the S phase in a medium containing excess thymidine (3mM), and the specific antibody production rate was examined. Figure 2 shows the time–courses for the viable cell and antibody concentrations in the medium containing 3mM thymidine and in a normal RDF medium. In the 3mM-thymidine culture, cell growth was stopped as shown in Fig. 2 because DNA synthesis was inhibited by the excess thymidine. However, the time–course for the antibody concentration in the 3mM-thymidine culture is similar to that in the normal RDF medium. Therefore, the specific antibody production rate in the 3mM–thymidine culture was twice as high as that in a normal RDF medium (Table I).

The DNA distribution in the 3mM–thymidine and normal media are each shown in Figure 2. In the 3mM–thymidine culture, the cells in the G_1 phase increase. Thus, an increase in the G_1-phase cell fraction might have caused the increase in the specific antibody production rate. Each fraction of the cell–cycle phase was closely related to the specific growth rate. In the continuous culture, it was possible to control the specific growth rate by changing the dilution rate and to control the cell–cycle phase distribution. Hence the relationship between the cell–cycle phase fraction and specific growth rate in the continuous culture was investigated.

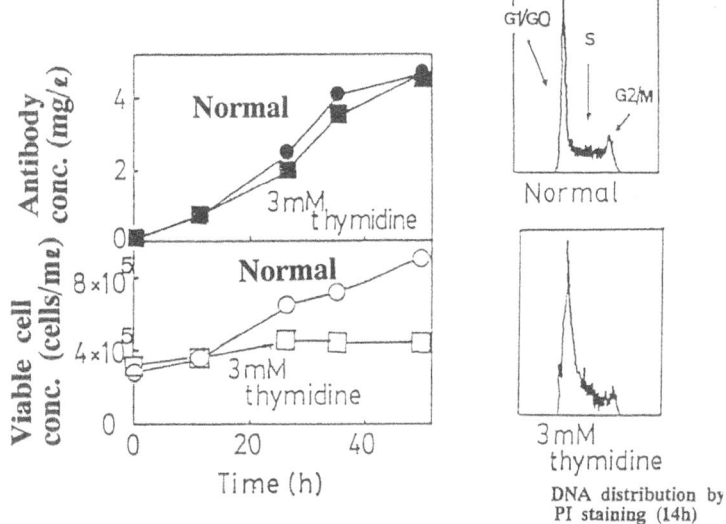

Figure 2. Comparison between culture in normal RDF and 3mM–thymidine media

Table I The comparison of specific antibody production rate

	Normal medium	3mM thymidine medium
Specific antibody production rate $(pg \cdot cell^{-1} \cdot h^{-1})$	0.170	0.308

3.4 GLUCOSE–LIMITED CONTINUOUS CULTURE

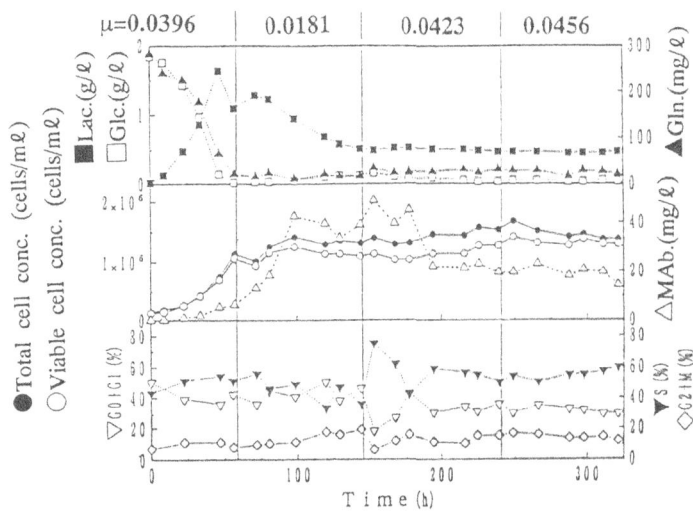

Figure 3. Time–courses for the glucose–limited continuous culture

Figure 3 shows an example of the time–courses for the glucose–limited continuous culture, in which the specific growth rates of 0.0181, 0.0423 and 0.0456 h^{-1} were achieved by changing the dilution rate. The batch culture was started with a glucose concentration of 1.0 $g \cdot \ell^{-1}$, and when the glucose had been completely consumed, the feed medium (glucose concentration = 1.0 $g \cdot \ell^{-1}$) was fed into the spinner flask. In the continuous culture, as the glucose concentration in spinner flask was almost zero, it was considered to be a glucose–limited condition. The cell concentration during the continuous operation was almost constant in spite of the changes in specific growth rate. The G_1 phase ratio was increased at a low specific growth rate compared with that of the higher specific growth rate. The kinetic parameters for this cultivation are shown in Table II.

Table II Kinetic parameters for the glucose–limited continuous culture

	Batch	Continuous		
Specific growth rate (h^{-1})	0.0396	0.0181	0.0423	0.0456
Specific glucose consumption rate $(10^{-10}g \cdot cell^{-1} \cdot h^{-1})$	1.23	0.103	0.262	0.298
Specific glutamine consumption rate $(pg \cdot cell^{-1} \cdot h^{-1})$	1.46	0.367	0.862	0.894
Cell yield $(cells \cdot g\text{-}glucose^{-1})$	3.2×10^8	17.5×10^8	16.2×10^8	15.5×10^8
Specific antibody production rate $(pg \cdot cell^{-1} \cdot h^{-1})$	0.345	0.830	0.510	0.598

The G_1–phase cell fraction was increased at the lower specific growth rate, and accordingly, the specific antibody production rate also increased at the lower specific growth rate, this result being the same as that in the previous synchronous culture experiments. From this result, the total antibody production rate was higher under the lower specific growth rate, because the cell concentration was constant during cultivation. The relationship between the specific antibody production rate and specific growth rate is shown in Figure 4.

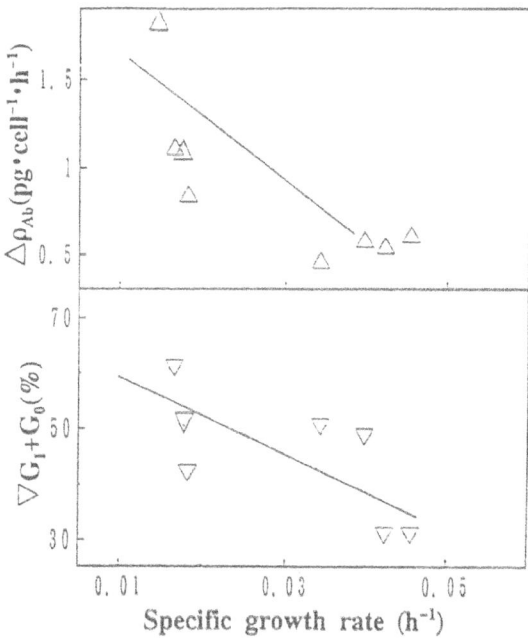

Figure 4. Relationship between the specific growth rate and specific antibody production and the fraction of G_1+G_0 phase in the glucose–limited continuous culture

This figure shows the average G_1+G_0 phase fraction and specific antibody production against the specific growth rate. The results show that it was possible to change the fraction of the cell–cycle phase and to increase the specific antibody production rate by changing the dilution rate. Hence, an increasing fraction of G_1–phase cells is important for increasing the specific antibody production rate.

In conclusion, in order to maximize antibody production, it was proved important to increase the fraction of G_1–phase cells. Therefore, it is necessary to investigate further the operating conditions which affect the cell–cycle distribution, for example, nutrient limitations, growth factor limitations, etc., and to construct a cell–cycle model based on the experimental data from continuous cultivation. Furthermore, it is desirable to apply this model to a high–cell density culture, i.e., perfusion and hollow fiber cultivation, to optimize the cultivation conditions.

4. REFERENCES

1. Watanabe, S., Yagi., Y. and Pressman, D. (1973) 'Immunoglobulin production in synchronized cultures of human hematopoietic cell lines', J. Immunol. 111, 797–804.

2. Turner, K.J., Plozzza, T.M., and Holt, P.G. (1985) 'Cell cycle–dependent fluctuations in IgE secretion in human myeloma lines', Clin. Immunol. Immunopathol. 36, 212–216.

3. Omasa, T., Higashiyama, K., Shioya, S., and Suga, K. (1992) 'Effects of lactate concentration on hybridoma culture in lactate–controlled fed–batch operation', Biotechnol. Bioeng.39, 556–564.

4. Omasa, T., Ishimoto, M., Higashiyama, K., Shioya, S., and Suga,K. (1992) 'The enhancement of specific antibody production rate in glucose– and glutamine–controlled fed–batch culture', Cytotechnology 8, 75–84.

5. Omasa, T., Kobayashi, M., Ishimoto, M., Higashiyama, K., Shioya, S., and Suga, K. (1992) 'The effects of glutamine and glucose concentration on hybridoma cell growth and antibody productivity', in S. Furusaki, I. Endo and R. Matsuno (eds.), Biochemical Engineering for 2001, Springer–Verlag, Tokyo, pp. 323–326.

PHYSIOLOGICAL ASPECTS OF SCALING UP AND MONITORING

C.A.M. VAN DER VELDEN-DE GROOT[1,] J.M.COCO MARTIN[1], D.E.MARTENS[2] AND E.C. BEUVERY[1]
[1] National Institute for Public Health and Environmental Protection (RIVM), PO Box 1, 3720 BA Bilthoven, The Netherlands
[2] Department of Food Science, Food and Bioprocess Engineering Group, Agricultural University, Bomenweg 2, 6703 HD Wageningen, The Netherlands.

ABSTRACT. In order to optimize the production of monoclonal antibodies several homogeneous culture systems are studied. Data are presented on cultivation of hybridoma cell line MN12 in a continuous culture system with cell retention. A computerized measurement and control system was used for monitoring and control of process parameters as pH, dissolved oxygen concentration (DO), stirrer speed and temperature. Various analytical methods were used to study parameters as cell growth, viability, metabolism, oxygen consumption, monoclonal antibody (Mab) production, integrity and carbohydrate moiety of the monoclonal antibodies.
During cultivation a change from aerobic to a more anaerobic metabolism was observed which not only affected the levels of nutrients and metabolites but also the glycosylation pattern of the antibodies produced. The antibody production and the number of antibody producing cells remained unchanged.

1. Introduction

For many applications of animal cell culture large numbers of animal cells are required. As scaling up via multiples of small cultures is not feasible enlarging the size of the culture vessel seems attractive. However as the size increases several parameters change for instance degree of environmental control and the means of maintaining the correct physiological conditions. Understanding of fundamental physiological processes forces to study growth kinetics, metabolic activity and product formation/excretion. Factors influencing these parameters are: specific growth rate, uptake of nutrients, amino acid utilization, oxygen transfer, hydrodynamic effects, osmotic condition and agitation. All these parameters together present the black box of the cultivation of animal cells.
The study we present here is of a multidisciplinary approach and undertaken to increase productivity and to explain failures. The disciplines involved are Bioprocess Technology, Cell Biology and Bio- and Immunochemistry. The principle of homogeneous continuous perfusion allowes for relatively high cell concentrations with the drawback of insufficient oxygen supply. The type of bioreactor and mode of control of environmental conditions influences the physiology of the cell. Characteristics of the homogeneous continuous perfusion system are stable cell-, product-, nutrient- and metabolite-concentrations. However oxygen limitation greatly influences the metabolism. During cultivation of cell line Mn12 a number of in-line and off-line analysis were performed to obtain more insight in the production process in order to optimize product formation, both in quantity and quality.

S. Kaminogawa et al. (eds.), Animal Cell Technology: Basic & Applied Aspects, Vol. 5, 551–556.
© 1993 *Kluwer Academic Publishers.*

2. Materials and methods

2.1 Cultivation of Hybridoma Cell line MN12

The cell line MN12, producing IgG2a antibodies against the outer membrane protein PI.16 of Neisseria meningitidis (kindly provided by Dr.J.Poolman of the RIVM, Bilthoven, The Netherlands) was cultivated in a 3 liter glass bioreactor connected to an CF 500 computerized measurement and control system (Applikon Dependable Instruments, Schiedam, The Netherlands) [1]. Culture conditions were: pH 7.2, Temperature 37^{0}C, DO 25 % air saturation and stirrer speed control between 100 and 200 rpm. Besides the DO concentration in the culture fluid also the Dissolved Oxygen Controller Output (DOCO) was measured which is in case of a constant Oxygen transfer coefficient (K_La) a measure for oxygen demand of the cells.The cultivation medium consisted of Iscove's modified Dulbecco's medium (Gibco Life Technology, Grand Island, NY) supplemented with 0.25 % Primatone RL (Sheffield Products, Norwich, NY), 3% fetal calf serum (Sanbio, Uden, The Netherlands) and antibiotics. The medium perfusion rate was one culture volume per day.

2.2 Analytical Methods

-   Cell number and viability were determined by the Trypan Blue exclusion method.
-   Glucose, Lactate and Ammonia were analyzed using enzymatic kits (Boehringer Mannheim, Mannheim, Germany).
-   Murine IgG was measured using isotype specific Elisa techniques [2].
-   The number of antibody secreting cells was determined by an isotype specific spot-Elisa [3] and by Flowcytometer analysis [4].
-   The Glycosylation pattern was determined by Protein A purification of the IgG followed by determination of carbohydrate moieties by binding of Lectins specific for terminal sugar residues [5/6].

3. Results

3.1 Cell Growth and Metabolism.

The cultivation of cell line MN12 in a homogeneous continuous perfusion system was repeated several times and appeared to be reproducible with a correlation coefficient of $\geq$ 90%. The cultures were characterized by two exponential growth phases and two steady states. The cell density in the first steady state was 7×10^6 cells/ml and in the second steady state 13×10^6 cells/ml. The viability remained ±85% (Fig.1).

cell number ($^*10^4$ cells/ml)

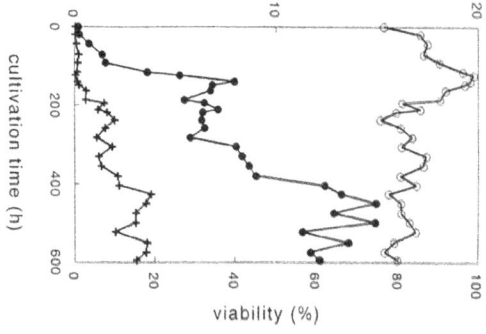

Figure 1. Growth curve and viability.
(●) viable cells,
(+) dead cells,
(O) viability.

viability (%)

The DOCO showed a similar trend as the curve of the viable cell concentration with a correlation coefficient of 0.94 (Fig.2). However during the second steady state the oxygen demand of the cells was higher than the the maximum DOCO which could be registered.

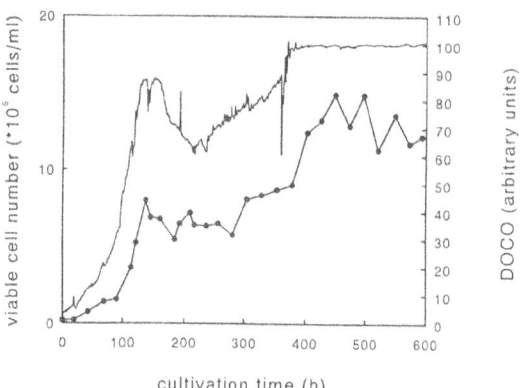

Figure 2A. *Growth curve and course of the dissolved oxygen controller output (DOCO)*

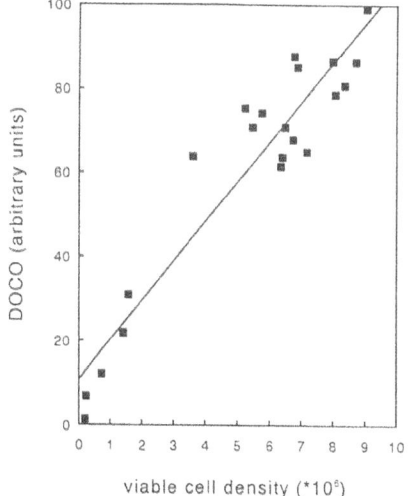

Figure 2B. *Correlation between DOCO and viable cells.*

554

The trend analysis of the measurement and control unit CF 500 showed a stable pH with a tendency towards a lower pH during the second part of the culture. The DO decreased from 25% to 3% during the second steady state, indicating more or less anaerobic conditions (Fig.3).

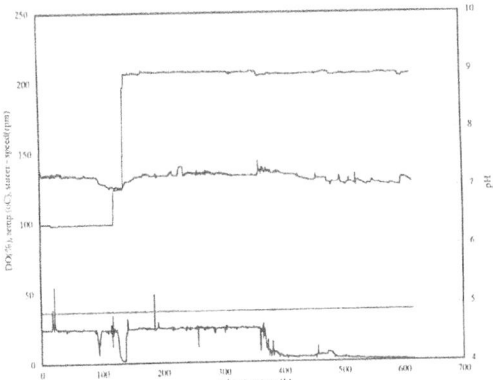

Figure 3. *Course of pH, DO, temperature and stirrerspeed during cultivation of hybridoma cell line Mn12 in a homogeneous continuous perfusion system.*

The concentration of glucose decreased rapidly from 20 to 1 mM during the first growth phase while during the first and second steady state the levels were 5 and 2 mM respectively (Fig.4A). The lactate concentration showed a reversed curve: a sharp increase during the first growth phase and rather stable levels of around 10 and 20 mM during the first and second steady state (Fig.4B).

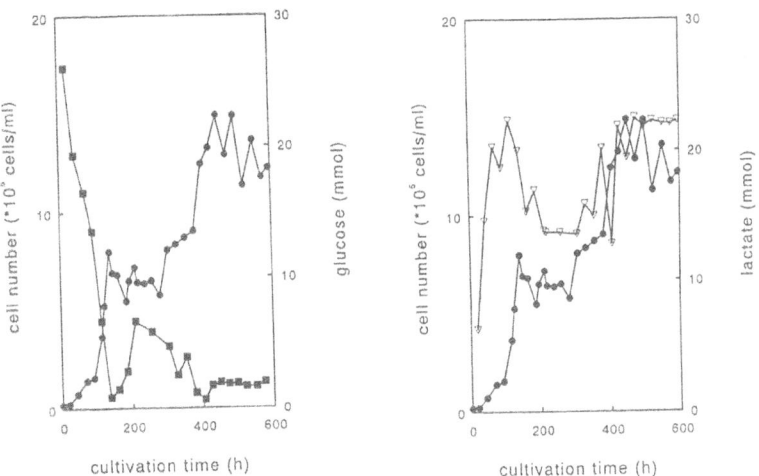

Figure 4. *Growth curve and course of glucose (A) and lactate (B).*

The increase of the molar ratio lactate to glucose indicates a less efficient consumption of glucose in the second stage of the culture. This can be explained by the fact that glucose is degraded in three steps: the digestion (outside the cells) where the glucose is split in subunits, the glycolysis where a limited amount of ATP is produced (2 mol ATP/1 mol glucose)and the complete oxydation where a lot of ATP is produced (8 mol ATP/1 mol glucose). Only for this last step molecular oxygen is required. During the second stage of the culture a low DO was observed and consequently not enough oxygen was present for complete oxydation. As the lactate production is a function of the glucose concentration and transport of glucose is increased at high glucose concentrations which again influences pH and bicarbonate consumption control of glucose at a low level is advisable.

The level of the ammonia concentration during cultivation was low and stable at 3 mM.

3.2 Antibody Production and Integrity

The curve of the antibody production showed the same pattern as the curve of the viable cell concentration with levels of 35 and 75 ug/ml IgG respectively in the first and second steady state (Fig.5). The percentage of antibody producing cells, determined by Flowcytometric analysis and by spot-Elisa remained constant and high at a level of $\geq$90%. The calculated specific antibody production per cell remained constant at 6 μg/10^6 cells/day.

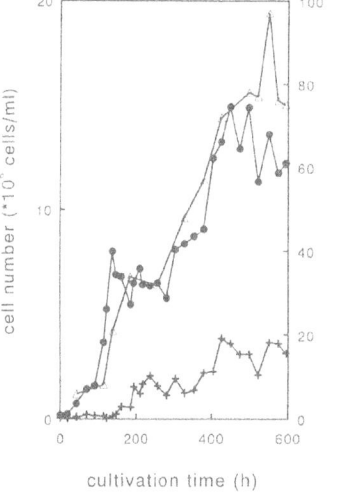

Figure 5. *Growth curve and antibody production.*
(●) viable cells,
(+) dead cells,
(△) antibody concentration.

cultivation time (h)

The glycosylation pattern of the secreted antibodies was determined by lectins binding specifically to terminal sugar residues. During cultivation the glycosylation pattern of most terminal sugars remained constant except for an additional band with a lectin specific for mannose during the second steady state. As production method and/or culture conditions can dramatically affect the glycosylation profile of the secreted antibodies [7] probably the oxygen limitation or the glucose starvation caused degradation of oligosaccharides. As little is known about extracellular factors influencing intracellular glycosylation events validation of the production process with respect to lot to lot consistency and a reproducible glycosylation pattern is essential. The lectin method only gives global information. For more insight in carbohydrate structures ion-spray MS and High Resolution NMR Spectroscopy are methods of choice.

Conclusion

In this paper we studied the cultivation of hybridoma cell line MN12 in a homogeneous continuous perfusion system. The use of a sophisticated computerized measurement and control system allowed us to get information about the condition and metabolic state of the cells. The culture showed two exponential phases and two steady states. For this phenomena which we saw also in other cultures we have no explanation. During the second steady state the high cell concentration caused DO-limitation which caused a shift in metabolism. This did not affect the viability of the cells, the antibody production and the specificity of the antibody. However a change in glycosylation pattern of the antibody was observed which might influence the pharmacokinetic properties. These results indicate the importance of well controlled conditions in the bioreactor during cultivation of animal cells and validation of the production process.

5. References

[1] Wieten et al.G. (1989) 'The CF 1000: second generation instrumentation for monitoring and control of bioreactors' in R.E.Spier, J.B.Griffiths, J.Stephenne and P.J.Crooy (eds.), Advances in Animal Cell Biotechnology and Technology for Bioprocesses, Butterworth and & Co. LTD., Sevenoaks, 408-411

[2] Van der Velden-de Groot C.A.M., Coco Martin J.M. and Beuvery, E.C. (1990) New developments in the cultivation of hybridoma cells in homogeneous continuous culture systems. Develop. Biol. Standard. 71, 45-54.

[3] Coco-Martin, J.M., Koolwijk, P., Van der Velden- de Groot, C.A.M. and Beuvery, E.C. (1991a) An isotype-specific spot-Elisa for the enumeration of antibody-secreting hybridoma cells in homogeneous continuous culture systems. J.Immunol. Methods 145, 11-18.

[4] Coco-Martin, J.M., Oberink, J.W., van der Velden-de Groot, C.A.M. and Beuvery, E.C. (1991b) Methods for studying the stability of antibody expression by hybridoma cells in homogeneous continuous perfusion culture systems. Anal. Chim. Acta 249, 257-262.

[5] Haselbeck, A., Schickaneder, E., von der Eltz, H. and Hosel, W. (1990) Structural characterization of glycoprotein carbohydrate chains by using digoxigenin-labeled lectins on blots. Anal. Biochem. 191, 25-30.

[6] Coco-Martin, J.M., Brunink, F., van der Velden-de Groot, C.A.M. and Beuvery, E.C. (1992) On the glycosylation of two mouse IgG2a monoclonal antibodies. J.Immunol.Methods, 155, 242-248.

[7] Goochee, C.F. and Monica, T. (1990) Environmental effects on glycosylation. Bio/technology 8, 421-427.

A COMPARATIVE STUDY OF THE IMMUNOPOTENTIATING ACTIVITY OF ADJUVANT FORMULATIONS CONTAINING GLYCOPEPTIDES

Grubhofer, N. and Cooper, P.M.

C-C Biotech Corporation
16766 Espola Road, Poway, CA 92064, USA.

Bacterial cell walls are known to contain glycopeptides which enhance the immune response. The minimal characteristic subunit has been established as N-Acetylmuramyl-L-alanyl-D-isoglutamine "Adjuvant Peptide" (MDP). [1]

In the course of investigation of the cell wall peptidoglycan from *L. bulgaricus* it was shown that a glucosaminyl derivative of MDP possessed more potent biological activity. N-Acetyl-glucosaminyl-Beta(1-4)-N-acetylmuramyl-L-alanyl-D-isoglutamine (GMDP). [2] Besides enhanced adjuvant activity, some other remarkable biological properties such as hypnogenic and antitumor activity and the stimulation of nonspecific resistance to bacterial and viral infections were discovered.

This study was initiated to determine the optimum dosage and formulation for GMDP to be used as an adjuvant in aqueous phase, without the potential harmful effects of Freund's emulsion.

GMDP and MDP at the correct dose level are effective in aqueous solution without oil emulsion. At the dose related optimum they surpassed Freund's adjuvant in stimulating the production of antibodies in different host animals. GMDP was shown to be better than MDP. The results of these studies are presented and have resulted in the development of a new, aqueous phase adjuvant, available under the name "Gerbu Adjuvant".

Figure 1

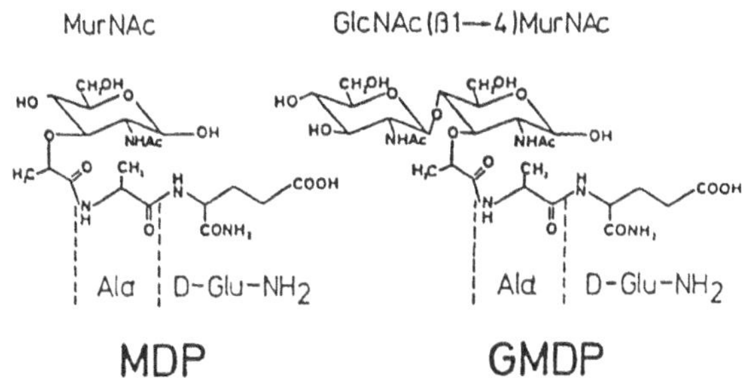

MurNAc   GlcNAc(β1—4)MurNAc

MDP   GMDP

A Schematic Presentation

557

S. Kaminogawa et al. (eds.), Animal Cell Technology: Basic & Applied Aspects, Vol. 5, 557–564.
© 1993 *Kluwer Academic Publishers.*

GMDP has been the subject of extensive study in recent years and many publications have appeared in the Russian literature. Andronova and Ivanov have published a detailed account of the synthesis and biological properties of the glucosaminylmuramyl peptides and this provides a good summary of the Russian research. [3]

One of the first studies to determine the optimum dose level of GMDP was performed by Man'ko et al. [4]

TABLE 1

The effect of GMDP on the accumulation of SRBC-specific antibodies in the spleens of mice, relative to GMDP and antigen doses and the genetype of immunized animals, during the primary immune response.

Strain of mouse	SRBC Number of cells	Stimulation Index of SRBC specific antibodies at varying GMDP levels ug/animal				
		0.01	1.0	10	100	1000
C57BL/6	5×10^6	2.35 (21)	3.61 (22)	2.71 (20)	4.14 (27)	5.21 (27)
	2×10^7	–	3.03 (14)	–	2.51 (21)	1.73 (21)
	2×10^8	–	–	–	1.02 (11)	0.73 (12)
CBA	5×10^6	2.68 (18)	3.26 (20)	3.06 (18)	1.83 (18)	1.89 (16)
	2×10^7	–	1.68 (11)	–	0.75 (14)	1.00 (16)
(CBA x C57BL/6)F₁	5×10^6	2.59 (13)	2.31 (20)	3.98 (26)	4.13 (26)	5.13 (28)
	2×10^7	–	1.38 (14)	–	1.50 (21)	1.49 (23)

Three different strains of mice were used. The numbers in parenthesis designate the number of mice. The stimulation index is the factor by which the antibody titer increased against a control without GMDP. The GMDP was injected four days after intraperitoneal immunization.

The stimulation effect shows correlation to the dose level although with CBA mice the effect was less obvious and declined above 10ug per mouse. At high antigen dose levels the immunostimulating effect is lessened.

Andronova *et al* investigated the stimulation of circulating antibodies to ovalbumin with GMDP in BALB/c mice. [3]

TABLE 2

Stimulation of circulating antibodies to ovalbumin.
OVA (25ug) with adjuvant in Freund's Incomplete (IFA) was
injected i/p in BALB/c mice. The sera was assayed on day 42
by ELISA

Adjuvant	Dose ug/Kg	Antibody Titer $-\log_2(AD_{492}=1.0)$	Actual Dilution	Stimulation Index
IFA	—	10.0	1024	1.0
IFA + MDP	0.5	9.0	512	0.5
	5.0	11.7	3326	3.25
IFA + GMDP	0.05	9.6	776	0.76
	0.5	11.0	2048	2.0
	5.0	13.0	8192	8.0

This data shows good stimulation with dose levels between 0.02 and 0.2ug per mouse, but it is not clear whether maximum stimulation was achieved. The stimulation effect of GMDP did persist over the period of 42 days after only a single injection.

Further data was presented by Noscov *et al.*, where varying dose levels of GMDP were used with mice immunized with BSA and inactivated influenza virus (H_3N_2). [5]

TABLE 3

The adjuvant activity of GMDP in saline. The ratio of
antibody titers in the presence and absence of GMDP.

Antigen	Dose ug/Mouse	Stimulation Index	Administration route
BSA	150	10.2	s/c
	150	5.7	i/p
	1500	8.3	oral
H_3N_2	3	2.2	i/v
	7.6	2.4	i/p
	7.6	2.0	oral

Subsequent work showed the injection route did not significantly alter the effectiveness of GMDP. [6]

Andronova and Ivanov were able to show significant protective
effects of GMDP and MDP in mice with experimental infections.
Mice were immunized with lethal doses of bacteria, but preceded
by prophylactic injections of GMDP and MDP 24-48 hours before. [3]

		TABLE 4					
Antigen	Dose	GMDP			MDP		
	ug	1	2	4	1	2	4*
E.coli	0.1	25	20	70	10	20	50
	1.0	25	40	40	50	40	40
	10.0	25	60	100	50	50	70
	50.0	40	50	100	70	50	60
	100.0	60	70	100	70	70	60
	200.0	60	70	90	60	70	70
P.aeurginosa	0.1	50	80	80	70	70	90
	1.0	20	70	60	50	80	100
	10.0	40	50	70	50	80	90
	50.0	40	60	70	40	60	80
	100.0	40	60	60	70	80	80
	200.0	50	70	60	70	60	90
S,aureus	0.1	0	0	0	0	0	0
	1.0	0	0	0	0	0	0
	10.0	0	0	10	10	0	0
	100.0	0	0	0	0	0	0

The numbers show the % survival. In the control group the
viability was 0%. * Number of injections.

These studies show that both GMDP and MDP produce preventative and
protective action. This action is dependent on dose level and
injection frequency. Multiple injections are more effective. However,
neither GMDP nor MDP was effective with the *Staphylococcus* infection.

With samples of GMDP provided by the Shemiakin Institute, Moscow,
researchers at Stanford University Medical School, studied the
idiotype vaccination against murine B cell lymphoma. [7] (Fig 2).
In this case the GMDP was combined with a carrier emulsion
comprising of squalene, Pluronic L 121 and Tween 80. This
formulation was first described by Allison *et al.* [8] [9]

At Syntex Research Laboratories, Palo Alto, California, there has been
extensive research on adjuvant peptides, especially the
N-acetylmuramyl-L-threonyl-D-isoglutamine, an analog of MDP. In this
case the immunopotentiator was combined with the squalene carrier.

Since much of the data summarised does not clearly define the most effective dose levels for GMDP, we systematically undertook an evaluation of the immunopotentiating possibilities for GMDP in antisera production. The preliminary data is presented below, but further work is planned.

RESULTS

In cooperation with the Shemiakin Institute we directly established the effectiveness of GMDP as an immunopotentiator in mice.

TABLE 5

Immune response in mice against ovalbumin.

Adjuvant	ug/mouse	Stimulation Index		
		Exp 1	Exp 2	Exp 3
FCA	–	8.87	8.57	–
IFA	–	2.22	2.60	2.46
ISAF-1	–	4.10	1.15	2.93
ISAF-1 + MDP	20	3.60	5.30	–
ISAF-1 + GMDP	20	2.70	1.70	–
ISAF-1 + GMDP	10	–	–	3.19
ISAF-1 + GMDP	1	–	–	6.60

FCA Freund's Complete Adjuvant. IFA Incomplete Freund's Adjuvant.
ISAF-1 Squalene, Pluronic L 121, Tween 80 formulation. [8] [9]
The control was antigen in water.

This data began to show GMDP was effective at lower than expected concentrations and may even have immunosuppressive effects at higher concentrations. At 1ug per mouse, GMDP was showing similar immunostimulation as FCA. However, at this level of concentration, there was no detectable toxicity or side effects on the animals.

Independently we continued experiments using human lambda light chain as antigen to confirm the initial data from Shemiakin Institute.

TABLE 6

Stimulation of antibody response to human lambda light chain.
Reciprocal ELISA titer.

GMDP Dose ug/mouse	Day 21 Bleed	Day 42 Bleed
20	1050	1960
10	960	3060
5	920	4000
1	860	5060

Four groups of 5 mice each were used and immunized with 25ug of human lambda light chain. The mice were immunized twice by subcutaneous injection – day 1 and day 21. GMDP and antigen was administered in water.

The findings again confirmed that the best response was at the
1ug/mouse level. Because this experiment terminated at 42 days, we
decided to extend this period in an additional experiment. We also
included in this experiment other commercially available adjuvants.

TABLE 7

*Stimulation of antibody response to DNP-BSA in mice. Control
was antigen in water.*

Adjuvant Used	GMDP Dose ug/mouse	Day 42 Bleed	Day 65 Bleed	Day 84 Bleed
None	–	1.0	1.0	1.0
IFA	–	1.9	2.0	1.2
FCA	–	3.8	3.4	2.0
ISAF-1	–	2.5	2.2	1.6
GMDP	5.0	4.4	3.7	2.6
GMDP	1.0	4.6	4.5	3.2
GMDP	0.5	3.6	2.7	4.7
GMDP	0.1	1.4	1.5	0.7
GMDP	0.05	3.5	2.5	2.3
GMDP	0.01	3.7	2.3	1.8
RIBI	–	1.3	2.3	0.7
TiterMax	–	2.9	4.9	3.3

12 Groups of 5 mice each were used. The antigen was injected
twice - on day 1 and day 21. 10ug was used per mouse.
GMDP was administered in water.
FCA Freund's Complete Adjuvant. IFA Freund's Incomplete.
ISAF-1 squalene Pluronic L 121 Tween 80.
RIBI Ribi Immunochemical Research, Hamilton MT 59840.
TiterMax CytRx Corporation, Norcross, GA 30092.

With this data we concluded that the GMDP administered in water only
is an excellent adjuvant for antibody stimulation and the optimum dose
is at 1ug/mouse. Approximately 30ug/Kg of animal weight.
A graphic representation of the above results is given in the form of
titer curves obtained by ELISA. Fig 2.

Figure 2

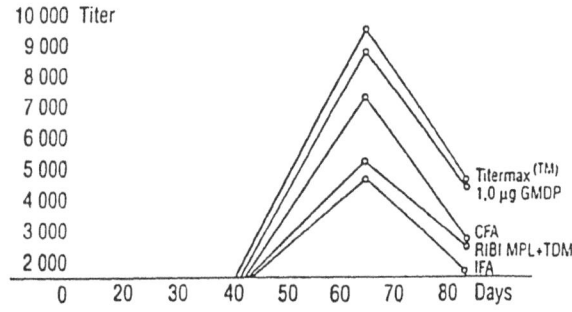

DISCUSSION

Experimental work is currently continuing with larger animal species
and preliminary results appear to confirm our findings with mice. The
particular advantage of GMDP is that no carrier emulsion is needed and
it is therefore very simple to use. The toxicity and pyrogenicity have
been well established and at the optimum doses indicated in our work,
no adverse effects were detected in any animals.

GMDP was better than Freund's and when administered in water, showed
none of the typical adverse effects of Freund's. Therefore, because of
increasing pressure to avoid inflicting undue stress on the animals,
GMDP shows distinct promise of becoming a valuable adjuvant in the
future.

Acknowledgements. The authors would like to express their appreciation
to Dr. V. Ivanov and Dr. T. Andronova of the Shemiakin Institute of
Bioorganic Chemistry for their assistance. We also wish to thank Dr.
S. Sisson, Cascade Immunology Corporation, Tryon, N.C.

REFERENCES

1. Ellouz, F. Adam, A., Ciorbaru, R. and Lederer, E. 1974.
   Minimal structural requirements for adjuvant activity of bacterial
   peptidoglycan derivatives. *Biochemical Biophysical Communications*
   59:1317.
2. Bogdanov, I.G., Dalev, P.G., Gurevich, A.I., Kolosov, M.N.,
   Mal'kova, V.P., Plemyannikova, L.A. and Sorokina, I.B. 1975
   Antitumor glycopeptides from *Lactobacillus bulgaricus* cell wall.
   FEBS Letters, 57:259.
3. Andronova, T. and Ivanov, V. 1991 In Press.
   The structure and immunomodulating function of glucosaminylmuramyl
   peptides. *Soviet Medical Reviews, Immunological Reviews*
   (R.V. Petrov Ed.)
4. Man'ko, V.M., Skvortsov, V.Yu., Masternak, T.B., Ivanova, A.S.,
   Larin, A.S., Andronova, T.M., Ivanov, V.T. and Zabanova, E.V.
   188. Study of the immunomodulating effect of the
   glucosaminylmuramyl dipeptide (GMDP). Effect on the hmoral
   respnse. *Immunoloiya*, 6:34.
5. Noskov, F.S., Nikitina, L.E., Maslennikova, L.K. and Fridman, E.A.
   1984. *Immunostimulating and anti-infectious activity of synthetic
   analogs of bacterial cell wall glycopeptides*. Immunolgiya, 4:53.
6. Kalina, M.G., Elkina, S.I., Moshiashvili, I.Ya. and Sergeev, V.V.
   1983. Adjuvant effects of synthetic glycopeptides on humoral
   response to a thymus-dependent antigen. *Zhurnal Mikrobiologii,
   Epidemiologii i Immunobiologii,* 6:97.
7. Campbell, M.J., Esserman, L., Byars, N.E., Allison, A.C. and
   Levy, R. 1990. *Idiotype vaccination against murine B cell lymphoma.*
   J. Immunology, 145:1029.
8. Allison, A.C., Byars, N.E. and Waters, R.V. 1986. Immunologic
   adjuvants: efficacy and safety considerations.
   Advances in Carriers and Adjuvants for Veterinary Biologics.
   R.M. Nervig et al. Ed. The Iowa State University Press, Ames, Iowa
   p19.
9. Allison, A.C. and Byars, N.E. 1986. An adjuvant formulation that
   selectively elicits the formation of antibodies of protective
   isotypes and of cell-mediated immunity. *J. Immunol. Methods* 95:157.

ESTABLISHMENT AND CHARACTERIZATION OF MONOCLONAL ANTIBODIES AGAINST FISH VIRUS (*Iridoviridae*)

Hiroshi Oda,[1] Tadakazu Tamai,[1] Kazunari Tsujimura,[2] Nobuyuki Sato,[1] Shoji Kimura,[1] Sanetaka Shirahata,[2] and Hiroki Murakami[2]

[1]*Central Research Institute, Taiyo Fishery Co. Ltd, 16-2 Wadai, Tsukuba, Ibaraki 300-42, Japan;* [2]*Graduate School of Genetic Resources Technology, Kyushu University, Fukuoka 812, Japan*

Abstract

Fish virus of the family *Iridoviridae* often seriously damages the aquaculture of such fish as the red sea bream (*Pagrus major*) and striped jack (*Caranx delicatissimus*). The virus causes symptoms like severe anemia, cell hypertrophy in some tissues, and mortality; however, the virus has not been thoroughly characterized. For the purpose of early diagnosis and healing, we tried to establish murine monoclonal antibodies (MAbs) against the virus. We first prepared a crude viral preparation from a spleen extract of diseased striped jacks. To ascertain that the virus belonged to the family *Iridoviridae*, PCR primers were designed according to the conserved sequences between insect and frog iridoviruses. A specific DNA fragment was amplified by using the virus DNA as a template. The amino acid sequence deduced from the nucleotide sequence of the PCR products was 40% and 47% identical to those of the insect and frog iridoviruses, respectively, without counting gaps, suggesting that the virus causing the irido-symptom belonged to the family *Iridoviridae*. Six MAbs (IgM, κ) reactive to the virus were established by immunizing mice with the crude virus preparation. The reactivities of MAbs against sera derived from the striped jack and red sea bream suffering from the iridovirus diseases were examined by ELISA. Sera from healthy striped jack and red sea bream showed no reactivity to the MAbs, but a strong correlation was found between the results of ELISA and the appearance of lesion. These results suggest that the MAbs established here will be useful for detecting the viral infection of fish in aquaculture.

Introduction

In 1990, a new epizootic causing severe mortality was reported by some aquafarms of the red sea bream on Shikoku Island, Japan. During the last two years, it has also been reported for more species, e.g., striped jack and yellow tail. This epizootic seriously damaged the aquaculture of these fish. From pathological and virological studies, Inoue *et al.* have demonstrated that the epizootic was a viral infectious disease [3]. They also identified that the virus belonged to the family *Iridoviridae*. Diseased fish became dark in color, less active, and showed anemia. In the spleen, heart, kidney, liver, and gill, characteristic cell hypertrophy was often observed. Thus far, the diagnostic procedures have been dependent on the appearance of lesions or on the histological demonstration of hypertrophic cells. The difficulty of diagnosis by these techniques, especially in the case of low-grade infections, lead us to develop a more sensitive and specific diagnostic method using monoclonal antibodies (MAbs). We report here the establishment and characterization of MAbs against the virus.

Materials and Methods

Preparation of the crude virus solution

Ten spleens were collected from diseased red sea bream and striped jack showing the irido-

565

S. Kaminogawa et al. (eds.), Animal Cell Technology: Basic & Applied Aspects, Vol. 5, 565–571.
© 1993 *Kluwer Academic Publishers.*

symptom. After microscopically observing cell hypertrophy, they were disrupted with a glass homogenizer. The homogenates were centrifuged at 15,000 x g for 10 min to remove debris, and the supernatants were filtered through a 0.45-μm membrane. These filtrates were used as a crude virus preparation and stored frozen until needed for use. Controls (virus-free spleen extracts) were prepared from healthy red sea bream and striped jack showing none of the irido-symptom.

Polymerase chain reaction

A genomic template was prepared from the crude virus preparation by the SDS-phenol method [5]. The PCR conditions are summarized in Fig. 2.

DNA sequencing

The PCR products was sequenced by the Sanger method [6].

Production of hybridomas

Six-week-old BALB/c mice were immunized with the crude virus preparation (100 μg of protein in each injection) at approximately 3-week intervals. The first injection was administered intradermally with complete Freund's adjuvant, while the second and third (final) injections were made intrapenitoneally without the adjuvant. Three days after the final injection, serum samples from each mouse were taken and assayed for the fish-iridovirus antibody by ELISA.

The fusion protocol was a modification of the PEG method [2]. Spleen cells were fused with a HAT-sensitive mouse myeloma line of P3-X63-AgU-U1 cells at a ratio of 10:1 with PEG 1500. The fused cells were resuspended in a growth medium (GM) consisting of ERDF supplemented with 15 % fetal calf serum. Aliquots of 0.1 ml of the cell suspension were put in each well of 96-well microtitre plates. On the first day of the test, 0.1 ml of GM supplemented with hypoxanthine (0.1 mM), aminopterin (4 x 10^{-4}mM) and thymidine (1.6 x 10^{-2}mM) (GM-HAT) was added to each well. The cultures were maintained by feeding every 2 days with GM-HAT for 8 days. The supernatants were tested periodically for the fish iridovirus-specific antibody by ELISA. Hybridomas producing specific antibodies were expanded and cloned at least twice by the limiting dilution method. The established hybridomas were cultured in ERDF medium supplemented with insulin (10 μg/ml), transferrin (35 μg/ml), ethanolamine (20 μM) and selenium (25 nM) (ITES)[4].

Reactivity of MAbs against sera of fish in aquafarms

Fifteen blood samples from striped jack and 42 from red sea bream were collected from aquafarms on Kyushu Island and Amami Island, Japan. These blood samples were held at room temperature for 1 h and centrifuged at 12,000 x g for 10 min to remove clots. The reactivity of MAbs against these sera samples was examined by ELISA.

Results

PCR amplification of the fish iridovirus genome

To ascertain that the virus belonged to the family *Iridoviridae*, PCR was performed by using a DNA solution prepared from the crude virus preparation. The design of PCR primers was based on the conserved sequences between insect and frog iridoviruses as shown in Fig. 1.

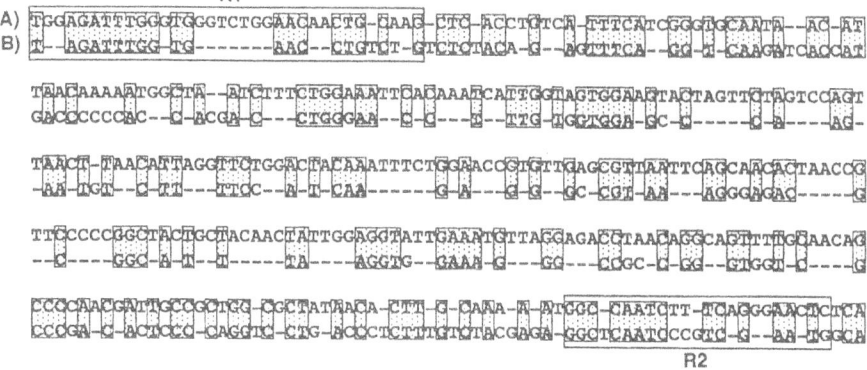

Fig. 1. Comparison of nucleotide sequences between (A) insect iridescent virus type 6 and (B) frog virus 3. PCR primers 1 and 2 were based on highly homologous regions R1 and R2, respectively.

Primer 1 : 5'-TAGATTTGGTGAACCTGTCTG-3'   Primer 2 : 5'-CATTCGACGGGATTGAGCC-3'

Composition of the Reaction Mixture

Template DNA	13 μg
d NTP s	each 20 n mol
Primer 1	20 p mol
Primer 2	20 p mol
Taq DNA polymerase	25 U
Total vol.	100 ml

Cycle Parameters

Denature	94 °C	30 sec
Annealing	37 °C	2 min
Extension	72 °C	1 min

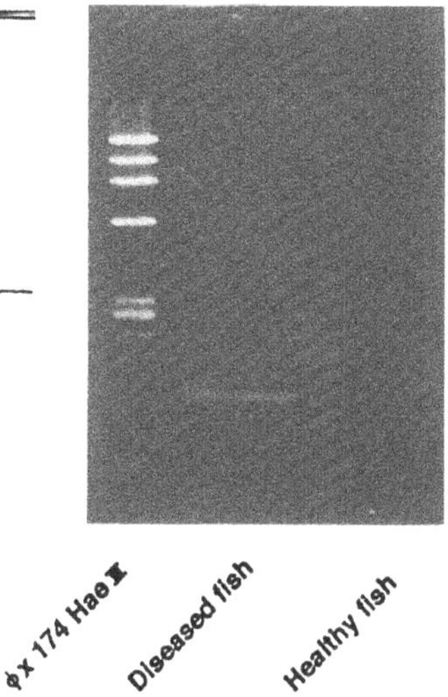

Fig. 2. PCR conditions and amplification of a specific fragment from the crude virus preparation of diseased fish. This fragment was not found with DNA prepared from healthy fish.

The upper sequence (A) was derived from the insect iridescent virus type 6 genome [1], and the lower sequence (B) was derived from the frog virus 3 genome [7]. Primers 1 and 2 were based on highly homologous regions R1 and R2, respectively.

A pair of primers 1 and 2 gave a specific DNA fragment of 180 bp, the length being thought to be reasonable compared with other iridovirus genomes (Fig. 2). This specific PCR fragment was not observed with DNA extracted from the spleens of healthy striped jack.

568

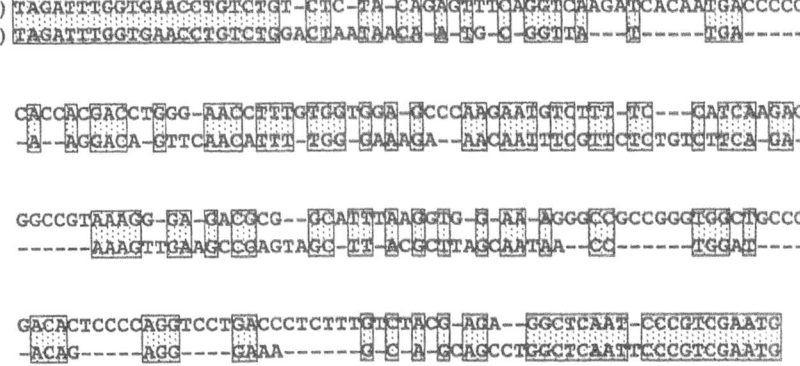

Fig. 3. Comparison of nucleotide sequences between (A) frog virus 3 and (B) PCR products. These nucleotide sequences show 54% homology, and the amino acid sequences deduced from these nucleotide sequences show 47% homology. The homology of the nucleotide and amino acid sequences between insect iridescent virus type 6 and the PCR products is 44% and 40%, respectively.

Sequence analysis of the PCR products

To clarify the primary structure of the amplified DNA fragment, it was subcloned into a pUC 119 sequencing vector and sequenced by the Sanger method. The nucleotide sequences of frog virus 3 and the PCR products showed 54% homology (Fig. 3). The amino acid sequences deduced from these sequences showed 47% homology. On the other hand, the homology of the nucleotide and deduced amino acid sequences between insect iridescent virus type 6 and the PCR products was 44% and 40%, respectively.

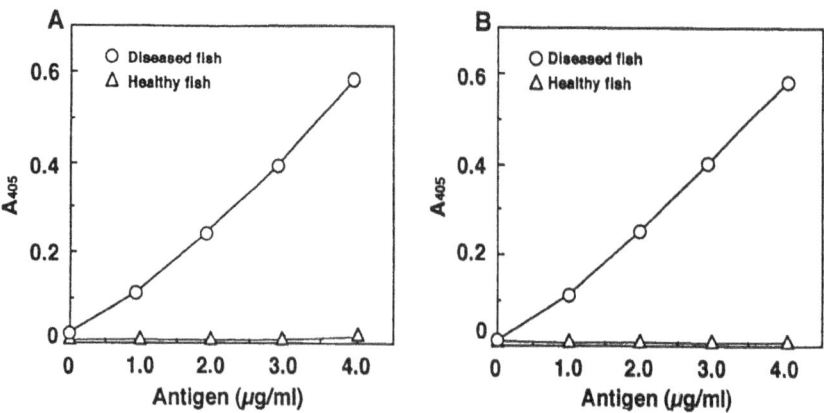

Fig. 4. Reactivity of SM-803 (A) and -807 (B) against the spleen extract of diseased red sea bream. They were reactive not only to the spleen extract of diseased striped jack, but also to that of diseased red sea bream. They were not reactive to the spleen extract of healthy red sea bream in the same manner as that of healthy striped jack. Other MAbs (SM-804, -805, -808 and -809) had the same criteria.

Table 1. Reactivity of MAbs against sera from striped jack

Striped jack No.	Appearance of lesion*	Monoclonal antibodies**					
		SM-803	SM-804	SM-805	SM-807	SM-808	SM-809
O 1-1	advanced	+	+	+	+	+	+
O 2-1	moderate	+	+	+	+	+	+
O 3-1	no lesion	+	-	-	-	-	-
O 4-1	no lesion	-	-	-	-	-	-
O 5-1	slight	-	+	-	-	-	-
N 1-1	no lesion	-	-	-	-	-	-
N 2-1	no lesion	-	-	-	-	-	-
N 3-1	slight	-	-	+	-	+	-
K 1-1	no lesion	-	-	-	-	-	-
K 1-2	no lesion	-	-	-	-	-	-
K 2-2	no lesion	-	-	-	-	-	-
S 2-1	slight	-	+	-	-	+	-
S 2-2	slight	-	-	+	-	+	-
S 4-1	no lesion	-	-	-	-	-	-
S 5-1	no lesion	-	+	+	-	-	-

*The appearance of lesion was judged from anemia, anorexia, abnormal swimming, and dark coloration of the body.
**++ , results from ELISA (A405) > 0.4 ; + , A405 > 0.1 ; - , no detectable absorbance

Production of hybridomas

To confirm the production of antibodies against the virus, antibody titres of BALB/c mice immunized with the crude virus preparation were determined by ELISA. After the fusion experiment described in the materials and methods section, six positive hybridomas were established. All of them produced MAbs (SM-803, -804, -805, -807, -808 and -809) reactive to the spleen extract of diseased striped jack, but not reactive to that of healthy fish. All of these MAbs were demonstrated to be IgM and to possess κ as the light chain. These MAbs also showed high reactivity to the spleen extract of diseased red sea bream, but were not reactive to that of healthy fish. The reactivity of SM-803 and SM-807 is shown in Fig. 4.

Reactivity of MAbs against the sera of fish in aquafarms

The reactivity of these six MAbs against serum samples was examined in several aquafarms. Before collecting blood, the appearance of lesion judged from anemia, anorexia, abnormal swimming, and dark coloration of the body was checked. As shown in Table 1, the sera of striped jack which had advanced or moderate lesion had high reactivity against all MAbs. The sera of striped jack which had slight or no lesion showed no detectable reactivity. The results of an examination using the sera of red sea bream were similar to those using the sera of striped jack (Table 2).

Discussion

For the purpose of an early diagnosis of the irido-symptom, we tried to establish MAbs against fish iridovirus. As it was difficult to obtain purified virus particles, we used the crude

Table 2. Reactivity of MAbs against sera from red sea bream

Red sea bream No.	Appearance of lesion*	Monoclonal antibodies**					
		SM-803	SM-804	SM-805	SM-807	SM-808	SM-809
O 2-1	advanced	++	++	++	++	++	++
O 3-1	advanced	++	+	++	++	++	++
O 4-1	moderate	+	+	+	+	+	+
O 5-1	advanced	+	+	+	+	+	+
K 1-2	no lesion	-	-	-	-	-	-
K 1-3	advanced	+	+	+	+	+	+
K 1-4	no lesion	-	-	-	-	-	-
K 1-5	no lesion	-	-	+	+	-	-
K 2-1	no lesion	-	-	+	-	-	-
K 2-2	no lesion	-	-	-	-	-	-
K 2-3	no lesion	-	-	-	-	-	-
K 2-4	slight	-	+	+	-	-	-
K 2-5	slight	-	-	-	+	-	-
K 4-1	no lesion	-	-	-	-	-	-
K 7-1	no lesion	-	-	-	-	-	-
K 8-1	moderate	+	+	+	+	+	+
S 1-1	slight	-	-	-	-	-	-
S 2-1	no lesion	-	-	-	-	-	-
S 2-2	moderate	+	+	+	+	++	++
S 2-3	no lesion	-	-	-	-	-	-
S 3-1	no lesion	-	-	-	-	-	-
S 5-1	no lesion	-	-	-	-	-	-
S 5-2	slight	+	+	+	+	+	+
S 5-3	no lesion	-	-	-	-	-	-
T 1-1	moderate	-	-	-	-	-	-
T 1-2	advanced	+	+	+	+	+	+
T 2-1	moderate	+	+	+	+	+	+
T 3-1	advanced	+	+	+	+	+	+
T 4-1	advanced	++	++	+	++	+	+
T 4-2	advanced	+	+	+	+	+	+
T 5-1	advanced	++	++	++	++	++	++
T 5-2	moderate	-	-	+	-	+	-
G 1-1	no lesion	-	-	-	-	-	-
G 2-1	slight	+	-	-	+	+	+
G 3-1	slight	-	-	-	-	-	-
G 4-1	slight	+	-	-	+	+	+
A 2-1	no lesion	-	-	-	-	-	-
A 2-3	no lesion	-	-	-	-	-	-
A 2-5	advanced	+	+	+	+	+	+
A 3-1	no lesion	-	-	-	-	-	-
A 3-3	slight	-	-	-	-	-	-
A 3-4	no lesion	-	-	+	-	-	-

*The appearance of lesion was judged from anemia, anorexia, abnormal swimming, and dark coloration of the body.

**++, results from ELISA (A405) > 0.4 ; +, A405 > 0.1 ; -, no detectable absorbance

spleen extract of diseased striped jack as an antigen in this study. Since this crude extract contained many contaminating compounds other than virus particles, a crude extract of healthy striped jack which would not contain virus particles was prepared as a control.

In PCR, a unique DNA fragment of 180 bp was amplified only when the crude DNA solution from diseased striped jack was used as a template. The nucleotide sequence of this fragment showed about 50% homology with other iridoviruses. These results suggest that crude virus solution from diseased striped jack contained virus particles belonging to the family *Iridoviridae*.

BALB/c mice were immunized with this crude virus solution, and six positive hybridomas were cloned. All MAbs produced by these hybridomas were reactive to the spleen extracts of diseased striped jack and red sea bream. However, they were not reactive to the spleen extracts of healthy striped jack and red sea bream. Although their species were different, these MAbs were only reactive to the spleen extracts of diseased fish. It is suggested that these MAbs will recognize the virus particles.

When the reactivity of these MAbs against serum samples was examined in several aquafarms, a strong correlation was found between the results from ELISA and the appearance of the irido-symptom. Thus far, fish iridovirus has not been reported in peripheral blood. However, these MAbs could react to serum samples with such correlation, suggesting that the virus particles would also be present, probably at a low level, in peripheral blood.

Although the epitopes of these MAbs have not been characterized because of difficulty in virus amplification, the results presented here strongly suggest that the MAbs that were established will be useful for detecting the iridovirus infection of fish in aquaculture. We are now attempting to obtain purified virus particles to elucidate the viral epitopes.

References

1. Fischer, M., Schnitzler, P., Scholz, J., Wolff, A. R., Delius, H., and Darai, G. (1988) 'DNA nucleotide sequence analysis of the *Pvu*II DNA fragment L of the genome of insect iridescent virus type 6 reveals a complex cluster of multiple tandem, overlapping, and interdigitated repetitive DNA elements,' Virology **167**, 497-506.

2. Galfre, G., and Milstein, C. (1981) 'Preparation of monoclonal antibodies, strategies and procedures,' Methods Enzymol. **73**, 3-46.

3. Inouye, K., Yamano, K., Maeno, Y., Nakajima, K., Matsuoka, M., Wada, Y., and Sorimachi, M. (1992) 'Iridovirus infection of cultured red sea bream, *Pagrus major*,' Gyobyo Kenkyu **27**, 19-27.

4. Murakami, H., Masui, H., Sato, G. H., Sueoka, N., Chow, T. P., and Kano-Sueoka, T. (1982) 'Growth of hybridoma cells in serum-free medium, ethanolamine is an essential component,' Proc. Natl. Acad. Sci. U.S.A. **79**, 1158-1162.

5. Sambrook, J., Fritsch, E. F., and Maniatis, T. (1989) 'Extraction of bacteriophage λ DNA,' Ford, N. and Nolan, C. (eds.), Molecular Cloning 2nd ed., Cold Spring Harbor Laboratory Press, New York, pp. 2.80-2.81.

6. Sanger, F., Nicklen, S., and Coulson, A. R. (1977) 'DNA sequencing with chain-terminating inhibitors,' Proc. Natl. Acad. Sci. U.S.A. **74**, 5463-5467.

7. Willis, D., Foglesong, D., and Granoff, A. (1984) 'Nucleotide sequence of an immediate-early frog virus 3 gene,' J. Virol. **53**, 905-912.

THE PRODUCTION OF RECOMBINANT ANTIBODIES USING THE GLUTAMINE SYNTHETASE (GS) SYSTEM

J.R. BIRCH, C.R. BEBBINGTON, R. FIELD, G. RENNER, H. BRAND AND
H. FINNEY.
Celltech Limited
216 Bath Road
Slough
Berks
SL1 4EN
United Kingdom

ABSTRACT. There is increasing interest in the use of genetically engineered, especially 'humanised' antibodies for a range of therapeutic applications. Efficient expression systems and production methods are required to manufacture these antibodies in the large (multikilogram) quantities which are likely to be required. We have developed a very efficient expression system based on glutamine synthetase (GS) selection. Antibody genes are transfected into Myeloma or CHO cells in vectors which also contain the GS gene and appropriate promoter sequences to ensure efficient expression. Transfected cells are selected by their ability to grow in glutamine free culture medium. Using the GS system we have been able to develop and optimise a large scale, serum free batch culture process using airlift fermenters. High product levels (up to one gram per litre) have been achieved for monoclonal antibodies and antibody fragments. Production is stable over many cell generations.

Introduction

There has been a rapid growth of interest in the use of recombinant DNA technology to produce monoclonal antibodies for therapeutic applications. Particular attention has been paid to reducing the immunogenicity of rodent antibodies by grafting the antigen binding regions of the protein into an essentially human molecule. It follows that if genetically engineered antibodies prove to be useful in therapy they are likely to be required in large quantities; tens and perhaps hundreds of kilograms per year. The provision of these quantities requires both an effective expression system and a suitable manufacturing process. A number of antibody expression systems using mammalian cells have been described, mostly based on Chinese hamster ovary (CHO) or rodent myeloma cells (see for example Wood et al 1990, Page et al 1991, Gillien et al 1989, Dorai and Moore, 1987). A particularly efficient system based on the use of the glutamine synthetase gene as a selectable marker has been described for use in CHO cells (Bebbington 1991) and myeloma cells (Bebbington et al, 1992). Alongside the development of the GS expression system progress has been made in developing large scale serum free cultivation of myeloma cells in airlift fermenters (Rhodes & Birch 1988). In this paper we describe our current experience with the productivity of myeloma and CHO cells producing recombinant antibodies using the GS expression system.

S. Kaminogawa et al. (eds.), Animal Cell Technology: Basic & Applied Aspects, Vol. 5, 573–577.
© 1993 Kluwer Academic Publishers.

Results

USE OF GS AS A SELECTABLE MARKER

GS catalyses the synthesis of glutamine from glutamate and ammonia and provides the only pathway in mammalian cells for glutamine synthesis. If glutamine is not added to the culture medium, GS becomes an essential enzyme. Whilst some established cell lines (such as CHO) have endogenous GS others (such as some myeloma lines) lack the enzyme and are therefore dependent on glutamine in the culture medium. Transfection of a GS gene into GS deficient myeloma cells confers the ability to grow in glutamine-free medium and this is the basis for using GS as a selectable marker in such cells (Bebbington et al, 1992). Amplification of genetic vectors containing GS can be achieved by selecting cells which are resistant to the specific inhibitor of GS, methionine sulphoximine (MSX). In the case of cell lines which have endogenous GS (eg. CHO), MSX may also be used to inhibit this endogenous enzyme and allow selection of cells expressing increased levels of enzyme as a result of transfection with the GS gene. (Bebbington 1991). For antibody expression, vectors have been designed which contain the GS gene driven by an SV40 early promoter and the genes for antibody light and heavy chains driven by hCMV-MIE promoters (Bebbington et al 1992).

VECTOR AMPLIFICATION

For the mouse NS0 myeloma cell, vector amplification is achieved by treating cells with MSX at a concentration of 10-100μM. Unlike other amplification systems such as dhfr there is no necessity to carry out more than one round of amplification and high levels of expression can be achieved even in non-amplified cell lines. This is a significant advantage in shortening development timescales and it is possible to progress from cDNA to cell lines expressing 200-300mg antibody/l in shake flasks in 4-5 months. The copy number in NS0 cells typically increases from 1 in non amplified transfectants to 4-10 copies after one round of amplification.

CELL LINE STABILITY

It is important that productivity of cell lines is stably maintained through the manufacturing process. In our case this process is based on a 2000 litre airlift reactor and in practice this dictates that the cell line must be stable for at least 60 generations beyond the Manufacturers Working Cell Bank. Our experience to date has been that most cell lines based on NS0 are stable when grown in glutamine free medium, regardless of whether the vector is amplified and even in the absence of the selective drug MSX. In the case of CHO cells it has been necessary to retain MSX as a selective pressure to ensure stability (Brown et al in press). It is assumed that because CHO has endogenous GS, the use of glutamine-free medium may be insufficient to maintain selection.

PROCESS PERFORMANCE

Recombinant CHO and myeloma cell lines are grown in serum free culture in airlift reactors, the operation of which has been described by Birch et al (1987). Recombinant antibody yields of hundreds of milligrams/litre have been achieved with both CHO and NS0 cell lines (Bebbington 1991, Bebbington et al 1992). These yields result both from the efficiency of the GS expression system and from extensive process optimisation. An example of the results of such optimisation for a particular amplified NS0 cell line is given in Fig. 1.

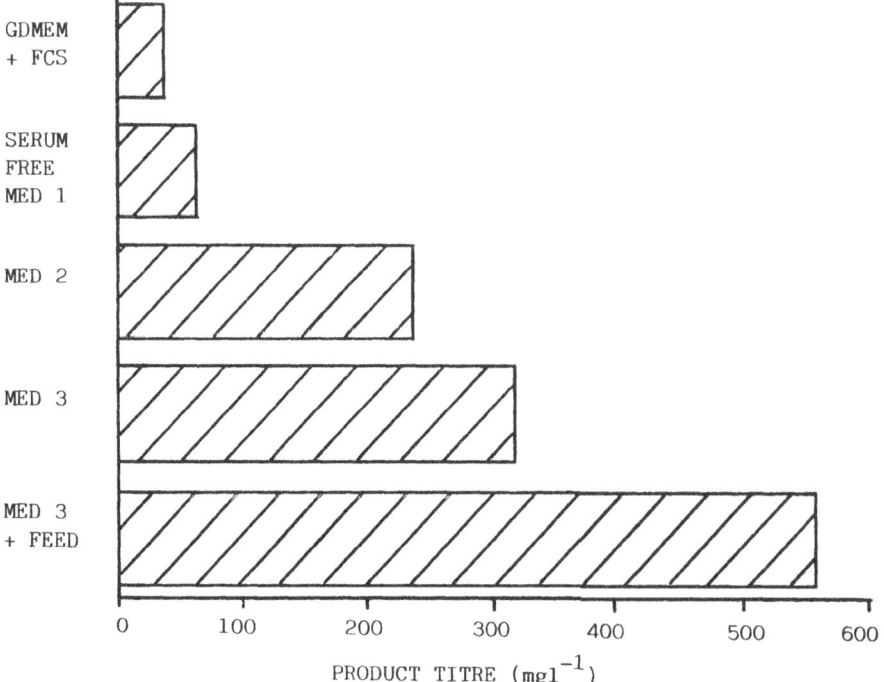

FIG 1.   PROCESS IMPROVEMENT; RECOMBINANT MAB 1
          PRODUCTION IN GS-NSO CELL LINE.

Titres of 40-50mg/l were achieved in GDMEM + 10% FBS in shake flask culture in 250-300 hours. An iterative programme of medium development increased productivity to 300mg/l in fermenters. The application of a small volume of key nutrients as a feed during the growth phase increased titres to 560-585 mg/l in a 300 hour batch culture. Yields as high as one gram/litre have now been achieved (Fig. 2), and productivity of hundreds of mgs per litre are obtained for a range of antibodies (examples given in Table 1).

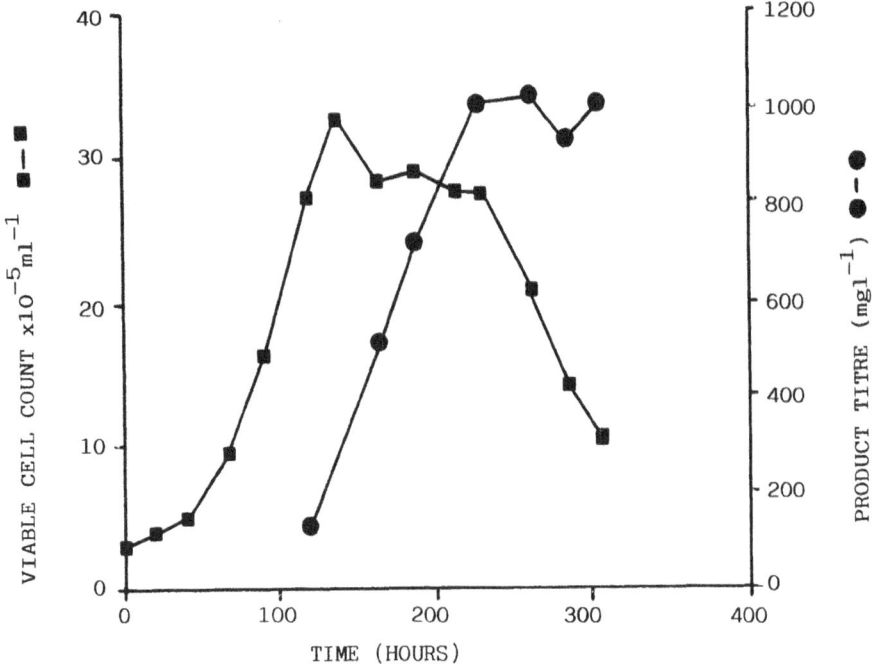

FIG 2.   PRODUCTION OF RECOMBINANT MAB 5 FROM GS—NSO
         CELL LINE IN SERUM FREE FERMENTATION.

Table 1. The GS System - Yields in Suspension Culture (fermenters)

Expression System	Product	Antibody Titre (mg/l)
GS-CHO non amp	MAb 1	110
GS-CHO amp	MAb 1	195
GS-CHO amp	FAb	580
GS-NS0 amp	MAb 1	585
GS-NS0 amp	MAb 2	900
GS-NS0 amp	MAb 3	895
GS-NS0 non amp	MAb 4	440
GS-NS0 non amp	MAb 5	180
GS-NS0 amp	MAb 5	1026

Conclusions

The GS system has provided a reproducible and efficient technology for recombinant antibody production. In combination with process optimisation it has been possible to achieve yields of antibody up to 1 gram per litre in fed batch airlift fermentations. This makes feasible the economic production of multikilogram quantities of therapeutic product.

References

Bebbington, C.R. (1991) 'Expression of recombinant genes in non-lymphoid mammalian cells'. Methods: A companion to Methods in Enzymology 2, 135-145.

Bebbington, C.R., Renner, G., Thomson, S., King, D., Abrams, D. and Yarranton, G.T. (1992) 'High level expression of a recombinant antibody from myeloma cells using a glutamine synthetase gene as an amplifiable selectable marker'. BioTechnology 10, 169-175.

Birch, J.R., Lambert, K., Thompson, P.W., Kenney, A.C. and Wood, L.A. (1987) 'Antibody Production with Airlift Fermentors in Lydersen, B.K. (ed) Large Scale Cell Culture Technology, Hanser Publishers, Munich, Vienna, New York pp 1-20.

Brown, M.E., Renner, G., Field, R.P. and Hassell, T. (1992) 'Process development for the production of recombinant antibodies using the glutamine synthetase (GS) system'. Cytotechnology, In press.

Dorai, H., and Moore, G.P., (1987) 'The effect of dihydrofolate reductase-mediated gene amplification on the expression of transfected immunoglobulin genes'. J. Immunol. 139, 4232-4241.

Gillies, S.D., Dorai, H., Wesolowski, J., Majeau, G., Young, D., Boyd, J., Gardner, J. and James, K. (1989) 'Expression of human anti-tetanus toxoid antibody in transfected murine myeloma cells'. BioTechnology 7, 799-804.

Page, M.J. and Sydenham, M.A. (1991) 'High level expression of the humanized monoclonal antibody Campath-1H in Chinese hamster ovary cells'. BioTechnology 9, 64-68.

Rhodes, P.M. and Birch, J.R., (1988) 'Large-scale production of proteins from mammalian cells'. BioTechnology 6, 518-523.

Wood, C.R., Dorner, A.J., Morris, G.E., Alderman, E.M., Wilson, D. O'Hara. R.M., and Kaufman, R.J. (1990) 'High level synthesis of immunoglobulins in Chinese hamster ovary cells'. J. Immunol 145, 3011-3016.

GENERATION OF SUPRANATURAL ANTIBODIES BY MODIFYING THE L CHAINS: CELL HYBRIDIZATION AND GLYCOTECHNOLOGY

Hirofumi TACHIBANA, Sanetaka SHIRAHATA, Ken-ichi NAGAMINE*, and Hiroki MURAKAMI

Graduate School of Genetic Resources Technology, Kyushu University, Fukuoka, 812, Japan
NICHIREI Corporation, General Research Institute

Abstract
The antigen-binding of antibodies was altered by manipulating the light chains. For this purpose, we used human hybridoma HB4C5 that produces antibodies reactive to lung adenocarcinoma cells. Complementarity-determining region 1 in the λ light chain was N-glycosylated, and hybrid antibody molecules were obtained by generating hybrid hybridomas from fusion of the HB4C5 cell line with the SU-1 hybridoma line to produce anti-double-stranded DNA antibody. The hybrid antibody molecules had light chains derived only from SU-1, and had higher antigen-binding strength than that of the original antibodies. We examined the effect on antigen-binding of altering the structure of the carbohydrate chain located on the light chain. Some molecular weight variants of the light chain were produced by concanavalin A-resistant types of C5TN, which was a subclone of HB4C5, to give antibodies cross-reactive with carboxypeptidase A, *Candida* cytochrome C and double-stranded DNA. The variant antibodies significantly altered the original cross-reactivitiy with each of these antigens or lost the ability to bind the antigen.

Introduction

Human monoclonal antibodies (MAbs) have many uses in immunotherapy, immunoassay and immunodiagnosis [1]. The development of the *in vitro* immunization technique has allowed to an MAb specific to an antigen to be obtained [2]. However, this technique has not yet provided the ability to generate MAbs having more specific and higher affinity to antigens yet. The work reported in this paper is directed at generating MAbs with altered antigen-binding by modifications to the light chain. We used human hybridoma cell line C5TN producing MAb cross-reactive to bovine carboxypeptidase A (Cpase), *Candida* cytochrome C (Cyt C) and double-stranded DNA (ds DNA), the light chain of the antibody being N-glycosylated in the variable region [4]. We then generated the antibodies by modifying the light chains in two ways: 1) cell hybridization and 2) glycotechnology. Cell hybridization was used to generate antibodies with heterogeneous light chains by biological

S. Kaminogawa et al. (eds.), Animal Cell Technology: Basic & Applied Aspects, Vol. 5, 579–583.
© 1993 Kluwer Academic Publishers.

recombination of the heavy and light chains. Glycotechnology was applied to obtain cells with varied degrees of glycosylation, the variant cells being expected to produce antibodies having a light chain with an altered carbohydrate chain located on the antigen binding site.

Materials and Methods

Cells and Cell Culture

C5TN is a subclone of human hybridoma HB4C5 [3], and produces MAb (IgM, λ) cross-reactive to Cpase, Cyt C and ds DNA. SU-1 is a human hybridoma producing anti-ds DNA MAb (IgM, λ and κ)[4]. Hybrid hybridomas were produced by fusing the G418-resistant clone of HB4C5, HT2, and the mycophenolic acid-resistant clone (SU-1-D2) of SU-1 [5]. Variant clones of glycosylation were isolated for resistance to cytotoxic effect of concanavalin A (Con A) [6].

Glycosidase Digestion

Antibodies were treated with various glycosidases in a 0.1M sodium citrate buffer (pH 4.5) at 37°C for 18h. These glycosidases were 1) neuraminidase (0.1 unit); 2) a mixture of neuraminidase (0.1 unit) and β-galactosidase (0.1 unit); and 3) a mixture of neuraminidase (0.1 unit), β-galactosidase (0.1 unit) and β-N-acetylhexosaminidase (0.05 unit).

Antibody Binding Assay

Antigen binding of antibodies was assayed by the ELISA method. In brief, an immunoplate was coated with Cpase, Cyt C, or ds DNA immobilized with protamine sulfate. Antibodies were reacted with the antigens for 1 h, and the antibodies bound to the antigens were detected by the peroxidase-conjugated (POD) anti-human μ chain antibody. The peroxidase activity was assayed by measuring the absorbance at 405 nm of a colored substance produced by the enzyme reaction.

Western Blotting.

Antibodies were applied to SDS-PAGE and transffered to a nitrocellulose membrane as described elsewhere [3]. The blotted membrane was incubated with POD anti human λ or κ light chain antibody. The color reaction was developed using 4-chloro-1-naphthol as substrate.

Results and Discussion

Hybrid antibodies with increased antigen-binding by association with heterogeneous light chains [5]

Seven hybrid hybridoma clones were generated by fusing the HT2 and SU-1-D2 cell lines. Six of the resulting hybrid hybridomas produced hybrid antibodies with higher reactivity to Cpase and ds DNA compared with the parental antibodies. Among them, the E5H8H1 clone was used for further analysis. One hybrid hybridoma (E10D3) an produced antibody reactive to Cpase but not to ds DNA. Table 1 shows the reactivity and classification of the light chains of the antibodies produced by hybrid hybridomas E5H8H1 and E10D3, and by parental cells HT2 and SU-1-D2 . The E10D3 antibody had slightly higher reactivity to Cpase than that of parental HT2 antibody . The E5H8H1 antibody reacted with Cpase with higher affinity than the E10D3 antibody and had a higher binding strength to ds DNA than did not HT2 and SU-1-D2 antibodies. To examine the differences in reactivity of the hybrid antibodies, the light chain class of each antibody was determined. The light chain of the E10D3 antibody consisted only of the κ chain derived from the SU-1-D2 antibody. On the other hand, the light chains of the E5H8H1 antibody were κ and λ chains derived from the SU-1-D2 antibody. No light chain derived from the HT2 antibody was detected in either of the hybrid antibodies. These results indicat that the immunoglobulin molecule consisted of the k chain, as the light chain (i.e., the E10D3 antibody) had slightly higher binding to Cpase but not to ds DNA. The higher antigen binding of the E5H8H1 antibody to Cpase and to ds DNA was due to association of the heavy chain derived from the HT2 antibody with the λ light chain from the SU-1-D2 antibody. These findings suggest that hybrid antibodies with higher affinity can be generated by the heterogeneous association of heavy and light chains.

Generation of specificity-variant antibodies by modifying the carbohydrate chain on the light chain [6]

A subclone of HB4C5, C5TN, produced an antibody cross-reactive to Cpase, Cyt C, and ds DNA. The light chain of the antibody was N-glycosylated at complementarity-determining region 1. We attempted to clarify the role of the carbohydrate on the light chain regarding antigen binding. Variant cell clones that produced antibodies with altered molecular weights of the light chains were isolated for their resistance to Con A. The antibodies produced by the variant cells showed a dramatic alteration in cross-reactivity to the antigens or lost the ability of antigen binding (Table 2). The variant antibodies consisting of light chains with an altered carbohydrate chain were also generated by trimming the carbohydrates with glycosidases. The light chain of the C5TN antibody consisted of complex or hybrid carbohydrate, as the carbohydrate was sensitive to a neuraminidase treatment. As shown in Table 2, the reactivity of the C5TN antibody was significantly increased by removing the sialic acid residues. Additional digestion of the antibody by β- galactosidase and N-acetylhexosaminidase induced higher affinity. A variant antibody (variant A) having no reactivity to the antigens acquired reactivity by a

TABLE 1. Characterization of the antibodies from hybrid hybridomas
and parental hybridomas

	Parental cells		Hybrid hybridoma cells	
	HT2	SU-1-D2	E10D3	E5H8H1
L chain	λ†	λ, κ	κ	λ, κ
Anti-Cpase (OD 405)	0.150	0.005	0.200	>0.5
Anti-ds DNA (OD 405)	0.100	0.150	0.005	>0.3

†The λ light chain contained the N-linked carbohydrate chain.

TABLE 2. Effect of glycosidase digestion on Ig reactivity.

				Digested with		
			Control Ig	1. Neurami-nidase	2. Neurami-nidase+β-galactosi-dase	3. Neurami-nidase+b-ga-ractosidase +b-N-acetyl-hexosamini-dase
Ig	M.W. of L chain	Coated substance				
C5TN	32 KD	Cpase	0.490	1.020	1.060	1.090
		Cyt C	0.024	0.056	0.069	0.071
		ds DNA	0.203	0.753	0.874	0.824
Variant A	32.5 KD	Cpase	0.003	0.136	0.231	0.264
		Cyt C	0.000	0.046	0.076	0.081
		ds DNA	0.000	0.000	0.000	0.000
Variant B	28 KD	Cpase	0.260	0.272	0.300	N.D.
		Cyt C	0.330	0.336	0.362	N.D.
		ds DNA	0.000	0.000	0.000	N.D.

Reactivity was determined by ELISA, and these values represent the
absorbance at 405 nm.

neuraminidase treatments. This acquired reactivity was progressively enhanced by β-galactosidase and N-acetylhexosaminidase treatment. Thus, the carbohydrate structure affected the affinity and specificity of antigen binding. On the other hand, another variant antibody (variant B) consisting of a light chain resistant to these glycosidases had no change in antigen binding by these treatments, indicating that the carbohydrate chain on the constant region of the heavy chain was not associated with the change in reactivity, but that the carbohydrate on the light chain had a critical role in antigen binding.

Conclusions

We have described the results of our work to generatie supranatural human antibodies. Several antibodies with high affinity or altered specificity were produced by substituting the heterogenous light chain *in vivo* and by altering the carbohydrate chain on the light chain. In addition, we have already reported the generation of a bispecific human antibody by the association of the heterogeneous heavy and light chains [7]. The findings from these studies suggest that modification of the light chain is a useful technique for generating functional human antibodies, including bispecific antibody.

References

1. Hashizume, S., Mochizuki, K., Murakami, H., Yano, T., Yasumoto, K. and Nomoto, K. Serodiagnosis of cancer, using porcine antigens recognized by human monoclonal antibody, HB4C5. *Biotherapy* 1989, 1: 109-115.
2. Kawahara, H., Shirahata, S., Tachibana, H. and Murakami, H. In vitro immunization of human lymphocytes with human lung cancer cell line A549. *Hum. Antibod. Hybridomas* 1992, 3: 8-13.
3. Tachibana, H., Shirahata, S., and Murakami, H. Alteration of reactivity of human monoclonal antibodies produced by concanavalin A-resistant hybridomas. In Animal Cell Technology: Basic & Applied Aspects. 1992, Vol. 4, pp.547-551.
4. Hashizume, S., Murakami ,H., Kamei, M., Hirose, S., Shirai, T., Yamada, K., and Omura, H. Specificity of anti-polynucleotide monoclonal antibodies from human-human hybridomas. *In Vitro Cell. Develop. Biol.* 1987, 23: 53-56.
5. Tachibana, H., Shirahata, S., Kawahara ,H., and Murakami, H. Increased antigen binding strengths of hybrid antibodies produced by human hybrid hybridomas. *Cytotechnology* 1991, 7:1-6.
6. Tachibana, H., Shirahata, S., and Murakami, H. Generation of specificity-variant antibodies by alteration of carbohydrate in light chain of human monoclonal antibodies. *Biochem. Biophy. Res. Commun.* 1992, 189: 625-632.
7. Tachibana, H., Shirahata ,S., and Murakami, H. Human bifunctional antibody generated by heterologous association of heavy and light chains. *Hum. Antibod. Hybridomas* 1993, 4: 42-46.

TWO-STAGE CHEMOSTAT STUDIES OF HYBRIDOMA GROWTH, NUTRIENT UTILISATION, AND MONOCLONAL ANTIBODY PRODUCTION

David C. Venables, *Robert C. Boraston and Michael E. Bushell.
Microbial Physiology Group, School of Biological Science, University of Surrey, Guildford Surrey, GU2 5XH and *Celltech Biologics Plc., 216 Bath Road, Slough, Berks, SL1 4DY

ABSTRACT.

Batch and single-stage and two-stage chemostat studies have been carried out on a mouse x mouse hybridoma, NB1, producing IgM against the B subgroup of red blood cells. Positively growth associated production kinetics were observed during stirred batch culture. However, negatively growth associated monoclonal antibody production was observed in equivalent single-stage chemostat culture. During two-stage, non-supplemented chemostat culture, increased titre in the second stage was observed. Utilisation of residual substrate from the first stage was also evident. The viable cell count was higher in stage two except at the lowest dilution rates. The greatest rate of antibody production was observed in the second stage when the calculated growth rate was negative. This increased production occurred despite the observation that several essential amino acids were exhausted from the media.

INTRODUCTION.

Several types of production kinetics of antibody in batch culture have been described by various investigators, these have been classified into three groups by Merten et al. (1987). Type two and type three kinetics are the most often observed, usually referred to as growth dissociated and growth associated production respectively. Antibody production kinetics similar to those observed in batch culture have also been observed in single-stage chemostat culture. These include positively growth associated antibody production (Low et al.,1987; MacMichael, 1989), negatively growth associated production (Boraston et al., 1984, Reuveny et al., 1986, Miller et al., 1988; and Frame and Hu, 1991), as well as growth dissociated production (Ray et al., 1989, Hiller et al., 1991).

In batch culture and single-stage chemostat culture negatively growth associated antibody production is most frequently observed. The usefulness of single-stage chemostat culture in the study of negatively growth associated antibody production is limited because steady-states can become destabilised at very low dilution rates and, because growth is a prerequisite for steady-state operation, such a system is inappropriate for the study of metabolites produced by non-growing cells. The use of multiple-stage chemostats provides a convenient system for the study of negatively growth associated production phenomena because the first stage provides a continuous inoculum for subsequent stages. Growth is therefore only required in the first stage and subsequent stages can be operated at very low, zero and negative growth rates and therefore used for the study of non growth associated product formation. Conditions optimal for growth and antibody production are often different, two-stage chemostat culture therefore allows the operation of a first stage which generates viable biomass while a second stage can be optimised for maximising antibody production.

S. Kaminogawa et al. (eds.), Animal Cell Technology: Basic & Applied Aspects, Vol. 5, 585–594.
© 1993 Kluwer Academic Publishers.

MATERIALS AND METHODS

The cell line used in this study was the mouse x mouse hybridoma NB1 (Sacks and Lennox, 1981). The antibody produced was an IgM which has agglutinating activity against the group B antigen of human red blood cells.

Cell stocks were maintained in Dulbecco's modified Eagle's medium (DMEM) (Gibco) supplemented with 5% foetal calf serum (Imperial Laboratories).

Stocks of cells were maintained in a spinner culture consisting of a 250 ml bottle containing a magnetic flea placed on a stirrer base and incubated in a 5% carbon dioxide air atmosphere at 36.5^0C.

Cell concentration was determined using a Fuchs-Rosenthal heamocytometer. The culture samples were first diluted in 0.04% w/v solution of erythrosin B stain (Sigma) in saline. Dead cells stained red while live cells remained colourless. All counts were performed in duplicate.

Batch and chemostat experiments were performed using an LH 2000 series fermenter fitted with a two litre working volume glass vessel. Temperature was controlled at a set-point of 36.5^0C by passing water through a finger from a thermostatically controlled supply. The culture was stirred at a rate of 60 rpm by a top driven agitator fitted with a single Rushton turbine impeller. Dissolved oxygen tension (DOT) was measured using an autoclavable polarographic oxygen electrode (Ingold). The DOT was controlled at 40% of air saturation by the sparging of air into the culture when required. Culture pH was monitored by a pH electrode (Ingold) and controlled between setpoints of 7.0 - 7.2 by the addition of carbon dioxide or 1M NaOH.

A single stage chemostat was set up by using a peristaltic pump to feed medium from a refrigerated reservoir into the culture vessel. The culture volume was kept constant by removing culture via a weir.

A two-stage chemostat consisted of two LH 2000 fermenters in series where the overflow from the first stage was fed into the second stage (Figure A).

Medium feed rates were measured by a pipette fitted to the feed line via a 'T' piece connector.

Steady state culture parameters were measured over many culture doubling times at each steady state over the dilution rate range 0.01 to $0.05h^{-1}$

During the course of a fermentation samples were removed from the culture and, after centrifugation to remove cells and cellular debris, were dispensed into 1.5ml vials and stored at -20^0C. These samples were used for biochemical analyses and determination of antibody concentration.

Glucose and lactate concentrations were measured using a GM7 analyser (Analox Instruments Ltd.). Glutamine analysis was carried out using the Boehringer Mannheim glutamic acid assay kit. This was adapted for the measurement of L-glutamine by the addition of asparaginase (Grade V Asparaginase. Sigma) converting L-glutamine into L-glutamate. Ammonia concentration was determined using a Sigma Diagnostics ammonia determination assay.

The amino acid concentrations in culture supernatant were analysed using a Waters PICO.TAG Amino Acid Analyser system. The samples were derivatised with phenylisothiocyanate (PITC) to produce phenylcabamyl (PTC) amino acids which were then separated using a C18 column and their elution detected at 254nm.

The concentration of IgM monoclonal antibody in cell culture supernatant was determined by double-antibody sandwich ELISA.

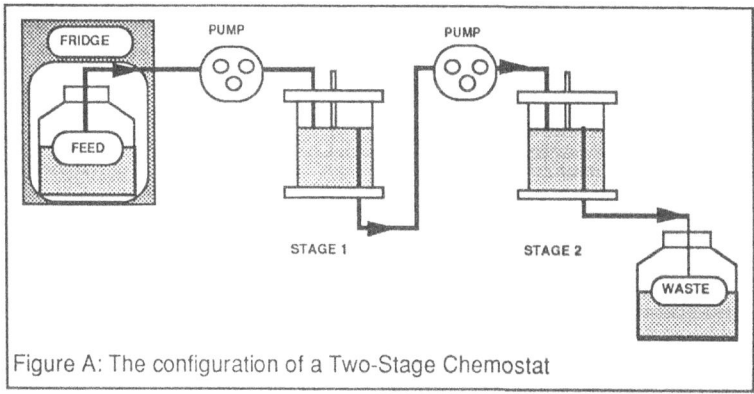

Figure A: The configuration of a Two-Stage Chemostat

RESULTS and DISCUSSION

BATCH CULTURE

Batch Culture. Increasing viable cell concentration was accompanied by decreasing glucose and glutamine concentrations and increasing lactate, ammonia and antibody. The point at which the culture entered the decline phase coincided with the point at which glutamine became depleted. After this point there was no further increase in total cell concentration, glucose utilisation, lactate or ammonia production. Also, there was very little further increase in antibody production (Figure 1a and 1b).

This result indicates that glutamine was essential for cell growth and antibody production. Antibody production is growth associated as production and growth cease simultaneously. This may be because antibody is produced only by a dividing cell population, or a nutrient essential for both growth and production has become exhausted, or a toxin has accumulated which inhibits growth and production.

Glutamine Supplemented Batch Culture. Extra glutamine was added to a batch culture of NB1 hybridomas to assess whether this would result in increased growth and product formation, and whether any further antibody production was growth associated. The results are shown in Figure 1c and 1d. The extra glutamine was added as a single "shot" during the batch culture (at the end of "exponential phase") to reduce the amount lost through spontaneous decomposition into pyrrolidone-carboxylic acid and ammonia (Tritsch and Moore, 1962). Rather than entering a rapid decline phase immediately after the growth phase as in the non-supplemented culture, the viable cell population stabilised in a prolonged stationary phase. During the stationary phase the extra glutamine was utilised, as was the glucose to a lesser extent. Antibody production continued through the stationary phase in what appeared to be a growth dissociated manner as the viable cell population was stationary. However, the increase in total cell concentration indicates that although the viable cell population was stationary, individual cells were still dividing. Antibody production was therefore still associated with a growing viable cell population.

It is likely that the antibody is being actively secreted by the viable cell population rather than being released by dead cells breaking up as no increase in production was observed in the non-pulsed culture when the cell population entered the decline phase.

588

Renard et al (1988) related antibody production kinetics to the integral of the viable cell curve in batch culture to provide evidence that antibody is secreted by viable cells rather than released by dead cells. Their hypothesis

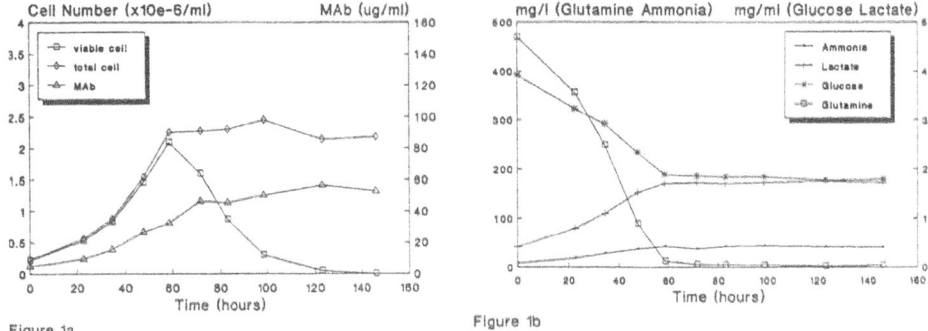

Figure 1a   Figure 1b

A typical batch culture of NB1 hybridoma cells showing (1a) Viable cell, Total cell and Monoclonal antibody (MAb) produced, and (1b) utilisation of Glutamine and Glucose, and production of Ammonia and Lactate.

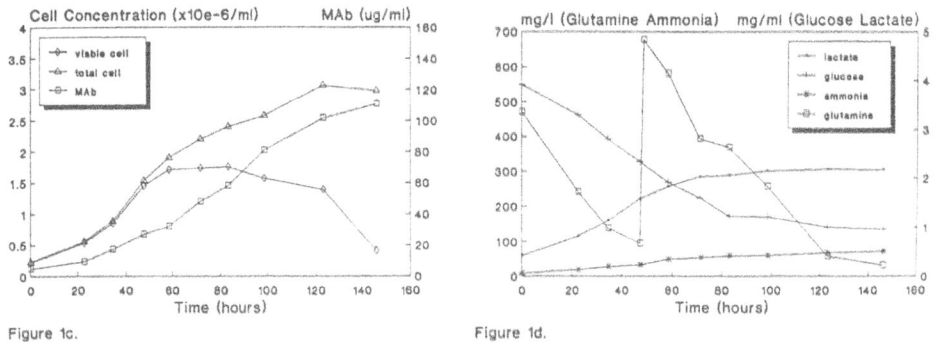

Figure 1c.   Figure 1d.

Glutamine Pulsed Batch Culture showing (1c) Viable cell, Total cell and Monoclonal antibody (MAb) produced, and (1d) utilisation of Glutamine and Glucose, and production of Ammonia and Lactate.

states that the titre of antibody in the supernatant at any time should be proportional to the time integral of viable cells. Thus, the production curve and the integral curve should show analogous profiles and regression of one over the other should be linear. This was shown to be true for the glutamine-pulsed culture, and also for the non glutamine-pulsed culture before glutamine became exhausted from the medium, once again demonstrating the requirement of glutamine for both growth and production.

SINGLE -STAGE CHEMOSTAT CULTURE

Cell Yield. In single stage studies steady state total cell concentration remained fairly constant across the dilution rate range, while viable cell concentration decreased with decreasing dilution rate (Figure 2a).

There was an increase in residual glutamine with increasing growth rate while the ammonia concentration remained relatively constant.The stability in ammonia concentration over the dilution range was not surprising considering that the change in residual glutamine was only 25 mg/l from a supply of 584 mg/l.

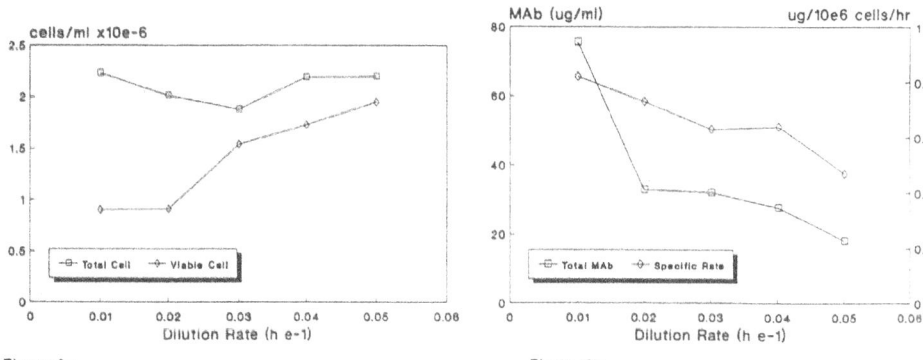

Figure 2a.   Figure 2b.

Single-Stage Chemostat Culture of NB1 Hybridoma Cells: (2a) Total and Viable cell concentration, (2b) Total Monoclonal Antibody (MAb) produced and the Specific Rate of production.

Residual glucose, in contrast to glutamine, showed a general trend of increasing concentration with decreasing growth rate, and, as would be expected, lactate production showed the reverse. These results would tend to suggest that the use of glucose and glutamine as an energy source is not interchangeable as it would appear to be for some cell lines.

Monoclonal Antibody Production. Monoclonal antibody concentration increased with decreasing dilution rate as did the specific rate of antibody production (Figure 2b). This is typical of inversely growth associated production and is in contrast to the growth associated production kinetics observed in batch culture.

The results of the batch and chemostat experiments indicate that viable cell concentration and growth rate control each affect volumetric antibody productivity significantly. Single-stage chemostat culture results in either high viability and high growth rate or low viability and low growth rate while maximum productivity requires a high viable cell concentration and low growth rate. The feasibility of achieving this with two-stage chemostat culture was therefore explored.

TWO-STAGE CHEMOSTAT CULTURE

A two stage chemostat with no extra feed to the second stage was set up as shown in the Materials and Methods (Fig. A).

Cell Concentration. At the highest dilution rates nutrients remaining from stage one were utilised in stage two resulting in higher total cell concentration (Figure 3a), while at lower dilution rates nutrients were fully utilised in stage one resulting in decreased viable cell concentration in stage two (Figure 3b). Viable cell number was higher in stage two only at the highest dilution rate.

In a single-stage chemostat, when a culture is 100% viable then the growth rate, (u) is equal to the dilution rate, (D). However, if the culture is not 100% viable then the specific growth rate exceeds the dilution rate (because the viable cell population is responsible for generating the non-viable as well as the viable cells) according to the equation (Pirt, 1975):

$$u = D/B$$

where B is the fraction of the culture which is viable (Figure 3c).

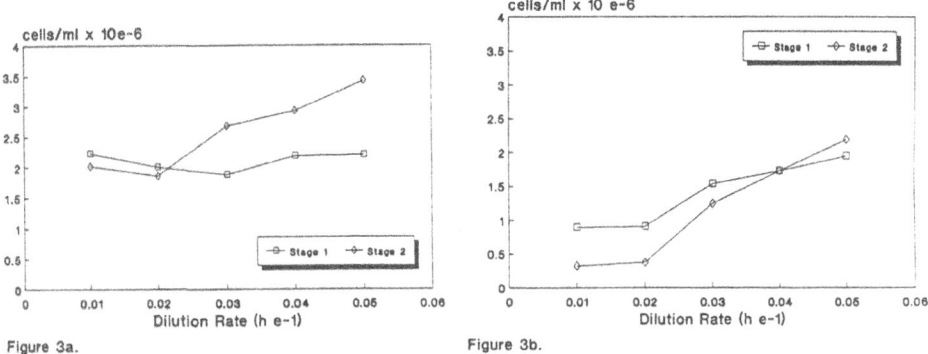

Figure 3a.　　　　　　　　　　　　　　　　　Figure 3b.

Two-Stage Chemostat Culture of NB1 Hybridoma Cells:
(3a) Total cell concentration in stage one compared to stage two
(3b) Viable cell concentration in stage one compared to stage two

Specific growth rate in the second stage of a two-stage chemostat (u_2) with no extra feed to the second stage are calculated according to the following equation (Pirt, 1975)

$$u_2 = D_2(X_2 - X_1)/X_2$$

where D_2 is the dilution rate in stage two; X_2 is the viable cell concentration in stage two and X_1 is the viable cell concentration in stage one.

Figure 3c shows a plot of growth rate calculated in this way and indicates that a positive growth rate is only observed in stage two at $D = 0.05$ h^{-1}, all other dilution rates showing a zero or negative growth rate.

However, although the viable cell population may show a negative growth rate there is no reason to suppose that individual cells are not dividing. This would typically be shown by an increase in the total cell concentration between stage one and stage two, unless the dead cells were being broken up into small fragments. At dilution rates of 0.05, 0.04, and 0.03 h^{-1} there was an increase in total cell concentration between stages one and two while only at D=0.05 h^{-1} was there an increase in viable cell concentration. At dilution rates of 0.02 and 0.01h^{-1} there was a decrease in total cell concentration.

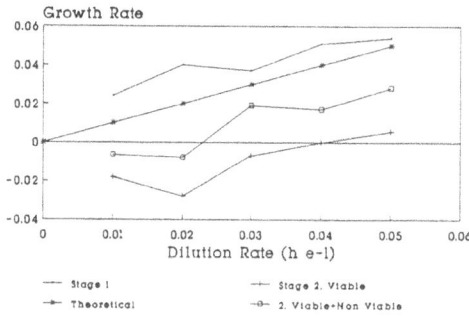

Growth Rates Observed in Two-Stage Chemostat Culture. Theoretical: assuming 100% viability u = D in stage-one. Stage 1: calculated accounting for the generation of non-viable cells. Stage 2 Viable: calculated for viable cell concentration only. 2 Viable + Non-Viable: calculated accounting for generation of non-viable cells.

Figure 3c.

The equation of Pirt (1975) has been adapted to account for the generation of non-viable cells as follows:

$$u_2 = D_2 (X_{t2} - Xt_1)/X_{v2}$$

where X_{t2} is the total cell concentration of stage two, Xt_1 is the total cell concentration of stage one, and X_{v2} is the viable cell concentration of stage two. This equation is analogous to D/B for a single-stage chemostat.Based on this calculation a more accurate representation of growth rate in stage two is obtained (Figure 3c) The calculation of growth rates in the second stage of two stage chemostat culture indicates that the cells may have a minimum growth rate in a two-stage system as observed in single stage culture both by us and by other workers (Tovey and Brouty-Boye, 1976; Boraston et al, 1984; Miller et al, 1988).

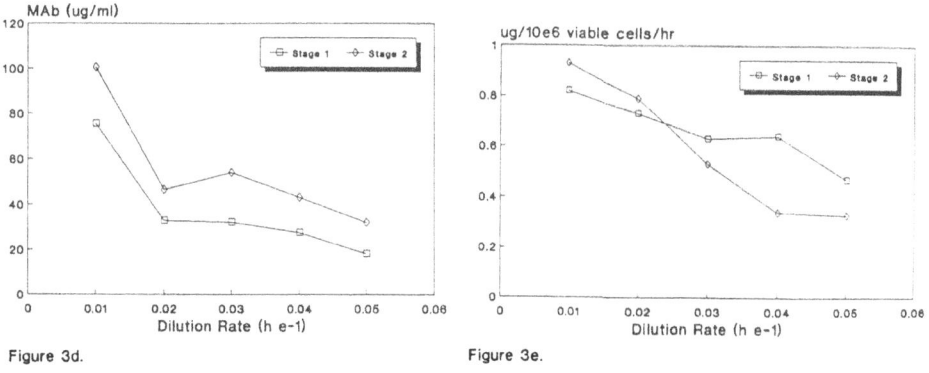

Figure 3d.   Figure 3e.

Two-Stage Chemostat Culture of NB1 Hybridoma Cells:
(3d) Total antibody produced in stage-one compared to stage-two
(3e) Specific rate of antibody production in stage-one compared to stage-two

Monoclonal Antibody Production. The specific rate of antibody production, as well as the volumetric titre (steady state concentration), increased with decreasing dilution rate in stage one and stage two (Figure 3d, and 3e).

At high dilution rates the specific production rate was greater in stage one, but at lower dilution rates was greater in stage two.

592

The results for specific production at high dilution rates were consistent with observations in glutamine depleted batch culture, that is, antibody production stopped when glutamine ran out, therefore we see reduced production in stage two where glutamine is depleted. However, the increase in production rate seen in stage two compared to stage one at a low dilution rate is not consistent with batch culture. This may be partly due to the fact that batch culture results in nutrient depletion, while chemostat culture results in nutrient limitation.

There are several other possible reasons why antibody production was greater in stage two at low dilution rates.

1. Antibody production was better able to compete with growth for common precursors or pathways because either a nutrient essential for growth but not for production was limiting; or toxins produced by the cells may be inhibiting growth but not antibody production. This is supported by the fact that at dilution rates of 0.02 and $0.01h^{-1}$ when the growth rate is negative specific antibody production is greater, indicating that when no cell growth is occuring production is increasing.

2. Cells increase production of some proteins when they are subject to stress; the cells are likely to be very stressed at low dilution rates which may result in greater antibody production rates.

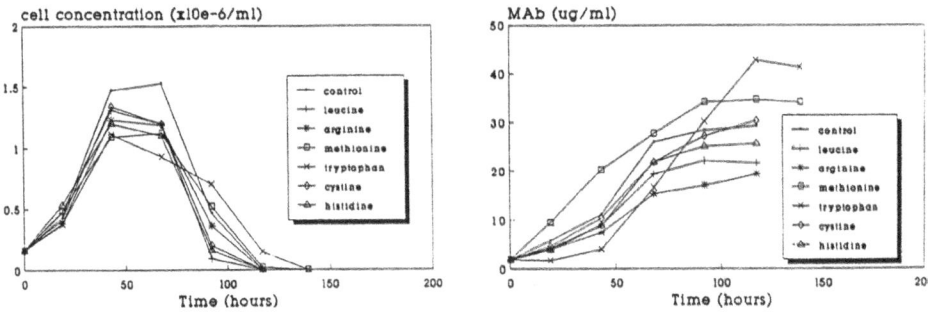

Figure 4a.   Figure 4b.

Batch Culture of NB1 Hybridoma Cells Supplemented with an Increased Concentration of Potentially Limiting Amino Acids: (4a) Viable cell concentration, (4b) Total antibody produced

Amino Acid Utilisation: From amino acid data for two-stage chemostat culture it would appear that glutamine was the growth limiting substrate in this set of experiments, however, with a reduction in dilution rate the amino acids methionine, cystine, tryptophan and leucine also became depleted. In non glutamine-pulsed batch culture no other amino acids apart from glutamine were depleted, however in the glutamine-pulsed batch culture methionine and cystine were.

Batch experiments were set-up in which depleted amino acids, chosen from the chemostat and batch results, were added in excess amounts (Figure 4a and 4b). None of the amino acids added had a positive effect on viable cell concentration, while tryptophan, methionine and cystine had a positive effect on final antibody titre. However, in batch culture it is not possible to discern whether the amino acids had a direct effect on antibody production, or caused an increase in production indirectly by inhibiting cell growth, or encouraging prolonged viability.

The results suggest that four interrelated factors are affecting the final antibody titre obtained, these are:

1. Maximum viable cell concentration
2. Growth rate
3. Maintenance of the viable cell concentration.
4. Specific rate of antibody synthesis.

The relationship between the addition of extra nutrients and productivity is being further studied in two-stage chemostat culture in which constant nutritional conditions and growth rate can be maintained, and growth and production can be optimised in two separate stages.

CONCLUSIONS

In batch culture antibody appears to be produced by the viable cell population in a growth associated manner, while in chemostat culture antibody production is inversely growth associated.

Cell growth appears to be limited by glutamine in single-stage chemostat culture and glutamine and several other amino acids in the second stage of a two-stage system. The specific rates of synthesis for the energy catabolites ammonia and lactate reflect the rates of glutamine and glucose utilisation respectively.

In batch culture glutamine appears to be an essential nutrient for both growth and antibody production, when glutamine runs out growth and production cease. In contrast to batch culture, glutamine depletion in two-stage chemostat culture does not appear to result in a decline in antibody production at low growth rates. It is possible that other medium components replace the glutamine requirement at low growth rates resulting in increased antibody production.

Death of a cell population in DMEM is due to nutrient depletion, not inhibitor accumulation, demonstrated by continued growth in batch culture when extra glutamine were added. However the drop in viability during extended stationary phases may be due to stress or inhibitor accumulation.

Growth and antibody production can be extended and increased in batch by supplementing the culture with extra glutamine and amino acids.

Two-stage chemostat culture has been shown to be a practical possibility in the culture of animal cells, and is a technique which can be further used in the study of factors involved in inversely growth associated antibody production.

ACKNOWLEDGEMENTS

David Venables was supported in his work by an SERC grant and Celltech Biologics Plc.
The assistance of Sarah Boork, Kabi Pharmacia, with amino acid analysis was much appreciated.

REFERENCES

Boraston R., Thompson P.W., Garland S. and Birch J.R. (1984) 'Growth and Oxygen Requirements of Antibody Producing Mouse Hybridoma Cells in Suspension Culture'. Develop. Biol. Standard. 55: 103-111.

594

Frame K.K. and Hu W-S. (1991) 'Kinetic Study of Hybridoma Cell Growth in Continuous Culture. 2. Behaviour of Producers and Comparison to Non-Producers'. Biotech. Bioeng. 38: 1020-1028.

Hiller, G.W., Aeschlimann, A.D., Clark, D.S., and Blanch, H.W. (1991) 'A Kinetic Analysis of Hybridoma Growth and Metabolism in Continuous Suspension Culture on Serum-Free Medium'. Biotech. Bioeng. 38: 733-741.

Low K.S., Harbour C., and Barford J.P. (1987) 'A Study of Hybridoma Cell Growth and Antibody Production Kinetics in Continuous Culture'.Biotech. Lett. 4: 239-244.

MacMichael G.J. (1989) 'The Use of Continuous Culture to Enhance Monoclonal Antibody Production'. Hybridoma 8: 117-126.

Merten O-W., Palfi G.E., Klement G. and Steindl F. (1987) 'Specific Kinetic Patterns of Production of Monoclonal Antibodies in Batch Cultures and Consequences on Fermentation Processes'. Spier and Griffiths (Eds) Modern Approaches to Animal Cell Technology p381-396.

Miller W.M.; Blanch H.W. and Wilke C.R. (1988) 'A Kinetic Analysis of Hybridoma Growth and Metabolism in Batch and Continuous Suspension Culture: Effect of Nutrient Concentration, Dilution Rate, and pH'. Biotech. Bioeng. 32: 947-965.

Pirt, S.J. (1975) Principles of Microbe and Cell Cultivation. Blackwell Scientific Publications, Oxford, U.K.

Ray N.G.; Karkare S.B. and Runstadler P.W. (1989) 'Cultivation of Hybridoma Cells in Continuous Cultures: Kinetics of Growth and Product Formation'. Biotech. Bioeng. 33: 724-730.

Renard, J.M., Spagnoli, R., Mazier, C., Salles, M.F., and Mandine, E. (1988) 'Evidence That Monoclonal Antibody Production Kinetics is Related to the Integral of the Viable Cell Curve in Batch Systems'.Biotech. Letts. 10: 91-96.

Reuveny S.; Velez D.; Macmillan J.D. and Miller L. (1986) 'Factors Affecting Cell Growth and Monoclonal Antibody Production in Stirred Reactors'. J. Immunol. Methods 86: 53-59.

Sacks S.H. and Lennox E.S. (1981) 'Monoclonal Anti-B as a New Blood-Typing Reagent'. Vox Sang.40: 99-104.

Tovey M. and Brouty-Boye D. (1976) 'Characteristics of the Chemostat Culture of Murine Leukemia L 1210 Cells'. Exp. Cell Res. 101: 346-354.

Tritsch, G.L. and Moore, G.E. (1962) 'Spontaneous Decomposition of Glutamine in Cell Culture Media'. Exp. Cell Res. 28: 360-364.

THE USE OF ARTIFICIAL INTELLIGENCE TO ENHANCE THE ACCURACY OF HYBRIDOMA GROWTH AND METABOLISM

P.C. FU and J.P. BARFORD
Department of Chemical engineering
Sydney University, NSW 2006
Australia

ABSTRACT. Structured models and detailed simulations can be utilized to provide useful insights into the complicated mechanisms in cell growth and metabolism. The integration of artificial intelligence is able to enhance the accuracy of the model predictions when kinetic information about cell metabolic networks is incomplete and imprecise. In this paper we have yielded a combination of numeric calculation and intelligent decision-making in the framework of hybrid modelling. It features a parallel arrangement of a numeric program in Fortran and decision module in Prolog, and an interactive information transfer process. Simulation result has been shown to demonstrate its advantage in accurate modelling.

INTRODUCTION

To describe hybridoma growth and metabolism, mathematical models can be developed on different levels and with differing degrees of complexity (1). Most of the models presented in the literature are unstructured, and unable to simulate experiments carried out at widely diverse operating conditions. It is therefore interesting to develop structured, kinetic models to analyze systematically the dynamics of the biochemical process.

Development of such a model depends on a good understanding of the rationale of cell metabolism, and successful identification of critical parameters and relationships. Due to the nonlinearity and simultaneous multicomponent nature of the pathway interactions, it is difficult to derive detailed and precise information for the model formulation. To solve this problem, new modelling skills are needed for the complex situations.

In this paper, a structured mathematical model for hybridoma growth and metabolism has been expanded to encompass the qualitative reasoning mechanism to treat the nonlinear and time-varying nature. A separate intelligent reasoning module was designed to process the parameter errors of the existing mathematical model, and to identify and compensate the variation of the model. The emphasis is placed on the solution of complex simulation problem by knowledge acquisition and utilization. The framework has proven to be able to produce more accurate model prediction.

S. Kaminogawa et al. (eds.), Animal Cell Technology: Basic & Applied Aspects, Vol. 5, 595–600.
© 1993 Kluwer Academic Publishers.

MODEL ARCHITECTURE AND FUNCTION

Formulation of the structured mathematical model for cellular metabolism in hybridoma growth have been discussed in detail in the literature (2,3). Here we concentrate mainly on the architecture and function of the decision-making module as shown in Figure 1. Domain knowledge for the judgement of the modelling discrepancy and the production of the compensation for the model parameters is encoded in a knowledge base, with the aid of an experimental database which stores actual observations from literature and our experiments. In order to interpret the error information in an appropriate way, we have developed an approach based on Order-of-Magnitude (denoted O[M]) reasoning (4). The O[M] method offers an intermediate abstraction level between qualitative and quantitative modelling. It can reason not only with directionality(i.e. +, 0 or -), but also with the relative magnitudes of variables as well (5).

Having received model prediction from the mathematical model equations in Fortran procedure, this Prolog program will access the experimental database for the records at the same time interval. Then the O[M]-relations, as illustrated in Tables 1 and 2, between the simulated and actual values of the bioprocess are used to translate the information into a qualitative representation. When the translation process is finished, the module will infer from the relationship between the numeric model variables and its parameters by invoking the knowledge base which consists of O[M]-relations, numeric information, algebraic constraints and if-then rules. As a result, decision of model parameter modifications will be established and transferred to the Fortran procedure for the model to make more precise predictions.

SIMULATION RESULT

The experimental data were from batch cultivation of hybridoma cell lines (6). The decision-making and module communication procedure can be summarized as follows:

> 1. Numeric module transfers computed values of the system variables to Decision module;
> 2. Comparing the model predictions with the measurements stored in the experimental database;
> 3. Organizing modelling variations by means of quantitative knowledge in Table 1 and other supplementary qualitative knowledge in knowledge base;
> 4. Reasoning from a set of first type of frames with O[M] relations and other encoded domain knowledge in knowledge base, obtaining the decision of compensation for the modelling disparity;
> 5. Sending the information for modification of the model parameters back to Numeric module.

Both the Fortran module and the Prolog module are installed in a time-shared minicomputer system DEC Micro VAX II in our department. There is a "soft switch" in the

Fortran program to turn "on" or "off" the involvement of the Prolog program. Therefore, we are able to select running the numeric module only, or running the hybrid model. Figures 2 to 4 show the comparisons of the experimental data, simulations without AI participation, and the model predictions with the knowledge utilization. It can be seen from the figures that the results of the hybrid model with certain degree of intelligence are closer to the experimental data and more adaptive to the changing environment, comparing to the outputs of the numeric model.

CONCLUSION

A hybrid model which is able to provide more accurate simulation of sophisticated, nonlinear and time-dependent hybridoma culture process has been developed. It introduced an artificial intelligence framework in the modelling and simulation of the biochemical process and has proven to be a real aid to solve the complex structured modelling problem. The methodology may be a useful alternative for the conventional programming technique in the treatment of modelling and control of biochemical systems. It needs to be pointed out that a well-organized knowledge base and an available experimental database are prerequisites for the successful application of the system.

REFERENCES

1.   Schügerl, K. (1985) 'Modelling of biotechnological processes', in the proceedings of IFAC Modelling and Control of Biotechnological processes, Noordwijkerhout, pp.13-31.

2.   Barford J.P., Phillips P.J. and Harbour C. (1992) 'Enhancement of productivity by yield improvement using simulation techniques'. in Animal Cell Technology: Basic and Applied Aspects. Murukami, H., Shirahata, S. and Tachibana, H. (eds.), Vol.4, Kluwer Academic Publishers, Dordrecht, pp.397--403.

3.   Barford J.P., Phillips P.J. and Harbour C. (1993) 'Simulation of animal cell metabolism', CYTOTECHNOLOGY (in press).

4.   Mavrovouniotis M.L. and Stephanopoulos G. (1988) 'Formal Order-of-Magnitude reasoning in process engineering', Comput. Chem. Eng., 12, pp.867--880.

5.   Quantrille, T.E. and Lin, Y.A. (1991) Artificial intelligence in chemical engineering, Academic Press, San Diego

6.   Koh B.T. (1991) personal communication

TABLE 1: DEFINITION OF ERROR STRENGTH LEVEL

VARIABLE	ERROR STRENGTH LEVEL				
	0	1	2	3	4
Glucose (mm/L)	< .5	.5 -- 1	1 -- 2	2 -- 4	>4
Lactate (mm/L)	< 1	1 -- 2	2 -- 3	3 -- 5	>5
Pyruvate (mm/L)	< .1	.1 -- .2	.2 -- .4	.4 --.6	>.6
Glutamine(mm/L)	< .2	.2 -- .4	.4 -- .6	.6 -- 1	>1
NH_3 (mm/L)	< .2	.2 -- .3	.3 -- .5	.5 -- 1	>1
Growthrate(1/H)	<.01	.01--.015	.015--.02	.02--.03	>.03
Total cell level (10^6 /L)	<100	100 -- 200	200 -- 400	400 -- 800	>800
Viable cell level (10^6 /L)	<50	50 -- 100	100 -- 150	150 -- 300	>300
MAb (mg/L)	<1	1 -- 2	2 -- 5	5 -- 8	>8

TABLE 2: SEMIQUANTITATIVE KNOWLEDGE OF THE MODELLING ERRORS

O[M]-RELATION	STRENGTH LEVEL	VERBAL EXPLANATION
r_1: S << M	-4	S is much smaller than M
r_2: S < M	-3	S is smaller than M
r_3: S -< M	-2	S is moderately smaller than M
r_4: S ~< M	-1	S is slightly smaller than M
r_5: S ~= M	0	S is generally equal to M
r_6: S >~ M	+1	S is slightly larger than M
r_7: S >- M	+2	S is moderately larger than M
r_8: S > M	+3	S is larger than M
r_9: S >> M	+4	S is much larger than M

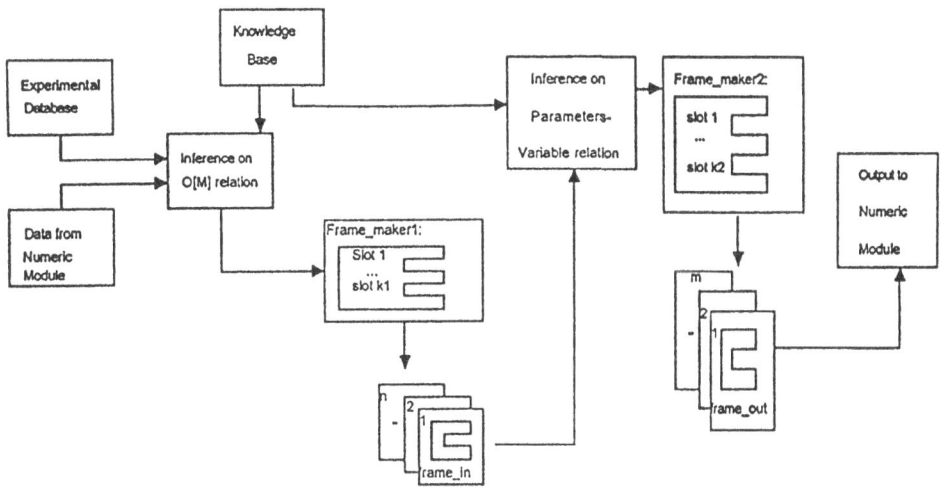

Figure 1: INFERENCE PROCESS WITHIN THE FRAMEWORK OF KNOWLEDGE MODULE

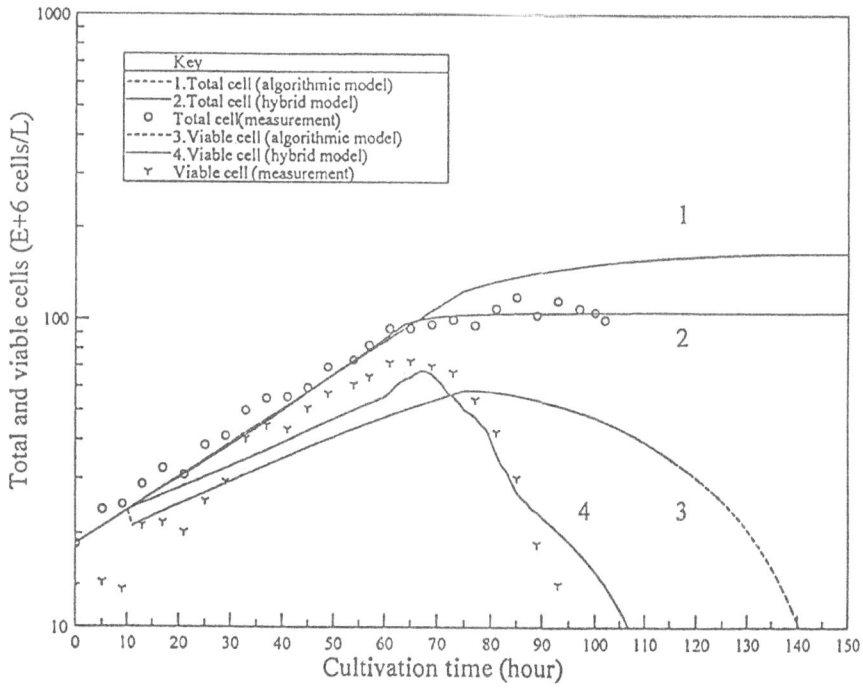

Figure 2: Total/Viable cell levels during batch culture

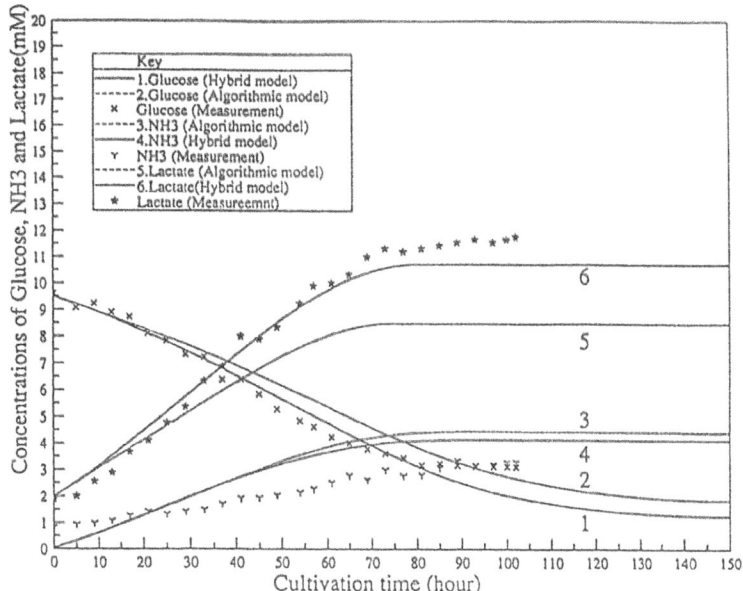

Figure 3: Substrate/product concentrations during batch culture

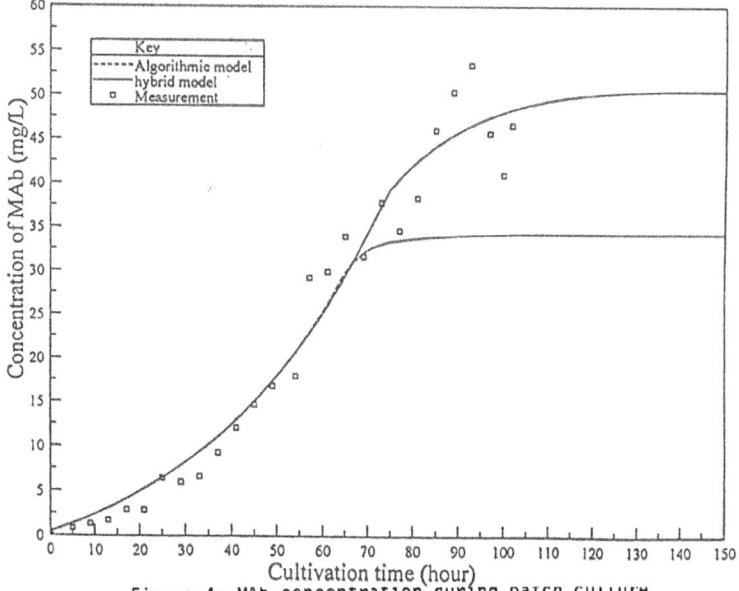

Figure 4: MAb concentration during batch culture

IMMUNOGLOBULIN PRODUCTION STIMULATING FACTOR (IPSF) DERIVED FROM NAMALWA CELLS

Takuya Sugahara,[1] Hiroto Nakajima,[1] Sanetaka Shirahata,[1]
Ken-ichi Nagamine,[2] and Hiroki Murakami[1]

[1] Graduate School of Genetic Resources Technology, Kyushu University,
6-10-1 Hakozaki, Higashi-ku, Fukuoka 812, Japan
[2] Nichirei Corporation, General Research Institute, Higashimurayama-shi,
Tokyo 189, Japan

ABSTRACT. Two immunoglobulin production stimulating factors (IPSFs) were found in human Burkitt's lymphoma Namalwa cells. These IPSFs were IPSF-IIα and IPSF-IIβ, respectively. They enhanced IgM production of human-human and mouse-mouse hybridoma cells under serum-free conditions. IPSF-IIα was purified from a Namalwa cell lysate and estimated to be a 112 kDa protein composed of one 40 kDa and two 36 kDa subunits. The active subunit of IPSF-IIα (a 36 kDa peptide) was identified as glyceraldehyde-3-phosphate dehydrogenase (GPD; EC 1.2.1.12) on the basis of the N-terminal amino acid sequence and enzymic activity. As a result of evaluating the of IPSF activity of GPD, the enzyme showed as much IPSF activity as IPSF-IIα.

The mode of action of GPD was then examined. Iodine-labeled GPD retained IPSF activity in spite of the complete loss of its enzymic activity. Therefore, the IPSF activity of this enzyme did not descend from its enzymic activity. Actinomycin D, a transcriptional inhibitor, could not suppress the enhancing effect of GPD on the antibody production of hybridomas. The enzyme stimulated the translation activity of a cell-free translation system made from HB4C5 cells. These results suggest that GPD stimulated monoclonal antibody production by enhancing the translation process of protein synthesis in hybridoma cells.

IPSF-IIβ was also purified and identified as enolase α-chain (EC 4.2.1.11), an enzyme in the glycolytic pathway. The IPSF activity of enolase was not derived from its enzymic activity, like GPD. Moreover, inhibition of the transcription process in hybridoma cells with actinomycin D did not suppress the IPSF activity of enolase. As the result of an investigation on the mode of action of enolase, it seems that enolase stimulated monoclonal antibody production by the same mechanism as that for GPD.

1. INTRODUCTION

There are three basic methods for producing monoclonal antibodies in large quantities. They are large-scale culture, high-density culture, and enhancemed cellular productivity. With the first two methods, the production of a large quantity of protein can be achieved as a consequence of culturing a large number of cells. To accomplish enhanced cellular productivity under serum-free conditions, we have searched for cellular protein factors which could enhance the productivity of the immunoglobulin of hybridomas. Several factors, named immunoglobulin production stimulating factor (IPSF), were found. IPSF-IIα and IPSF-IIβ were derived from the cell lysate of human Burkitt's lymphoma Namalwa cells. IPSF-IIα and IPSF-IIβ were

S. Kaminogawa et al. (eds.), Animal Cell Technology: Basic & Applied Aspects, Vol. 5, 601–607.
© 1993 Kluwer Academic Publishers.

purified and identified as glyceraldehyde-3-phosphate dehydrogenase (GPD) and human enolase α-chain, respectively [1,2,3]. Both IPSF-IIα and IPSF-IIβ enhanced the IgM production of both human-human and mouse-mouse hybridomas in a serum-free medium. The hybridomas immediately started to enhance IgM production after adding these IPSFs to the serum-free medium, and the enhanced secretion of IgM was maintained for 72 h without any growth-stimulating activity for the hybridomas. We now report the modes of action of GPD and enolase on monoclonal antibody production.

2. MATERIALS AND METHODS

2-1.   Assay of IPSF activity

To assay IPSF activity, a human-human hybridoma line of HB4C5 cells was used. The HB4C5 cells were maintained in ITES-ERDF medium and inoculated at 1×10^5 cells/ml in a 96-well plate containing 200 μl of ITES-ERDF medium supplemented with GPD or enolase. The amount of secreted immunoglobulin (IgM) during 6h of cultivation was measured by ELISA.

2-2.   Assay of the enzymic activity of GPD

GPD catalyzes the conversion of glyceraldehyde-3-phosphate to 1,3-bisphosphoglycerate in the presence of NAD^+ and arsenate. The enzymic reaction could be quantitatively monitored by spectrophotometrically measuring the quantity of NADH produced at 340 nm.

2-3.   Cell-free translation

A cell-free translation system was made from the HB4C5 cell lysate. HB4C5 cells cultured in 5 % FCS/ERDF medium were collected and washed with ice-cold phosphate-buffered saline (PBS). The cells (5×10^8) were washed twice with ice-cold 146 mM NaCl/35 mM Tris (pH 7.6) and then suspended in 2 volumes of 10 mM KCl/1.5 mM magnesium acetate/10 mM Tris (pH 7.5). After 5 min, the cell suspension was homogenized in an ice-bath with a Teflon homogenizer for 2 or 3 min. A one-third volume of ice-cold 125 mM KCl/5 mM magnesium acetate/6 mM 2-mercaptoethanol/30 mM Tris at pH 7.5 was then added to the lysate. The suspension was centrifuged at 15,000 x g for 30 min, and the supernatant was used for cell-free translation. The HB4C5 cell-free translation mixture was composed of 10 mM ATP, 1 mM GTP, 2 ng/ml of creatinkinase, 50 mM creatinphosphate, 1mM amino acid mixture (methionine-free) and 100 μg/ml of the HB4C5 cell lysate. The translation reaction was initiated by adding ^{35}S-methionine and mRNA derived from HB4C5 with or without GPD at 30 °C. The translation reaction was terminated to assay the protein synthesis by adding 1 M NaOH/5 % H_2O_2. After incubating for 15 min, the protein was precipitated by 25 % trichloroacetic acid. The precipitate was collected by filtration through glass-fiber discs (Whatman GF/C), washed three times with 8% trichloroacetic acid and dried. The radioactivity on the filters was counted by using a toluene scintillant (PACKARD 2250CA TRI-CARB).

3. RESULTS

3-1.   Relationship between the enzymic activity and IPSF activity

Although iodine-labeled GPD completely lost its enzymic activity, the IPSF activity of the enzyme was not affected at all by the labeling with iodine (Fig. 1). These results make it clear that the IPSF activity of GPD was unrelated to the enzymic activity of the enzyme.

3-2.   Effect of an actinomycin D treatment of the hybridomas on IPSF activity

Human-human hybridoma line HB4C5 cells were treated with 4 µg/ml of actinomycin D for 1 h to suppress the transcription activity.   After this treatment, these cells were inoculated to examine the IPSF effect of GPD or enolase on the transcription-suppressed HB4C5 cells (Table 1).   The actinomycin D treatment caused a decline in the antibody productivity of HB4C5 cells, but GPD and enolase enhanced the relative antibody productivity of actinomycin D-treated HB4C5 cells as much as that of non-treated HB4C5 cells.   These results suggest that the antibody production-stimulating activity of both GPD and enolase was not derived from any enhancement of the transcription process for protein synthesis, and that the enzymes affected other processes involved in protein synthesis.

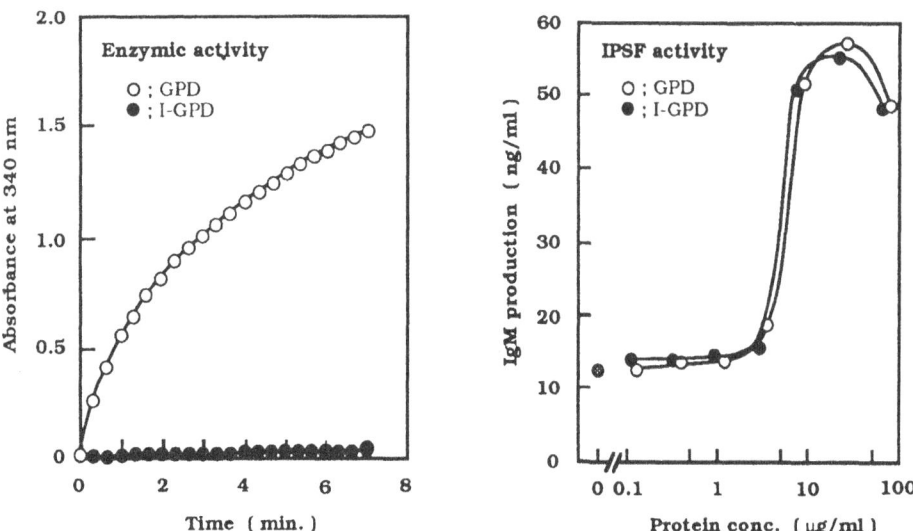

Fig. 1. Enzymic and IPSF activities of iodine-labeled GPD.

Table 1. Effect of actinomycin D treatment on IPSF activities of IPSFs.

Actinomycin D	Relative IgM production		
	None	GPD	Enolase
Non-treated	1.0	4.5	5.2
Treated	1.0	4.9	5.1

3-3. Effect of GPD on the cell-free translation system

Since it was found that GPD did not participate in the transcription process to enhance antibody synthesis, the effect of the enzyme on the translation process was examined. GPD was added to the cell-free translation system derived from HB4C5 cells. The addition of 40 μg/ml of the enzyme stimulated the incorporation of ^{35}S-methionine into protein (Fig. 2). The effect of GPD was particularly apparent at the beginning of the reaction. These results support the possibility that the enzyme stimulated antibody production by enhancing the translation process of protein synthesis.

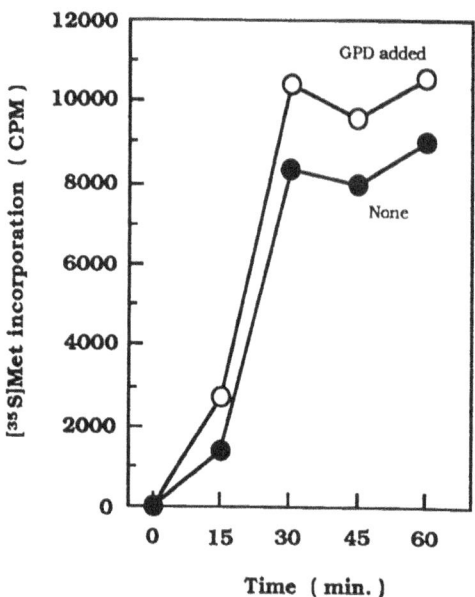

Fig. 2. Effect of GPD on HB4C5 cell-free translation system. The mRNA extracted from HB4C5 cells was added to the *in vitro* translation system using HB4C5 cell lysate. Translation activity was assessed by determination of the incorporation of ^{35}S-Met into translated protein.

4. DISCUSSION

We have previously reported that immunoglobulin production stimulating factor IIα (IPSF-IIα) was purified from an extract of human Burkitt's lymphoma Namalwa cells, and that IPSF-IIα was composed of two 36 kDa and one 40 kDa subunits. The 36 kDa subunit, which exclusively retained IPSF activity, was identified as glyceraldehyde-3-phosphate dehydrogenase, a key enzyme in the glycolytic pathway. This enzyme from other origins stimulated IgM production of human-human and mouse-mouse hybridomas under serum-free conditions.

The mode of action of the enzyme as IPSF-IIα was then investigated. The enzyme completely lost its enzymic activity with the iodine-binding reaction, but retained as much IPSF activity as the native enzyme. These results prove that the IPSF activity of the enzyme was irrelevant to its enzymic activity or its reaction products, and that it was derived from another function of the enzyme.

The IPSF effect of the enzyme was observed as soon as the enzyme was added to the medium. Enhanced IgM production was recognized within 1 h after an inoculation, and the enhanced productivity was maintained for 3 days until the condition of the hybridomas became worse due to deficient nutrition and the accumulation of cytotoxic metabolites. GPD must have affected the antibody productivity of the hybridomas, because the effect was not accompanied by cell growth of the hybridomas. These results suggest that GPD acted on the protein synthesis of the cell. To investigate the participation of the enzyme in protein synthesis, hybridoma line HB4C5 was acted on by the enzyme after treating HB4C5 with actinomycin D. The enzyme stimulated the IgM production of HB4C5 whose transcription was suppressed with actinomycin D. If the enzyme acted on the transcription process to enhance antibody production, the IPSF effect of GPD on transcription-suppressed hybridomas should not have observed. In addition, the immediate IPSF effect of the enzyme proves that the effect was not derived from stabilizing mRNA for the antibody peptide. Therefore, these results support the assumption that the enzyme stimulated the translation of peptides.

GPD was added to a rabbit reticulocyte cell-free translation system, translation of the antibody peptide being enhanced by the addition of GPD (data not shown). A cell-free translation system derived from the HB4C5 lysate enhanced the incorporation of ^{35}S-methionine into the protein translated by GPD. This stimulating effect of the enzyme on cell-free translation was much lower than that on the productivity of the cells. It can be inferred that there are several reasons for the lower activity in cell-free translation. One reason is derived from essential features of a cell-free translation system, which is inferior to in vivo translation and has a limit to its translation activity. It is expected that these features of cell-free translation may have suppressed the IPSF effect of the enzyme. The other expected reason is that the enzyme may be a precursor of the IPSF active substance, and that a change is necessary to act as IPSF. It is assumed that GPD is an inactive or low-activity form that would be activated during incorporation by the hybridomas.

GPD has been studied by many workers for its associating property with cytoplasmic membrane proteins and cellular organelles such as the erythroid cell membrane [4,5,6,7], plasma membrane of the intact human red blood cell [8], microtubule of a human colon tumor cell line [9], the rod outer segment of the retinal rod photoreceptor cell [10], and calsequestrin of skeletal muscle [11]. The enzyme has a specific affinity for highly acidic regions of certain proteins, and the association between the enzyme and these organelles is inhibited by physiological ionic strength [5,11]. We, however, investigated the possibility of receptors for the enzyme on the surface of the hybridoma, but no evidence for this was obtained (data not shown). This result presumes that glyceraldehyde-3-phosphate dehydrogenase would associate with the hybridoma in the same way as with cytoplasmic membranes or cellular organelles. According to this presumption, the decrease in the amount of enzyme associated with the hybridomas must have been caused by the high ionic strength of the ERDF medium. GPD has other properties besides those already mentioned. The enzyme associates with single-strand DNA [12], mRNA [13] and mono- and polyribosomes [14]. Moreover, the enzyme is one of the three major RNA-binding proteins of rabbit reticulocytes [13]. We assume that these features of this enzyme are concerned with its IPSF activity, because some DNA-binding proteins such as histone H1, H2A and H2B have IPSF activity. In addition, lactate dehydrogenase, which associates with single-strand DNA [15] also has IPSF activity.

We found another IPSF, IPSF-IIβ, in Namalwa cells. After purification, IPSF-IIβ was estimated to be a monomeric peptide of 46 kDa by gel filtration and SDS gel electrophoresis [3]. The partial amino acid sequence for 26 residues of the 46 kDa peptide completely coincide with human enolase α-chain, an enzyme in the glycolytic pathway similar to GPD. It is expected that enolase would act as IPSF in the same way as GPD, because the IPSF activity of enolase by an actinomycin D treatment of hybridomas. Both GPD and enolase are identified as heat-shock proteins of *Xenopus laevis* and yeast, respectively [16,17]. This common feature may be concerned with the IPSF activity of these enzymes. Uncovering the mechanism for the action of these IPSFs will contribute to the enhanced cellular productivity of the MAbs of hybridomas.

5. REFERENCES

[1] Sugahara, T., Shirahata, S., Yamada, K., and Murakami, H. (1991) 'Purification of immunoglobulin production stimulating factor-IIα derived from Namalwa cells,' *Cytotechnology* **5**, 255-263.

[2] Sugahara, T., Shirahata, S., Akiyoshi, K., Isobe, T., Okuyama, T., and Murakami, H. (1991) 'Immunoglobulin production stimulating factor-IIα (IPSF-IIα) is glyceraldehyde-3-phosphate dehydrogenase like protein,' *Cytotechnology* **6**, 115-120.

[3] Sugahara, T., Nakajima, H., Shirahata, H., and Murakami, H. (1992) 'Purification and characterization of immunoglobulin production stimulating factor-IIβ derived from Namalwa cells,' *Cytotechnology* (in press).

[4] Allen, R. W., Trach, K. A., and Hoch, J. A. (1987) 'Identification of the 37-kDa protein displaying a variable interaction with the erythroid cell membrane as glyceraldehyde-3-phosphate dehydrogenase,' *J. Biol. Chem.* **262**, 649-653.

[5] Kliman, H. J. and Steck, T. L. (1980) 'Association of glyceraldehyde-3-phosphate dehydrogenase with the human red cell membrane,' *J. Biol.. Chem.* **255**, 6314-6321.

[6] Kant, J. A. and Steck, T. L. (1973) 'Specificity in the association of glyceraldehyde 3-phosphate dehydrogenase with isolated human erythrocyte membranes,' *J. Biol. Chem.* **248**, 8457-8464.

[7] Shin, B. C., and Carraway, K. L. (1973) 'Association of glyceraldehyde 3-phosphate dehydrogenase with the human erythrocyte membrane,' *J. Biol.,Chem.* **248**, 1436-1444.

[8] Rogaliski, A. A., Steck, T. L., and Waseem, A. (1989) 'Association of glyceraldehyde-3-phosphate dehydrogenase with the plasma membrane of the intact human red blood cell,' *J. Biol. Chem.* **264**, 6438-6446.

[9] Launay, J. F., Jellali, A., and Vantier, M. T. (1989) 'Glyceraldehyde-3 phosphate dehydrogenase is a microtubule binding protein in a human colon tumor cell line,' *Biochimica et Biophysica Acta* **996**, 103-109.

[10] Hsu, S. C. and Molday, R. S. (1990) 'Glyceraldehyde-3-phosphate dehydrogenase is a major protein associated with the plasma membrane of retinal photoreceptor outer segments,' *J. Biol. Chem.* **265**, 13308-13313.

[11] Caswell, A. H., and Corbett, A. M. (1985) 'Interaction of glyceraldehyede-3-phosphate ehydrogenase with isolated microsomal subfractions of skeletal muscle,' *J. Biol. Chem.* **260**, 6892-6898.

[12] Perucho, M., Salas, J., and Salas, M. L. (1977) 'Identification of the mammalian DNA-binding protein P8 as glyceraldehyde-3-phosphate dehydrogenase,' *Eur. J. Biochem.* **81**, 557-562.

[13] Ryazanov, A. G. (1985) 'Glyceraldehyde-3-phosphate dehydrogenase is one of the three major RNA-binding proteins of rabbit reticulocytes,' *FEBS letters* **192**, 131-134.

[14] Ryazanov, A. G., Ashmarina, L. I., and Muronetz, V. I. (1988) 'Association of glyceraldehyde-3-phosphate dehydrogenase with mono- and polyribosomes of rabbit reticulocytes,' *Eur. J. Biochem.* **171**, 301-305.

[15] Grosse, F., Nasheuer, H. P., Scholtissek, S., and Schomburg, U. (1986) 'Lactate
      dehydrogenase and glyceraldehyde-phosphate DNA-polymerase-α-primase
      complex,' *Eur. J. Biochem.* **160**, 459-467.
[16] Nickells, R. W., and Browder, L. W. (1988) 'A role for glyceraldehyde-3-phosphate
      dehydrogenase in the development of thermotorerance in *Xenopus laevis*
      embryos,' *J. Cell Biology* **107**, 1901-1909.
[17] Iida, H., and Yahara, I. (1985) 'Yeast heat-shock protein of Mr 48,000 is an
      isoprotein of enolase,' *Nature* **315**, 688-690.

ONE-STEP PURIFICATION OF F(ab')₂ FRAGMENTS OF MOUSE MONOCLONAL ANTIBODIES (IMMUNOGLOBULINS G1 AND M) BY HYDROPHOBIC INTERACTION HPLC.

Kuniyo INOUYE(1) and Koichi MORIMOTO(2)
(1) Department of Food Science and Technology,
Faculty of Agriculture, Kyoto University, Kyoto
606, and (2) Biotechnology Research Laboratories,
TOSOH Corporation, Ayase, Kanagawa 252, Japan

ABSTRACT. Hydrophobic interaction HPLC, using TSKgels Phenyl-5PW or Ether-5PW, was applicable to one-step purification of F(ab')₂ fragments from pepsin digests of mouse monoclonal antibodies (MABs) of the IgG1 and IgM classes. The pepsin digests were applied to a gel equilibrated with phosphate-buffered saline containing 1M ammonium sulfate. F(ab')₂ fragments were adsorbed onto the gel, and eluted by reducing the ammonium sulfate concentration to zero. The fraction containing F(ab')₂ was homogeneous (>98% purity). The molecular sizes of the fragments from IgG1 and IgM were 110 and 144 kD, respectively. Each F(ab')₂ fragment fully maintained the immunoreactivity of the original MAb. The method shown here is convenient and suitable for the large-scale preparation of F(ab')₂ fragments.

INTRODUCTION

F(ab')₂ fragments of monoclonal antibodies (MAbs) are currently of great interest as both diagnostic and therapeutic agents [1,2]. They are more useful than the original MAbs, because they do not retain any biological effects due to Fc regions, and their interaction with non-specific proteins is less. Having a smaller molecular size than the original MAb is another advantage.

The preparation of F(ab')₂ fragments by pepsin digestion of mouse IgG1 MAbs has been reported [3,4]. The methods are time-consuming, and cannot afford sufficient purity and recovery. On the other hand, in the case of IgM MAbs, their size (approx. 900kD) limits the number of applications. In particular, for immunohistochemical studies, IgM cannot sufficiently penetrate tissues. In enzyme immunoassays, IgM has been generally avoided because of its low solubility and difficulty in handling.

609

S. Kaminogawa et al. (eds.), Animal Cell Technology: Basic & Applied Aspects, Vol. 5, 609–616.
© 1993 Kluwer Academic Publishers.

Therefore, suitable methods are required for producing active fragments of IgM MAbs. Although some papers have reported the fragmentation of IgM [5-8], the sizes of the fragments varied, and the method has not been well established.

We have recently reported a new method for preparing F(ab')₂ fragments from pepsin digests of mouse MAbs of the IgM and IgG1 classes by hydrophobic interaction HPLC [9,10]. In this paper, we describe the one-step purification and characterization of F(ab')₂ fragments of IgG1 and IgM MAbs.

MATERIALS AND METHODS

Ten IgG1 MABs and 12 IgM MAbs were used. The hybridomas secreting them were all established in our laboratory by fusing spleen cells from an antigen-immunized Balb/c mouse with NS-1 or SP/2-Ag14 myeloma cells [11]. The hybridoma cells were injected into pristane-primed Balb/c mice, and were grown in ascites. The ascites were collected and centrifuged at 3000 x g for 20 min to remove the cells, and the supernatants were passed through Millipore filters (0.8 μm pore size, AA type). Ammonium sulfate (solid) was added to each filtrate to give a 50% saturation for IgG1 and a 60% saturation for IgM, this being followed by centrifugation at 10,000 x g for 20 min after standing for 2 h. The precipitates were dissolved in phosphate-buffered saline (PBS, pH 7.4), and ammonium sulfate was added to the solution to give the requied 50% or 60% saturation. The precipitates of IgG1 and IgM were collected by centrifugation and dissolved in 0.1 M citrate buffer at pH 3.5 (buffer A) and at pH 4.2 (buffer B), respectively, giving partially-purified MAbs.

These partially-purified MAbs were then digested by porcine pepsin (Sigma, 3900 U/mg), the starting concentration of the IgG1 MAbs being 10-20 mg/ml in buffer A, and that of the IgM MAbs being 0.5-5 mg/ml in buffer B. Pepsin was added to the MAb solutions at weight ratios of 1:100 for IgG1 and of 1:200 for IgM. Digestion proceeded while gently stirring at 37°C for 2 h, and was stopped by adding 10 volumes of 3 M Tris to give a pH of around 7.

The HPLC apparatus consisted of a CCPM solvent-delivery system, a UV-8010 UY monitoring system, an FC-8000 fraction collector and an SC-8010 computer control system was from TOSOH (Tokyo). The elution was monitored by absorbance at 280 nm, and fractions of 1 ml were collected.

Hydrophobic interaction HPLC was performed in TSKgel Phenyl-5PW and Ether-5PW columns (7.5 mm innner diameter x 75 mm; TOSOH) for the pepsin digests of IgG1 and IgM MAbs, respectively. The pepsin-digested MAbs were salted out with 60% saturated ammonium sulfate, and the precipitates

were immediately dissolved in PBS containing 1 M ammonium
sulfate (pH 7.4). This solution was then applied to the
appropriate column equilibrated with the same buffer, and
eluted with a linear gradient of ammonium sulfate from 1 to
0 M in PBS over 30 min, which was generated 10 min after
the sample injection, at a flow-rate of 1 ml/min at room
temperature. Gel-filtration HPLC was performed in a TSKgel
G4000SWxL column (7.8 mm inner diameter x 30 cm) with PBS
(pH 7.4) at a flow-rate of 1 ml/min.

SDS-PAGE was done according to Laemmli, and protein
concentrations were determined by the Lowry-Folin method,
using BSA as the standard. Binding of F(ab')₂ and MAb to
human C1q complements [12] and their antigen-binding
activity [9,10] were measured by a solid-phase enzyme
immunoassay. Neutral sugars bound to proteins were
estimated by the phenol-sulfuric acid method.

RESULTS AND DISCUSSION
The time dependence for the pepsin digestion of MAbs
was monitored by SDS-PAGE. In the case of IgG1 MAbs, a
160-kD band corresponding to IgG1 disappeared gradually
during the reaction and had vanished completely after a 2-h
digestion. On the other hand, a 110-kD band, which is
considered to be F(ab')₂, appeared during the progress of
the reaction. Under reducing conditions, it was shown that
the heavy (H) chain (50 kD) disappeared in the reaction,
and a new band of 30 kD appeared. The light (L) chain
looked not to have been degraded. After digesting for 2 h,
the H-chain band disappeared completely and only 30- and
28-kD bands remained.

By SDS-PAGE under reducing conditions, IgM showed two
bands of 75 and 27 kD, corresponding to H (μ) and L (k)
chains, respectively. The H chain disappeared and a new
47-kD band appeared during the pepsin digestion. The L
chain was not digested. After a 2-h digestion, the H-chain
band disappeared completely, and only 47- and 27-kD bands
were observed. This new 47-kD band must have been derived
from cleavage of the H (μ) chain.

The pepsin digests of IgG1 and IgM MAbs were separated
by hydrophobic interaction HPLC, using TSKgels Phenyl-5PW
and Ether-5PW, respectively (Figs. 1 and 2). The pepsin
digests were obtained by digesting MAb CU204 (IgG1; anti-
carcinoembryonic antigen (CEA)) and MAb TS7M (IgM; anti-
human myoglobin) for 0, 15, and 120 min.

IgG1 MAb was eluted at 34 min, and as the digestion
progressed, the MAb peak decreased (Fig. 1). The peak had
disappeared completely after digesting for 2 h, and
inversely, a new peak corresponding to F(ab')₂ appeared at
30 min. Fractions from 28 to 31 min were collected for
further analyses. As shown in Fig. 2, IgM was eluted at

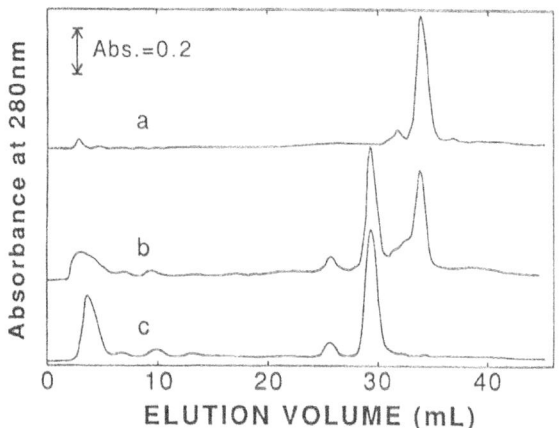

Fig. 1. Purification of the F(ab')₂ fragments from pepsin
digests of MAb CU204 (IgG1) by hydrophobic interaction HPLC
with a TSKgel Phenyl-5PW column. Time for pepsin
digestion: a, 0 min; b, 15 min; c, 120 min.

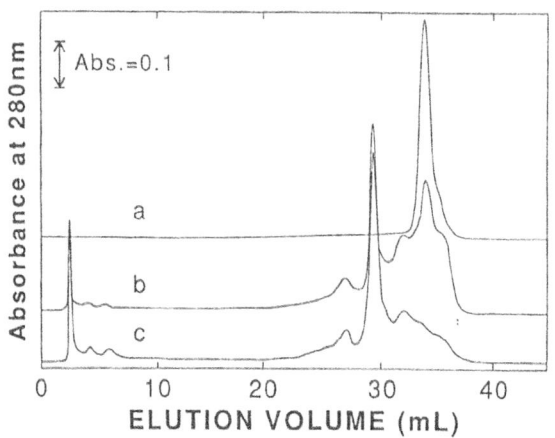

Fig. 2. Purification of the F(ab')₂ fragments from pepsin
digests of MAb TS7M (IgM) by hydrophobic interaction HPLC
with a TSKgel Ether-5PW column. Time of pepsin digestion:
a, 0 min; b, 15 min; c, 120 min.

33 min. With the progress of digestion, the peak
decreased, and a new peak of F(ab')₂ appeared at 29 min.
Fractions from 28 to 30 min were collected. The other IgG1
and IgM MAbs examined showed almost the same chromatograms
as those in Figs. 1 and 2, respectively.

The fractions of F(ab')₂ collected by hydrophobic interaction HPLC were then applied to gel-filtration HPLC. The proteins were eluted as a single peak, showing that the purity of the fragments was more than 98%. SDS-PAGE of the fractions showed a single band under non-reducing conditions, and only two bands under reducing conditions. F(ab')₂ (110 kD) prepared from IgGl is considered to have been composed of two sets of H (30 kD) and L (28 kD) chains. Similarly, F(ab')₂ (144 kD) from IgM is suggested to have been made of two sets of H (47 kD) and L (27 kD) chains.

When the total MAbs (160 kD IgGl; 900 kD IgM) were entirely converted to F(ab')₂ fragments, the weight ratio of the fragments to MAb should be 110/160=0.69 for IgGl and 144x5/900=0.80 for IgM. In case of TS7M (IgM), 70 mg of MAb was applied to pepsin digestion. Therefore, the theoretical quantity of F(ab')₂ was calculated to be 70x0.80=56 mg. As 32 mg of F(ab')₂ was obtained from hydrophobic interaction HPLC (Fig. 2), the recovered yield was 32/56=0.57 (57%). The recovered yields for IgGl MAbs were in the range of 42-58%, and those for IgM MAbs were 50-66%. The cycle time for the HPLC was 45 min, and in one cycle, F(ab')₂ of up to 2200 mg was purified from IgGl, and of up to 98 mg from IgM.

The immunoreactivity of MAb CU204 and its F(ab')₂ against specific antigen CEA is shown in Fig. 3. Each well of a microtiter plate was coated with 0.1 ml of a CEA solution at various concentrations in a 0.1 M carbonate buffer (pH 9.5). After blocking with BSA, 0.1 ml of 12 nM MAb or F(ab')₂ was added. Their immunoreactivity against CEA was in good agreement, suggesting that the immunoreactivity of MAb was fully maintained in the F(ab')₂. The immunoreactivity of MAb TS7M and its F(ab')₂ against human myoglobin is shown in Fig. 4. The slopes for MAb and F(ab')₂ are almost the same, indicating that the MAb immunoreactivity was fully recovered in F(ab')₂. The absorbance at 415 nm observed at the antigen concentration of zero was due to the non-specific binding of the peroxidase-labeled antibody to the plate and/or to the antibody coated onto the plate. The absorbance of non-specific binding for TS7M was 0.19, and that for F(ab')₂ was 0.01, the non-specific binding being greatly reduced by using F(ab')₂ in place of IgM MAb. This suggests that the accuracy of the immunoassay could be greatly improved by using F(ab')₂ instead of MAb.

All IgM MAbs showed binding activity to human Clq complements; however, every F(ab')₂ completely lacked this activity. The sugar content of each MAb (IgM) was in the range of 5.8-11.0%, and that of each F(ab')₂ was between 3.7 and 4.2%. In every case, the sugar content was decreased by fragmentation of IgM to F(ab')₂.

614

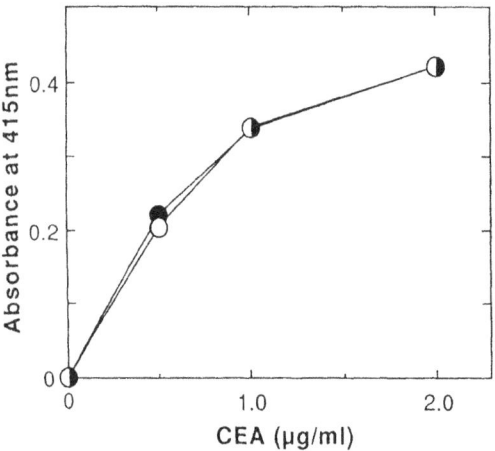

Fig. 3.   Immunoreactivity of MAb CU204 (O) and its F(ab')₂
fragment (●) against carcinoembryonic antigen (CEA).

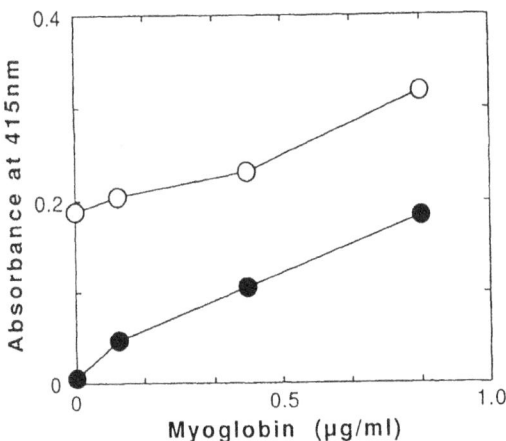

Fig. 4.   Immunoreactivity of MAb TS7M (O) and its F(ab')₂
fragment (●) against human myoglobin.

Although there have already been some trials in the
fragmentation of IgM [7,8,10], the proteolytic conditions
applied varied, the products were diverse and, even if the
products are referred to F(ab')₂ fragments, their molecular
sizes were rather different.   It has recently been shown
that pepsin digestion of IgM at low temperature (4°C)
produced F(ab')₂ of 134 kD [8].   The digestion was done for
24 h with a considerably high ratio of pepsin to MAb (1:25)
at pH 4.0.   We have shown here that digestion at 37°C and

pH 4.2 with a ratio of 1:200 for 2 h gave a homogeneous F(ab')₂ of 144 kD. With both methods, as the L chains remained intact, the cleaved position in the H (μ) chain may have been different according to the conditions for pepsin digestion.

Difficulty in the application of IgM has resulted in many efforts to select IgG-producing hybridomas and to screen even class-switch variants from IgM-producing hybridomas. Most of the MAbs so far developed against antigens containing sugars, such as cancer-related gangliosides, fall into the IgM class. In the present paper, a new and convenient method for preparing F(ab')₂ fragments from IgM and IgG1 MAbs is proposed. This method could develop new applications of MAbs, especially those in the IgM class.

REFERENCES

1. Mage, M. and Lamoyi, E. (1987) 'Preparation of F(ab) and F(ab')₂ fragments from monoclonal antibodies', in L. B. Schook (ed.), Monoclonal Antibody Production Techniques and Applications, Marcel Dekker, New York, pp. 79-97.
2. Goding, J. W. (1986) Monoclonal Antibodies: Principles and Practice, 2nd Edn., Academic Press, London, pp. 125-133.
3. Lamoyi, E. and Nisonoff, A. (1983) 'Preparation of F(ab')₂ fragments from mouse IgG of various subclasses', J. Immunol. Methods 56, 235-243.
4. Parham, P. (1983) 'On the fragmentation of monoclonal IgG1, IgG2a, and IgG2b from Balb/c mice', J. Immunol. 131, 2895-2902.
5. Beale, D. and Van Dort, T. (1982) 'A comparison of proteolytic fragmentation of immunoglobulin M from several different mammalian species', Comp. Biochem. Physiol. 71B, 475-482.
6. Johnstone, A. and Thorpe, R. (1987) Immunochemistry in Practice, 2nd Edn., Blackwell, Oxford, pp. 65-73.
7. Maillet, T., Roche, A.-C., Therain, F. and Monsigny, M. (1985) 'Time course localization of immunoglobulin M monoclonal antibody and its fragments in leukemic tumor-bearing mice', Cancer Immunol. Immunother. 19, 177-182.
8. Pascual, D. W. and Clem, L. W. (1992) 'Low temperature pepsin proteolysis. An effective procedure for mouse IgM (Fab')₂ fragment production', J. Immunol. Methods 146, 249-255.
9. Morimoto, K. and Inouye, K. (1992) 'Single-step purification of F(ab')₂ fragments of mouse monoclonal antibodies (immunoglobulins G1) by hydrophobic interaction high-performance liquid chromatography

using TSKgel Phenyl-5PW', J. Biochem. Biophys. Methods 24, 107-117.

10. Inouye, K. and Morimoto, K. (1993) 'Single-step purification of F(ab')$_{2\mu}$ fragments of mouse monoclonal antibodies (immunoglobulins M) by hydrophobic interaction high-performance liquid chromatography using TSKgel Ether-5PW', J. Biochem. Biophys. Methods, 26, 27-39.

11. Kohler, G. and Milstein, C. (1975) 'Continuous cultures of fused cells secreting antibody of predefined specificity', Nature 256, 495-497.

12. Hardy, R. R. (1986) 'Complement fixation by monoclonal antibody-antigen complexes', in D. W. Weir (ed.), Handbook of Experimental Immunology, Vol. 1, Blackwell, Oxford, pp. 40.1-40.12.

APPLICATION OF ANTI-TETANUS HUMAN MONOCLONAL ANTIBODIES

M. MATSUDA[1], M. KAMEI[2], N. SUGIMOTO[1] and S. HASHIZUME[2]
[1] Department of Tuberculosis Research , Research Institute for Microbial Diseases, Osaka University 3-1, Yamadaoka, Suita, Osaka, 565 Japan
[2] Morinaga Institute of Biological Science, 2-1-1 Shimosueyoshi, Tsurumi-ku, Yokohama, Kanagawa 230, Japan

ABSTRACT. To develop anti-tetanus human monoclonal antibodies (MAbs) applicable for clinical use, five hybrid cell lines (human lymphocytes-a mouse/human hetero-myeloma SHM-D33 or RF-S1) stably producing human MAbs with high toxin-neutralizing activity in large amounts in serum-free medium were established. The MAbs they produced were the IgG type, had very high affinity (Ka ca. 10^{11} M^{-1}) and were directed against one of the three domains [A], [B] or [C] of the tetanus toxin molecule. Their protective effects were studied in comparison with that of polyclonal antibody by observing the effects on progress of tetanus symptoms of mice. Mice injected im with toxin pretreated in vitro with the individual MAbs above sufficient doses survived, although at doses below the sufficient doses, delayed intoxication and progress of symptoms over 96 h and delayed death of the mice were observed. Injection iv of the individual MAbs or their mixtures at over doses equivalent to 0.03~0.1 IU per mouse protected from subsequent challenge with 20 MLD of toxin. Mice injected im with 4 MLD of toxin could be protected by iv injection of individual MAbs or mixture at doses equivalent to 0.03 IU per mouse, even 10 h after injection of toxin. Thus these human MAbs should be useful for prevention and treatment of human tetanus.

1. Introduction

Since the discovery of antitoxin therapy, horse polyclonal antitoxin serum preparations have long been used, and then homologous human polyclonal antitoxin antibody in the form of an immunoglobulin preparation is currently usually used for prevention and therapy of human tetanus. However, the supply of the human antitoxin preparations is limited and use of donor blood plasma as a source involves the risk of possible infection with viruses such as HIV. Therefore, hybrid cell lines that secrete human monoclonal antibody against tetanus in vitro should be new and better sources of antibody for clinical use.

If monoclonal antibodies (MAbs) are to be used clinically, at least all the following requirements must be satisfied for the hybrid cell lines obtained: (1) They must secrete human MAbs, and (2) produce antibodies stably, (3) their antibodies must have a high neutralizing activity, comparable with that of human polyclonal antitoxin currently used clinically, (4) they must produce at least the same amount of antibody in serum-free medium that they produce in serum containing medium and in sufficiently high amounts for large scale production in continuous culture and (5) the antibodies must be readily

S. Kaminogawa et al. (eds.), Animal Cell Technology: Basic & Applied Aspects, Vol. 5, 617–623.
© 1993 Kluwer Academic Publishers.

purified. We established five hybrid cell lines that meet all of these requirements for the first time (Kamei et al., 1990).

Here we described the preparation of the hybrid cell lines and further characteristics of these human MAbs and in detail the protective effects of the invidual MAbs and their mixtures by in vitro neutralizaiton test and by in vivo preventive and curative tests in mice, in comparison with those of anti-tetanus human polyclonal antibody. Here we also propose a reproducible method of evaluating actual protective activity of a MAb on the basis of characteristics of kinetic patterns of neutralization by the individual MAbs.

2. Material and Methods

2.1. FUSION

Hybrid cell lines G1, G2, G3, G4 and G6 producing anti-tetanus human MAbs were generated as described previously (Kamei et al., 1990). Briefly 50% polyethylene glycol was used for fusing peripheral blood lymphocytes from humans hyperimmunized with tetanus toxoid, with a mouse/human heteromyeloma SHM-D33 (ATCC CRL 1668, American Type Culture Collection, Rockville, Md.) or RF-S1, prepared from a human myeloma line RPMI 8226 and a mouse myeloma line FO (Kamei et al., 1990). 10 to 15 days after fusion, MAb-positive wells were screened by ELISA. Hybridomas secreting anti-tetanus human MAb were cloned at least three times by the limiting dilution method.

2.2. MONOCLONAL ANTIBODIES

MAbs were produced at 37 °C in serum-free RDF medium (Kamei et al., 1990) and purified by affinity chromatography on a Protein A-Cellulofine (Seikagaku Kogyo, Co., Ltd., Tokyo) column and ion-exchange chromatography on a Mono S column (Pharmacia LKB Biotechnology, Uppsala, Sweden) as described previously (Kamei et al., 1990).

2.3. TETANUS TOXIN AND ITS FRAGMENTS

Tetanus toxin was prepared and purified from culture filtrates (Matsuda and Yoneda, 1975) and fragments [A-B] and [C] were separated and purified from mildly papain-treated toxin as described previously (Ozutsumi et al., 1989), and used for immunoblotting with or without reduction with ß-mercaptoethanol to determine localization, in the tetanus toxin molecule, of epitopes to which the MAbs were directed.

2.4. BINDING AFFINITY

The binding affinities of the individual MAbs to purified tetanus toxin were estimated by a solid phase method using enzyme-linked immunosorbent assay (ELISA) and Ka was determined from the slopes of Scatchard plots (Kimoto, 1985).

2.5. IN VITRO NEUTRALIZATION

In vitro toxin-neutralization was performed by injecting im into OF1 mice (22~26 g) 0.5 ml of the mixture of the standard test toxin (Lot. TA-4B from the National Institute of Health, Tokyo) 20 MLD (0.1 ml) and the antibodies (0.4 ml) diluted $10^{0.5}$-fold with PBS containing 0.2% gelatin after incubating at 37 °C for 1 h.

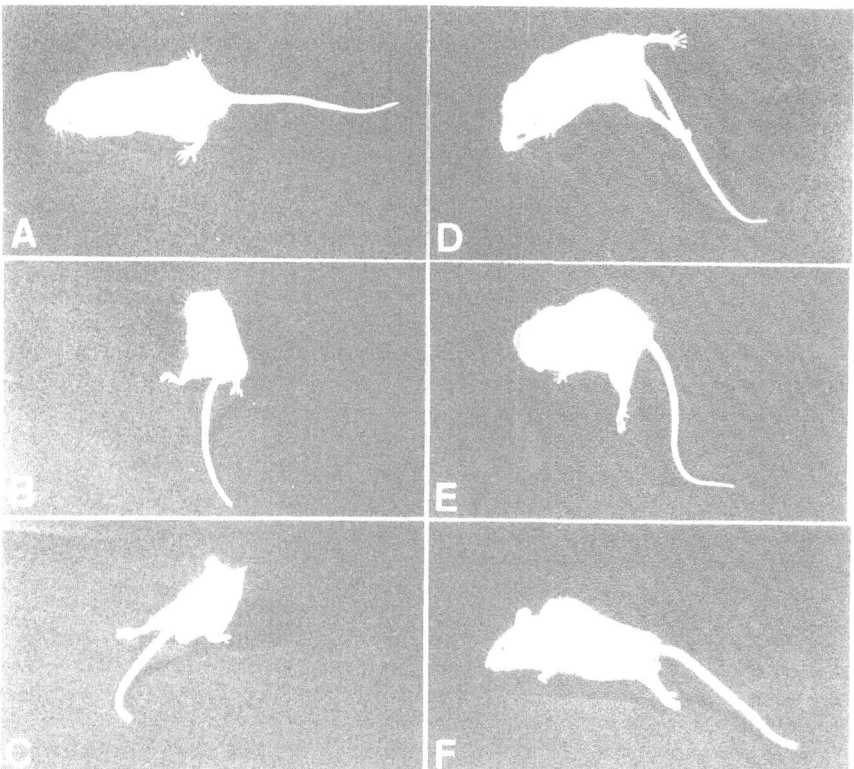

Figure 1. Scores of the symtpoms of tetanus-intoxicated mice. Scores: 0, no symptoms; 0.5, slight stiffness in injected limb barely visible only when the mouse was suspended by its tail; 1, obvious stiffness in injected limb visible only when the mouse was suspended by its tail; 1.5, slight limp of the injected limb, foot of the injected limb was slightly outward; 2, obvious limp, though injected limb was still used effectively for walking; 2.5, obvious limp, foot of the injected limb was ca. 90° outward from the sagittal line; 3, injected limb still movable, but non-functional, foot-pad of the injected limb was completely turned to the dorsal position, tail was near the sagittal line; 3.5, injected limb rigid and was ca. 45° outward from the sagittal line, tail stiff and bent slightly towards the injected limb, toes immovable; 4, injected limb was rigid and stretched ca. 90° outward from the sagittal line, tail stiff and bent notably to the injected limb; 4.5, injected limb rigid, animal became very irritable, slight stimulation readily induced either a convulsive run or convulsion, easily induced to supine position; 5, animal showed generalized convulsion, and was very ill; 5.5, animal often showed generalized convulsions, backbone notably elevated at the neck, eyes closed, body temperature was low, animal became motionless; 6, dead. For intermediate symptoms between two of those described above, scores of 0.25 were added to that of the lower grade of symptoms. A, Score 1; B, Score 2; C, Score 3; D, Score 3.5; E, Score 4 and F, Score 5.

2.6. IN VIVO PROTECTION

For examination of the preventive effect of the antibodies an antibody diluted $10^{0.5}$-fold
with PBS containing 0.2% gelatin was injected iv into mice and ca. 1 h later the mice
were challenged im with 20 MLD of the standard test toxin. For examination of the
curative effect of the antibodies, mice were treated im or iv with 4 MLD of tetanus toxin
and then after various times (0~30 h), 0.1 ml of the antibody solution containing 0.03 IU
equivalent per mouse was iv injected.

2.7. SCORES OF THE SYMPTOMS OF TETANUS-INTOXICATED MICE

To examine the protective effects, the symptoms of mice were observed and scored as
originally reported by van Heyningen and Mellanby (1971) for tetanus intoxicated mice
with a modification (Matsuda et al., 1982) as described in Figure. 1, from score 0, no
symptoms to score 6, death.

3. Results

3.1. PRODUCTION OF HUMAN ANTI-TETANUS MONOCLONAL ANTIBODIES

The hybridomas produced relatively large amounts of human anti-tetanus MAbs in the
culture supernatants. The amounts produced by the cells lines, G1, G2, G3, G4 and G6
were 12, 12, 11, 20 and 12 µg IgG/ml respectively. The production rates were 6.5, 6.4,
12, 23 and 8.4 pg/cell/day, respectivity. The hybrid cell lines grew as well in serum-free
medium as in medium containing serum and produced similar amounts (G2 for example
about 12 µg IgG/ml) of anti-tetanus antibody in 5 days. Cell line G2 produced MAb with
stirring culture in a 500-ml bioreactor: the concentration of MAb increased to a plateau
(about 7 µg IgG/ml in the spent medium) 6 days after the start of the culture with a
perfusion rate of 800 ml per day. Thus 5 to 6 mg of MAb per day in 500-ml bioreactor
was obtained.

TABLE1. Characterization of monoclonal antibodies

Antibody	Domain	Ig class		Affinity (Ka) $(\times 10^{10}$ M$^{-1})$
		H-chain	L-chain	
MAb-G4	[A]	γ	κ	14
MAb-G1	[B]	γ	λ	8.3
MAb-G3	[B]	γ	λ	2.9
MAb-G6	[B]	γ	κ	10
MAb-G2	[C]	γ	λ	11

3.2. CHARACTERIZATION OF MONOCLONAL ANTIBODIES

The results of the experiments on characterization of the MAbs are summarized in Table 1. They are human MAbs of the IgG1 type. Western blotting using purified toxin, fragments of the toxin and dissociated toxin showed that MAb-G1, G3 and G6 recognized fragment [B], G2 recognized fragment [C] and G4 recognized fragment [A] of the tetanus toxin molecule. The MAbs had very high affinity ca. $10^{11}M^{-1}$, hough MAb-G3 showed a slightly low affinity (Table 1). They were purified readily by affinity column chromatography on a Protein A-Cellulofine column and further by ion-exchange column chromatography on a Mono S column. They have high neutralizing (protective) activity as described below in detail.

3.3. PROTECTIVE EFFECTS

3.3.1. *In vitro Toxin-Neutralization* . Mice died about 26 h after injection of the test toxin (20 MLD) alone. Preincubation of the toxin with increasing amounts of the MAb before its injection resulted in a progressive decrease in the rate of progress of the symptoms in the case of MAb-G4 which recognizes fragment [A] or suppression of progress of the symptoms in the case of MAb-G3 which recognizes fragment [B] and in the case of MAb-G2 which recognizes fragment [C]. Decrease in the rate and suppression of progress of the symptoms were observed in the case of MAb-G6 which recognizes fragment [B]. Thus at lower doses than those sufficient to rescue mice, MAbs suppressed the development and/or slowed the rate of progress of symptoms for over 96 h and delayed sudden death of the mice occurred. Sufficiently high doses of all the individual MAbs rescued the mice from death from the toxin effect and resulted in final disappearance of all symptoms. Complete healing took about 50 days after the injection.

The kinetic patterns of progress of symptoms with the individual MAbs differed each other and different from that of anti-tetanus human polyclonal antibody. In sharp contrast to results with MAbs, administration of toxin pretreated with human polyclonal antibody, if lethal, resulted in death within about 96 h after its injection. A mixture of the MAbs had the same effect on progress of tetanus symptoms as polyclonal antibody.

3.3.2. *Prophylactic Effects.* Intravenous injection of individual MAbs or their mixtures at doses over 0.03~0.1 IU per mouse protected mice from subsequent challenge with 20 MLD of tetanus toxin.

3.3.3. *Curative Effects.* Mice could be rescued by iv injection of individual MAbs, at doses equivalent to 0.03 IU per mouse, even 10 h after im injection of 4 MLD of tetanus toxin. Similar results were obtained for all MAbs and mixture of MAbs also for human polyclonal antibody.

3.3.4. *A Reliable Method of Evaluating Actual Neutralizing Activity of a Monoclonal Antibody.* The results of the experiments on the method of evaluating neutralizing activity for a MAb are summarized in Table 2. The results show the reliability of the present method based on the chatacteristics of kinetic patterns of neutralization of toxin by the MAbs.

4. Discussion

There have been numerous reports on anti-tetanus human MAb (Simpson et al., 1990).

TABLE 2. Neutralizing activities of human anti-tetanus monoclonal antibodies

Antibody	Minimum survival * dose (µg IgG)	IU per 70 µg IgG	Ratio of titers ** (Routine/ Present method)
MAb-G4	2.8	0.3	1
MAb-G1	0.89	1	3
MAb-G3	2.8	0.3	3
MAb-G6	0.028	30	30
MAb-G2	0.089	10	9
Mixture of MAbs [G1,2,3,4,6]	0.028	30	1

*Minimum survival dose (MSD) against 20 MLD of toxin.
**Present method determines the neutralizing activities of the MAbs in terms of international units by comparing their MSD with that of the polyclonal antibody of known international units. The assay method routinely used for polyclonal antitoxin determines the neutralizing activity using the endpoint of 96 h.

However neutralizing activities so far reported on human anti-tetanus MAbs were all overestimated or only qualitatively estimated and the neutralizing antivities so far reported were low or unreliable in terms of complete protection. Human MAbs directed against domains [C] and [A-B] have been reported but not MAbs directed against domains [A] and [B].

The present results show that we obtained a set of hybridomas producing MAbs directed against the three domains [A], [B] and [C] respectively and that these MAbs meet all the requirements at least to be satisfied for practical application . The toxin-neutralizing activities of these MAbs estimated by the reliable method proposed in our study were very high and those of the mixture of appropriate combination of two or three kinds of the MAbs were comparable with the neutralizing activity of polyclonal IgG antitoxin preparation currently used clinically on the basis of toxin-specific IgG content. Therefore only 1/50~1/100 of IgG protein of that for polyclonal antitoxin is required to achieve the same neutralizing potency for the MAb preparation. The present results showed the complete protecion from tetanus by the MAbs contrarily to the pessimistic view with the murine MAbs that showed only partial neutralizing effect that resulted in delayed intoxication process and delayed death but no complete healing. With the MAbs reported here we could show the preventive and curative effects of tetanus antitoxin in vivo which have been controversial so far in treatment of human tetanus. Thus the above results clearly show that these MAbs should be useful for prevention and treatment of human tetanus, in place of the currently used human polyclonal antitoxin IgG preparations.

5. References

Kamei, M., Hashizume, S., Sugimoto, N., Ozutsumi, K. and Matsuda, M.(1990) 'Establishment of stable mouse/human-human hybrid cell lines producing large amounts of anti-tetanus human monoclonal antibodies with high neutralizing activity', Eur. J. Epidemiol. 6, 386-397.

Kimoto, H. (1985) 'A method for estimating the affinity of monoclonal antibodies a solid phase method', Saibo-kogaku (Cell Technology) 4, 792-796 (in Japanese).

Matsuda, M., Sugimoto, N. and Ozutsumi, K. (1982) 'Acute botulinum-like intoxication by tetanus toxin in mice and the localization of the acute toxicity in the N-terminal papain-fragment of the toxin', in Proceedings of the 6th International Conference on Tetanus, Fondation Marcel Merieux, Lyon, pp. 21-32.

Matsuda, M. and Yoneda, M. (1975) 'Isolation and purification of two antigenically active, "complementary" polypeptide fragments of tetanus neurotoxin,' Infect. Immun. 12, 1147-1153.

Ozutsumi, K., Lei, D-L., Sugimoto, N. and Matsuda, M. (1989) 'Isolation and purification by high performance liquid chromatography of a tetanus toxin fragment (Fragment [A-B]) derived from mildly papain-treated toxin,' Toxicon 27, 1055-1057.

van Heyningen, W.E. and Mellanby, J.H. (1971) 'Tetanus toxin.' in Microbial Toxins. Vol. 3, S. Kadis, T.C. Montie and S.J. Ajl (eds.), Academic Press, New York, pp. 69-108.

GENETICALLY ENGINEERED ANTIBODIES

James W. Larrick MD PhD and Robert Balint PhD
Palo Alto Institute of Molecular Medicine
2462 Wyandotte Street
Mountain View, CA 94043, USA
TEL: 415--694-4996
FAX: 415--851-3959

ABSTRACT. The development of PCR technology for rapid cloning and sequencing of antibody variable regions has provided the basis for recombinant antibody technology. Murine antibodies can be reshaped and the entire immunoglobulin repertoire can be immortalized. Recent advances using phage display and combinatorial approaches demonstrate that entirely synthetic monoclonal antibodies can be generated without *in vivo* immunization and cell fusion technology.

Human monoclonal antibodies (HuMABs) have enormous potential as human therapeutics (Larrick 1989, 1990). However, despite advances in the *in vitro* immunization of human B cells (Borrebaeck et al., 1988) and the development of immunodeficient mice (McCune et al., 1988) for the reconstitution of the human immune system *ex vivo*, immortalization of antigen-specific human B cells remains the limiting step in the generation of these molecules. Previously we (Larrick et al, 1992) and others (Masuho et al., 1990) have used Epstein-Barr virus transformation followed by subcloning, confirmation of antigen binding and hybridization of the B lymphoblasts to suitable fusion partners such as mouse-human heteromyelomas. This general approach is effective and widely used, however it is time consuming and erratic immortalization occurs. For this reason, we and others have devised methods to obtain directly the variable regions from small numbers of human B cells (Larrick et al, 1987).

1. Rapid direct cloning of antibody variable regions.

Recent developments suggest that recombinant DNA technology can replace cell fusion as a means of generating monoclonal antibodies (Chiswell and McCafferty, 1992). This type of immunoengineering technology has been made possible by gene amplification technology [i.e. polymerase chain reaction]. The concept is quite simple: a mixture of oligomer primers in the 5' leader sequences or framework I region combined with 3' constant region primers permits the amplification of any immunoglobulin variable region from very small numbers of

625

S. Kaminogawa et al. (eds.), Animal Cell Technology: Basic & Applied Aspects, Vol. 5, 625–639.
© 1993 *Kluwer Academic Publishers.*

cells (Larrick et al, 1989a, 1989b, Chiang et al., 1989). Primer design and other methodological details can be found in our previous papers (Larrick and Fry, 1991a, 1991b). The PCR fragments can be directly sequenced and/or ligated into expression vectors. The method has been used to obtain variable regions of both heavy and light chains from single human B lymphocytes. It is also possible to obtain sequences from individual B cells deposited by the fluorescence activated cell sorter (FACS) into microtiter wells containing carrier RNA and guanidinium isothiocyanate. Complementary DNA can be synthesized and amplified by PCR for sequencing. Thus the variable region genes of B cells can be obtained from *in vitro* antigen expanded cultures or from peripheral blood on a suitable day post-immunization. As described in detail below others have reported success using the PCR to obtain antibody variable regions for construction of rMABs (Orlandi et al, 1989, Roux et al, 1990) directly from libraries of phage combining heavy and light chains together artificially *in vitro*. Thus, the stage has been set for a new era of rapid progress in understanding and using antibodies.

2.   Genetically engineered chimaeric monoclonal antibodies.

Antibodies are composed of disulfide-linked heavy and light chains each comprised of variable and constant domains. It is thought that the most immunogenic portion of antibodies is the species conserved constant regions. For this reason several laboratories have used recombinant DNA technology to construct chimeric rodent-human monoclonal antibodies by attaching human constant regions to the rodent variable regions--see table 1 (Morrison et al., 1984; Boulianne et al., 1984), for reviews see Morrison and Oi, 1989). Because the antibody combining site is localized within the variable regions these molecules maintain their combining affinity for the antigen and acquire the function of the substituted constant regions (Steplewski et al. 1988; Bruggemann et al., 1987).

Although some laboratories have linked variable regions or Fab fragments by biochemical means to human Fc regions (Hamblin et al., 1987) most mouse monoclonal antibodies have been successfully chimerized using recombinant DNA technology (see Table 1). When therapeutic use was intended, most investigators have used the IgG1 constant region because of its serum half-life, and its capacity to fix complement and bind to Fc receptors. Details of the therapeutic chimeric antibodies summaried in table 1 can be found in Larrick and Fry, (1991a).

Summary of work with therapeutic chimeric MABs.

Many of the above antibodies are in various stages of pre-clinical development. How successful has this first generation of rMABs been? In all cases the chimeric MAB retained the antigen binding characteristics of the parental mouse MABs and in most cases the levels of expression were mid-range for hybridomas. In many cases the chimeric MAB had superior activity in ADCC and other functional activities using human effector cells. At the present very little is known about the immunogenicity of chimeric MABs, although evidence from rodents (using chimeric rodent MABs) indicates recipients can still recognize these molecules as non-self. Bruggeman et al. (1989) immunized mice with model xenogeneic (both the VH frameworks and the CH domains of human origin), chimeric (human VH frameworks only), or self MAB, and the antiantibody responses were dissected. Only the self MAB did not elicit an immune response. A strong response was elicited by the most xenogeneic MAB with approximately 90% against the constant domains

Table 1: Therapeutic Recombinant Chimaeric Antibodies

Targets	**Reference**
Tetanus	Larrick et al. (1992)
Hepatitis surface antigen	Li et al. (1990)
Human immunodeficiency virus antigen gp 120	Liou et al, (1989) Liu et al. (1987)
CD7 T lymphocyte antigen	Heinrich et al. (1989)
CD4	Centocor
Cancer antigens	
Carcinoembryonic antigen (CEA)	Beidler et al. 1989; Neumaier et al. 1990; Koga et al.1990; Hardman et al. 1989
Ganglioside GD2	Mueller et al. 1990
Common acute lymphocytic antigen	Nishimura et al. 1987; Saga et al. 1987; Yokoyama et al. 1987
Multiple drug transporter, P170	Hamada et al. 1990
Colorectal antigen 17-1A	Sun et al. 1987; Shaw et al. 1987; Fogler et al, 1989.
Melanoma (Nrml-05)	Marchitto et al, 1989
Tumor associated glycoproteins (eg. B72.6, L6)	Whittle et al. 1987; Liu et al. 1987
Ovarian cancer	Gallo et al. 1988
Transferrin receptor	Hoogenboom et al. 1990
Misc. cancer cells	Hank et al. 1990; Sahagan et al. 1986
BR96	Bristol Myers Squibb

and approximately 10% against the V domain. The anti-V response was not attenuated in the chimeric antibody, demonstrating that foreign VH frameworks can be sufficient to lead to a strong anti-antibody response. The magnitude of this xenogeneic anti-VH response was similar to that of the allotypic response elicited by immunizing mice of the Igha allotype with an Ighb antibody. Thus, although chimerization can diminish anti-antibody responses, there is reason to believe that chimeric MABs will be immunogenic in immunocompetent human patients. Recent data from LoBuglio et al. suggests that some V regions are immunogenic [?possess helper T cell epitopes?] whether they are in the original mouse MAB or chimerized or humanized.

3. Reshaped or composite antibodies.

The laboratory of Winter has pioneered a more sophisticated approach for construction of human antibodies from rodent monoclonals by splicing the rodent hypervariable--complementarity determining regions (CDRs) onto human variable framework sequences (Jones et al, 1986). Short of deriving a human monoclonal antibody from an immune human B cell this is about as "humanized" as a rodent monoclonal can become using rDNA technology (see table 2). This is feasible because the antibody combining site is constructed from several hypervariable regions held together to form the antigen binding cleft by a beta-sheet comprised of framework sequences. The first of these 'composite' monoclonals was constructed by grafting the CDRs from the heavy-chain variable region of mouse antibody B1-8, which binds the hapten NP-cap (4-hydroxy-3-nitrophenacetyl caproic acid; K_D = 1.2 uM), onto a human myeloma protein (Jones et al., 1986). In combination with the B1-8 mouse light chain, the new antibody acquired the hapten affinity of the B1-8 antibody (K_D = 1.9 uM). The affinity of a second composite MAB was less than the parent murine MAB (Verhoeyen et al., 1988).

The rat anti-CAMPATH-1 monoclonal (Reichmann et al., 1988) recognizes a glycoprotein (CDw52) expressed on virtually all human lymphocytes and monocytes, but is absent from the hematopoietic stem cells. Depletion of cells bearing this antigen appears to be an important therapeutic approach for control of graft-versus-host disease in bone marrow transplantation, prevention of bone marrow and other organ rejection episodes and for treatment of various lymphoid malignancies (Waldmann et al, 1988). The six hypervariable regions from the heavy and light-chain variable region domains of the rat antibody grafted onto the framework regions of a human IgG1 antibody yielded a 'reshaped' human monoclonal antibody with effector functions equal to (complement fixation) or better than (cell-mediated lysis of human lymphocytes) the parent CAMPATH-1 monoclonal. In the initial clinical trials this pioneer reshaped antibody eliminated large numbers of tumor cells, resulting in disease remission for patients with non-Hodgkin lymphoma. Significantly, there was no antiglobulin response in these patients (Hale et al, 1988).

The Mr 55,000 interleukin 2 receptor peptide (Tac; CD25) is not expressed by normal resting T-cells but is markedly up-regulated in adult T-cell leukemia and other malignancies, as well as on T-cells activated in normal immune, autoimmune, allograft, and graft-versus-host settings. Anti-Tac is a mouse monoclonal antibody directed against the Tac peptide which inhibits proliferation of T cells by blocking IL2 binding. Early attempts to use this MAB in humans for antitumor therapy and immune regulation were limited by weak recruitment of effector functions and neutralization by antibodies to mouse immunoglobulins. Queen et al. (Queen et al.,1989; Junghans et al., 1990) humanized the anti-Tac antibody using human framework and constant regions. The human framework regions were chosen to maximize homology with the anti-Tac antibody sequence. A computer model of murine anti-Tac was used to identify several amino acids which, while outside the CDRs, were likely to interact with the CDRs or antigen. These mouse amino acids were also retained in the humanized antibody. The composite anti-Tac antibody was shown to have an affinity for p55 of 3×10^9 M^{-1}, which is about 1/3 that of murine anti-Tac. Furthermore, the composite Tac rMAB blocked T-cell activation and facilitated ADCC with human effector cells.

Table 2: **Humanized murine MABs**

Concept	Reference
NIP	Jones et al., 1989
Lysozyme	Verhoeyen et al., 1988

Immunomodulatory

Campath-1 (CDW52)	Reichman et al., 1988
CD3 (OKT3)	Woodle et al., 1992
CD4	Gorman et al,, 1991
IL6	Tsuchiya et al., 1992
CD18 (60.3)	Hsiao et al., 1992
ICAM-1 (CD54)	Miglietta et al., 1992
IL2-receptor (Tac)	Queen et al., 1989; Junghans et al., 1990

Cancer

CEA	Gussow and Seemann, 1991
HER2 (p185)	Carter et al., 1992
17-1a	Centocor
B72-3	NIH
EGR receptor	Kettleborough et al., 1991
CD33	Co et al., 1992
Lewis Y (SDZ ABL 364)	Loibner et al. 1992

Anti-infectious

E5 (anti-LPS)	Xoma
Hepatitis	Scotgen
Respiratory syncytial virus	Tempest et al., 1991
Cytomegalovirus	Scotgen
Herpes simplex virus	Co et al., 1991
HIV	Maeda et al., 1991

Respiratory syncytial virus (RSV) is a major cause of acute respiratory morbidity and mortality particularly among young children. When Tempest et al.(1991)

directly transferred the CDRs of a neutralizing anti-RSV murine MAB into a human IgG1 framework binding activity was lost. Binding activity and neutralizing capacity were restored when murine amino acids 91-94 were used to replace the corresponding human framework amino acids.

Two major approaches to humanization have emerged from this work. The first case pioneered by Queen et al. at Protein Design Labs relies on choosing human framework regions most homologous to the murine sequences. Murine amino acids that contact the CDRs are also transferred into the human frameworks. In addition unusual amino acids in the human frameworks are replaced with consensus human amino acids. In an alternative approach taken by Tempest et al (1991) at Scotgen, a particular human framework is used as the basis to reshape all MABs.

It should be noted that even fully humanized murine MABs may be immunogenic. Although limited studies have demonstrated that chimeric mouse-human antibody 17-1A was less immunogenic in humans than the parent mouse monoclonal, more studies will be required to determine how much of a problem the human anti-idiotype response will be. In principle, the idiotype of a reshaped recombinant monoclonal could be changed by altering the CDRs or framework regions. However, grafting the CDRs into several cassettes might focus the immune response onto the combining site. This might be one method to potentiate development of effective anti-idiotype vaccines.

4.   Immortalization of the immunoglobulin repertoire using rDNA technology.

Ward et al. (1989) used PCR primers flanking the V regions to construct libraries of VH genes from spleen genomic DNA of mice immunized with either lysozyme or keyhole-limpet haemocyanin (KLH). From these libraries, VH domains were expressed and secreted from *E. coli*. Binding activities were detected against both antigens, and two VH domains were characterized with affinities for lysozyme in the 20 nM range. These isolated single domain antibodies were called "dAbs". The immortalization of an entire antibody repertoire laid the groundwork for an enormous technical advance, the construction of whole synthetic antibodies independent of hybridomas.

5.   Recombinatorial antibody libraries.

As noted above, the capacity of PCR to amplify essentially any V region permits the simultaneous amplification and subsequent cloning of an entire library of heavy and/or light chain variable regions. This revolutionary finding means that rMABs can be constructed without resort to hybridoma technology. Several groups have produced recombinant libraries in *E. coli* using phage. In one case heavy chains were amplified by PCR primers flanking the V regions and demonstrated to possess antigen binding activity in the absence of light chains (Gussow et al, 1989).

Huse et al. (1990) described a technique for the generation of recombinant libraries encoding the entire antibody repertoire. PCR primers flanking the variable regions are used to amplify variable regions combined with expression of F(ab) fragments in *E. coli*. Heavy and light chains can be expressed in separate vectors and recombined artificially *in vitro*. The recombinants release F(ab) fragments into the periplasmic space. Hence the recombinants can be screened directly for antigen-binding fragments in the same manner as a conventional lambda gtll library is screened with antibody.

Mullinax et al. (1990) were the first to apply this technology to a clinically relevant human antibody. They immunized volunteers with tetanus toxoid. mRNA was prepared from lymphocytes harvested six days post-immunization. The mRNA was converted to cDNA using light or heavy chain primers. PCR primers were then used to amplify immunoglobulin H or L chain sequences with sets of primers hybridizing to conserved leader sequences in the 5' ends and to the 3' end of the light chain (full length) or just 3' to the first cysteine codon in the hinge exon of the H chain. The product results in an Fd fragment of the IgG1 isotype with conservation of the H-L disulfide bond. These fragments were digested with different restriction enzymes and ligated into linearized Lambda Zap vectors. These vectors were constructed to have a ribosome binding site and a pelB leader sequence for secretion. The ligated recombinant phage DNA was then packaged. These vectors are constructed to permit co-ligation and generation of heavy-light chain recombinatorial libraries. Prior to immunization the frequency of B cells producing anti-tetanus antibody was <1/500,000. After immunization this rose to as high as 1/3000. In the library that was screened with radioiodine-labelled tetanus toxoid approximately 0.2% of the clones were positive. On further examination several of these showed an apparent affinity of 9×10^8 M^{-1} for tetanus. Improvements in this basic technology have been developed by Stratcyte Corporation, La Jolla, CA in the form of the Surf-ZAP vectors that combine the packaging function of lambda phage with phage display (see below).

The immunoexpression approach combined with *in vitro* recombination of heavy and light chains permits the generation of wholly synthetic antibodies. When the libraries are combined with expression systems a very high number of clones can be screened in a short period of time. Highly conserved antigens eg. human antibody fragments, autoantigens or tumor-associated self-antigens and antigens from toxic or dangerous organisms can be used to screen libraries to generate therapeutic human rMABs using this technology. In principle it should be possible to immortalize the entire antigen combining repertoire as well as many novel recombinants not present in B cells, i.e. heavy and light chains not normally associated in vivo. Libraries can be generated at different time points, from different lymphoid organs and after different immunization strategies. This important advance will facilitate studies of the antibody network and immunoglobulin development, the immune response, B lymphoma carcinogenesis etc.

6.   Phage Antibody Libraries: wholly synthetic MABs

Libraries in which antibodies are displayed on the surface of filamentous bacteriophage offer a number of important advantages over the *E. coli* expression libraries discussed above (see Table 3). Smith (1985) and coworkers (Parmley and Smith, 1988; Scott and Smith, 1990) pioneered the expression of peptide "epitope" libraries on the surface of fd phage by genetically engineering random peptides into the N-terminal domain of the phage gene III protein. Several copies of this protein located at the tip of the phage mediate its attachment to the *E. coli* F pilus, whereupon infection is initiated. Parmley and Smith (1988) showed that fusion of heterologous domains to the N-terminus of the gene III protein does not significantly impair its function, and furthermore, that such domains are accesible to exogenous ligands. Populations of phage expressing as many as 1×10^8 different epitopes can be generated and "panned" against immobilized ligands to enrich for desired binding specificities. Bound phage may be eluted, amplified in *E. coli*, and subjected to successive rounds of panning until maximum affinity epitopes have been identified. Such libraries have been used to identify MAB epitopes (eg., Cwirla et al., 1990), peptide mimics of non-peptide epitopes (Devlin et al., 1990; Oldenburg et al., 1992; Scott et al., 1992), alternative ligands for receptors (Scott and Smith, 1990), and peptide protease inhibitors (Roberts et al., 1992).

Recently, McCafferty et al. (1990) greatly enlarged the versatility of this technology by demonstrating that antibody fragments encoded at the N-terminus of the gene III protein could be displayed on the surface of fd phage with full epitope-binding activity. This discovery opened a novel route for the isolation of monoclonal antibodies. Repertoires of antibody V-region genes could be amplified by PCR, as discussed above, and cloned into fd or M13 phage to be expressed as gene III fusions, thus creating large libraries of phage, each displaying a specific antibody. By panning such libraries against the antigen of interest binding specificities as rare as one in 10^7 could be isolated. Repertoires from a variety of sources have been used with success. For example, spleens or peripheral blood lymphocytes (PBLs) from immunized or unimmunized donors may be used. Alternatively, specific V-region genes may be used as templates for artificially creating diversity by error-prone PCR or oligonucleotide-directed randomization. Antibody domains may be expressed as single-chain V-regions (ScFv) fused to the gene III protein, in which heavy and light chain V-regions are tethered together by a flexible linker, or as Fab fragments, in which the heavy chain Fd fragment is fused to the gene III protein (Hoogenboom et al., 1991). For the latter, both chains are expressed separately from the same cistron as signal peptide fusions directing them to the periplasmic space of *E. coli* where they typically accumulate in concentrations exceeding the association constant of the heterodimer, thereby allowing appropriate assembly of the two-chain antibody structure on the phage surface. V-region repertoire cloning has recently been improved by the introduction of phagemid vectors in place of the original phage genomic vectors (Kang et al., 1991). The greater transforming efficiency of phagemids permits the construction of larger libraries and their greater stability insures the production of monospecific phage.

Phage libraries expressing repertoires of antibody V-regions obtained from hyperimmunized mouse spleens have yielded monoclonal antibodies with affinities comparable to those of traditional MABs (Clackson et al., 1991), while phage libraries from unimmunized donors have yielded affinities in the sub-micromolar range, which is comparable to primary response

Table 3: Therapeutic Phage-derived MABs

Concept	Reference
NIP/pOX	Hoogenboom and Winter, 1992; Clackson et al, 1991;
Lysosome, BSA etc.	Marks et al., 1991
Blood group antigens (B,D,E etc.)	Hoogenboom et al, 1992

Immunomodulatory

C5a	Ames et al., 1992
Tumor necrosis factor	Hoogenboom et al., 1992

Cancer

CEA	Hoogenboom et al., 1992
Mucins	Hoogenboom et al., 1992
EGF receptor	Kettleborough et al., 1993

Anti-infectious

Tetanus toxin	Persson et al, 1991
Hepatitis B virus	Zebedee et al, 1992
Respiratory syncytial virus	Barbas et al., 1992a
HIV-1	Burton et al. 1991, Barbas et al., 1992b
Influenza	Caton and Koprowski, 1990

affinities (Marks et al., 1991). The technology has also been used to rescue antibody V regions from immunized Hu-SCID mice (Duchosal, 1992). Recently, Winter and coworkers (Marks et al., 1992) described the generation of antibodies with nanomolar affinities using phage libraries constructed from a non-immunized human repertoire, thus demonstrating the ability of phage technology to produce human antibodies with therapeutically useful affinities without immunization. They constructed a phage library from an unimmunized human PBL repertoire expressed in single chain form, from which they isolated a low-affinity antibody to a hapten. They then recombined the VH domain of this molecule with a VL repertoire from the same donor, and then isolated a higher affinity molecule from the resulting library. The increased affinities generated by light chain shuffling were attributed to decreased dissociation rate ('off' rate) rather than increased 'on' rates. This was accomplished by pre-loading phage with biotinylated antigen and then diluting into excess unlabelled antigen for varying times prior to capture on streptavidin-coated paramagnetic beads. The gene fragment encoding VL plus VH CDR3 from the highest affinity phage arising from light-chain shuffling was then recombined with a repertoire of VH minus CDR3 from the same donor, since

CDR3 usually makes the most extensive contacts with antigen. From this library they isolated 90 clones with higher affinities than the parent, the best of which had a Kd of 1.1 nM, 320-fold lower than that of the initial antibody. Thus, using chain shuffling alone, Winter and co-workers were able to mimic affinity maturation *in vitro*. However, in view of the known differences in immunogenicity between haptens and proteins, it remains to be shown that such an approach can produce high-affinity antibodies to therapeutic targets.

Random mutagenesis of V-regions *in vitro* has also been used to generate phage libraries from which improved affinities could be isolated. Winter and co-workers (Hawkins et al., 1992) achieved a four-fold improvement of a hapten-binding antibody by limited randomization (~1.7 bases per V_H) using error-prone PCR. Using oligonucleotide-directed randomization Barbas, et al. (1992) constructed a Fab phage library containing 5×10^7 heavy-chain CDR3 variants of a human anti-tetanus toxoid antibody. This library rivals in size the naive mouse repertoire, which can recognize a seemingly unlimited number of antigens. From this library they isolated fluorescein-binding Fabs with 100-1000-fold greater affinity than the parent antibody. Thus, from the successes that have been achieved thus far using phage selection technology along with chain shuffling and random mutagenesis to enhance natural antibody diversity, it is likely that therapeutic huMABs produced entirely *in vitro* from naive human repertoires. This should include huMABs directed against 'self' or other non-immunogenic antigens. Preliminary work in this direction has been reported by Hoogenboom and Winter (1992). Phage selection technology should also be able to facilitate humanization of therapeutically promising murine MABs as well as facilitate alterations of specificity and improvement of selectivity of other therapeutically promising MABs.

7. References

Ames RS, Tornetta MA, Tsui P. 1992. Isolation of anti-C5a monoclonal antibodies froma filamentous phage Fab display library. Proc. 3rd Intl. Conf. Antibody Engineering, San Diego, CA.

Barbas CF, Crowe JE Jr, Cababa D, Jones TM, Zebedee SL, Murphy BR, Chanock RM, Burton DR. 1992a. Human monoclonal Fab fragments derived from a combinatorial library bind to respiratory syncytial virus F glycoprotein and neutralize infectivity. Proc. Natl. Academy of Sci.(USA) 89:10164-8.

Barbas CF, Bjorling E, Chiodi F, Dunlop N, Cababa D, Jones TM, Zebedee SL, Persson MA, Nara PL, Norrby E et al. 1992b. Recombinant human Fab fragments neutralize human type 1 immunodeficiency virus *in vitro*. Proc. Natl. Acad. Sci. (USA) 89:9339-43.

Barbas CF, Bain JD, Hoekstra DM, Lerner RA (1992) Semisynthetic combinatorial antibody libraries: A chemical solution to the diversity problem. Proc. Natl. Acad. Sci. USA 89:4457-4461.

Beidler, C.B., J.R. Ludwig, J. Cardenas, J. Phelps, C.G. Papworth, E. Melcher, M. Sierzega, L.J. Myers, B.W. Unger, M. Fisher, et al. 1988. Cloning and high level expression of a chimeric antibody with specificity for human carcinoembryonic antigen. J Immunol 141:4053-60.

Borrebaeck, C.A.K., L. Danielsson, S.A. Moller. 1988. Human monoclonal antibodies produced by primary *in vitro* immunization of peripheral blood lymphocytes. Proc Natl Acad Sci (USA) 85:3995-3999.

Boulianne, G.L., N. Hozumi, M.J. Shulman. 1984. Production of functional chimaeric mouse/human antibody. Nature 312: 644-646.

Bruggemann, M., G. Winter, H. Waldmann, M.S. Neuberger. 1989. The immunogenicity of chimeric antibodies. J Exp Med 170:2153-7.

Burton DR, Barbas CF, Persson MAAA, Koenig S, Chanock RM, Lerner RA. 1991. A large array of human monoclonal antibodies to type 1 human immunodeficiency virus from combinatorial libraries of asymptomatic individuals. Proc. Natl. Acad. Sci. (USA) 88:10134-10139.

Carter P, Prestal L, Gorman CM, Ridgway JB, Henner D, Wong WL, Rowland AM, Kott C, Shepard HM (1992) Humanization of an anti-p185[HER2] antibody for human cancer therapy. Proc Natl Acad Sci (USA) 89:4285-4290.

Caton AJ, Koprowski H. Influenza virus hemagglutinin-specific antibodies isolated from a combinatorial expression library are closely related to the immune response of the donor. Proc. Natl. Acad. Sci. (USA) 87: 6450-6455.

Chiang, Y.L., R. Dong, and J.W. Larrick. 1989. Enzymatic amplification and direct cloning of rearranged immunoglobulin cDNA. Biotechniques 7:360-366.

Chiswell DJ, McCafferty J. 1992. Phage antibodies: Will new 'coliclonal' antibodies replace monoclonal antibodies. Trends Biotechnology 10:80-84.

Clackson T, Hoogenboom JR, Griffiths AD, Winter G (1991) Making antibody fragments using phage display libraries. Nature 352:624-628.

Co MS, Deschamps M, Whitley RJ, Queen C. 1991. Humanized antibodies for antiviral therapy. Proc Natl Acad Sci (USA) 88:2869-2873.

Co MS, Audalovic NM, Caron PC, Audalovic MV, Scheinberg DAO, Queen C. 1992. Chimaeric and humanized antibodies with specificity for the CD33 antigen. J Immunol 148:1149-1154.

Cwirla SW, Peters EA, Barrett RW, Dower WJ (1990) Peptides on phage: A vast library of peptides for identifying ligands. Proc. Natl. Acad. Sci. USA 87:6378-6382.

Devlin JJ, Panganiban LC, Devlin PE (1990) Random Peptide Libraries: A Source of Specific Protein Binding Molecules. Science 249:404-406.

Duchosal MA; Eming SA; Fischer P; Leturcq D; Barbas CF 3d; McConahey PJ; Caothien RH; Thornton GB; Dixon FJ; Burton DR. 1992. Immunization of hu-PBL-SCID mice and the rescue of human monoclonal Fab fragments through combinatorial libraries. Nature 355:258-62.

Fogler, W.E., L.K. Sun, M.R. Klinger, J. Ghrayeb, P.E. Daddona. 1989. Biological characterization of a chimeric mouse-human IgM antibody directed against the 17-1A antigen. Cancer Immunol Immunother 30:43-50.

Foote J, Winter G. 1992. Antiboy framework residues affecting the conformation of the hypervariable loops. J Mol Biol 224:487-499.

Gallo, M.G., V.K. Chaudhary, D.J. Fitzgerald, M.C. Willingham, I. Pastan. 1988. Cloning and expression of the H chain V region of antibody OVB3 that reacts with human ovarian cancer. J Immunol 141:1034-40.

Gascoigne, N.R., Goodnow, C.C., K.I. Dudzik, V.T. Oi, M.M. Davis. 1987. Secretion of a chimeric T-cell receptor-immunoglobulin protein. Proc Natl Acad Sci USA 84:2936-40.

Gorman SD, Clark MR, Routledge EG, Cobbold ST, Waldmann H. 1991. Reshaping a therapeutic CD4 antibody. Proc Natl Acad Sci (USA) 88:4181-4185.

Gussow, D., E.S. Ward, A.D. Griffiths, P.T. Jones, G. Winter. 1989. Generating binding activities from E. coli by expression of a repertoire of immunoglobulin variable domains. Cold Spring Harb Symp Quant Biol 54:265-72.

Gussow D, Seemann G. 1991. Humanization of monoclonal antibodies. Methods Enzymol. 203:99-121.

Hale, G., M.J.S. Dyer, M.R. Clark, J.M. Phillips, R. Marcus, L. Riechmann, G. Winter, H. Waldmann, et al. 1988. Remission induction in non-hodgkin lymphoma with reshaped human monoclonal antibody CAMPATH-1H. Lancet ii:1394-1399.

Hamada, H., K. Miura, K. Ariyoshi, Y. Heike, S. Sato, K. Kameyama, Y. Kurosawa, T. Tsuruo. 1990. Mouse-human chimeric antibody against the multidrug transporter P-glycoprotein. Cancer Res 50:3167-71.

Hamblin, T.J., A.R. Cattan, M.J. Glennie, M.R. MacKenzie, F.K. Stevenson, H.F. Watts, G.T. Stevenson. 1987. Initial experience in treating human lymphoma with a chimeric univalent derivative of monoclonal anti-idiotype antibody. Blood 69:790-7.

Hardman, N., L.L. Gill, R.F. De Winter, K. Wagner, M. Hollis, F. Businger, D. Ammaturo, F.

636

Buchegger, J.P. Mach, C. Heusser. 1989. Generation of a recombinant mouse-human chimaeric monoclonal antibody directed against human carcinoembryonic antigen.Int J Cancer 44:424-33.

Hawkins RE, Russell SJ, Winter G (1992) Selection of phage antibodies by binding affinity. Mimicking affinity maturation. J. Mol. Biol. 226:889-896.

Heinrich, G., H. Gram, H.P. Kocher, M.H. Schreier, B. Ryffel, A. Akbar, P.L. Amlot, G. Janossy. 1989. Characterization of a human T cell-specific chimeric antibody (CD7) with human constant and mouse variable regions. J Immunol 143:3589-97.

Hoogenboom, H.R., J.C. Raus, G. Volckaert. 1990. Cloning and expression of a chimeric antibody directed against the human transferrin receptor. J Immunol 144:3211-7.

Hoogenboom HR, Griffiths AD, Johnson KS, Chiswell DJ, Hudson P, Winter G (1991) Multi-subunit proteins on the surface of filamentous phage: methodologies for displaying antibidy (Fab) heavy and light chains. Nucl. Acids Res. 19:4133-4137

Hoogenboom HR, Marks JD, Griffiths AD, Winter G. 1992. Building antibodies from their genes. Immun Rev 130:41-68.

Hoogenboom HR, Winter G. 1992. By-passing immunisation. Human antibodies from synthetic repertoires of germline VH gene segments rearranged in vitro. J Mol Biol 227:381-8.

Hsiao Ku-Chuan, Bajorath J, Harris LJ. 1992 Humanization of anti-CD18 mAB 60.3. Proc. 3rd Intl. Conf. Antibody Engineering, San Diego, CA, Engineering, San Diego, CA.

Huse, W., L. Sastry, S.A. Iverson, A.S. Kang, M Alting-Mees, D.R. Burton, S. J. Benkovic, R.A. Lerner. 1989. Generation of a large combinatorial library of the immunoglobulin repertoire in phage lambda. Science 246:1275-1281.

Jones, P.T., P.H. Dear, J. Foote, M.S. Neuberger, G. Winter. 1986. Replacing the complementarity-determining regions in a human antibody with those from a mouse. Nature 321:522-525.

Junghans, R.P., T.A. Waldmann, N.F. Landolfi, N.M. Avdalovic, W.P. Schneider, C. Queen. 1990. Anti-Tac-H, a humanized antibody to the interleukin 2 receptor with new features for immunotherapy in malignant and immune disorders. Cancer Res 50:1495-502.

Kang AS, Barbas CF, Janda KD, Benkovic SJ, Lerner RA (1991) Linkage of recognition and replication functions by assembling combinatorial antibody Fab libraries along phage surfaces. Proc. Natl. Acad. Sci. USA 88:4363-4366

Kettleborough CA, Saldanha J, Heath VJ, Morrison CJ, Bendig MM. 1991. Humanization of a mouse monoclonal antibody by CDR-grafting: the importance of framework residues on loop conformation. Protein Eng 4:773-780.

Kettleborough CA, KH Ansell, RW Allen, D Gussow, MM Bendig. 1993. Use of pahge display libraries to isolate novel anti-EGF receptor antibodies. abstract in *Pharmaceutical Design using Epitope Selection Technologies*, Palo Alto, CA.

Koga, H., H. Kanda, M. Nakashima, Y. Watanabe, K. Endo, T. Watanabe. 1990. Mouse-human chimeric monoclonal antibody to carcinoembryonic antigen (CEA): *in vitro* and *in vivo* activities. Hybridoma 9:43-56.

Larrick JW, Chiang YL, Sheng-Dong R, Senyk G, Casali P. 1987. Generation of specific human monoclonal antibodies by *in vitro* expansion of human B cells. In: *International Symposium on In Vitro Immunization in Hybridoma Technology*. Tylösand, Sweden: Elsevier.p. 231.

Larrick, J. W., L. Danielson, C. Brenner, E. Wallace, M. Abrahamson, K.E. Fry, C. Borrebaeck. 1989a. Rapid direct cloning of rearranged immunoglobulin genes from small number of human hybridoma cells. Biotechnology. 7:934-938.

Larrick, J. W., L. Danielsson, C.A. Brenner, M. Abrahamson, K.E. Fry, C. Borrebaeck. 1989b. Cloning of rearranged immunoglobulin genes from small numbers of anti-HIV human hybridoma cells. Biochem. Biophys. Res. Comm. 160:1250-1256.

Larrick, J.W. 1989. Antibody inhibition of the Immunoinflammatory Cascade. J. Critical Care 4:211.

Larrick, J. 1990. Potential of monoclonal antibodies as pharmacological agents. Pharmacological Reviews 41:539-557.

Larrick J, Fry KE. 1991a. Recombinant antibodies. Human Antibodies Hybridomas 2:172-189.

Larrick JW and KE Fry. 1991b. PCR Amplification of Antibody Genes in RA Lerner and DR Burton (eds.) *METHODS--a companion to Methods in Enzymology Volume: New techniques in antibody generation.* 2:106-110.

Larrick JW, Wallace EF, Coloma MJ, Bruderer U, Lang AB, Fry KE. 1992. Therapeutic human antibodies derived from PCR amplification of B-cell variable regions. Immunol Rev 130:69-85.

Li, Y.W., D.K. Lawrie, P. Thammana, G.P. Moore, C.W. Shearman. 1990. Construction, expression and characterization of a murine/human chimeric antibody with specificity for hepatitis B surface antigen. Mol Immunol 27:303-11.

Liou, R.S., E.M. Rosen, M.S. Fung, W.N. Sun, C. Sun, W. Gordon, N.T. Chang, T.W. Chang. 1989. A chimeric mouse-human antibody that retains specificity for HIV gp120 and mediates the lysis of HIV-infected cells. J Immunol 143:3967-75.

Liu, A.Y., R.R. Robinson, K.E. Hellstrom, E.D. Murray Jr, C.P. Chang, I. Hellstrom. 1987. Chimeric mouse-human IgG1 antibody that can mediate lysis of cancer cells. Proc Natl Acad Sci (USA) 84:3439-43.

Loibner H, Baker J, Bednarik K, Janzek E, Neruda W, Plot R, Co MS. 1992. Generation and characterization of humanized anti-Lewis Y antibodies. Proc. 3rd Intl. Conf. Antibody

Maeda H, Matsushita S, Eda Y, Kimachi K, Tokiyshi SO, Bendig MM. 1991. Construction of reshaped human antibodies with HIV neutralizing activity. Hum Antib Hybrid 2:124-135.

Marchitto, K.S., W.R. Kindsvogel, P.L. Beaumier, S.K. Fine, T. Gilbert, S.D. Levin, C.S. Woodhouse, A.C. Morgan, Jr. 1989. Characterization of a human-mouse chimeric antibody reactive with a human melanoma associated antigen. Prog Clin Biol Res 288:101-5.

Marks JD, Griffiths AD, Malmquist M, Clackson TP, Bye JM, Winter G (1992) Bio/Technology 10:779-783.

Marks JD; Hoogenboom HR; Bonnert TP; McCafferty J; Griffiths AD; Winter G. 1991. By-passing immunization. Human antibodies from V-gene libraries displayed on phage. J. Mol Biol 222:581-97.

Marks JD, Hoogenboom HR, Bonnert TP, McCafferty J, Griffiths AD, Winter G (1991) By-passing immunization. Human antibodies from B-gene libraries displayed on phage. J. Mol. Biol. 222:581-597.

Masuho Y, Y-I Matsumoto, T. Sugano, T. Tomiyama, S. Sasaki, T. Koyama. 1990. Development of a human monoclonal antibody against cytomegalovirus with the aim of passive immunotherapy. in CAK Borrebaeck and JW Larrick (eds). *Therapeutic Monoclonal Antibodies.* Stockton Press. pp.187-207.

McCafferty, J., A.D. Griffiths, G. Winter, D.J. Chiswell. 1990. Phage antibodies: filamentous phage displaying antibody variable domains. Nature 348:552-554.

McCune, J.M., R. Namikawa, H. Kaneshima, et al. 1988. The SCID-hu mouse: Murine model for the analysis of human hematolymphoid differentiation and function. Science 241:1632-1639.

Miglietta J, Shrutkowski A, Farrell T, Kishimoto K, Brown M, Kehry M, Morrison S, Griffin J. 1992. Alteration of framework residues modulate binding of a CDR-grafted anti-human ICAM-1 (CD54). Proc. 3rd Intl. Conf. Antibody Engineering, San Diego, CA,

Morrison, S.L., M.J. Johnson, L.A. Herzenberg, and V.T. Oi. 1984. Chimeric human antibody molecules: Mouse antigen-binding domains with human constant region domains. Proc. Natl. Acad. Sci. (USA) 81:6851-6855.

Morrison, S.L. and V.T. Oi. 1989. Genetically Engineered Antibody Molecules. Adv. Immunol. 44:65-92.

Mueller, B.M., C.A. Romerdahl, S.D. Gillies, R.A. Reisfeld. 1990. Enhancement of antibody-dependent cytotoxicity with a chimeric anti-GD2 antibody. J Immunol 144:1382-6.

Mullinax, R.L., E.A. Gross, J.R. Amberg JR, B.N. Hay, H.N. Hogrefe, M.M. Kubitz, A. Greener, M. Alting-Mees, D. Ardourel, J.M. Short, J.A. Sorge, B. Shopes. 1990. Identification of human antibody fragment clones specific for tetanus toxoid in a bacteriophage γ immunoexpression library. Proc. Natl. Acad. Sci. (USA) 87:8095-8099.

Neumaier, M., L. Shively, F.S. Chen, F.J. Gaida, C. Ilgen, R.J. Paxton, J.E. Shively, A.D. Riggs. 1990. Cloning of the genes for T84.66, an antibody that has a high specificity and affinity for carcinoembryonic antigen, and expression of chimeric human/mouse T84.66 genes in myeloma and Chinese hamster ovary cells. Cancer Res 50:2128-34.

Nishimura, Y., M. Yokoyama, K. Araki, R. Ueda, A. Kudo, T. Watanabe. 1987. Recombinant

638

human-mouse chimeric monoclonal antibody specific for common acute lymphocytic leukemia antigen. Cancer Res 47:999-1005.

Oldenburg KR, Loganathan D, Goldstein IJ, Schultz PG, Gallop MA. Peptide ligands for a sugar-binding protein isolated from a random peptide library. Proc. Natl. Acad. Sci. USA 89:5393-5397

Orlandi, R., D.H. Gussow, P.T. Jones, G. Winter. 1989. Cloning immunoglobulin variable domains for expression by the polymerase chain reaction. Proc Natl Acad Sci (USA) 86:3833-7.

Parmley SF, Smith GP (1988) Antibody-selectable filamentous fd phage vectors: affinity purification of target genes. Gene 73:305-318.

Persson MA; Caothien RH; Burton DR. 1991. Generation of diverse high-affinity human monoclonal antibodies by repertoire cloning. Proc. Natl. Acad. Sci(USA) 88:2432-6.

Queen, C., W.P. Schneider, H.E. Selick, P.W. Payne, N.F. Landolfi, J.F. Duncan, N.M. Avdalovic, M. Levitt, R.P. Junghans, T.A. Waldmann. 1989. A humanized antibody that binds to the interleukin 2 receptor. Proc Natl Acad Sci USA 86:10029-33.

Riechmann, L., M. Clark, H. Waldmann, G. Winter. 1988. Reshaping human antibodies for therapy. Nature 332:323-327.

Roberts BL, Markland W, Ley AC, Kent RB, White DW, Guterman SK, Ladner RC Directed evolution of a protein: Selection of potent neutrophil elastase inhibitors displayed on M13 fusion phage. Proc. Natl. Acad. Sci. USA 89:2429-2433

Roux, K.H., P. Dhanarajan. 1990. A strategy for single site PCR amplification of dsDNA: priming digested cloned or genomic DNA from an anchor-modified restriction site and a short internal sequence. Biotechniques 8:48-57.

Saga, T, K Endo, M. Koizumi, Y. Kawamura, Y. Watanabe, J. Konishi, R. Ueda, Y. Nishimura, M. Yokoyama, T. Watanabe. 1990. In vitro and in vivo properties of human/mouse chimeric monoclonal antibody specific for common acute lymphocytic leukemia antigen. J Nucl Med 31:1077-83.

Sahagan BG, H Dorai, J. Saltzgaber-Muller, F. Toneguzzo, C.A. Guindon, S.P. Lilly, K.W. McDonald, D.V. Morrissey, B.A. Stone, G.L. Davis, et al. 1986. A genetically engineered murine/human chimeric antibody retains specificity for human tumor-associated antigen. J Immunol 137:1066-74.

Scott J.K. and G. P. Smith. 1990. Searching for peptide ligands with an epitope library. Science 249:386-90.

Scott JK, Loganathan D, Easley RB, Gong X, Goldstein IJ A family of concanavalin A-binding peptides from a hexapeptide epitope library. Proc. Natl. Acad. Sci. USA 89:5398-5402

Scott JK, Smith GP (1990) Searching for Peptide Ligands with an Epitope Library. Science 249:386-390

Shaw, D.R., M.B. Khazaeli, L.K. Sun, J. Ghrayeb, P.E. Daddona, S. McKinney, A.F. LoBuglio. 1987. Characterization of a mouse/human chimeric monoclonal antibody (17-1A) to a colon cancer tumor-associated antigen. J Immunol 138:4534-8.

Smith GP, (1985) Filamentous Fusion Phage: Novel Expression Vectors That Display Cloned Antigens on the Virion Surface. Science 228:1315-1317

Steplewski, Z., L.K. Sun, C.W. Shearman, et al. 1988. Biological activity of human-mouse IgG1, IgG2, IgG3 and IgG4 chimeric monoclonal antibodies with antitumor specificity. Proc Natl Acad Sci (USA) 85:4852-4856

Sun, L.K., P. Curtis, E. Rakowicz-Szulczynska, J. Ghrayeb, N. Chang, S.L. Morrison, H. Koprowski. 1987. Chimeric antibody with human constant regions and mouse variable regions directed against carcinoma-associated antigen 17-1A. Proc Natl Acad Sci U S A 84:214-8.

Tempest, P.R., P. Bremner, M. Lambert, G. Taylor, J.M. Furze, F.J. Carr, W.J. Harris. 1991. Reshaping a human monoclonal antibody to inhibit human respiratory syncytial virus infection in vivo. BioTechnology 9:266-271.

Tsuchiya M, Sato K, Saldanha J, Tsunenari T, Koishihara Y, Ohsugi Y, Kishimoto T, Bendig MM. 1992. The humanization of two mouse antibodies that inhibit IL-6-dependent tumor cell growth. Proc. 3rd Intl. Conf. Antibody Engineering, San Diego, CA,

Verhoeyen, M., C. Milstein, G. Winter. 1988. Reshaping Human Antibodies: Grafting an Antilysozyme Activity. Science 239:1534-1536.

Waldmann, H., G. Hale, M. Clark, et al. 1988. Monoclonal antibodies for immunosuppression. Prog.

Allergy 45:16-30.

Ward, E.S., D. Gussow, A.D. Griffiths, P.T. Jones, G. Winter. 1989. Binding activities of a repertoire of single immunoglobulin variable domains secreted from Escherichia coli. Nature 341:544-6.

Whittle, N., J. Adair, C. Lloyd, L. Jenkins, J. Devine, J. Schlom, A. Raubitschek, D. Colcher, M. Bodmer. 1987. Expression in COS cells of a mouse-human chimaeric B72.3 antibody. Protein Eng (ENGLAND) 1:499-505.

Woodle ES, Thistlewaite JR, Jolliffee LK, Zivin RA, Coltins A, Adair JR, Bodmer M, Athwal D, Alegre ML, Bluestone JA. 1992. Humanized OKT3 antibodies: successful transfer of immune modulating properties and idiotype expression. J Immunol 148:2756-2763.

Yokoyama, M., Y. Nishimura, T. Watanabe. 1987. Suppression of tumor growth by in vivo administration of a recombinant human-mouse chimeric monoclonal antibody. Jpn J Cancer Res (JAPAN) 78 (11) p1251-7.

Zebedee SL, Barbas CF, Hom Y.-L, Caothien RH, Graff R, Degaw J, Pyati J, LaPolla R, Burton DR, Lerner RA, Thornthon GA. 1992. Human combinatorial antibody libraries to hepatitis B surface antigen. Proc. Natl. Acad. Sci. (USA) 89:3175-3180.

APPLICATIONS OF HUMAN MONOCLONAL ANTIBODIES TO VARIOUS CLINICAL USES

S. HASHIZUME[1], S. SATO[1], M. KATO[1], M. KAMEI[1], K. MOCHIZUKI[1],
K. KURODA[1], K. KUSAKABE[2], K. KANAYA[2], K. YASUMOTO[3],
K. NOMOTO[4] and H. MURAKAMI[5]
[1]Morinaga Institute of Biological Science, 2-1-1
Shimosueyoshi, Tsurumi-ku, Yokohama 230, Japan
[2]Department of Radiology, Tokyo Women's Medical College,
8-1 Kawada-cho, Shinjuku-ku, Tokyo 162, Japan
[3]Kitakyushu Municipal Medical Center, 2-1-1 Bashaku,
Kokurakita-ku, Kitakyushu, Fukuoka 802, Japan
[4]Medical Institute of Bioregulation, Kyushu University,
Fukuoka 812, Japan
[5]Graduate School of Genetic Resources Technology, Kyushu
University, Fukuoka 812, Japan

Abstract. Human monoclonal antibodies, produced by hybridomas which
were obtained by fusing regional lymph node B cells from cancer
patients with the fusion partners of high fusion efficiency, were
applied to serodiagnosis, cytodiagnosis and radioimmunoimaging of lung
cancer. Early stages of lung cancer could be detected by ELISA using
an animal antigen (histone H2B) recognized by a lung cancer-specific
human monoclonal antibody HB4C5. Cancer cells in sputa, pleural
effusions and washings were reactive with the antibodies HB4C5 and/or
H-6. Clear images of human lung cancer cell lines transplanted into
nude mice were obtained with ^{125}I-labelled monoclonal antibody HB4C5.
From these data, human monoclonal antibodies described above are
considered to be of potential use in external and internal diagnoses.

1. Introduction

For the application of monoclonal antibodies (MoAbs) to clinical uses,
including external and internal diagnoses as well as therapy, it is
required that the MoAbs have high specificities and adequate
antigen-binding activities. Since few human MoAbs have met these
requirements as compared with those of mouse origin, mouse MoAbs have
been the major vehicle for the clinical uses described above.
However, the mouse MoAbs administered into human bodies inevitably
give rise to human antibodies against the mouse MoAbs (Schroff et al.
(1985)), which would cause adverse effects in the subsequent doses
such as anaphylaxis and rapid clearance of the mouse MoAbs from

641

S. Kaminogawa et al. (eds.), Animal Cell Technology: Basic & Applied Aspects, Vol. 5, 641–646.
© 1993 *Kluwer Academic Publishers.*

circulating blood (Seccamani et al. (1989)). Although chimera or humanized mouse MoAbs made by the gene manipulation can reduce these adverse effects (Winter and Milstein (1991)), the use of human MoAbs would be most preferable for the clinical purposes (Borrebaeck et al. (1990) and Persson et al. (1991)).

We have generated many hybridomas secreting human MoAbs with varied specificities and high antigen-binding affinities. In this proceedings, the application of these human MoAbs to various clinical uses are shown.

2. Materials and Methods

2.1. GENERATION OF HYBRIDOMAS SECRETING HUMAN MONOCLONAL ANTIBODIES

The generation of hybridomas was carried out according to the method detailed for HB4C5 (Murakami et al. (1985)). Briefly, lymphocytes, obtained from the regional lymph nodes which were dissected from lung cancer patients, were fused with fusion partners such as NAT-30 and RF-S1 (Kamei et al. (1990)) in 50% polyethylene glycol.

2.2. MONOCLONAL ANTIBODIES

All human MoAbs were produced by culturing the hybridomas at 37 C in serum-free medium, and the antibodies of IgM type were purified by the method as described elsewhere (Yano et al. (1988)). MoAbs of IgG type in the spent media were purified by the affinity chromatography on protein A-Cellulofine (Seikagaku Kogyo Co., Ltd., Tokyo) as described previously (Kamei et al. (1990)).

2.3. CYTODIAGNOSIS OF CANCER

Cellular specimens of sputa, pleural effusions and washings were obtained from patients with lung cancer. The immunoperoxidase staining was carried out as described previously (Hirose et al. (1991)).

2.4. SERODIAGNOSIS OF CANCER USING HISTONE H2B RECOGNIZED BY HUMAN MONOCLONAL ANTIBODY HB4C5

The serodiagnosis of cancer was carried out by ELISA of anti-histone H2B antibody as described previously (Kamei et al. (1992)). In brief, calf thymus histone H2B (Boehringer Mannheim GmbH, Germany) was dissolved to a concentration of 50 ug/ml in 43 mM sodium carbonate buffer, pH 9.6, containing 0.05% fetal calf serum (Armour Pharmaceutical Co., USA), and wells of immunoplates were coated with this histone H2B solution. All subsequent procedures of ELISA were performed according to the conventional method.

2.5. RADIOIMMUNOIMAGING

Xenografted nude mice after the i.v. injection with ^{125}I-labelled human MoAb were subjected to the gamma-scintigraphy on day 5 of post-injection using a gamma-scintillation camera (Model Sigma 410, Ohio Nuclear, USA) (Hashizume et al. (1990)).

3. Results and Discussion

3.1. DETECTION OF LUNG CANCER CELLS IN CELLULAR SPECIMENS USING HUMAN MONOCLONAL ANTIBODIES HB4C5 and H-6

Cellular specimens from 13 lung cancer patients were immunostained with human MoAbs HB4C5 and H-6 by the avidin-biotin-peroxidase method. As shown in TABLE 1, HB4C5 reacted with cancer cells in 7 of the 9 cellular specimens from patients with lung adenocarcinoma.

TABLE 1. Reactivity of human MoAb HB4C5
with cellular specimens from patients
with lung adenocarcinoma

Specimen	No. positive/No. examined
Sputum	3/4
Pleural effusion	1/2
Washing	3/3

Human MoAb H-6 was reactive with cancer cells in all specimens from 4 patients with lung squamous cell carcinoma (data not shown). These data suggest that a mixture of human MoAbs HB4C5 and H-6 would be valuable for the detection of lung cancer cells in cellular specimens such as sputa, pleural effusions and washings.

3.2. SERODIAGNOSIS OF CANCER USING HISTONE H2B

An ELISA method for the serodiagnosis of cancer was developed by using histone H2B which is recognized by a lung cancer-specific human MoAb HB4C5. This diagnosis can be performed by measuring anti-histone H2B antibody levels in sera. By this method, the early diagnosis of lung cancer at stages I and II was successfully performed at a high detection rate of more than 70% (Figure 1), though the rate for total lung cancer patients with stages covering I to IV was 37%. With these data, it would be possible that cancer at early stages can be

644

diagnosed by measuring anti-tumor antibody instead of tumor markers such as CEA and SCC, since the antibody generated in the patients could permit a highly sensitive assay even though the antigen concentrations in sera are still too low to be detected.

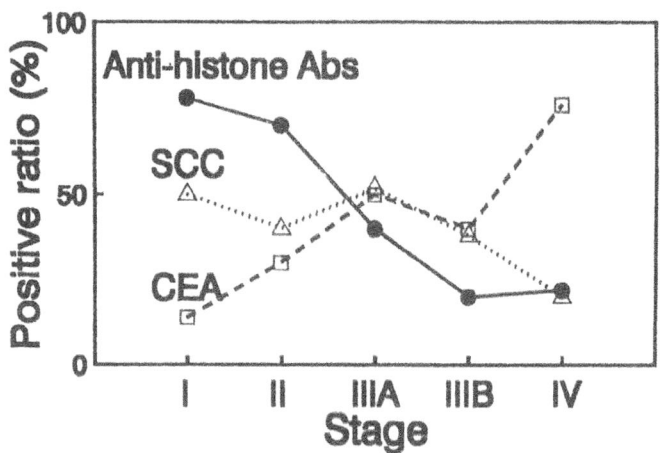

Figure 1. Serodiagnosis on lung cancer patients at various stages

TABLE 2. Radioimmunoimaging of cancer in nude mice bearing various human lung cancer cell lines by using ^{125}I-labelled human MoAb HB4C5

Cell line	Imaging
Adenocarcinoma	
A-549	+
L-2	+
Calu-3	−
Calu-6	−
Squamous cell carcinoma	
LC-6	+
PC-10	+

+: Clear image of cancer was obtained.
−: Clear image of cancer was not obtained.

3.3. RADIOIMMUNOIMAGING OF CANCER

Clear images of 4 kinds of cancer xenograft were obtained in nude mice bearing 6 different human cancer cell lines on day 5 of post-administration with ^{125}I-labelled human MoAb HB4C5 (TABLE 2).
    These data suggest that human MoAb HB4C5 is widely useful for radioimmunoimaging of human lung cancer.
    Thus, human MoAbs are considered to be of potential usefulness in external and internal diagnoses.

4. References

Borrebaeck C.A.K., Danielsson L., Ohlin M., Carlsson J. and
    Carlsson R. (1990) 'The use of in vitro immunization, cloning
    of variable regions, and SCID mice for the production of human
    monoclonal antibodies', in C.A.K. Borrebaeck and J.W. Larrick
    (eds.), Therapeutic Monoclonal Antibodies, Stockton Press, New
    York, pp. 1-15.

Kamei M., Kato M., Mochizuki K., Kuroda K., Sato S., Hashizume
    S., Yasumoto K., Murakami H. and Nomoto K. (1992)
    'Serodiagnosis of cancers by ELISA of anti-histone H2B
    antibody', Biotherapy 4, 17-22.

Kamei M., Hashizume S., Sugimoto N., Ozutsumi K. and Matsuda M.
    (1990) 'Establishment of stable mouse/human-human hybrid cell
    lines producing large amounts of anti-tetanus human monoclonal
    antibodies with high neutralizing activity', Eur. J. Epidemiol.
    6, 386-397.

Hashizume S., Sato S., Matsuyama M., Tamaki S., Hanada K.,
    Murakami H., Yasumoto K., Nomoto K., Nakano K. and Kusakabe K.
    (1990) 'Accumulation of ^{125}I-labelled human monoclonal antibody
    (HB4C5), specific to lung cancer, into transplanted human lung
    cancer in nude mouse', in H. Murakami (ed.), Proceedings of the
    Second Annual Meeting of Japanese Association for Animal Cell
    Technology, Kodansha, Tokyo, pp. 167-172.

Murakami H., Hashizume H., Ohashi H., Shinohara K., Yasumoto K.,
    Nomoto K. and Omura H. (1985) 'Human-human hybridomas secreting
    antibodies specific to human lung carcinoma', In Vitro Cell.
    Develop. Biol. 21, 593-596.

Hirose H., Sato S., Tai H., Okano H., Yasumoto K., Murakami H.,
    Nomoto K., Matsuyama M., Tamaki S. and Hashizume S. (1991)
    'Detection of lung cancer in clinical specimens using a human
    monoclonal antibody HB4C5-clone 3', Hum. Antibod. Hybridomas 2,
    200-206.

Persson M.A.A., Caothien R.H. and Burton D.R. (1991) 'Generation of diverse high-affinity human monoclonal antibodies by repertoire cloning', Proc. Natl. Acad. Sci. U.S.A. 88, 2432-2436.

Schroff R.W., Foon K.A., Beatty S.M., Oldham R.K. and Morgan Jr. A.C. (1985) 'Human anti-murine immunoglobulin responses in patients receiving monoclonal antibody therapy', Cancer Res. 45, 879-885.

Seccamani E., Tattanelli M., Mariani M., Spranzi E., Scassellati G.A. and Siccardi A.G. (1989) 'A simple qualitative determination of human antibodies to murine immunoglobulins (HAMA) in serum samples', Nucl. Med. Biol. 16, 167-170.

Winter G. and Milstein C. (1991) 'Man-made antibodies', Nature 349, 293-299.

Yano T., Yasumoto K., Nagashima A., Hashizume S., Murakami H. and Nomoto K. (1988) 'Immunohistological characterization of human monoclonal antibody against lung cancer', J. Surg. Oncol. 39, 108-113.

REACTIVITY OF HUMAN MONOCLONAL ANTIBODY BD9-D12 TO CANCER CELLS AND TISSUES

Kiyohiko Seki (1), Hiroharu Kawahara (1), Susumu Sato (2), Hirofumi Tachibana (1), Isao Kido (1), Asako Hashimoto (1), Shuichi Hashizume (2), and Hiroki Murakami (1)

(1) Graduate School of Genetic Resources Technology, Kyushu University, Fukuoka, Japan
(2) Morinaga Institute of Biological Science, Yokohama, Japan

ABSTRACT

A human monoclonal antibody (MAb), BD9-D12 (IgG/κ), has been generated by an *in vitro* immunization technique using a lung cancer cell line, A549, as the antigen. The MAb reacted to various human cancer cell lines, especially with lung cancer and breast cancer cell lines, but did not react to normal cell lines. BD9-D12 reacted to cancer cells in various lung cancer tissues from patients, but did not react to normal cells around the cancer tissues. The MAb specifically reacted to 13 of 14 different specimens of lung cancer tissues. These results suggest that the MAb recognizes a lung cancer-related antigen.

INTRODUCTION

Monoclonal antibodies (MAbs) are expected to be useful as drugs for the diagnosis and therapy of various kinds of human diseases, especially cancer [1,2]. We have established several kinds of human parent cells showing high fusion efficiency such as NAT-30 [3], HO-323 [4] and A4H12 [5], which enabled to be established many human-human hybridomas secreting cancer-specific MAbs. Recently, a new *in vitro* immunization method for human lymphocytes has been developed in our laboratory [5]. A human MAb, BD9-D12, was established by the *in vitro* immunization method, using lung cancer cell line A549 as the antigen. This paper reports the reactivity of BD9-D12 MAb to various kinds of cancer cell lines and cancer tissues.

S. Kaminogawa et al. (eds.), Animal Cell Technology: Basic & Applied Aspects, Vol. 5, 647–652.
© 1993 Kluwer Academic Publishers.

MATERIALS AND METHODS

Cells and cell culture

Human parent cell line A4H12 [5] is an HAT-sensitive mutant derived from human leukemic T cell line Molt4 [6]. Human lung cancer cell lines A549, PC-9 and QG-90, human breast cancer cell lines MCF-7 and HBC5, stomach cancer cell lines MKN-28 and MKN-45, colon cancer cell line Colo 201, and normal fibroblast cell line WI-38 were cultured in ERDF medium supplemented with 10% fetal bovine serum (FBS; HyClone Laboratory, Logan, UT, U.S.A.) in a humidified atmosphere of 5% CO_2/95% air.

Generation of human hybridomas

Muramyl peptides (MDP) were purchased from Calbiochem, La Jolla, CA, U.S.A., and recombinant human IL-2 and IL-6 were obtained from Genzyme, Boston, MA, U.S.A. Human MAb BD9-D12 was generated by the in vitro immunization technique, as described previously [5]. Briefly, human peripheral blood lymphocytes were separated by density-gradient centrifugation from several normal donors. A549 cells were inoculated into each well of a 24-well culture plate and cultured for 36 h, before human peripheral blood lymphocytes were seeded over the A549 cells. MDP (10 μg/ml), IL-2 (100 U/ml) and IL-6 (10 U/ml) were added as immuno-activating reagents. Culture for an in vitro immunization was performed for 4 days. To immortalize the lymphocytes immunized in vitro, fusion partner A4H12 was fused with the sensitized lymphocytes by using polyethylene glycol (PEG). The hybridomas were selected in 15% FBS-ERDF medium containing 100 mM hypoxanthine, 0.4 mM aminopterin, and 16 mM thymidine (HAT). The hybridomas were further selected by checking the reactivity of MAbs to A549 cells with a cellular enzyme-linked immunosorbent assay (ELISA). The selected hybridomas were cloned two times by limiting dilution.

Reactivity of human MAb produced by hybridomas to various human cancer cell lines

Lung, breast, stomach and colon cancer cell lines, and normal cell lines were cultured to confluency in 96-well microplates [3]. These cells were fixed with 0.05% glutaraldehyde, washed and then incubated in a 3% bovine serum albumin (BSA) solution. The cells were next reacted with BD9-D12 MAb, and after incubating at 37°C for 1hr, MAb combined with the cancer cells was incubated with peroxidase-conjugated goat anti-human IgG (1:2000, TAGO). The cells were incubated in the substrate solution (0.1M citrate buffer [pH 4.0] containing 0.003% H_2O_2 and 0.3 mg/ml of p-2.2'-azino-di (3-ethylbenzothiazoline-6-sulfonic acid) diammonium salt) for

20min. The absorbancy at 405 nm of the reaction mixture was measured by an ELISA reader.

Reactivity of human MAb BD9-D12 to various cancer tissues

The following steps for staining were carried out at room temperature, unless otherwise indicated: Biopsy tissues were fixed in 10% formalin, and embedded in paraffin blocks [7]. 4-µm sliced sections of the biopsy specimens were deparaffinized in xylene and rinsed in acetone. To inactivate the endogenous peroxidase activity, the sections were incubated in methanol containing 0.3% hydrogen peroxide for 30 minutes. Cellular specimens destained as already described were directly immersed in this methanol solution. After being treated with 1.8% normal goat serum in PBS(-), the sections were blocked to endogenous IgG by incubating with the 1% Fab fragment of goat anti-human IgG and 1.8% goat serum in PBS(-). After being blocked with 1.8% goat serum in PBS(-) again, the section were treated with biotinylated MAb BD9-D12 (2µg/ml) overnight at 4°C. The sections of cellular specimens were then washed with PBS and soaked in a 3,3'-diaminobenzidine solution (DAB) containing 0.01% hydrogen peroxidase for about 2 minutes to develop an immunoreaction. The sites bound to MAb BD9-D12 were browned. Finally, the sections were counterstained with hematoxylin.

RESULTS AND DISCUSSION

Generation of BD9-D12 MAb by in vitro immunization

Peripheral blood lymphocytes from several normal donors were fused with human parent cells A_4H_{12}. After fusing four times and screening several hundred clones, five human-human hybridoma clones secreting MAbs reactive to lung cancer cell line A549 were established. Two of them were IgG producers and others were IgM producers. These MAbs were also reactive with other cancer cell lines such as breast, stomach and colon cancer (data not shown). Among them, BD9-D12 was an IgG producer, and the IgG showed high reactivity to A549 cells.

The production-efficiency of MAbs reactive to the cancer cell lines was as high as five clones for about every four hundred clones in the case of in vitro immunization. On the other hand, the efficiency was as low as one clone for several thousands of clones in the case of using lymph node lymphocytes derived from cancer patients. Furthermore, it was very difficult to obtain IgG producers in the latter case.

Reactivity of BD9-D12 human MAb to various cancer cell lines

The reactivity of MAb BD9-D12 was examined to various cancer cell lines and a normal cell line by ELISA (Table 1). BD9-D12 reacted to breast cancer cell line MCF-7 strongly, as well as to lung adenocarcinoma cell line A549, which was used as an antigen for *in vitro* immunization (Table 1). Among the lung cancer cell lines, this MAb reacted strongly to large-cell carcinoma PC-13, and moderately to squamous cell carcinoma QG-56 and small-cell carcinoma QG-90. The MAb also showed weak reactivity to Adenocarcinoma PC-9, and showed moderate reactivity to stomach cancer and colon cancer cell lines. However, the MAb showed no reactivity to normal fibroblast WI-38. These results suggest that the MAb recognized some cancer-related antigens in the cancer cell lines examined. Since BD9-D12 MAb reacted to the lung cancer and breast cancer cell lines, the MAb is expected to be useful for the diagnosis and therapy of these cancers.

Reactivity of BD9-D12 MAb to various lung cancer tissues

It is desirable to examine the reactivity of human BD9-D12 MAb to cancer tissues derived from patients, because cancer cells grow in tissues of the body and the environment of cancer cells in tissues will be different from those of cultured cells. Furthermore, cancer-relating antigens may be different from those in tissues due to the change in nature of the cultured cells.

We examined the reactivity of BD9-D12 MAb to various types of lung cancer tissues derived from patients (Table 2). The MAb reacted strongly to all adenocarcinoma tissues derived from 6 patients, but did not react to the normal cells around the cancer cells. The MAb could clearly stain cancer cells in the tissues. Large and small carcinoma cells were also reactive to the MAb, showing little reactivity to the normal cells. The MAb also reacted weakly to squamous cell carcinoma cells. These results strongly suggest that BD9-D12 would be very useful for diagnosing lung cancer such as by RI-imaging or serological diagnosis.

Table 1.   Reactivity of BD9-D12 MAb to Various Cell Lines.

Cell line	Reactivity
Lung cancer	
Adenocarcinoma	
A549	0.839
PC-9	0.101
Squamous cell carcinoma	
QG-56	0.432
Large cell carcinoma	
PC-13	0.530
Small cell carcinoma	
QG-90	0.302
Breast cancer	
MCF-7	1.585
HBC5	0.693
Stomach cancer	
MKN-28	0.500
MKN-45	0.250
Colon cancer	
Colo 201	0.256
Normal fibroblast	
WI-38	0.049

The cells cultured to confluency were fixed 0.05 % glutaraldehyde and used for ELISA.   BD9-D12 MAb reacted to various cancer lines and especially reacted strongly the adenocarcinoma line A549 and the breast cancer cell line MCF-7.   This MAb did not react to normal fibroblast line WI-38, originating from lung tissue.

Table 2.   Reactivity of BD9-D12 MAb to Various Lung Cancer Tissues by Immunoperoxidase Staining.

Tissue		Reactivity	
		cancer	normal
Adenocarcinoma			
	Patient 1	+++	-
	Patient 2	+++	-
	Patient 3	++	-
	Patient 4	++	-
	Patient 5	+++	-
	Patient 6	+++	+
Squamous cell carcinoma			
	Patient 7	+	-
	Patient 8	+	-
	Patient 9	+	-
Large cell carcinoma			
	Patient 10	+	-
	Patient 11	+++	+
Small cell carcinoma			
	Patient 12	++	-
	Patient 13	-	-
	Patient 14	+++	-

Tissues sliced with a thickness of 4 μm on slide glass were fixed with 10 % formaldehyde and used for the assay.   Degree of staining intensity was scored as follows: +++,strongly positive; ++,moderately positive; +,weakly positive; -,negative.   Histologically normal tissues adjacent to the lung cancer of the same patient whose cancer tissues were assayed.

REFERENCES

1. Hashizume, S., Sato, S., Matsuyama, M., Tamaki, S., Hanada, K., Murakami, H., Yasumoto, K., Nomoto, K., Nakano, K., Kusakabe, K. (1990) Accumulation of 125I-labelled human monoclonal antibody (HB4C5), specific to lung cancer, into transplanted human lung cancer in nude mouse. In: Murakami H (ed), *Trends in Animal Cell Culture Technology*, Kodansha, VCH, 167-172

2. Hashizume, S., Mochizuki, K., Murakami, H., Yano, T., Yasumoto, K., Nomoto, K. (1989) Serodiagnosis of cancer, using porcine antigens recognized by human monoclonal antibody, HB4C5. *Biotherapy*, 1, 109-115

3. Murakami, H., Hashizume, S., Ohashi, H., Shinohara, K., Yasumoto, K., Nomoto, K., Omura, H. (1985) Human-human hybridomas secreting antibodies specific to human lung carcinoma. *In Vitro Cell. Develop. Biol.*, 21, 593-596.

4. Ohashi, H., Hashizume, S., Murakami, H., Aihara, K., Shinohara, K., Omura, H. (1986) HO-323, a human B-lymphoblastoid cell line useful for making human-human hybridomas. *Cell Biology International Reports*, 10, 77-83

5. Kawahara, H., Shirahata, S., Tachibana, H., Murakami, H. (1992) *In vitro* immunization of human lymphocytes with human lung cancer cell line A549. *Hum. Antibod. Hybridomas*, 3, 8-13

6. Minowada, J., Ohmura, T., Moore, GE. (1972) Rosette-forming human lymphoid cell lines. I. Establishment and evidence for origin of thymus-derived lymphocytes. *J. Natl. Cancer. Inst.*, 49, 891-895

7. Yano, T., Yasumoto, K., Nagashima, A., Murakami, H., Hashizume, S., Nomoto, K. (1988) Immunohistological characterization of human monoclonal antibody against lung cancer, *J. Surg. Oncol.*, 39, 108-113.

CELL CULTURE IN EUROPE: A REVIEW OF TODAY'S MARKET FOR CELL CULTURE PRODUCTS AND SERVICES

T.J.M. CLARK
European Business Associates
22 rue Dernier Sol
L-2543 Luxembourg
Grand Duchy of Luxembourg

ABSTRACT. There is already today a substantial European biotechnology industry utilising cell culture in production processes. As a result usage of cell culture consumables, reagents and growth systems is increasing. Numbers of plants using cell culture are growing and cell culture derived therapeutics are becoming increasingly important.

1. Introduction

In Europe there are in 1992 at least 350 specialised commercial biotechnology organisations including 90 which utilise cell culture technology in their production processes. 32 are vaccine producers, 47 produce monoclonal antibodies, 5 specialise in cytokines and a further 6 concentrate in other areas such as hormones, TPA and urokinase.

In addition several thousand academic and industrial laboratories, clinical diagnostic laboratories and quality control laboratories also use cell culture techniques.

2. Growth Potential

Continuing developments in cell biology and cell culture are leading to increased demands both for improved methods and for new equipment in almost every sector. However, as Table 1 shows, the rate of growth in various application areas varies significantly.

2.1. Clinical Diagnostics

Historically cell substrates have been an important tool for the virologist in detecting the presence of viruses in

S. Kaminogawa et al. (eds.), Animal Cell Technology: Basic & Applied Aspects, Vol. 5, 653–659.
© 1993 Kluwer Academic Publishers.

a biological sample. Cell substrates are also used in confirming chlamydial infections and in a number of indirect fluorescence antibody staining techniques. Routine immunoassay testing is today becoming more popular and new DNA-probe based tests are rapidly appearing on the European market. The speed of changeover from traditional cell culture will largely depend on relative costs of new techniques and on their acceptance for reimbursement by national authorities.

2.2. Quality Control Laboratories

The use of cell cultures as a viable alternative to laboratory animals in toxicity testing procedures has been adopted more slowly than anticipated. This is because of the inordinately long time taken by some regulatory authorities to accept new techniques in official test procedures. For non-regulatory applications however, cell cultures offer lower costs, more reproduceable and more humane methodologies. They are being increasingly used in toxicity testing for drugs, packaging materials, cosmetics etc. as well as in process control of therapeutic manufacturing processes.

2.3. Human Therapeutics and Vaccines

The production of human vaccines was the earliest industrial application of cell culture technology and is still a very important sector. Although rationalisation of human vaccine production in Europe has reduced the total number of producers, manufacture of newly developed vaccines continue to produce steady overall growth. In addition production of human growth hormone, cytokines, growth factors and other therapeutic products are moving from research through pilot plant stages to full production. This sector which is today growing by 15% annually, may well accelerate to a rate of 25-30% by the end of the decade.

2.4. Veterinary Therapeutics and Vaccines

The veterinary vaccine market, also, is growing rapidly, both for reasons of increasing demand and also of newly developed vaccines especially in the small animal sector. However the veterinary vaccine market is much more price conscious than many other cell culture sectors. Although production is growing rapidly, technologies are increasingly cost driven.

2.5. Mab-based Therapeutics

Therapeutic and in-vivo imaging products based on monoclonal antibodies are beginning to move through to commercial marketing and actual use. Centoxin from Centocor and Xomen E5 from Xoma are the first Mabs for treatment of septic shock and are already now generally on the market in Europe. OKT3 from Ortho for cell adhesion therapy has been around for a little longer. Mabs for in-vivo imaging from Centocor and Cytogen are beginning to appear. Of at least 60 companies worldwide involved in therapeutic Mab development, 19 are European. The value of European sales in this sector in 1991 was only $5 million, but it is projected to increase to over $1 billion by the end of the century. Cancer therapeutics are expected to be the most important area, followed by AIDS treatments and a rapidly growing autoimmune disease sector.

2.6. Mab-based Diagnostics

Today a substantial proportion of clinical immunoassays use monoclonal antibodies for reasons of sensitivity and specificity. While some assays use only one Mab, in other systems as many as three different Mabs may be used, directed against different epitopes on the target antigen. At least 50 European companies are producing and marketing immunoassays using monoclonal antibodies. The infectious disease assay sector is increasing fastest, providing an overall 15% per year growth rate.

2.7. OTC Diagnostics

The self-testing diagnostics sector, which consists largely of pregnancy tests, and to a lesser extent fertility tests, has already largely been converted to the use of Mab technology. Demands for greater sensitivity levels and faster test results have resulted in over 95% of all pregnancy tests now using monoclonal antibodies. Heavy promotional programmes are continuing to increase the overall European market size at around 10% each year.

2.8. Industrial Purification

A small but important use for monoclonal antibodies is in industrial purification systems. Undoubtedly the greatest part of production of Mabs for this purpose will be in-house. Such products are unlikely to be offered for sale commercially since the producers will be well aware of the high value of such products in their own production processes.

2.9.In Vitro Fertilization

The use of Assisted Procreation Techniques in humans, and
especially In Vitro Fertilization (IVF) is growing rapidly
worldwide and especially in Europe. The increasing tendency
of couples to delay planning families until in their mid-
30's, the extended use of oral contraceptives and an
increasing incidence of sexually transmitted diseases are
all producing greater fertility problems. IVF techniques
using feeder cell monolayers and specially developed media
are as a result becoming a rapidly growing niche market
sector.

3. Cell Culture Consumables

Large quantities of specially treated polystyrene flasks,
multiwell plates tubes and petri dishes are used in
research and clinical laboratories in Europe. In 1991 over
16 million units were used with over 80% produced in
Europe. About one million roller bottles, widely used in
vaccine production, were sold together with about 1000
kilos of microcarriers, mostly for use in the same sector.
Including special laboratory glassware for cell culture
use, this sector was valued at $30 million last year.

4. Cell Culture Reagents

Of 450 000 litres of animal sera used annually in European
cell culture, about 100 000 are foetal bovine serum.
Currently this single product constitutes over 40% of the
value of cell culture reagents used (Table 2). Foetal
bovine serum (FBS) is an "accidental" product. The
availability depends on the number of cows slaughtered in
those few countries where cattle are not closely
supervised. As a result, farmers do not know whether cows
are in calf or not. Sizes of cattle herds in these areas
(USA, Canada, Australia, New Zealand etc) will vary greatly
with climatic and economic conditions so that in turn the
availability and price of FBS also changes. For cell
culture scientists such an uncertain supply situation is
most unsatisfactory. Some laboratories have been able to
convert cell growth systems to other sera, while others
have concentrated on developing serum-free or low-serum
growth systems. Overall, however, an over reliance on using
FBS is still evident in too many laboratories. Effective
multi-purpose substitutes are greatly needed.

For industrial cell culture organisations the spread of
bovine spongiform encephalitis (BSE) from Britain to
mainland Europe is also creating problems, with increasing
proportions of all sera being imported from outside Europe.
Disease-free supply sources are becoming a major problem
worldwide. It is anticipated that FBS usage will slowly
reduce in Europe, other serum usage remain largely static,
and serum-free medium usage grow substantially over the
next 5-6 years. 10.5 million litres of liquid media and
powder media equivalents were used in 1991 in Europe. This
quantity is projected to increase to 16 million litres by
1996.

5. Cell Culture Growth Systems

Many cell culturists work on a relatively small scale using
traditional flasks, tubes and plates very effectively. For
large scale research and industrial producers, however, a
wide range of alternative systems are available in Europe
and are in practical use.

For the growth of attachment dependent cells, roller
bottles are still the most widely used method. New expanded
surface bottles have improved production economies and
robotic handling systems have reduced contamination levels.
A number of large producers are obtaining excellent results
with bioreactor systems and microcarriers, while plastic
stacked tray systems, packed columns and ceramic core
bioreactors are also in use.

Hybridoma production is the most important in the
suspension culture sector. The most generally used systems
are bioreactors with operating volumes of up to 1000 litres
or more. These use a wide variety of control and monitoring
equipment, much of which is designed for use in specific
growth systems. Although a large number of hollow fibre
bioreactors are also in use, most of these are in small and
medium sized production units.

6. Conclusions

The acceptance and use of newer more efficient production
methodologies has largely been restricted to new
manufacturing units. Relatively few pre-existing European
production facilities have actually made far-reaching
changes in their basic methods. This is because many
existing end products are subject to strict validation
procedures. Major changes in methodologies would incur
substantial financial and managerial expenses which may be
difficult to justify. New product developments and new

production processes are, however, expected to result in a substantial number of new production facilities over the next five years. Today there are 141 major production units in Europe, in 90 separate facilities culture methodology (many use several different systems for differing end products). It is anticipated that this will have increased to well over 200 by the end of the decade.

TABLE 1

COMMERCIAL APPLICATIONS
FOR CELL CULTURE -
GROWTH POTENTIAL

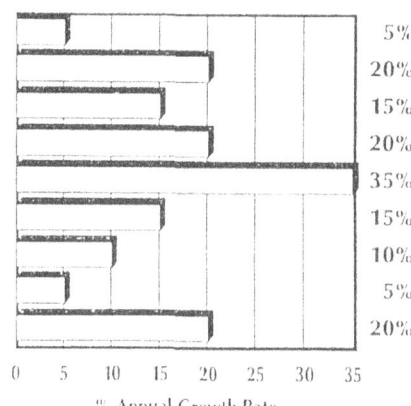

Traditional Clinical Diagnostics	5%
Quality Control Labs.	20%
Human Therapeutics/Vaccines	15%
Veterinary Therapeutics/Vaccines	20%
Mab-based Therapeutics	35%
Mab-based Clinical Diagnostics	15%
Mab-based OTC Diagnostics	10%
Industrial Purification	5%
In Vitro Fertilisation	20%

% Annual Growth Rate

Source: EUROPEAN BUSINESS ASSOCIATES

TABLE 2

CELL CULTURE REAGENTS
EUROPE - 1991

	Value $ million	Volume litres
Foetal Bovine Serum	33.0	100,000
Other Sera	15.0	350,000
Liquid media/supplements	18.5	1.5 million
Powder media/chemicals	8.5	9.0 million
Others	3.5	
Cell Cultures	2.5	
	81.0	

Source: EUROPEAN BUSINESS ASSOCIATES

Author Index

Subject Index

The manufacturer's authorised representative in the EU is Springer
Nature Customer Service Centre GmbH, Europaplatz 3, 69115 Heidelberg,
Germany. If you have any concerns regarding our products, please
contact ProductSafety@springernature.com

Printed and bound by CPI Group (UK) Ltd, Croydon, CR0 4YY

24/04/2026

02096308-0015